MW01489501

Molecular Bacteriology:
Protocols and Clinical Applications

METHODS IN MOLECULAR MEDICINE™

John M. Walker, SERIES EDITOR

15. **Molecular Bacteriology**: *Protocols and Clinical Applications,* edited by *Neil Woodford and Alan P. Johnson,* 1998

14. **Tumor Marker Protocols,** edited by *Margaret Hanausek and Zbigniew Walaszek,* 1998

13. **Molecular Diagnosis of Infectious Diseases,** edited by *Udo Reischl,* 1998

12. **Diagnostic Virology Protocols,** edited by *John R. Stephenson and Alan Warnes,* 1998

11. **Therapeutic Application of Ribozymes,** edited by *Kevin J. Scanlon,* 1998

10. **Herpes Simplex Virus Protocols,** edited by *S. Moira Brown and Alasdair MacLean,* 1998

9. **Lectin Methods and Protocols,** edited by *Jonathan M. Rhodes and Jeremy D. Milton,* 1998

8. *Helicobacter pylori* **Protocols,** edited by *Christopher L. Clayton and Harry L. T. Mobley,* 1997

7. **Gene Therapy Protocols,** edited by *Paul D. Robbins,* 1997

6. **Molecular Diagnosis of Cancer,** edited by *Finbarr Cotter,* 1996

5. **Molecular Diagnosis of Genetic Diseases,** edited by *Rob Elles,* 1996

4. **Vaccine Protocols,** edited by *Andrew Robinson, Graham H. Farrar, and Christopher N. Wiblin,* 1996

3. **Prion Diseases,** edited by *Harry F. Baker and Rosalind M. Ridley,* 1996

2. **Human Cell Culture Protocols,** edited by *Gareth E. Jones,* 1996

1. **Antisense Therapeutics,** edited by *Sudhir Agrawal,* 1996

METHODS IN MOLECULAR MEDICINE™

Molecular Bacteriology:
Protocols and Clinical Applications

Edited by

Neil Woodford
and
Alan P. Johnson

Central Public Health Laboratory, London, UK

Humana Press ✳ **Totowa, New Jersey**

© 1998 Humana Press Inc.
999 Riverview Drive, Suite 208
Totowa, New Jersey 07512

This publication is printed on acid-free paper. ∞
ANSI Z39.48-1984 (American Standards Institute) Permanence of Paper for Printed Library Materials.

Cover illustration: Fig. 3 from "Biochemical and Enzyme Kinetic Applications for the Characterization of ß-Lactamases" by David J. Payne and Tony H. Farmer.

Cover design by Jill Nogrady.

For additional copies, pricing for bulk purchases, and/or information about other Humana titles, contact Humana at the above address or at any of the following numbers: Tel.: 973-256-1699; Fax: 973-256-8341; E-mail: humana@humanapr.com; Website: http://humanapress.com

Printed in the United States of America. 10 9 8 7 6 5 4 3 2 1

Library of Congress Cataloging-in-Publication Data

Molecular Bacteriology: Protocols and Clinical Applications/edited by Neil Woodford and Alan B. Johnson
    p. cm — (Methods in molecular medicine; 15)
    Includes index.
    ISBN 0-89603-498-4 (alk. paper)
    1. Diagnostic bacteriology. 2. Bacterial diseases–Molecular diagnosis. 3. Molecular microbiology. I. Woodford, Neil. II. Johnson, Alan P. (Alan Patrick), 1951– . III. Series.
    [DNLM: 1. Bacteriological Techniques. 2. Genetic Techniques. 3. Bacterial Infections–diagnosis.
    QY 100 M718 1988]
    QR67.2.M65 1998
    616.07'581–DC21

Preface

The enormous advances in molecular biology that have been witnessed in recent years have had major impacts on many areas of the biological sciences. Not least of these has been in the field of clinical bacteriology and infectious disease. *Molecular Bacteriology: Protocols and Clinical Applications* aims to provide the reader with an insight into the role that molecular methodology has to play in modern medical bacteriology.

The introductory chapter of *Molecular Bacteriology: Protocols and Clinical Applications* offers a personal overview by a Consultant Medical Microbiologist of the impact and future potential offered by molecular methods. The next six chapters comprise detailed protocols for a range of such methods. We believe that the use of these protocols should allow the reader to establish the various methods described in his or her own laboratory. In selecting the methods to be included in this section, we have concentrated on those that, arguably, have greatest current relevance to reference clinical bacteriology laboratories; we have deliberately chosen not to give detailed protocols for certain methods, such as multilocus enzyme electrophoresis that, in our opinion, remain the preserve of specialist laboratories and that are not currently suited for general use. We feel that the methods included in this section will find increasing use in diagnostic laboratories and that it is important that the concepts, advantages, and limitations of each are thoroughly understood by a wide range of workers in the field. To assist in this, the subsequent chapters in the volume describe the application of these and other methods to the investigation of a variety of bacterial pathogens, diseases, and antimicrobial resistances. Our aim is that by cross-referring between chapters, the reader should become conversant with both the practical and theoretical aspects of the topics covered.

We believe that *Molecular Bacteriology: Protocols and Clinical Applications* will provide a valuable source of information for workers in both clinical and academic settings. In particular, we feel the Notes sections included at the ends of most of the chapters should prove to be of particular interest as they often include "tricks of the trade," that the various contributors have learned through personal experience.

Neil Woodford
Alan P. Johnson

Contents

Preface v

Contributors ix

1. Impact of Molecular Methods on Clinical Bacteriology
    Robert C. George 1
2. Genomic DNA Digestion and Ribotyping
    J. Zoe Jordens 17
3. Pulsed-Field Gel Electrophoresis
    Mary Elizabeth Kaufmann 33
4. Plasmid Analysis
    Alan P. Johnson and Neil Woodford 51
5. DNA Amplification: *General Concepts and Methods*
    Nick A. Saunders and Jonathan P. Clewley 63
6. Arbitrarily Primed PCR Methods for Studying Bacterial Diseases
    David Ralph and Michael McClelland 83
7. Genomic Fingerprinting by Application of rep-PCR
    Anne M. Ridley 103
8. Molecular Approaches to the Identification of Streptococci
    Robert E. McLaughlin and Joseph J. Ferretti 117
9. Pneumococcal Diseases
    Anthony M. Smith and Keith P. Klugman 139
10. Molecular Approaches in *Mycobacterium tuberculosis* and Other Infections
    Caused by *Mycobacterium* Species
    Madhu Goyal and Douglas B. Young 157
11. Diagnosis and Epidemiology of Diptheria
    Androulla Efstratiou, Kathryn H. Engler, and Aruni de Zoysa 191
12. Diagnosis and Epidemiology of Infections Caused by *Legionella* spp.
    Norman K. Fry and Timothy G. Harrison 213
13. Molecular Methods for *Haemophilus influenzae*
    Mark A. Herbert, Derrick Crook, and E. Richard Moxon 243
14. The Impact of Molecular Techniques on the Study of Meningococcal Disease
    Martin C. J. Maiden 265
15. Gonorrhea
    Catherine A. Ison 293
16. Chancroid
    Stephen A. Morse, David L. Trees, and Patricia A. Totten 309

17. Mycoplasma/Ureaplasma Infection
    Claire B. Gilroy and David Taylor-Robinson **335**

18. Application of Molecular Methods to the Study of Infections
    Caused by *Salmonella* spp.
    E. John Threlfall, Mike D. Hampton, and Anne M. Ridley **355**

19. Cholera
    Timothy J. Barrett and Daniel N. Cameron **369**

20. Diagnosis and Investigation of Diarrheagenic *Escherichia coli*
    James P. Nataro and Juan Martinez **387**

21. *Campylobacter* Infections: *Species Identification and Typing*
    Janet R. Gibson and Robert J. Owen **407**

22. Detection *and* Typing of *Helicobacter pylori*
    Robert J. Owen and Janet R. Gibson **419**

23. Nosocomial Infections Caused by Staphylococci
    Ariane Deplano and Marc J. Struelens **431**

24. Application of Molecular Techniques to the Study of Nosocomial Infection
    Caused by Enterococci
    *Teresa M. Coque, Prema Seetulsingh, Kavindra V. Singh,
    and Barbara E. Murray* **469**

25. Molecular Approaches for the Detection and Identification
    of ß-Lactamases
    David J. Payne and Christopher J. Thomson **495**

26. Biochemical and Enzyme Kinetic Applications for the Characterization
    of ß-Lactamases
    David J. Payne and Tony H. Farmer **513**

27. ß-Lactam Resistance Mediated by Changes in Penicillin-Binding Proteins
    Christopher G. Dowson and Tracey J. Coffey **537**

28. The Application of Molecular Techniques for the Study
    of Aminoglycoside Resistance
    *Karen J. Shaw, Frank J. Sabatelli, Linda Naples, Paul Mann,
    Roberta S. Hare, and George H. Miller* **555**

29. Molecular Investigation of Glycopeptide Resistance
    in Gram-Positive Bacteria
    Neil Woodford and Jill M. Stigter **579**

30. Quinolone Resistance
    Janice C. Brown and Sebastian G. B. Amyes **617**

31. Resistance to Tetracyclines, Macrolides, Trimethoprim,
    and Sulfonamides
    Marilyn C. Roberts **641**

Index **665**

Contributors

SEBASTIAN G. B. AMYES • *Department of Medical Microbiology, University of Edinburgh Medical School, Edinburgh, UK*

TIMOTHY J. BARRETT • *Foodborne and Diarrheal Diseases Branch, Division of Bacterial and Mycotic Diseases, National Center for Infectious Diseases, Centers for Disease Control and Prevention, Atlanta, GA*

JANICE C. BROWN • *Department of Medical Microbiology, University of Edinburgh Medical School, Edinburgh, UK*

DANIEL N. CAMERON • *Foodborne and Diarrheal Diseases Branch, Division of Bacterial and Mycotic Diseases, National Center for Infectious Diseases, Centers for Disease Control and Prevention, Atlanta, GA*

JONATHAN P. CLEWLEY • *Molecular Biology Unit, Central Public Health Laboratory, London, UK*

TRACEY J. COFFEY • *Biological Sciences, University of Warwick, Coventry, UK*

TERESA M. COQUE • *Center for the Study of Emerging and Re-Emerging Pathogens, Division of Infectious Diseases, Department of Internal Medicine, Univeron, UK*

DERRICK CROOK • *Oxford Public Health Laboratory, Oxford, UK*

ARIANE DEPLANO • *Unite dí Epidemiologie et díHygiene Hospitaliere, Laboratoire de Microbiologie, Hospital Erasma, Bruxelles, Belgium*

ARUNI DE ZOYSA • *Respiratory and Systemic Infection Laboratory, Central Public Health Laboratory, London, UK*

CHRISTOPHER G. DOWSON • *Biological Sciences, University of Warwick, Coventry, UK*

ANDROULLA EFSTRATIOU • *Respiratory and Systemic Infection Laboratory, Central Public Health Laboratory, London, UK*

KATHRYN H. ENGLER • *Respiratory and Systemic Infection Laboratory, Central Public Health Laboratory, London, UK*

TONY H. FARMER • *Department of Molecular Microbiology (UP1345), SmithKline Beecham Pharmaceuticals, Collegeville, PA*

JOSEPH J. FERRETTI • *Department of Microbiology and Immunology, University of Oklahoma Health Science Center, Oklahoma City, OK*

NORMAN K. FRY • *Respiratory and Systemic Infection Laboratory, Central Public Health Laboratory, London, UK*

ROBERT C. GEORGE • *Division of Hospital and Respiratory Infection, Central Public Health Laboratory, London, UK*

JANET R. GIBSON • *Laboratory of Enteric Pathogens, Central Public Health Laboratory, London, UK*

CLAIRE B. GILROY • *MRC STD Research Group, Jefferiss Trust Laboratories, St Mary's Hospital Medical School, London, UK*

MADHU GOYAL • *Imperial College School of Medicine, St. Mary's Campus, London, UK*

MIKE D. HAMPTON • *Laboratory of Enteric Pathogens, Central Public Health Laboratory, London, UK*

ROBERTA S. HARE • *Schering-Plough Research Institute, Kenilworth, NJ*

TIMOTHY G. HARRISON • *Respiratory and Systemic Infection Laboratory, Central Public Health Laboratory, London, UK*

MARK A. HERBERT • *Paediatric Infectious Diseases Group, Institute of Molecular Medicine, John Radcliffe Hospital, Oxford, UK*

CATHERINE A. ISON • *Department of Infectious Diseases and Microbiology, St. Mary's Campus, London, UK*

ALAN P. JOHNSON • *Antibiotic Reference Unit, Laboratory of Hospital Infection, Central Public Health Laboratory, London, UK*

J. ZOE JORDENS • *Oxford Public Health Laboratory, John Radcliffe Hospital, Oxford, UK*

MARY ELIZABETH KAUFMANN • *Laboratory of Hospital Infection, Central Public Health Laboratory, London, UK*

KEITH P. KLUGMAN • *MRC Pneumococcal Disease Research Unit, South African Institute for Medical Research and the University of Witwatersand, Johannesburg, South Africa*

MARTIN C. J. MAIDEN • *Division of Bacteriology, National Institute for Biological Standards and Control, Hertfordshire, Hatfield, UK*

PAUL MANN • *Schering-Plough Research Institute, Kenilworth, NJ*

JUAN MARTINEZ • *Center for Vaccine Development, University of Maryland, School of Medicine, Baltimore, MD*

MICHAEL MCCLELLAND • *Sidney Kimmel Cancer Center, San Diego, CA*

ROBERT E. MCLAUGHLIN • *Department of Microbiology and Immunology, University of Oklahoma Health Science Center, Oklahoma City, OK*

GEORGE H. MILLER • *Schering-Plough Research Institute, Kenilworth, NJ*

STEPHEN A. MORSE • *Division of AIDS, STD and TB Laboratory Research, National Center for Infectious Diseases, Centers for Disease Control and Prevention, Atlanta, GA*

E. RICHARD MOXON • *Paediatric Infectious Diseases Group, Institute of Molecular Medicine, John Radcliffe Hospital, Oxford, UK*

BARBARA E. MURRAY • *Center for the Study of Emerging and Re-Emerging Pathogens, Division of Infectious Diseases, Departments of Internal Medicine, Microbiology, and Molecular Genetics, University of Texas Medical School, Houston, TX*

LINDA NAPLES • *Schering-Plough Research Institute, Kenilworth, NJ*

JAMES P. NATARO • *Center for Vaccine Development, University of Maryland, School of Medicine, Baltimore, MD*

ROBERT J. OWEN • *Laboratory of Enteric Pathogensity of Texas Medical School, Houston, TX*

DAVID J. PAYNE • *Anti-Infective Research (UP1345), SmithKline Beecham Pharmaceuticals, Collegeville, PA*

DAVID RALPH • *Urocor Inc., Oklahoma City, OK*

ANNE M. RIDLEY • *Laboratory of Enteric Pathogens, Central Public Health Laboratory, London, UK*

MARILYN C. ROBERTS • *Department of Pathobiology SC-38, School of Public Health and Community Medicine, University of Washington, Seattle, WA*

FRANK J. SABATELLI • *Schering-Plough Research Institute, Kenilworth, NJ*

NICK A. SAUNDERS • *Molecular Biology Unit, Central Public Health Laboratory, London, UK*

PREMA SEETULSINGH • *Department of Microbiology, Queen Mary's University Hospital, London , UK*

KAREN J. SHAW • *Schering-Plough Research Institute, Kenilworth, NJ*

KAVINDRA V. SINGH • *Center for the Study of Emerging and Re-Emerging Pathogens, Division of Infectious Diseases, Department of Internal Medicine, University of Texas Medical School, Houston, TX*

ANTHONY M. SMITH • *MRC Pneumococcal Disease Research Unit, South African Institute for Medical Research and the University of Witwatersand, Johannesburg, South Africa*

JILL M. STIGTER • *Department of Biochemistry, University of Cambridge, Cambridge, UK*

MARC J. STRUELENS • *Unite díEpidemiologie et díHygiene Hospitaliere, Laboratoire de Microbiologie, Hospital Erasma, Bruxelles, Belgium*

DAVID TAYLOR-ROBINSON • *MRC STD Research Group, Jefferiss Trust Laboratories, St Mary's Hospital Medical School, London, UK*

CHRISTOPHER J. THOMSON • *Department of Medical Microbiology, University of Edinburgh Medical School, Edinburgh, UK*

E. JOHN THRELFALL • *Laboratory of Enteric Pathogens, Central Public Health Laboratory, London, UK*

PATRICIA A. TOTTEN • *Division of Infectious Diseases, University of Washington, Harborview Medical Center, Seattle, WA*

DAVID L. TREES • *Division of AIDS, STD and TB Laboratory Research, National Center for Infectious Diseases, Centers for Disease Control and Prevention, Atlanta, GA*

NEIL WOODFORD • *Antibiotic Reference Unit, Laboratory of Hospital Infection, Central Public Health Laboratory, London, UK*

DOUGLAS B. YOUNG • *Imperial College School of Medicine, St. Mary's Campus, London, UK*

1

Impact of Molecular Methods on Clinical Bacteriology

Robert C. George

1. Introduction

The impact of molecular (nucleic acid-based) methods on the basic science of medical microbiology is undeniable. Indeed, microbiologists have been at the forefront of the molecular biology revolution that has had such a dramatic effect on our understanding of biological science. Although the foregoing is indisputable, have these techniques yet found an appropriate, cost-effective, and quality-assured place in the clinical bacteriology laboratory? Are patients and the infections from which they may be suffering managed more effectively and efficiently through the application of molecular methods? This introduction seeks to explore these issues from the perspective of a clinical bacteriologist. Detailed theoretical and practical guidance on the application of these techniques to the diagnosis, management, and epidemiology of a wide range of infections is provided in the succeeding chapters.

In very broad terms, the functions of a clinical bacteriology laboratory are twofold: first, the examination of biological samples (and the organisms isolated from, or detected in them), to determine the etiological diagnosis, specific treatment, and control of bacterial infections; and second, the formulation of specific and general advice, guidance, and policy for the management, control, and prevention of bacterial infections in individuals and communities. In considering the impact of molecular methods on clinical bacteriology, this chapter will concentrate more on the former than the latter functions of the diagnostic laboratory. However, new insights provided by these novel technologies—in particular, rapid and simple methods for microbial "fingerprinting" and/or detection of particular antimicrobial resistance or virulence genes for epidemiologic purposes—may be expected to have a significant impact on infection control policies and their implementation.

From: *Methods in Molecular Medicine, Vol. 15: Molecular Bacteriology: Protocols and Clinical Applications*
Edited by: N. Woodford and A. P. Johnson © Humana Press Inc., Totowa, NJ

As noted above, the impact of molecular methods on our understanding of the basic science of clinical bacteriology has been significant and will become increasingly so. However, at present many of these techniques are of greater relevance and use to the reference or other specialist laboratory than to the clinical laboratory. Some significant exceptions to this general statement are the following: laboratories serving tertiary referral hospitals that with their particular patient populations and clinical specialities, can make cost-effective use of molecular methods in diagnosis, therapy, and epidemiology, and, of course, clinical virology laboratories. The latter have embraced rapid molecular diagnostic technologies far more speedily and comprehensively than their bacteriology counterparts. There are several reasons for this; in particular, the specialized, skill-dependent, and retrospective nature of many conventional virological methods and the increasing number of viral infections amenable to specific antiviral or immunomodulation therapy. As a consequence, rapid, sensitive, and specific viral diagnostic and therapeutic monitoring methods are required for the effective use of these therapies. It is noteworthy that commercial suppliers of molecular diagnostics have targeted this market far more aggressively and successfully than clinical bacteriology.

2. Areas of Potential Impact on Clinical Bacteriology

The molecular methods actually or potentially applicable in clinical bacteriology laboratories include the polymerase chain reaction (PCR) or other DNA amplification techniques and/or gene probing methods for the identification of bacteria and specific virulence or antimicrobial resistance genes (either in cultures isolated by conventional methods or directly in clinical material), and genomic analysis by one or more of a range of techniques for bacterial "fingerprinting" and typing for epidemiologic purposes.

The uptake and impact of molecular methods will, in part, be dictated by the clinical necessity or epidemiologic requirement for a truly rapid or otherwise unachievable result and the implication of that result for the individual patient and health-care staff. For certain infections, particularly those acquired in hospitals or those of wide and general public health significance, a positive result may have widespread ramifications. Increased speed and sensitivity in achieving that result—whether it is an etiological diagnosis, the detection of a specific virulence determinant or antimicrobial resistance gene(s), or the definition of the degree of relatedness of isolates from episodes of presumed hospital or community crossinfection—allows the implementation of appropriate therapeutic and control measures more rapidly than might otherwise be possible. It is in these areas of clinical bacteriology that molecular methods may be expected to have the greatest impact on medical practice.

2.1. Impact on Laboratory Methods
for Diagnosis and Pathogen Identification

For the vast majority of common bacteriologic investigations undertaken in clinical laboratories on samples from immunocompetent individuals, biological amplification by overnight culture using simple agar or broth media is the method of choice and is likely to remain so. Notable exceptions include slow-growing or difficult-to-culture organisms (e.g., mycobacteria and chlamydiae) and infections in the immunocompromised, for which diagnostic accuracy and speed are essential and can be lifesaving.

The greatest scope for widespread application of molecular methods in routine bacteriology is in the further examination and identification of agar-grown pure cultures. The last 20 years have seen an ever-increasing acceptance and use in the clinical bacteriology laboratory of a wide range of commercially produced test "kits" for these purpose. Such kits have simplified and standardized phenotypic testing. It is therefore likely that conveniently packaged, competitively priced, and quality-assured DNA-based identification and other test systems will find a ready market.

2.1.1. Identification and Characterization of Isolated Bacteria

In essence, these will be new ways of doing old tests on agar-grown pure cultures. Speciation by DNA amplification methods and/or gene probing may replace biochemical or other phenotypic identification procedures. For certain organisms, in addition to speciation, it is also necessary to determine their pathogenic potential by demonstrating the presence or absence of certain factors (e.g., diphtheria toxin in isolates of *Corynebacterium diphtheriae*). A positive result will substantiate the diagnosis and may define the course of clinical and epidemiologic management. In such circumstances, speed of detection may be very important and molecular methods have much to offer over conventional phenotypic tests.

As such, tests will be undertaken with large amounts of target DNA obtained from bacterial colonies. Crosscontamination of reagents and equipment are of perhaps slightly less concern than for the application of molecular diagnostic methods, such as PCR, directly to clinical samples where target DNA may be present in vanishingly small amounts. Laboratory managers who have to ensure quality assurance and control of all aspects of the work undertaken will almost certainly wish to use commercial kits with built-in controls and validation steps. Determining factors in any widespread successful application of these methods will be the total costs of reagents, dedicated equipment, and facilities, as well as the training, skill base, and number of staff required to operate them. Clearly, there is ample scope for cost-beneficial automation of such test systems with colorimetric, fluorimetric, or other machine-readable endpoints. For the majority of potential applications in this general sphere of activity, speed of testing is perhaps slightly less relevant, because many conventional phenotypic test kits for bacterial identification and characterization already give same-day results.

2.1.2. Detection of Pathogens in Clinical Samples

Molecular methods offer the promise of rapid and direct detection of bacterial pathogens in clinical material and, for a few infections, this promise has begun to be realized. Researchers, both in the commercial and public sectors, have concentrated their attention on slow-growing or difficult-to-culture organisms, such as *Mycobacterium tuberculosis* and *Chlamydia trachomatis*. Semiautomated commercial systems utilizing DNA amplification are available for the diagnosis of these latter infections and are increasingly utilized. The advantages, particularly in speed of diagnosis over conventional culture methods for slow-growing and difficult-to-culture organisms, are obvious and offer new opportunities for early clinical and epidemiologic interventions in the management of both individual patients and communities. However, rapid microbial evolution in response to ecological pressures, such as antibiotic use and advances in medical care, is occurring constantly in organisms of relevance to clinical bacteriology. Therefore, it is difficult to envisage whether a non-culture method will ever provide the same actual or potential information as a bacterial isolate.

Any relevant literature search on this subject will reveal numerous publications. However, a close analysis reveals that many of these published studies are technical evaluations of the potential of these methods, using artificially "spiked" samples or retrospective analyses rather than real-time, clinical outcome-based studies. As a consequence, and with certain specific exceptions, considerably more work is required before such techniques are likely to replace conventional methods. As always in consideration of any new diagnostic method, issues of sensitivity and specificity are paramount and, if nonculture molecular methods are to replace rather than complement standard culture techniques, they will need to be at least as sensitive and specific. Sensitivity, which can usually be improved through various technical manipulations of the sample(s) and test conditions, is ultimately unlikely to be a limiting factor. Specificity is rather more problematic and a recently published example of misdiagnosis by PCR of cerebral nocardia infection in a renal transplant patient with suspected cerebral toxoplasmosis is illustrative *(1)*. Primers for the P30 gene of *Toxoplasma gondii* as target gave positive results with material from a cerebral abscess, apparently confirming the clinical diagnosis. However, conventional culture of the abscess material revealed *Nocardia asteroides* and subsequent PCR with the *T. gondii* P30 gene primers, and DNA from the

N. asteroides yielded an amplicon of the expected size; an example of unrelated, but clinically significant, crossreactivity leading to misdiagnosis. A subsequent publication *(2)* by the originators of the *T. gondii* P30 gene PCR pointed out that the primer sequences for this gene were unique to *T. gondii* according to published data at the time of their original publication in 1990. Furthermore, crossreactivity studies also showed amplicons of the expected size with *Plasmodium* spp. and *M. tuberculosis*, in addition to *N. asteroides*. It is self-evident that primers can only be selected for specificity according to what is published at the time of primer selection and that this body of knowledge is expanding at an exponential rate. Therefore, in addition to crossreactivity and specificity studies with species related to the target pathogen(s), the originators of new molecular diagnostic tests must consider organisms likely to be found in the same or similar anatomical sites and clinical conditions.

An important and expanding application of molecular methods is in the diagnosis of partially treated infections in which conventional culture has been compromised by prior antimicrobial therapy (e.g., meningococcal meningitis treated with penicillin in advance of hospital admission and diagnostic sample collection). Confirmation of a clinical diagnosis of meningococcal infection by, for example, PCR is desirable both for the individual patient and also in view of the public health control measures that are necessary to prevent further cases. Similarly, DNA amplification techniques offer considerable promise in the diagnosis of pneumococcal infections, for which conventional culture is often negative in patients presenting to hospitals after partial treatment in the community. In both of these examples, molecular methods offer not just a rapid diagnosis, but also the potential to make a specific etiological diagnosis that would not otherwise be possible. As new vaccines are developed and used widely it will become increasingly important to diagnose meningococcal and pneumococcal infections accurately and specifically, in order to define and characterize anticipated changes in their epidemiology.

Another important area for consideration in the design and application of molecular methods to primary diagnostic specimens is the type of sample being examined. The detection of single pathogens in normally sterile site specimens of relatively standard composition (e.g., pneumococci or meningococci in blood or cerebrospinal fluid; CSF) presents fewer technical and specificity problems than searching for evidence of one or more of several potential pathogens in complex and variable nonsterile site samples (e.g., *Legionella pneumophila* or *Mycoplasma pneumoniae* in sputum). All diagnostic DNA amplification methods applied directly to clinical samples should include internal controls for each sample to ensure that inhibition of amplification, which could result in false-negative results or significant reduction in sensitivity, is detected. In addition, and in contrast to the application of molecular methods to pure cultures of conventionally isolated bacteria, the preparation and handling of samples, reagents, and equipment (as well as the rooms in which the work is undertaken) must be rigorously monitored and controlled to avoid crosscontamination, which may result in false-positive results. These considerations will add significantly to the overall costs of such methods as applied to diagnostic samples and, in the short to medium term, will favor the use of commercial and quality-assured test systems and protocols in a small number of dedicated or specialist laboratories as opposed to widespread use in all diagnostic bacteriology laboratories. Careful consideration of the total costs of these methods in comparison with the clinical and public health benefit resulting from them is required, and this will vary according to the particular infection to be diagnosed and local arrangements for specimen transport and processing.

2.2. Impact on Therapy

Antimicrobial susceptibility testing of bacteria isolated from clinical specimens is one of the primary functions of the clinical bacteriology laboratory. Quality assurance and control are imperative because the consequences of errors in such tests (e.g., a resistant organism being incorrectly reported as susceptible) may be devastating for the patient, in the worst possible scenario resulting in an avoidable death. In addition, the reputation of the laboratory with its user community and the trust placed in its work may be seriously compromised if such errors occur.

For organisms and antimicrobials where the genetic basis of resistance is clearly understood and the underlying DNA sequences are known, molecular methods may be used to detect and/or amplify specific antibiotic resistance genes as a supplement or alternative to conventional susceptibility testing methods. Speed of reporting is often, but not always, critical, and, as for primary diagnostic work, certain slow-growing, difficult-to-culture, or technically demanding organism/antimicrobial combinations may be more suited to molecular methodologies than others. In addition, such techniques offer new opportunities for the study of structure–function relationships in antimicrobial action (and consequently for the design of new antimicrobials), because DNA encoding both antimicrobial susceptible and resistant targets may be amplified for sequencing and further study.

Unfortunately, the selective pressure exerted by the widespread use of antimicrobials in both human and veterinary medicine results in the rapid emergence of novel resistance determinants and continuing evolution of existing ones. A variety of naturally occurring genetic transfer mechanisms ensures that under the appropriate selection, pressure resistance determinants will disseminate to other species and genera. Occasionally, entirely new and unexpected resistance mechanisms emerge, the most notable in recent years being vancomycin resistance in previously susceptible Gram-positive species. Because the application of DNA amplification and/or probing technology to antibiotic resistance gene detection can (on the whole) only seek for what is already known and novel resistance mechanisms will arise continuously under the selective pressure of antibiotic use, there will be a continued requirement for phenotypic testing. Conventional antimicrobial susceptibility testing methods will therefore need to be maintained in addition to molecular methods, thus leading to considerable duplication of effort and additional expense.

It must always be borne in mind that the vast majority of all antimicrobial prescribing is undertaken on a best-guess basis, in the absence of a relevant laboratory report at the time of prescription. It is, and will remain, one of the principal functions of the diagnostic laboratory to ensure that the clinicians it serves are aware of the likely pathogens in particular clinical conditions and of the expected antimicrobial susceptibility of those pathogens. Such data, which must be updated and disseminated regularly, will be based on locally gathered and analyzed surveillance data, irrespective of the methodologies used to undertake the susceptibility tests.

For the majority of infections, surrogate susceptibility testing by detection of antimicrobial resistance genes should be regarded more as a rapid screening test, allowing earlier modification of empirically prescribed therapy and/or institution of infection control interventions than might otherwise be possible, as opposed to a definitive result. In some clinical and epidemiologic situations, such rapid results are highly desirable (e.g., infections caused by methicillin-resistant *Staphylococcus aureus* [MRSA] and *M. tuberculosis*). These two examples are explored below as models that may be useful in consideration of the wider application of molecular methods to susceptibility testing.

2.2.1. Detection of Methicillin Resistance in Staphylococcus aureus

MRSA is probably the major nosocomial infection problem of the current decade and is often resistant to several other relevant antimicrobials in addition to methicillin. The increased prevalence of methicillin and multiply resistant *S. aureus* strains in hospitals compromises empirical and definitive therapeutic options for presumed or proven staphylococcal sepsis, leading to an increased use of glycopeptide antimicrobials (such as vancomycin) that, at present, are the only agents for which susceptibility may be assumed. Screening of patients and staff for MRSA carriage and colonization is undertaken regularly in addition to diagnostic sampling in individual patients. Depending on the local prevalence of MRSA and infection control policies and procedures, detection of an MRSA may have very significant consequences for management of the individual patient (treatment, side room isolation, and barring of transfer to other units and hospitals), staff members (treatment to eradicate carriage/colonization, transfer to other duties, temporary lay-off), and the hospital (ward or unit closure to new admissions, cohort nursing, enhancement of infection control activities). As a consequence, it is important to get the right result the first and every time, and as rapidly and reliably as possible. Unfortunately, and in contrast to many other organism/antimicrobial combinations, detection of methicillin resistance in *S. aureus* by conventional methods is methodologically complex. Indeed, there is a great deal of literature devoted to variables affecting its detection by conventional in vitro methods. Some recommendations suggest incubating plates for up to 48 h before reporting a final result, and no single conventional susceptibility testing procedure can be relied on to yield 100% accuracy. In contrast to the complexities of phenotypic testing, molecular methods for MRSA, in particular DNA amplification, are directed at detection of a single target gene, *mecA*, that encodes the production of penicillin-binding protein 2' (PBP 2'), which seems to be a requirement for methicillin resistance in most staphylococci. Thus, in this example, detection of the presence or absence of the *mecA* gene by DNA amplification technology offers the possibility of same day, unequivocal results with 100% positive and negative predictive values. Even for the diehard septic this has to be considered a major advance and with the positive spinoffs for individual patients (treatment options), infection control staff (more rapid institution of appropriate infection control measures), and hospital managers (infection control costs and consequences and pharmacy budgets) should find wide acceptance. Simple, inexpensive kit tests for detection of *mecA* in conventionally isolated cultures with built-in QA/QC would find very wide application in diagnostic laboratories. Methicillin resistance in coagulase-negative staphylococci is very common and may also be mediated by *mecA*. Consequently, direct detection of MRSA in clinical or screening samples (that may also contain methicillin-resistant coagulase-negative staphylococci) in advance of conventional culture results presents a number of additional technical problems. However, the clinical and infection control advantages of such a technology becoming available suggest that these difficulties can and will be overcome.

2.2.2. Detection of Antibiotic Resistance Genes in M. tuberculosis

Despite recent advances and increasing use of radiometric or other monitoring methods with commercial broth culture-based systems, conventional culture and antimicrobial susceptibility testing of *M. tuberculosis* from clinical material remains a relatively slow process. In the absence of specialized test systems for mycobacterial culture, it can still take 6–12 wk (sometimes longer) from specimen collection to final culture and susceptibility test results. In a period of increased prevalence of tuberculosis and increasing concern over drug-resistant strains this is not satisfactory.

Although antituberculous chemotherapy will be prescribed on the basis of clinical and/or early microbiological findings and may be specifically tailored to take account of the suspicion of drug resistance on clinical or epidemiologic grounds, definitive therapy must await susceptibility testing results. Definitive therapy should be curative and, importantly, will reduce infectivity for those in contact with the patient. Ineffective or suboptimal therapy will prolong infectivity for others and, if the infecting strain is only susceptible to one of three or four drugs prescribed, will, in effect, be monotherapy. In such circumstances, there is a significant chance that the infecting strain will develop resistance to the previously effective antimicrobial. Patients suffering from infectious, multidrug-resistant TB may be isolated more strictly, for longer, and additional infection control precautions for health-care staff and others in contact with the patients are usually advised. A further complicating factor is the toxicity of many antitubercular antimicrobials; the possible inappropriate prescription of a more toxic agent on suspicion of drug resistance may expose the patient to an unnecessarily increased incidence of adverse reactions, some of which can be quite severe.

From the foregoing, it may be seen that rapid detection of antimicrobial resistance in *M. tuberculosis* using DNA amplification and amplicon analysis has much to offer to the individual patient, their actual and potential contacts, and health-care attendants, as well as those charged with the implementation of relevant public health interventions. In addition, specific resistance genes may be used as epidemiologic markers in the investigation of the spread of particular strains. The costs-to-benefit ratio of achieving resistance detection perhaps weeks to months in advance of conventional methods is strongly weighted in favor of molecular methods, particularly if they can be applied directly to clinical material. As noted above for other pathogens, it is unlikely that molecular determination of antimicrobial resistance in *M. tuberculosis* will ever completely substitute for phenotypic testing, because novel and previously uncharacterized resistance determinants will arise and must be detected. Rather, these techniques offer real scope for rapid screening for resistance with resulting improvements in the control and prevention of TB and limitation of further resistance development.

The two examples discussed above consider the impact of molecular methods on therapy for infections and/or antimicrobials for which conventional methods leave much to be desired. The rapid results made possible by these technologies can make a real difference to the care of patients and the control and prevention of infections. As new and user-friendly technologies are developed for resistance screening and detection, it is likely that similar approaches for other organisms and antimicrobials will become increasingly cost-effective and find a place in the diagnostic laboratory.

2.3. Impact on Epidemiology and Control Measures

Epidemiologic typing of bacteria isolated from infected or colonized humans or animals and environmental or food sources has long been the mainstay of our understanding of the sources, routes, and modes of transmission of bacterial infections of clinical and public health importance. The information gained has advanced the study of the epidemiology of many bacterial infections, permitted an understanding of outbreaks, and allowed the design of rational and targeted interventions to prevent or limit transmission.

Over the past decade, there has been an explosion of molecular methods for the intraspecies discrimination of bacterial isolates, so-called "molecular typing." The term "molecular fingerprinting" is perhaps more appropriate because, to date, internationally accepted and meaningful "molecular type" designations have remained elusive. A large part of this volume is devoted to molecular methods for "typing" and "fingerprinting" and the introduction does not seek to explore the pros and cons of any particular method, but rather aims to consider the context(s) in which the methods are, or may be, used and how this will impact on clinical bacteriology.

Traditionally, the definition of "types" of particular microorganisms by serologic classification of variable surface antigens (protein and polysaccharide-based serotyping) or susceptibility to lysis by particular bacteriophages (phage-typing) has been undertaken by reference or specialist laboratories that usually produce the relevant reagents "in house." Such production is expensive, may require the use of animals to raise specific antisera, and requires considerable experience and expertise, as well as extensive quality control in both production and use. With a few exceptions, most such reagents are not commercially available, limiting the application of conventional typing techniques to a small number of centers, often a single national reference laboratory in each country and sometimes not even that. Such laboratories are usually members of formal or informal international networks undertaking similar work, and in consequence, interlaboratory and international standardization and definition of "types" by exchange of strains and reagents is the rule rather than the exception. This is particularly important if the type antigen defined by a particular serotyping scheme is, or is likely to be, a component of a vaccine for widespread use or the techniques are being used to study the epidemiology of an international outbreak.

Bacteriologists in diagnostic laboratories seeking to discriminate among isolates of the same species for epidemiologic, outbreak investigation, and/or infection control purposes have traditionally referred such isolates for further investigation by classical typing techniques to the relevant reference or specialist laboratory. There are advantages and disadvantages to the traditional approach to microbial typing. On the "plus side" are inter- and intralaboratory consensus on what constitutes a "type" designation, an ability to identify widely dispersed, but low-level common source outbreaks if indistinguishable "types" are referred from geographically dispersed laboratories, and an overview of the dissemination of particular strain types between different hospitals and communities at regional, national, and international level. Timely communication of typing results and associated epidemologic information to both the originator and the wider customer base is essential. Disadvantages include the following:

1. The costs of maintaining and quality assuring classical typing schemes and associated reagent production.
2. In many countries an increasing requirement to limit the use of laboratory animals wherever possible (in this instance for antiserum production).
3. The extremely skill-dependent nature of many of the classical schemes not least in the interpretation of the results obtained.
4. The emergence or recognition of "new pathogens" for which development from scratch of classical sero- or phage-typing schemes is not a realistic or cost-effective option.
5. The slow response time of classical schemes to the emergence of new "types," resulting in nontypability unless significant time and expense is invested in defining the new "type."
6. An increasing recognition that classical typing schemes may not be sufficiently discriminatory for meaningful epidemiologic analysis.

For these reasons, many reference and specialist laboratories have enthusiastically embraced the new technologies, often using them in a hierarchical manner to subdivide classical types, or to devise and apply new schemes for previously "untypable" organisms.

With a relatively small investment in reagents, equipment, and staff training, molecular fingerprinting of bacterial isolates by any one of a wide range of techniques is now within the reach of almost any reasonably well-resourced laboratory. The widespread availability of microcomputer software packages for the analysis of molecular fingerprint banding patterns on agarose gels, photographs, or blots has considerably simplified the interpretation of the results obtained from such studies. However, it should be noted that virtually all such software packages still require training and experience for scientifically valid and efficient use. On occasion, it may be possible to produce relevant results more rapidly if such tests are undertaken locally rather than referring the isolates to a specialist laboratory. However, this will by no means always be the case, particularly if each new investigation requires reallocation of personnel and resources. The increasing availability and use of these new technologies has resulted in a huge increase in the number of publications describing their use. Not all such publications address with sufficient rigor the issues of reproducibility, standardization, discriminatory power, and clinical and epidemiologic significance of the molecular fingerprinting results obtained.

Should clinical bacteriologists invest staff time and other resources in developing these methods within their own laboratories? The answer to this question will vary from place to place and organism to organism, and in large part will depend on the current provision of, and financial arrangements for, regional or national reference/specialist typing laboratories. Where such reference provision is freely and speedily available there would seem to be little point, but where it is lacking, or restricted, or too slow to be clinically or epidemiologically relevant then local molecular fingerprinting to answer local questions has much to recommend it. It should be remembered that where appropriate reference facilities exist they usually encompass a particular and specialized knowledge and skill base for the organism or infection in question and an accumulation of expertise and experience in the investigation of outbreaks and incidents. This knowledge and expertise is an important but difficult to quantify aspect of reference work. Although bacterial molecular fingerprinting techniques can be undertaken in almost any laboratory, a sufficient level of organism, infection control, and epidemiologic-based knowledge to extract the maximum benefit from these new techniques may not be so widely available.

In summary, the positive impact of molecular methods on epidemiology and control measures for bacterial infections has been significant and is likely to become increasingly so. Unfortunately, to date we have not achieved definition of inter- and intralaboratory agreed molecular "types," and in consequence we will need to continue to deal in comparative fingerprinting as opposed to true typing. If an ever-increasing range of molecular fingerprinting techniques is applied in an ever larger number of different laboratories, then there is a distinct possibility of increasing fragmentation and mutual incomprehension in discussion and provision of molecular typing and fingerprinting data. This is of particular concern in considering the investigation of multicenter or international outbreaks. In the absence of defined and agreed molecular "types," this method-driven potential for confusion could result in an extremely negative impact for epidemiology and infection control. Fortunately, a number of groups at national and international levels are seeking to define rule sets and agree methodologies to avoid this potentially untoward and dangerous outcome for these new approaches to the characterization of bacterial isolates.

3. Quality Assurance and Quality Control

All those working in clinical microbiology laboratories will be aware of an increasing requirement for demonstrable quality indicators, as well as in many countries, strict accreditation requirements. Molecular microbiological methods are currently in a transition phase between research and routine diagnostic applications, and in consequence techniques that have performed well in the dedicated hands of researchers are now being undertaken much more widely. Quality issues may not have been considered in sufficient detail by all those seeking to introduce molecular diagnostic technology into routine use. For example, a MEDLINE search undertaken in May 1996 using the search term "PCR" produced 33,584 citations, but when the search terms "quality control" and/or "quality assurance" were added, the number of citations decreased to 43 (0.1%) and only a few of these were related to microbiological topics. Considering specific areas within diagnostic microbiology, MEDLINE found 294 publications on "*M. tuberculosis* and PCR" between 1992 and May 1996, declining to one when the search terms "quality control" and/or "quality assurance" were added. In this one publication, Noordhoek et al. reported on results obtained by seven different laboratories that examined 200 BCG seeded samples; the false-positivity rate was between 3 and 22% with one value of 77%, and the sensitivity varied widely between participants. For example, a positive result was reported for between 2 and 90% of samples with 10^3 mycobacteria *(3)*.

As noted earlier in this chapter, clinical virology laboratories have been undertaking diagnostic PCRs for longer than most clinical bacteriology laboratories and the number of publications on quality issues is greater. Unfortunately, the results of external quality assurance exercises have not been too encouraging. For example, in the detection of hepatitis B virus DNA by amplification methods, Quint et al. sent coded samples to 39 different laboratories; these generated 43 data sets, of which 19 data sets (44.2%) had false-negative and/or false-positive results *(4)*. In a similar exercise examining the detection of HIV DNA, seven laboratories following their own protocols examined two sets of 20 coded samples; false-positives and false-negatives were noted in all of the laboratories and the concordance with HIV serology ranged from 40 to 100% *(5)*. Direct examination of clinical samples presents the greatest quality challenge for molecular methods because using these techniques to study further pure cultures or undertake "fingerprinting" utilizes large amounts of readily available target DNA and the opportunities for error are thus reduced.

It is quite clear that much remains to be done in improving the overall quality and reliability of diagnosis by DNA amplification from clinical material in routine laboratories. The levels of false-positivity and possibly low sensitivity noted in published studies are not acceptable for routine diagnostic use. However, many of the problems and issues are now considerably better understood than they were even a few years ago and remedies and solutions are available for the majority. Physical separation of the various stages of examination of samples is clearly important to avoid crosscontamination of reagents and equipment. The minimum requirement must be for separate rooms for specimen handling and nucleic acid extraction, preparation of amplification "mixes" and addition of extracts and amplification, and analysis of amplicons. Careful consideration should be given to the inclusion of internal controls for both extraction and amplification. After nucleic acid extraction from clinical material, it is prudent to examine the extract by PCR for the human mitochondrial cytochrome gene or similar human DNA. A positive result will ensure that the original sampling was adequate and that nucleic acid in an amplifiable form has been extracted from the sample. In addition, a synthetic nucleic acid construct should be included with the reaction mix. In the absence of inhibitors of amplification, this would be expected to react with primers directed at the target nucleic acid in order to produce an amplicon of defined size in addition to, and distinct from, that obtained if the sample contains the target organism. The use of such an internal control is mandatory for the reliable detection of negative results.

From the foregoing, it may be seen that a considerable capital and staff training investment is required in order to bring molecular diagnostic methods to a sufficient level of sensitivity, specificity, and reliability for routine use in clinical bacteriology. The increasing availability of commercial DNA amplification tests kits with built-in controls, strict protocols, and well-defined sensitivity and specificity will overcome many of the problems associated with "in house" tests; this seems to be the route that most diagnostic laboratories will follow.

4. Conclusions

Molecular methods have had an enormous impact on the basic science of clinical bacteriology and antibiotic resistance. Molecular "fingerprinting" methods have much to offer, but at present their application is not well coordinated leading to possible confusion in definitions of "strains" and "types." Routine diagnostic microbiology is on the threshold of a molecular revolution but, with some notable exceptions, remains at the door. Quality issues remain a major consideration, as do the total costs of DNA amplification approaches to bacteriological diagnosis. Genes will continue to evolve and consequently new and previously undefined questions will require answers. Thus, for most infections it will be necessary to continue phenotypic tests in parallel, leading to duplication of effort and cost. The clinical necessity for, and implications of, a positive result and confidence in a negative finding (whatever the test), together with the speed and reliability of conventional methods will define the place of molecular biological tests in the diagnostic bacteriology laboratory. For a few indications, the advantages are already obvious. For many more, this remains to be proven.

References

1. McHugh, T. D., Ramsay, A. R. C., James, E. A., Mognie, R., and Gillespie, S. H. (1995) Pitfalls of PCR: misdiagnosis of cerebral nocardia infection. *Lancet* **346,** 1436.
2. Holliman, R. E., Patel, B., Johnson, J. D., Mangan, J., and Savva, D. (1996) Pitfalls of PCR: misdiagnosis of cerebral nocardia infection. *Lancet* **347,** 335,336.
3. Noordhoek, G. T., Kolk, A. H. J., Bjune, G., Catty, D., Dale, J. W., Fine, P. E. M., et al. (1994) Sensitivity and specificity of PCR for detection of *Mycobacterium tuberculosis*: a blind comparison study among seven laboratories. *J. Clin. Microbiol.* **32,** 277–284.
4. Quint, W. G. V., Heijtink, R. A., Schirm, J., Gerlich, W. H., and Niesters, H. G. M. (1995) Reliability of methods for hepatitis B virus DNA detection. *J. Clin. Microbiol.* **33,** 225–228.
5. Defer, C., Agut, H., Garbarg-Chenon, A., Moncany, M., Morinet, F., Vignon, D., et al. (1992) Multicentre quality control of polymerase chain reaction for detection of HIV DNA. *AIDS* **6,** 659–663.

2

Genomic DNA Digestion and Ribotyping

J. Zoe Jordens

1. Introduction

Bacteria may need to be characterized for a number of reasons. Newly discovered organisms are characterized to determine their taxonomic position; clinical isolates are characterized to provide an indication of pathogenic potential and likely antibiotic susceptibility. Such studies generally involve characterization to the genus and species level and provide an identification for the organism. Highly discriminatory methods of intraspecies characterization are required for epidemiologic purposes in order to establish sources and routes of infection to which control measures may be directed. For an individual patient, differentiation between relapsing infection or reinfection may aid treatment choice and clinical management. The pattern of DNA fragments produced by digestion of genomic DNA with a restriction endonuclease, directly or after hybridization, can be highly discriminatory and enable strain characterization, or typing, suitable for epidemiologic studies. The practice of obtaining these patterns (or profiles) for epidemiologic purposes is the subject of the present chapter, although the methods can also be applied to questions regarding identification at higher taxonomic levels.

It is appropriate to characterize and compare the DNA of bacteria as this molecule determines phenotype and hence, behavior. As it is not, at present, practically feasible to sequence and compare entire genomes, so techniques are employed that examine parts of the genome that are thought to be representative of the genome as a whole. The patterns resulting directly from restriction endonuclease digestion of genomic DNA (genomic DNA digestion, GDD) can be compared between organisms. With the choice of an enzyme appropriate for the species under investigation, profiles that are easy to read visually (i.e., that consist of ≤20 well-separated fragments) can be obtained with simple

From: *Methods in Molecular Medicine, Vol. 15: Molecular Bacteriology: Protocols and Clinical Applications*
Edited by: N. Woodford and A. P. Johnson © Humana Press Inc., Totowa, NJ

agarose gel electrophoresis using widely available equipment. The actual enzyme chosen is often determined empirically, but potentially useful ones can be mathematically predicted *(1)*. This method of GDD detects the distribution of restriction sites around a genome, but generally analyzes only a small proportion of the genome. More of the genome can be analyzed by pulsed-field gel electrophoresis (PFGE) of large fragments generated by digestion with restriction endonucleases that cut the DNA less frequently (*see* Chapter 3). With these techniques, fragments of similar size are assumed to represent identical parts of the genome, although this may not be the case, and the presence of plasmids may complicate the patterns. The use of good quality DNA and standardization of the quantities digested are essential for pattern clarity. GDD has been successfully used to characterize strains from a variety of species *(2)*, including those associated with outbreaks of hospital-acquired infection.

Some of the problems associated with GDD (e.g., too many fragments and unknown DNA content of the fragments) can be overcome with the use of probes for specific genes. To obtain patterns consisting of multiple bands and representing as much of the genome as possible, targets should ideally be present in multiple copies dispersed around the genome. For many bacterial species, suitable targets include ribosomal RNA (rRNA) operons (*rrn;* **Fig. 1**) or insertion sequences. The patterns of restriction fragments hybridizing with a probe for *rrn* are commonly termed ribotyping patterns. Probe-based methods such as ribotyping, detect variation within and flanking the probe sequence *(3)*. The choice of restriction endonuclease for ribotyping is again determined empirically or based on the expected frequency of cutting *(1)*. Various probes have been employed for ribotyping; ones containing as much of the operon as possible will detect the most variation *(3)*. 16S and 23S rRNA from *Escherichia coli* is widely used as a universal probe for many species *(4)*. Ribotyping has been successfully applied to the investigation of a range of hospital-acquired infections *(5)*.

For both GDD and ribotyping, the most appropriate enzyme(s) should be determined with a selection of epidemiologically unrelated strains shown to be different by established typing methods (preferably together with a few epidemiologically unrelated strains with similar characteristics). Once the appropriate enzyme(s) (and probe) have been determined, isolates of interest from putative outbreaks of infection may be characterized. All such investigations should include a selection of epidemiologically unrelated strains with similar characteristics (e.g., serotype, antibiotic susceptibility pattern, biotype, and so forth) to those under investigation, for comparison, in order to exclude the possibility of indistinguishable outbreak-associated isolates belonging to a common type (or clone) rather than representing recent spread of a single strain. Only if all these "controls" are distinguishable, can indis-

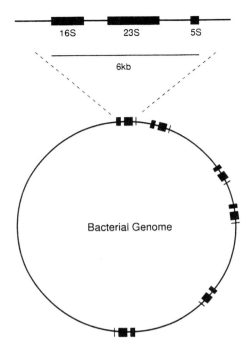

Fig. 1. Schematic representation of a ribosomal RNA operon *(rrn)* showing the relative positions of the genes encoding 16S, 23S, and 5S rRNA (top) and an example of a genome containing 6 *rrn* dispersed around the genome (bottom). The position of restriction sites (for the endonucleases used in ribotyping) within and flanking each copy determines the ribotyping pattern. Variation in these sites enable strains to be differentiated.

tinguishable patterns be interpreted to indicate that isolates represent a single strain and, hence, indicate an outbreak. Indistinguishable patterns obtained consistently with different characterization methods (that examine different and unrelated parts of the genome) provide evidence for a single strain. Knowledge of the population genetics structure of the species under investigation is, therefore, essential for appropriate interpretation of results. In order to check pattern reproducibility and aid intergel comparisons, one strain should be included in all experiments in addition to a molecular size standard. There are, at present, no agreed criteria for interpreting patterns, although many workers use differences in two or more bands to define different strains. Criteria for interpreting differences in patterns from PFGE of macrorestriction fragments of genomic DNA have been advocated such that differences in the positions of more than three bands probably indicate different strains *(6)* (*see* Chapter 3); this may be applicable to patterns produced

Extract DNA

↓

digest with restriction endonuclease

↓

separate fragments by agarose gel electrophoresis

↓   Stain & photograph
     = Genomic DNA Digestion

Transfer fragments to nylon membrane by
Southern blotting

↓

Hybridise with DNA probe labelled with digoxigenin

↓

Detect probe-target hybrids with
antibody-enzyme conjugate + substrate

↓

= Pattern of chromosomal DNA restriction fragments containing rRNA
genes (Ribotyping pattern)

Fig. 2. The experimental stages involved in obtaining patterns by GDD and ribotyping.

by GDD or ribotyping. However, the interpretation varies with the informativity of the system used, with fewer differences being significant if there is little heterogeneity. Estimates of strain relatedness for clustering purposes (i.e., population genetics studies) based on ribotyping have been shown to be discordant with those based on multilocus enzyme electrophoresis *(7)* and should therefore be avoided.

To obtain DNA profiles, GDD requires DNA extraction, digestion, and separation of resulting fragments by gel electrophoresis. In addition to these stages, ribotyping requires the subsequent steps of Southern blotting, hybridization with a probe, and detection of probe-target hybrids (**Fig. 2**). These methods are described in this chapter.

2. Materials

2.1. Rapid Extraction of Genomic (total cellular) DNA from Bacteria

1. Purity plate (heavy growth) of organism.
2. Tris-EDTA (TE) buffer: 10 mM Tris-HCl, pH 8.0, 1 mM Na$_2$EDTA (sterile).
3. Guanidium-EDTA-Sarkosyl (GES) lysing solution: 5 M guanidium thiocyanate, 100 mM EDTA, 0.5% v/v sarkosyl. Mix 60 g guanidium thiocyanate and 20 mL 0.5 M Na$_2$EDTA and leave at 65°C until dissolved, cool, then add 5 mL 10% v/v sarkosyl and make up to 100 mL with distilled water. Store at room temperature. *Note:* Guanidium thiocyanate is hazardous; gloves should be worn.
4. 7.5 M ammonium acetate (cold, +4°C).
5. Chloroform:isoamyl alcohol (24:1). *Note:* Isoamyl alcohol is harmful by inhalation.
6. Isopropanol (propan-2-ol) (cold, –20°C). *Note:* Skin contact should be avoided.
7. Absolute ethanol (cold, –20°C).
8. 70% ethanol.
9. Finely drawn-out pasteur pipets.

2.1.1. Determination of DNA Concentration and Quality by UV Spectroscopy

1. Ultraviolet (UV) spectrophotometer.
2. Matching quartz cuvets.
3. TE buffer (*see* **Subheading 2.1.**).

2.2. Restriction Endonuclease Digestion of DNA and Separation of Fragments by Agarose Gel Electrophoresis

2.2.1. DNA Digestion

1. Restriction endonuclease (5–10 U/µL).
2. 10X buffer (usually supplied with endonuclease).
3. Sterile distilled water (pharmacy or tissue culture quality).
4. Chromosomal DNA (of known concentration).

2.2.2. Agarose Gel Electrophoresis

1. Electrophoresis tank, gel frame, well-former, and tape as appropriate for system in use (e.g., DNA sub-cell GT electrophoresis cell, Bio-Rad, Hemel Hempstead, UK).
2. Power supply (for constant voltage in the range 20–150V).
3. UV transilluminator and camera. *Note:* Suitable eye and skin protection should be worn when working with UV.
4. Agarose: a molecular biology grade (e.g., UltraPure, Life Technologies, Paisley, UK).
5. Tris-borate-EDTA (TBE) buffer: 89 mM Tris, 89 mM borate, 2 mM Na$_2$EDTA. Stored as 10X stock solution (pH 8.0–8.5, not adjusted).
6. Digested DNA.
7. Marker: a 1 kb ladder (Life Technologies) or λ *Hin*dIII for genomic DNA digestion and digoxigenin-labeled λ *Hin*dIII (DIG marker II, Boehringer Mannheim, Germany) for ribotyping.

8. Loading buffer (6X): 0.25% bromophenol blue, 30% glycerol, 50 m*M* EDTA in 1X TBE (*see* **step 5**, above).
9. 1 µg/mL ethidium bromide (EB) solution. Store in a dark bottle as a 10 mg/mL stock solution. *Note:* EB is a mutagen and suspected carcinogen; avoid skin contact and inhalation.

2.3. Transfer of DNA to Nylon Membrane for Hybridization (Southern or Capillary Blotting)

1. Depurinating solution: 0.25 *M* HCl.
2. Denaturing solution: 1.5 *M* NaCl, 500 m*M* NaOH.
3. Neutralizing solution: 1 *M* Tris-HCl, pH 8.0, 1.5 *M* NaCl.
4. Transfer buffer: 10X SSC: Stored as 20X stock: 3 *M* NaCl, 300 m*M* tri-sodium citrate.
5. 2X SSC (*see* **step 4**, above).
6. Sandwich box.
7. Chromatography or filter paper: 1 sheet ~ 45 × 50 cm.
8. Glass or perspex plates.
9. Cling film or plastic wrap.
10. Nylon membrane (e.g., Hybond-N, Amersham, Little Chalfont, Bucks, UK).
11. Strong tissue paper or paper towels (or other flat, absorbent material) and newspapers.
12. Approximately 1 kg weight (e.g., a house brick or 1 L water).

2.4. Probe Preparation (Digoxigenin Labeling of cDNA from 16S + 23S rRNA)

1. Moloney murine leukaemia virus reverse transcriptase (M-MLV-RT), 200 U/µL (Life Technologies).
2. RT buffer (5X, supplied with M-MLV-RT).
3. Dithiothreitol (DTT, 10X, supplied with M-MLV-RT).
4. p(dN)$_6$ (random hexamer primers) (Boehringer Mannheim).
5. dNTPs: 5 m*M* in water, supplied as 100 m*M* stocks (Pharmacia Biotech, St. Albans, Herts, UK). Dilute each 1/20 and prepare a 10X ACGT mix with 10 µL dATP + 10 µL dCTP + 10 µL dGTP + 4 µL dTTP + 6 µL water.
6. Bovine serum albumin (BSA) 500 µg/mL (commercial stock solution is usually 10 mg/mL).
7. Digoxigenin-11-dUTP (Boehringer Mannheim).
8. RNase inhibitor (Boehringer Mannheim).
9. 500 m*M* EDTA.
10. 16S + 23S rRNA from *E. coli* (supplied at a concentration of 4 mg/mL, Boehringer Mannheim), diluted 1/10.

2.5. Hybridization

1. Hybridization oven, bottles, and meshes (e.g., Hybaid, Teddington, UK or Techne, Cambridge, UK) (*see* **Note 1**).
2. Blocking reagent (Boehringer Mannheim) 10% (w/v) in 100 m*M* maleic acid, 150 m*M* NaCl, pH 7.5, autoclaved (15 psi for 15 min) and stored at 4°C (*see* **Note 2**).

3. Prehybridization solution: 5X SSC, 0.1% (w/v) sarkosyl, 0.02% (w/v) SDS, 1% (w/v) blocking reagent (can be stored at −20°C). SSC is stored as a 20X stock solution (*see* **Subheading 2.3.**), SDS stored as 10% (w/v) stock, sarkosyl NL30 (BDH, Poole, Dorset, UK) is 30% (w/v). *Note:* SDS solid is an irritant; avoid eye and skin contact.
4. Probe (*see* **Subheading 3.4.**) (25 µL, approx 200 ng DNA).
5. Wash solution 1: 2X SSC; 0.1% (w/v) SDS (add SDS to diluted SSC or it will precipitate). Make up on day of use.
6. Wash solution 2: 0.1X SSC; 0.1% (w/v) SDS. Heat to hybridization temperature immediately prior to use.

2.6. Chemiluminescent Detection of Probe DNA-Target DNA Hybrids

1. Buffer 1: 100 mM Tris-HCl, 150 mM NaCl, pH 7.5. Stored as 10X stock.
2. Buffer 2: 1% (w/v) blocking reagent (*see* **Subheading 2.5.**) in buffer 1.
3. Buffer 3: 100 mM Tris-HCl, 100 mM NaCl, 50 mM MgCl$_2$, pH 9.5. Stored as two 10X stocks: 10X 3A: 1 M Tris-HCl, pH 9.5, 1 M NaCl, and 10X 3B: 500 mM MgCl$_2$.
4. TE buffer (*see* **Subheading 2.1.**).
5. Antibody-conjugate: antidigoxigenin-alkaline phosphatase (Fab fragments) (Boehringer Mannheim) diluted to 150 mU/mL (1:5000) in buffer 2. Diluted antibody-conjugate solutions are stable for only about 12 h at +4°C.
6. Chemiluminescent substrate: Lumigen PPD (Boehringer Mannheim) diluted 1:100 in buffer 3 (*see* **Note 3**).

3. Methods

Genomic DNA digestion and ribotyping require a series of consecutive, self-contained methods (**Fig. 2**) that can be used for other purposes. Therefore, each constituent method is given separately.

3.1. Rapid Extraction of Genomic (Total Cellular) DNA from Bacteria (Modified from ref. 8)

1. Using a dry sterile swab (or loop if preferred), harvest bacterial growth into 0.2 mL of TE buffer (*see* **Note 4**).
2. Add 0.45 mL of GES, mix by inversion, and leave at room temperature until lysed (approx 3 min) (*see* **Note 5**). If clearing occurs before this, store samples on ice.
3. Add 0.25 mL of cold 7.5 M ammonium acetate, mix by inversion, and store samples on ice for 10 min.
4. Add 0.25 mL of chloroform:isoamyl alcohol, mix thoroughly by hand (shake) for 3 min or until an even emulsion is formed.
5. Centrifuge for 20 min at 12,000g to separate the aqueous and organic phases.
6. Using a wide-bore tip, carefully transfer a measured volume (as much as possible without disturbing the interface—about 0.75–0.85 mL) of the upper aqueous layer to a fresh microcentrifuge tube.
7. Add 0.54X volume of cold isopropanol (i.e., 432 µL for 0.8 mL of supernatant), mix gently by inversion for 1 min or until DNA threads are visible (*see* **Note 6**).

8. Centrifuge at 12,000g for 2.5 min, then pour off the supernatant (*see* **Note 7**). Invert the open tubes on paper towel to drain off excess alcohol.

9. Add 0.5 mL of 70% ethanol, mix, centrifuge at 12,000g for 5 min and carefully discard the supernatant (*see* **Note 8**). Repeat twice more.

10. Briefly invert the open tubes on paper towel to remove excess fluid and remove the remainder by capillary action with a drawn-out pasteur pipet (avoid touching the pellet).

11. Resuspend the DNA pellet in 0.2 mL of TE buffer (*see* **Note 9**) and allow it to dissolve overnight at 4°C. Store the DNA at 4°C (*see* **Note 10**).

12. Determine the concentration and purity of the DNA by ultraviolet (UV) spectroscopy (*see* **Subheading 3.1.1.**).

3.1.1. Determination of DNA Concentration and Quality by UV Spectroscopy

1. Ensure that the DNA is completely dissolved (*see* **Note 10**). Prepare a 1/50 dilution in TE buffer.

2. Using TE buffer as a blank, measure the absorbance of the dilution at 260 and 280 nm.

3. Calculate the concentration of DNA. An OD of 1 at 260 nm corresponds to a DNA concentration of 50 µg/mL, therefore;
   DNA concentration (µg/µL) = (OD_{260}) × (dilution factor) × (50 µg DNA/1 OD_{260} U) / 1000 e.g., if for a 1/50 dilution of DNA the OD_{260} = 0.12,
   DNA concentration = [0.12 × 50 (dilution factor) × 50 (conversion factor)]/ 1000 = 0.3 µg/µL (i.e., for simplicity just multiply OD_{260} by 2.5). Therefore, when digesting 2 µg of DNA (*see* **Subheading 3.2.**), 2/0.3 = 6.7 µL (7 µL) of the extract is required.

4. Estimate the purity of each DNA preparation from the ratio: OD_{260}/OD_{280}. It should be approx 1.8 (*see* **Note 11**).

3.2. Restriction Endonuclease Digestion of DNA and Separation of Fragments by Agarose Gel Electrophoresis

3.2.1. DNA Digestion

1. Calculate the volume (in µL) of each DNA sample that contains 2 µg DNA (*see* **Subheading 3.1.1.** and **Note 12**). Calculate the volume of water necessary to make the total reaction volume 20 µL, assuming 2 µL of buffer per reaction and 1 µL of endonuclease per reaction (equivalent to 2.5–5 U of enzyme per 1 µg of DNA).

2. Combine sufficient restriction buffer and endonuclease for all reactions plus 10% extra (e.g., for 10 digests, mix 22 µL buffer + 11 µL endonuclease), mix well. Aliquot 3 µL per tube.

3. Add the appropriate amount of water (*see* **Note 13**) and the appropriate amount of DNA to each tube (calculated in **step 1**, above). Mix briefly (10–15 s) in a microcentrifuge.

4. Incubate overnight (or 4 h minimum) at 37°C (or at the appropriate temperature for the enzyme chosen). Store the digests at –20°C, if they are not to be immediately used.

3.2.2. Agarose Gel Electrophoresis

1. Prepare a 0.8% (w/v) agarose gel in 1X TBE (*see* **Note 14**): dissolve by boiling and allow to cool to 60°C. Pour into a gel tray with the ends sealed with tape and a well-former in position. Allow the gel to set (30–60 min). Remove the well-former and tape, place the gel in the electrophoresis tank, and submerge with 1X TBE.
2. Add 4 µL of loading buffer to each digest, mix briefly by centrifugation (*see* **Note 15**). Load marker (~ 1 µg) into at least one well and 12 µL of the digested DNA (i.e., half of the sample, equivalent to 1 µg of DNA) into the remainder. Store remaining digested DNA at –20°C.
3. Electrophorese until the blue dye front is about 3/4 of the way down the gel (or appropriate separation of fragments is achieved) (e.g., 1 V/cm overnight or 2.5 V/cm for 4–6 h). For a Bio-Rad sub-cell, 35V overnight separates fragments in the range ~0.5–10kb.
4. Stain the gel in 1 µg/mL EB for 20 min and destain in distilled water for 20 min. Photograph under UV transillumination (with a fluorescent ruler aligned with wells to estimate distances migrated and, hence, fragment sizes, if desired). If the gels are to be blotted, expose to UV for 1–2 min.

GDD patterns are obtained at this stage. For ribotyping, the following methods (*see* **Subheadings 3.3.** and **3.4.**) must also be performed.

3.3. Transfer of DNA to Nylon Membrane for Hybridization (Southern or Capillary Blotting; see Note 16)

1. Expose the gel to UV for 1–2 min (i.e., slightly longer than necessary for photographing the gel). *Note: see* **Subheading 2.2.2.**
2. Soak the gel in 0.25 *M* HCl for 15 min (depurination) on an orbital shaker.
3. Soak the gel in denaturing solution for 30–40 min to denature DNA.
4. Soak the gel for 30–40 min in neutralizing solution.
5. Transfer the DNA to neutral nylon membrane (*see* **Note 17**) overnight (**Fig. 3**). Put ~250 mL of 10X SSC in a sandwich box. Form a wick by wetting two pieces of filter paper (~3 cm wider than the gel and long enough to rest in transfer buffer) in transfer buffer and position them flat on a perspex plate placed across the box, so that both ends of the filter paper are immersed in the transfer buffer. Place the gel on this (trim if necessary) (*see* **Note 18**) and place cling film around it to prevent the buffer short-circuiting. Place a membrane (~1 cm larger than the gel) on the gel, ensuring that there are no air bubbles between the gel and membrane (*see* **Note 19**). Place two pieces of dry filter paper (~2 cm larger than gel) over this followed by flat, absorbant material (soft paper towels, then newspapers) to a depth of ~2 cm. Place another perspex plate on the top with a ~1 kg weight. Leave overnight (*see* **Note 16**).
6. Remove materials above the membrane without disturbing it. Note the position of any air bubbles (mark the limits of the bubble on the nylon if in an important area of the gel). Mark well positions, label blot with pencil or marker, and note which side DNA is on. Remove the membrane and rinse in 2X SSC for 5 min.

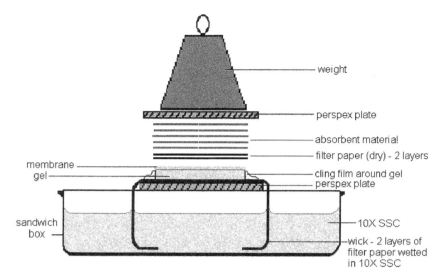

Fig. 3. Diagram illustrating the transfer of DNA from agarose gels to membranes by Southern (or capillary) blotting.

Stain the posttransfer gel in EB (*see* **Subheading 3.2.2.**) to check the efficiency of transfer (there should be little or no DNA left in it). Drain off excess SSC from the membrane with blotting paper, then wrap the membrane in cling film and fix the DNA to the nylon either by exposing it (DNA side down) to UV on a transilluminator for 3 min or by baking at 80°C for 2 h. The membrane can be immediately hybridized or stored in the dark for a short while (<3 d). To store for longer periods (weeks or months), remove the cling film, place the nylon between two pieces of filter paper, wrap this in aluminium foil, and store at room temperature.

3.4. Probe Preparation (Digoxigenin Labeling of cDNA from 16S + 23S rRNA). (Modified from ref. 9; see Notes 20 and 21).

1. Boil 5 μL of diluted *E. coli* rRNA (i.e., 2 μg) + 7 μL of water for 5 min.
2. Place on ice and add, in the following order, 4 μL of ACGT mix, 10 μL of RT buffer, 5 μL of DTT, 10 μL of random primers, 1 μL of RNAase inhibitor, 5 μL of BSA, 1 μL of dig-11-dUTP, and 2 μL of M-MLV-RT. Mix briefly by centrifugation.
3. Incubate for 1 h at 37°C, cool on ice, add 1 μL of EDTA and mix well to stop the reaction. Store at −20°C (stable for months).

3.5. Hybridization (see Note 22)

1. Prehybridize the membranes in hybridization bottles or bags with ~50 mL of prehybridization solution (prewarmed to 65°C) (*see* **Note 23**) at 65°C for at least 2 h.
2. Boil the probe (made in **Subheading 3.4.**) for 5 min immediately before use and place on ice.

3. Prepare the hybridization mix by adding the freshly denatured (boiled) probe to 10 mL of prehybridization solution. This gives a final probe concentration of ~20 ng/mL.
4. Discard the prehybridization solution from the bottles or bags and replace with the hybridization mix. Note that filters should not become dry when exchanging the prehybridization with hybridization mix.
5. Incubate overnight at 65°C (*see* **Note 23**).
6. Pour off the hybridization solution (store this at –20°C, *see* **Note 24**), add ~200 mL of wash solution 1 and mix at room temperature for 5 min. Discard, replace with fresh wash solution 1, and again wash at room temperature for 5 min. Wash twice with wash solution 2 at 65°C (*see* **Note 23**) for 15 min.
7. Detect hybrids immediately or air-dry the membrane on filter paper, place between two sheets of filter paper, and wrap in aluminium foil to store (for weeks or months) for later detection.

3.6. Chemiluminescent Detection of Probe DNA-Target DNA Hybrids (see Note 25)

The following steps are all performed at room temperature with shaking or mixing. The volumes of the solutions are calculated for a filter size of 100 cm^2, and represent the minimum volumes required. Larger volumes may be used.

1. Wash the membrane briefly (for 1 min) in buffer 1.
2. Incubate for 30 min with about 100 mL of buffer 2 (blocking step).
3. Incubate the membrane for 30 min with about 20 mL of freshly diluted antibody-conjugate solution.
4. Remove the unbound antibody-conjugate by washing twice for 15 min with 100 mL of buffer 1.
5. Equilibrate the membrane for 2 min with 20 mL of buffer 3.
6. Incubate the membrane in ~10 mL of chemiluminescent substrate for 5 min, mixing thoroughly (fast on orbital shaker; for colorimetric detection, *see* **Note 25**).
7. Let excess liquid drip off the membrane (hold edges with plastic forceps—metal ones damage it and produce high background), blot briefly on filter paper, and wrap in Saran wrap™ (some clingfilms produce background). Incubate at 37°C for ~1 h, then expose to X-ray film (e.g., Kodak X-OMAT AR) until bands are clearly visible. Re-expose as necessary. The used substrate can be stored at 4°C or –20°C in the dark and reused within 1 wk. Chemiluminescent detection enables stripping and reprobing of membranes (*see* **Note 26**).
8. Compare banding patterns (*see* **Note 27**).

4. Notes

1. If using meshes, they must be thoroughly washed in boiling distilled water between uses as dirty meshes can inhibit hybridization. A shaking water bath and polythene bagging for pre-hybridization and hybridization and a sandwich box for the stringency washes works extremely well.

2. Blocking reagent powder (Boehringer Mannheim) can be used directly, but needs to be prepared in buffer 1 about 1 h in advance and incubated at 56°C to dissolve. 10X concentrated liquid blocking reagent is available from Genosys, Cambridge, UK).

3. A 1:1000 dilution also works well.

4. Overnight 10 mL broth cultures (~10^{10} organisms) may be used, in which case bacteria should be harvested by centrifugation, the supernatant discarded and the pellet resuspended in 1 mL of TE buffer. This can then be transferred to a micro-centrifuge tube, centrifuged to pellet the cells (~2 min at 12,000g; avoid excess centrifugation or the organisms will be difficult to resuspend), the supernatant discarded and the rice-grain sized pellet thoroughly resuspended in 0.2 mL of TE buffer (use a vortex mixer or pipet tip; thorough resuspension is essential for efficient lysis). The method then continues from **step 2**.

5. This will lyse most Gram-negative bacteria. Gram-positive bacteria will need appropriate enzymic pretreatment before the addition of GES . For *Staphylococcus aureus*, harvest in 0.2 mL of TES (10 mM Tris-HCl, pH 8.0, 1 mM EDTA, 100 mM NaCl) and add lysostaphin to a final concentration of 200 μg/mL (a 1 mg/mL stock solution should be made by dissolving to a concentration of 2 mg/mL in 75 mM NaCl and adding an equal volume of glycerol; it is stored at –70°C). Incubate at 37°C for 30 min. For coagulase-negative staphylococci, harvest in TES and add lysostaphin to 100 μg/mL and lysozyme to 200 μg/mL and incubate at 37°C for 30 min. For enterococci, add lysozyme to 50 mg/mL and incubate at 37°C for 30 min. For *Streptococcus pneumoniae* harvest in TEG (100 mM Tris-HCl, pH 7.0, 10 mM EDTA, 25% w/v glucose), add mutanolysin to 20 μg/mL and incubate at 37°C for 5–10 min. For pyogenic streptococci, incubate for 1 h. Continue from **step 2**. If extracts fail to digest, *see* **Note 12**.

6. If no precipitate is visible, store at –20°C for 15–30 min.

7. This spin should be as brief as possible to collect the precipitate without impacting the DNA. Over-centrifugation makes the pellet difficult to dissolve. If the pellet is not well-defined (i.e., it looks "sticky"), respin and/or add more isopropanol, mix and respin. Sticky pellets may result in poor quality DNA that may not digest well.

8. These pellets are not as firm as isopropanol or absolute-ethanol precipitates and may float. Respin or carefully discard ethanol with a pipet tip if the pellet is mobile.

9. If the final pellet is very small, or not visible, resuspend in less TE (i.e., 50 μL).

10. Mix thoroughly, but do not vortex. If not thoroughly dissolved, incubate for 15–30 min at 56°C and add more TE if necessary.

11. The OD$_{260}$ is only a reliable means of determining DNA concentration if the extract is pure. A ratio of 2 indicates the presence of excessive RNA (extracts can be treated with 50 μg/mL RNAase at 37°C for 5 min), whereas values of less than 1.7 indicate protein contamination. In practice, extractions usually result in a consistent yield and ODs are unnecessary. Initial digests can be set up with the same volumes of DNA (e.g., 2 μL), which can be adjusted for subsequent digestions, based on the results of the initial gel.

12. The amount of DNA to be digested depends on the application and the restriction endonuclease used. Ribotyping usually requires less DNA than direct analysis of the genomic DNA digestion. DNA at the top of the gel tracks should generally be horizontal, not curved. Curved bands indicate too much DNA, unless large amounts are necessary to obtain distinct small fragments. Complete digestion is essential for both methods. If partial digests are obtained, increase the digestion time to overnight, double the amount of endonuclease or decrease amount of DNA. If "partials" persist, clean up the DNA extract (or half of it) as follows. Add an equal volume of phenol:chloroform:isoamyl alcohol (25:24:1), mix and centrifuge at 12,000*g* for 10 min. Transfer the upper layer to a clean tube and add two volumes of cold (–20°C) absolute ethanol (or an equal volume of cold isopropanol). NaCl and spermidine may be added to 300 and 0.1 m*M*, respectively (stock solution of NaCl is 5 *M*, stored at room temperature, and that of spermidine is 5 m*M*, stored at –20°C) to aid precipitation. Mix and store at –20°C for at least 30 min. Centrifuge at 12,000*g* for 2–3 min, discard all ethanol, and resuspend the pellet in half the original volume of TE buffer (as some DNA will be lost in the process). *Note:* Phenol is extremely toxic and causes severe burns; appropriate precautions should be taken.
13. If the concentration of DNA in all samples is consistent (and, hence, the volume required is consistent ± 1 µL), add water to the bulk reaction mix and aliquot all three reagents together.
14. 0.8–1% (w/v) agarose is generally used, but this can be increased (up to 2%) for separation of smaller fragments or decreased (to 0.5%) for separation of larger fragments.
15. Alternatively, place 2 µL spots of loading buffer onto parafilm and mix 10 µL of each digest with a spot and load the well. This is useful if digestion may be incomplete as the remainder could be reincubated, whereas adding loading buffer to the tubes will stop digestion.
16. Vacuum blotting is faster (~ 2 h) and works well.
17. Positively-charged membranes can be used but, in the author's experience, these may result in high backgrounds.
18. Removal of excess gel (leaving 5 mm either side of tracks containing DNA) reduces membrane waste. As membrane is often supplied 30-cm wide and many gels are 15-cm wide, cutting 0.5 cm off each side of the gel enables a 15-cm wide membrane to be used to cover the gel and maximizes the use of expensive membrane.
19. Do not handle the membrane; mark on backing paper with pencil and cut. Handle edges of membrane only, wearing rinsed gloves, to minimize background. To remove air bubbles from between plate and wick, or wick and gel, gently roll a clean 10 mL pipet over the surface.
20. This is a modification of a method described originally for biotinylated probes *(9)*.
21. Amplification products (e.g., from 16S and 23S rDNA) can be labeled with digoxigenin. Use 1 µL of PCR product as target in a second round of amplification. Prepare a PCR reaction mix as usual, but replace 35% of the dTTP with dig-11-dUTP. Check incorporation of label by comparing the electrophoretic

mobilities of unlabeled target and labeled product; the labeled product is retarded. Probes containing entire 16S and 23S genes are generally preferable to oligonucleotide probes (which must be end-labeled) as they detect more variation within and flanking the operon *(3)*. Plasmid-based probes should be labeled by random priming (e.g., DIG High Prime, Boehringer Mannheim).

22. A useful booklet "The DIG system user's guide for filter hybridization" is available from Boehringer Mannheim, on request.

23. 65°C is highly stringent and is appropriate for Gram-negative organisms (i.e., those closely related to the probe source, *E. coli*). For less-related organisms (e.g., Gram-positives), prehybridization, hybridization, and stringency washes should be performed at 50°C. These may also require longer detection and, hence, less marker.

24. Hybridization mixtures can be stored at −20°C and reused several times. Boil before reuse and use to replace prehybridization solution.

25. Hybrids may also be colorimetrically detected. For this, perform all steps up to and including **step 5**. Incubate membrane in the dark in a sealed plastic bag together with about 10 mL of freshly prepared color solution (45 µL of nitroblue tetrazolium solution and 35 µL of 5-bromo-4-chloro-3-indolylphosphate solution [NBT/BCIP, Life Technologies] are added to 10 mL of buffer 3). The color precipitate starts to form within a few minutes and the reaction is usually complete after 1 d. Do not shake or mix while the color is developing. When the desired bands are detected, stop the reaction by washing the membrane for 5 min with 50 mL of TE buffer. Drain off excess solution on filter paper and blot gently. Document the results by photocopying the membrane or by photography. The dry membrane may be stored between two sheets of filter paper, in the dark, at room temperature. Membranes detected in this way cannot be easily stripped and reprobed.

26. To strip, rinse membranes briefly in water, incubate twice for 10 min in 0.2 *M* NaOH, 0.1% SDS at 37°C. Rinse in 2X SSC and prehybridize as before.

27. Patterns can be compared visually or with commercially available computer software (*see* Chapter 3).

Acknowledgments

The author is grateful to N. Leaves and D. Griffiths for help with preparation of the figures.

References

1. Forbes, K. J., Bruce, K. D., Jordens, J. Z., Ball, A., and Pennington, T. H. (1991) Rapid methods in bacterial DNA fingerprinting. *J. Gen. Microbiol* **137,** 2051–2058.

2. Owen, R. J. (1989) Chromosomal DNA fingerprinting—a new method of species and strain identification applicable to microbial pathogens. *J. Med. Microbiol* **30,** 89–99.

3. Jordens, J. Z. and Leaves, N. I. (1997) Source of variation detected in ribotyping patterns of *Haemophilus influenzae*: Comparison of traditional ribotyping, PCR-ribotyping and rRNA restriction analysis. *J. Med. Microbiol.* **46,** 763–772.

4. Grimont, F. and Grimont, P. A. D. (1986) Ribosomal ribonucleic acid gene restriction patterns as potential taxonomic tools. *Ann. Inst. Pasteur/Microbiol.* **137B,** 165–175.
5. Bingen, E. H., Denamur, E., and Elion, J. (1994) Use of ribotyping in epidemiological surveillance of nosocomial outbreaks. *Clin. Microbiol. Rev.* **7,** 311–327.
6. Tenover, F. C., Arbeit, R. D., Goering, R. V., Mickelsen, P. A., Murray, B. E., Persing, D. H., and Swaminathan, B. (1995) Interpreting chromosomal DNA restriction patterns produced by pulsed-field gel electrophoresis: criteria for bacterial strain typing. *J. Clin. Microbiol.* **33,** 2233–2239.
7. Leaves, N. I. and Jordens J. Z. (1994) Development of a ribotyping scheme for *Haemophilus influenzae* type b, *Eur. J. Clin. Microbiol. Infect. Dis.* **13,** 1038–1045.
8. Pitcher, D. G., Saunders, N. A., and Owen, R. J. (1989) Rapid extraction of bacterial genomic DNA with guanidium thiocyanate. *Lett. Appl. Microbiol.* **8,** 151–156.
9. Pitcher, D. G., Owen, R. J., Dyal, P., and Beck, A. (1987) Synthesis of a biotinylated DNA probe to detect ribosomal RNA cistrons in *Providencia stuartii, FEMS Microbiol. Lett.* **48,** 283–287.

3

Pulsed-Field Gel Electrophoresis

Mary Elizabeth Kaufmann

1. Introduction

Pulsed-field gel electrophoresis (PFGE) was first described by Schwartz and Cantor *(1)*. It is now an umbrella term for the alternating of an electric field in more than one direction through a solid matrix to achieve the separation of DNA fragments. The method requires the preparation of unsheared DNA, digestion of the DNA using a rare-cutting restriction endonuclease, separation of fragments by PFGE, and the visualization and interpretation of banding patterns (**Fig. 1**). Conventional agarose gel electrophoresis employs a static field and can resolve DNA fragments up to 50 kilobases (kb), although in practice, fragments larger than 20 kb co-migrate under the conditions usually applied. By introducing a pulse or change in the direction of the electric field, fragments as large as 10 megabases (Mb) can be separated. The time required by DNA fragments of different sizes to reorientate to the new electric field is proportional to their molecular weight and it is this factor that allows the separation and focusing of DNA fragments.

The main applications of PFGE in medical microbiology are as described in Chapter 2. With the implication in infections of an increasing variety of species for which phenotypic typing schemes are not available, PFGE provides a potentially universal method for fingerprinting and comparing isolates. However, it is not a typing method as, with the ability to produce an infinite variety of banding patterns, a PFGE profile can rarely be used to identify a documented bacterial strain, particularly when the technique is performed in different laboratories. Despite this, as a means of comparing isolates to determine whether they are the same or different, PFGE is arguably the most discriminating method currently available *(2)*.

From: *Methods in Molecular Medicine, Vol. 15: Molecular Bacteriology: Protocols and Clinical Applications*
Edited by: N. Woodford and A. P. Johnson © Humana Press Inc., Totowa, NJ

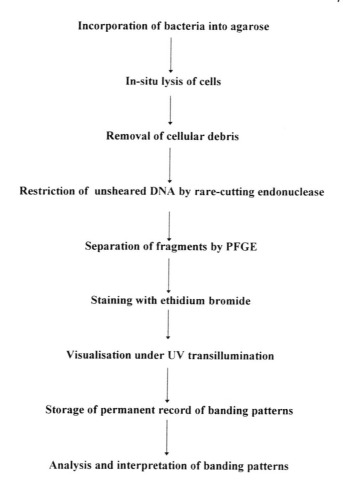

Fig. 1. Practical steps involved in PFGE.

The most commonly used equipment is of two types; in one type the electric fields pulse forwards and backwards through the gel as in field-inversion gel electrophoresis (FIGE) *(3)*, whereas the other employs the contour-clamped homogeneous electric field (CHEF) *(4)*. FIGE equipment is simple, using a conventional electrophoresis tank, powerpack, and a switching unit. To achieve net forward migration in FIGE, the forward pulse time is longer than the backward pulse time, usually in a constant ratio. A development of FIGE is zero-integrated field electrophoresis (ZIFE) in which the forward voltage gradient is higher than the reverse, but the product of voltage and time is only slightly higher in the forward direction. As the difference in the two fields is only small, long run times are required, but the resulting bands are of high resolution.

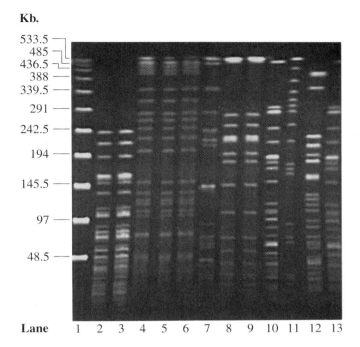

Kb.

Lane 1 2 3 4 5 6 7 8 9 10 11 12 13

Fig. 2. PFGE of *Xba*I chromosomal digests from various Gram-negative species. Lane 1, Lambda concatamer; Lanes 2 and 3, *Pseudomonas aeruginosa*; Lanes 4–7, *Klebsiella* spp.; Lanes 8–13, *Enterobacter* spp.

The main advance in PFGE was the production of a homogeneous electric field by the use of multiple electrodes and differential voltages *(4)*. The equipment for CHEF uses 24 electrodes arranged in a hexagon with the electric field intersecting at an angle of 120° in all parts of the gel. This gives rise to well-focused bands in straight lanes with DNA in all parts of the gel being subjected to the same conditions (**Fig. 2**). A further advance, programmable autonomously controlled electrodes (PACE), provides computer linkage to the CHEF apparatus and gives enhanced versatility to the system. This allows variable angles of intersection, multiple electric fields, different voltages, and a combination of programs within a single run. This system is marketed as the CHEF Mapper by Bio-Rad Laboratories (Richmond, CA) and as the Gene Navigator by Pharmacia Biotech (Uppsala, Sweden).

1.1. Parameters Affecting DNA Separation

Each aspect of the method will influence the final result and great care, therefore, needs to be given to the standardization of the method if gel-to-gel comparisons of banding patterns are to be made.

1.1.1. DNA Quality and Concentration

The main advantage of PFGE is its ability to separate large fragments of DNA following the restriction of unsheared chromosomes with rare-cutting endonucleases. DNA extracted as an aqueous solution for conventional electrophoresis is subjected to mechanical shearing resulting in DNA fragments of <500 kb—which is smaller than many of the fragments generated by rare-cutters—it is, therefore, not suitable for PFGE. Unsheared DNA is achieved by the incorporation of bacterial cells into agarose, followed by the lysis and disruption of the cells and the washing away of cellular debris.

The concentration of DNA affects its mobility and, hence, the clarity of bands and also the ease with which comparisons of banding patterns can be achieved both by eye and by computer-assisted analysis. DNA concentrations over 70 µg/mL reduce mobility and bands become increasingly diffuse. It is not possible to estimate the concentration of DNA embedded in agarose, so the concentration of cells embedded has to be carefully controlled to give the required end result.

1.1.2. Agarose Concentration

As in conventional gel electrophoresis, increases in agarose concentration retard the mobility of DNA fragments. For most PFGE applications, a concentration of 1% (w/v) is recommended, although increased resolution can be achieved with concentrations of 1.2–1.5% (w/v), whereas to separate fragments over 3 Mb, lower percentage gels are necessary, with a corresponding reduction in the clarity of bands. The quality of agarose used will produce noticeable differences in the mobility of fragments larger than 2 Mb. It is usually recommended that special PFGE agarose is used for these very large fragments, whereas molecular grade agarose can be used for those smaller than 2 Mb.

1.1.3. Voltage and Pulse Times

In general, the mobility of DNA fragments smaller than 2 Mb increases with field strength, but this may be accompanied by a reduction in band clarity. The voltage most commonly used in PFGE is 6 V/cm distance between electrodes, whereas lower voltages, 2 V/cm, are required to separate DNA fragments over 2–3 Mb.

The most important parameter influencing the separation of large DNA fragments is the switching (or pulse) time. Simplistically, as pulse times increase, so does the size of DNA fragments that can be separated, provided that all other conditions remain the same. It is usual to ramp (change linearly) the pulse times from short to long during a run to increase the effectiveness of DNA separation. State of the art equipment allows the rate of ramping to be nonlinear, thereby improving the separation of the lower or higher molecular weight fragments and by combining periods with different ramp ratios, almost any range of fragment sizes can be spread out to optimize and facilitate analysis. If a wide range of fragment sizes is expected, it may be necessary to divide the run into several different periods, varying the voltage applied as well as the range of pulse times.

1.1.4. Buffer Strength and Temperature

The mobility of DNA fragments through agarose increases with temperature, with large fragments being more susceptible to this influence than small fragments. Conversely, the clarity of focused bands decreases as the temperature rises. It is important to obtain a balance between speed of separation and clarity of banding patterns, and it is generally accepted that a realistic optimum temperature for PFGE is 14°C. Electrophoresis at 6 V/cm will generate a considerable amount of heat that, without active cooling, will raise the buffer temperature to an unacceptable level. Running equipment in a cold-room with the buffer circulating through ice will keep the temperature in the range of 10–14°C, but to maintain a constant monitored temperature, special cooling apparatus is required (available from most distributors of electrophoretic equipment). Reducing buffer concentration to 0.5X buffer instead of using 1X buffer also helps to reduce the generation of heat and to keep the temperature low. However, fragments move faster in 0.5X buffer necessitating a shorter run time.

Two buffers can be used for PFGE; Tris-borate EDTA buffer (TBE) and Tris-acetate EDTA buffer (TAE). In general, fragments move faster with TAE than with TBE, but the buffering capacity of TAE is lower, and for long runs the buffer may become exhausted before the completion of the run.

1.2. Interpretation of Banding Patterns

Banding pattern interpretation will depend on the purpose of, and the question to be answered by, the PFGE. One of the most common uses of PFGE in medical microbiology is the comparison of epidemiologically related bacterial isolates in the investigation of outbreaks of infections in hospitals or in the community. Comparison of banding patterns by eye may be sufficient to decide which isolates are involved in the outbreak and which are different. To facilitate interpretation and to provide quantitation of the level of relatedness within and between groups of isolates, there are a number of computer programs available. These include GelCompar™ (Applied Maths, Kortrijk, Belgium), MVSP (Korvach Computing Services, Anglesey, UK), GelManager™ (BioSystematica, Tavistock, Devon, UK), Lane Manager™ (TDI), Phoretix 1D & 2D (Phoretix International Ltd., Newcastle, UK), and Dendron® (Solltech Inc., Oakdale, CA).

1.3. Applications of PFGE

1.3.1. Comparison of Epidemiologically Related Isolates

The main application of PFGE to clinical microbiology is in the determination of the relatedness of bacterial isolates from epidemiologic incidents. For some species, traditional typing schemes (serotyping, phage-typing, biotyping, and bacteriocin-typing) are available and, although of long standing, still provide a means of strain identification that is cheap to run, quick to provide, and reproducible. Reagents for these schemes are not, however, universally available nor are schemes available for all species. Chromosomal DNA is a fundamental molecule of all bacteria, and provides a focus for molecular methods that can be applied to all species. PFGE employs >80% of the chromosome to produce banding patterns, making it potentially the most discriminating method available today.

1.3.2. Investigation of Specific Regions of DNA Following Southern Blotting and Hybridization

The principles behind hybridization of PFGE gels are as described in Chapter 2. Difficulties can occur in the transfer of large DNA fragments and extra depurination incubation may be required. Transfer to nylon membrane takes longer for PFGE gels than conventional gels as they are thicker. Capillary blotting takes at least 24 h, whereas vacuum blotting takes about 4 h. Once the DNA has been transferred, the methods for hybridization are as described in Chapter 2.

1.3.3. The Separation and Sizing of Large Plasmids (5)

Large plasmids can be separated and sized by PFGE, without the removal of chromosomal DNA. Bacteria embedded in agarose are subjected to lysis followed by incubation with S1 nuclease, which converts supercoiled plasmids to full-length linear molecules. The migration of linear plasmid DNA is consistent with commonly used molecular weight markers allowing accurate plasmid sizing.

1.3.4. Physical Chromosome Mapping by Two-Dimensional PFGE (6)

Two-dimensional PFGE (2-D PFGE), in which partially digested DNA fragments are separated in the first direction followed by completion of digestion and separation of all fragments in the second direction, allows assembly of a chromosomal map based on the fact that bands liberated from a single partial band must be adjacent to each other on the chromosome. 2-D PFGE in which completely digested DNA is separated in the first direction, followed by complete digestion with a second enzyme in the second direction, can also be used to produce a genomic map. However, hybridization with known fragments may be required to complete the map by this method.

1.3.5. Comparison of Yeast Isolates

Methods for the preparation of DNA from yeast cells have been described *(7)* and used successfully to produce clear banding patterns. A word of caution has been expressed by Bostock et al. *(8)* when comparing *Candida albicans* isolates, however, as the high level of genetic switching that is characteristic of this species may result in two genetically similar isolates appearing different because of genetic rearrangement.

2. Materials
2.1. Preparation of Unsheared DNA
2.1.1. Incorporation of Bacteria into Agarose

1. Bacterial culture.
2. Low-gelling agarose.
3. Casting mold.
4. SE buffer: 75 mM NaCl, 25 mM EDTA Na$_2$ (filter sterilized). Store at 4°C.
5. Waterbath at 50–56°C.

2.1.2. Lysis of Bacteria

1. Appropriate lysis buffer:
   Gram-negative bacteria: 1% (w/v) N-lauroyl sarcosine, 500 mM EDTA Na$_2$ pH 9.5 (*see* **Note 1**). Filter sterilize and store at room temperature. *See* **Table 1** for alternative lysis buffers.
   Gram-positive bacteria: 6 mM Tris-HCl, 100 mM EDTA Na$_2$, 1 M NaCl, 0.5% (w/v) Brij 58, 0.2% (w/v) sodium deoxycholate, 0.5% (w/v) lauroyl sarcosine, pH 7.5, stored at room temperature (*see* **Table 1**).
2. Proteinase K (Sigma, St. Louis, MO, P-0390). Stock solution 50 mg/mL, stored at –20°C.
3. Lysostaphin (for staphylococci only). For best results, using lysostaphin from Aplin and Barrett Ltd, Trowbridge, Wilts, UK, is recommended. Stored at 4°C as 3000 U/mL in 50 mM Tris-HCl, 150 mM NaCl, pH 7.5.
4. Lysozyme (*Acinetobacter* spp. and Gram-positive bacteria only). Store as powder at –20°C and add the appropriate amount to lysis buffer when required.
5. RNase. Stored at –20°C.
6. Proteolysis buffer for Gram-positive bacteria (same composition as Gram-negative lysis buffer, *see* **step 1**, above).
7. TE buffer: 10 mM Tris-HCl, 10 mM EDTA Na$_2$, pH 7.5. Filter sterilize and store at 4°C.

2.2. DNA Restriction

1. Rare-cutting restriction endonuclease (*see* **Note 2**).
2. 10X reaction buffer (usually supplied with endonuclease).
3. Sterile distilled water.
4. Agarose block with incorporated DNA.

Table 1
Lysis Buffers used in PFGE

Species	Lysis buffer	Additives	Reference
Staphylococcus aureus	6 m*M* Tris-HCl, 100 m*M* EDTA Na$_2$, 1 *M* NaCl, 0.5% Brij-58, 0.2% sodium deoxycholate, 0.5% lauroyl sarcosine, pH 7.5	30–50 U/mL lyzostaphin 500 µg/mL lysozyme	*9*
Coagulase negative staphylococci	As for *S. aureus*	As for *S. aureus*, may need to increase lysozyme concentration	Author's unpublished observations
Micrococci	As for *S. aureus*	2 mg/mL lysozyme	Author's unpublished observations
Enterococci	As for *S. aureus*	1 mg/mL lysozyme 20 µg/mL RNase	*10*
Listeria monocytogenes	100 m*M* EDTA Na$_2$, 1% Sarcosyl		*11*
Mycobacterium tuberculosis	Incubate in 10 m*M* Tris-HCl, 1 m*M* EDTA Na$_2$ prior to incorporation into agarose plug	2 mg/mL deoxycholic acid 2.5 mg/mL lysozyme 2 mg/mL lysozyme	*12*
Bordetella pertussis	As for *S. aureus*	1 mg/mL lysozyme 20 mg/mL RNase	*13*
Acinetobacter spp	May require *S. aureus* lysis buffer	1 mg/mL lysozyme	Author's unpublished observations

2.3. Separation of DNA Fragments

1. Electrophoresis tank, power source, and switching unit suitable for PFGE (*see* **Subheading 1.**).
2. Cooling apparatus (*see* **Subheading 1.1.4.**).
3. Gel cassette and comb (*see* **Note 3**).
4. UV transilluminator and camera or gel documentation system. *Note:* UV-protective face masks should be worn when working with a UV transilluminator as UV radiation can seriously damage the cornea and activate *Herpes simplex*.
5. Molecular grade agarose.
6. 0.5X TBE buffer: 44.5 m*M* Tris base, 44.5 m*M* boric acid, 1 m*M* EDTA Na$_2$, pH 8.0–8.5. Store as 5X TBE stock solution at room temperature. The pH should not require adjustment.
7. Agarose block containing restricted DNA.
8. Molecular weight marker (e.g., lambda phage concatamers, *Escherichia coli* MG1655 digested with *Not*I *[10]*).
9. 1 µg/mL ethidium bromide (EB). Store as 10 mg/mL aqueous solution at room temperature. *Note:* EB is a mutagen and carcinogen, avoid skin contact and inhalation and discard as toxic waste.

3. Methods
3.1. Preparation of Unsheared DNA

Gloves should be worn at all stages of the method.

3.1.1. Incorporation of DNA into Agarose

1. From plate cultures suspend sufficient growth in 1 mL of SE buffer to give an optical density of McFarland's tube 4 (*see* **Note 4**).
2. Prepare 2% low gelling agarose (LGA) in SE buffer. Dissolve by boiling and keep at 50–56°C until ready to use.
3. Assemble the mold.
4. Warm the bacterial suspensions to 50–56°C and allow any clumps to settle.
5. Mix 500 µL of LGA with an equal volume of bacterial suspension, mix gently but thoroughly. Do not vortex.
6. Dispense the mixture into the mold and allow to solidify at 4°C. Many blocks can be made per isolate as, following lysis, blocks can be kept for many months in TE buffer provided the buffer is changed regularly.

3.1.2. Lysis of Bacteria

3.1.2.1. Gram-Negative Bacteria

1. Prepare Gram-negative lysis buffer containing 500 µg/mL proteinase K.
2. Dispense sufficient lysis buffer into bijoux to cover the number of blocks made per isolate (2–3 mL is sufficient for five blocks).
3. Incubate overnight at 56°C (*see* **Note 5**).

4. Wash three times for 30 min each in TE buffer at 4°C.
5. Store the blocks in TE buffer at 4°C.

3.1.2.2. GRAM-POSITIVE BACTERIA

A more rigorous lysis is required to disrupt the Gram-positive bacterial cell wall, and different chemicals have been recommended in the literature for different genera. Some recommended cocktails are shown in **Table 1**.

1. Prepare the appropriate Gram-positive lysis buffer.
2. Dispense sufficient lysis buffer into bijoux to cover the number of blocks made per isolate (2–3 mL is sufficient for five blocks).
3. Incubate at 37°C overnight (*see* **Note 5**).
4. Replace the lysis buffer with proteolysis buffer containing 100 µg/mL proteinase K.
5. Incubate at 56°C overnight.
6. Wash and store the blocks as described for Gram-negative bacteria.

3.2. DNA Restriction

1. Cut a portion of an agarose block containing DNA and transfer to a 500 µL reaction tube.
2. Cover with 100 µL of 1X reaction buffer and keep at 4°C for 30 min.
3. Replace reaction buffer with 100 µL of 1X reaction buffer containing 20 U of restriction endonuclease.
4. Incubate at the temperature recommended by manufacturer for 4 h (*see* **Note 2**).

3.3. Separation of DNA Fragments

1. Prepare a 1–1.2% agarose gel in 0.5X TBE buffer. Dissolve agarose by boiling, allow to cool to 60°C before pouring into a gel-forming cassette (*see* **Note 6**) with a comb for wells in position. Keep 5 mL of agarose at 56°C for sealing wells.
2. Leave the gel for at least 30 min to set. Remove the comb and carefully load the agarose plugs containing digested DNA into the wells, ensuring that they are lined up against the leading edge of each well (*see* **Note 3**). If banding patterns are to be analyzed by computer, load molecular weight markers in every fourth, fifth, or sixth well.
3. Seal each well with the reserved agarose, making sure that no bubbles are trapped.
4. Remove the gel from cassette and place into the electrophoresis tank. Secure the gel in position (*see* **Note 7**).
5. Cover the gel with 0.5X TBE (usually about 2 L) and allow to cool for at least 30 min by circulating the buffer through the cooling apparatus.
6. Set the required parameters for electrophoresis (*see* **Note 8**).
7. When electrophoresis is complete, remove the gel from the tank and stain in EB for 1 h.
8. Discard the tank buffer as it cannot be reused.
9. Destain the gel in distilled water for 1–2 h, changing the water every 30 min.

10. Visualize the banding patterns under UV transillumination. Keep a permanent record of the gel for analysis (*see* **Subheading 3.4.**).
11. Discard gels as toxic waste.

3.4. Interpretation of Banding Patterns

3.4.1. Visual Interpretation

The level of analysis required to interpret banding patterns is in part dependent on the question to be answered. It is often possible to group isolates giving identical or very similar banding patterns and to distinguish between markedly different profiles at a glance, thereby determining quickly whether an outbreak is represented, whether cross-infection has occurred, or whether there is no sharing of isolates between patients.

Determination of the number of band differences between banding patterns of isolates from suspected outbreaks of infection has been advocated by Tenover et al. *(14)* for the interpretation of profiles having more than 10 bands. Each difference, indicated by the presence or absence of a band, is considered to be significant and four categories of relatedness have been proposed:

1. Indistinguishable, no band differences;
2. Closely related, two to three band differences;
3. Possibly related, four to six band differences; and
4. Different, seven or more band differences.

It should be noted that the larger the number of bands present, the more likely one is to see changes in the banding patterns, and although these criteria hold true for many species, some produce more variation within a strain than these criteria suggest, whereas others may be more clonal. Some authors equate a single band difference with the acquisition or loss of DNA by a fragment and its consequent apparent change in position because of change in molecular size *(15)*. However, this situation is considered by Tenover et al. *(14)* to represent two band differences. If band differences are to be used as the criteria for clustering and differentiating between isolates, it is advisable to consider first the number of bands found between strains, and then see how many bands need to be accommodated within a strain. Whereas this number will usually be within the criteria set down by Tenover et al. *(14)*, this will not always be so, and one will gain an understanding of the diversity to be expected within different species.

3.4.2. Computer Analysis

Many computer-assisted analysis packages are now available. These range in sophistication from those where the presence and absence of bands have to be keyed in for analysis (e.g., MVSP) to those employing image files produced from scanned photographs or from gel documentation systems. Such analysis packages calculate coefficients of similarity and represent these as dendrograms (usually using unweighted pair group matching by an arithmetic averages algorithm, UPGMA, **Fig. 3**), similarity matrices, or evolutionary trees. Two commonly used coefficients are the Jaccard and Dice coefficients, both of which use the number of bands in common between profiles, and the total number of possible band positions to calculate the percentage similarity between the isolates. The formula for the Jaccard coefficient is:

$$n_{AB}/n_A + n_B - n_{AB}$$

where n_A and n_B are the numbers of bands in the profiles of isolates A and B, and n_{AB} is the number of shared bands. The Dice coefficient gives more weight to matching bands and is represented as:

$$2n_{AB}/n_A + n_B.$$

A third coefficient offered by analysis packages is Pearson's correlation coefficient. One advantage of this calculation is that the whole densitometric trace is employed and specific band positions do not have to be defined. However, the quality of banding patterns affects the calculation, with variations in background staining being included. Hence, isolates with identical band positions that calculate as 100% similar by Jaccard or Dice coefficients may be calculated as only 80–90% similar by Pearson's coefficient. Irrespective of the coefficient used to calculate percentage similarity, the interpretation of the data has to be related to the epidemiologic situation when deciding which isolates are representatives of the "outbreak" strain and which are not. Figures of 80–85% similarity have been quoted to define a strain *(15,16)*. However, just as the rules for number of band differences are not always applicable, so too do rules of percentage similarity sometimes fall down. It is, however, usual to find that there is a definite gap (more than 20%) between the percentage at which similar isolates cluster and the percentage at which isolates are clearly different, with only a few isolates falling between the two.

3.5. Troubleshooting

It is not uncommon for an electrophoretic procedure to be followed faithfully without obtaining the excellent results that one would expect. In techniques like PFGE, which are lengthy and contain many steps, it may be only when looking expectantly at the stained gel that one realizes that something has gone wrong, and it is often difficult to determine at what stage error occurred. This section contains some of the more common problems that may occur and details how to detect the cause.

1. There are faint bands, diffuse bands, or no bands present on the gel.
   a. A few species of bacteria produce endogenous endonucleases that will

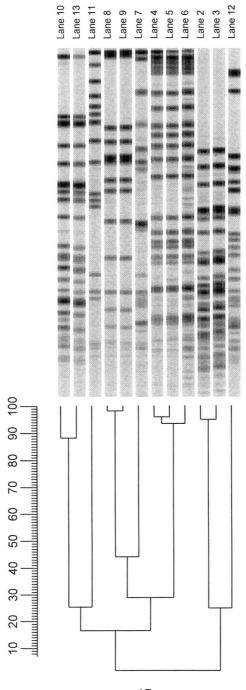

Fig. 3. Dendrogram of percentage relatedness of isolates in Fig. 2, as calculated by Pearson's correlation coefficient represented by UPGMA.

degrade digested DNA into fragments which will pass through the gel during PFGE. Incubating the bacterial suspension in 10% formalin for 1 h followed by three washes in saline prior to incorporation in agarose has been found to inhibit DNase activity *(17)*.

   b. Some Gram-negative bacteria lyse if kept at 56°C for a long time (more than 30 min) prior to incorporation into agarose. This may result in shearing of DNA and a reduction in DNA concentration to a level that produces faint bands, diffuse bands, or no bands. If the DNA concentration is too high, bands may appear to bulge and not be clearly separated.

   c. DNA digestion at an inappropriate temperature may result in incomplete or no restriction. As a result, a single heavy band of DNA, a smear of DNA running the length of the lane, or DNA only in the wells may be seen.

   d. The parameters set for the electrophoresis may be inappropriate for the sizes of fragments present (*see* **Note 8**).

   e. The concentration and temperature of the running buffer affects the mobility of DNA fragments. If buffer concentration is high, the buffer temperature will rise during electrophoresis and may result in unresolved bands.

   f. If the buffer circulation pump is set too fast, the passage of buffer over the gel can cause the gel to vibrate, which results in fuzzy bands. Similarly, if the gel is not firmly secured before buffer is added, the gel may float either totally or partially with the result that bands are not well-defined.

   g. EB that has been used to stain PFGE gels more than three times becomes too dilute and will fail to bind sufficiently with DNA to be visualized. Conversely, bands are sometimes obscured by excessive background staining when using freshly made EB. Extra washing in distilled water with gentle shaking will remove unbound EB and bands can be visualized.

2. Bands are well-defined, but the lanes are bent, are of unequal length, or appear to be double.

   a. Provided the gel has not moved during the run, the most likely cause of lane distortion is that one or more of the electrodes has broken during electrophoresis. Electrodes are subject to wear and tear as a result of the continuous switching of field direction, and the life of electrodes is considerably reduced if pulse times shorter than 5 s are used.

   b. The hour glass-like appearance of lanes that sometimes occurs can be reduced by equilibrating the gel in cold tank buffer for longer than suggested in **Subheading 3.3.**

   c. Bubbles accumulating under the gel during electrophoresis will result in differential cooling across the gel, which will cause the lanes in the center of the gel to be bowed, whereas the outer lanes remain straight.

   d. If the agarose block containing digested DNA breaks on loading, the DNA of each portion will run as separate lanes, with a gap between.

3. Bands are present, but are not evenly spaced along the gel lane.

   This is usually because the parameters used are inappropriate for the range of fragment sizes to be separated. Some banding patterns have a high concentra-

tion of bands within a small range of fragment sizes. It may be necessary to apply nonlinear ramping (CHEF mapper) or a series of linked short runs (CHEF DR-II or DR-III) to open up areas of dense bands.
4. Bands are very distorted.
    a. It is very important that bacterial suspensions are uniform for incorporation into agarose. Some species aggregate or clump in suspension and, if these are incorporated into agarose, variable densities of cells will be present, and lysis will be inefficient. If lysis is achieved, DNA concentrations will not be uniform and DNA mobilities will vary across the lane causing distortion.
5. The stained gel appears speckled.
    a. Laboratory gloves may contain powder that fluoresces under UV transillumination. This will be detected by the photographic equipment. Analysis of speckly gels is very difficult by eye and nearly impossible by computer.
    b. Agarose that has not been boiled for long enough and that is, therefore, not fully dissolved will have visible specks within the matrix. These will appear as spots in photographs.

4. Notes

1. To achieve the required pH, it is necessary to add approx 10 mL of 10 M NaOH per 100 mL of lysis buffer before adjusting the pH dropwise.
2. The choice of a rare-cutting endonuclease, which is an enzyme which recognizes DNA sites occurring infrequently in the genome, is dependent on the G + C ratio of the species under investigation, and the recognition site of the enzyme. Complex mathematical formulae are available for predicting the cutting frequency of an endonuclease *(18)*, but as a general rule enzymes recognizing mostly G and C will be rare cutters of AT-rich genomes and vice versa. Longer recognition sites will tend to appear less frequently in the chromosome. Endonucleases that recognize target sites that include the sequence CTAG are also usually rare cutters (**Table 2**). The majority of restriction endonucleases require incubation at 37°C, however, there are some that have different requirements. The length of time that an incubated endonuclease remains active varies considerably, and for some it may be necessary to use two periods of 2-h incubation with fresh buffer, rather than a single 4-h incubation.
3. It is important that the agarose blocks containing digested DNA are loaded carefully into the wells. Broken inserts will distort the banding pattern. If some inserts are loaded onto the leading edge and others are placed on the back edge of the well, identical banding patterns will not align. An alternative to loading inserts into wells is to place them onto the well-forming comb and carefully pour the gel around them. Once the gel has set, the comb can be removed and the holes filled with cool molten agarose. It is worth practicing this method before using it for an investigation because the inserts will swim away from the comb if the gel is poured too rapidly.
4. Some bacterial species grow better in liquid culture. Optical density readings of liquid cultures can be calibrated to viable counts for estimation of cell concentra-

Table 2
Commonly used Rare-Cutting Restriction Endonucleases

Restriction endonuclease	Recognition site	Bacterial groups for which it acts as a rare cutter
*Apa*I	GGGCC↓C	*Acinetobacter* spp., Enterococci, *Listeria monocytogenes, Streptococcus* spp.
*Asn*I	AT↓TAAT	Enterobacteriaceae, "Pseudomonads," *Mycobacterium* spp., *Bordetella pertussis.*
*Dra*I	TTT↓AAA	Enterobacteriaceae, "Pseudomonads," *Mycobacterium* spp., *Bordetella pertussis*
*Not*I	GC↓GGCCGC	*Escherichia coli* MG1655, *Staphylococcus aureus, Listeria monocytogenes*
*Sma*I	CCC↓GGG	*Staphylococcus aureus*, Enterococci, coagulase-negative staphylococci, *Listeria monocytogenes, Streptococcus* spp.
*Spe*I	A↓CTAGT	Enterobacteriaceae, "Pseudomonads," *Mycobacterium* spp., *Bordetella pertussis*
*Ssp*I	AAT↓ATT	Enterobacteriaceae, "Pseudomonads," *Mycobacterium* spp., *Bordetella pertussis*
*Sst*II	CCGC↓GG	*Staphylococcus aureus*, coagulase negative-staphylococci
*Xba*I	T↓CTAGA	Enterobacteriaceae, "Pseudomonads," *Mycobacterium* spp, *Bordetella pertussis*

tion. However, following reading, the cells need to be spun and washed three times before resuspending in SE buffer and incorporation into agarose. This may result in the loss of and disruption of cells, with corresponding changes in DNA concentration.

5. Many authors have published faster lysis regimens. For example, Lindhardt et al. *(19)* incubated *Staphylococcus aureus* and coagulase-negative staphylococci for 1 h in lysis buffer followed by 2 h in proteolysis buffer. If speed of result is the driving force, then a quick lysis step is indicated. However, from personal experience, the quality of the end product may not be guaranteed.

6. Commercially available tanks are supplied with gel-casting molds. It is possible to run gels of any size that will fit within the arrangement of the electrodes. With the largest possible gel sizes, some lane distortion may occur in the extremities.

7. During electrophoresis, buffer is circulated to maintain an optimum temperature. If the gel is not firmly secured, it will float and move with the current resulting in distortion of bands or no bands at all. Most commercial tanks are supplied with a means of anchoring compatible gels. If gels of different sizes are used, it may be necessary to secure these with small pieces of plastic (e.g., cut off ends of tips or reaction tubes) and waterproof tape.

8. Electrophoresis parameters that produce the best separation of banding patterns will need to be found empirically. Some general guidelines can, however, be given.
   a. In general, short switching times separate small fragments, whereas longer switches are required to separate larger fragments.
   b. Longer gels require longer run times (e.g., banding patterns that spread evenly along the lane in a 15-cm gel in 18 h, may need 30–35 h in a 20 cm gel and 40–48 h in a 25-cm gel).
   c. Composite runs may be required to separate fragments clearly if fragments differ only slightly in size.
   d. A useful set of conditions from which to start experimenting are:

Initial pulse	5 s
Final pulse	35 s
Ratio	1:1
Run time	18 h for 15-cm gel.

References

1. Schwartz, D. C. and Cantor, C. R. (1984) Separation of yeast chromosome-sized DNAs by pulsed field gradient gel electrophoresis. *Cell* **37,** 67–75.
2. Grundman, H., Schneider, C., Hartung, D., Daschner, F. D., and Pitt, T. L. (1995) Discriminatory power of three DNA-based typing techniques for *Pseudomonas aeruginosa*. *J. Clin. Microbiol.* **33,** 528–534.
3. Carle, G. C., Frank, M., and Olson, M. V. (1986) Electrophoretic separation of large DNA molecules by periodic inversion of the electric field. *Science* **232,** 65.
4. Chu, G., Vollrath, D., and Davis, R. W. (1986) Separation of large DNA molecules by contour clamped homogeneous electric field. *Science* **234,** 1582.
5. Barton, B. M., Harding, G. P., and Zuccarelli, A. J. (1995) A general method for detecting and sizing large plasmids. *Anal. Biochem.* **226,** 235–240.
6. Römling, U., Grothues, D., Heuer, T., and Tümmler, B. (1992) Physical genome analysis of bacteria. *Electrophoresis* **13,** 626–631.
7. Khattak, M. N., Burnie, J., Matthews, R. C., and Oppenheim, B. (1992) Clamped homogeneous electric field gel electrophoresis typing of *Torulopsis glabrata* isolates causing nosocomial infections. *J. Clin. Microbiol.* **30,** 2211–2215.
8. Bostock, A., Khattak, M. N., Matthews, R., and Burnie, J. (1993) Comparison of PCR fingerprinting, by random amplification of polymorphic DNA, with other molecular typing methods for *Candida albicans*. *J. Gen. Microbiol.* **139,** 2179–2184.
9. Goering, R. V. and Duensing, T. D. (1990) Rapid field inversion gel electrophoresis in combination with an rRNA gene probe in the epidemiological evaluation of staphylococci. *J. Clin. Microbiol.* **28,** 426–429.
10. Murray, B. E., Singh, K. V., Heath, J. D., Sharma, B. R., and Weinstock, G. M. (1990) Comparison of genomic DNAs of different enterococcal isolates using restriction endonucleases with infrequent recognition sites. *J. Clin. Microbiol.* **28,** 1059–1063.
11. Brosch, R., Buchrieser, C., and Rocourt, J. (1991) Subtyping of Listeria monocytogenes serovar 4b by use of low-frequency-cleavage restriction endonucleases and pulsed-field gel electrophoresis. *Res. Microbiol.* **142,** 667–675.

12. Zhang, Y., Mazurek, G. H., Cave, M. D., Eisenach, K. D., Pang, Y., Murphy, D. T., and Wallace, R. J. (1992) DNA polymorphisms in strains of mycobacterium tuberculosis analysed by pulsed-field gel electrophoresis: a tool for epidemiology. *J. Clin. Microbiol.* **30,** 1551–1556.

13. Khattak, M. N., Matthews, R. C., and Burnie, J. P. (1992) Is *Bordetella pertussis* clonal? *BMJ* **304,** 813–815.

14. Tenover, F. C., Arbeit, R. D., Goering, R. V., Mickelsen, P. A., Murray, B. E., Persing, D. H., and Swaminathan, B. (1995) Interpreting chromosomal DNA restriction patterns produced by pulsed-field gel electrophoresis: Criteria for bacterial strain typing. *J. Clin. Microbiol.* **33,** 2233–2239.

15. Goering, R. V. (1993) Molecular epidemiology of nosocomial infection: Analysis of chromosomal restriction fragment patterns by pulsed-field gel electrophoresis. *Infect. Control. Hospital. Epidemiol.* **14,** 595–600.

16. Struelens, M J., Deplano, A., Godard, C., Maes, N., and Serruys, E. (1992) Epidemiological typing and delineation of genetic relatedness of methicillin-resistant *Staphylococcus aureus* by macrorestriction analysis of genomic DNA by using pulsed-field gel electrophoresis. *J. Clin. Microbiol.* **30,** 2599–2605.

17. Gibson, J. R., Sutherland, K., and Owen, R. J. (1994) Inhibition of DNAse activity in PFGE analysis of DNA from *Campylobacter jejuni. Lett. Appl. Microbiol.* **19,** 357–358.

18. Forbes, K. J., Bruce, D. B., Jordens, Z., Ball, A., and Pennington, T. H. (1991) Rapid methods in bacterial DNA fingerprinting. *J. Gen. Microbiol.* **137,** 2051–2058.

19. Linhardt, F., Ziebuhr, W., Meyer, P., Witte, W., and Hacker, J. (1992) Pulsed-field gel electrophoresis of genomic restriction fragments as a tool for the epidemiological analysis of *Staphylococcus aureus* and coagulase-negative staphylococci. *FEMS Microbiol. Lett.* **95,** 181–186.

4

Plasmid Analysis

Alan P. Johnson and Neil Woodford

1. Introduction

Bacterial plasmids are extra-chromosomal, covalently-closed circular (CCC) molecules of DNA that are capable of autonomous replication *(1)*. Plasmids may contain genes for a variety of phenotypic traits, such as antibiotic resistance, virulence, or metabolic activities, although some plasmids comprise genes conferring no detectable phenotype and are said to be "cryptic." Some plasmids (referred to as selftransferable or conjugative plasmids) have the ability to transfer copies of themselves to other bacterial strains or species; this trait is encoded by *tra* genes. In addition, some other plasmids are incapable of selftransfer, but are able to utilize the *tra* functions of conjugative plasmids present in the same bacterial cell to ensure that they are also passed to other strains and species; such plasmids are said to be mobilized.

Determination of the plasmid content of bacterial isolates has been used in medical bacteriology for a number of years. This reached a peak during the early and mid 1980s when "plasmid typing" was very much in vogue as plasmid content was used to investigate possible relationships between epidemiologically related groups of isolates. In these studies banding patterns on agarose gels, produced either by the intact CCC plasmids or by digestion of the total plasmid content with restriction endonucleases, were used as fingerprints for interisolate comparisons. However, for the purpose of bacterial typing, analysis of plasmids has been largely superseded by methods based upon analysis of chromosomal DNA, such as ribotyping (*see* Chapter 2), pulsed-field gel electrophoresis (*see* Chapter 3), or various PCR-based typing approaches (*see* Chapters 6 and 7). In part, the decreased usage of plasmid analysis is based upon the fact that some plasmids may be relatively unstable characteristics of a strain (in comparison with the bacterial chromosome); they can be horizontally

From: *Methods in Molecular Medicine, Vol. 15: Molecular Bacteriology: Protocols and Clinical Applications*
Edited by: N. Woodford and A. P. Johnson © Humana Press Inc., Totowa, NJ

transferred between unrelated bacterial strains or species, or they may be lost from bacterial strains in the absence of selective pressure. For these reasons, in modern medical bacteriology, plasmids are usually investigated because of the phenotypes they confer on the bacteria carrying them. In such studies, plasmid analysis is often combined with Southern blotting (*see* Chapter 2) and hybridization with specific DNA probes in order to determine the location of genes of interest (e.g., antibiotic resistance genes, toxin genes, and so forth).

Several different methods have been developed for the extraction, purification, and analysis of plasmids from bacteria. In the earliest method, bacteria were lysed and plasmid DNA was separated from chromosomal DNA by ultra-centrifugation using a cesium chloride-ethidium bromide density gradient *(2)*. However, this method is time-consuming, not readily applicable to the analysis of large numbers of samples, and has been superseded by the rapid, small-scale methods that are now used routinely in many laboratories. In this chapter, three such methods are described: a modification of the alkaline lysis method of Birnboim and Doly *(3)*, a modification of the method of Kado and Liu *(4)*, and the Brij lysis method *(5)*. In the authors' laboratory, they find the first of these methods suitable for extracting plasmids from a range of gram-negative and gram-positive bacteria, whereas they use the second and third methods to extract plasmids from various gram-negative bacteria and from coagulase-positive and -negative staphylococci, respectively. In addition, methods for the preparation of plasmids suitable for subsequent digestion with restriction endonucleases are also presented, along with methods for the analysis of either undigested or digested plasmid DNA by agarose gel electrophoresis.

The reader should be aware that there are numerous commercially-available kits for the extraction of plasmid DNA. Many of these kits are based upon the principles of the alkaline lysis extraction method described in this chapter (*see* **Subheading 3.1.1.**) and, after lysis of bacterial cells, use cartridges or columns to bind the released plasmid DNA. This DNA is washed on the column and then eluted for subsequent analysis or manipulation. The main reason for the profusion of these kits has been the rapid expansion of recombinant DNA technology into diverse laboratory settings. Accordingly, many of the kits were developed to allow rapid extraction of small, multicopy recombinant plasmids or cosmids, from standard laboratory strains of *Escherichia coli*. In the authors' experience, such kits have been less reliable for extracting plasmids of a range of sizes from clinical isolates of *E. coli* or other species. However, new products appear regularly, and this situation will hopefully improve. Cost is also a concern in most laboratories. In a small laboratory with occasional requirements for plasmid analysis, it may well be worth considering a commercial kit and perfecting its use for the species of interest. However, for larger laboratories that require rapid turnaround of large numbers of isolates belonging to diverse bacterial species, the authors feel that the use of the "in house" methods described in this chapter offers the most cost-effective approach.

2. Materials

2.1. Extraction of Plasmid DNA

2.1.1. Method I

1. Tris-EDTA (TE) buffer: 10 mM Tris-HCl, 1 mM EDTA, pH 8.0.
2. Solution I: TE buffer containing 25% sucrose and 10 mg/mL lysozyme (*see* **Note 1**).
3. Solution II: 0.2 M NaOH, 1% sodium dodecyl sulfate (SDS) (*see* **Notes 2** and **3**).
4. Solution III: 3 M potassium acetate, pH 4.8, stored at 4°C.
5. Phenol:chloroform:isoamyl alcohol (25:24:1), stored at 4°C (*see* **Note 4**).
6. 100% ethanol stored at –20°C.
7. Ribonuclease (RNase): 1 mg/mL stock solution (*see* **Note 5**).

2.1.2. Method II

1. Suspending solution: 50 mM Tris-HCl, 1 mM EDTA, pH 8.0.
2. Lysis solution: 50 mM Tris base, 3% SDS, pH 12.5 (*see* **Note 6**).
3. Phenol:chloroform:isoamyl alcohol (25:24:1), stored at 4°C *(see* **Note 4**).

2.1.3. Method III

1. Suspending solution: 2.5 M NaCl, 50 mM EDTA, pH 7.5.
2. Lysostaphin solution: 200 U/mL in 20 mM sodium acetate stored at –20°C.
3. Lysozyme solution: 100 mg/mL in sterile distilled water.
4. Lysis solution: 50 mM Tris-HCl, 50 mM EDTA, 1% Brij-58, 0.4% sodium deoxycholate, pH 8.0.
5. Phenol:chloroform:isoamyl alcohol (25:24:1), stored at 4°C (*see* **Note 4**).
6. 100% ethanol, stored at –20°C.
7. Protease: 1 mg/mL in sterile distilled water, stored at –20°C.
8. RNase: 1 mg/mL in sterile distilled water, stored at –20°C (*see* **Note 5**).

2.2. Analysis of Plasmids Using Restriction Endonucleases

2.2.1. Gram-Negative Bacteria

1. Suspending solution : 50 mM Tris-HCl, 1 mM EDTA, pH 8.0.
2. Lysis solution: 50 mM Tris base, 3% SDS, pH 12.5 (*see* **Note 6**).
3. Solution III: 3 M potassium acetate, pH 4.8 stored at 4°C.
4. 100% ethanol, stored at –20°C.
5. TE buffer: 10 mM Tris-HCl, 1 mM EDTA, pH 8.0.
6. Phenol:chloroform:isoamyl alcohol (25:24:1), stored at 4°C (*see* **Note 4**).
7. 7.5 M ammonium acetate, stored at 4°C.
8. RNase: 1 mg/mL stock solution, stored at –20°C (*see* **Note 5**).
9. Restriction endonucleases (usually supplied at concentrations of 5–10 U/µL).
10. Reaction buffers for restriction endonuclease enzymes (supplied with the enzymes as 10X concentrated stocks).

2.2.2. Enterococci or Gram-Negative Bacteria

1. All reagents and solutions are those used in **Subheading 2.1.1.** above.
2. Restriction endonucleases (usually supplied at concentrations of 5–10 U/µL).
3. Reaction buffers for restriction endonucleases enzymes (supplied with the enzymes as 10X concentrated stocks).

2.2.3. Staphylococcus aureus

1. Suspending solution: 2.5 M NaCl, 50 mM EDTA, pH 7.5.
2. Lysostaphin solution: 200 U/mL in 20 mM sodium acetate, stored at –20°C.
3. Lysis solution: 0.5% cetyltrimethylammonium bromide (CTAB), 0.25% sarkosyl.
4. 0.5% CTAB.
5. TE buffer: 10 mM Tris-HCl, 1 mM EDTA, pH 8.0.
6. RNase: 1 mg/mL stock solution, stored at –20°C (*see* **Note 5**).
7. Chloroform:isoamyl alcohol (24:1) stored at 4°C.
8. 100% ethanol, stored at –20°C.
9. Restriction endonucleases (usually supplied at concentrations of 5–10 U/µL).
10. Reaction buffers for restriction endonucleases enzymes (supplied with the enzymes as 10X concentrated stocks).

2.3. Examination of Plasmid DNA by Agarose Gel Electrophoresis

1. Loading buffer: 50 mM EDTA, pH 8.0 containing 25% Ficoll and 0.25% bromophenol blue.
2. Tris-borate-EDTA (TBE) electrophoresis buffer (5X concentrated): 445 mM Tris base, 445 mM boric acid, 10 mM EDTA pH 8.0 (*see* **Note 7**).
3. 0.7 or 0.8% (w/v) agarose gels (*see* **Note 8**).
4. Ethidium bromide (EB): 5–10 mg/mL stock solution (*see* **Note 9**).

3. Methods

3.1. Extraction of Plasmid DNA

3.1.1. Method I (for Gram-positive and Gram-negative bacteria; adapted from **ref. 3**)

1. Grow bacteria overnight on agar plates (*see* **Note 10**).
2. Resuspend a loopful of growth in 100 µL of solution I in a sterile 1.5-mL microfuge tube.
3. Incubate suspensions of Gram-negative bacteria at room temperature for 5 min, but incubate suspensions of enterococci and other Gram-positive bacteria at 37°C for 30 min (*see* **Notes 11** and **12**).
4. Add 200 µL of solution II and mix by inverting the tube gently until the cells lyse (*see* **Note 13**).
5. Add 150 µL of cold solution III and mix by shaking vigorously (*see* **Note 14**).
6. Spin in a microcentrifuge for 5 min at maximum speed. Carefully pour the supernatant to a fresh tube and then add 400 µL of phenol:chloroform:isoamyl alcohol. Vortex and spin in a microfuge for 5 min.

7. Carefully transfer 200 μL of the upper aqueous phase to a fresh tube.
8. Add 400 μL of ice-cold ethanol (100%), shake and leave at room temperature for 2 min (*see* **Note 15**).
9. Spin in a microcentrifuge for 5 min, then pour off the ethanol. Residual ethanol may be removed by using a pipet.
10. Resuspend the pellet in 39 μL of distilled water and add 1 μL of RNase (*see* **Note 16**).
11. Add 5 μL of loading buffer and load 20 μL aliquots onto a 0.7% agarose gel.
12. Separate plasmids by agarose gel electrophoresis at 90 V for 2–3 h (*see* **Subheading 3.3.**).

3.1.2. Method II (for Gram-negative bacteria; adapted from **ref. 4**)

1. Grow bacteria overnight on agar plates (*see* **Note 10**).
2. Suspend a loopful of growth in 20 μL of suspending buffer. Vortex briefly to ensure suspensions are uniform (*see* **Note 12**).
3. Add 100 μL of lysis buffer and vortex briefly.
4. Incubate at 56°C for 40 min.
5. Add 100 μL of phenol:chloroform:isoamyl alcohol, mix well and spin for 10–30 min in a microcentrifuge at maximum speed (the aqueous and organic phases must separate totally; *see* **Note 17**).
6. Transfer 35 μL of the upper aqueous phase to a fresh tube.
7. Add 5 μL of loading buffer and load 20 μL aliquots onto a 0.7% agarose gel.
8. Separate plasmids by agarose gel electrophoresis at 90 V for 2–3 h (*see* **Subheading 3.3.**).

3.1.3. Method III (for staphylococci; adapted from **ref. 5**)

1. Grow bacteria overnight on agar plates.
2. Suspend a loopful of growth in 250 μL of suspending buffer in a microfuge tube.
3. For cultures of *S. aureus*, add 12 μL of lysostaphin solution, and incubate at 37°C for 20 min. For coagulase-negative staphylococci, add 38 μL of lysostaphin solution, 12 μL lysozyme solution, and incubate at 37°C for 1 h.
4. Add 400 μL of lysis buffer, mix by inverting gently and spin on a microcentrifuge for 1 h.
5. Very slowly, pipet (or pour) 100 μL of the supernatant into a clean microfuge tube, taking great care not to decant the partially lysed bacteria that often consist of a gelatinous "blob."
6. Add 10 μL of protease and incubate at 37°C for 30 min.
7. Add 200 μL of ice-cold ethanol and precipitate DNA at –20°C for at least 30 min (*see* **Note 15**).
8. Spin in a microcentrifuge for 5 min to pellet the DNA, then pour off most of the alcohol and remove excess with a pipet.
9. Resuspend the DNA in 39 μL sterile distilled water and add 1 μL of RNase (*see* **Note 16**).
10. Add 5 μL of loading buffer and load 10–20 μL aliquots onto a 0.7% agarose gel.
11. Separate plasmids by agarose gel electrophoresis at 90 V for 2 to 3 h (*see* **Subheading 3.3.**).

3.2. Analysis of Plasmids Using Restriction Endonucleases

The three miniprep methods described in **Subheading 3.1.** yield plasmid preparations that contain residual chromosomal DNA. Although this does not pose a problem for the assessment of bacterial plasmid content by agarose gel electrophoresis, the presence of such chromosomal DNA in preparations to be digested will result in production of linear DNA that may not easily be differentiated from that of plasmid origin. In the authors' laboratory, plasmids to be digested are prepared using the methods described below. These have been developed to yield preparations that contain less chromosomal DNA.

3.2.1. Gram-Negative Bacteria

1. Grow bacteria overnight on blood or nutrient agar plates.
2. Reuspend a loopful of growth in 40 μL of suspending buffer (*see* **Note 12**).
3. Add 400 μL of lysis buffer, mix and incubate at 56°C for 30 min.
4. Add 300 μL of cold solution III, mix well and incubate on ice for 20 min.
5. Spin on a microcentrifuge for 5 min. Pour the supernatant into a fresh microfuge tube, through a piece of medical gauze (*see* **Note 18**).
6. Precipitate DNA by adding 2 vol of ice-cold ethanol (approx 800 μL) and leave at room temperature for 2 min (*see* **Note 15**).
7. Pellet the DNA in a microcentrifuge for 10 min, then pour off the alcohol. Remove residual alcohol using a pipet.
8. Resuspend the pellet in 200 μL of TE buffer and add 200 μL of phenol:chloroform:isoamyl alcohol. Mix well and spin on a microcentrifuge for 5 min.
9. Remove 200 μL of the upper aqueous phase and precipitate the DNA by adding 100 μL of 7.5 *M* ammonium acetate and 600 μL of ice-cold ethanol. Store at –20°C for at least 1 h (*see* **Note 15**).
10. Collect the DNA by centrifugation for 5 min in a microcentrifuge. Pour off the alcohol and dry the pellet at 37°C for at least 1 h.
11. Dissolve the pellet in 24 μL of sterile distilled water. Add 1 μL of RNase, 3 μl of the appropriate (10X) reaction buffer and 2 μL of restriction endonuclease (10–20 U).
12. Incubate at 37°C (or the appropriate temperature for the particular enzyme) for at least 4 h (but usually left overnight).
13. Add 5 μL of loading buffer and load 15 μL of the sample onto a 0.8% agarose gel (the remaining sample may be stored at –20°C for subsequent reuse; *see* **Note 19**).
14. Separate linear fragments by agarose gel electrophoresis at 90 V for 2–3 h (*see* **Subheading 3.3.**).

3.2.2. Enterococci or Gram-Negative Bacteria (from **ref. 6**)

1. Grow bacteria overnight on agar plates.
2. Reuspend a loopful of growth in 100 μL of solution I and incubate at 37°C for 30 min (for enterococci) or at room temperature for 5 min (for gram-negative bacteria) (*see* **Note 12**).
3. Add 200 μL of solution II, mix, and incubate at 56°C for 1 h.

4. Add 150 µL of cold solution III, mix by shaking vigorously, and spin in microcentrifuge for 5 min.
5. Pour the supernatant into a fresh tube, through a piece of medical gauze (*see* **Note 18**).
6. Add 400 µL of phenol:chloroform:isoamyl alcohol, mix well, and spin in a microcentrifuge for 5 min.
7. Carefully transfer 200 µL of the upper aqueous phase to a fresh tube, add 400 µL of ice-cold ethanol, mix well, and leave at room temperature for 2 min (*see* **Note 15**).
8. Spin in a microcentrifuge for 5 min, pour off the alcohol, and dry the pellet by incubation at 37°C for at least 1 h.
9. Resuspend the pellet in 22 µL of sterile distilled water. Add 1 µL of RNase, 3 µL of the appropriate (10X) reaction buffer, and 4 µL of restriction endonuclease (30–40 U).
10. Incubate at 37°C (or the appropriate temperature for the particular enzyme) overnight.
11. Add 5 µL of loading buffer and load 15 µL of the sample onto a 0.8% agarose gel (the remaining sample may be stored at –20°C for subsequent reuse) (*see* **Note 19**).
12. Separate linear fragments by agarose gel electrophoresis at 90 V for 2–3 h (*see* **Subheading 3.3.**).

3.2.3. Staphylococcus aureus (adapted from **ref. 7**)

1. Grow bacteria overnight on nutrient agar plates.
2. Resuspend a loopful of growth in 90 µL of suspending solution. Add 10 µL of lysostaphin solution, mix well, and incubate for 20 min at 37°C.
3. Add 500 µL of lysis solution and gently invert the tube for 2–3 min. Complete cell lysis by heating at 56°C for 10–30 min.
4. Centrifuge in a microcentrifuge at maximum speed for 15 min and carefully decant the supernatant into a clean tube.
5. Add 500 µL of 0.5% CTAB and leave at room temperature for 10–30 min. At this stage a sparse, white precipitate should be produced.
6. Centrifuge in a microcentrifuge on the "low speed setting" (~ 5000 rpm) for 10 min. Discard the supernatant and resuspend the pellet in 25 µL of suspending solution.
7. Add 50 µL of TE buffer and 5 µL of RNase, and incubate at 37°C for 30 min.
8. Add 100 µL of chloroform:isoamyl alcohol, invert several times and then centrifuge in a microcentrifuge on the 'low speed setting' (~ 5000 rpm) for 10 min.
9. Remove 50–75 µL of the upper aqueous phase to a clean tube.
10. Precipitate the DNA with 150 µL of ice-cold ethanol and collect by spinning in a microcentrifuge for 5 min at maximum speed (*see* **Note 15**).
11. Remove all traces of ethanol (pour and then use a pipet tip) and complete drying at 37°C for at least 30 min.
12. Resuspend the dried pellet in 14 µL of distilled water and add 2 µL of the appropriate (10X) reaction buffer and 4 µL of restriction endonuclease (30–40 U).
13. Incubate at 37°C (or the appropriate temperature for the particular enzyme) overnight.
14. Add 5 µL of loading buffer and load 15–20 µL of the sample onto a 0.8% agarose gel.
15. Separate linear fragments by agarose gel electrophoresis at 90 V for 2–3 h (*see* **Subheading 3.3.**).

3.3. Examination of Plasmid DNA by Agarose Gel Electrophoresis

Under alkaline conditions, DNA molecules have a net negative charge and will migrate toward a positive electrode (anode) under the influence of an electric field. When subjected to electrophoresis in alkaline buffer and using an agarose gel, DNA molecules of different sizes are separated on the basis of molecular weight. This is the basis for separating mixtures of CCC plasmids or linear DNA fragments resulting from digestion.

1. Fill an electrophoresis tank with 0.5X TBE buffer.
2. Remove the comb from the solidified agarose gel (*see* **Note 8**) to reveal a series of wells. Remove the tape from the ends of the gel tray, while supporting the gel, place the tray together with the gel in the electrophoresis tank. Ensure that the gel is completely covered by the TBE buffer.
3. Load the samples of either undigested or digested plasmid DNA into the wells using a pipet. Also load appropriate markers (*see* **Subheading 3.4.**).
4. Connect the electrophoresis tank to a power pack and carry out electrophoresis at a constant voltage of about 90–100 V until the dye front reaches the end of the gel (2–3 h) or at 15–20 V overnight.
5. Remove the gel from the electrophoresis tank and place it (still on its tray) in a solution of EB (1 μg/mL) to stain the DNA (*see* **Notes 9** and **20**).
6. After staining for 15–20 min, the gel is placed on a transilluminator and DNA bands are detected by irradiation with UV light. The position of DNA bands on the gel are recorded either by photography, or by using an image analyzer (*see* **Note 21**).

3.4. Determining the Sizes of DNA Molecules Separated by Agarose Gel Electrophoresis

The molecular size of plasmids or linear fragments of DNA (expressed either as megadaltons, MDa, or kilobases, kb) may be determined by comparing their electrophoretic mobilities in agarose gels with those of DNA molecules of known size. The electrophoretic mobility of the molecular size standards are assessed by measuring, on gel photographs or images, the distance the DNA bands have migrated, and a calibration curve may be produced by plotting the $\log_{10}$ of the molecular size (MDa or kb) against the distance traveled. The size of the test plasmids or DNA fragments may then be determined from this curve. However, CCC molecules of plasmid DNA exhibit a different electrophoretic mobility from linear DNA fragments of the same size. Thus the size of CCC plasmids may only be accurately estimated if CCC plasmids are used as standards. Similarly, the size of linear fragments of DNA may only be determined by comparison of their mobilities with linear DNA standards.

For determining the sizes of CCC plasmids, a number of bacterial strains containing plasmids of known size are available. In the authors' laboratory, they routinely use two strains of *E.coli;* strain 39R861 (**8**, NCTC 50192) and strain V517 (*9*, NCTC 50193). These two strains contain a number of plasmids of known size within the range 1–100 MDa (approx 1.5–150 kb). For determination of the size of linear fragments of DNA, standard linear DNA fragments (such as digests of phage lambda DNA) are used. These linear DNA markers may be prepared "in house" or obtained commercially.

Within limits, there is a linear relationship between the $\log_{10}$ of the size of DNA molecules and the distance moved through the gel matrix. The size range of DNA molecules falling within the linear part of the curve may be adjusted by altering the agarose concentration of the gel. An agarose concentration of 0.7–0.8% (w/v) allows reasonable estimation of the sizes of plasmids of between 1 and 100 MDa. Smaller plasmids require the use of higher concentrations of agarose in the gel (e.g., 1% w/v), whereas larger molecules require the use of lower concentrations (e.g., 0.5% w/v).

4. Notes

1. To prepare solution I, dissolve 25 g sucrose in TE buffer to a final volume of 100 mL. This solution is stored at room temperature. Lysozyme powder (stored at –20°C) is added on the day of use.
2. To prepare Solution II, add 1 mL of a stock solution of 2 *M* NaOH to 8 mL of distilled water, then add 1 mL of 10% SDS.
3. Stock solutions of 10% SDS are kept at room temperature. However, if the ambient temperature falls, the SDS may precipitate. It may be redissolved by incubation at 37°C.
4. As phenol oxidises (becomes pink) following exposure to light, all solutions containing it should be stored at 4°C in a dark bottle or in a bottle wrapped in foil.
5. Dissolve ribonuclease in distilled water at a concentration of 1 mg/mL and heat the solution at 80°C for 10 min to destroy any contaminating DNase activity. Cool, then store aliquots at –20°C.
6. To prepare lysis solution, dissolve 0.6 g Tris base in 75 mL of distilled water and add 3 g of SDS. Heat at 37°C to dissolve the SDS, then add 4.1 mL of 2 *M* NaOH and adjust the final volume to 100 mL. Check that pH = 12.5 (pH paper may be used). Store at room temperature.
7. To prepare 5X concentrated TBE buffer pH 8.0, dissolve 54 g Tris base in 800 mL of distilled water, then add 3.72 g EDTA (disodium salt). Finally, dissolve 27.83 g boric acid and adjust the final volume to 1 L. Check that the solution is alkaline using pH paper. Dilute this stock solution 10-fold with distilled water prior to use to give working strength of 0.5X concentration.
8. To prepare a 0.7% (or 0.8%) agarose gel, add 0.35 g (or 0.4 g) of agarose to 45 mL of distilled water in a 100 mL conical flask and add 5 mL of 5X concentrated TBE buffer (*see* **Note 7**). Cover the flask with foil to prevent evaporation and boil to dissolve the agarose. Allow the agarose to cool to about 60°C, then pour into a gel tray, the ends of which have been sealed with tape. Position a comb with the required number of teeth to form wells and allow the gel to set.

9. Stock solutions of ethidium bromide are stored at room temperature in a dark bottle or in a bottle wrapped in silver foil, to prevent photoinactivation. A working solution of ethidium bromide (0.5–1 μg/mL) is prepared by diluting 100 μL of stock solution in 500 mL distilled water.

10. Blood or nutrient agars may be used, but in the authors' experience, bacteria grown on MacConkey agar tend not to give good plasmid preparations.

11. In the alkaline lysis procedure, it is the initial step that can be modified to allow use of the technique with a wide variety of bacteria. If, on addition of solution II, adequate lysis is not achieved, the concentration of lysozyme used may be increased, incubation times may be extended or alternative cell wall-weakening enzymes may be used.

12. Certain bacteria (e.g., some strains of *Klebsiella* spp. and other Enterobacteriaceae) are mucoid because they produce excessive capsular polysaccharide, and this may hinder extraction of plasmid DNA. Such strains may be incubated with snail acetone powder (extracted from *Helix pomatia*). The powder is dissolved to a concentration of 10 mg/mL in solution I (*see* **Subheading 2.1.1.**) by vigorous vortexing and microcentrifuged briefly to remove undissolved particles. Bacterial growth is suspended in 100 μL of this solution and incubated for 15 min at room temperature *(10)*. Bacteria are recovered by brief microcentrifugation, the supernatant is discarded, and extraction of plasmid DNA from the pellet may proceed.

13. Addition of solution II causes cell lysis. Ideally this should not be total as this results in contamination of the preparations with excessive chromosomal DNA. Thus, the solution will become opalescent (as opposed to the turbidity of the suspension prior to treatment). Do not vortex the preparations at this point (*see* **Note 16**). The high pH denatures DNA.

14. Addition of the acetate causes selective precipitation of chromosomal DNA (which is linearized during extraction and totally denatured on exposure to solution II) rather than plasmid DNA (which is not totally denatured and can return to its CCC form). RNA-SDS and protein-SDS complexes (formed when cells are lysed by solution II) also precipitate at this point.

15. When 100% ethanol has been added to solutions to cause precipitation of DNA, the DNA is stabilized. Such solutions may be stored for extended periods at –20°C (hours, days, or weeks) before extractions are completed. This provides a convenient point at which many plasmid extraction techniques may be paused.

16. DNA pellets should be resuspended by gently flicking the tube and must not be vortexed. Intact plasmids are required as CCC molecules. Excessive physical and chemical stress during extraction will introduce nicks into the DNA strands. Nicking a strand of a plasmid at a single point introduces a point of free rotation and the supercoiled CCC form will untwist into its open-circular (OC) form. Importantly, OC forms migrate through agarose gels at rates different from their CCC counterparts. If nicks are introduced into both strands of a plasmid, the molecule becomes linearized. Hence, care should be exercised whenever intact plasmids are extracted in order to minimize loss of CCC molecules into their alternative molecular forms.

17. Occasionally, the two phases are not clearly separate after centrifugation (e.g., if excessive bacterial growth was used and a thick pellet of debris masks the upper phase). If this occurs, add 25 µL of suspending solution to the tube, vortex briefly, respin, and continue the procedure.

18. Cut 1 cm² double-layered gauze squares, hold them over the clean microfuge tube, and carefully pour the preparation through it. If poured too quickly, the solution will "roll" over the surface of the gauze and fall onto the bench.

19. Unused digested DNA (i.e., linear fragments) may be stored at –20°C for running on another gel at a later date. Many restriction endonucleases have asymmetric cutting sites and leave "sticky ends" (3' or 5' overhanging bases) on digested fragments. On cooling digested DNA, there is the theoretical possibility of random association and hydrogen bond formation between fragments. This might affect a repeated gel. To avoid this, frozen samples of digested DNA should be rapidly thawed and heated at 65°C for 5 min prior to loading a gel. This heating step will break any H-bonds formed. The authors do not recommend storage of completed preparations of intact, CCC plasmids (*see* **Note 15**). These should be analyzed on the day of extraction and excess should be discarded. Storage at either 4 or –20°C with subsequent thawing will each increase the proportion of OC plasmid molecules (*see* **Note 16**) in the preparation and may lead to confusing gels.

20. In some protocols, ethidium bromide is incorporated into the agarose gel or is included in the electrophoresis buffer (at concentrations ~ 1 µg/mL). Whereas these practices are acceptable for the analysis of linear DNA fragments (including digested plasmids), they are inappropriate for the analysis of CCC plasmids. When ethidium bromide is intercalated into DNA, the molecules become susceptible to nicking by UV light. Whereas this will have negligible effects on the migration of linear DNA, the resulting alterations in the molecular form of CCC plasmids should be avoided.

21. If after visualization of DNA under UV, it is discovered that linear fragments are poorly separated, the gel may be returned to the electrophoresis tank and the voltage restored. However, this is not possible for gels of CCC plasmids (*see* **Note 20**).

References

1. Grinsted, J. and Bennett, P. M. (1986) Introduction in *Methods in Microbiology*, vol. 21, Plasmid Technology (2nd ed.) (Grinsted, J. and Bennett, P. M., eds.), Academic Press, London, pp. 1–10.
2. Sambrook, J., Fritsch, E. F., and Maniatis, T. (1989) *Molecular Clonings: A Laboratory Manual* (2nd ed.). Cold Spring Harbor Laboratory, Cold Spring Harbor, NY.
3. Birnboim, H. C. and Doly, J. (1979) A rapid alkaline extraction procedure for screening recombinant plasmid DNA. *Nucleic Acid Res.* **7**, 1513–1523.
4. Kado, C. I. and Liu, S. T. (1981) Rapid procedure for detection and isolation of large and small plasmids. *J. Bacteriol.* **145**, 1365–1373.
5. Naidoo, J. (1984) Interspecific co-transfer of antibiotic resistance plasmids in staphylococci in vivo. *J. Hyg. (Camb)* **93**, 59–66.

6. Woodford, N., Morrison, D., Cookson, B., and George, R. C. (1993) Comparison of high-level gentamicin-resistant *Enterococcus faecium* isolates from different continents. *Antimicrob Agents Chemother* **37**, 681–684.

7. Townsend, D. E., Ashdown, N., Bolton, S., and Grubb, W. B. (1985) The use of cetyltrimethylammonium bromide for the rapid isolation from *Staphylococcus aureus* of relaxable and non-relaxable plasmid DNA suitable for *in vitro* manipulation. *Letts. Appl. Microbiol.* **1**, 87–94.

8. Threlfall, E. J., Rowe, B., Ferguson, J. L., and Ward, L. (1986). Characterization of plasmids conferring resistance to gentamicin and apramycin in strains of *Salmonella typhimurium* phage type 204c isolated in Britain. *J. Hyg. (Camb)* **97**, 419–426.

9. Macrina, F. L., Kopecko, D. J., Jones, K. R., Ayers, D. J., and McCowen, S. M. (1978) A multiple plasmid-containing *Escherichia coli:* a convenient source of plasmid size reference molecules. *Plasmid* **1**, 417–420.

10. Hibbert-Rogers, L. C. F., Heritage, J., Gascoyne-Binzi, D. M., Hawkey, P. M., Todd, N., Lewis, I. J., and Bailey, C. (1995) Molecular epidemiology of ceftazidime resistant Enterobacteriaceae from patients on a paediatric oncology ward. *J. Antimicrob. Chemother.* **36**, 65–82.

5

DNA Amplification

General Concepts and Methods

Nick A. Saunders and Jonathan P. Clewley

1. Introduction

The polymerase chain reaction (PCR) was first described in 1985 *(1)*, although its theoretical roots go back beyond that time *(2)*. It is the most versatile of the amplification methods; the others (*see* **Subheading 4.**) are more or less confined to diagnostic applications. For example, the product of a PCR, often referred to as an amplicon, can be readily sequenced for diagnostic, typing, fingerprinting, or molecular epidemiologic reasons. PCR is now taking its place in diagnostic microbiology laboratories as an adjunct to culture and serologic tests. PCR tests are available in kit form under the AMPLICOR™ name (Roche Diagnostic Systems, Basel, Switzerland).

2. The Design of PCR Assays
2.1. The PCR Process

PCR requires two synthetic oligonucleotide primers, each of about 20 bases, that are used in an amplification reaction with a thermostable DNA polymerase (e.g., *Taq*). The reaction is incubated using a heating block that cycles between a high denaturing temperature (e.g., 95°C) to melt double-stranded DNA, a lower annealing temperature (e.g., 50–62°C) to allow the primers and target sequences to anneal, and an extension temperature (usually 72°C) to allow the *Taq* DNA polymerase to synthesize the product amplicon. This product amplicon can be any size from around 50 bases to up to 30 kilobases (*see* **Note 1**). For sensitive diagnostic PCR, amplicons are usually in the size range 100–1000 bases.

From: *Methods in Molecular Medicine, Vol. 15: Molecular Bacteriology: Protocols and Clinical Applications*
Edited by: N. Woodford and A. P. Johnson © Humana Press Inc., Totowa, NJ

PCR is susceptible to false-positive results through contamination of the reaction with products of previous reactions (dubbed "carryover" contamination). Problems can also be caused by cross-contamination of clinical specimens at the time of their collection or processing. For these reasons it is essential that extreme care is used at all times during PCR, including the use of physical separation of the preparation, amplification, and analysis stages. For diagnostic purposes, laboratories need to be modified or specially designed if they are to be used for diagnostic PCR *(3–5)*. Additionally, chemical anticontamination measures can be used *(6, 7)*, of which the most common is the incorporation of dUTP instead of TTP in the reaction mixture. The amplicons subsequently made are then susceptible to degradation with the enzyme uracil *N*-glycosylase (UNG), in distinction to the natural target, which will not contain UTP *(see* **Note 2**). This form of enzymatic contamination prevention is part of the AMPLICOR system.

2.2. Design of Primers

It is important that the primers to be used in a PCR are chosen with care *(5)*. They need to give sensitive and specific amplification with no formation of artifactual products through interaction with one another or with other sequences *(8)*. The means whereby a pair of primers is chosen will depend on the intended purpose of the PCR, and the extent of sequence knowledge of the target. When there is only one available sequence of the intended target, selection of the primers is straightforward and can be achieved either empirically by eye, or with the help of a computer program *(9)*. Many commercial DNA analysis software packages offer primer design algorithms (e.g., OLIGO from National Bioscienses, Plymouth, MN; Primer Premier from Biosoft International, Palo Alto, CA; Lasergene from DNAStar Madison, WI; GeneJockey from Biosoft, Cambridge, UK; MacVector from Oxford Molecular, UK); there are also noncommercial programs available (e.g., Primer, Amplify) through the World Wide Web at sites that offer access to many useful programs *(see* **Note 3**).

When choosing the primers by eye for a diagnostic PCR, they should be sufficiently far apart to produce an amplicon that can be resolved and visualized easily on a gel; for example, smaller amplicons (<1000 bp) require specialized agarose *(see* **Note 4**) for satisfactory migration, whereas larger amplicons (>1000 bp) can be resolved on standard (SeaKem) agarose *(see* **Subheading 3.1.1.**).

The primers themselves should be about 18–22-mers chosen to have, as far as possible, an approximately equal representation of the four bases, with no self-complementary sequences (hairpins), or complementarity with each other (which may lead to primer dimer formation). There are different schools of thought about the most appropriate 3' base. C-G pairs are held together by three hydrogen bonds and, hence, have a more stable local structure than A-T pairs (which interact via two hydrogen bonds). However, a 3' T may be tolerated by the polymerase if there is a mismatch, whereas other bases will not be *(10)*. If the PCR is intended to detect a group of organisms for which only a single sequence example of the gene target is known, but for which there may be sequence variability, a codon with minimal degeneracy (*see* **Note 5**) could be chosen as a suitable location for the 3' primer terminus. If, however, several sequences from the target are available (e.g., from different strains or species of the organism), then a multiple alignment should be done (e.g., using Clustal) and the primers chosen from conserved regions.

2.3. PCR Optimization

The optimal pH, Mg^{2+} ion concentration and other reaction conditions for any particular primer pair and template can most easily be determined by using one of the commercially available optimisation kits (e.g., from Invitrogen or Stratagene).

2.4. Hot-Start PCR

The component mixes for PCR are most conveniently assembled at room temperature. However, during this procedure the *Taq* polymerase may start to synthesize unwanted products. These arise from interactions between the primers themselves (primer dimers), between the primers and the source DNA containing the target sequences (mispriming), and from nicks in the source DNA. A consequence of this can be a background of nonspecific bands visible when the products are examined by gel electrophoresis. When this occurs, a "hot-start" protocol may be employed to minimize the problem. The components of the reaction are only all brought together at high temperature, usually greater than 80°C. This can be achieved by using wax beads to separate one or more of the ingredients (e.g., Mg^{2+} ions) from the bulk of the reaction. At high temperature the wax melts and the reaction is initiated. Alternatively, anti-*Taq* antibody can be used to prevent the *Taq* from working until the antibody is denatured. A recent development is the use of a modified form of *Taq* polymerase (Amplitaq Gold, Perkin Elmer), which is essentially inactive prior to treatment at elevated temperature.

2.5. Touch-Down PCR

A progressive decrease in the annealing temperature (Ta) over several cycles until the predicted optimal Ta is reached can also be used to reduce nonspecific amplification *(11)* (computer programs can be used to predict the Ta and Tm, e.g., PrimerSelect in the Lasergene, DNASTar package; *see* **Note 3**). The final temperature that amplifies the fragment of interest has first to be determined. A touchdown PCR protocol is then programmed in the thermal cycler. This could start 5–10 degrees higher than the optimal Ta and decrease one degree per cycle. This helps to ensure that only the specific amplicon is made during the initial PCR cycles and that any nonspecific product made at the lowest Ta is outcompeted.

2.6. Nested or Heminested PCR

PCR is a very efficient process that can result in the production of microgram scale quantities of the target sequence. However, there are constraints upon the system that limit the degree of amplification that can be achieved. These can be considered under three categories:

1. The exhaustion of PCR mix components (primers, deoxynucleotides and polymerase);
2. The accumulation of competing DNAs that, in some cases, are synthesized more rapidly than the desired product (spurious amplicons, primer artifacts, incomplete PCR products, and nicked DNA); and
3. The loss of product caused by hydrolysis at elevated temperature.

Poorly designed primers can be rapidly incorporated into primer dimers, causing their loss from the reaction. However, the polymerase is usually the component of the reaction that is most vulnerable to exhaustion. Although the polymerases used in the PCR are highly thermostable, loss of activity occurs particularly during the denaturation step. This loss is cumulative so that it may be significant by the end of 30 or more cycles.

The accumulation of competing DNAs and hydrolysis of the required PCR product are generally the crucial factors limiting the efficiency of amplification. Careful design and optimization (*see* **Subheading 2.3.**) of the PCR conditions can minimize the effects of these factors. However, it is often not practical to produce, in a standard PCR mixture, sufficient quantities of PCR product for subsequent analyses. This especially applies when the target copy number is very low, or when large quantities of nontarget DNA are present in the specimen.

A commonly used approach to improve the yield and purity of the PCR product is to apply a second PCR to the material produced in the initial amplification. In order to avoid the problem presented by the accumulation of competing DNAs that would rapidly overwhelm a fresh round of PCR using the same primers, it is necessary to employ either one (heminested PCR) or two (nested PCR) new primers that anneal to internal sites on the first-round PCR product (**Fig. 1**).

Generally, only a small aliquot (typically 1 μL for a 50 μL mix) of the initial PCR is added to the heminested or nested amplification. Nested PCR has

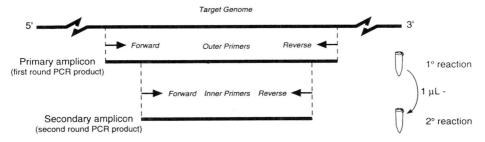

Fig. 1. Nested PCR. The outer primers in the 1^0 reaction mix give a PCR product 1 µL of which is transferred to the 2^0 reaction tube mix containing the inner primers.

proven to be a very successful approach for the amplification of low copy number or otherwise difficult targets. In addition, the specificity of the process is enhanced by the use of nested primers as the probability of products arising because of the amplification of nontarget sequences in both rounds of PCR is remote. The specificity of nested PCR can be further increased by having a 5–10-fold greater concentration of primers in the second round compared with the first (e.g., 25 pmol vs 5 pmol). This is known as booster PCR *(12)*.

The main drawbacks of using nested PCR are that it is more laborious, and that great care must be taken to avoid cross-contamination during transfer from the first round to the second. To overcome these difficulties, several authors have designed nested PCRs in which the primers for the first and second round PCRs are included in a single reaction mix *(13,14)*. The primers for the first round PCR are designed to anneal at a temperature that is too high for binding of the second round primers. The temperature cycle conditions are then varied so that the nested primers are not able to anneal to the template during the early annealing steps. This has been referred to as "drop in, drop out nested PCR" *(15)*. The difficulty of this method is that conditions must be carefully optimized to ensure that the outer primer pair do not continue to monopolize the template at the reduced annealing temperature. This can be done by ensuring that a low concentration of outer primers is used, and that the ratio of the inner to outer primers is high, as for booster PCR *(see above)*. An alternative approach to one-tube nested PCR is to suspend the primers (together with fresh polymerase and deoxynucleotides) in an aqueous drop within a silicon oil overlay *(16)*. After several cycles with a relatively low temperature annealing step, the sealed tube is centrifuged briefly to mix the first and second round PCR mixes, and is then returned to the thermal cycler for further rounds of amplification at a higher annealing temperature. Since the outer primers are designed to anneal only at the lower temperature, they should play no part in the nested reaction. A drawback of the suspended droplet approach is that a relatively large quantity of oil and nested primer mix are required to prevent premature mixing of the primers. Besides being relatively expensive in terms of reagents, the large volume requires a longer dwell time for temperature equilibration resulting in slower tests and some loss of specificity.

3. The Evaluation of PCR End-Points

An optimized PCR should be rapid, specific, and highly sensitive. Ideally, the product will be predominantly an amplicon of the desired DNA sequence. Usually this is the case if one or more nested primers is used. However, in some cases additional products are synthesized, and in a few instances, these may be major components of a mixture. In addition, products of the predicted size may occasionally be produced in reactions that do not include the target sequence. Consequently, on its own, electrophoretic analysis of the PCR product to ascertain that a DNA molecule of the expected size has been produced, although useful, cannot be relied upon and additional evidence must be provided for diagnostic PCR.

PCR amplicons can be resolved by gel electrophoresis, either on agarose or acrylamide, and detected directly by staining with ethidium bromide or Sybr Green (Molecular Probes, Eugene, OR), or indirectly via the incorporation of radioactivity, biotin, or digoxigenin. A gel picture shows the size of the product, how much is made, how many other products are formed (primer dimers, related targets, false priming, and so forth), and is, therefore, a very valuable way of assessing the performance of the amplification. When a new PCR is setup, the identity of the intended target band should be confirmed by a sequence-based assay (Southern or oligomer hybridization, or restriction enzyme digestion). Whether this is then routinely undertaken will depend on the purpose of the PCR. An optimized and validated PCR may be transferred to a solid-phase, microtitre plate-based detection system (*see* **Subheading 3.3.3.**).

3.1. Gel Electrophoresis

3.1.1. Agarose Gels

Large amplicons (0.5–40 kilobases) can be resolved on agarose gels in TAE or TBE buffer *(17)* (*see* Chapters 2 and 4). Smaller amplicons (50–1000 bases) can be run on NuSieve GTG or Metaphor agarose (*see* **Note 4**).

3.1.2. Acrylamide Gel Electrophoresis

Polyacrylamide gel electrophoresis (PAGE) offers higher resolution (down to a single base) than the best agarose gels, and details can be found in standard manuals *(17)*. PAGE is, however, generally less convenient than agarose gel electrophoresis and is, therefore, best suited for specialized amplicon analyses. As an example, heteroduplexes can be resolved by PAGE *(18–20)*. The

Pharmacia PhastSystem overcomes most of the disadvantages associated with acrylamide gels by using small precast gels and semiautomated equipment (*see* **Note 6**).

3.2. Identification of PCR Products by Restriction Fragment Length Polymorphism (RFLP)

Restriction endonucleases have sequence-specific sites of DNA cleavage. It is, therefore, good evidence that the desired PCR product has been synthesized when the fragmentation pattern generated by the action of a particular enzyme matches the profile predicted by examination of the sequence. The degree of confidence in the identity of a PCR product provided by a matching restriction fragment profile depends upon the resolution of the electrophoretic system used for analysis of the fragments. The number of sites and recognition sequence of the enzyme employed are also critical factors.

3.2.1. Choice of Enzyme

The first step in designing a system for the RFLP analysis of a PCR product is to identify enzymes with recognition sites within the sequence amplified. This can be rapidly accomplished using a suitable computer program (e.g., DNASIS from Hitachi America, Brisbane, CA; Lasergene from DNASTAR or the GCG package: *see* **Note 3**). Two criteria need to be considered in making a final selection. First, the restriction fragments generated by the enzyme of choice should produce a pattern of well-resolved bands on electrophoresis. Second, in general, the quality of the evidence for the identity of a PCR product provided by a restriction endonuclease depends upon the number of bases within its recognition site. Thus it is advantageous to choose a restriction endonuclease with a hexanucleotide recognition site (e.g., *Eco*RI, *Hind*III, *Bam*HI, and so on), rather than one with a tetranucleotide site (e.g., *Hae*III, *Sau*3A, *Msp*I). Whatever enzyme is used, its activity in the presence of the components of a PCR (dNTPs, primers, *Taq,* and so forth) should be determined before using it with an unpurified amplicon. If the enzyme is inhibited by the PCR mix, the amplicon must be purified before digestion (*see* **Subheading 3.4.**). The possibility that the restriction site can be lost by mutation of the template or *Taq* misincorporation should also be considered.

3.2.2. Electrophoretic Analysis

All of the gel formats discussed above are suitable for amplicon restriction fragment analysis. A high resolution system is advantageous since more accurate size determination of the restriction fragments decreases the probability of misidentification of the amplicon. However, in most cases, the convenience and reproducibility are more important considerations in the choice of a gel system. The use of appropriate molecular size markers that migrate as close as possible to the bands derived from the PCR product is of great importance. A wide range of suitable markers are commercially available.

3.2.3. Differentiation of Homologous PCR Products by RFLP

In some circumstances it may be desirable to differentiate between strains on the basis of interstrain differences in the base sequence of a PCR product. RFLP may be applicable to such cases when the sequence differences coincide with a restriction site in a subset of strains. However, synthetic oligonucleotide probes are generally more flexible and convenient tools for the comparison of PCR amplicon variants.

3.3. Internal Probes for Identification of PCR Products

Oligonucleotide probes are powerful tools for the identification of specific base sequences in nucleic acids, and they have several features that make them particularly useful in practice. First, they are chemically defined synthetic molecules that can be labeled with a range of useful reporter molecules. Second, they hybridize predictably and have excellent specificity. It may be calculated that the probability of a randomly chosen 20 base sequence occurring in a circular target genome is $0.25^{20}2N$ (where N is the size of the genome in base pairs). So for *Escherichia coli* with a genome of about 5×10^6 bp, a random 20-bp oligonucleotide has only approx 5×10^{-6} chance of matching. This specificity rapidly decays when mismatches are allowed and can be calculated using the general formula for binomial probabilities:

$$(n!/r!(n-r)!)0.25^r(1-0.25)^{n-r}2N$$

where n is the length of the oligonucleotide in bases and r is the number of matches.

For a 20 base probe, the chance of 18 bases matching in the *E. coli* genome is approx 1×10^{-2}. This level of similarity is significant since, in practice, the authors have found that two mismatches in a 20-base oligonucleotide represents the cut-off for significant hybridization to target DNA under nonoptimised conditions. Thus, although the probability of nonspecific hybridization to a chromosomal target is low it remains significant. Genome-specific factors, such as the G + C content and the occurrence of both common and rare base runs, increase the probability that a probe designed to hybridize to a specific sequence will also bind nonspecifically. However, even a PCR with relatively poor specificity produces a restricted range of molecules with a greatly reduced sequence complexity compared with the input DNA. Thus, the probability of nonspecific hybridization of an oligonucleotide probe to the product of a PCR is insignificant. Hybridization to a probe based on a sequence known to be located within the target sequence is, therefore, excellent evidence of the identity of the PCR product and has been widely used.

3.3.1. Choice of an Internal Sequence for Use as a Probe

Oligonucleotides to be used as probes can be complementary to either DNA strand of the amplicon and, in general, the precise choice of sequence is arbitrary. However, certain structural features of oligonucleotides are strongly associated with poor hybridization and these should be screened out.

1. To maintain specificity, the probe should not overlap either of the primer sequences used in the amplification.
2. Probes able to form a stable internal hairpin structure with a stem comprised of ≥3 bases may be inaccessible for hybridization and should be avoided.
3. Oligonucleotides selected as probes should have a (G + C)/(A + T) ratio as close to 50% as possible and should not include single base runs of four or more bases.

Beyond these factors, it is probably not worthwhile to try to predict the relative performance of different probes from their sequences. The length of the probe and its %G + C content will determine its dissociation temperature (Td) and specificity. A reasonable guide to the actual experimental Td is to add 4°C for each G or C residue and 2°C for A or T residues *(21)*. Thus 20 base probes comprising 50% G + C might be expected to have a Td of approx 60°C and, as shown above, to have high sensitivity. Probes of this size are widely used and are compatible with all of the hybridization assay formats described herein. Finally, it is worth searching the nucleotide sequence databases using a computer program such as BLAST to check for the presence of coincidental matches and near matches *(22)*. Although the sequences available in the databases are only a sample of the sequences present in nature, such a search may serve to identify particularly common sequence motifs so that they can be avoided.

3.3.2. Oligonucleotide Labeling

Many methods for labeling oligonucleotides are now available. Reporter molecules may be positioned at either end (5' or 3'), or internally in the probe and the range includes ^{32}P, biotin, fluorescein, rhodamine, coumarin, digoxigenin, and horse-radish peroxidase. The most convenient method of incorporation is to add the reporter to the 5' end of the completed chain during the final cycle on an automatic oligonucleotide synthesizer and the range of amidites available for this purpose now includes biotin, digoxigenin, and various fluorescent molecules. The major advantage of this approach is that the efficiency of labeling achieved is close to 100%. Clearly, the choice of label will depend upon the format of the hybridization assay adopted (*see* **Note 7**).

3.3.3. Hybridization Assay Formats

The classic method used for the identification of a PCR product is Southern blotting *(23)*, which has the advantage that both the molecular size and an internal sequence are used for recognition of the amplicon. The PCR product is electrophoresed through an agarose gel, transferred to a solid support, then denatured and hybridized to a suitably labeled probe (*see* Chapter 2). The main disadvantage of Southern blotting is that it is relatively time-consuming to perform. An alternative method is to denature the product and hybridize the probe prior to the electrophoretic analysis, and then to detect the hybridized probe directly by autoradiography of the frozen gel *(24)*. This method is relatively rapid as it does not involve blotting or the extensive washing of the solid supports, but it does require radioactive labeling of the probe, and the band image is poorly resolved owing to "flaring," which is a particular problem because of the gap between the film and the hybridized probe.

Currently, many methods for identification of PCR products rely on dot-blotting or reverse dot-blotting in which either the PCR product or the probe, respectively, are immobilized on a solid support. These methods, used on their own, do not confirm the size of the PCR amplicon of interest but, as discussed above, they are sufficiently specific to guarantee an accurate identification when the oligonucleotide probe is applied to PCR-derived DNA rather than to genomic DNA.

Dot-blotting relies upon binding of the denatured PCR product to a solid support, which is usually either a nitrocellulose or nylon membrane. The probe is then added and allowed to hybridize to the target sequence before appropriate detection. The dot-blotting method works well, but requires that the PCR products are bound to the filter for each batch of tests, and has now largely been replaced by the more flexible reverse dot-blot method in which the PCR product is hybridized to a surface-bound capture probe. Probes with a poly dT 3' tail may be bound to nitrocellulose filters and stored dry ready for use *(25)*. In order to use microtitre plates as the solid support, aminolinked probes may be directly conjugated with a protein and passively absorbed to the wells. Alternatively, biotin-labeled probes may be immobilized on commercially available, streptavidin-coated plates (LabSystems, Basingstoke, UK). The major advantages of basing the test on microtitre plates are ease of handling, excellent signal-to-noise ratio (SNR), and that the end-point may be assessed objectively using a microtitre plate reader. For simplicity, the strand of the PCR product complementary to the capture probe may be labeled. This can be achieved by using a tagged primer or by incorporation of a labeled nucleotide such as digoxigenin-dUTP. The haptens carried by hybridized material can then be colorimetrically detected, via a suitable enzyme linked to a hapten-specific antibody. The general approach is illustrated in **Fig. 2**. Alternatively, a

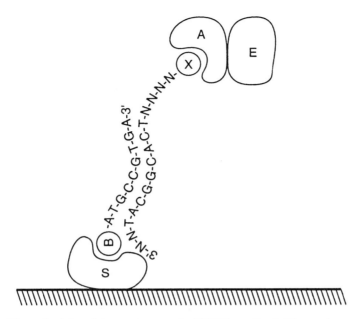

Fig. 2. The principle of the capture probe ELISA method. The capture probe has a 5' biotin moiety *(B)* that binds to streptavidin or avidin *(S)* that has been attached to the surface of a microtitre plate well. The complementary sequence from the denatured PCR amplicon, which is labeled with a hapten *(X)* recognized by the antibody *(A)*, is hybridized to the capture probe. Finally, hapten remaining bound to the well following washing is detected using the antihapten *(A)* enzyme *(E)* conjugated with an appropriate colored or chemiluminescent substrate.

hapten-labeled probe complementary to a second internal sequence may be included in the hybridization mix. The enhanced specificity afforded by the additional probe may be a significant advantage in some systems.

The power of capture probe systems for the identification of PCR products can be stretched further to allow genotyping by including a series of capture oligonucleotides that identify variations in the sequence of the PCR amplicon. For example, a single primer pair based on conserved regions on either side of a variable sequence of the small subunit rRNA gene can be used to amplify the corresponding amplicon from a wide range of bacterial species. In order to confirm the identity of a strain, a sample of DNA from it is amplified using the conserved sequence primers to form a PCR product that is hybridized to a series of capture probes based on the variable sequence. The strain can then be identified from the pattern of hybridization. The product of a single PCR reaction can be split and hybridized to a series of microtitre wells carrying many different capture probes.

A recent advance in the identification of PCR amplicons that also allows the synthesis of product to be assessed without interrupting the PCR process is the "Taq-man" system *(26)*. An oligonucleotide complementary to an internal sequence of the target, is labeled with a different fluorescent dye at either terminus. The dyes are selected so that when they are held relatively close together on the oligonucleotide their fluorescence is quenched. When this probe is added to the PCR it hybridizes to the specific PCR product, and is cleaved close to its 5' terminus, thus severing the link between the dye moieties and reducing the quenching effect. Cleavage is caused by an activity of the 5'-3' exonuclease activity of *Taq* polymerase, which encounters the free 5'-end of the probe during extension of the nascent DNA strand. Equipment supplied by Applied Biosystems is used to quantitate the increase in fluorescence within the PCR reaction during the cycling process.

3.4. Sequencing for Identification of PCR Products

The most direct and accurate method for confirmation of the identity of a PCR product is to determine its sequence *(27)*. Sequencing also provides the best data for comparison of strains at the genetic level as, for example, in studies on pathogenic mechanisms or for epidemiology. However, until recently, sequencing was considered to be too expensive and laborious to justify its use for the analysis of PCR products, except in comparative studies and in a few exceptional cases. This perception is changing because of the availability of automated sequencing equipment and the development of protocols for performing sequencing reactions directly on PCR products purified using rapid methods. At present, in most instances, sequence analysis still cannot compete with the use of RFLP or probes in terms of either convenience or speed, but a general outline of the method follows.

The most appropriate method for sequencing PCR products is by linear cycle sequencing. Labeling the extended sequencing primer for analysis on an automated sequencer is achieved either by the incorporation of 5'-modified primers, or by using modified dideoxynucleotides (dye terminators). Dye terminator sequencing allows greater flexibility since unlabeled primers can be used, and fewer manipulations are required *(27)*. Only one sequencing tube is needed for each reaction compared with the four used with the labeled primer method.

PCR amplicons need to be purified free of excess dNTPs, primers, and *Taq* before sequencing. Removal of these components can be achieved easily *(27)* by membrane filtration (e.g., Centricon 30, Amicon, Danvers, MA) or gel exclusion (e.g., Bio-Spin 30 columns, Bio-Rad, Richmond, CA). Recently, an enzymatic method for the removal of residual primers and deoxynucleotides has been described (Sequenase PCR Product Sequencing Kit protocol booklet, 1994, Amersham, Bucks, UK), which uses the enzymes shrimp alkaline phosphatase and exonuclease I. The main advantage of this approach is the greater potential for automation.

When the PCR amplicon of interest is not the only significant product of the PCR it must be purified by gel electrophoresis prior to sequence analysis. Recovery of DNA fragments from the gel can be achieved by using one of a plethora of different techniques, for example, using GenElute spin columns (Supelco, Poole, UK) (*see* **Note 8**).

4. Alternative Amplification Methods

PCR was originally conceived of as a means of detecting the sickle cell anemia mutation *(1)*, but was soon applied to pathogen detection, for example, HIV *(28)*. Prior to the development of PCR, there had been much speculation over the possibility of nucleic acid hybridization methods becoming as ubiquitous and as commercially profitable as MAb-based diagnostic protocols (e.g., ELISAs). There were perceived to be two main problems: many pathogens have a hit-and-run lifestyle such that their genome DNA or RNA has disappeared by the time symptoms and, hence, a need for diagnosis, becomes apparent, and direct DNA hybridization is insufficiently sensitive for diagnosis; the limit of detection of the most sensitive direct hybridization method, based on ^{32}P, is around 10^5 molecules.

The first obstacle to the use of pathogen detection by DNA hybridization is effectively insurmountable and, therefore, indirect methods based on the detection of an antibody response (IgM and IgG) are more useful. The second obstacle was tackled by many laboratories, but mainly from the angle of trying to increase the signal following the hybridization of a specific probe. A few protocols were described in which the target was amplified by an initial culture step (e.g., growth in tissue culture or on solid media), but this was not perceived as a universal solution. Hence, when in vitro amplification of the target by PCR was described, attention switched to it as the solution to the sensitivity problem of DNA diagnostics. Commercial exploitation was limited, however, because of the complicated ownership of patents and this, and the intellectual freedom provided by the knowledge that in vitro target amplification was possible, spurred the development of alternative methods to achieve the same ends *(29–31)*.

4.1. Alternative Target Amplification Methods
4.1.1. NASBA

The first alternative to PCR was the transcription-based amplification system (TAS) *(32)*. Variants of this technique have since been described by other workers, for instance, the self-sustained sequence (synthetic) reaction or "3SR" *(33,34)* and nucleic acid sequence based amplification (NASBA) *(35)*. The patent rights have been acquired by Organon Teknika, and diagnostic kits under the NASBA trademark are available for HIV-1 and *Mycobacterium tuberculosis*. The NASBA technique essentially involves the conversion, by reverse

transcriptase and RNase H, of the RNA template into a double-stranded complementary DNA molecule carrying a promoter for the T7 RNA polymerase. This
molecule acts as template for the RNA polymerase and RNA copies (both sense
and antisense strands) of the target sequence are made. These copies can in turn
be converted into more copies of the double-stranded cDNA. This is accomplished in a reaction containing the three enzymes and two template-specific
primers (each carrying the T7 RNA polymerase promoter sequence close to its 5'
terminus) at a constant temperature of 41°C. The product of this reaction is single-
stranded RNA, unlike PCR, which produces a double-stranded DNA product.
Like PCR, the RNA product can be detected by gel electrophoresis and hybridization. Unlike PCR, this method is not much used in independent laboratories; it
is more complex to set up, and not as robust as PCR. If Organon Teknika can
deliver a "black box" kit that is reliable and easy to use, it will be adopted in
diagnostic laboratories, if not it is likely only to occupy a specialist niche.

4.1.2. Strand Displacement Assay

The strand displacement assay (SDA) is an isothermal amplification technique
that uses a restriction enzyme (*Hinc*II) and DNA polymerase to give a similar
degree of sensitivity as PCR (10^8-fold). However, it cannot efficiently amplify
target sequences longer than about 120 nucleotides. It is being developed by Becton
Dickinson and has been described for the detection of *M. tuberculosis (36)*.

4.1.3. Ligase Chain Reaction

Since its initial description *(37)* the ligase chain reaction (LCR) has been applied
to the detection of both bacterial and viral sexually transmitted diseases: *Chlamydia
trachomatis* and HIV *(38–40)*. It has been commercialized by Abbott Laboratories.
LCR is similar to PCR in requiring successive cycles of denaturation, primer hybridization, and enzymatic manipulation to replicate the DNA sequence of interest. Unlike
PCR, LCR depends on two pairs of primers, one for each strand of DNA, and a
thermostable DNA ligase enzyme. The DNA ligase joins each pair of primers to
form a product that is the sum of their individual sizes. Although a thermostable
DNA polymerase may be used to fill in short gaps between the ends of the primers,
LCR does not offer the same opportunity for molecular epidemiology as PCR. The
LCR does not amplify the target except as represented by the added oligonucleotides,
or only a few bases between their 3' and 5' ends. Therefore, it can be thought of as a
signal amplification procedure. It is also called the ligation amplification reaction
(LAR) and the ligation detection reaction (LDR).

4.2. Signal Amplification Methods

Amplification of the signal generated by a DNA or RNA probe can be
approached in at least three ways:

1. As a problem similar to the enhancement of the signal from an ELISA, via an enzyme cascade or chemical reaction that generates a flash of light;
2. By sticking as much probe on the target as possible; or
3. By enzymatically amplifying the probe.

The first solution, the enhancement of colorimetric or chemiluminescent detection, is not dealt with here.

4.2.1. Branched DNA

The second solution has been called the "Xmas tree" or compound probes, and involves a cascade or network of probes all hybridizing to one another, thus amplifying the signal. It is available commercially for hepatitis C virus and HIV as the direct quantification branched DNA (bDNA) assay developed by Chiron Corporation. The bDNA assay is based on direct quantification of the target, rather than amplification of it. This is achieved by hybridization of a complex of probes (the branches) to the target, and detection of them by further hybridization with enzyme-linked probes. The enzyme catalyzes a colorimetric reaction in a similar fashion to the detection methods used in target amplification methods and ELISAs. It has been developed for HCV *(41)* and sexually transmitted diseases *(42)*.

4.2.2. Q-Beta Replicase-Directed Amplification

The Q-beta replicase-directed amplification method, based on the RNA bacteriophage Q-beta, was in gestation in 1983 before the reinvention of PCR in 1985 *(43,44)*. It has been described for the detection of *M. tuberculosis (45)*, and is being commercialized by Vysis, Inc. An RNA probe is used to detect the target sequence, rather than DNA. This probe has engineered into it the recognition sequence for Q-beta RNA replicase. After hybridization of the probe to the target, unbound probe is washed away and the remaining, specifically bound probe is enzymatically amplified in vitro with Q-beta RNA replicase. The major problem with this approach is removal of all nonspecifically bound probe; if this is not achieved, high background noise will obscure the signal causing false-positive results. Methods are being developed to overcome this problem *(29)*.

5. Notes

1. The rate of incorporation of dNTPs by *Taq* polymerase is maximal at 72°C and it is usual to allow 30–60 s extension time per kilobase. The enzyme also has significant activity at lower temperatures. The upper limit for amplicon size is usually accepted as being between 4–5 kilobases. This appears to be because of misincorporation of bases leading to the formation of products that are mismatched with the template at the 3'-end, and which are, therefore, extended very inefficiently. Larger amplicons can be produced by employing thermostable DNA polymerases

(such as *Pwo* or *Tth* polymerase) that have proofreading activity (3'-exonuclease). These enzymes have been used on their own or mixed with *Taq* polymerase. For very long PCR amplicons >10 kilobases, changes are also necessary in the reaction buffer and cycling conditions to reduce the level of depurination and subsequent hydrolysis of the template that occurs at elevated temperatures and lower pH values. Mixtures of glycerol (10%) and DMSO (5%) that reduce the temperature required for template denaturation are commonly used in conjunction with a modified buffer system.

2. TTP is replaced by UTP at the same molarity in all standard PCR reactions so that PCR amplicons are rendered sensitive to UNG. Subsequently, any PCR amplicons that are carried over into PCR mixtures can be inactivated (hydrolyzed) by pretreatment with UNG.

3. Volumes in the series, "Methods in Molecular Biology," that discuss the Computer Analysis of Sequence Data contain much useful information about DNA analysis programs. A good WWW starting point is the PCR Jump Station at http://apollo.co.uk/a/pcr. Other sites include http://www.ebi.ac.uk/biocat/biocat.html or http://www.bchs.uh.edu/Server; http://www.chemie.uni-marburg.de/~becker/welcome.html; and http://www.applepi.com.

4. Analysis of PCR products on NuSieve agarose: Prepare a 4% NuSieve GTG (FMC BioProducts, Rockland, ME) agarose gel in Tris-borate EDTA (TBE) buffer (89 mM Tris base, 89 mM boric acid, 2 mM EDTA, pH 8.3). Combine 10–20 µL of the amplification reaction with 4 µL Orange-G/Ficoll loading dye (20% Ficoll, 0.25% Orange G, in 10 mM Tris-HCl, pH 7.4, 1 mM EDTA) in the wells of a microtitre plate. Include as molecular size markers 500 ng of 1 kb, 123 bp or mass ladder (Gibco-BRL), or other suitable markers. With the gel in the electrophoresis tank add buffer to the level of the gel, but do not submerge it. Pipet the reaction products/Orange-G/Ficoll into the dry wells and run the Orange-G 0.5 cm into the gel at low power. Top up the tank with TBE buffer so the gel is submerged and turn the power up. Electrophorese at 125 V for 30–45 min or until the Orange-G dye reaches the end of the gel. Stain the gel with ethidium bromide (0.5 µg/mL in H$_2$O) for 30–60 min, and detect the DNA by UV transillumination. FMC BioProducts supply technical manuals that are helpful in choosing the optimal combination of agarose, buffer, and electrophoresis conditions for resolution of DNA and RNA on gels.

5. The amino acid tryptophan is encoded only by UGG, whereas arginine is encoded by CGU, CGC, CGA, CGG, AGA, AGG. This is a consequence of the redundancy of the genetic code. Thus, when a protein sequence is back-translated to DNA, only TGG specifies Trp, whereas any of six codons may specify Arg. However, organisms differ in the way in which they use the codons available for any particular amino acid, and this information has been collected into codon usage tables for those for which there is sufficient sequence data. These tables can be used to provide a best guess of the most likely DNA sequence coding for an amino acid sequence.

6. Analyses with the Pharmacia PhastSystem: Electrophorese the amplicons on 4–15% or 8–25% gradient gels using the PhastSystem *(20)* (Pharmacia LKB Bio-

technology AB, Uppsala, Sweden). The buffer system in the gels is 112 mM acetate and 112 mM Tris-HCl, pH 6.4. The electrode buffer contains 880 mM L-alanine and 250 mM Tris-HCl, pH 8.8. The separation of homoduplexes and heteroduplexes is achieved with a prerun of 250 V, 4 mA, 1 W, 15°C for 140 Vh. The samples are applied to the gel at 250 V, 0.8 mA, 1 W, 15°C for 5 Vh, and run at 250 V, 4 mA, 1 W, 15°C for a total of 150 Vh (4–15% gradient gel) or 200–300 Vh (8–25% gradient gel). The DNA bands are visualized with the PhastGel Silver Kit (Pharmacia Biotech) optimized for native-PAGE.

7. Companies supplying custom oligonucleotides vary in the range of 5' modifications offered. Labeling at the 5' end of the oligonucleotide is advantageous since the chemistry need not support the addition of additional residues. Thus, suitable amidites are relatively simple and cost-effective to produce.

8. GenElute spin columns for purification of PCR products from agarose gel: The spin column is prepared by adding 100 μL TE, placed on a microfuge tube and centrifuged for 5 s in a microcentrifuge (12,000g). The agarose slice containing the DNA fragment to be recovered is placed in the column on a fresh microfuge tube and centrifuged for 10 min (12,000g). DNA solution collected in the microfuge tube is concentrated by ethanol precipitation.

References

1. Saiki, R. K., Scharf, S., Faloona, F., Mullis, K. B., Horn, G. T., Erlich, H. A., and Arnheim, N. (1985) Enzymatic amplification of beta-globin sequences and restriction site analysis for diagnosis of sickle cell anemia. *Science* **230,** 1350–1354.
2. Kleppe, K., Ohtsuka, E., Kleppe, R., Molineux, I., and Khorana, H. G. (1971) Studies on polynucleotides XCVI. Repair replication of short synthetic DNAs as catalyzed by DNA polymerases. *J. Mol. Biol.* **56,** 341–361.
3. Dragon, E. A., Spadoro, J. P., and Madej, R. (1993) Quality control of the polymerase chain reaction, in *Diagnostic Molecular Microbiology: Principles and Applications* (Persing, D. H., Smith, T. F., Tenover, F. C., and White, T. J., eds.), American Society for Microbiology, Washington DC, pp. 160–168.
4. McCreedy, B. J. and Callaway, T. H. (1993) Laboratory design and workflow, in *Diagnostic Molecular Microbiology: Principles and Applications* (Persing, D. H., Smith, T. F., Tenover, F. C., and White, T. J., eds.), American Society for Microbiology, Washington DC, pp. 149–159.
5. Dieffenbach, C. W., Dragon, E. A., and Dveksler, G. S. (1995) Setting up a PCR laboratory, in *PCR Primer: A Laboratory Manual* (Dieffenbach, C. W. and Dveksler, G. S., eds.), Cold Spring Harbor Laboratory, New York, pp. 7–16.
6. Persing, D. H. and Cimino, G. D. (1993) Amplification product inactivation methods, in *Diagnostic Molecular Microbiology: Principles and Applications* (Persing, D. H., Smith, T. F., Tenover, F. C., and White, T. J., eds.), American Society for Microbiology, Washington DC, pp. 105–121.
7. Hartley, J. L. and Rashtchian, A. (1995) Enzymatic control of carryover contamination in PCR, in *PCR Primer: A Laboratory Manual* (Dieffenbach, C. W. and Dveksler, G. S., eds.), Cold Spring Harbor Laboratory, New York, pp. 23–29.

8. Rychlik, W. (1995) Priming efficiency in PCR. *Biotechniques* **18**, 84–90.

9. Dieffenbach, C. W. and Dveksler, G. S. (1995) Computer software for selecting primers, in *PCR Primer: A Laboratory Manual* (Dieffenbach, C. W. and Dveksler, G. S., eds.), Cold Spring Harbor Laboratory, New York, pp. 681–686.

10. Kwok, S., Chang, S.-Y., Sninsky, J. J., and Wang, A. (1995) Design and use of mismatched and degenerate primers, in *PCR Primer: A Laboratory Manual* (Dieffenbach, C. W. and Dveksler, G. S., eds.), Cold Spring Harbor Laboratory, New York, pp. 143–155.

11. Don, R. H., Cox, P. T., Wainwright, B. J., Baker, K., and Mattick, J. S. (1991) 'Touchdown' PCR to circumvent spurious priming during gene amplification. *Nucleic Acids Res.* **19**, 4008.

12. Ruano, G., Fenton, W., and Kidd, K. K. (1989) Biphasic amplification of very dilute DNA samples via 'booster' PCR. *Nucleic Acids Res.* **17**, 5407.

13. Wilson, S. M., McNerney, R., Nye, P. M., Godfrey-Faussett, P. D., Stoker, N. G., and Voller, A. (1993) Progress toward a simplified polymerase chain reaction and its application to diagnosis of tuberculosis. *J. Clin. Microbiol.* **31**, 776–782.

14. Wilson, P. A., Phipps, J., Samuel, D., and Saunders, N. A. (1996) Development of a simplified polymerase chain reaction-enzyme immunoassay for the detection of *Chlamydia pneumoniae. J. Appl. Bacteriol.* **80**, 431–438.

15. Erlich, H. A., Gelfand, D., and Sninsky, J. J. (1991) Recent advances in the polymerase chain reaction. *Science* **252**, 1643–1651.

16. Trka, J., Divoky, V., and Lion, T. (1995) Prevention of product carry-over by single tube two- round (ST-2R) PCR: application to BCR-ABL analysis in chronic myelogenous leukemia. *Nucleic Acids Res.* **23**, 4735–4737.

17. Sambrook, J., Fritsch, E. F., and Maniatis, T. (1989) *Molecular Cloning: A Laboratory Manual.* Cold Spring Harbor Laboratory, New York.

18. Delwart, E. L., Shpaer, E. G., Louwagie, J., McCutchan, F. E., Grez, M., Rubsamen-Waigmann, H., and Mullins, J. I. (1993) Genetic relationships determined by a DNA heteroduplex mobility assay: analysis of HIV-1 env genes. *Science* **262**, 1257–1261.

19. Delwart, E. L., Sheppard, H. W., Walker, B. D., Goudsmit, J., and Mullins, J. I. (1994) Human immunodeficiency virus type 1 evolution *in vivo* tracked by DNA heteroduplex mobility assays. *J. Virol.* **68**, 6672–6683.

20. Novitsky, V., Arnold, C., and Clewley, J. P. (1996) Heteroduplex mobility assay for subtyping HIV-1: improved methodology and comparison with phylogenetic analysis of sequence data. *J. Virol. Meth.* **59**, 61–72.

21. Suggs, S. V., Hirose, T. M., Kawashima, E. H., Johnson, M. J., Itakura, K., and Wallace, R. B. (1981) Use of synthetic oligonucleotides for the isolation of specific cloned DNA sequences, in *Development Biology Using Purified Genes* (Brown, D. and Fox, C. F., eds.), Academic, New York, pp. 683–693.

22. Altschul, S. F., Boguski, M. S., Gish, W., and Wootton, J. C. (1994) Issues in searching molecular sequence databases. *Nature Genet.* **6**, 119–129.

23. Southern, E. M. (1975) Detection of specific sequences among DNA fragments separated by gel electrophoresis. *J. Mol. Biol.* **98**, 503–517.

24. Gibson, K. M., McLean, K. A., and Clewley, J. P. (1991) A simple and rapid method for detecting human immunodeficiency virus by PCR. *J. Virol. Meth.* **32,** 277–286.

25. Stuyver, L., Rossau, R., Wyseur, A., Duhamel, M., Vanderborght, B., Van Heuverswyn, H., and Maertens, G. (1993) Typing of hepatitis C virus isolates and characterization of new subtypes using a line probe assay. *J. Gen. Virol.* **74,** 1093–1102.

26. Holland, P. M., Abramson, R. D., Watson, R., and Gelfand, D. H. (1991) Detection of specific polymerase chain reaction product by utilizing the 5' to 3' exonuclease activity of *Thermus aquaticus* DNA polymerase. *Proc. Natl. Acad. Sci. USA* **88,** 7276–7280.

27. Arnold, C. and Clewley, J. P. (1996) From ABI Sequence Data to LASERGENE'S EDITSEQ, in *Sequence Data Analysis Guidebook* (Swindell, S. R., ed.), (Walker, J. M., series ed.), Humana Press, Totowa, NJ, pp. 65–74.

28. Kwok, S., Mack, D. H., Mullis, K. B., Poiesz, B., Ehrlich, G., Blair, D., Friedman-Kien, A., and Sninsky, J. J. (1987) Identification of human immunodeficiency virus sequences by using *in vitro* enzymatic amplification and oligomer cleavage detection. *J. Virol.* **61,** 1690–1694.

29. Persing, D. H. (1993) In vitro nucleic amplification techniques, in *Diagnostic Molecular Microbiology: Principles and Applications* (Persing, D. H., Smith, T. F., Tenover, F. C., and White, T. J., eds.), American Society for Microbiology, Washington DC, pp. 51–87.

30. Landegren, U. (1993) Molecular mechanics of nucleic acid sequence amplification. *Trends Genet.* **9,** 199–204.

31. Dieffenbach, C. W. and Dveksler, G. S. (1995) Alternative amplification technology, in *PCR Primer: A Laboratory Manual* (Dieffenbach, C. W. and Dveksler, G. S., eds.), Cold Spring Harbor Laboratory, New York, pp. 623–630.

32. Kwoh, D. Y., Davis, G. R., Whitfield, H. L., Chappelle, L., DiMichele, L. J., and Gingeras, T. R. (1989) Transcription-based amplification system and detection of amplified human immunodeficiency virus type 1 with a bead-based sandwich hybridization format. *Proc. Natl. Acad. Sci. USA* **86,** 1173–1177.

33. Gingeras, T. R., Prodanovich, P., Latimer, T., Guatelli, J. C., Richman, D. D., and Barringer, K. J. (1991) Use of self-sustained sequence replication amplification reaction to analyze and detect mutations in zidovudine-resistant human immunodeficiency virus. *J. Infect. Dis.* **164,** 1066–1074.

34. Gingeras, T. R., Biery, M., Goulden, M., Ghosh, S. S., and Fahy, E. (1995) Optimization and characterization of 3SR-based assays, in *PCR Primer: A Laboratory Manual* (Dieffenbach, C. W. and Dveksler, G. S., eds.), Cold Spring Harbor Laboratory, New York, pp. 653–666.

35. Kievits, T., van Gemen, B., van Strijp, D., Schukkink, R., Dircks, M., Adriaanse, H., Malek, L., and Sooknanen, R. (1991) NASAB™ isothermal enzymatic in vitro nucleic acid amplification optimized for the diagnosis of HIV-1 infection. *J. Virol. Meth.* **35,** 273–286.

36. Spargo, C. A., Haaland, P. D., Jurgensen, S. R., Shank, D. D., and Walker, G. T. (1993) Chemiluminescent detection of strand displacement amplified DNA from

species comprising the Mycobacterium tuberculosis complex. *Mol. Cell. Probes*
7, 395–404.

37. Wu, D. Y. and Wallace, R. B. (1989) The ligation amplification (LAR)-amplification of specific DNA sequences using sequential rounds of template-dependent ligation. *Genomics* **4**, 560–569.

38. Bassiri, M., Hu, H. Y., Domeika, M. A., Burczak, J., Svensson, L. O., Lee, H. H., and Mardh, P. A. (1995) Detection of Chlamydia trachomatis in urine specimens from women by ligase chain reaction. *J. Clin. Microbiol.* **33**, 898–900.

39. Frenkel, L. M., Wagner, L. E., Atwood, S. M., Cummins, T. J., and Dewhurst, S. (1995) Specific, sensitive, and rapid assay for human immunodeficiency virus type 1 pol mutations associated with resistance to zidovudine and didanosine. *J. Clin. Microbiol.* **33**, 342–347.

40. Wiedmann, M., Barany, F., and Batt, C. A. (1995) Ligase chain reaction, in *PCR Primer: A Laboratory Manual* (Dieffenbach, C. W. and Dveksler, G. S., eds.), Cold Spring Harbor Laboratory, New York, pp. 631–652.

41. Chan, C. Y., Lee, S. D., Hwang, S. J., Lu, R. H., Lu, C. L., and Lo, K. J. (1995) Quantitative branched DNA assay and genotyping for hepatitis C virus RNA in Chinese patients with acute and chronic hepatitis C. *J. Infect. Dis.* **171**, 443–446.

42. Urdea, M. S., Kolberg, J., Clyne, J., Running, J. A., Besemer, D., Warner, B., and Sanchez-Pescador, R. (1989) Application of a rapid non-radioisotopic nucleic acid analysis system to the detection of sexually transmitted disease-causing organisms and their associated antimicrobial resistances. *Clin. Chem.* **35**, 1571–1575.

43. Miele, E. A., Mills, D. R., and Kramer, F. R. (1983) Autocatalytic replication of a recombinant RNA. *J. Mol. Biol.* **171**, 203–209.

44. Lizardi, P. M., Guerra, C. E., Lomeli, H., Tussie-Luna, I., and Kramer, F. R. (1988) Exponential amplification of recombinant-RNA hybridisation probes. *Biotechnology* **6**, 1197–1202.

45. Shah, J. S., Liu, J., Buxton, D., Hendricks, A., Robinson, L., Radcliffe, G., King, W., Lane, D., Olive, D. M., and Klinger, J. D. (1995) Q-beta replicase-amplified assay for detection of *Mycobacterium tuberculosis* directly from clinical specimens. *J. Clin. Microbiol.* **33**, 1435–1441.

6

Arbitrarily Primed PCR Methods for Studying Bacterial Diseases

David Ralph and Michael McClelland

1. Introduction

The most common application of the polymerase chain reaction (PCR) is to exponentially amplify a specific known and predictable sequence from a complex mixture of nucleic acids. This chapter describes a technique that, in contrast, uses PCR with arbitrary primers, to generate a fingerprint of PCR products from complex mixtures of nucleic acids and identify sequence polymorphisms and other differences that distinguish them. This robust and reproducible technique can amplify discrete sets of DNA fragments. When genomic DNAs from different individuals are compared, these differences represent point mutations and variously sized deletions and insertions. When different sources of RNA are examined from an isogenic source, differences in the particular set of DNA fragments amplified represent differentially expressed genes.

The technique described in this chapter has many subtle variations that have been given many different names in the literature. The common feature that unifies the techniques described in all of these publications is that they use oligonucleotides of arbitrarily chosen sequence to prime PCR on complex templates such as denatured genomic DNA or total cell RNA (**Fig. 1**). This is done at low stringency, which permits the oligonucleotide to anneal to many sites. The sequences of these annealing sites imperfectly complement the oligonucleotides. There are usually several mismatches between the oligonucleotides and the annealing sites. Nonetheless, the best annealing sites are specific for each oligonucleotide, and complex nucleic acid mixture examined. A subset of these annealing sites are of such quality that DNA polymerases are capable of extending the oligonucleotides and synthesizing copy DNAs. These copy DNAs can be separated from their templates by heating. The temperature of the

From: *Methods in Molecular Medicine, Vol. 15: Molecular Bacteriology: Protocols and Clinical Applications*
Edited by: N. Woodford and A. P. Johnson © Humana Press Inc., Totowa, NJ

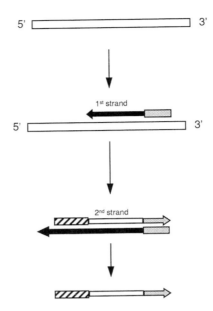

Fig.1. Schematic representation of the arbitrarily primed PCR method (**step 1** uses *Taq* polymerase [Stoffel fragment] for DNA or reverse transcriptase for RNA).

mixture can then be cooled permitting the oligonucleotide of arbitrarily chosen sequence to anneal under similar low stringency conditions to the copy DNA. Again, the oligonucleotide of arbitrary sequence anneals with low affinity, but high specificity to many sites along the copy DNAs. As before, a subset of these annealing sites on the copy DNA are of sufficient quality to permit DNA polymerases to extend the oligonucleotide to make a second-strand copy DNA. As shown in **Fig. 1**, completion of second-strand copy DNA synthesis results in a DNA fragment that has an arbitrarily chosen oligonucleotide at its 5' end, and the complement of an arbitrarily chosen oligonucleotide at its 3' end. Between these two oligonucleotide sequences is a sequence that has been sampled and selected from the original complex mixture of nucleic acids. These second-strand copy DNA fragments, flanked by the oligonucleotides, can be amplified by standard high stringency PCR. The resulting PCR products are separated and displayed by gel electrophoresis. The image of such an electrophoretic gel can be considered a fingerprint of the complex mixture of nucleic acids from which it was derived. Each band on such a fingerprint represents a different sampled fragment of nucleic acid from the original complex mixture. Which DNA fragments are selected for PCR amplification is dependent upon the sequence of the arbitrarily chosen oligonucleotide primer(s). Therefore, different fingerprints or different samples of the original complexity will be obtained for each chosen oligonucleotide of arbitrary sequence. Many different fingerprints of the same original template can be constructed by using various oligonucleotides of arbitrary sequence to extensively sample the original nucleic acid mixture.

A commonly used application of this technique is to compare genomic DNAs from different individuals to identify sequence polymorphisms that distinguish between them. Several kinds of sequence polymorphisms can be detected. A single nucleotide substitution in genomic DNA at an annealing site is likely to lower the affinity of the oligonucleotide ever further, resulting in a failure of DNA polymerases to extend the oligonucleotide at that site. The fingerprint of the genomic DNA from an individual with this nucleotide substitution would then be missing a band that was present in individuals that did not have that particular nucleotide substitution. Other single nucleotide polymorphisms will result in the creation of new annealing sites and their resulting bands on fingerprints. Deletions and insertions can also change the pattern of amplified PCR products. Obviously, if an annealing site is deleted, the relevant PCR product will not be made. Similarly, an insertion can disrupt an annealing site. Some insertions may create a new annealing site, or move one in close proximity to another, thereby permitting a new PCR product to be produced. Finally, small deletions and insertions or frameshift mutations can be detected as length polymorphism in the PCR products derived from the same annealing sites.

The two original publications that used this technique were by Welsh and McClelland *(1)* and Williams et al. *(2)*. Welsh and McClelland called the technique AP-PCR for arbitrarily primed-polymerase chain reaction. Williams et al. called the resulting products RAPD for random amplified polymorphic DNA. These two publications have been cited in over 1500 publications that have used PCR primed with oligonucleotides of arbitrary sequence annealed at low stringency to sample complex mixtures of nucleic acids. In this chapter, these and other variations of this technology will be discussed. This discussion will include both practical descriptions of how the technique is applied and theoretical consideration of experimental design.

2. Materials

1. Multipipettor for 5 to 200 µL vol.
2. TE buffer: 10 m*M* Tris-HCl, pH 8.0, 1 m*M* EDTA.
3. 10% SDS.
4. Proteinase K: 20 mg/mL.
5. 5 *M* NaCl.
6. CTAB/NaCl: 10 g hexadecyltrimethyl ammonium bromide, 4.1 g NaCl in H_2O to a final volume of 100 mL.
7. Chloroform:isoamyl alcohol (24:1).
8. Phenol:chloroform:isoamyl alcohol (25:24:1).
9. Isopropanol.

10. 70% ethanol.
11. Stocks of all four dNTPs: 5 mM.
12. Stocks of primers: 100μM (e.g., from Genosys, Woodlands, TX).
13. [α-^{32}P] dCTP: 3000 Ci/mmol.
14. 2X *Taq* polymerase mixture: 20 mM Tris-HCl, pH 8.3, 20 mM KCl, 8 mM MgCl$_2$.
15. Ampli*Taq* polymerase (Perkin-Elmer-Cetus, Norwalk, CT).
16. Ampli*Taq* Stoffel fragment (Perkin-Elmer-Cetus, Norwalk, CT).
17. GeneAmp PCR System 9600 thermocycler (Perkin-Elmer-Cetus).
18. Formamide dye solution: 96% formamide, 0.1% bromophenol blue, 0.1% xylene cyanol, 10 mM EDTA.
19. 40% Acrylamide stock solutions (19:1 ratio of acrylamide:bisacrylamide).
20. 8 M Urea.
21. Ammonium persulfate: fresh 10% solution.
22. TEMED.
23. 10X TBE buffer: 900 mM Tris-Borate, 20 mM Na$_2$EDTA, pH 8.3.

3. Methods

There are several variables that must be considered when designing a PCR fingerprinting experiment that uses oligonucleotide primers of arbitrary sequence. These include length of primer(s), whether to use one or two primers, MgCl$_2$ concentration, type of electrophoretic gel on which to display the sampled and amplified PCR products, and the thermocycling parameters, including the temperature at which the low stringency annealing is performed. The following is a generic protocol for fingerprinting microbial genomic DNAs that encompasses features of many published accounts.

3.1. Preparation of Template

The quality of a fingerprint is directly dependent on the quality of the genomic DNA used to generate it. Reasonable effort should be made to isolate genomic DNAs that are as free from contaminants as possible. It is important that each DNA is of similar quality. Optimum protocols for isolating and purifying genomic DNAs vary with the type of micro-organism being examined. It may be necessary to modify or independently develop a DNA isolation protocol that is specific for organism type. A protocol that works well for many applications is shown below and is based on that described by Ausubel et al. *(3)*.

1. Pellet 1.5 mL of a bacterial culture in a microcentrifuge
2. Resuspend the cell pellet in 567 μL of TE buffer.
3. Gently mix 30 μL of 10% SDS and 3 μL of a 20 mg/mL solution of Proteinase K into the resuspended cells and then incubate for 1 h at 37°C. The solution should become significantly less turbid and significantly more viscous. If turbidity and viscosity remain unchanged, another 15 μL of 10% SDS may be added. Mix gently and incubate for another 30 min.

4. Add 100 mL of 5 M NaCl and 80 mL of CTAB/NaCl. Incubate for 10 min at 65°C.
5. Extract the solution twice with organic solvents, the first time with chloroform:isoamyl alcohol (24:1) and the second time with phenol:chloroform:isoamyl alcohol (25:24:1). Microcentrifuge for 5 min after each extraction.
7. Precipitate the DNA with 0.6 vol of isopropanol. Wash with 70% ethanol. Resuspend in TE buffer.

3.2. Arbitrarily Primed PCR

It is very strongly recommended that all fingerprinting experiments are set up in duplicate or triplicate with differing concentrations of template being sampled in each replicate.

1. For bacterial genomic DNA, make up dilutions of the templates that are 1.0, 3.0, and 9.0 ng of DNA per µL of water. If only duplicates are desired, 2.0 and 8.0 ng per µL may be used. Dispense 10 µL of each solution into its own prelabeled thin-walled PCR tube (*see* **Note 1**).
2. The number of PCR reactions per arbitrary primer will be two or three times the number of genomic DNAs to be examined. Dispense 10 µL of the 2X PCR reaction mix into each PCR tube containing 10 µL of diluted genomic DNA for a 20 µL final reaction containing 10 mM Tris pH 8.3, 10 mM KCl, 4 mM MgCl$_2$, 200 µM of each dNTP, 2 µM of one or two primers, 1 µCi [α^{32}P]-dCTP and 0.4 U of Ampli*Taq* polymerase (*see* **Note 2**).
3. For arbitrary oligonucleotides that are 17 to 20 mers, the cycling parameters are: heat to 94°C for 2 min; perform two cycles of 94°C for 45 s, 40°C for 5 min, 72°C for 5 min; then 40 cycles of 94°C for 45 s, 55°C for 1 min, 72°C for 2 min; finally, hold at 72°C for 5 min.
4. After the PCR is completed, electrophorese an aliquot. If an agarose gel is preferred, then 10 µL of the final PCR reaction should be used. The amplified products can be visualized by staining with ethidium bromide (EB).
5. Alternatively, sequencing-style denaturing polyacrylamide gels may be desired because of their much superior resolution. This requires that [α^{32}P]-dCTP is incorporated into the PCR products during the reaction. In this case, dilute 2.5 µL of the labeled PCR products with 7.5 µL of sequencing-style formamide tracking dye. Heat to 80°C for 2 min. Chill on ice. Load 2.5 µL into each well of the polyacrylamide gel. Electrophorese at 100 W at constant power for about 3 h or until the xylene cyanol reaches the bottom of the gel. The gel can be dried in a gel dryer. After drying, the gel can be exposed to X-ray film. The PCR products are then visualized by autoradiography. If the gel is not dried, the X-ray film can be exposed to the gel in a frozen state at –80°C.

An example of an autoradiogram of a series of genomic fingerprints is shown in **Fig. 2**. Bands that appear in all lanes of all fingerprints derived from the same arbitrary oligonucleotide represent genetic loci that are identical or conserved between individuals. Bands that occur in all of the lanes of fingerprints of the same genomic DNA, but which vary between the various different

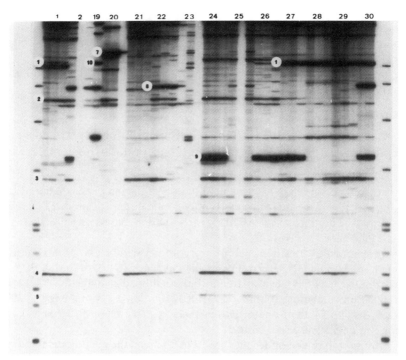

Fig. 2. Arbitrarily primed PCR of *Streptococcus pyogenes* genomes. DNA from various surface antigen types was fingerprinted at three concentrations (100, 50, and 25 µg per 20 µL) using the M13 reverse sequencing primer and *Taq* polymerase. Numbers above the three lanes indicate strain numbers. Numbers within the fingerprint indicate bands scored in a population study *(58)*.

genomic DNAs represent DNA sequences that are polymorphic between the sampled genomes.

3.3. Technical Considerations

3.3.1. Sensitivity of Arbitrarily Primed PCR Determines Appropriate Application

As with the application of any technology, it is important to match arbitrarily primed PCR fingerprinting with applications for which it is most appropriate. A significant attribute of arbitrarily primed fingerprinting to be considered when choosing an appropriate application, is the very high sensitivity of the technique. From the very first uses of this technique, the most successful applications have been to identify differences that distinguish genomes that were really quite similar. Examples of this earlier work include the detection of genetic differences that distinguish inbred strains of the same species of plants or animals *(1,2,4)*. The exquisite sensitivity of this technique to identify genetic polymorphisms within strains of a single species has been extended to many studies of microorganisms. One exceptional example has been the identification of genetic polymorphisms that distinguish various strains of *Escherichia coli* K-12 that have arisen in the 50 yr since K-12 was originally isolated and characterized *(5)*.

Attempts to study organisms with significantly more divergent genomes using arbitrarily primed PCR fingerprinting have proved less satisfactory. Two publications by the authors illustrate this point *(6, 7)*. These studies both examined a variety of strains and species from two genera of spirochetes. In a study that successfully related species, Welsh et al. *(6)* examined members of the genus *Borrelia*. The *Borrelia* species examined share very significant sequence similarity. Arbitrarily primed PCR fingerprinting was able to divide the examined *Borrelia* into three phylogenetic taxa based on shared polymorphic genetic traits as visualized by shared bands on their respective fingerprints. The strengths of this study were twofold. First, there were enough shared traits within each group to link firmly the members within each group to each other. Second, there were sufficient traits shared by all of the examined strains to link all of the examined spirochete strains into the common taxa of the genus *Borrelia*. This combination of shared and polymorphic traits permitted a phylogenetic analysis using the PAUP *(8)* and PHYLIP *(9)* software packages. Not only was it possible to divide the examined strains into three taxonomic groups, it was possible to determine the phylogenetic relationships of the groups to each other.

In contrast, in the study by Ralph et al. *(7)*, where members of the genus *Leptospira* were examined, similarities in fingerprinting patterns clearly grouped the examined strains into seven species; however, there were not enough shared traits to link the species to each other. The species of *Leptospira* were too dissimilar to determine their phylogenetic relationships by this method. Instead, relationships among species had to be determined by analysis of restriction site polymorphisms within their rRNA encoding genes. In this example, fingerprinting genomic DNAs by arbitrarily primed PCR was an effective means to identify a bacterial strain as belonging to a particular species, but was an inappropriate technology to investigate the genus *Leptospira* as a whole. Similar results have been reported by Corney et al. *(10)* who used fingerprinting with arbitrarily primed PCR to identify leptospiral strains isolated from bovids. These studies with spirochetes illustrate the point that the appropriateness of using arbitrarily primed PCR fingerprinting to examine a group of organisms depends both on the relative similarities of the genomes of the organisms involved and the information about the organisms that is desired. If simple strain identification is required, fingerprinting is a robust, reproducible, and highly sensitive technique. If a higher order of taxonomic or phylogenetic information is desired, then a more critical evaluation of the appropriateness of the technology is required.

3.3.2. Comparison of Arbitrarily Primed PCR Fingerprinting with Restriction Fragment Length Polymorphism

Arbitrarily chosen oligonucleotides and restriction endonucleases sample genomic DNAs based on the same kinds of genetic variations. Both methods are very sensitive to genetic alterations in their annealing or recognition sites. Both methods similarly detect insertion and deletion mutations. Data obtained from both methods can be similarly analyzed with a variety of genetic analysis software packages. The question to be considered is when is it appropriate to choose one method over the other. The answer to this question has to do with the complexity of the information that can be generated by both methods. Arbitrarily primed PCR fingerprinting results in an unbiased sampling of the whole genome resulting in 10–200 distinct bands or genetic loci. Digestion of most bacterial genomes with most restriction endonucleases results in the generation of many hundreds to many thousands of DNA fragments. Whereas such a digestion contains vast genetic information, this information is generally inaccessible by simple electrophoretic display because of the extreme complexity of the resulting pattern.

To reduce the complexity of the pattern, two strategies are employed. The first is to limit the survey for genetic variability to a small portion of the bacterial genome. This can be done by either hybridizing a Southern blot of a restriction digest with a probe specific for a small region of the genome and examining the resulting autoradiogram for restriction site polymorphisms, or by PCR amplifying a small portion of the genome, digesting the PCR products and displaying the fragments on an electrophoretic gel. Both of these methods can be very effective. Some have used a combination of restriction analysis and arbitrarily primed PCR fingerprinting to analyze bacterial groups. The problem with the Southern blot is that it is very labor intensive, time consuming, and uses up at least an order of magnitude more genomic DNA than would a similarly informative series of fingerprints made with arbitrary PCR primers.

These problems with Southern blots can be overcome by amplifying a small section of the bacterial genome using specific PCR and subjecting only the amplified portion to restriction analyses. The problem with this and the previously mentioned technique is that they only sample a small portion of the genome for genetic polymorphisms. The genetic diversity of a particular portion of a genome may not always be representative of the genome as a whole. In some contexts, this would be a significant disadvantage, whereas in other contexts, this could be very beneficial. A beneficial context occurs when it is desired to limit a survey for genetic diversity to a region of the genome that is believed to be more informative than the genome as a whole. For example, if the genomes to be compared are highly variable, it is frequently more informative to examine portions of the respective genomes that vary less than average. Such a region might be the rRNA genes that are well-characterized and easy to amplify by PCR. If the genomes to be compared are very similar, then PCR amplification and restriction digestion of a genomic region known to be hypervariable has been found by some to be most informative. Whereas these techniques may be useful in some contexts, they make the assumption that the PCR amplified region is representative in some way of the genome as a whole. This may not be true.

Another limitation of PCR amplification of a specific genomic region is that it requires that the investigator knows enough about the genomes to be examined to allow design of specific PCR primers. In contrast, no previous information is required to fingerprint genomes with arbitrary primers. In addition, by limiting the sample to a small portion of the genome, large-scale insertions and deletions will not be observed. These differences and the contributions to variability contributed by plasmids and other episomes will be entirely missed by a sample limited to a small portion of a genome. Arbitrarily primed PCR fingerprinting provides an unbiased sampling of the entire genetic complexity of the genome being examined and displays it in an accessible format.

The other strategy to reduce the complexity of restriction digest patterns is to use restriction enzymes with large recognition sequences and display the resulting very large DNA fragments by pulsed-field gel electrophoresis (PFGE; *see* Chapter 3). The information in such a display can be very great and very taxonomically informative. Like arbitrarily primed PCR fingerprinting, this strategy also examines the entire bacterial genome. Hirschl et al. *(11)*, Chachaty et al. *(12)*, Liu et al. *(13)*, van Belkum et al. *(14)* and Kersulyte et al. *(15)* have used this technique and arbitrarily primed PCR to examine the genetic diversity of strains of *Helicobacter* spp., *Clostridium difficile, Burkholderia cepacia, Staphylococcus aureus* and *Pseudomonas* spp., respectively. These investigations showed that the information available from a single PFGE experiment and a single arbitrarily primed PCR fingerprint are roughly equal. The disadvantages of PFGE are that sample preparation is time consuming and relatively tedious, and that significantly more DNA is consumed in a PFGE experiment than is typical for a fingerprinting experiment. In addition, there are a limited number of restriction enzymes with recognition sites that are sufficiently rare to be informative in PFGE experiments. There is a very large number of possible informative arbitrary primers. Another consideration to be taken into account is that the samples to be examined by PFGE are usually prepared from live organisms. This may not always be desirable. An advantage of PFGE is that it is very sensitive to large-scale genomic rearrangements that may be difficult to detect by other means (*see* for example **refs.** *16* and *17*).

In general, arbitrarily primed PCR fingerprinting is the technique of first choice when the genomes to be compared share significant similarity and when an unbiased sample of the entire genome is desired. Restriction digestion and examination of small regions of the genome may be most appropriate if the genomes being examined are significantly dissimilar. PFGE may be the technique of first choice when detection of large scale rearrangements are desired.

3.4. Applications of Fingerprinting using Oligonucleotides of Arbitrary Sequence

3.4.1. Taxonomy and Epidemiology

The most common application of arbitrarily primed fingerprinting to genomic DNA from micro-organisms has been for the purpose of identifying and distinguishing organisms that are very closely related to each other. A good example of this application is the work of Makino et al. *(18)* who examined over 100 isolates of an emerging form of cholera, strain 0139, that was first observed in India in 1992. Primary analysis by other means had indicated that 0139 was similar to the strain 01 of *Vibrio cholerae*. However, fingerprinting with arbitrary primers showed that 0139 was more closely similar to a pandemic form of the El Tor strain of *V. cholerae*. This information will be useful in tracking the spread of strain 0139 as part of an attempt to curtail its effects as a possibly pandemic strain. Strains of *Listeria monocytogenes* that were otherwise found to be identical or near identical, by other methods, could be distinguished by arbitrarily primed PCR *(19–22)*. In one study, it was found that a patient with a *L. monocytogenes* infection relapsed with the same strain of the bacterium some time after what had appeared to be an effective antibiotic treatment *(23)*. Strains of other micro-organisms that have been discriminated by fingerprinting with arbitrary primers include *Staphylococcus* spp. *(1,14)*, *Leptospira* spp. *(7,10)*, *E. coli (24–26)*, *Chlamydia trachomatis (27)*, the bovine pathogen *Haemophilus somnus (28)*, the chicken pathogen *Mycoplasma galliseptium (29)*, the mycoplasma-like plant pathogens *(30)*, *Pseudomonas aeruginosa* from patients with cystic fibrosis *(15)* or corneal ulcers *(31)* and isolates of *Aspergillus fumigatus (32)*. An interesting variation on this application was the use of arbitrarily primed PCR fingerprinting to identify taxa-specific genomic sequences *(33)*. These species- or taxa-specific sequences can then be used as probes to identify organisms by either hybridization or specific PCR.

As indicated by the studies cited above as well as many others, the high sensitivity of PCR fingerprinting with arbitrary primers makes this an excellent technique for epidemiologic studies. In one such study, isolates of *Neisseria meningitidis* from an outbreak of the disease at the University of Connecticut were compared with each other and with isolates from asymptomatic individuals at the same location, and with a variety of strains isolated across the US *(34)*. The disease isolates from the outbreak were indistinguishable from each other and from most of the isolates from asymptomatic individuals at the same location. The examined isolates from other locations were genetically diverse. The authors proposed that an outbreak involves "sporadic invasive progression by a strain that also frequently causes asymptomatic colonization." This ability to identify asymptomatic carriers of a pathogenic organism is of great value in an epidemiologic study because it can determine whether different patients are victims of the same epidemic, and because it can identify reservoirs of the pathogen not associated with disease.

Results similar to those reported by Woods et al. *(34)* were found by Makino et al. *(35)* who investigated an outbreak of Izuma fever in Japan. Izuma fever is caused by *Yersinia pseudotuberculosis*. This study demonstrated that the outbreak of Izuma fever was caused by a clonal population of the 5a serotype of this pathogen. Carriers and a possible environmental source were also identified. In addition, 10 *Y. pseudotuberculosis* type strains from various locations were examined and found to yield distinct fingerprints.

The ability of arbitrarily primed PCR fingerprinting to identify natural reservoirs of infectious agents has been used in other epidemiologic investigations. One of these was a study of patient-derived and environmental strains of the fungus *Cryptococcus neoformans* in Nagasaki Prefecture, Japan *(36)*. This study demonstrated that some strains of *C. neoformans* isolated from weathered bird excrement were identical to geographically-related patient specimens, indicating that both sets of isolates were derived from the same population of the pathogen.

In recent years, there have been a number of reports in which fingerprinting genomic DNAs with arbitrary primers has been used to examine the epidemiology of nosocomial infections in hospital wards. The central questions in these investigations are whether patients in hospitals who become ill with an infectious disease acquired their infections after entering the hospital and, if so, from what source within the hospital these infections were acquired. Answers to these questions have immediate impact on improving the quality of patient care. PCR fingerprinting of genomic DNAs with arbitrary primers provides a sufficiently sensitive means to distinguish between genetically distinct strains of pathogens that might otherwise appear identical. The authors demonstrated the value of molecular epidemiologic analysis in a nosocomial methicillin-resistant *S. aureus* outbreak *(37)*. Two studies examined strains of *Clostridium difficile*, a significant causative agent of antibiotic-associated diarrhea and colitis. In one *(38)*, 15 of 20 AIDS patients who had occupied the same hospital ward for various periods over 12-mo were all found to have acquired infections with the same strain of *C. difficile*. Some of these infections were probably nosocomial, and some of these probably involved environmental sources rather than other patients. In the other study *(39)*, strains of *C. difficile* isolated from neonates were examined. This study found that the neonates had acquired their infections from the hospital environment and not from their mothers. Frequently, a particular source within a hospital can be implicated as the reservoir of the pathogen involved in a nosocomial outbreak. One example involves a nosocomial outbreak of lower respiratory infections caused by *Enterobacter cloacae (40)*. In this case, several environmental sources were identified, but only two, isolated from equipment used to care for the affected patients, produced fingerprints that matched those of the patient-derived isolates.

Sometimes arbitrarily primed PCR fingerprinting can be used to show that a cluster of cases of a disease in a particular hospital are unrelated. Although this does not mean that the affected patients acquired their infections before entering the hospital, it clearly indicates that the infections were acquired from different sources. In one study, it was found that nine cases of *Stenotrophomonas maltophilia* from burn patients were unrelated *(41)*. Two examples of similar studies involving fungal pathogens were reported by Bart-Delabesse et al. *(42)*, and Leenders et al. *(43)* who examined clusters of cases of *Candida* septicemia and invasive aspergillosis, respectively. The patient isolates examined in these studies also indicated that most or all of the patient cases were unrelated.

As indicated in a previous section, when fingerprints contain a combination of shared and unshared bands that link and distinguish the various examined strains, respectively from each other, it is possible to treat these bands as phylogenetic characters. Sometimes it has been informative to construct phylogenetic hypotheses that predict the phylogenetic relationships of the examined organisms. An important epidemiologic question that can be answered by such an analysis is whether pathogenic and nonpathogenic strains of the same organism are phylogenetically distinct. The answer is important because it determines if organisms derived from a clonal lineage are switching between a pathogenic and nonpathogenic form. Two examples of studies involving pathogenic single-celled eukaryotes were reported by Mackenstedt and Johnson *(44)* and Guo and Johnson *(45)*. These investigations examined pathogenic and avirulent strains of *Entamoeba histolytica* and *Toxoplasma gondii*, respectively. It was found for both of these organisms that pathogenic and avirulent strains were phylogenetically distinct. An alternative hypothesis, that these organisms may exist as a phylogenetically indistinct community of benign symbionts that are occasionaly mobilized or activated to a pathogenic form was discounted by these data. However, this is not always the case. An example of such a study involved a phylogenetic analysis of over 100 isolates of *Porphyromonas gingivalis,* an organism associated with periodontal disease *(46)*. All of the strains isolated from humans and old world monkeys were phylogenetically related and distinct from isolates obtained from other animals. The evidence suggested that *Porphyromonas gingivalis* is a commensal that occasionaly, when conditions are permissive, becomes an opportunist pathogen. The evidence also suggests that the host determined which phylogenetic group of this organism colonized the oral cavity. These results are similar to a study of *Vibrio* spp. isolated from a wide variety of marine niches near Newfoundland *(47)*. In this study, phylogenetic analyses of fingerprints made by arbitrarily primed PCR showed that regionally isolated strains were distinct from strains isolated elsewhere. They also demonstrated that the phylogenetic group to which a strain of *Vibrio* belonged, correlated strongly with the type of other organism with which it was associated. For example, strains isolated from seaweed were more closely related to each other than they were to strains isolated from scallops.

3.4.2. Fingerprinting RNA with Arbitrary Primers to Identify Differentially Expressed Genes

Arbitrarily primed PCR fingerprinting is a sensitive means of identifying differences that distinguish populations of nucleic acids. The previous sections have described the application of this technique to identifying genetic differences that distinguish the genomes of relatively closely related micro-organisms. Another commonly used application of PCR primed with arbitrarily chosen oligonucleotides has been to identify differentially expressed genes by reverse transcribing mRNA, and then fingerprinting the resulting cDNA. In this application, differences in the populations of nucleic acids being sampled are derived from mRNAs whose abundances vary. This is particularly useful when examining genes that are differentially regulated in alternative developmental or regulatory pathways. Usually, mRNA populations are sampled from organisms representing a single genetically homogeneous strain that have been treated in some way to induce differential mRNA expression. By examining different regulatory states in a genetically homogeneous strain, differences in the fingerprints derived from arbitrarily primed PCR are caused by differentially expressed genes, and not by interstrain genetic sequence variations. An important characteristic of using arbitrarily primed PCR fingerprinting for identifying genes whose mRNA abundances vary, is that it samples the repertoire of expressed genes without prejudice and identifies those mRNAs that are differentially regulated. By eliminating prejudice in the survey methodology, the investigator is more likely to discover regulatory pathways that are unanticipated by previous results. Such unanticipated observations can result in new insights into the regulatory or developmental pathways being investigated.

Using arbitrarily primed PCR fingerprinting of RNA to identify differentially expressed genes was first developed to identify such genes in mammalian systems. The first two published accounts of the reduction to practice of this concept were by Liang and Pardee *(48)*, who called the technique differential display PCR (ddPCR), and a group including the authors *(49)*, who called the technique RNA fingerprinting by arbitrarily primed PCR or RAP-PCR. Both versions of this application are grounded in the earlier work of Williams et al. *(2)* and Welsh et al. *(1)*. Whereas both ddPCR and RAP-PCR have been used successfully to identify differentially expressed genes in a wide variety of mammalian systems, only the version of Welsh et al. *(50)* is applicable to non-eukaryotic systems because the version of Liang and Pardee *(48)* uses oligo-dT based primers to prime reverse transcription before PCR amplification with arbitrarily chosen oligonucleotide primers. Most microbial systems lack the polyA tails to which these oligo-dT based primers anneal. The structure of genetic information into operons in many microbial systems also makes these oligo-dT primers unsatisfactory. Welsh et al. *(49)* used oligonucleotides of arbitrarily chosen sequence to prime first strand cDNA synthesis and the following PCR.

To date, arbitrarily primed PCR fingerprinting of RNA has been used to investigate two microbial developmental systems. One was a model system for the induction of pathologically relevant genes in *Salmonella typhimurium* by hydrogen peroxide *(51)*. The other system involved the various developmental stages of trypanosomes *(52–54)*. An unusual variation on the use of arbitrarily primed PCR fingerprinting to identify differentially regulated genes has been reported by Pogue et al. *(55)* who examined a different group of trypanosomes in the genus *Leishmania*. The experiments reported in this study first used arbitrarily primed PCR fingerprinting to identify genomic DNA fragments that occurred differentially among the examined strains. These identified genomic fragments were then used on Northern blots to examine relative differences in mRNA abundances. They found that some of their genomic fragments encoded genes were differentially expressed. All of these reports demonstrate that arbitrarily primed PCR fingerprinting can be employed to identify differentially regulated genes in microbial systems, and indicate the likely utility of this approach in other microbial systems.

In conclusion, in investigations of microbial ecology, epidemiology or phylogenetics, it is frequently necessary or useful to identify genetic differences that discriminate between closely related and highly similar micro-organisms. Many of the techniques used in these types of studies either lack the sensitivity to discriminate between highly similar strains or are inconvenient to use. This chapter has described a technique that uses oligonucleotides of arbitrarily chosen sequence to prime PCR amplification of a sample of genetic material from a variety of sources. Upon display in an appropriate electrophoretic system, the PCR amplified DNA fragments constitute a fingerprint composed of an unbiased sampling of the DNA being examined. The technique, called many names by many different investigators, is robust, reproducible, and highly sensitive to small genetic differences. In studies involving micro-organisms, this technique has been most frequently used to examine genomic DNAs to identify genetic differences that distinguish closely related strains of micro-organisms. In numerous studies, many of which are cited herein, this technique has demonstrated its superior utility in this application. In a variety of eukaryotic systems, arbitrarily primed PCR fingerprinting has been used to construct genetic maps and to identify differentially expressed genes. These applications await further development in microbial systems.

4. Notes

1. Frequently, large quantities of genomic DNAs are not available. This will limit the number of fingerprints that may be attempted. This situation could arise when the organisms under investigation are fastidious, slow-growing, or difficult to obtain. It could also arise if the organism is pathogenic, making very large-scale propagation imprudent. Under these circumstances where there is limited starting genomic material, it is advisable that the investigator first make a large number of fingerprints on a single abundant template that is related to the template of interest using a large array of oligonucleotide primers, because some oligonucleotides yield better fingerprints than others. It is not clear why this occurs, but there may be effects because of genome composition. For example, oligonucleotides that are predominantly guanine and cytosine may give poor fingerprints on genomes that are predominantly adenine and thymine. A subset of all examined oligonucleotides will give significantly more robust fingerprints than average. By using only these superior oligonucleotides to fingerprint the rarer and more valuable genomic DNAs, the maximum amount of information can be obtained from these more precious *(21,56)*.

2. Selecting Oligonucleotides of Arbitrary Sequence for Fingerprinting. In the example of a basic protocol given in **Subheading 3.** above, the arbitrarily chosen oligonucleotides are described as 17–20 mers. Oligonucleotides of almost any length may be used. Many researchers have used 10-mers that may be purchased commercially (e.g., from Genosys Biotechnologies, The Woodlands, Texas, or Operon, Alameda, CA). If oligonucleotides of arbitrary sequence of less than 15 nucleotides are chosen for fingerprinting experiments, the protocol is the same as described above, except that the annealing temperature is not raised after the first two cycles of PCR. The number of PCR products can be influenced by the polymerase used. In particular, it is often desirable to use the truncated "Stoffel" form of *Taq* polymerase when the primers are shorter than about 15 bases because this enzyme generally produces a richer fingerprint *(57)*.

Acknowledgments

This work was supported by grants from the US National Institute of Allergy and Infectious Diseases, AI 34829 (McClelland), AI 32644 (Welsh), and AI40042 (Ralph).

References

1. Welsh, J. and McClelland, M. (1990) Fingerprinting genomes using PCR with arbitrary primers. *Nucleic Acids Res.* **18,** 7213–7218.

2. Williams, J. G., Kubelik, A. R., Livak, K. J., Rafalski, J. A., and Tingey, S. V. (1990) DNA polymorphisms amplified by arbitrary primers are useful as genetic markers. *Nucleic Acids Res.* **18,** 6531–6535.
3. Ausubel, F., Brent, R., Kingston, R., Moore, D., Seidman, J., Smith, J., and Struhl, K. (1992) *Short Protocols in Molecular Biology,* 2nd ed., John Wiley and Sons, New York.
4. Welsh, J., Petersen, C., and McClelland, M. (1991) Polymorphisms generated by arbitrarily primed PCR in the mouse: application to strain identification and genetic mapping. *Nucleic Acids Res.* **19,** 303–306.
5. Brikun, I., Suziedelis, K., and Berg, D. E. (1994) DNA sequence divergence among derivatives of *Escherichia coli* K-12 detected by arbitrary primer PCR (random amplified polymorphic DNA) fingerprinting. *J. Bacteriol.* **176,** 1673–1682.
6. Welsh, J., Pretzman, C., Postic, D., Saint Girons, I., Baranton, G., and McClelland, M. (1992) Genomic fingerprinting by arbitrarily primed polymerase chain reaction resolves *Borrelia burgdorferi* into three distinct phyletic groups. *Int. J. Syst. Bacteriol.* **42,** 370–377.
7. Ralph, D., McClelland, M., Welsh, J., Baranton, G., and Perolat, P. (1993) *Leptospira* species categorized by arbitrarily primed polymerase chain reaction (PCR) and by mapped restriction polymorphisms in PCR-amplified rRNA genes. *J. Bacteriol.* **175,** 973–981.
8. Swofford, D. (1991) PAUP: phylogenetic analysis using parsimony, version 3.0. *Illinois Natural History Survey, Champaign.*
9. Felsenstein, J. (1988) Phylogenies from molecular sequences: inference and reliability. *Annu. Rev. Genet.* **22,** 521–565.
10. Corney, B. G., Colley, J., Djordjevic, S. P., Whittington, R., and Graham, G. C. (1993) Rapid identification of some *Leptospira* isolates from cattle by random amplified polymorphic DNA fingerprinting. *J. Clin. Microbiol.* **31,** 2927–2932.
11. Hirschl, A. M., Richter, M., Makristathis, A., Pruckl, P. M., Willinger, B., Schutze, K., and Rotter, M. L. (1994) Single and multiple strain colonization in patients with *Helicobacter pylori* associated gastritis: Detection by microrestriction DNA analysis. *J. Infectious Diseases* **170,** 473–475.
12. Chachaty, E., Saulnier, P., Martin, A., Mario, N., and Andremont, A. (1994) Comparison of ribotyping, pulsed-field gel electrophoresis and random amplified polymorphic DNA for typing *Clostridium difficile* strains. *FEMS. Microbiol. Lett.* **122,** 61–68.
13. Liu, P. Y. -F., Shi, Z. -Y., Lau, Y. -J., Hu, B. -S., Shyr, J. -M., Tsai, W. -S., Lin, Y. -H., and Tseng, C. -Y. (1995) Comparison of Different PCR approaches for characterization of *Burkholderia (Pseudomonas) cepacia* Isolates. *J. Clin. Microbio.* **33,** 3304–3307.
14. van Belkum, A., Kluytmans, J., van Leeuwen, W., Bax, R., Quint, W., Peters, E., Fluit, A., Vandenbroucke Grauls, C., van den Brule, A., Koeleman, H., and et al., (1995) Multicenter evaluation of arbitrarily primed PCR for typing of *Staphylococcus aureus* strains. *J. Clin. Microbiol.* **33,** 1537–1547.
15. Kersulyte, D., Struelens, M. J., Deplano, A., and Berg, D. E. (1995) Comparison of arbitrarily primed PCR and macrorestriction (pulsed-field gel electrophoresis)

typing of *Pseudomonas aeruginosa* strains from cystic fibrosis patients. *J. Clin. Microbiol.* **33,** 2216–2219.

16. Honeycutt, R. J., McClelland, M., and Sobral, B. W. (1993) Physical map of the genome of *Rhizobium meliloti* 1021. *J. Bacteriol.* **175,** 6945–6952.

17. Liu, S. L. and Sanderson, K. E. (1995) Rearrangements in the genome of the bacterium *Salmonella typhi. Proc. Natl. Acad. Sci. USA* **92,** 1018–1022.

18. Makino, S., Kurazono, T., Okuyama, Y., Shimada, T., Okada, Y., and Sasakawa, C. (1995) Diversity of DNA sequences among *Vibrio cholerae* O139 Bengal detected by PCR-based DNA fingerprinting. *FEMS. Microbiol. Lett.* **126,** 43–48.

19. MacGowan, A. P., O'Donaghue, K., Nicholls, S., McLauchlin, J., Bennett, P. M., and Reeves, D. S. (1993) Typing of *Listeria* spp. by random amplified polymorphic DNA (RAPD) analysis. *J. Med. Microbiol.* **38,**322–327.

20. Czajka, J. and Batt, C. (1994) Verification of causal relationships between *Listeria monocytogenes* isolates implicated in food-borne outbreaks of Listeriosis by Random Amplified Polymorphic DNA patterns. *J. Clin. Micro.* **32,** 1280–1287.

21. Farber, J. M. and Addison, C. J. (1994) RAPD typing for distinguishing species and strains in the genus *Listeria. J. Appl. Bacteriol.* **77,** 242–250.

22. Boerlin, P., Bannerman, E., Ischer, F., Rocourt, J., and Bille, J. (1995) Typing *Listeria monocytogenes:* a comparison of random amplification of polymorphic DNA with 5 other methods. *Res. Microbiol.* **146,** 35–49.

23. Levett, P. N., Bennett, P., O'Donaghue, K., Bowker, K., Reeves, D., and MacGowan, A. (1993) Relapsed infection due to *Listeria monocytogenes* confirmed by random amplified polymorphic DNA (RAPD) analysis [letter]. *J. Infect.* **27,** 205–207.

24. Heuvelink, A. E., van de Kar, N. C., Meis, J. F., Monnens, L. A., and Melchers, W. J. (1995) Characterization of verocytotoxin-producing *Escherichia coli* O157 isolates from patients with haemolytic uraemic syndrome in Western Europe. *Epidemiol. Infect.* **115,** 1–14.

25. Leroy Setrin, S., Lesage, M. C., Chaslus Dancla, E., and Lafont, J. P. (1995) Clonal diffusion of EPEC-like *Escherichia coli* from rabbits as detected by ribotyping and random amplified polymorphic DNA assays. *Epidemiol. Infect.* **114,**113–121.

26. Madico, G., Akopyants, N. S., and Berg, D. E. (1995) Arbitrarily primed PCR DNA fingerprinting of *Escherichia coli* O157:H7 strains by using templates from boiled cultures. *J. Clin. Microbiol.* **33,** 1534–1536.

27. Scieux, C., Grimont, F., Regnault, B., Bianchi, A., Kowalski, S., and Grimont, P. A. D. (1993) Molecular typing of *Chlamydia trachomatis* by random amplification of polymorphic DNA. *Res. Microbiol.* **144,** 395–404.

28. Myers, L., Silva, S., Procunier, J., and Little, P. (1993) Genomic fingerprinting of *"Haemophilus somnus"* isolates by random-amplified polymorphic DNA analysis. *J. Clin. Microbiol.* **31,** 512–517.

29. Geary, S. J., Forsyth, M. H., Aboul Saoud, S., Wang, G., Berg, D. E., and Berg, C. M. (1994) *Mycoplasma gallisepticum* strain differentiation by arbitrary primer PCR (RAPD) fingerprinting. *Mol. Cell Probes.* **8,** 311–316.

30. Chen, K. H., Credi, R., Loi, N., Maixner, M., and Chen, T. A. (1994) Identification and grouping of mycoplasma-like organisms associated with grapevine yel-

lows and clover phyllody diseases based on immunological and molecular analyses. *Appl. Environ. Microbiol.* **60**, 1905–1913.

31. Bukanov, N., Ravi, V. N., Miller, D., Srivastava, K., and Berg, D. E. (1994) *Pseudomonas aeruginosa* corneal ulcer isolates distinguished using the arbitrarily primed PCR DNA fingerprinting method. *Curr. Eye Res.* **13**, 783–790.

32. Anderson, M., Gull, K., and Denning, D. (1996) Molecular typing by random amplification of polymorphic DNA and M13 hybridization of related paired isolates of *Aspergillus fumigatus. J. Clin. Micro.* **34**, 87–93.

33. Martinez Murcia, A. J. and Rodriguez Valera, F. (1994) The use of arbitrarily primed PCR (AP-PCR) to develop taxa specific DNA probes of known sequence. *FEMS. Microbiol. Lett.* **124**, 265–269.

34. Woods, J. P., Kersulyte, D., Tolan, R. W., Jr., Berg, C. M., and Berg, D. E. (1994) Use of arbitrarily primed polymerase chain reaction analysis to type disease and carrier strains of *Neisseria meningitidis* isolated during a university outbreak. *J. Infect. Dis.* **169**, 1384–1389.

35. Makino, S., Okada, Y., Maruyama, T., Kaneko, S., and Sasakawa, C. (1994) PCR-based random amplified polymorphic DNA fingerprinting of *Yersinia pseudotuberculosis* and its practical applications. *J. Clin. Microbiol.* **32**, 65–69.

36. Yamamoto, Y., Kohno, S., Koga, H., Kakeya, H., Tomono, K., Kaku, M., Yamazaki, T., Arisawa, M., and Hara, K. (1995) Random amplified polymorphic DNA analysis of clinically and environmentally isolated *Cryptococcus neoformans* in Nagasaki. *J. Clin. Microbiol.* **33**, 3328–3332.

37. Fang, F. C., McClelland, M., Guiney, D. G., Jackson, M. M., Hartstein, A. I., Morthland, V. H., Davis, C. E., McPherson, D. C., and Welsh, J. (1993) Value of molecular epidemiologic analysis in a nosocomial methicillin-resistant *Staphylococcus aureus* outbreak [see comments]. *JAMA* **270**, 1323–1328.

38. Barbut, F., Mario, N., Meyohas, M. C., Binet, D., Frottier, J., and Petit, J. C. (1994) Investigation of a nosocomial outbreak of *Clostridium difficile*-associated diarrhoea among AIDS patients by random amplified polymorphic DNA (RAPD) assay. *J. Hosp. Infect.* **26**, 181–189.

39. Martirosian, G., Kuipers, S., Verbrugh, H., van Belkum, A., and Meisel Mikolajczyk, F. (1995) PCR ribotyping and arbitrarily primed PCR for typing strains of *Clostridium difficile* from a Polish maternity hospital. *J. Clin. Microbiol.* **33**, 2016–2021.

40. Riain, U. N., Cormican, M. G., Flynn, J., Smith, T., and Glennon, M. (1994) PCR based fingerprinting of *Enterobacter cloacae. J. Hosp. Infect.* **27**, 237–240.

41. Chatelut, M., Dournes, J. L., Chabanon, G., and Marty, N. (1995) Epidemiological typing of *Stenotrophomonas* (*Xanthomonas*) *maltophilia* by PCR. *J. Clin. Microbiol.* **33**, 912–914.

42. Bart Delabesse, E., van Deventer, H., Goessens, W., Poirot, J. L., Lioret, N., van Belkum, A., and Dromer, F. (1995) Contribution of molecular typing methods and antifungal susceptibility testing to the study of a candidemia cluster in a burn care unit. *J. Clin. Microbiol.* **33**, 3278–3283.

43. Leenders, A., van Belkum, A., Janssen, S., de Marie, S., Kluytmans, J., Wielenga, J., Lowenberg, B., and Verbrugh, H. (1996) Molecular epidemiology of apparent

outbreak of invasive Aspergillosis in a hematology ward. *J. Clin. Microbiol.* **34,** 345–351.

44. Mackenstedt, U. and Johnson, A. M. (1995) Genetic differentiation of pathogenic and nonpathogenic strains of *Entamoeba histolytica* by random amplified polymorphic DNA polymerase chain reaction. *Parasitol. Res.* **81,** 217–221.

45. Guo, Z. G. and Johnson, A. M. (1995) Genetic characterization of *Toxoplasma gondii* strains by random amplified polymorphic DNA polymerase chain reaction. *Parasitology.* **111,** 127–132.

46. Menard, C. and Mouton, C. (1995) Clonal diversity of the taxon *Porphyromonas gingivalis* assessed by random amplified polymorphic DNA fingerprinting. *Infect. Immun.* **63,** 2522–2531.

47. Martin Kearley, J., Gow, J. A., Peloquin, M., and Greer, C. W. (1994) Numerical analysis and the application of random amplified polymorphic DNA polymerase chain reaction to the differentiation of *Vibrio* strains from a seasonally cold ocean. *Can. J. Microbiol.* **40,** 446–455.

48. Liang, P. and Pardee, A. B. (1992) Differential display of eukaryotic messenger RNA by means of the polymerase chain reaction [*see* comments]. *Science* **257,** 967–971.

49. Welsh, J., Chada, K., Dalal, S. S., Cheng, R., Ralph, D., and McClelland, M. (1992) Arbitrarily primed PCR fingerprinting of RNA. *Nucleic Acids Res.* **20,** 4965–4970.

50. Shroyer, A. L., Marshall, G., Warner, B. A., Johnson, R. R., Guo, W., Grover, F. L., and Hammermeister, K. E. (1996) No continuous relationship between Veterans Affairs hospital coronary artery bypass grafting surgical volume and operative mortality. *Ann. Thorac. Surg.* **61,** 17–20.

51. Wong, K. K. and McClelland, M. (1994) Stress-inducible gene of Salmonella typhimurium identified by arbitrarily primed PCR of RNA. *Proc. Natl. Acad. Sci. USA* **91,** 639–643.

52. Murphy, N. B. and Pelle, R. (1994) The use of arbitrary primers and the RADES method for the rapid identification of developmentally regulated genes in trypanosomes. *Gene* **141,** 53–61.

53. McClelland, M., Mathieu Daude, F., and Welsh, J. (1995) RNA fingerprinting and differential display using arbitrarily primed PCR. *Trends. Genet.* **11,** 242–246.

54. Mathieu Daude, F., Stevens, J., Welsh, J., Tibayrenc, M., and McClelland, M. (1995) Genetic diversity and population structure of *Trypanosoma brucei*: clonality versus sexuality. *Mol. Biochem. Parasitol.* **72,** 89–101.

55. Pogue, G. P., Koul, S., Lee, N. L., Dwyer, D. M., and Nakhasi, H.L (1995) Identification of intra- and interspecific *Leishmania* genetic polymorphisms by arbitrary primed polymerase chain reactions and use of polymorphic DNA to identify differentially regulated genes. *Parasitol. Res.* **81,** 282–290.

56. Yates-Silata, K., Sanders, D., and Keath, E. (1995) Genetic diversity in clinical isolates of the dimorphic fungus *Blastomyces dermatitidis* detected by a PCR-based random amplified polymorphic DNA assay. *J. Clin. Microbiol.* **33,**2171–2175.

57. Sobral, B. W. S. and Honeycutt, R. (1993) High output genetic mapping of polyploids using PCR-generated markers. *Theor. Applied Genet.* **86,** 105–112.

58. Welsh, J. and McClelland, M. (1996) Fingerprinting using arbitrarily primed PCR: application to genetic mapping, population biology, epidemiology, and detection of differentially expressed RNAs, in *The Polymerase chain reaction.* (Mullis, K., Ferre, F., and Gibbs, R., eds.), Birkhauser, Boston, pp. 295–303.

7

Genomic Fingerprinting by Application of rep-PCR

Anne M. Ridley

1. Introduction

Several families of short repetitive DNA sequences, widely distributed in the genome, have been identified in bacteria *(1)*. They have an intercistronic location, are not translated, and their function is unclear, although they may be involved in transcription termination, mRNA stability or chromosomal organization. Repetitive extragenic palindrome (REP) elements *(2)*, also known as palindromic units (PU) *(3)*, and enterobacterial repetitive intergenic consensus (ERIC) sequences *(4)* are the best characterized of these elements and were initially identified in *Salmonella typhimurium* and *Escherichia coli,* respectively. The REP consensus sequence was formulated through DNA sequence comparisons of intercistronic regions of different operons *(3,5)* and comprises a 38 nucleotide palindromic sequence that can form a stable stem-loop structure with a 5-bp variable loop in the central region *(2)*. There may be 50–1000 copies of the REP sequence in the genome, frequently present in complex clusters *(2)*, with each cluster comprising as many as 10 copies *(6)*. REP sequences have been located between genes within an operon or at the end of an operon, in different orientations and in tandem arrays, and in operons distributed throughout the genome *(2,3)*. The REP sequence has been identified in intergenic regions within operons from different bacterial species *(7,8)*. REP-like sequences have been shown to exist throughout the eubacterial kingdom, although the consensus sequences may differ among different bacteria *(9–11)*. The precise function of these elements has not been determined, but several functions have been postulated, including roles in gene regulation and retroregulation *(2,12,13)*. These functions may be a consequence of a stem loop structure in a specific chromosomal location *(1)*. Like their REP counterparts, the larger 126-bp ERIC elements have an extragenic location and contain

From: *Methods in Molecular Medicine, Vol. 15: Molecular Bacteriology: Protocols and Clinical Applications*
Edited by: N. Woodford and A. P. Johnson   © Humana Press Inc., Totowa, NJ

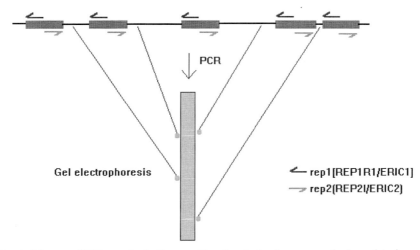

Fig. 1. The rep-PCR method. Outwardly directed primers are designed to be complementary to the rep consensus sequences. For REP-PCR this is a 38-bp sequence, however, for ERIC-PCR the primers are based on the central inverted repeat of the 126 bp consensus. Under the conditions of PCR, the primers amplify between adjacent repetitive elements within the limits of *Taq* polymerase extension. The PCR products are electrophoresed and the resultant bands generate a fingerprint. Differences in band sizes represent polymorphisms in the distance between adjacent repetitive elements.

a highly conserved central inverted repeat *(4,14)*. However, despite similarities in structure, the ERIC elements have no sequence homology to the REP consensus *(4)*.

The term rep-PCR refers to a general method that utilizes oligonucleotide primers matching interspersed repetitive DNA sequences to yield DNA fingerprints of individual bacterial isolates *(11)*. The use of degenerate, repetitive sequence oligonucleotides as primers in PCR-based amplification was described by Versalovic et al. *(11)*. It was hypothesized that repetitive DNA sequences were dispersed in the *E.coli* chromosome in different orientations and separated by various distances. As such, these dispersed repetitive sequences could be used as primer binding sites (**Fig. 1**). As the rep primers are targeted at endogenous interspersed repetitive sequences in the genomes of numerous bacterial species, the rep-PCR method enables a rational approach to primer design, such that the same primer set can generate genomic fingerprints for bacteria that contain these sequences. Outwardly directed primer sets, based on the REP or ERIC consensus sequences and designed to match inverted repeat sequences, can be used to generate bands clearly resolvable by agarose gel electrophoresis after PCR amplification using template genomic DNA from species that contain these sequences *(11)*. The bands represent amplification of DNA between adjacent repetitive elements (**Fig. 1**) within the approximate 5-kb limitation of the *Taq* polymerase extension. Band patterns provide DNA fingerprints that discriminate bacterial strains *(11,15–18)*. Differences in band sizes represent polymorphisms in the distances between repetitive sequence elements in different genomes. The method has been used to type bacterial pathogens, investigate infectious disease outbreaks, and identify clonal or sporadic outbreak isolates *(11,18–23)*. Time-consuming genomic DNA extraction, purification, and quantitation may be avoided by the use of cell lysates as the template for PCR amplification *(19,24)*. Although this increases the attractiveness of the method for investigations that require high-throughput strain typing, it is not suitable for all strains.

This chapter describes a PCR-based method for typing bacterial strains using outwardly directed primer sets based on REP and ERIC consensus sequences. Use of cell lysates is preferable for rapidity and simplicity. However, DNA template preparation methods have been included for both Gram-positive *(25)* and Gram-negative *(26)* bacteria.

2. Materials
2.1. Preparation of Crude Cell Lysates for PCR

1. Sterile distilled water.
2. 200 mM NaCl wash buffer.
3. TE buffer: 10 mM Tris-HCl, 1 mM EDTA, pH 8.0.

2.2. Preparation of DNA Template for PCR
2.2.1. Gram-Positive Bacteria

1. Lysozyme: 10 mg/mL in sterile distilled water (freshly prepared).
2. Lysis buffer: 5 M guanidine isothiocyanate, 0.1 M EDTA, pH 7.0, store at room temperature.
3. Ammonium acetate: 7.5 M.
4. Chloroform:isoamyl alcohol (24:1): store at 4°C.
5. Isopropanol: store at 4°C.
6. 70% ethanol: store at 4°C.
7. TE buffer: 10 mM Tris-HCl, 1 mM EDTA, pH 8.0 (autoclave at 120°C for 20 min and store at room temperature).

2.2.2. Gram-Negative Bacteria

1. TE buffer: 10 mM Tris-HCl, 1 mM EDTA, pH 8.0 (autoclave at 120°C for 20 min and store at room temperature).
2. Sodium dodecyl sulfate (SDS): 10% (w/v), store at room temperature.
3. Proteinase K: 20 mg/mL, store in small single-use aliquots at –20°C.
4. Sodium chloride: 5 M.

5. Hexadecyl trimethyl ammonium bromide (CTAB)/NaCl solution: 10% (w/v) CTAB in 0.7 M NaCl.
6. Chloroform:isoamyl alcohol (24:1): store at 4°C.
7. Phenol:chloroform:isoamyl alcohol (25:24:1): store at 4°C.
8. Isopropanol: store at 4°C.
9. 70% ethanol: store at 4°C.

2.3. Oligonucleotide Primers

The REP primers are based on the 38-bp REP consensus (*11*; **Fig. 1**) (*see* **Note 1**) and the ERIC primers are based on the central inverted repeat *(11)* of the published ERIC consensus sequence *(4)*. These primers may be obtained commercially (*see* **Notes 2 and 3**).

1. REP1R-I: 5'-IIIICGICGICATCIGGC-3' (I denotes the nucleotide inosine).
2. REP2–I: 5'-ICGICTTATCIGGCCTAC-3'.
3. ERIC1R: 5'-ATGTAAGCTCCTGGGGATTCAC-3'.
4. ERIC2: 5'-AAGTAAGTGACTGGGGTGAGCG-3'.

2.4. Buffers and Stock Solutions for PCR

All components of PCR reactions should be prepared in sterile deionized water dedicated for PCR. This can be obtained commercially (e.g., Sigma Chemical Co., St. Louis, MO).

1. PCR buffer (10X stock): 200 mM Tris-HCl, pH 8.4, 500 mM KCl. Commercially available from *Taq* polymerase manufacturer (*see* **Note 4**).
2. MgCl$_2$: 50 mM stock diluted for use at a concentration of 2 mM.
3. Deionized distilled water.
4. dNTPs: a mixture of all four dNTPs at 10 mM each in sterile, deionized water. Dilute from commercially supplied, buffered stock solutions (100 mM). Dispense into 20–50 µL aliquots (sufficient for 40–100 reactions) and store frozen. Minimize freeze thawing (*see* **Notes 5 and 6**).
5. Primers: 100 µM stock solutions of each primer in sterile deionized water. Store frozen in 100 µL aliquots (*see* **Note 6**).
6. *Taq* DNA polymerase: 5 U/µL, usually obtained commercially (*see* **Note 4**).
7. Mineral oil: commercially available.
8. Agarose: use a Molecular Biology quality grade suitable for the preparation of high concentration gels on which to separate fragment range of 100 bp to 5.5 kb (e.g., UltraPure from Life Technologies Ltd. [Gibco-BRL, Gaithersburg, MD]).
9. TBE electrophoresis buffer (10X concentrate): 0.9 M Tris base, 0.9 M boric acid, 0.02 M EDTA.
10. DNA dilution buffer (TE): 10 mM Tris-HCl, 1 mM EDTA, pH 8.0.
11. Sample loading buffer (5X concentrate): 60% sucrose, 0.025% (w/v) bromophenol blue in TE buffer.
12. 1-kb and/or 100-bp lambda DNA markers: available commercially.

13. Ethidium bromide (EB): 5 mg/mL in water. *Caution:* EB is a powerful mutagen and potential carcinogen. Always wear gloves when handling the solid or liquids containing the chemical, and dispose of appropriately. EB solutions can be purchased from commercial sources; this represents the safest way of preparing the stock solution.

3. Methods

3.1. Preparation of Cell Lysates for PCR

1. To prepare a crude cell lysate from a pure overnight broth culture, transfer 200 µL of the culture to a clean microcentrifuge tube and pellet by centrifugation at $10,000g_{max}$ for 4 min.
2. Discard the supernatant, taking care not to disturb the pellet, and resuspend in 500 µL of 200 mM NaCl to wash the cells.
3. Pellet the cells by centrifugation, ensuring that as much as possible of the 200 mM NaCl is removed and resuspend in 100–200 µL of water or TE buffer to an approximate cell density equal to an A_{600} value of 0.5 (*see* **Note 7**).
4. Heat to 95°C on a heating block or block of thermal cycler for 10 min to encourage cell disruption.
5. Centrifuge briefly (~15 s) before using 5 µL as template for PCR.

3.2. Preparation of DNA Template for PCR

3.2.1. Gram-Positive Bacteria

1. Harvest bacterial cells into a 1.5-mL microcentrifuge tube and pellet by centrifugation at $10,000g_{max}$ for 3 min. The recommended pellet volume is 50–100 µL. Volumes of reagents, however, may be proportionately scaled up or down as required.
2. Resuspend the cells by vortexing in 100 µL of lysozyme solution and incubate at room temperature for 30 min.
3. Lyse the cells by addition of 200 µL of 5 M guanidine isothiocyanate lysis buffer and mix by inversion. After the solution has cleared add 150 µL of ammonium acetate and mix gently.
4. Add 450 µL of chloroform:isoamyl alcohol and emulsify the lysate by vigorous shaking or vortexing.
5. Separate the phases by centrifugation at $10,000g_{max}$ for 2 min and transfer the upper aqueous phase to a fresh tube.
6. Add 0.54 vol of isopropanol to the aqueous phase and mix by inversion to precipitate the high molecular weight DNA.
7. Immediately separate the precipitated DNA, which forms a dense stringy precipitate within a few seconds of mixing, from the liquid phase with a pipet tip (*see* **Note 8**).
8. Wash the DNA twice in 1 mL of ethanol and dry under vacuum.
9. Dissolve the dried pellet in TE buffer (*see* **Note 9**).

3.2.2. Gram-Negative Bacteria

1. Harvest bacterial cells by centrifugation as for Gram-positive bacteria.
2. Resuspend the bacterial pellet in 567 µL of TE buffer by repeated pipeting.
3. Add 30 µL of 10% SDS and 3 µL of 20 mg/mL proteinase K to give a final concentration of 100 µg/mL proteinase K in 0.5% SDS. Mix thoroughly and incubate for 1 h at 37°C (*see* **Note 10**).
4. Add 100 µL of 6 *M* NaCl and mix well (*see* **Note 11**).
5. Add 80 µL of CTAB/NaCl solution. Mix and incubate at 65°C for 10 min.
6. Add 800 µL of chloroform:isoamyl alcohol. Mix and centrifuge 5 min at $10,000g_{max}$ to remove the CTAB-protein/ polysaccharide complexes, which appear as a white interface after centrifugation.
7. Remove the supernatant to a fresh microcentrifuge tube, taking care to avoid disrupting the interface.
8. Add 750 µL of phenol:chloroform:isoamyl alcohol, mix well, and separate phases by centrifugation.
9. Transfer the upper aqueous phase to a fresh microcentrifuge tube and repeat phenol:chloroform:isoamyl alcohol extractions until the interface is clear.
10. Add 0.6 vol (~450 µL) of isopropanol to precipitate the nucleic acids and invert the tube until a stringy clump becomes visible (*see* **Note 8**).
11. Transfer the pellet to a fresh tube containing 70% ethanol. Wash the pellet twice in ethanol to remove residual CTAB.
12. Dry under vacuum and resuspend pellet in an appropriate volume of TE buffer (*see* **Note 9**).

3.3. Polymerase Chain Reaction

3.3.1. Preparation of Reaction Mixes

1. Use 25 µL reaction volumes in small microfuge tubes of a size and shape compatible with the thermal cycler in use.
2. Prepare a master mix for the PCR reaction. The volumes of each component are dependent on the number of individual reactions to be prepared (*see* **Note 12**). The following components are added in order, to a separate tube (volumes given are per reaction):14.75 µL of sterile distilled water, 2.5 µL of 10X PCR buffer, 1.25 µL of 50 m*M* MgCl$_2$, 0.5 µL of dNTP mix, 0.5 µL of each primer (REP1R-I and REP2–I or ERIC1R, and ERIC2). Replace cap and store on ice.
3. Aliquot each template in a total volume of 5 µL to a fresh labeled microfuge tube (*see* **Note 13**).
4. Add *Taq* polymerase to a concentration of 1–2 U per reaction to the reaction mix. Mix well by inversion.
5. Immediately aliquot the reaction mix to each tube of template taking great care to avoid cross-contamination. The simplest way is to use a separate pipet tip for each transfer (*see* **Note 5**).
6. Overlay the reaction mixture with mineral oil and pulse to ensure no reaction mix is left on the walls of the tube.

3.3.2. Thermal Cycling

1. When using REP primers, program an initial denaturing step of 95°C for 7 min, followed by 30 cycles of 90°C for 30 s; 42°C for 1 min, and 65°C for 8 min with a final cycle of 90°C for 30 s, 40°C for 1 min and 65°C for 16 min. Program the machine to maintain the samples at 4°C until the PCR products can be analyzed by gel electrophoresis. When using ERIC primers, the annealing temperature is increased to 52°C, but all times and the temperatures of denaturing and extension steps are identical to those used for the REP primers.
2. For reactions involving the use of cell lysates as template, it may be necessary to add EDTA to a concentration of 50 m*M* as soon as possible after completion of thermal cycling (*see* **Note 14**).

3.3.3. Analysis of PCR Products

1. Prepare a 1.5% agarose gel containing 1X TBE electrophoresis buffer.
2. Add 5 µL of loading dye to the finished PCR. There is no need to remove the mineral oil; add the dye by pipeting it onto the inside wall of the tube and centrifuge briefly to mix it into the aqueous phase.
3. Load an appropriate volume of the sample into the well of the gel (e.g., 10 µL for a 2 mm well). Load DNA markers to at least one well, flanking the sample wells.
4. Run the gel in 1X TBE at 10 V/cm until the bromophenol blue dye has migrated approx 4/5 the length of the gel.
5. Stain the gel in 600 ng/mL EB in water. *Caution:* Wear gloves whenever handling the gel.
6. Photograph the gel using a UV transilluminator to visualize the DNA bands.

3.4. Evaluation of Results

The band pattern following agarose gel electrophoresis of products of rep-PCR amplification is specific for the test bacterium and primer combination utilized (*see* **Note 15**). A typical gel photograph is shown in **Fig. 2**. Amplification product sizes resulting from rep-PCR generally range from <100 bp – 5 kp (the latter is the limit of *Taq* polymerase extension). Fingerprints should be visually inspected. Profiles should be considered highly similar when all visible bands from the test isolates have the same apparent migration distance. Variations in intensity or shape of bands should not be used to justify designation of a separate pattern type, particularly when cell lysates are used as templates, as this observation often results from differences in the amount of DNA in the reaction available for amplification. The profile obtained for a sample constitutes a fingerprint for the test micro-organism, and can be compared with those obtained for other bacteria under the same reaction conditions.

Failure to generate a rep-PCR profile may be caused by several factors. These include

1. Sequences homologous to the REP or ERIC consensus sequences may not be present in the target organism;

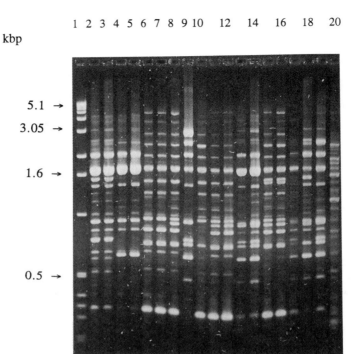

Fig. 2. REP-PCR profiles of *Listeria monocytogenes,* including six epidemiologically related pairs, generated using primers REP1R-I/REP2–I. Lane 1, 1 kb λ ladder molecular weight marker; lane 2, serovar 1/2a; lane 3, serovar 1/2a; lane 4, serovar 1/2c; lane 5, serovar 1/2c; lane 6, serovar 1/2a; lane 7, serovar 4bX; lane 8, serovar 1/2a; lane 9, serovar 1/2a; lane 10, serovar 1/2a; lane 11, serovar 1/2b; lane 12, serovar 4bX; lane 13, serovar 1/2a; lane 14, serovar 4b; lane 15, serovar 1/2b; lane 16, serovar 4b; lane 17, serovar 1/2b; lane 18, serovar 1/2a; lane 19, serovar 1/2a; lane 20, serovar 1/2a.

2. Incomplete cell lysis (*see* **Note 7**) or concentration of target DNA is too low;
3. Inhibition of PCR by a factor(s) introduced with the template (*see* **Note 5**); and
4. A component(s) in the reaction is limiting (*see* **Notes 16** and **17**) .

3.4.1. Computer-Assisted Analysis of Fingerprints to Evaluate Similarity Among Amplification Profiles

The patterns may be recorded by a scanner or densitometer linked to a computer. The digitized, optimized density values are transfered to a computer and are stored as densitograms of patterns or as a bitmap image file. Using a software package such as Gel Compar (Applied Maths, Belgium) or Dendron (Solltech, Oakdale, CA) the tracks are further processed. This involves normalization of tracks, generation of databases, and grouping or identification of tracks by quantification of their resemblance. However, this is only feasible if optimization of the PCR reactions, ensuring reproducibility between runs, has been undertaken.

4. Notes

1. Primer pairs REPIR-I plus REP2–I *(11)* are based on the 38-bp REP consensus and contain the nucleotide inosine at ambiguous positions. Inosine contains the base hypoxanthine and can form Watson–Crick base pairs with any of the four nucleotide bases (A,C,G, and T). Weaker than A:T base pairs, however, they form the least destabilizing, but most discriminatory pairs.

2. Primers should be COP-cartridge or even HPLC purified. This appears to enhance reproducibility between batches of primers (author's unpublished observation).

3. A fluorophore-enhanced rep-PCR (FERP) technique has been recently developed *(27)* that combines DNA amplification and fluoresence detection for the analysis of bacterial pathogens. If this system is desired, a covalently linked fluoresent dye should be added to the 5' end of each of the primers described in **Subheading 2.** This is useful for the laser detection of specific PCR products when the amplicons are resolved by polyacrylamide gel electrophoresis (PAGE).

4. Commercially available *Taq* polymerases (e.g., Life Technologies [Gibco BRL, Gaithersburg, MD] or *AmpliTaq* [Perkin Elmer, Norwalk, CT]) are supplied with 10X PCR buffer. This may contain 500 mM MgCl$_2$, or this component may be supplied separately, allowing for easy optimization. Some manufacturers supply a 1% solution of the detergent W-1, which should be added at a final concentration of 0.05% (v/v), and is thought to improve the thermostability of the enzyme. The experience of the author is that this component does not enhance the appearance of the amplification products. It is advisable to use cloned *Taq* polymerase, licensed for PCR, as negative control lanes were clear of contamination, in contrast to corresponding control reactions generated using uncloned *Taq* polymerase.

5. Contamination of PCR must be minimized. Reaction products must not contaminate stock reaction components or they will act as templates in any reactions set up using the contaminated component solutions yielding misleading results. This can be minimized using clean, sterile pipet tips for each reaction component, establishing separate areas for handling of reaction components and products, using designated pipets or positive-displacement pipets and tips to prevent aerosols, using commercially obtained sterile distilled water, and by using aliquotted batches of dNTPs, primers, and PCR reaction buffers to appropriate volumes.

6. For long term storage of stocks of primers, dNTPs, or DNA templates (including crude lysates), it is advisable to store concentrated solutions in small aliquots at $-20°C$. Solutions of DNA are susceptible to degradation by nucleases at higher temperatures.

7. Differences in amplification profiles obtained using cell lysates compared with purified DNA templates have been observed by the author. These mainly consist of greater numbers and intensities of small (<200 bp) amplicons when generated from cell lysates, whereas larger fragments tend to predominate when DNA is

used as a template. Inhibition of the PCR may result from the use of a cell lysate containing too many cells. It is easier to standardize cell concentration when liquid cultures are used. Although an OD reading provides a good approximation of cell concentration, it is not appropriate for a large number of samples. In this case, an OD reading could be obtained for a single sample, and the remainder diluted to match. The volumes of cells used can be adjusted accordingly. For this reason it is advisable to optimize conditions of cell lysis, taking into account the buffers and detergents (if any) used, cell concentration, and whether centifugation of the lysate is required.

8. If no stringy DNA precipitate forms with the addition if isopropanol using either of the DNA extraction methods outlined, it is likely that cell lysis was incomplete, or that the DNA has sheared into relatively low molecular weight pieces. DNA should not be collected by centrifugation because this results in the formation of a compacted pellet, which is difficult to resuspend. In addition, a cloudy precipitate, probably consisting mainly of RNA and protein, forms in some preparations and this is collected with the DNA if centrifugation is used.

9. The nucleic acid concentration should be spectrophotometrically determined. A 1 mg/mL solution of double-stranded DNA has an absorbance of 20 OD U at 260 nm (i.e., an OD_{260} of 1 is equivalent to 50 µg/mL of dsDNA). Alternatively, a small aliquot (0.2–1 µL) may be run on 0.6% agarose gel with a sample of known concentration run in adjacent well to give an estimate of the DNA concentration in samples. The DNA should be stored frozen in aliquots at –20°C at 50–100 ng/µL and diluted in TE before use.

10. As the detergent lyses the bacterial cell wall, the solution should become viscous. If cell lysis is poor, it may be necessary to predigest the bacterial cell wall with lysozyme.

11. To ensure that nucleic acids are retained in solution, thereby allowing CTAB to form polysaccharide complexes, it is necessary to add concentrated salt solution. The protein/polysaccharide CTAB complexes are removed by precipitation.

12. For multiple PCR reactions encompassing a number of different bacterial lysates and/or DNA templates, a master PCR reaction mix should be prepared, allowing for control reactions and pipetting inaccuracies.

13. When the volume of cell lysate or DNA to be added to each PCR reaction is greater than 1 µL, the volume of water added to the reaction mix should be adjusted accordingly.

14. Some bacteria contain nucleases that will degrade DNA products amplified from cell lysates; this is especially noticable for *Salmonella* spp. For this reason, it is advisable to add EDTA to a final concentration of 50 mM to the reaction immediately after completion of the PCR.

15. Optimization of reaction conditions should be undertaken for different templates. The factors most likely to cause variation of rep-PCR profiles include magnesium ion concentration and annealing temperature. Many workers use the annealing temperature (40°C) reported by Versalovic et al. *(11)* for use with these primers, although temperatures as high as 45°C have

been reported to generate reproducible patterns with a sensible number of amplicons for the purpose of typing *(24)*. Magnesium ion concentration has a significant effect on the complexity of the profiles, as has the addition of dimethylsulfoxide *(24)*. PCR-amplified fragments of a size close to the limit of *Taq* polymerase extension (4–5 kb) are not always reproducible, and should not be considered when interstrain comparisons are made.

16. In general, more discriminatory DNA fingerprints are provided by rep-PCR when the REP primers (REP1R-I/REP2–I), rather than the ERIC primers (ERIC1R/ ERIC2) are employed. The level of discrimination afforded using the REP primers has been observed to be at least as high as ribotyping *(24)* (author's unpublished observation), but the method has the advantage of speed and simplicity.

17. Some strains may be observed to generate profiles consisting of smears, with or without few visible bands. These strains may contain large amounts of substances that can inhibit PCR, such as lipopolysaccharides, or may release nucleases that degrade DNA released from lysed cells. In this case, it is preferable to use a purified DNA template. However, it should be noted that small differences occur between profiles generated from DNA, and cell lysates and analyses should be performed with the same template preparation method.

References

1. Lupski, J. R. and Weinstock, G. M. (1992) Short, interspersed repetitive DNA sequences in prokaryotic genomes. *J. Bacteriol.* **174,** 4525–4529.
2. Stern, M. J., Ames, G. F. L., Smith, N. H., Robinson, E. C., and Higgins, C. F. (1984) Repetitive extragenic palindromic sequences: a major component of the bacterial genome. *Cell* **37,** 1015–1026.
3. Gilson, E., Clement, J. M., Brutlag, D., and Hofnung, M. (1984) A family of dispersed, repetitive, extragenic palindromic DNA sequences in *E.coli. Embo J.* **3,** 1417–1421
4. Hulton, C. S. J., Higgins, C. F., and Sharp, P. M. (1991) ERIC sequences: a novel family of repetitive elements in the genomes of *Escherichia coli, Salmonella typhimurium* and other enterobacteria. *Mol. Microbiol.* **5,** 825–834.
5. Higgins, C. F., Ames, G. F. L., Barnes W. M., Clement, J. M., and Hofnung, M. (1982) A novel intercistronic regulatory element of prokaryotic operons. *Nature* (London) **298,** 760–762.
6. Makino, K., Kim, S. K., Shinagawa, H., Amemura, M., and Nakata, A. (1991) Molecular analysis of the cryptic and functional *phn* operons for phosphonate use in *Escherichia coli* K-12. *J. Bacteriol.* **173,** 2665–2672.
7. Dahl, M. K., Francoz, E., Saurin, W., Boos, W., Manson, M. D., and Hofnung, M. (1989) Comparison of sequences from the *malB* regions of *Salmonella typhimurium* and *Enterobacter cloacae* with *Escherichia coli* K12: a potential new regulatory site in the interoperonic region. *Mol. Gen. Genet.* **218,** 199–207.
8. Higgins, C. F., McLaren, R. S., and Newbury, S. F. (1988) Repetitive extragenic palindromic sequences, mRNA stability and gene expression: evolution by gene conversion? A review. *Gene* **72,** 3–14.

9. Dimri, G. P., Rudd K. E., Morgan, M. K., Bayat, H., and Ames, G. F. L. (1992) Physical mapping of repetitive extragenic palindromic sequences in *Escherichia coli* and phylogenetic distribution among *Escherichia coli* strains and other enteric bacteria. *J. Bacteriol.* **174,** 4583–4593.

10. Gilson, E., Perrin, D., and Hofnung, M. (1990) DNA polymerase I and a protein complex bind specifically to E. coli palindromic unit highly repetitive DNA: implications for bacterial chromosome organisation. *Nucleic Acids Res.* **18,** 3946–3952.

11. Versalovic J., Koeuth, T., and Lupski, J. R. (1991) Distribution of repetitive DNA sequences in eubacteria and application to fingerprinting of bacterial genomes. *Nucleic Acids Res.* **19,** 6283–6831.

12. Newbury S. F., Smith, N. H., Robinson, E. C., Hiles, I. D., and Higgins, C. F. (1987) Stabilization of translationally active mRNA by prokaryotic REP sequences. *Cell* **48,** 297–310.

13. Stern, M. J., Prossnitz, E., and Ames, G. F. -L. (1988) Role of the intercistronic region in post-transcriptional control of gene expression in the histidine transport operon of *Salmonella typhimurium*: involvement of REP sequences. *Mol. Microbiol.* **2,** 141–152.

14. Sharples, G. J. and Lloyd, R. G. (1990) A novel repeated DNA sequence located in the intergenic regions of bacterial chromosomes. *Nucleic Acids Res.* **18,** 6503–6508.

15. de Bruijn, F. J. (1992) Use of repetitive (repetitive extragenic palindrome and enterobacterial repetitive intergenic consensus) sequences and the polymerase chain reaction to fingerprint the genomes of *Rhizobium melitoti* isolates and other soil bacteria. *Appl. Environ. Microbiol.* **58,** 2180–2187.

16. Georghiou, P. R., Wright, C. E., Versalovic, J., Koeuth, T., Watson, D., Hamill, R., and Lupski, J. R. (1992) Molecular epidemiology of *Enterobacter aerogenes* using plasmid profile restriction enzyme analysis (REA) and repetitive element polymerase chain reactions (rep-PCR). in *Program Abstr. 32nd Intersci. Conf. Antimicrob. Agents Chemother.,* Abstr. 1415.

17. Versalovic, J., Koeuth, T., Zhang, Y.-M., McCabe, E. R. B., and Lupski, J. R. (1992) Quality control for bacterial inhibition assays: DNA fingerprinting of microorganisms by rep-PCR. *Screening* **1,** 175–183.

18. Woods, C. R., Versalovic, J., Koeuth, T., and Lupski, J. R. (1992) Analysis of relationships among isolates of *Citrobacter diversus* by using DNA fingerprints generated by repetitive sequence-based primers in the polymerase chain reaction. *J. Clin. Microbiol.* **30,** 2921–2929.

19. Woods, C. R., Versalovic, J., Koeuth, T., and Lupski, J. R. (1993) Whole-cell repetitive element sequence based polymerase chain reaction allows rapid assessment of clonal relationships of bacterial isolates. *J. Clin. Microbiol.* **31,** 1927–1931.

20. van Belkum, A., Bax, R., Peerbooms, P., Goessens, W. H. F., van Leeuwen, N., and Quint, W. G. V. (1993) Comparison of phage typing and DNA fingerprining by polymerase chain reaction for discrimination of methicillin resistant *Staphylococcus aureus* strains. *J. Clin. Microbiol.* **31,** 798–803.

21. van Belkum, A., Streulens, M. J., and Quint, W. G. V. (1993) Typing of *Legionella pneumophila* strains by polymerase chain reaction-mediated DNA fingerprinting. *J. Clin. Microbiol.* **31,** 2198–2200.

22. Giesendorf, B., van Belkum, A., Koeken, A., Stegeman, H., Henkens, M., van der Plas, J., Niesters, B., and Quint, W. (1993) Development of species-specific DNA probes for *Campylobacter jejuni, Campylobacter coli* and *Campylobacter lari* by PCR fingerprinting. *J. Clin. Microbiol.* **31,** 1541–1546.

23. Streulens, M. J., Carlier, E., Maes, N., Serruys, S., Quint, W. G. V., and van Belkum, A. (1993) Nosocomial colonisation and infection with multiresistant *Acinetobacter baumannii*: outbreak delineation using DNA macrorestriction analysis and PCR fingerprinting. *J. Hosp. Infect.* **25,** 15–32.

24. Snelling, A. M., Gerner-Smidt, P., Hawkey, P. M., Heritage, J., Parnell, P., Porter, C., Bodenham, A. R., and Inglis, T. (1996) Validation of use of whole-cell repetitive extragenic palindromic sequence-based PCR (REP-PCR) for typing strains belonging to the *Acinetobacter calcoaceticus-Acinetobacter baumannii* complex and application of the method to the investigation of a hospital outbreak. *J. Clin. Microbiol.* **34,** 1193–1202.

25. Ridley, A. M. and Saunders, N. A. (1993) Restriction fragment length polymorphism (RFLP) analysis for epidemiological typing of *Listeria monocytogenes,* in *New Techniques in Food and Beverage Microbiology* (Kroll, R. G., et al. eds.), Technical Series, Blackwell Scientific, Oxford, UK, pp. 231–249.

26. Wilson, K. (1987) Preparation of genomic DNA from bacteria, in *Current Protocols in Molecular Biology*, (Ausubel, F. M., et al. eds.), Wiley, New York, pp. 2.4.1.,2.4.2.

27. Versalovic, J., Kapur, V., Koeuth, T., Mazurek, G. H., Whittam, T. S., Musser, J. M., and Lupski, J. R. (1995) DNA fingerprinting of pathogenic bacteria by fluorophore-enhanced repetitive sequence-based polymerase chain reaction. *Arch. Pathol. Lab. Med.* **119,** 23–29.

8

Molecular Approaches
to the Identification of Streptococci

Robert E. McLaughlin and Joseph J. Ferretti

1. Introduction

The ability to identify rapidly organisms to the species, and at times subspecies level, is an important step in the treatment of bacterial infections and for monitoring the spread of microorganisms. Conventional identification of streptococci relies on the isolation and culturing of bacterial cells, and then submitting the culture to a battery of biochemical tests. Whereas these panels are useful and have a fairly high degree of accuracy, they can suffer from preparation time and problems with the identification of nutritionally variant strains (mainly with the viridans streptococci). Biochemical classification also lacks the ability to type species to the level of a particular clonal population or strain. Although not as important from the diagnostic perspective, the ability to type bacteria to the clonal level is important for epidemiologic studies of disease outbreaks.

Molecular typing techniques for the identification of microorganisms have been used to study the epidemiology of infectious agents. These techniques employ amplification of specific genes, arbitrarily primed polymerase chain reaction (PCR) profiling *(1)*, restriction fragment length polymorphism (RFLP), and multilocus enzyme electrophoresis (MLEE) *(2)*. In addition, there are numerous variations on the aforementioned techniques. In this chapter, the authors describe several methods for the identification of streptococci based on DNA methodology, pointing out the advantages and disadvantages of each.

2. Materials

The following are common laboratory reagents used in the isolation of DNA, and are used in one or more of the techniques described in this chapter.

From: *Methods in Molecular Medicine, Vol. 15: Molecular Bacteriology: Protocols and Clinical Applications*
Edited by: N. Woodford and A. P. Johnson © Humana Press Inc., Totowa, NJ

1. Brain Heart Infusion Broth (BHI): 30 g per L (Difco, Detroit, MI). Add 15 g/L of agar for solid medium and autoclave for 15 min at 121°C.
2. Todd-Hewitt Yeast Extract Broth (THY): 30 g per L (Difco) plus 0.1% yeast extract. Add 15 g/L of agar for solid medium and autoclave for 15 min at 121°C.
3. Proteinase K: dissolve in distilled water to a final concentration of 20 mg/mL. Aliquot and store at –20°C.
4. Mutanolysin: suspend contents of lyophilized mutanolysin (Sigma, St. Louis, MO) in distilled water to give a final concentration of 5000 U/mL. Divide into small aliquots (~50–100 μL) and store at –20°C.
5. Lysozyme: dissolve in distilled water or TE buffer to a final concentration of 50 mg/mL. Aliquot in 500 μL volumes and store at –20°C.
6. Buffer-saturated phenol (pH 8.0).
7. Buffer-saturated phenol (pH 8.0):chloroform:isoamyl alcohol (25:24:1).
8. Chloroform:isoamyl alcohol (24:1).
9. TE buffer: 10 mM Tris-HCl, pH 7.5, 1 mM EDTA.
10. 10X TBE: 0.9 M Tris base pH 8.3, 0.9 M sodium borate, 25 mM EDTA.
11. 20X SSC: 3.0 M NaCl, 0.3 M sodium citrate.
12. 5 M sodium chloride.
13. 7.5 M ammonium acetate.
14. 1 M sodium acetate.

2.1. Materials Required for Specific Techniques

The following list is comprised of specific materials required for a particular technique. The technique and section number for the technique is given in the heading.

2.1.1. Isolation of Purified DNA by SDS Lysis (**Subheading 3.1.1.**)

1. Wash buffer: 10 mM Tris-HCl, pH 7.8, 5 mM EDTA.
2. SDS lysis solution: 20% SDS in 50 mM Tris-HCl, pH 7.8, 20 mM EDTA.

2.1.2. Isolation of Purified DNA by Guanidinium Thiocyanate Lysis (**Subheading 3.1.2.**)

1. GES: 5 M guanidinium thiocyanate, 100 mM EDTA pH 8.0, 0.5% sarkosyl (*see* **Note 1**). Filter through a 0.45-μm filter and store at room temperature. Handle guanidinium solutions with gloves.

2.1.3. Preparation of Crude DNA for PCR (**Subheading 3.1.3.**)

1. Lysozyme solution: 20% (w/v) sucrose in [10 mM Tris-HCl, pH 8.0, 100 mM NaCl, 1 mM EDTA] containing 25 mg/mL lysozyme.
2. Lysis solution: 50 mM KCl, 10 mM Tris-HCl, pH 8.3, 0.1% gelatin, 0.45% Nonidet P-40, and 0.45% Tween-20.

2.1.4. Isolation of Intact DNA
*for Long-PCR and Electrophoresis (**Subheading 3.1.4.**)*

1. Suspension buffer: 1 M NaCl, 10 mM Tris-HCl, pH 7.6.
2. Lysis solution I: 6 mM Tris-HCl, pH 7.6, 1 M NaCl, 100 mM EDTA, pH 8.0, 0.5% Brij 58, 0.2% deoxycholate, 0.5% N-lauryl sarcosine containing 1 mg/mL lysozyme and 20 μg/mL RNase (*see* **Note 2**).
3. Lysis solution II: 0.5 M EDTA, pH 8–9, 0.5% N-lauryl sarcosine, 1 mg/mL proteinase K.

2.1.5. Identification of Streptococci by RFLP (**Subheading 3.3.**)

1. PC2 buffer (final concentration): 20 mM Tris-HCl, pH 8.55, 150 μg/mL bovine serum albumin, 16 mM (NH$_4$)$_2$SO$_4$, 3.5 mM MgCl$_2$.

2.1.6. Identification of Streptococci Based on Hybridization
*to 23S or 16S rRNA genes (**Subheading 3.4.1.**)*

1. Prehybridization solution: 2X SSC, 5X Denhardt's solution, 50% formamide containing 200 μg/mL denatured salmon sperm DNA.
2. Hybridization solution: 2X SSC, 5X Denhardt's solution, 3% dextran sulfate, 50% formamide containing 50 μg/mL denatured salmon sperm DNA.

2.1.7. Differentiation of Group A Streptococci
*by Reverse Dot/Line Blot Hybridization of Specific
Probes with M-Protein (emm) genes (**Subheading 3.4.2.,3.4.3.**)*

1. Prehybridization solution: 6X SSC, 0.5% SDS, 5X Denhardt's solution containing 100 μg/mL denatured salmon sperm DNA.

2.1.8. Identification of Streptococci Based
*on DNA-DNA Hybridization (**Subheading 3.5.**)*

1. Coating buffer: phosphate-buffered saline (pH 7.2) containing 0.1 M MgCl$_2$.
2. Prehybridization solution: 2X SSC, 5X Denhardt's solution, 50% formamide containing 200 μg/mL denatured salmon sperm DNA.
3. Hybridization solution: 2X SSC, 5X Denhardt's solution, 3% dextran sulfate, 50% formamide containing 50 μg/mL denatured salmon sperm DNA.

3. Methods
3.1. Extraction of DNA from Streptococci

Several methods for the extraction of DNA from streptococcal cells are presented in this review. These procedures can be scaled to the desired cell volume as needed. There are numerous other methods and modifications in the literature, all of which may work equally well for isolation of DNA suitable for PCR analysis. The primary concern is that the DNA should be relatively intact (not highly sheared) and that it is in a final solution of low ionic strength containing few or no Mg^{2+} ions. Significant amounts of Mg^{2+} ions in the template DNA will need to be compensated for in subsequent PCR reactions by adjusting the amount of extra $MgCl_2$ added to the reaction. This adjustment is necessary as changes in Mg^{2+} ion concentration can alter the specificity of a given primer pair.

3.1.1. Isolation of Purified DNA by SDS Lysis (3–5)

1. Grow cells in 1.5 mL of medium (BHI or THY) overnight at 37°C until late log-phase is reached.
2. Pellet cells and wash with 1 mL of wash buffer,
3. Resuspend cell pellet in 350 μL of wash buffer and add 25 μL of lysozyme (*see* **Note 2**).
4. Incubate cells for 30 min at 37°C.
5. Add 20 μL of lysis solution and 3 μL of proteinase K.
6. Incubate cells at 37°C for 1 h.
7. Add 200 μL of 5 *M* NaCl to samples, mix by inversion several times and microfuge at maximum speed for 10 min. Transfer the supernatant fluid to a new microfuge tube (*see* **Note 3**).
8. Add an equal volume of saturated phenol:chloroform to the supernatant and mix by inversion several times. Microfuge at maximum speed for 5 min and transfer the aqueous phase (top layer) to a new microfuge tube.
9. Repeat **step 8** extracting the aqueous phase with chloroform:isoamyl alcohol.
10. Add 30 μL of 1 *M* sodium acetate to the aqueous phase from **step 9** and mix by inversion.
11. Precipitate the DNA from solution by addition of 2.5 vol of 95% ethanol and store at −20°C for at least 2 h (*see* **Note 4**).
12. DNA is pelleted by centrifugation (maximum speed) for 5 min at 4°C and washed with 500 μL of 70% ethanol.
13. Dry the final DNA pellet under vacuum or let air dry. Suspend the pellet in 30 μL of TE buffer (this volume may vary depending on the yield) (*see* **Note 5**).

3.1.2. Isolation of Purified DNA by Guanidinium Thiocyanate Lysis (6)

1. Grow cells in 1.5 mL of medium (BHI or THY) overnight at 37°C until late log-phase is reached.
2. Suspend cells in 100 μL of TE buffer containing lysozyme (50 mg/mL) (*see* **Note 2**).
3. Incubate for 30 min at 37°C .
4. Add 500 μL of GES to sample and mix by inversion or gently vortexing. Lysis should occur in 5–10 min (*see* **Note 6**).
5. Chill samples on ice for 10 min.
6. Add 250 μL of ice-cold 7.5 *M* ammonium acetate and mix by inversion several times.
7. Incubate samples on ice for 10 min.
8. Add 500 μL of chloroform:isoamyl alcohol and mix by inversion several times.
9. Microfuge at maximum speed for 10 min.

10. Transfer the aqueous phase (top layer) to a new microfuge tube and precipitate DNA by addition of 0.54 vol of 2–propanol.
11. Collect the fibrous DNA precipitate by centrifugation for 30 s.
12. Wash the DNA pellet five times with 70% ethanol (*see* **Note 7**).
13. Dry the final DNA pellet under vacuum or let air dry. Suspend in 100 µL of TE buffer (this volume may vary depending on the yield) (*see* **Note 5**).

3.1.3. Preparation of Crude DNA for PCR (7)

1. Grow cells in 1.5 mL of medium (BHI or THY) overnight at 37°C until late log-phase is reached (*see* **Note 8**).
2. Pellet the cells and suspend in 1 mL of lysozyme solution (*see* **Note 2**).
3. Incubate samples for 30–60 min at 37°C.
4. Pellet the cells and suspend in 1 mL of lysis solution supplemented with 100 µg/mL proteinase K.
5. Incubate samples at 60°C for 60 min followed by 95°C for 10 min (*see* **Note 9**).
6. Transfer samples directly to an ice-water bath until ready for use or store at 4°C for several days.

3.1.4. Isolation of Intact DNA
for Long-PCR and Electrophoresis (1,8,9)

1. Grow cells in 5 mL of medium (BHI or THY) overnight at 37°C until late log-phase is reached (*see* **Note 10**). Pellet cells.
2. Suspend cells in 2 mL/(50 µL) of suspension buffer.
3. Mix cell suspension with 1 mL/(50 µL) of 1.5–2.0% low-melting temperature agarose in suspension buffer (InCert Agarose; FMC, Rockland, ME) and pipet into a plug mold (Bio-Rad, Gaithersburg, MD).
4. After solidification, place plugs in 10 mL/(150 µL) of lysis solution I (*see* **Note 11**).
5. Incubate samples overnight at 37°C.
6. Add 10 mL/(150 µL) of lysis solution II.
7. Incubate the samples overnight at 50°C.
8. Wash plugs three times in 10 mL/(150 µL) TE buffer for approx 2 h each time and store samples at 4°C (*see* **Note 12**).

3.2. Identification of Streptococci by PCR Amplification Techniques

This section presents several methods for the identification and differentiation of streptococci based on PCR, including hybridization and electrophoretic analysis. In addition to these methods, there are numerous other methods for the identification of streptococci that have been used. There are several commercial kits for the identification of streptococci based on biochemical and physiological characterization. It is generally recommended to identify clinical isolates with one of the commercial methods in addition to using molecular methods. These panels are also useful for typing strains to a particular group, thereby simplifying the choice as to which molecular method of identification to employ.

PCR-based protocols provide a rapid method for the identification of streptococci. As with all PCR protocols, certain precautions need to be followed to ensure accurate amplification of the target DNA. Contaminating DNA, excess salt (especially Mg^{2+} ions), or high levels of proteins in the template preparation can all effect the outcome of a PCR reaction. All solutions for template preparation and PCR reagents should be made in high quality deionized water. Such care should eliminate the possibility of excess salts sometimes present in distilled water supplies and should also eliminate foreign DNA that may be present in the water. Aliquot all reagents in small volumes (including primers) and store at –20°C. Pipeting should be done using barrier tips, with a new tip used for each step to avoid cross-contamination of samples.

The use of thin-walled PCR tubes, although not essential, can increase the specificity of some reactions by increasing the rate of heat transfer, thereby making it possible to shorten steps in the amplification cycle. The use of master mixes (*see* **Note 13**), as opposed to individual reactions, should limit the number of times access is made into the reagents, and ensures a consistent reaction mixture for all samples of a given set. For routine analysis, a 25–50 μL reaction is usually sufficient. An example of a representative 25 μL reaction is given below:

Volume	Component	Final Conc.
4 μL	dNTP solution (1.25 mM each)	200 μM each
2.5 μL	primer 1 (5–10 μM stock)	0.5–1 μM
2.5 μL	primer 2 (5–10 μM stock)	0.5–1 μM
2.5 μL	10X PCR Buffer	[10 mM Tris-HCl (pH 8.3), 50 mM KCl]
1–4 μL	MgCl$_2$ (25 mM)	1–4 mM

add deionized water to bring the total reaction to 25 μL, taking into account the volume of template DNA to be added

0.2 μL	*Taq* (5 U/μL)	1U
1–2 μL	Template DNA	

A list of primers and PCR conditions that have been used successfully for the identification and/or differentiation of streptococci is presented in **Table 1**. The concentration of Mg^{2+} ions in the reaction mixture is also given. The concentration of other components should be within the limits given above. The cycling conditions and reagent concentrations given should be used as an initial guide, however, experimental conditions may vary slightly with the quality of the template and ionic strength of the reaction mixture.

3.2.1. Setting Up the PCR Reaction

1. Program the thermal cycler to match the parameters for amplification primers selected from **Table 1**.

Table 1
Primers for PCR Amplification of Streptococcal DNA

Target gene	Primer sequences (5'–3')	Product size (bp)	Cycle conditions	Cycles	Specificity	Reference	Application
mf	ATGAATCTACTTGGATCAAGA GAGTAGGTGTACCGTTATGG	808	94°C, 40 c/55°C, 90 s/72°C, 60 s 1.5 mM Mg²⁺	30	Group A >99%	(1)	Direct PCR
speA	CACCGAGAATGTGAAATCT GGTCCATTAGTATATAGTTGC	397	94°C, 1 m/38°C, 1 m/72°C, 1 m 4.0 mM Mg²⁺	30	Group A 13–44%	(1,2)	Direct PCR
speB	GGCGATTTCAGAATTGATGGC TAGTGCGTCAAGACGGAAG	342	94°C, 1 m/48°C, 1 m/72°C, 1 m 4.0 mM Mg²⁺	30	Group A >99%	(24, 25)	Direct PCR
speC	CCGAAATGTCTTATGAGGCC CCCTTCATTGGTGAGTC	382	94°C, 1 m/48°C, 1 m/72°C, 1 m 4.0 mM Mg²⁺	30	Group A 35–71%	(24, 25)	Direct PCR
slo	AGAACACAATATACTGAATCAATGGGT ACTTTTCGCCACCATTCCCAAGC	868	94°C, 40 s/52°C, 90 s/72°C, 60 s 1.5 mM Mg²⁺	30	Group A, >99%1	(24)	Direct PCR
tRNA^iMeta tRNA^Phe intergenic region	AAGGTCGTAGGTTCAAATCC ACCAATGAGCTACCGAGCC	<100	94°C, 1 m/60°C, 1 m/72°C, 2 m or 94°C, 30 s/60°C, 30 s/72°C, 60 s	40	Various (see Fig. 2, Note 16)	(23)	Direct PCR
tRNA genes	AGTCCGGTGCTCTAACCAACTGAG AGGTCGCGGGTTCGAATCC	Multiple	94°C, 30 s; 50°C, 30 s; 72°C, 2 m 1.5 mM Mg²⁺	40	Various	(23, 26)	PCR fingerprinting
16S rRNA gene	CCAAGCTTGCTCAGGAGAACGCT CGGGATCCGCCGCCGGGAACGTATTCAC	1400–1500	93°C, 90 S/56°C, 90 S/75°C, 90 S 1.5 mM Mg²⁺	35	Various	(3,4)	PCR-RFLP
16S rRNA gene V2 region	GAGAGTTTGATCCTGGCTCAGGA TTACCGCGGCTGCTGGCACGT	529	92°C, 2 m; 55°C, 1 m; 72°C, 1.5 m 1.5 mM Mg²⁺	30	Various	(5)	PCR-hybridization
23S rRNA gene	AAGGGGCGCACGGTGGATGCCTG ATCTTATCACTCGCAGTCTG	950	92·C, 2 m/55°C, 60 s/72°C, 90 s 1.5 mM Mg²⁺	30	Various	(12)	PCR-hybridization
vir	AAACCGTATCTTTGACGCACTCG AGGACAATTTGCGAGATTAG AGACATGAGCTCAATGGCAAGTT ATCAAATGGTAATTTTTG		94°C, 10 s/60°C, 2 m/68°C, 6 m	25	Group A	(1)	PCR-RFLP
emm genes^b	GGGGGGGGGATCCATAAGGAGCATAAAA ATGGCT GGGGGGGAATTCAGCTTAGTTTTTCTTCT TTGCG	~1–2 kb range	94°C, 1 m/50°C, 1 m/72°C, 2.5 m (Dot Blot) 94°C, 1 m/55°C, 1 m/72°C, 2.5 m (Line Blot)	30	Group A	(13, 27)	Dot/line blot
emm genes	TATTG/CGCTTAGAAAATTAA GCAAGTTCTTCAGCTTGTT	~1000	Conditions for PCR cycling were not given		Group A	(28)	PCR-sequence

[a]Amplification can also occur with some strains of group C and G streptococci as well as *Bacillus cereus* and *Enterococcus faecalis* (1).
[b]Underlined bases are restriction sites and clamp regions used for force cloning of the PCR product. If these sequences are deleted the amplification conditions may need to be modified.

2. Make a master mix of all components except template DNA (*see* **Note 13**).
3. Add template DNA to each reaction tube, include known positive, negative and nontemplate controls. One to two microliters of DNA from any of the procedures listed in **Subheading 3.1.** above is sufficient in most cases.
4. Preheat the thermal cycler to 95°C at this time (*see* **Note 14**).
5. Add the appropriate amount of *Taq* polymerase to the master mix. Mix by pipeting up and down several times.
6. Add master mix to each reaction tube (the volume added is dependent on the amount of template added and the final volume of the PCR reaction). Mix by gently pipeting several times.
7. Overlay each reaction with approx 50 µL of mineral oil and transfer to the thermal cycler.

3.2.2. Direct Analysis of PCR Products

The identification of streptococci by direct analysis of PCR products is based on the presence (or absence) and size of the unique region amplified from the chromosome of a particular species. This method has the advantage of being rapid and having the fewest manipulation steps, however, accurate identification of only a limited number of species is possible.

Primer pairs for use in what is termed direct PCR give products that can be resolved by electrophoresis through 1–2% agarose gels in 1X TBE buffer at 100 V for 2–4 h (time varies depending on the size of expected product, length of gel and percentage of agarose used). For exceptionally small fragments, or when high resolution is required, electrophoresis through 3% NuSeive agarose gels (FMC), or a similar product may be required. For exact size determination, the products can be amplified using a radiolabeled primer pair and resolved on a standard denaturing sequencing gel with a sequencing reaction being used as the size standard (*see* **Note 15**). Primers suitable for direct PCR are presented in **Table 1**, along with the expected size of the amplification product and the specificity of the primers (when known).

3.2.3. Differentiation of Group A Streptococci by Direct Cycle-Sequencing of emm-Genes (10)

M-proteins have been used for many years as a means of differentiating group A streptococci. The traditional method involves typing streptococcal isolates with a panel of M protein antisera. Whereas this method has been widely accepted, accurate typing is limited by the availability of antisera. In addition, newly emerging M-types are often missed because of the lack of specific antisera or cross-reactivity with multiple antisera.

Recently, new methods for the differentiation of group A strains based on M-type have been presented. These methods rely on direct analysis of the M-protein *(emm)* genes by PCR sequencing or hybridization. These methods have the advantage over immunologic typing in that laboratories with the ability to sequence DNA can perform the analysis in a relatively short time. They may also be less susceptible to "mistyping" as the analysis is being performed directly on the DNA and not the protein, thereby eliminating problems with cross-reactivity, interference, or lack of a particular antiserum.

The direct cycle sequencing of M-protein *emm* genes is performed as follows: Target DNA is amplified with the appropriate primers (*see* **Table 1** and **Subheading 3.2.2.**). The PCR products are purified using Wizard columns (Promega, Madison, WI) or similar products. Approximately 60 ng of the purified product is subjected to cycle-sequencing using primer 1 as the sequencing primer (*see* **Note 17**). Identification of M-type is based on homology with known *emm* gene sequences found in the GenBank/EMBL databases.

3.3. Identification of Streptococci by RFLP

Identification of streptococci by RFLP is based on analysis of restriction fragments generated by digestion of PCR products (or intact chromosomal DNA, *see* **Subheading 3.1.4.**) amplified from a conserved region within the chromosome of a particular genus or species. This type of analysis allows the identification of a wider range of species compared with direct PCR. The major disadvantage of this procedure is the additional manipulation required of the samples and the time involved in the steps subsequent to PCR. Resolution of similar size DNA fragments may also interfere with data interpretation, and may require fragment separation on a denaturing sequencing gel with a sequencing reaction being used as the size standard (*see* **Note 15**). The following procedure is for the identification of streptococcal species based on RFLP analysis of 16S ribosomal DNA *(3,4)* (*see* **Note 18**).

1. Amplify target DNA with appropriate primers (*see* **Table 1** and **Subheadings 3.1.** and **3.2.**)
2. Extract PCR reactions with chloroform to remove the mineral oil overlay from sample.
3. Add 2 vol of 95% ethanol and precipitate DNA at –20°C for approx 2 h.
4. Microfuge at maximum speed for 10 min to pellet DNA.
5. Suspend the pellet in 25–50 µL of distilled water.
6. Transfer approx 5 µL (amount may vary) of amplification product to each of a series of microfuge tubes and digest with the appropriate restriction enzymes (*see* **Table 2**) for 2 h at 37°C in a total reaction volume of 20 µL.
7. Electrophorese of the entire digest as described above (*see* **Subheading 3.2.2.**) and determined the size of each fragment. Species identification should be determined by comparison of the fragment patterns with the patterns produced from known strains. Alternatively, some species may be identified by comparison of the fragment sizes obtained with the data in **Table 2** and **Fig. 1**.

RFLP analysis of the *vir* regulon allows identification of group A streptococci from other species, and differentiation of group A strains *(1)*. This tech-

Table 2
Identification of Streptococcal Species Based
on Fragment Sizes Generated by Digestion of rDNA
Amplification Products with the Appropriate Enzyme

| Species[a] | Restriction Enzyme | |
	RsaI[b]	MspI[c]
S. uberis	200	
S. parauberis	200	
S. agalactiae	270	+570
S. dysgalactiae	270	+490
S. mitis	930	+100/+130/+330
S. salivarius	930	+490
S. equinus	930	−100/−13/−330/−490

[a]All listed species produce a characteristic *Hha*I pattern consisting of three bands with sizes of 280, 550, and 600 bp.

[b]Digestion of the amplification product with *Rsa*I results in multiple fragments, but bands of the given sizes are specific for the particular species. In addition, *S. uberis* and *S. parauberis* may be further differentiated from each other by the absence of a 370-bp product in *S. parauberis* and the presence of a 240-bp product in *S. uberis.*

[c]The presence (+) or absence(−) of bands of the given size are used in conjunction with data from the *Rsa*I digestion to differentiate the species further (e.g., *S. agalactiae* is distinguished from *S. dysgalactiae,* which has the same specific *Rsa*I fragment, by the presence or absence of the indicated *Msp*I fragments).

nique does not allow identification of other streptococcal species. Differentiation is based on comparison of *Hae*III restriction digest patterns of *vir* region amplification products as follows:

1. Amplify target DNA in a 50 µL reaction consisting of 250 µ*M* of each dNTP, 20 pmol of each primer (*see* **Table 1**), 1 µL of template DNA (*see* **Subheading 3.1.4.**) and 0.4 µL of KlentaqLA-16 *(11)* (*see* **Note 19**) in PC2 buffer. Follow the amplification cycle information presented in **Table 1**.
2. Verify amplification product and concentration by electrophoresis of 5 µL of PCR product through a 0.8% agarose gel.
3. Transfer approx 0.5 µg of PCR Product (8–25 µL) to a microfuge tube and digest with 2 U of *Hae*III for 1 h.
4. Separate fragments by electrophoresis through a 1.5% agarose gel.
5. Visualize DNA bands (*see* **Subheading 3.2.2.**) and determine the size of each fragment. Strain identification is determined by comparing the restriction digest patterns.

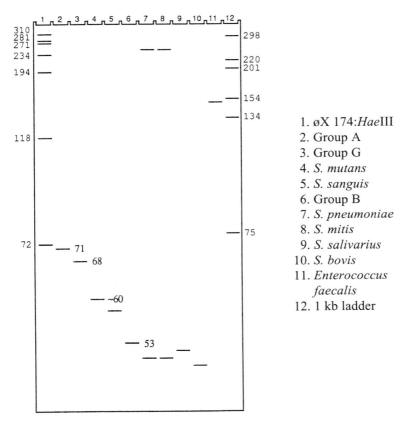

Fig. 1. Computer representation of tRNA gene intergenic regions amplified from several streptococcal species *(23)*. Sizes of amplification product are given when known, the other products are approximate, based on gel migration

3.4. Identification of Streptococci
by Hybridization Analysis of PCR Products

Hybridization of specific primers to PCR products amplified from conserved regions of the streptococcal genome can be used for identification of some strepto-coccal species. This technique has the advantage of increased sensitivity over direct PCR and RFLP (the amount of template and product required can be substantially less) and also being relatively quick to perform. The major disadvantage is the need for numerous species-specific primers for hybridization. Hybridization pro-tocols will vary depending on the type of probe labeling (isotopic versus nonisotopic) and the particular manufacturer's product being used (*see* **Note 15**). The main criterion for successful hybridization is high-efficiency probe labeling to reduce the amount of time necessary for hybridization and detection.

3.4.1. Identification of Streptococci Based on Hybridization to 23S or 16S rRNA genes (5,12)

1. The 23S or 16S rRNA gene is amplified using the appropriate primer pair (*see* **Table 1**) that amplifies a conserved region of the streptococcal genome (*see* **Subheadings 3.1.** and **3.2.**).
2. Verification of product size should be done by agarose gel electrophoresis (*see* **Subheading 3.2.2., Table 1**).
3. Amplified products (~1 μg) can be denatured by heating to >95°C for 5 min (this can be done in the thermal cycler) and directly applied to a positively charged nylon membrane (results may vary depending on the type of membrane used) using a dot- or slot-blot apparatus.
4. DNA is fixed to the membrane by baking at 80°C for 30 min (a vacuum oven is not required for nylon membranes) or by UV crosslinking the DNA to the membrane.
5. Label the appropriate probe (*see* **Table 3**) using an isotopic or nonisotopic method (*see* **Note 15**).
6. Prehybridize the membrane for 1 h at 42°C followed by hybridization for 1.5–2 h at 42°C (*see* **Notes 20** and **21**).
7. Perform two high stringency posthybridization washes with 1X SSC, 0.1% SDS at the appropriate wash temperature (*see* **Table 3**, **Note 20**).
8. Detect hybridization signal (*see* **Note 22**).
9. Identity of the streptococcal species is based on hybridization with the specific probe (*see* **Table 3**).

3.4.2. Differentiation of Group A Streptococci by Reverse Dot Blot Hybridization of Specific Probes with M-Protein (emm) Genes (13)

The M-protein is one of the major factors for the differentiation of group A streptococci. Whereas most typing is performed by laboratories using M-type specific antisera, other methods for M-typing are being developed. The following methods rely upon the variability of the *N*-terminal region of the *emm*-genes for identification.

As with other procedures of this nature, differentiation is limited by the number and availability of specific oligonucleotide probes. However, with the reduced cost of oligonucleotide synthesis and the availability of numerous *emm*-gene sequences deposited in GenBank, this method is becoming more practical. One potential problem associated with this method is the inability of a single oligonucleotide probe to detect all the *emm*-genes of a single M-type. This problem has been shown with some of the M-type 1 and 49 isolates *(14,15)*. Therefore, all results should be verified using an alternate method.

1. Amplify the *emm* gene from the target DNA (*see* **Subheadings 3.1.** and **3.2.**) using the appropriate primer pair (*see* **Table 1**), incorporating a nonisotopic label into the product (*see* **Note 23**).
2. Verify amplification product by running a sample of the reaction on an agarose gel.

Table 3
Hybridization Probes used for the Differentiation
of Streptococci Following rRNA Gene Amplification

Probe	Sequence	Specificity	Stringency wash, temperature (°C)
23S rRNA gene amplification *(12)*			
Su1	GAAGTGGGACATAAAGTTAATA	*S. uberis*	42
Su2	TTGACTTTAGCCCTAGCAGT	*S. uberis*	50
Sp1	GACGTGGGATCAAATACTATA	*S. parauberis*	50
Sp2	TCACCTTTCCCTCACGGTACTGG	*S. parauberis*	47
All hybridizations were done at 42°C.			
16S rRNA gene amplification *(5)*			
BB9967	ACTAACATGTGTTAATTACTCTTA	*S. agalactiae*	42
BB9969	AGTACATGAGTACTTAATTGTCA	*S. parauberis*	42
BB9970	GGTACATGTGTACCCTATTGTC	*S. uberis*	42
All hybridizations were done at 42°C (*see* **Note 21**).			

3. Set up a dot blot panel with a serial dilution of each specific oligonucleotide probe (*see* **Note 24, Table 4**).
4. Bake blot at 80°C for 30 min or UV crosslink to bind oligonucleotides to the membrane.
5. Prehybridize the blot at 55°C for 1 h.
6. Add approx 10 µL of the labeled PCR product (denature prior to addition) directly to the prehybridization solution.
7. Incubate blots for at least 1 h at 50°C.
8. Wash the blots at room temperature with 2X SSC, 0.5% SDS for 5 min followed by 2X SSC, 0.1% SDS for 10 min.
9. Wash the blots once for 20 min at 50°C with 0.1X SSC, 0.1% SDS.
10. Perform chemiluminescent detection (colorimetric detection can also be used) of the captured probe according to the manufacturer's instructions (Boehringer Mannheim, Germany).

3.4.3. Differentiation of Group A Streptococci by Reverse-Line Blot Hybridization of Specific Probes with M-Protein (emm) Genes *(13)*

This procedure is similar to that given for reverse-dot blot hybridization, however, it has the advantage of being able to process multiple samples on a single blot. The *emm*-specific oligonucleotides are covalently linked to activated Biodyne C membrane. This probe acts to capture the labeled *emm*-gene fragment.

1. Amplify the *emm* gene from the target DNA (*see* **Subheadings 3.1.** and **3.2.**) using the appropriate primer pair (*see* **Table 1**), incorporating a nonisotopic label into the product (*see* **Note 25**).
2. Verify amplification product by running a sample of the reaction on an agarose gel.

Table 4
Oligonucleotide Probes Used
for the Identification of *emm* Specific Genes (13,15)

Probe	Nucleotide Sequence
*emm*1	TTC TAT AAC TTC CCT AGG ATT ACC ATC ACC
*emm*2	TGC TTC TTT TTT GAC AGG GAC AGG GTT CTT
*emm*3	CAT GTC TAG GAA ACT CTC CAT TAA CAC TCC
*emm*4	CCA CGC TGA ATC AGC CTG AGG CTT TTT AAT
*emm*5	CGG GTC ATT TAT TGT ACC CCT AGT CAC GGC
*emm*6	TGC TTT GTC CGG GTT TTC TAC CGT CCC CCT
*emm*8	TCG TTA TTA GAA ATA CTA TGA GAT TTT GGG
*emm*11	GCG TCA CGT TTG TAC CTT TAG GAG CGC TTT
*emm*12	ACG TTG TTT TTC TGC GAC TAA ATC ACT ATG
*emm*18	CGT CTT TAT TGT CTG CTG TAG CTC GAG TAA
*emm*19	ATC TTC TGG CGT ATG CCT AGT ATA ACG CAC
*emm*24	TTC TTG TAC TTT TTC CAG AGT ATC TGT CTG
*emm*41	GTA AAG CTT CTT CTT GGG CTT GTG CTA ATC
*emm*49H	ATT CTC CGC AAC CTC AAC TTT AGC CTC AAC
*emm*49K	ATT CTC CGC AGC CTC AAC
*emm*49HK1	TTT TTC TCT TCT TGC AAC GCT AGA CAC GTT
*emm*49HK2	TGT AAG ATC GGC GAT TTG GTC GTA TAG CTC
*emm*52	TTC GGT ATA TCG ATG GTG ATC AAC AGG TTG
*emm*57	CAA AAT AGG CGT CAT CGA AGT AAT ATC GCT

3. Activate the Biodyne C membrane (Pall Biosupport, NJ) by incubation for 10 min in 16% 1-ethyl-3-(3-dimethyl-aminopropyl) carbodiimide *(16)*.
4. Rinse the membrane in distilled water and place in a Line blot apparatus (Immunetics, ITK Diagnostics, Uithoorn, The Netherlands).
5. Fill each slot with a different *emm*-specific 5' amino-modified oligonucleotide (an amino group is added during synthesis) (150 µL of 0.3 µ*M* probe in 500 m*M* NaHCO$_3$, pH 8.4).
6. Incubate the membrane for 1 min at room temperature and then remove excess probe by aspiration.
7. Remove the blot from the apparatus and treat with 100 m*M* NaOH for 10 min to inactivate membrane.
8. Wash membrane for 5 min at 50°C with 2X SSPE (360 m*M* NaCl, 20 m*M* NaH$_2$PO$_4$, 2 m*M* EDTA, pH 7.2), 0.1% SDS.
9. Replace the blot in the line blot apparatus in a position 90° rotated relative to the original orientation.
10. Fill the slots of the blotting apparatus with heat-denatured biotin-labeled PCR products (30 µL of PCR product in 120 µL 2X SSPE, 0.1% SDS).
11. Incubate the blot at 50°C for 45 min.
12. Remove excess probe by aspiration and wash with 2X SSPE, 0.5% SDS at 50°C for 10 min.

13. Incubate blot with streptavidin-peroxidase conjugate (Boehringer Mannheim or other) diluted to the appropriate concentration in 2X SSPE, 0.5% SDS for 10 min at 42°C .

14. Wash the blot in 2X SSPE, 0.5% SDS at 42°C for 10 min and rinse briefly in 2X SSPE.

15. Perform chemiluminescent detection (colorimetric detection can also be used) of the captured probe according to the manufacturer's instructions (Boehringer Mannheim, Germany).

3.5. Identification of Streptococci
Based on DNA-DNA Hybridization (17–20)

DNA-DNA hybridization as a means of species identification relies of the fact that closely related species will have more homologous DNA than more distantly related organisms. A panel of DNA from known strains is used as the target for probing with labeled DNA isolated from the strain to be identified. Identity is determined based on which of the standards has the strongest hybridization. This technique is a rapid and sensitive method for the identification of unknown species, however it does have several limitations. The first and foremost is the need for DNA from numerous species to act as the target. It is also possible to identify incorrectly an isolate if the homologous species is not present among those of the test panel. In addition, as the identification is based on a quantitative measurement of the hybridization signal, care must be taken to ensure the same amount of target DNA and probe DNA is present in each assay.

The following method has been successfully used to differentiate species of viridans group streptococci *(7)*. The assay is performed in a microtiter plate assay format using nonisotopic hybridization. This procedure can be easily adapted to include other streptococcal species.

1. Isolate DNA from large volumes (1.5 L) of known species of streptococci. The procedures presented in **Subheading 3.1.** can be scaled up to accommodate the volume of cells or an alternate procedure can be used.

2. Determine the concentration of each stock DNA solution based on $OD_{260/280}$ ratio. Aliquot the DNA stocks into small volumes and store at 4°C (30 d) or –20°C (prolonged storage).

3. Dilute the DNA to 10 µg/mL in coating buffer and heat denature at 100°C for 5 min.

4. Dispense 100 µl of the denatured DNA solution into the wells of a microplate (Immunoplate, Nalgene Nunc, Rochester, NY; or similar). Dispense denatured salmon sperm DNA (10 µg/mL) into one well to serve as the negative control (*see* **Note 26**).

5. Seal the microplate(s) and incubate at 30°C for 12 h.

6. Isolate DNA from the strain(s) to be identified using one of the methods presented in **Subheading 3.1.**

7. Add 200 µL of prehybridization solution to each well and incubate at 37°C for 1 h.

8. Remove the prehybridization solution and add 10–50 ng of photobiotinylated DNA (*see* **Note 27**) from the test species (**step 6**) in 100 µL hybridization solution to each well.
9. Seal the microtiter plate(s) and incubate for 2 h at 55°C.
10. Wash the wells four times with 300 µL of 2X SSC buffer.
11. Add 100 µL of streptavidin-horseradish peroxidase in phosphate buffered saline (PBS) to each well (*see* **Note 28**).
12. Incubate the plates at 37°C for 10 min.
13. Wash the wells three times with 1X SSC.
14. Add 100 µL of substrate solution to each well (50 µL of tetramethyl-benzidine [400 µg/mL] and 50 µL of 3% H_2O_2) (*see* **Note 29**).
15. Monitor the optical density of the wells at 650 nm. The assay is considered complete when the ratio of the maximal color intensity and the intensity of the negative control is greater than 1.9. The value for the maximum color intensity is taken as 100% and the control is taken as 0%.
16. The test strain is identified as the plate-bound species when none of the other wells have greater then 70% maximum color intensity.

3.6. Differentiation of Streptococci by Gel Electrophoresis

Pulsed-field gel electrophoresis (PFGE) (*see* Chapter 3) can be used as a means of typing particular strains or clonal populations of streptococci. This technique provides a means to study the molecular epidemiology of streptococcal disease outbreaks *(4)* it is not a reliable method, however, for typing or verifying unknown streptococcal isolates as there can be considerable differences in PFGE patterns even between strains within a given species (*see* **Fig. 2**). The presence and number of phage present within a given strain can also greatly alter the electrophoretic pattern of otherwise similar strains. The use of PFGE is also the most time-consuming method for differentiating streptococci. The procedure below assumes the use of a Bio-Rad CHEF electrophoresis unit.

1. Isolate DNA from streptococcal cells as described in **Subheading 3.1.4.**
2. Soak the DNA-agarose plugs (or a portion of the plug) in 1X restriction buffer for 30–60 min.
3. Transfer the gel slice to 250 µL of 1X buffer and add 25–50 U of the desired restriction enzyme (*Sma*I and *Sfi*I work well for streptococcal DNA).
4. Incubate samples for >3 h at the temperature appropriate for the restriction enzyme.
5. Make a 1–1.2% agarose gel (Fast Lane or Sea Plaque GTG; FMC) in 0.5X TBE buffer.
6. Carefully remove gel slices from restriction buffer and place in the wells of the gel.
7. Overlay the sample gel slice with agarose (~45°C) to seal the wells.
8. Electrophoresis of the samples is performed at 180–200 V for 22 h with a pulse rate starting at 6 s and ramping to 30 s at the end of the run.
9. Visualize gels by staining with ethidium bromide (EB) (150 µg/mL) for 30–40 min followed by destaining with several changes of distilled water for 30–60 min.

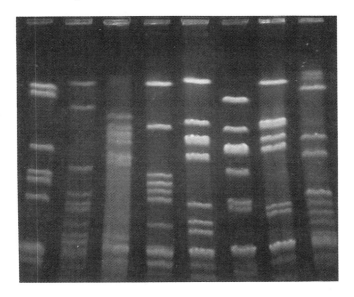

Fig. 2. PFGE pattern of several clinical isolates of *S. pyogenes* digested with *Sma*I.

4. Notes

1. To make GES solution, mix 60 g of guanidinium thiocyanate, 20 mL of 0.5 M EDTA (pH 8.0) and 20 mL of distilled water. Heat to 65°C until dissolved. After cooling, add 5 mL of 10% sarkosyl and dilute to 100 mL.

2. The addition of mutanolysin (50 U/mL final concentration) will improve the subsequent cell lysis step. This can be omitted, but the ultimate yield of DNA may be reduced.

3. The addition of 5 M NaCl allows for the precipitation and removal of proteins and cell debris from the samples. Chilling the samples on ice for 5 min prior to centrifugation will allow precipitation of more protein from particularly dirty samples. It is important to remove as much protein as possible at this stage to maintain high purity in the DNA preparation.

4. The DNA will precipitate out of solution as a filamentous mass in samples with a high yield of intact DNA. In these instances, the samples can be immediately pelleted (by microfuging at room temperature for 30 s) and washed with 70% ethanol.

5. Allow the DNA to remain in solution for several minutes or longer as needed to allow the DNA to go into solution. The sample can be mixed by tapping the bottom of the microfuge tube. Avoid pipeting the sample or vortexing as this can cause shearing of the DNA. Samples can also be heated to 50°C for 10 min to accelerate the DNA into solution.

6. If lysis is not readily apparent, samples can be heated in a 65°C water bath for 5–10 min.

7. The excessive washing steps are necessary to remove all the guanidinium and ammonium acetate from the samples. Failure to remove all the salt may interfere with subsequent use of the DNA.

8. Addition of glycine to the medium (0.5–1.75%), such that the growth rate is reduced by approx 25%, will increase the cells lysis in subsequent steps because of cell wall weakening. This will ultimately increase the overall yield of DNA from most streptococcal species.

9. This step allows for the inactivation of proteinase K and any other proteinase that might be present in the sample and is necessary to prevent subsequent digestion of *Taq* polymerase during PCR amplification. In addition, this step acts as an initial DNA denaturing step prior to PCR.

10. This procedure can be modified for isolation of smaller amounts of DNA suitable for Long-PCR. For this modification, a loopful of cells from overnight growth is used. All steps are carried out in a microtiter plate. Values in parenthesis should be used with this modification.

11. For Long-PCR template production, add solution to the microtiter wells for incubation, do not remove plugs from the wells.

12. Template for Long-PCR is obtained by melting an agarose plug in 450 µL of TE buffer at 65°C for 10 min. One microliter of this solution is used for subsequent amplification. Alternatively, a portion of the agarose plug can be melted in a correspondingly reduced volume of TE buffer. Unused portions may be solidified and reused, however, deterioration of template quality may occur with repeated processing.

13. A master mix can be made by multiplying each component by the number of reactions to be performed plus one (don't forget positive, negative, and no-template controls); maintain the mixture on ice. Add the appropriate amount of water to the tube first, followed by the remaining components. The *Taq* polymerase is added to the mixture just prior to use.

14. An initial 95°C denaturing step of 30 s to 2 min prior to initiating the cycling may improve the overall yield of the amplification reaction.

15. The procedures for labeling probes and performing sequencing reactions is beyond the scope of this chapter. The researcher should refer to general molecular laboratory manuals for information, or follow the manufacturer's recommended procedure, which accompanies labeling and sequencing kits.

16. This method is a rapid means of differentiating streptococci from other genera. However, it has not been sufficiently tested to determine if the sizes are consistant throughout a particular species. It is also possible that the intergenic region may not differ enough between species to be useful for species differentiation without another set of primers being used.

17. Sequencing is done with primer emmseq 2 [TATT(G/C)GCTTAGAAAATT-AAAAACAGG]. If using an ABI 373 automated system, the procedure for sequencing with dye termination mix is performed as per the manufacturer's instruction (Applied Biosystems, Foster City, CA), the cycle parameters are 25 cycles of 96°C for 30 s, 55°C for 1 s and 60°C for 4 min. It is also possible to use manual systems for cycle-sequencing with ^{32}P-labeled primer.

18. A similar identification strategy has also been devised based on fragment length polymorphism analysis of tRNA genes *(15,20)*. This procedure uses a primer pair combination (*see* **Table 1**) to amplify conserved tRNA gene sequences to produce a fingerprint that should be unique to the particular species. The amplification is performed using radiolabeled primers, and separation is accomplished on a 5% denaturing sequencing gel. Identification is made by comparison of the fingerprint pattern with the pattern produced from known species (*see* **Fig. 1**).

19. KlentaqLA-16 is made-up as follows: Mix 15/16 μL of Klentaq1 (AB Peptides, St. Louis, MO) with 1/16 μL *pfu* DNA polymerase (Stratagene, La Jolla, CA). Klentaq1 is first standardized to a working concentration such that 0.25 μL effectively (whereas 0.12 μL less effectively) amplified a 2.0-kb target from 10 ng of plasmid template. This can be done using a standard cloning vector containing an ~2-kb insert and the M13F and M13R primer sites by amplifying for 24 cycles (1 cycle = 95°C for 30 s [denaturation] and 65°C for 7 min [combined annealing and extension]).

20. The procedure described for identification of streptococci based on hybridization to the amplified variable region of the 16S rRNA gene is primarily based on the use of a magnetic capture system and microtiter hybridization *(3)*. This procedure is more involved than most laboratories are currently equipped to carry out, so the authors have listed the probes for use with standard dot/slot blot hybridization. The wash temperature presented should be used as a starting point, but may have to be increased if cross reactivity occurs. A list of potential oligonucleotides for the identification of other streptococci is also presented *(5)*. These oligonucleotides have not been examined for specificity, however, and, therefore, should be used with caution.

21. Longer hybridization times are acceptable, though excessive hybridization can lead to increased nonspecific binding. However, if less than 1 μg of amplification product is used, longer hybridization times may be required.

22. Follow the manufacturer's directions for the detection procedure appropriate for the type of labeled probe used. Detection of probe can be accomplished by chemiluminescence (greatest sensitivity and speed), colorimetric detection or autoradiography. In most instances, colorimetric detection is sufficiently sensitive to obtain strong signals within 60 min, although color development can proceed for several hours without significant increases in background levels.

23. The nonisotopic label can be incorporated into the product by the addition of the labeled nucleotide (digoxigenin-UTP), directly to the amplification reaction.

24. Poly dT-tailing of the oligonucleotides may improve binding to the nylon membrane and signal intensity. Incubate 200 pmol of the oligonucleotide for 60 min at 37°C in a 100 μL reaction containing 4 μL of 25 m*M* CoCl$_2$, 100 nmol of dTTP, 60 U of terminal deoxyribonucleotidyltransferase (Boehringer Mannheim) and 20 μL of the supplied 5X enzyme buffer. Stop the reaction by addition of 100 μL of 10 m*M* EDTA and 100 μL of TE buffer.

25. In this method, a biotin molecule is incorporated at the 5' end of the downstream oligonucleotide during synthesis. The amplification reaction is setup as follows. Four microliters of crude lysate is used as the template in a 50 μL reaction con-

taining *Tth* buffer (10 m*M* Tris·HCl, 50 m*M* KCl, 1.5 m*M* MgCl$_2$, 0.01% gelatin, 0.1% Triton X-100 pH 9.0), 200 μ*M* of each dNTP, 20 pmol of each primer and 0.5 U of *Tth* DNA polymerase. Overlay reactions with paraffin oil and heat for 4 min at 94°C before initiation of cycling.

26. Covalink microtiter plates (Nalgene Nunc) can be used instead of the immuno-plates, however, different coating and washing procedure should be followed *(20,21)*. The use of Covalink plates allows for the covalent binding of DNA to the microplate wells. Use of these plates may improve the sensitivity, reduce the background of the hybridization, and allow for the repeated probing of the assay plates without the need for recoating with target DNA *(20)*.

27. Procedure for photobiotinylation of DNA *(17,18,22)*:
    a. Mix 5 μL of photobiotin (Vector Laboratories, Burlingame, CA) with an equal volume containing approx 5 μg of denatured DNA in a microfuge tube.
    b. Irradiate with a 500 W sunlamp at a distance of approx 10 cm for 15 min.
    c. Dilute labeled DNA to 100 μL with 0.1 *M* Tris-HCl (pH 9.0).
    d. Extract twice with an equal volume of 2–butanol to remove free photobiotin.
    e. Use immediately for hybridization experiments.

28. Dilute streptavidin-horseradish peroxidase to a final working concentration according to the manufacturer's specifications. Alkaline-phosphatase based conjugates can also be used to detect the bound label. The substrate and wavelength of the microplate reader will need to be changed to accommodate the alternate conjugate.

29. If a microplate luminometer is available, then the bound probe can be detected using a chemiluminescent substrate. This will increase the sensitivity of the assay and decrease the overall hybridization time required *(20)*.

References

1. Gardiner, D., Hartas, J., Currie, B., Mathews, J., Kemp, D., and Sriprakash, K. (1995) Vir typing: A long-pcr typing method for group A streptococci. *PCR Methods Applic.* **4**, 288–293.

2. Musser, J., Kapur, V., Peters, J., Hendrix, C., Drehner, D., Gackstetter, G., Skalka, D., Fort, P., Maffei, J., Li, L. -L., and Melcher, G. (1994) Real-time molecular epidemiologic analysis of an outbreak of *Streptococcus pyogenes* invasive disease in US Air Force trainees. *Arch. Pathol. Lab. Med.* **118**, 128–133.

3. Jayarao, B., Doré, J., Baumbach, G., Matthews, K., and Oliver, S. (1991) Differentiation of *Streptococcus uberis* from *Streptococcus parauberis* by polymerase chain reaction and restriction length polymorphism analysis of 16S ribosomal DNA. *J. Clin. Microbiol.* **29**, 2774–2778.

4. Jayarao, B., Doré, J., and Oliver, S. (1992) Restriction fragment length polymorphism analysis of 16S ribosomal DNA of *Streptococcus* and *Enterococcus* species of bovine origin. *J. Clin. Microbiol.* **30**, 2235–2240.

5. Bentley, R. and Leigh, J. (1995) Development of PCR-based hybridization protocol for identification of streptococcal species. *J. Clin. Microbiol.* **33**, 1296–1301.

6. Pitcher, D., Saunders, N., and Owen, R. (1989) Rapid extraction of bacterial genomic DNA with guanidium thiocyanate. *Lett. Appl. Microbiol.* **8**, 151–156.

7. Hynes, W., Ferretti, J., Gilmore, M., and Segarra, R. (1992) PCR amplification of streptococcal DNA using crude cell lysates. *FEMS Microbiol. Lett.* **94**, 139–142.
8. Suvorov, A. and Ferretti, J. (1994) Heterogeneity of group A streptococcal M1 chromosomal RFLP patterns as determined by PFGE, in *Pathogenic Streptococci: Present and Future.*, (Totolian, A., ed.), Lancer, St. Petersburg, Russia, pp. 21–23.
9. Gordillo, M., Singh, K., Baker, C., and Murray, B. (1993) Typing of group B streptococci: Comparison of pulsed-field gel electrophoresis and conventional electrophoresis. *J. Clin. Microbiol.* **31**, 1430–1434.
10. Beall, B., Facklam, R., and Thompson, T. (1996) Sequencing *emm*-specific PCR products for routine and accurate typing of group A streptococci. *J. Clin. Microbiol.* **34**, 953–958.
11. Barnes, W. (1994) PCR amplification of up to 35-kb DNA with high fidelity and high yield from λ bacteriophage templates. *Proc. Natl. Acad. Sci. USA* **91**, 2216–2220.
12. Harland, N., Leigh, J., and Collins, M. (1993) Development of gene probes for the specific identification of *Streptococcus uberis* and *Streptococcus parauberis* based upon large subunit rRNA gene sequences. *J. Appl. Bacteriol.* **74**, 526–531.
13. Kaufhold, A., Podbielski, A., Baumgarten, G., Blokpoel, M., Top, J., and Schouls, L. (1994) Rapid typing of group A streptococci by the use of DNA amplification and non-radioactive allele-specific oligonucleotide probes. *FEMS Microbiol. Lett.* **119**, 19–26.
14. Penney, T., Martin, D., Williams, L., Malmanche, S. D., and Bergquist, P. (1995) A single emm gene-specific oligonucleotide probe does not recognise all members of the *Streptococcus pyogenes* M type 1. *FEMS Microbiol. Lett.* **130**, 145–150.
15. Kaufhold, A., Podbielski, A., Johnson, D., Kaplan, E., and Lütticken, R. (1992) M protein gene typing of *Streptococcus pyogenes* by nonradioactively labeled oligonucleotide probes. *J. Clin. Microbiol.* **30**, 2391–2397.
16. Zhange, Y., Coyne, M., Will, S., Levenson, C., and Kawasaki, E. (1991) Single-base mutational analysis of cancer and genetic diseases using membrane bound modified oligonucleotides. *Nucleic Acids Res.* **19**, 3929–3933.
17. Ezaki, T., Hashimoto, Y., Takeuchi, N., Yamamoto, H., Liu, S. -L., Miura, H., Matsui, K., and Yabuuchi, E. (1988) Simple genetic method to identify viridans group streptococci by colorimetric dot hybridization and fluorometric hybridization in microdilution wells. *J. Clin. Microbiol.* **26**, 1708–1713.
18. Ezaki, T., Hashimoto, Y., and Yabuuchi, E. (1989) Fluorometric deoxyribonucleic acid-deoxyribonucleic acid hybridization in microdilution wells as an alternative to membrane filter hybridization in which radioisotopes are used to determine genetic relatedness among bacterial strains. *Int. J. Sys. Bacteriol.* **39**, 224–229.
19. Kikuchi, K., Enari, T., Totsuka, K. -I., and Shimizu, K. (1995) Comparison of phenotypic characteristics, DNA-DNA hybridization results, and results with a commercial rapid biochemical and enzymatic reaction system for identification of viridans group streptococci. *J. Clin. Microbiol.* **33**, 1215–1222.
20. Adnan, S., Li, N., Miura, H., Hashimoto, Y., Yamamoto, H., and Ezaki, T. (1993) Covalently immobilized DNA plate for luminometric DNA-DNA hybridization to identify viridans streptococci in under 2 hours. *FEMS Microbiol. Lett.* **106**, 139–142.

21. Rasmussen, S., Larsen, M., and Rasmussen, S. (1991) Covalent Immobilization of DNA onto polystyrene micrwells: The molecules are only bound at the 5' end. *Anal. Biochem.* **198**, 138–142.

22. Gebeyechu, G., Rao, P., SooChan, P., Simms, D., and Klevan, L. (1987) Novel biotinylated nucleotide-analogs for labeling and colorimetric detection of DNA. *Nucleic Acids Res.* **15**, 4513–4534.

23. McClelland, M., Petersen, C., and Welsh, J. (1992) Length polymorphisms in tRNA intergenic spacers detected by using polymerase chain reaction can distinguish streptococcal strains and species. *J. Clin. Microbiol.* **30**, 1499–1504.

24. Yutsudo, T., Okumura, K., Iwasaki, M., Hara, A., Kamitani, S., Minamide, W., Igarashi, H., and Hinuma, Y. (1994) The gene encoding a new mitogenic factor in a *Streptococcus pyogenes* strain is distributed only in group A streptococci. *Infect. Immun.* **62**, 4000–4004.

25. Chaussee, M., Liu, J., Stevens, D., and Ferretti, J. (1996) Genetic and phenotypic diversity among isolates of *Streptococcus pyogenes* from invasive infections. *J. Infect. Dis.* **173**, 901–908.

26. Welsh, J. and McClelland, M. (1991) Genomic fingerprints produced by PCR with consensus tRNA gene primers. *Nucleic Acids Res.* **19**, 861–866.

27. Podbielski, A., Melzer, B., and Lütticken, R. (1991) Application of the polymerase chain reaction to study the M protein (-like) gene family in beta-hemolytic streptococci. *Med. Microbiol. Immunol.* **180**, 213–227.

28. Whatmore, A. and Kehoe, M. (1994) Horizontal gene transfer in the evolution of group A streptococcal *emm*-like genes: gene mosaics and variation in Vir regulons. *Mol. Microbiol.* **11**, 363–374.

9

Pneumococcal Diseases

Anthony M. Smith and Keith P. Klugman

1. Introduction

The pneumococcus is now the most important bacterial pathogen causing pneumonia, meningitis, and otitis media. Traditional approaches to the diagnosis of *Streptococcus pneumoniae* as the etiologic agent causing these diseases include colony morphology, microscopy, optochin sensitivity, bile solubility, and immunologic reaction with type-specific antisera *(1)*. The rationale for developing a molecular approach to diagnosis includes the very low yield of pneumococci isolated from blood culture in presumed pneumococcal pneumonia, and the need for a rapid diagnostic test that defines susceptibility to antibiotics, particularly in the setting of meningitis

2. Diagnosis

The basis of molecular diagnosis makes use of the gene probe, and more recently the polymerase chain reaction (PCR).

2.1. DNA Probing

The identification of the pneumococcus via DNA probe analysis has been based on genes coding for pneumococcal autolysins and ribosomal RNA (rRNA). The *lytA* gene *(2)* codes for the major autolysin of the pneumococcus, an *N*-acetylmuramoyl-L-alanine amidase *(3,4)*. The cell wall autolytic activity of the amidase has an absolute requirement for choline residues, which are predominantly found in the cell walls of pneumococci *(5,6)*. Pozzi and coworkers *(7)* analyzed the chromosomal DNA of 60 pneumococcal isolates, via dot-blot hybridization analysis with radioactively labeled plasmid DNA housing the 1.2-kb *lytA* gene, and obtained positive reactions for all isolates. No hybridization occurred with any of the other streptococci or related spe-

From: *Methods in Molecular Medicine, Vol. 15: Molecular Bacteriology: Protocols and Clinical Applications*
Edited by: N. Woodford and A. P. Johnson © Humana Press Inc., Totowa, NJ

cies tested, which included strains of *S. sanguis*, *S. mutans*, *S. pyogenes*, *S. agalactiae*, *S. bovis*, *Enterococcus faecalis*, and *E. faecium*. Fenoll and co-workers *(8)* later constructed a modified version of the autolysin DNA probe that contained only a 650-bp region of the *lytA* gene coding for the amino-terminal part of the amidase. This construction deleted the region involved in the specific recognition of cell wall choline residues, therefore, reducing the possibility of the probe hybridizing with nonpneumococcal genes coding for proteins that could contain homologous choline-binding domains. The probe demonstrated high specificity, with negative results obtained for the related streptococcal species *S. oralis*, which contains choline in its cell wall. The suitability of the *lytA* gene probe for diagnostic use was confirmed by positive hybridization results with 27 pneumococcal isolates previously identified as atypical by presumptive physiological methods.

It has been shown that the genes coding for autolysin and pneumolysin (discussed later) can be deleted in laboratory strains of pneumococci, resulting in autolysin-deficient and pneumolysin-deficient mutants *(9,10)*. One must bear in mind that these deletions could, therefore, also occur in clinical isolates.

The AccuProbe Culture Identification Test (Gene-Probe, San Diego, CA) is a commercially available DNA probing kit for pneumococci, and has been evaluated in at least two separate studies *(11,12)*. This system utilizes a chemi-luminescent-labeled DNA probe that hybridizes with target rRNA released in solution following lysis of bacterial cells. Denys and Carey *(11)* reported 100% sensitivity and specificity for this system after examining pure colonies of 172 pneumococcal isolates and 204 nonpneumococcal isolates that included viridans streptococci. Similar results have been obtained using this system on positive blood culture pellets of bacteria *(12)*.

2.2. Polymerase Chain Reaction

The diagnosis of pneumococci in clinical specimens by PCR has made use of a number of target genes that, to date, include the pneumolysin gene, autolysin gene, DNA polymerase 1 gene, penicillin-binding protein (PBP) 2B gene, and rRNA genes.

An accurate diagnosis of pneumococcal disease requires the culture of *S. pneumoniae* from sterile body fluids such as blood, pleural fluid, or cerebrospinal fluid (CSF), which takes at least 12–24 h. Also, cultures may only be positive in less than 30% of patients with severe pneumococcal pneumonia, often because patients have received antibiotic treatment prior to sampling *(13)*. A number of studies have, therefore, evaluated PCR diagnosis of pneumococci from whole blood and blood culture products.

Rudolph and coworkers *(14)* evaluated nested primer PCR (*see* Chapter 5) for pneumococcal DNA detection in the blood of patients with invasive pneumococcal disease, by targeting genes coding for autolysin and pneumolysin. Pneumolysin is a sulfhydryl-activated cytolysin that acts by binding cholesterol in host cell membranes *(15)*. Primers 1+2 and 3+4 (**Table 1**), define 649-bp and 559-bp regions of the pneumolysin gene *(16)*, whereas primers 5+6 and 7+8 (**Table 1**), define 553-bp and 445-bp regions of the *lytA* gene *(2)*. These primer pairs were able to detect a minimum of 10 fg (1×10^{-14}g) or 4.3 genome equivalents of purified pneumococcal DNA, whereas 100% specificity was shown upon analysis of DNAs from isolates of 20 different pneumococcal serotypes and 41 isolates of nonpneumococcal bacteria and fungi. The detection of bacterial DNA from whole blood or buffy coat fractions maintained an acceptable level of specificity (93%), however sensitivity showed a marked decrease, with whole-blood and buffy coat fractions demonstrating sensitivities of 38% and 63–75%, respectively. The decrease in sensitivity was attributed either to a low concentration of target DNA in the presence of high levels of background human DNA, or to the incomplete removal of DNA polymerase inhibitors during the treatment of specimens.

Zhang and coworkers *(17)* more recently described a PCR-based assay for the detection of pneumococci in whole blood that may be more sensitive. It targets the PBP 2B gene of penicillin-resistant and susceptible pneumococci, and is able to detect as little as four colony forming units (cfu) per mL of whole blood. This system is based on the amplification of a conserved 86-bp region of the PBP 2B gene *(18)* using primers 9+10 (**Table 1**), followed by detection of the PCR product with a probe (oligonucleotide 11, **Table 1**) via liquid hybridization-gel retardation analysis. Thirty-six pediatric blood specimens evaluated by culture and PCR demonstrated a PCR-sensitivity of 80% (four of five culture-positive specimens were PCR-positive) and a PCR specificity of 84% (26 of 31 culture-negative specimens were PCR-negative). It was not mentioned whether the five culture-negative and PCR-positive specimens were considered to be true positives. The specificity of the assay was further confirmed by the inability to support amplification from a series of nonpneumococcal and yeast genomic DNAs. The increased sensitivity of this PCR compared with that of Rudolph and coworkers *(14)* probably resulted from a superior extraction method for whole blood, which produced a high yield of good quality DNA (*see* **Note 1**). Rudolph and coworkers *(14)* used an extraction procedure based on the lysis of blood cells with sodium dodecyl sulfate (SDS) and proteinase-K, followed by organic extraction and ethanol precipitation, whereas Zhang and coworkers *(17)* mechanically lyzed specimens by vortexing with glass beads, followed by a high-efficiency purification (4–5 µg/ 0.5 mL) of DNA from whole blood lysates via QIAamp columns (QIAgen, Chatsworth, CA). The PCR method (PBP 2B gene target) of Zhang and coworkers *(17)* was further used to detect pneumococcal DNA in the CSF of

**Table 1
Sequences of Oligonucleotide Primers
and Probes used in the PCR Diagnosis of Pneumococci**

	Target gene	Sequence of oligonucleotide (5'-3')	Reference
1.	Pneumolysin	AATAATGTCCCAGATAGAATGCAGTAT	*14*
2.	Pneumolysin	GATACAACTCTGATTCCAATGTCGAAT	*14*
3.	Pneumolysin	TGGAACAACTCAAGGTCAAGTTTGGTT	*14*
4.	Pneumolysin	AATGCACTGTTACATCAACGCTGGAAA	*14*
5.	Autolysin	AGAATGAAGCGGATTATCACTGGCGGA	*14*
6.	Autolysin	TATATGCTTGCAGACCGCTGGAGGAAG	*14*
7.	Autolysin	AACGGTTGCATCATGCAGGTAGGACCT	*14*
8.	Autolysin	AAAATCAATGGCACTTGGTACTACTTT	*14*
9.	PBP 2B	ATGCAGTTGGCTCAGTATGTA	*17*
10.	PBP 2B	CACCCAGTCCTCCCTTATCA	*17*
11.	PBP 2B	CAAATAATGGTGTTCGTGTGGCTCCTCGTA	*17*
12.	PBP 2B	ATCAATTCTTGGTATACTCAGG	*20*
13.	PBP 2B	AGTAGATTCATCTGGTAGGTC	*20*
14.	PBP 2B	CTAGGCCAATGCCGATTACG	*20*
15	PBP 2B	AGTAGATTCATCTGGTAGGTC	*20*
16.	PBP 2B	CCAAACCTTAACAGATCAGC	*20*
17.	PBP 2B	AGTAGATTCATCTGGTAGGTC	*20*
18.	Autolysin	TGAAGCGGATTATCACTGGC	*20*
19.	Autolysin	GCTAAACTCCCTGTATCAAGCG	*20*
20.	DNA Polymerase I	GTCAAGATATGCTGAGTGAAGAG	*21*
21.	DNA Polymerase I	CCATAAAGACTAGCGATGGTCGC	*21*
22.	Autolysin	ATGGAAATTAATGTGAGTA	*23*
23.	Autolysin	AGGTCTCAGCATTCCA	*23*
24.	Autolysin	TTACTTCGCCTAATAGTGACC	*23*
25.	Autolysin	GGAGTAGAATATGGAAATTAATGT	*24*
26.	Autolysin	GCTGCATAGGTCTCAGCATTCCAA	*24*
27.	Pneumolysin	ATTTCTGTAACAGCTACCAACGA	*25*
28.	Pneumolysin	GAATTCCCTGTCTTTTCAAAGTC	*25*
29.	Pneumolysin	CCCACTTCTTCTTGCGGTTGA	*25*
30.	Pneumolysin	TGAGCCGTTATTTTTTCATACTG	*25*
31.	16S rRNA	AGGAGGTGATCCAACCGCA	*29*
32.	16S rRNA	AACTGGAGGAAGGTGGGGAT	*29*
33.	16S rRNA	AACTGAGACTGGCTTTAAGAGATTA	*29*
34.	16S rRNA	AACT(C/A)CGTGCCAGCAGCCGCGGTAA	*28*
35.	16S rRNA	AAGGAGGTGATCCA(G/A)CCGCA(G/C)(G/C)TTC	*28*
36.	16S rRNA	GTACAACGAGTCGCAAGC	*28*
37.	16S-23s rRNA spacer	AGGATAAGGAACTGCG	*32*
38.	16S-23s rRNA spacer	CTTATTTTCTGACCTTTCA	*32*

patients with culture-proven bacteremia and meningitis *(19)*. CSF specimens were prepared by sonication for 60 s and boiling for 10 min.

The PBP 2B gene was further used as a PCR target by Ubukata and coworkers *(20)* to identify penicillin resistance in pneumococcal isolates. Penicillin resistance is primarily the result of mutations in the PBP 2B gene, with these altered genes falling into two major mutated classes, A and B *(18)*. Three sets of primers were designed to amplify:

1. 240-bp fragment (primers 12+13, **Table 1**) of the PBP 2B gene of penicillin-susceptible isolates;
2. A 215-bp fragment (primers 14+15, **Table 1**) of a class-A mutated gene present in penicillin-resistant isolates (MICs ≥0.1 µg/mL); and
3. A 286-bp fragment (primers 16+17, **Table 1**) of a class-B mutated gene present in penicillin-resistant isolates (MICs ≥0.1 µg/mL).

In addition, the amplification of a 273-bp fragment (primers 18+19, **Table 1**) of the *lytA* gene was applied in combination with the above to identify positively pneumococci from colony lysates.

Friedland and coworkers *(21)* demonstrated how different blood anticoagulants can effect the activity of the DNA polymerase enzyme that catalyzes the PCR. The target DNA sequence was a 322-bp segment (primers 20+21, **Table 1**) of the pneumococcal DNA polymerase I gene *(22)*, which demonstrated excellent sensitivity and specificity. When PCR reactions were performed incorporating 1-µL aliquots of blood culture media, positive PCRs were only obtained with lithium heparin, sodium heparin, and sodium polyanetholesulfonate (SPS)-anticoagulated blood, whereas the addition of EDTA and citrate-anticoagulated blood inhibited the PCR assay (*see* **Note 2**).

Hassan-King and coworkers *(23)* described a PCR that was sensitive enough to identify pneumococci from blood cultures that were negative on culture. Their PCR targeted a 247-bp region of the *lytA* gene (primers 22+23, **Table 1**). This PCR combined with a confirmatory radioactively-labeled probe (oligonucleotide 24, **Table 1**) was able to detect femtogram quantities of purified DNA, whereas 100% specificity was shown for a number of pneumococcal isolates and nonpneumococcal bacteria that included, *S. oralis, S. mitis, S. sanguis,* and *S. parasanguis.* DNA was initially extracted from positive blood culture samples using organic reagents (phenol:chloroform), but this resulted in negative PCR results, probably because of the inadequate removal of DNA polymerase inhibitors. This method was replaced by processing blood culture samples using the Micro Tubergen DNA extraction and purification kit (Invitrogen, San Diego, CA). Positive PCRs were then obtained that showed concordance with positive blood cultures, whereas PCR also identified pneumococci in blood cultures demonstrating false-negative culture. The *lytA* gene has proven to be the most popular target for the molecular diagnosis of pneumococci. Gillespie and coworkers *(24)* described yet another primer pair (primers 25+26, **Table 1**) targeting the *lytA* gene, resulting in the amplification of a 263-bp product. This PCR assay also demonstrated high sensitivity and specificity, and was employed to detect pneumococci in sputum samples, following the preparation of sputum samples via heat inactivation (at 80°C for 20 min) and chloroform extraction.

Salo and coworkers *(25)* demonstrated excellent sensitivity and specificity for PCR detection of bacterial DNA extracted from the sera of patients with acute pneumococcal pneumonia. Serum specimens were treated with a solution of NaOH, NaCl, and SDS, to lyze bacterial cells, followed by organic extraction and ethanol precipitation of DNA. The sensitivity required for the PCR was gained via nested primer PCR that targeted the pneumolysin gene to amplify a 348-bp fragment (primers 27+28, **Table 1**), followed by an internal 208-bp fragment (primers 29+30, **Table 1**). This nested primer PCR assay was also used to assist in the diagnosis of pneumococci from middle ear fluids from children with acute otitis media *(26)*.

Evolutionarily conserved DNA, such as the gene encoding 16S rRNA *(27)*, lends itself to the design of universal PCR primers for conserved regions of the gene, which would amplify a product from all species of pathogenic bacteria. Further analysis of the PCR product via techniques such as nested primer PCR *(28)*, DNA probing *(29)*, restriction endonuclease analysis *(30)*, and single-strand conformation polymorphism *(31)*, can then determine the exact bacterial origin of the PCR product. Greisen and coworkers *(29)* described the amplification of a 333-bp fragment of the 16S rRNA gene from 124 species of bacteria using universal primers 31+32 (**Table 1**). Positive reactions following Southern blot hybridization with a pneumococcal DNA probe (oligonucleotide 33, **Table 1**), determined that the PCR product originated from *S. pneumoniae*. Radstrom and coworkers *(28)* also targeted the 16S rRNA gene in their diagnosis of bacterial meningitis by examination of CSF. A seminested primer PCR was performed on aliquots of CSF that were treated by boiling for 15 min to lyze bacterial cells and to denature proteinaceous material. The first PCR consisted of an amplification of a 1032-bp fragment of the 16S rRNA gene with universal primers 34+35 (**Table 1**). This identified a number of bacterial pathogens including, *S. pneumoniae, S. agalactiae, Neisseria meningitidis,* and *Haemophilus influenzae*. The second PCR (primers 36+35, **Table 1**) defined a 295-bp fragment that was specific for *S. pneumoniae* and *S. agalactiae*. This PCR assay demonstrated a high specificity on analysis of DNA from 28 different species of bacteria, whereas analysis of 304 clinical CSF specimens demonstrated a sensitivity of 94% and a specificity of 96%. Saruta and coworkers *(32)* identified pneumococcal DNA, with a 100% sensitivity and specificity, from CSF specimens, via a PCR assay that targeted the spacer region between the 16S and 23S rRNA genes; an area of the ribosomal DNA that is highly variable between species, but conserved within a species. CSF specimens were centrifuged at 15,000g for 30 min, and DNA was extracted from CSF pellets by cell lysis with SDS and proteinase-K, phenol extraction, and ethanol precipitation. Aliquots of extracted CSF DNA resulted in the PCR amplification of a 247-bp fragment (primers 37+38, **Table 1**).

3. Epidemiologic Investigation

The following typing techniques have been documented in the epidemiologic analysis of pneumococcal disease: multilocus enzyme electrophoresis, PBP gene fingerprinting, ribotyping, BOX-fingerprinting, pulsed-field gel electrophoresis, and PCR-based fingerprinting (REP-PCR, AP-PCR). The authors recommend that a minimum of two techniques are used for molecular typing of pneumococcal strains.

3.1. Multilocus Enzyme Electrophoresis (MLEE)

Although the technique of MLEE cannot be classified as molecular, it has been used with much success in determining clonal relationships between pneumococcal isolates and, therefore, deserves mention. In MLEE, the electrophoretic mobilities of a selected set of metabolic enzymes are compared between isolates *(33)*. Bacterial cells are lyzed by sonication, and following centrifugation, the supernatant is electrophoresed in a nondenaturing starch gel matrix. The gel is then sliced into many replicates, and each gel slice is stained with a specific colorimetric substrate to reveal the position of a particular enzyme. Variations in the electrophoretic mobility of an enzyme (electromorphs), identify allelic variations in the chromosomal genes encoding the enzyme. A combination of particular electromorphs are referred to as an electrophoretic type. MLEE is thought to be a reliable and reproducible phenotypic typing technique. However, it is labor-intensive and the colorimetric substrates are expensive. Nevertheless, MLEE has played a major role in determining the clonal relationships of pneumococci, particularly of the multiresistant serotype 23F pneumococci that were originally found only in Spain, but are now isolated worldwide *(34–37)*. In 1990, multi-resistant serotype 23F strains were isolated from children attending a day care center in Cleveland, OH *(34)*. These isolates were found to be identical to Spanish 23F isolates with respect to MLEE at 14 enzyme-encoding loci and gene fingerprint profiles of PBP 2B and 2X, therefore demonstrating the intercontinental spread of serotype 23F pneumococci from Spain to the US. Later, MLEE and ribotyping analysis of multiresistant serotype 23F strains isolated in 17 US cities demonstrated the dispersal of this 23F clone across the US *(35)*. Of 22 isolates analyzed, 16 revealed

Table 2
Summary of Results Obtained in the PCR Diagnosis of Pneumococci

Target gene	Type of test	Oligonucleotides[a]	Specimen	Sensitivity[b]	Specificity[c]	Reference
Pneumolysin	Nested PCR	1+2, 3+4	whole blood	38%	93%	*14*
pneumolysin	Nested PCR	1+2, 3+4	buffy coat	75%	93%	*14*
Autolysin	Nested PCR	5+6, 7+8	whole blood	38%	93%	*14*
Autolysin	Nested PCR	5+6, 7+8	buffy coat	63%	93%	*14*
PBP 2B	PCR+probing	9+10, 11	whole blood	80%	84%	*17*
PBP 2B	PCR+probing	9+10, 11	CSF, blood	ND	ND	*17*
PBP 2B[d]	Standard PCR	12+13	colony lysate	99%	ND	*20*
PBP 2B[e]	Standard PCR	14+15	colony lysate	ND	ND	*20*
PBP 2B[f]	Standard PCR	16+17	colony lysate	ND	ND	*20*
Autolysin	Standard PCR	18+19	colony lysate	ND	ND	*20*
DNA polymerase 1	Standard PCR	20+21	blood culture	100%	100%	*21*
Autolysin	PCR+probing	22+23, 24	blood culture	100%	100%	*23*
Autolysin	Standard PCR	25+26	sputum	93%	100%	*24*
pneumolysin	Nested PCR	27+28, 29+30	serum	100%	100%	*25*
pneumolysin	Nested PCR	27+28, 29+30	middle ear fluid	91%	96%	*26*
16S rRNA	PCR+probing	31+32, 33	purified DNA	ND	ND	*29*
16S rRNA	Seminested PCR	34+35, 36+35	CSF	94%	96%	*28*
16S-23S rRNA spacer	Standard PCR	37+38	CSF	100%	100%	*32*

[a] Oligonucleotides are described in Table 1.
[b,c] ND, these data were not documented in the article.
[d] Primers were designed to amplify only PBP 2B genes from penicillin-susceptible isolates (MICs, ≤0.06 μg/mL).
[e] Primers were designed to amplify only class-A mutated PBP 2B genes from penicillin-resistant isolates.
[f] Primers were designed to amplify only class-B mutated PBP 2B genes from penicillin-resistant isolates.

MLEE allelic profiles at 20 enzyme loci that were identical to a Spanish 23F isolate. MLEE and gene profiles of PBP 2B, 2X, and 1A have further identified isolates of the same Spanish 23F clone in the UK *(36)* and South Africa *(37)*. Besides its popularity for the analysis of serotype 23F pneumococci, MLEE has also been widely used to analyze the genetic diversity of antibiotic-resistant and susceptible pneumococcal strains (of all serotypes) isolated in different parts of the world *(38–44)*.

3.2. PBP Gene Fingerprinting

Restriction enzyme analysis (REA) or DNA fingerprint analysis of PBP 2B, 2X, and 1A genes are often used to assist in the epidemiologic analysis of pneumococci. PBP gene fingerprinting is only applicable to β-lactam resistant pneumococci, since the resistance, which is caused by altered forms of PBPs, is encoded by highly divergent "mosaic" genes consisting of regions containing extensive nucleotide substitutions alternating with regions identical to sequences of genes from susceptible strains *(45)*. The first step to gene fingerprinting involves the isolation of the PBP genes from the chromosomal DNA using PCR. The 1.5-kb PBP 2B gene is amplified with primers 1+2, **Table 3** *(45)*, the 2-kb PBP 2X gene is amplified with primers 3+4, **Table 3** *(34)*, and the 2.4-kb PBP 1A gene is amplified with primers 5+6, **Table 3** *(36)*. The genes are then purified, digested with restriction enzymes (*Sty*I or *Hinf*I for the PBP 2B gene, *Mse*I-*Dde*I or *Hinf*I for the PBP 2X and 1A genes), radioactively end-labeled, fractionated on a 6% polyacrylamide gel, and autoradiographed to reveal the gene fingerprint patterns. In the previous paragraph the authors have already referred to the use of PBP gene fingerprinting, together with MLEE and ribotyping, in tracking the intercontinental spread of multiresistant Spanish serotype 23F pneumococci. Furthermore, PBP gene fingerprinting is widely used to assist in the analysis of the genetic diversity of β-lactam-resistant pneumococcal strains isolated worldwide *(40,43,46)*.

3.3. Ribotyping

Ribotyping refers to the DNA fingerprint pattern obtained on Southern blots of chromosomal DNA digests, following probing with 23S, 16S, and 5S rRNA or their cDNA *(47)* (*see* Chapter 2). For ribotyping of pneumococci, the restriction enzyme *Hin*dIII has proven the most popular enzyme for digesting chromosomal DNA *(35,43,48,49,50)*. However, digestion with *Eco*RI *(40,51)* and *Pvu*II *(52)* has also been documented. Generally, two ribotyping probes are used: radio-labeled *Escherichia coli* 16 and 23S rRNA *(43)* or cold-labeled (digoxigenin-labeled) cDNA, reverse transcribed from *E.coli* 16 and 23S rRNA *(48)*. In particular, ribotyping analysis (*Hin*dIII digestion) was used together with MLEE to establish the presence of Spanish multiresistant serotype 23F

Table 3
Sequences of Oligonucleotide Primers and Probes
used in the Epidemiological Investigation of Pneumococci

Target gene	Sequence of Oligonucleotide (5'-3')	Reference
1. PBP 2B	GATCCTCTAAATGATTCTCAGGTGG	*45*
2. PBP 2B	CAATTAGCTTAGCAATAGGTGTTGG	*45*
3. PBP 2X	CGTGGGACTATTTATGACCGAAATGG	*34*
4. PBP 2X	GGCGAATTCCAGCACTGATGGAAATAAACATATTA	*34*
5. PBP 1A	CGGCATTCGATTTGATTCGCTTCT	*36*
6 PBP 1A	CTGAGAAGATGTCTTCTCAGG	*36*
7. repetitive DNA element	ACAACCTCAAAACAGTGTTT	*52*
8. repetitive DNA element	IIINCGNCGNCATCNGGC	*65*
9. repetitive DNA element	NCGNCTTATCNGGCCTAC	*65*
10. arbitrary DNA target	AAGTAAGTGACTGGGGTGAGCG	*52*
11. arbitrary DNA target	CGGTGGCGAA	*
12. arbitrary DNA target	TTGACAACTG	*
13. arbitrary DNA target	CCTGCGAGCGTAGGCGTCGG	*
14. BOX repeat	ATACTCTTCGAAAATCTCTTCAAAC	*56*

* McGee, Lee, Friedland, and Klugman, unpublished results.

strains in the US *(35)*, and recently assisted in determining that the emergence of extended-spectrum cephalosporin resistance in pneumococci, has occurred on several occasions in different clonal populations of pneumococci in the United States *(43)*. Ribotyping (*Eco*RI digestion) has also demonstrated considerable diversity amongst penicillin-resistant pneumococci isolated in Nairobi, Kenya (11 profiles among 23 isolates) *(40)*. Hermans and coworkers *(52)* identified seven ribotypes (*Pvu*II digestion) among 28 pneumococcal isolates from the Netherlands, whereas, other genotypic techniques yielded 14 to 19 subtypes. They, therefore, suggested that ribotyping should only be used to investigate long-term epidemiologic events, such as determining the genetic relatedness of pneumococci isolated from different parts of the world. Although the authors agree with this interpretation of their data, they also believe that more ribotypes might have been observed if they had probed with nucleic acid sequences for both the 16S and 23S rRNA, instead of using only a 16S rRNA gene probe.

3.4. BOX-Fingerprinting

Highly conserved, repeated DNA sequences exist interspersed in bacterial genomes. Repetitive extragenic palindromic (REP) elements and enterobacterial repetitive intergenic consensus (ERIC) sequences are two families of repetitive elements that have been well-described in bacteria *(53,54; see* Chapter 7*)*. BOX elements are conserved repetitive DNA sequences located within intergenic regions of the pneumococcal genome. Approximately 25 BOX elements exist within the pneumococcal genome *(55)*. Probes containing nucleic acid sequences for these BOX elements can, therefore, be used to type Southern blots of chromosomal DNA, similar to the principle of ribotyping (*see* Chapter 2). Hermans and coworkers *(52)* demonstrated BOX-fingerprinting of pneumococcal chromosomal DNA (*Pvu*II digestion) using a radio-labeled probe (oligonucleotide 7, **Table 3**) and identified 19 types among 28 isolates from the Netherlands. Although ribotyping and BOX-fingerprinting are relatively easy techniques to perform, many time-consuming steps are involved and, therefore, results can only be obtained after 2–3 d. A BOX repeat PCR assay for typing of pneumococci has recently been described *(56)*.

3.5. Pulsed-Field Gel Electrophoresis (PFGE)

When chromosomal DNA is digested with rare-cutting restriction enzymes, a limited number (5 to 30) of extremely large DNA fragments are obtained, that can only be electrophoretically separated by the technique of PFGE *(57)* (*see* Chapter 3). PFGE is a popular technique and has been widely used to compare genetically pneumococci isolated worldwide. Lefevre and coworkers *(58)* evaluated PFGE of pneumococci and found that the technique was excellent in differentiating between epidemiologically unrelated pneumococcal strains, with digestion of chromosomal DNA (*Apa*I or *Sma*I) producing 10–19 fragments and revealing 22 different fingerprint patterns amongst 23 unrelated isolates. PFGE of chromosomal DNA (*Apa*I or *Sma*I digestion) from antibiotic-resistant pneumococci isolated in Iceland, Portugal, and France has demonstrated the clonal spread of Spanish multiresistant strains to these countries *(59,60–62)*. PFGE has also identified multiple clones of antibiotic-resistant pneumococci amongst strains isolated in the US *(43,63)*, the Czech Republic and Slovakia *(49)*, Croatia *(64)*, the Netherlands *(52)*, and the UK *(44)*. Disadvantages of PFGE include the long and delicate procedures that are required for isolating and digesting high molecular weight genomic DNA, the need for expensive sophisticated equipment and the need for extended electrophoresis times (*see* **Note 3**).

3.6. REP-PCR

PCR-based fingerprinting is by far the easiest and quickest typing method that can be used to assess genomic variability between bacterial strains. In REP-PCR, repetitive elements serve as primer binding sites in the amplification of DNA sequences located between successive repetitive elements *(65)* (*see* Chapter 7). The number of PCR products is proportional to the number of REP elements, whereas the size of the products is proportional to the length of the spacer between successive REP elements. Since the numbers and positions of REP elements are variable, the numbers and sizes of PCR products will vary between strains. Chromosomal DNA is first isolated from pneumococcal bacteria using standard techniques. Nanogram quantities of DNA are then used as a template in a standard PCR incorporating REP primers 8+9, **Table 3** *(65)*. Following PCR, amplification products are separated by size by agarose or polyacrylamide gel electrophoresis, to reveal DNA fingerprint patterns. Forty-eight penicillin-resistant pneumococcal strains isolated in Houston, TX, were analyzed by REP-PCR and MLEE and revealed a heterogeneous array of 22 clonal genotypes, indicating that resistance has emerged on many occasions in this area *(41)*. Thirty-two pneumococci isolated from 11 HIV sero-positive patients were characterized by REP-PCR and ribotyping, and identified 14 distinguishable strains were identified, which enabled clinicians to determine whether relapse of pneumonia was because of the original strain or reinfection with a new strain *(51)*.

3.7. Arbitrarily Primed PCR (AP-PCR)

AP-PCR is the amplification of genomic DNA with one or more primers (typically 10–20 bases) of arbitrary/random nucleotide sequence, to target specific but unknown sites on the bacterial genome *(66,67)* (*see* Chapter 6). PCR is carried out at conditions of low stringency that favor binding of primer to multiple sites on the chromosomal DNA template. These conditions include a low primer annealing temperature of 35°C and high concentrations of *Taq* DNA polymerase, $MgCl_2$, and primers. This results in the amplification of multiple products of different sizes, that are then electrophoretically separated on agarose or polyacrylamide gels to reveal a unique DNA fingerprint pattern. To maintain reproducibility, it is vital that all steps and parameters in AP-PCR are kept constant. In particular, these include the method used to isolate chromosomal DNA, the concentration of PCR components (template DNA, $MgCl_2$, primer, dNTPs, *Taq* DNA polymerase), the make of *Taq* DNA polymerase, the times and temperatures of PCR cycles, and the type of DNA thermal cycler used. It is also preferable for all isolates that have to be typed by AP-PCR to be sent to a single reference laboratory where the technique can be performed by one individual (*see* **Note 4**). AP-PCR is a versatile technique that often gives interpretable results. Any primer that may be available in the laboratory can be tested in an AP-PCR. For example, primer 13 (**Table 3**) is the reverse primer used in the amplification of the IS*6110* insertion sequence of *Mycobacterium tuberculosis*, but it also generates suitable AP-PCR fingerprints of pneumococcal DNA (McGee, Lee, Friedland, and Klugman, unpublished results). Hermans and coworkers *(52)* demonstrated 15 subtypes among 28 pneumococci from the Netherlands using AP-PCR with primer 10 (**Table 3**). REP-PCR and independent AP-PCRs with primers 11, 12, and 13 (**Table 3**), demonstrated identical profiles for 13 Korean isolates of multiresistant serotype 23F pneumococci, which were also identical to profiles of the Spanish 23F clone, therefore, demonstrating the clonal spread of this strain from Spain to Korea (McGee, Lee, Friedland, and Klugman, unpublished results).

4. Future Prospects and Goals

A primary goal for the molecular diagnosis of pneumococcal infection would be the ability to identify a pneumococcal etiology of pneumonia by means of a molecular assay on sputum or blood. The confounding variable is the presence of DNA from colonizing strains of pneumococci that may colonize the nasopharynx in 5–20% of adults and more than 50% of children. It is possible that identification of mRNA from genes expressed only by invading strains of pneumococci and not by colonizing strains may provide this answer. Another essential goal is the ability to identify resistant strains rapidly. The elucidation of the molecular basis of resistance to the common antimicrobials used to treat pneumococcal infections may allow for rapid PCR-based assays that will reliably predict resistance.

The molecular epidemiology of the pneumococcus has dramatically expanded during the past few years. A future goal would be the standardization of methods and the development of a worldwide data bank of pneumococcal types based on the standardized method. This will allow the elucidation of patterns of pneumococci responsible for different types of invasive disease and give further insights into the spread of antibiotic-resistant pneumococci worldwide.

5. Notes

1. The use of PCR on blood requires high efficiency purification of DNA using reagents such as QIAamp columns (QIAgen, Chatsworth, CA) or purification kits (e.g., Micro Tubergen DNA extraction kit [Invitrogen, San Diego, CA]).
2. The anticoagulants used should be heparin or SPS rather than EDTA or citrate.
3. For pulsed-field electrophoresis of pneumococci, the problem of shearing of large DNA molecules can be reduced by using mutanolysin (Sigma, St. Louis, MO). Also remember to use a wide-necked pipet for manipulation of the DNA (*see* Chapter 3).
4. For AP-PCR standardized DNA extraction methods are essential even within individual laboratories (*see* Chapter 6).

References

1. Mufson, M. A. (1990) *Streptococcus pneumoniae*, in *Principles and Practice of Infectious Diseases,* 3rd ed. (Mandell, G. L., Douglas, R. D., and Bennet, J. E., eds.), Churchill Livingstone, New York, pp. 1539–1550.

2. Garcia, P., Garcia, J. L., Garcia, E., and Lopez, R. (1986) Nucleotide sequence and expression of the pneumococcal autolysin gene from its own promoter in *Escherichia coli*. *Gene*. **43**, 265–272.

3. Howard, L. V. and Gooder, H. (1974) Specificity of the autolysin of *Streptococcus (Diplococcus) pneumoniae*. *J. Bacteriol*. **117**, 796–804.

4. Mosser, J. L. and Tomasz, A. (1970) Choline containing teichoic acid as a structural component of pneumococcal cell wall and its role in sensitivity to lysis by an autolytic enzyme. *J. Biol. Chem*. **245**, 286–298.

5. Holtjie, J. V. and Tomasz, A. (1975) Specific recognition of choline residues in the cell wall teichoic acid by the *N*-acetyl-muramyl-L-alanine amidase of pneumococcus. *J. Biol. Chem*. **250**, 6072–6076.

6. Tomasz, A. (1968) Biological consequences of the replacement of choline by ethanolamine in the cell wall of pneumococcus: chain formation, loss of transformability and loss of autolysis. *Proc. Natl. Acad. Sci. USA* **59**, 86–93.

7. Pozzi, G., Oggioni, M. R., and Tomasz, A. (1989) DNA probe for identification of *Streptococcus pneumoniae*. *J. Clin. Microbiol*. **27**, 370–372.

8. Fenoll, A., Martinez-Suarez, J. V., Munoz, R., Casal, J., and Garcia, J. L. (1990) Identification of atypical strains of *Streptococcus pneumoniae* by a specific DNA probe. *Eur. J. Clin. Microbiol. Infect. Dis*. **9**, 396–401.

9. Berry, A. M., Lock, R. A., Hansman, D., and Paton, J. C. (1989) Contribution of autolysin to virulence of *Streptococcus pneumoniae*. *Infect. Immun*. **57**, 2324–2330.

10. Berry, A. M., Yother, J., Briles, D. E., Hansman, D., and Paton, J. C. (1989) Reduced virulence of a defined pneumolysin-negative mutant of *Streptococcus pneumoniae*. *Infect. Immun*. **57**, 2037–2042.

11. Denys, G. A. and Carey, R. B. (1992) Identification of *Streptococcus pneumoniae* with a DNA probe. *J. Clin. Microbiol*. **30**, 2725–2727.

12. Davis, T. E. and Fuller, D. D. (1991) Direct identification of bacterial isolates in blood cultures by using a DNA probe. *J. Clin. Microbiol*. **29**, 2193–2196.

13. Kalin, M. and Lindberge, A. A. (1983) Diagnosis of pneumococcal pneumoniae: a comparison between microscopic examination of expectorate, antigen detection, and cultural procedures. *Scand. J. Infect. Dis*. **15**, 247–255.

14. Rudolph, K. M., Parkinson, A. J., Black, C. M., and Mayer, L. W. (1993) Evaluation of polymerase chain reaction for diagnosis of pneumococcal pneumonia. *J. Clin. Microbiol*. **31**, 2661–2666.

15. Johnson, M. K., Geoffroy, C., and Alouf, J. E. (1980) Binding of cholesterol by sulfhydryl-activated cytolysins. *Infect. Immun*. **27**, 97–101.

16. Walker, J. A., Allen, R. L., Falmagne, P., Johnson, M. K., and Boulnois, G. J. (1987) Molecular cloning, characterization, and complete nucleotide sequence of the gene for pneumolysin, the sulfhydryl-activated toxin of *Streptococcus pneumoniae*. *Infect. Immun*. **55**, 1184–1189.

17. Zhang, Y., Isaacman, D. J., Wadowsky, R. M., Rydquist-White, J., Post, J. C., and Ehrlich, G. D. (1995) Detection of *Streptococcus pneumoniae* in whole blood by PCR. *J. Clin. Microbiol*. **33**, 596–601.

18. Dowson, C. G., Hutchison, A., Woodford, N., Johnson, A. P., George, R. C., and Spratt, B. G. (1990) Penicillin-resistant viridans streptococci have obtained altered

penicillin-binding protein genes from penicillin-resistant strains of *Streptococcus pneumoniae. Proc. Natl. Acad. Sci. USA* **87,** 5858–5862.

19. Isaacman, D. J., Zhang, Y., Rydquist-White, J., Wadowsky, R. M., Post, J. C., and Ehrlich, G. D. (1995) Identification of a patient with *Streptococcus pneumoniae* bacteremia and meningitis by the polymerase chain reaction (PCR). *Mol. Cell. Probes* **9,** 157–160.

20. Ubukata, K., Asahi, Y., Yamane, A., and Konno, M. (1995) Combinational detection of autolysin and penicillin-binding protein 2B genes of *Streptococcus pneumoniae* by PCR. *J. Clin. Microbiol.* **34,** 592–596.

21. Friedland, L. R., Menon, A. G., Reising, S. F., Ruddy, R. M., and Hassett, D. J. (1994) Development of a polymerase chain reaction assay to detect the presence of *Streptococcus pneumoniae* DNA. *Diagn. Microbiol. Infect. Dis.* **20,** 187–193.

22. Lopez, P., Martinez, S., Diaz, A., Espinosa, M., and Lacks, S. A. (1989) Characterization of the *PolA* gene of *Streptococcus pneumoniae* and comparison of the DNA polymerase I it encodes to homologous enzymes from *Escherichia coli* and phage T7. *J. Biol. Chem.* **264,** 4255–4263.

23. Hassen-King, M., Baldeh, I., Secka, O., Falade, A., and Greenwood, B. (1994) Detection of *Streptococcus pneumoniae* DNA in blood cultures by PCR. *J. Clin. Microbiol.* **32,** 1721–1724.

24. Gillespie, S. H., Ullman, C., Smith, M. D., and Emery, V. (1994) Detection of *Streptococcus pneumoniae* in sputum samples by PCR. *J. Clin. Microbiol.* **32,** 1308–1311.

25. Salo, P., Ortqvist, A., and Leinonen, M. (1995) Diagnosis of bacteremic pneumococcal pneumonia by amplification of pneumolysin gene fragment in serum. *J. Infect. Dis.* **171,** 479–482.

26. Virolainen, A., Salo, P., Jero, J., Karma, P., Eskola, J., and Leinonen, M. (1994) Comparison of PCR assay with bacterial culture for detecting *Streptococcus pneumoniae* in middle ear fluid of children with acute otitis media. *J. Clin. Microbiol.* **32,** 2667–2670.

27. Gray, M. W., Sankoff, D., and Cedergren, R. J. (1984) On the evolutionary descent of organisms and organelles: a global phylogeny based on a highly conserved structural core in small subunit ribosomal RNA. *Nucleic. Acids. Res.* **12,** 5837–5852.

28. Radstrom, P., Backman, A., Qian, N., Kragsbjerg, P., Pahlson, C., and Olcen, P. (1994) Detection of bacterial DNA in cerebrospinal fluid by an assay for simultaneous detection of *Neisseria meningitidis, Haemophilus influenzae,* and Streptococci using a seminested PCR strategy. *J. Clin. Microbiol.* **32,** 2738–2744.

29. Greisen, K., Loeffelholz, M., Purohit, A., and Leong, D. (1994) PCR primers and probes for the 16S rRNA gene of most species of pathogenic bacteria, including bacteria found in cerebrospinal fluid. *J. Clin. Microbiol.* **32,** 335–351.

30. Avaniss-Aghajani, E., Jones, K., Chapman, D., and Brunk, C. (1994) A molecular technique for identification of bacteria using small subunit ribosomal RNA sequences. *Biotechniques* **17,** 144–149.

31. Widjojoatmodjo, M. N., Fluit, A. C., and Verhoef, J. (1995) Molecular identification of bacteria by fluorescence-based PCR-single-strand conformation polymorphism analysis of the 1sRNA gene. *J. Clin. Microbiol.* **33,** 2601–2606.

32. Saruta, K., Matsunaga, T., Hoshina, S., Kono, M., Kitahara, S., Kanemoto, S., Sakai, O., and Machida, K. (1995) Rapid identification of *Streptococcus pneumoniae* by PCR amplification of ribosomal DNA spacer region. *FEMS. Microbiol. Lett.* **132,** 165–170.

33. Selander, R. K., Caugant, D. A., Ochman, H., Musser, J. M., Gilmour, M. N., and Whittam, T. S. (1986) Methods of multilocus enzyme electrophoresis for bacterial population genetics and systematics. *Appl. Environ. Microbiol.* **51,** 873–884.

34. Munoz, R., Coffey, T., Danials, M., Dowson, C. G., Liable, G., Casal, J., Hakenbeck, R., Jacobs, M., Musser, J. M., Spratt, B. G., and Tomasz, A. (1991) Intercontinental spread of a multiresistant clone of serotype 23F *Streptococcus pneumoniae. J. Infect. Dis.* **164,** 302–306.

35. McDougal, L. K., Facklam, R., Reeves, M., Hunter, S., Swenson, J. M., Hill, B. C., and Tenover, F. C. (1992) Analysis of multiply antimicrobial-resistant isolates of *Streptococcus pneumoniae* from the United States. *Antimicrob. Agents. Chemother.* **36,** 2176–2184.

36. Coffey, T. J., Dowson, C. G., Daniels, M., Zhou, J., Martin, C., Spratt, B. G., and Musser, J. M. (1991) Horizontal transfer of multiple penicillin binding protein genes and capsular biosynthetic genes in natural populations of *Streptococcus pneumoniae. Mol. Microbiol.* **5,** 2255–2260.

37. Klugman, K. P., Coffey, T. J., Smith, A. M., Wasas, A., Meyers, M., and Spratt, B. G. (1994) Cluster of an erythromycin-resistant variant of the Spanish multiply resistant 23F clone of *Streptococcus pneumoniae* in South Africa *Eur. J. Clin. Microbiol. Infect. Dis.* **13,** 171–174.

38. Munoz, R., Musser, J. M., Crain, M., Briles, D. E., Marton, A., Parkinson, A. J., Sorensen, U., and Tomasz. A. (1992) Geographic distribution of penicillin-resistant clones of *Streptococcus pneumoniae:* Characterization by penicillin-binding protein profile, surface protein A typing and multilocus enzyme electrophoresis. *Clin. Infect. Dis.* **15,** 112–118.

39. Sibold, C., Wang, J., Henrichsen, J., and Hakenbeck, R. (1992) Genetic relationships of penicillin-susceptible and-resistant *Streptococcus pneumoniae* strains isolated on different continents. *Infect. Immun.* **60,** 4119–4126.

40. Kell, C. M., Jordens, J. Z., Daniels, M., Coffey, T. J., Bates, J., Paul, J., Gilks, G., and Spratt, B. G. (1993) Molecular epidemiology of penicillin-resistant pneumococci isolated in Nairobi, Kenya. *Infect. Immun.* **61,** 4382–4391.

41. Versalovic, J., Kapur, V., Mason, E. O., Shah, U., Koeuth, T., Lupski, J. R., and Musser, J. M. (1993) Penicillin-resistant *Streptococcus pneumoniae* strains recovered in Houston: Identification and molecular characterization of multiple clones. *J. Infect. Dis.* **167,** 850–856.

42. Reichmann, P., Varon, E., Gunther, E., Reinert, R. R., Luttiken, R., Marton, A., Geslin, P., Wagner, J., and Hakenbeck, R. (1995) Penicillin-resistant *Streptococcus pneumoniae* in Germany: genetic relationship to clones from other European countries. *J. Med. Microbiol.* **43,** 377–385.

43. McDougal, L. K., Rasheed, J. K., Biddle, J. W., and Tenover, F. C. (1995) Identification of multiple clones of extended-spectrum cephalosporin-resistant *Strepto-*

coccus pneumoniae isolates in the United States. *Antimicrob. Agents. Chemother.* **39**, 2282–2288.

44. Hall, L. M. C., Whiley, R. A., Duke, B., George, R. C., and Efstratiou, A. (1996) Genetic relatedness within and between serotypes of *Streptococcus pneumoniae* from the United Kingdom : analysis of multilocus enzyme electrophoresis, pulsed-field electrophoresis and antimicrobial resistance patterns. *J. Clin. Microbiol.* **34**, 853–859.

45. Dowson, C. G., Hutchison, A., and Spratt, B. G. (1989) Extensive remodelling of the transpeptidase domain of penicillin-binding protein 2B of a South African isolate of *Streptococcus pneumoniae. Mol. Microbiol.* **3**, 95–102.

46. Smith, A. M., Klugman, K. P., Coffey, T. J., and Spratt, B. G. (1993) Genetic diversity of penicillin-binding protein 2B and 2X genes from *Streptococcus pneumoniae* in South Africa. *Antimicrob. Agents. Chemother.* **37**, 1938–1944.

47. Stull, T. L., LiPuma, J. J., and Edlind, T. D. (1988) A broad-spectrum probe for the molecular epidemiology of bacteria: ribosomal RNA. *J. Infect. Dis.* **157**, 280–286.

48. Harakeh, H., Bosley, G. S., Keihlbauch, J. A., and Fields, B. S. (1994) Heterogeneity of rRNA gene restriction patterns of multiresistant serotype 6B *Streptococcus pneumoniae* strains. *J. Clin. Microbiol.* **32**, 3046–3048.

49. Sa Figueiredo, A. M., Austrian, R., Urbaskova, P., Teixeira, L. A., and Tomasz, A. (1995) Novel penicillin-resistant clones of *Streptococcus pneumoniae* in the Czech Republic and in Slovakia. *Microbial Drug Resis.* **1**, 71–78.

50. Coffey, T. J., Daniels, M., McDougal, L. K., Dowson, C. G., Tenover, F. C., and Spratt, B. G. (1995) Genetic analysis of clinical isolates of *Streptococcus pneumoniae* with high-level resistance to expanded-spectrum cephalosporins. *Antimicrob. Agents. Chemother.* **39**, 1306–1313.

51. Zoe Jordens, J., Paul, J., Bates, J., Beaumont, C., Kimari, J., and Gilks, C. (1995) Characterization of *Streptococcus pneumoniae* from human immunodeficiency virus-seropositive patients with acute and recurrent pneumonia. *J. Infect. Dis.* **172**, 983–987.

52. Hermans, P. W. M., Sluijter, M., Hoogenboezem, T., Heersma, H., Van Belkum, A., and De Groot, R. (1995) Comparative study of five different DNA fingerprint techniques for molecular typing of *Streptococcus pneumoniae* strains. *J. Clin. Microbiol.* **33**, 1606–1612.

53. Stern, M. J., Ames, G. F. L., Smith, N. H., Robinson, E. C., and Higgins, C. F. (1984) Repetitive extragenic palindromic sequences: a major component of the bacterial genome. *Cell* **37**, 1015–1026.

54. Hulton, C. S. J., Higgins, C. F., and Sharp, P. M. (1991) ERIC sequences: A novel family of repetitive elements in the genomes of *Escherichia coli, Salmonella typhimurium* and other enterobcteria. *Mol. Microbiol.* **5**, 825–834.

55. Martin, B., Humbert, O., Camara, M., Guenzi, E., Walker, J., Mitchell, T., Andrew, P., Prudhomme, M., Alloing, G., Hakenbeck, R., Morrison, D. A., Boulnois, G. J., and Claverys, J.-P. (1992) A highly conserved repeated DNA element located in the chromosome of *Streptococcus pneumoniae. Nucleic Acids Res.* **20**, 3479–3483.

56 van Belkum, A., Sluÿter. M., de Groot, R., Verbrugh, H., and Hermans, P. W. M. (1996) Box repeat PCR assay for high-resolution typing of *Streptococcus pneumoniae* strains. *J. Clin. Microbiol.* **34**, 1176–1179.

57. Gardiner, K. (1991) Pulsed field gel electrophoresis. *Analytical Chem.* **63,** 658–665.
58. Lefevre, J. C., Faucon, G., Sicard, A. M., and Gasc, A. M. (1993) DNA finger-printing of *Streptococcus pneumoniae* strains by pulsed-field gel electrophoresis. *J. Clin. Microbiol.* **31,** 2724–2728.
59. Soares, S., Kristinsson, K. G., Musser, J. M., and Tomasz, A. (1993) Evidence for the introduction of a multiresistant clone of serotype 6B *Streptococcus pneumoniae* from Spain to Iceland in the late 1980s. *J. Infect. Dis.* **168,** 158–163.
60. Vaz Pato, M. V., De Carvalho, C. B., and Tomasz, A. (1995) Antibiotic suscepti-bility of *Streptococcus pneumoniae* isolates in Portugal. A multicenter Study between 1989 and 993. *Microbial Drug Resis.* **1,** 59–69.
61. Lefevre, J. C., Bertrand, M. A., and Faucon, G. (1995) Molecular analysis by pulsed-field gel electrophoresis of penicillin-resistant *Streptococcus pneumoniae* from Toulouse, France. *Eur. J. Clin. Microbiol. Infect. Dis.* **14,** 491–497.
62. Gasc, A.-M., Geslin, P., and Sicard, A. M. (1995) Relatedness of penicillin-resistant *Streptococcus pneumoniae* serogroup 9 strains from France and Spain. *Microbi-ology* **141,** 623–627.
63. Moreno, F., Crisp, C., Jorgensen, J. H., and Patterson, J. E. (1995) The clinical and olecular epidemiology of bacteremias at a university hospital caused by pneu-mococci not susceptible to penicillin. *J. Infect. Dis.* **172,** 427–432.
64. Tarasi, A., Sterk-Kuzmanovic, N., Sieradzki, K., Schoenwald, S., Austrian, R., and Tomasz, A. (1995) Penicillin-resistant and multidrug-resistant *Streptococcus pneumoniae* in a paediatric hospital in Zagreb, Croatia. *Microbial Drug Resis.* **1,** 169–176.
65. Versalovic, J., Koeuth, T., and Lupski, J. R. (1991) Distribution of repetitive DNA sequences in eubacteria and application to fingerprinting of bacterial genomes. *Nucleic Acids Res.* **19,** 6823–6831.
66. Welsh, J. and McClelland, M. (1990) Fingerprinting genomes using PCR with arbitrary primer. *Nucleic Acids Res.* **18,** 7213–7218.
67. Williams, J. G. K., Kubelik, A. R., Livak, K. J., Rafalski, J. A., and Tingey, S. V. (1990) DNA polymorphisms amplified by arbitrary primers are useful as genetic markers. *Nucleic Acids Res.* **18,** 6531–6535.

10

Molecular Approaches in *Mycobacterium tuberculosis* and Other Infections Caused by *Mycobacterium* Species

Madhu Goyal and Douglas Young

1. Introduction

The genus *Mycobacterium* consists of a diverse group of organisms that are ubiquitous and are believed to be some of the oldest bacteria on earth. They may exist as free-living commensals inhabiting soil and water, but they are also potentially pathogenic to man and other animals, being transmitted by airborne or droplet spread. At least 25 species of mycobacteria have been associated with human disease. Robert Koch in 1882, identified the acid-fast bacterium (AFB), *Mycobacterium tuberculosis* as the causative agent of tuberculosis (TB). TB is an ancient disease that remains a significant global health problem. With improved living standards and the introduction of chemotherapy in 1950s, the incidence of TB in most industrialized countries showed a progressive decline, with very little mortality by the mid 1980s. This pattern has changed over the last decade.

The magnitude of the global problem is enormous. According to a recent estimate, approx 90 million new cases of TB will occur world-wide during the decade 1990 through 1999 *(1)*. Among adults, tuberculosis is the world's foremost cause of death from a single infectious agent. If global control of TB remains at the 1990 level, 30 million people are expected to die of this disease by the year 2000 *(1)*.

In the United States and in developing countries, the increase in TB has been attributed to coinfection with the human immunodeficiency virus (HIV) *(2,3)*, deterioration of the public health infrastructure, and social disruption, including homelessness, drug abuse, and immigration. Multidrug-resistant TB (MDR-TB) has emerged as a major problem among HIV-infected persons.

From: *Methods in Molecular Medicine, Vol. 15: Molecular Bacteriology: Protocols and Clinical Applications*
Edited by: N. Woodford and A. P. Johnson © Humana Press Inc., Totowa, NJ

Because infection with HIV is associated with increased susceptibility to TB, or at least with accelerated progression and mortality of the disease, it is feared that the incidence of TB will increase even further as the HIV epidemic continues to develop. There is, therefore, an urgent need for new tools to assist in TB control. In the present chapter, the authors describe the progress in the development of different methods, including molecular techniques, used for the diagnosis and epidemiology of TB.

2. Diagnosis
2.1. Conventional Diagnostic Methods

The presumptive diagnosis of TB can be made on the basis of patient histories, clinical and radiological findings, and the presence of AFB in patient specimens. The clinical symptoms usually are persistent cough, mild fever, sweats, and weight loss. A chest X-ray with characteristic cavitating shadowing of the affected lobe or lobes provides additional valuable evidence of the disease. Further contributory information can be obtained by performing a traditional tuberculin skin test. The tuberculin skin test is a delayed-type hypersensitivity reaction to mycobacterial protein derivative, and is an indicator of prior exposure to mycobacterial species. Although currently available tuberculin skin tests are substantially less than 100% sensitive and specific for detection of *M. tuberculosis* (MTB) infection *(4)*, useful information is still provided by the test. In countries where BCG vaccination is not used, conversion to a positive skin test is used to identify recent infection requiring preventive therapy.

2.1.1. Microbiology

Microscopy of specimens such as sputa using auramine and Ziehl-Neelsen (ZN) stains followed by culture on Lowenstein Jensen (LJ) or in Middlebrook media form the basis of mycobacterial diagnosis. These techniques are one of the cornerstones of all anti-TB programs. However, both techniques have limitations. Microscopy, although quick and easy, has poor sensitivity, with 10^4 organisms per mL of sputum required for a 50% probability of a positive result. Smear-negative disease is thus a feature of half of all pulmonary TB infections, and is virtually the rule for infection at other sites, where smaller numbers of organisms cause clinical disease e.g., pleura and meninges. The diagnosis of primary TB in children is another such case.

The definitive diagnosis of TB continues to depend on culture of MTB from secretions or tissue from the infected host together with a compatible clinical picture of the disease. Cultures are traditionally prepared by treatment of samples with 2% NaOH for 15 min to kill other bacteria, followed by inoculation on egg-based solid LJ or Ogawa media, incubated at temperatures between 35 and 37°C in the presence of 5% CO_2 for up to 8 wk. Selective liquid media containing a cocktail of antibiotics to prevent growth of nonmycobacterial contaminants are often used in conjunction with solid media. Culture is several hundred times more sensitive than microscopy and can in practice detect 10–100 viable organisms per sample. However, the results are not available for many weeks or even months, thus limiting the usefulness of the technique. This is primarily because of the long generation time (12–18 h) of the organism. Data on drug susceptibility is available after a further 2–3 wk. These delays may compromise survival of patients infected with drug-resistant strains already commenced on seemingly appropriate therapy. Speciation of the infecting organism has become increasingly important in light of the profound impact of HIV on host susceptibility *(2)*. This has led to an increase in the number of nontuberculous infections, particularly by strains belonging to the *M. avium-M. intracellulare* (MAI) complex *(5)*, which are often resistant to the drugs used for treating TB. In addition to allowing initiation of appropriate treatment for the individual patient, early diagnosis of TB is crucial for TB control programs in preventing further spread of the organism to exposed contacts. At the 1992 conference "Meeting the Challenge of Multidrug-Resistant Tuberculosis," held at the Centers for Disease Control and Prevention (CDC), the use of rapid testing and reporting methods was recognized as essential for laboratories to achieve the goals to control TB.

2.1.2. Rapid Detection and Speciation

2.1.2.1 BACTEC

The radiometric BACTEC system (Becton Dickinson, Rutherford, NJ) has shortened the time required for isolation of MTB and atypical mycobacteria by sampling and automatically measuring CO_2 produced by mycobacterial metabolism of ^{14}C palmitic acid. It is also possible to identify presumptively MTB complex strains through the incorporation of *p*-nitro benzoic acid (PNB) or *p*-nitro-acetylamino-B-hydroxypropiophenone (NAP) in BACTEC (BACTEC-NAP) *(6)*, which inhibit the growth of MTB, but does not inhibit growth of nontuberculous mycobacteria.

2.1.2.2. MYCOBACTERIA GROWTH INDICATOR TUBE (MGIT)

A new method has been developed by Becton Dickinson Microbiology Systems for the detection of mycobacteria by using silicon rubber impregnated with a ruthenium metal complex as a fluorescence quenching based oxygen sensor (BBL, MGIT). The first reports of the primary isolation of mycobacteria from clinical and stock cultures and detection of MTB resistance to isoniazid (INH) and rifampicin using MGIT technique seem to be quite promising *(7,8)*.

2.1.2.3. HIGH PERFORMANCE LIQUID CHROMATOGRAPHY (HPLC)

HPLC can be used to detect the species-specific mycolic acids produced by mycobacteria *(9)*. When used on primary culture isolates, HPLC can provide definitive species identification for any of more than 50 *Mycobacterium* species in less than four h. More recently, Jost et al. *(10)* have developed and evaluated highly sensitive and automated HPLC that utilized fluorescence detection (HPLC-FL) of mycolic acid 6,7-dimethoxycoumarin esters to identify MTB and MAI directly from fluorochrome-stained smear-positive sputum specimens and young BACTEC cultures. HPLC-FL could identify MTB and MAI in 1 d from 56.8 and 33.3% smear-positive sputa. Rapid and sensitive identification of MTB and MAI from young BACTEC cultures was achieved in 99.0 and 94.3% of samples, respectively.

Capillary gas chromatography has been used to study the short-chain fatty acids and cleavage products of mycolic acids from mycobacteria with some success *(11)*, but differentiation within the MTB complex was not possible. Thus far, these methods have acquired a limited use because of expensive instrumentation, difficulties in standardization, and requirement for specialized expertise *(12)*.

Rapid confirmation of tuberculous meningitis has always been difficult. Recently, adenosine deaminase, a host enzyme produced by activated T cells and easily detected by a colorimetric procedure, was shown to increase in concentration during the active stages of tuberculous meningitis and to decrease to normal levels after effective antituberculous therapy. The presence of tuberculostearic acid in the spinal fluid or serum of patients has been detected using gas liquid chromatography (GLC), and mass spectrophotometry. The presence of this compound in patients with meningitis supports a tuberculous etiology *(13,14)*, but these techniques are not widely available.

2.2. Molecular Diagnostic Methods

2.2.1. Gene Probes

DNA probes have greatly aided in the rapid identification and detection of *Mycobacterium* species from cultures of clinical specimens. Both isotopic-labeled probes *(15)* and nonisotopic probes—the SNAP system (Syngena, San Diego, CA), which utilizes DNA probes labeled with horseradish peroxidase *(16)*, and the Accuprobe system (Gen-Probe, San Diego, CA), which employs acridinium ester as the probe label *(17)*—have been used to identify colonies isolated on solid media and *Mycobacterium* species directly from BACTEC cultures. Single-stranded DNA probes complementary to the ribosomal RNA (rRNA) of the target organism are used in the Gen Probe system (Gen-Probe). The Accuprobe system has been found to be as sensitive in identifying MTB and MAI complex organisms from colonies on solid media as the isotopically-labeled version of the probes *(17)*, but less sensitive than isotopic labeled probes in identifying *Mycobacterium* species from BACTEC cultures *(18)*. However, in combination with initial BACTEC isolation, the Accuprobe system can provide speciation within 2 wk. These probes are easy to use and rapidly provide results, but they cover a limited range of species and problems concerning specificity and sensitivity have not been resolved.

The sensitivity of nucleic acid hybridization assays can be dramatically increased by using branched DNA signal amplification *(19)* (*see* Chapter 5). A bifunctional oligonucleotide probe is used that contains one sequence specific for the target molecule, and another sequence to which a second branch oligonucleotide can bind. The second oligonucleotide in turn has many binding sites to which a third oligonucleotide conjugated to an enzyme can bind. The third oligonucleotide produces a detectable signal in the presence of substrates. Theoretically, such a procedure could amplify a hybridization signal 10–100-fold, which might improve the detection limit of the hybridization assays to as few as 100–1000 organisms per specimen.

Another sort of signal amplification system for detecting MTB in clinical specimens has been developed by Gene Trak *(20)*. It is based on reversible target capture of MTB 23S rRNA followed by amplification of a replicable detector probe with Q beta replicase *(21)* (*see* Chapter 5). The Q beta replicase assay is not inhibited by sputum, and results were in 100% agreement with those of culture, including detection of 10 culture-positive specimens.

2.2.2. Target Amplification (see Chapter 5)

Technological advances in amplifying and detecting a specific region of bacterial DNA or RNA provide the opportunity to make improvements in the laboratory diagnosis of TB. Several such procedures have been described for use with MTB, including polymerase chain reaction (PCR) amplification *(22)*, strand displacement amplification (SDA) *(23)* and ligase chain reaction (LCR) amplification *(24)*.

2.2.2.1. POLYMERASE CHAIN REACTION (PCR)

In the last few years, many investigators have described the use of PCR-based assays with a variety of clinical samples *(22,25,26)*, using a diversity of genetic elements as the target template (**Table 1**). Previously published protocols used PCR to detect mycobacterial gene sequences that code for rRNA *(27)* or proteins—including the 65 kDa *(28)* and 38 kDa *(29)* proteins and, MPB 64 *(30)* and MPB 70 *(31)*—or insertional elements such as IS*6110* *(22)* and *IS986* *(32)*. Other systems used repetitive sequences *(33)*, or specific restriction fragments like 1.5 kb *Eco*RI-*Bam*HI *(34)*.

The most frequently used PCR target has been the insertion element IS*6110* *(35–37)*. This element is specific for strains belonging to the MTB complex and it is usually present in multiple copies, thus enhancing sensitivity of the PCR in comparison with single-copy targets *(38)*. Using the IS*6110* sequence,

Table 1
Targets and Primer Sequences used in the PCR Assays for the Diagnosis of MTB

Target	Primers	Product Size	Detection	Ref
1. 65 kDa	5'GAGATCGAGCTGGAGGATCC3' 5'AGCTGCAGCCCAAAGGTGTT3'	383 bp	Agarose gel electrophoresis and hybridization	*28*
2. IS*6110*	5'CCTGCGAGCGTAGGCGTCGG3' 5'CTCGTCCAGCGCCGCTTCGG3'	123 bp	Agarose gel electrophoresis and hybridization	*37*
3. MTP 40	5'CAACGCGCCGTCGGTGG3' 5'CCCCCACGGCACCGC3'	396 bp	Agarose gel electrophoresis and hybridization	*134*
4. Repetitive seq contained on 375bp *Kpn*I-*Sma*I fragment	5'GCGGCTCGGGCGGCGTCGGTGGCTT3' 5'GCCAGAACCGACCAACCCGCCGATA3'	336 bp	Agarose gel electrophoresis and hybridization	*33*
5. 38 kDa protein Ag b (Pab)	5'ACCACCGAGCGGTTCGCCTGA3' 5'GATCTGCGGGTCGTCCCAGGT3' Probe Sequence 5''CGCTGTTCAACCTGTGGGGTCCGGCCTTTC3'	419 bp	Agarose gel electrophoresis and hybridization	*29*
6. MPB70	5'GAACAATCCGGAGTTGACAA3' 5'AGCACGCTGTCAATCATGTA3'	372 bp	Agarose gel electrophoresis	*31*
7. IS986	5'CGTGAGGGCATCGAGGTGGGC3' 5'GCGTAGGCGTCGGTGACAAA3'	205 bp	Agarose gel electrophoresis	*32*

Eisenach and colleagues *(39)* demonstrated the use of PCR for the diagnosis of TB from sputum samples. They tested 162 sputum samples; 51 of these were from patients who had either a positive smear, a positive culture, or both. Fifty of the 51 were PCR-positive. Of 68 specimens from patients with nontuberculous disease or no mycobacterial infection, only one false-positive was obtained. There are now many reports of the use of PCR for the detection of MTB from clinical specimens.

Although PCR-based diagnostic tests should have the greatest impact in detecting paucibacillary disease, few studies have included large numbers of smear-negative, culture-positive specimens in the evaluation of PCR based assays. Nolte et al. *(40)*, compared the PCR results using IS*6110* sequence with results of microscopy and conventional culture for the detection of MTB in 313 sputum specimens. PCR detected 105 of 110 (95%) of the smear-positive and 8 of 14 (57%) of the smear-negative specimens. There were no false-positive results by PCR (specificity 100%). Shawar et al. *(41)* using the same target for amplification detected 90% of smear-positive and 53% of smear-negative specimens. In a similar study of over 5000 sputum samples, Clarridge et al. *(42)* found that PCR could rapidly diagnose smear-negative, culture-positive disease in 33 of 52 (63%) cases, with only 5.7% of samples PCR-positive and culture-negative, giving overall sensitivity, specificity, and positive predictive values (all compared with culture) of 83.5, 99, and 94.2%, respectively.

Whereas PCR clearly provides exciting potential for improved diagnosis of TB, a series of drawbacks remain to be addressed. Walker et al. *(43)* used PCR to amplify IS*6110* from clinical samples obtained from 87 patients. IS*6110* DNA was identified in samples from all the six patients with active TB, from 15 of 18 patients with past TB, from five of nine contacts of patients with TB and from 9 of 54 patients with lung disease unrelated to TB. Similar results have been reported by Yuen et al. *(44)*, Schulger et al. *(45)*, and more recently, in a prospective evaluation of 103 patients with suspected TB, by Beige et al. *(46)*. A positive PCR result may be indicative of the presence of MTB, but may not always correlate with clinically significant disease.

There are also technical problems in implementing PCR assays. A report describing unacceptably high variation in sensitivity and specificity of PCR carried out in different laboratories underlines the need for rigorous quality control in this area *(47)*. The presence of inhibitors of the PCR that may cause false-negative results, and procedures for isolation of small amounts of DNA from samples from different sites in the body needs improvement. Studies using PCR have described lysis conditions using a combination of sonication *(48)*, heating at 55–100°C *(39,49,50)*, detergents *(39,28)*, proteases *(48,51)*, lysozyme *(49,26)*, freeze-thaw *(51)*, sodium hydroxide *(39,28)*, and hypertonic NaCl *(28)*. In some cases, these methods have been followed by adsorption to silica *(39)* or phenol:chloroform extraction *(33)*. Recently, Kolk et al. *(52)*, have described the construction of a strain of *M. smegmatis* containing a modified IS*6110* fragment integrated in the chromosome. This modified *M. smegmatis* strain can serve as an effective internal control to monitor efficacy of DNA extraction and to detect the presence of inhibitors of PCR assays necessary for a single clinical sample. The PCR product from the control DNA target can be distinguished from the MTB target by the difference in size. Contamination with amplicons from previous reactions can lead to false-positive results; the use of uracil-N glycosylase (UNG) and dUTP, instead of dTTP, has been suggested as a way of reducing this problem *(53)*.

2.2.2.2. COMMERCIAL KITS FOR PCR

1. Amplicor. Although simplified procedures have been investigated for PCR-based amplification methods *(54,55)*, they remain complex and insufficiently reliable for use in the routine clinical laboratory. The recent development of commercial test systems will perhaps circumvent these problems. One such system is the Amplicor assay developed by Roche Diagnostic Systems. It is a PCR kit used to amplify a target within the 16S rRNA of MTB. During testing of 2173 specimens in six different laboratories, the sensitivity of Amplicor test was 86%. A total of 95% of patients with culture-proven tuberculosis were diagnosed by the Amplicor test, whereas direct microscopy detected mycobacteria in only 72% of these patients. The specificity of the assay was 98% *(56)*. Recently Dilworth et al. *(57)* compared Amplicor PCR with IS*6110*-PCR and found that, though both the tests were equally sensitive (96%), Amplicor PCR was more specific (98%) than PCR using IS*6110* as a target (79%).

2. Amplified Mycobacterium Tuberculosis Direct Test (AMTD). The other commercially available test is the AMTD developed by Gen-Probe. It is a direct specimen assay for the identification of MTB from respiratory samples and can be completed within 4–5 h. In this method rRNA is amplified and the product is detected with a specific chemiluminescent probe. Vlaspolder et al. *(58)* used AMTD to diagnose tuberculosis from 550 respiratory and non-respiratory specimens from 340 patients and compared the results with a conventional culture method. They found a sensitivity of 98.4% and specificity of 98.9% and positive and negative predictive values of 93.8 and 99.7%, respectively.

   Direct comparisons of these two kits have been made in two different studies; each study concluded that both nucleic acid amplification methods are rapid, sensitive and specific for the detection of MTB in respiratory specimens *(59,60)*.

2.2.2.3. STRAND DISPLACEMENT ASSAY (SDA)

SDA is an isothermal in vitro DNA amplification method utilizing a DNA polymerase and a restriction enzyme to achieve exponential amplification (approx 10^7-fold) in 2 h. Species-specific SDA assays have been developed by

Becton and Dickinson Research Centre for detecting MTB, *M. avium,* or *M. kansasii (23)*. The *M. avium*-specific assay and MTB assay can be combined to generate a single assay to detect these two most common species in smear-positive specimens.

2.2.2.4. LIGASE CHAIN REACTION (LCR)

LCR is being developed by Abbott laboratories (Abbott Park, IL). Using this method, Leckie et al. *(24)* have shown excellent sensitivity and specificity for detecting MTB in sputum specimens, and the assay can be completed within 5 h after processing of the specimen. The performance of this test in the diagnostic laboratory has not yet been tested.

2.2.3. PCR and Speciation

The 16S rRNA gene is a suitable target for amplifying mycobacterial nucleic acids at the genus level, confirming correct amplification by genus-specific probes and differentiating at the species level with a set of species probes or by sequence analysis *(61)*. Each *Mycobacterium* sp. has a unique variable region that can be used as a species-specific hybridization probe to identify the mycobacterial species. These variable sequences are called signature sequences (*see* **Note 1**).

Using a variety of clinical samples, Kirschner et al. *(62)* have evaluated a molecular assay that targeted the 16S rRNA gene, allowing the detection of multiple mycobacterial pathogens in a single amplification reaction. The sensitivity and the specificity of PCR assay was 84.5 and 99.5%, respectively, and the method performed well in comparison with culture. This approach provides a convenient and cost effective way to identify all mycobacterial species instead of using multiple amplification assays developed to detect and identify single species of *Mycobacterium.*

2.3. Drug Resistance

Infection with drug sensitive strains of MTB can be efficiently cured with a combination of INH, rifampicin, and pyrazinamide. However, with the rise in the number of disease cases, there has been an increase in the recovery of MTB isolates resistant to one or more of the primary antimicrobial agents used as therapy *(63)*. Moreover, the emergence of MDR-TB strains, defined as those resistant to at least INH and rifampicin, has resulted in fatal outbreaks in many countries, including the US. Strains of MDR-TB, some of which are resistant to as many as seven drugs, are deadly to both HIV-negative and HIV-positive individuals *(64)*. The traditional method of assessing drug susceptibility is by testing the effect of the drug on bacterial growth. The extended time course associated with conventional drug susceptibility testing can have fatal consequences in the context of MDR-TB, and improvements are needed to yield accurate analysis in a shorter period. As with initial diagnosis, the BACTEC approach can be used to reduce the time required for susceptibility testing, and this strategy is recommended as routine procedure by the College of American Pathologists and CDC. Increasing drug resistance, the lack of effective new antituberculous drugs, and the need to design rapid strategies for susceptibility testing have led to the substantial research devoted to elucidating the molecular genetic basis of resistance *(65)*.

2.3.1. Molecular Basis of Drug Resistance

Rifampicin is a key component of therapeutic regimens. Patients infected with rifampicin-resistant organisms have a poor prognosis, particularly if this is associated with resistance to additional antituberculous drugs *(66)*. Resistance to rifampicin has been attributed to changes in RNA polymerase. Recently, the entire *rpoB* genes of *M. leprae (67)* and MTB *(68)* have been sequenced, and several mutations associated with rifampicin resistance have been identified in both species *(69, 70)*. In 96% of rifampicin-resistant MTB isolates, mutations are clustered in an 81-bp region of the *rpoB* gene *(65)*.

The molecular genetic basis of resistance to INH is less well understood. In 1954, Middlebrook *(71)* observed that highly-resistant strains of MTB lacked or had greatly decreased catalase activity. This finding, together with biochemical evidence implicating catalase-peroxidase in the antituberculous activity of INH *(72)*, led to the cloning and characterization of the *katG* gene encoding the catalase-peroxidase enzyme *(73,74)*. Initial studies reported deletion of the *katG* gene from the chromosome of a subset of highly INH-resistant strains of MTB *(75)*, with more detailed investigations revealing the widespread occurrence of point mutations and small deletions or insertions in the *katG* gene of resistant isolates *(76,77)*. A different approach led to cloning of a DNA fragment capable of conferring enhanced resistance to INH and ethionamide in *M. smegmatis (78)*. This fragment was found to contain the *inhA* gene, encoding an enzyme involved in mycolic acid biosynthesis, which may be a target of the activated drug. Sequence analysis of the *katG* gene and two genes within the *inhA* locus of 34 resistant and 12 susceptible isolates of MTB, suggests that mutations in these regions occur in around 84% of INH-resistant strains *(79)*. Similar findings have been reported in MDR-TB strains *(80)*. Another gene that may play a role in INH-resistance is *ahpC*, encoding an alkyl hydroperoxidase *(81)*.

Streptomycin, a broad-spectrum antibiotic, was the first drug that was used to treat tuberculosis and acts at the level of the ribosome, preventing translation of mRNA to protein. High-level resistance to streptomycin in MTB results predominantly from missense mutations in the *rpsL* gene, which encodes the ribosomal protein S12. The most common mutation in streptomycin-resistant clinical isolates involves an A to G transition in codon 43 that leads to the substitution of lysine by arginine *(82,83)*. A small minority have mutations in conserved loops of the 16S rRNA, which is encoded by the *rrs* gene. However, no mutations are found in the *rpsL* or *rrs* genes of about 30% of streptomycin-resistant clinical isolates of MTB studied *(82,84)*.

The emergence of resistance to the first-line antituberculous drugs has led to increased interest in the use of fluoroquinolones for TB treatment, with ciprofloxacin currently the most promising candidate. The target for these drugs is DNA gyrase, an enzyme involved in the supercoiling and unwinding of DNA, which is important for transcription and replication. DNA gyrase is composed of two A and two B subunits, encoded by the *gyrA* and *gyrB* genes; mutations in these genes confer a high-level of fluoroquinolone resistance in a range of bacterial species (*see* Chapter 30). Analysis of 12 resistant MTB strains found point mutations affecting one of four residues of *gyrA*, with half of these affecting residue 94 *(85)*. No mutations in *gyrB* were found. Recent observations suggest that low-level resistance to fluoroquinolones can be conferred by enhanced drug efflux, rather than alteration of the drug target *(86)*.

2.3.2. Identification of Resistant Isolates

Mutations can be identified either by nucleotide sequencing, or by use of other techniques suitable for detection of point mutations. Telenti et al. *(70)* used PCR followed by single-strand conformational polymorphism (SSCP) to detect mutations in the *rpoB* gene. PCR was performed using primers (5'TGCACGTCGCGGACCTCCA3') and (5'TCGCCGCGATCAAGGAGT3') in the presence of ^{32}P-dCTP to generate a radiolabeled 157-bp product (*see* **Note 2**). The diluted denatured PCR product was electrophoresed on a denaturing sequencing format gel and autoradiographed. They were able to identify 64 of 66 rifampicin-resistant MTB isolates using PCR-SSCP. PCR-SSCP is a promising technique for monitoring the emergence of drug resistance, although it is not without limitations. It is most effective (97% success) in detecting rifampicin resistance *(70)*, which is important because rifampicin resistance is a powerful surrogate marker for predicting resistance to other drugs. PCR-SSCP is less useful for detecting resistance to INH because, as described in **Subheading 2.3.1.**, multiple resistance mechanisms are involved and, therefore, multiple tests have to be employed.

A simplified protocol (Inno-Lipa Rif TB) for the detection of point mutations in the *rpoB* gene has recently been developed (Innogenetics, Belgium). In this method a 250-bp fragment of the *rpoB* gene of MTB is amplified by PCR (*see* **Note 3**), and allowed to hybridize with five wild-type and four mutant oligonucleotide probes immobilized on a membrane strip (**Table 2**). Presence

Table 2
Sequences of the Probes (Wild and Mutant) in the *rpoB* Gene used in the Inno-Lipa Method (135)

Target	Wild type Probes	Mutant probes
rpoB gene	S1 C AGC CAG CTG AGC CAA TTC ATG	
	S2 TTC ATG GAC CAG AAC AAC CCG CT	R2 AA TTC ATG GTC CAG AAC CCG
	S3 AAC AAC CCG CTG TCG GGG TTG ACC	
		R4A TTG ACC TAC AAG CGC CGA CTG TC
	S4 TTG ACC CAC AAG CGC CGA CTG TC	
		R4B TTG ACC GAC AAG CGC CGA CTG TC
	S5 CGA CTG TCG GCG CTG GGG C	R5 CGA CTG TTG GCG CTG GGG C

or absence of binding is detected by colorimetric reaction, allowing identification of resistant strains. Initial results look encouraging (**Fig. 1**; *136*).

Strategies that measure the effects of drugs on bacterial growth or bacterial metabolism overcome the need to design multiple tests for different drugs. Kawa et al. *(87)* used the rapid Diagnostic System for MTB (Gen-Probe, San Diego, CA) to monitor rRNA levels in MTB cultures that were treated with INH. They were able to distinguish INH-resistant strains from INH-sensitive strains in 3–5 d, as compared with the 21–28 d period required for conventional culture assays.

An alternative application of a molecular genetic approach to detect drug resistance in MTB is based on a reporter-phage system *(88)*. Genes encoding a luciferase enzyme *(lux)* are inserted into a mycobacterial phage in such a way that a chemiluminescent signal is generated when the phage infects a viable tubercle bacillus. Following a brief incubation in the presence of test drugs, bacterial cultures are mixed with the reporter phage. Phage infection of drug-resistant, viable bacteria results in light emission, in contrast to the lack of signal in drug-sensitive cultures. This type of assay has been shown to provide susceptibility results within 18–24 h of the propagation of culture. In its present format, it requires >10^6 bacteria, and has not yet been evaluated for use with clinical specimens.

An approach suitable for rapid detection of an already identified MDR-TB strain was developed in order to allow tracking of outbreaks caused by MTB strain W in the US *(89)*. This involved a multiplex PCR assay, which targeted a direct repeat of IS*6110* with a 556-bp intervening sequence (NTF-1) specific to this particular multiresistant isolate.

3. Epidemiologic Investigation

An important factor in the control of TB is the ability to identify outbreaks and track the transmission of particular strains of MTB. The typing of strains from infected individuals could play an important role in tracking the source of infection. In the past, proof of transmission of TB was based on the occurrence of cluster outbreaks *(90)*, transmission of strains with characteristic resistance patterns *(91)*, and tuberculin skin test conversion after exposure to an active case. Comparing antibiotic sensitivity patterns of isolates can be used to trace spread of an unusual pattern of resistance, but this approach is limited by the relatively low number of different sensitivity patterns possible, and by the fact that mycobacterial strains may change their patterns of resistance when they are exposed to antituberculous drugs. Phage typing can be used to differentiate strains, but is of limited value in the epidemiology of TB because only a few different phage types can be recognized *(92)*. Serotyping has been even less successful in differentiating strains of MTB *(93)*. Similarly, plasmid profiling

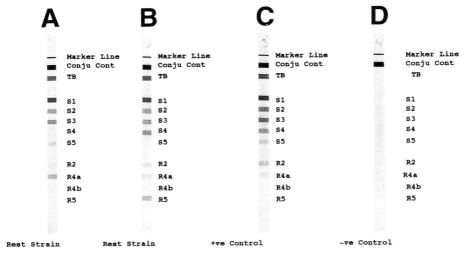

Fig. 1. Hybridization patterns of drug-resistant and drug-sensitive strains of MTB isolates using the Inno-Lipa kit (Innogenetics, Belgium). From top to bottom of the strip: Red Marker line, Conjugate control line (line 1), MTB complex specific probe (line 2), Wild type probes S1 to S5 (line 3 to 7) and Mutant probes R2, R4a, R4b, and R5 (line 8 to 11). Strip A shows S4 region mutation, strip B shows S5 region mutation, strip C shows no mutation (H37Rv DNA), and strip D is a negative control.

has not proved possible in MTB because of the low number or absence of plasmids *(94)*. Other techniques for typing mycobacteria have included electrophoresis of cell proteins *(93)* and biochemical heterogeneity *(95)*. These techniques suffer from problems of reproducibility and low numbers of different types. Multilocus enzyme electrophoresis (MLEE) has also been used to differentiate mycobacteria *(96)*, but the single MTB and *M. bovis* strains analyzed could not be differentiated by this method.

The availability of restriction endonucleases that recognize specific nucleotide sequences in DNA molecules has provided another approach for assessing relationships among closely related strains. Imeda *(97)* reported that the distribution patterns of the DNA fragments produced with various restriction endonucleases were indistinguishable among representative strains of MTB, but Collins and DeLisle *(98)* have successfully used this approach to differentiate strains of bovine TB in New Zealand. Gross restriction patterns of chromosomal DNA are difficult to analyze. A more sensitive technique for analyzing restriction fragment heterogeneity is hybridization with molecular probes consisting of cloned DNA fragments. Using this approach Eisenach et al. *(99)* showed that selected DNA probes were useful in detecting differences among MTB strains.

Recent studies using repetitive elements have revealed significant DNA polymorphisms in the chromosome of MTB. Two classes of repetitive DNA elements have been used for the molecular epidemiology of tuberculosis: Insertion sequence (IS) elements that have the capacity to move within the genome, and which have a size of about 1300 bp, and repetitive DNA sequences varying in size from 3 to 36 bp *(100)*. The DNA polymorphism associated with insertion elements is presumed to be caused by their inherent capacity to move to random locations on the genome. The nature of the genetic rearrangements driven by the small repetitive elements has been investigated only for the direct repeat (DR) region, and probably involves homologous recombination between DRs and rearrangements driven by the insertion element IS*6110*, which is present in the DR region of the majority of the MTB strains *(101)*. The DR locus contains multiple, well-conserved 36-bp DRs, interspersed by nonrepetitive spacer sequences, 34–41 bp in length. Strains vary in the number of DRs and in the presence or absence of particular spacers. It is likely that the same mechanism contributes to the DNA polymorphism associated with other small repetitive sequences, e.g., the polymorphic GC rich repeat sequences (PGRs), the major polymorphic tandem repeat (MPTR), and the GTG repeat *(102–104)*.

3.1. IS6110-*Restriction Fragment Length Polymorphism (RFLP)*

The insertion element IS*6110* is the most widely used marker for epidemiologic studies because of the high degree of discrimination obtained with this element. This approach takes advantage of the fact that MTB strains carry multiple copies of IS*6110* and that the precise locations of the IS*6110* elements in the MTB genome vary significantly from strain to strain, providing a unique DNA fingerprint for each MTB strain (**Fig. 2**). International consensus exists on a standardized method of IS*6110* fingerprinting, thus enabling the comparison of DNA types from different laboratories *(105)*. In brief, the method involves the extraction of whole genomic DNA from cultured bacteria followed by digestion with restriction enzyme *Pvu*II. The resulting DNA fragments are separated by gel electrophoresis, transferred to nylon membranes, and probed with a labeled probe directed against the insertion element IS*6110*. The IS*6110*-RFLP technique has been shown to be a reliable and reproducible method for differentiating MTB strains *(106,107)*. Fingerprint patterns for individual isolates are stable in laboratory and animal passage *(108)*, and show little or no change with time in chronic TB patients *(107)*. IS*6110*-based DNA fingerprints do not change during the development of drug resistance *(109)*.

IS*6110*-RFLP has been extensively used for epidemiologic studies of TB, including outbreak investigations, transmission in the community, and dissemination of multidrug-resistant clones *(110–113)*. It has also been used to detect

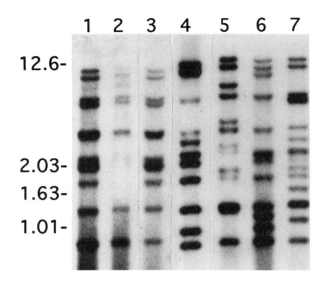

Fig. 2. IS*6110*-RFLP patterns of seven clinical isolates of MTB. The numbers on the left hand side of the figure indicate the sizes of the molecular marker (1 kb ladder).

suspected laboratory cross-contamination *(114)*. The method can be used to distinguish between a newly acquired infection and a relapse. Small et al. *(115)* demonstrated in AIDS patients in New York City that the patients were exogenously reinfected shortly after completing successful therapy. In another study, IS*6110*-RFLP was used to determine whether the infectivity of a particular clone of MTB was influenced by the HIV status of the host *(116)*. The results of the study showed an equal risk of infection with a defined MTB clone for HIV-positive and HIV-negative individuals.

Three population-based studies have been done using IS*6110*-RFLP analysis *(117–119)*. These studies showed that only a fraction of the main transmission routes of TB are disclosed by classical contact tracing practices, even in locations where an optimal infrastructure for TB control exists. Based on the assumption that clusters of infection with identical isolates are indicative of recent transmission, the two studies carried out in San Francisco and New York concluded that newly-acquired infection—not only reactivation of latent disease—was making a substantial contribution to increased TB rates *(117,118)*.

Patterns of *IS6110*-RFLPs show marked geographic variation. In general, strains from countries with a high prevalence of MTB, such as the Central African Republic, exhibit less DNA polymorphism than strains in countries with a low prevalence of infection, such as the Netherlands *(108)*. Analysis of the population structure of MTB strains from the People's Republic of China showed that vast majority belong to a genetically closely related group *(120)*.

The IS*6110*-RFLP typing method has several limitations. One major disadvantage is the time and labor required to perform this procedure. Because the method requires a sufficient amount of extracted DNA, the primary culture must have abundant growth or must be subcultured to yield enough organisms. After culture growth, the procedure involves DNA extraction, restriction endonuclease treatment, electrophoresis and Southern hybridization. Hence, the entire RFLP procedure may take 3–4 wk to yield interpretable results. Another limitation of this approach is the discovery in Asia of MTB strains with no copies of IS*6110* *(121,116)*.

Oligonucleotide $(GTG)^5$ has also been used to analyze DNA of MTB for polymorphism *(104)*. The method may be particularly useful for typing isolates that have few or no copies of the IS*6110* sequence.

3.2. Pulsed-Field Gel electrophoresis (PFGE)

A modification of the RFLP technique called PFGE (*see* Chapter 3) has also been used to type MTB isolates. This method uses restriction enzymes that cut genomic DNA infrequently to produce a small number of very large fragments. Zhang et al. *(122)* were able to separate 26 isolates of MTB into different types using PFGE. The results were in agreement with IS*6110*-RFLP. However, as with IS*6110*-RFLP, this technique also requires genomic DNA and, therefore, culture of the organism.

3.3. PCR-Based Typing

To overcome the problem of culturing the organism, many PCR-based typing methods have been developed by different laboratories using a variety of DNA targets in MTB (**Table 3**). Palittapongarnpim et al. *(123)* used arbitrarily primed (AP) PCR (*see* Chapter 6) to differentiate MTB isolates. They were able to show polymorphism in different strains of MTB and the degree of differentiation was almost the same as obtained by the IS*6110*-RFLP method. Random amplified polymorphic DNA analysis (RAPD) is a rapid and simple DNA fingerprinting method, which has been successfully applied to differentiate strains of MTB *(124)*. However, a large number of primers have to be tested before obtaining optimal results. Abed et al. *(125)* used the 16S-23S rRNA spacer region as RAPD template DNA for efficient discrimination of MTB strains with an arbitrary primer. They demonstrated that RAPD performed under these conditions is more efficient than with whole genomic DNA used as a target.

Mixed linker PCR *(126)* used oligonucleotide linkers ligated to restriction fragments and IS*6110* as primer targets for PCR-mediated amplification of

Table 3
Primers Used in the Various PCR-based
Typing Methods for MTB Complex Isolates

Method	Primers (5'-3')	Reference
1. Arbitrarily primed (AP) PCR	AATCGGCTG (arbitrary primer) CCGGGGCGGTTC (part of the IS) CCGCCGACCGAG (Arbitrary primer)	*123*
2. Random Amplified Polymorphic DNA method (RAPD)	CCTGCGAGCGTAGGCGTCGG GCGTAGGCGTCGGTGACAAA ACGCTCAACGCCAGAGACCA GATGAACCACCTGACATGAC	*124*
3. Mixed Linker PCR	First pair of primers TCGACTGGTTCAACCATCGCCG AGAACTGACCTCGACTCGCA For nested PCR ACCGTACTGCGGCGACGTC AGAACTGACCTCGACTCGCA	*126*
4. 16S-23S rRna RAPD	GAAGTCGTAACAAGG CAAGGCATCCACCGT PCR product obtained using above primers was fingerprinted using random primer- TGCATGCTGA	*125*
5. 2 kb *Kpn* I fragment present upstream of *katG* gene	TACCGCTGTTGAAGAAGCC TCGGCTTATTCGGCGTAAATCC	*129*
6. MPTR+IS*6110* PCR	(Target insertion seq and MPTR) First round primers TCGACTGGTTCAACCATCGCCG GGCAACACCGGCCTC Second round primers ACCAGTACTGCGGCGACGTC GGCAACACCGGCCTC Probe for hybridization TCTGATCTGAGACCTCAGC	*128*
7. DRE-PCR	Sequences from target IS*6110* GGCTGAGGTCTCAGATCAG ACCCCATCCTTTCCAAGAAC Sequences from target PGRS CCGTTGCCGTACAGCTG CCTAGCCGAACCCTTTG	*130*

discriminatory amplicons. A less complex PCR-based typing method was outlined by Ross and Dwyer *(127)*. They used primers that bind to the ends of IS*6110* to amplify DNA between closely spaced copies of this element. Plikaytis et al. *(128)* combined MPTR and IS*6110* to develop the IS*6110* ampliprinting typing method, which seems to be more convenient than the classical fingerprinting technique. Another variable sequence present upstream of INH resistance gene *(katG)* has also been used in a PCR based method to type MTB organisms *(129)*.

Another novel subtyping method utilized the amplification of MTB DNA segments located between two copies of repetitive elements IS*6110* and the PGRs. The rationale for the procedure is based on the fact that the distances between the repetitive elements and the copy numbers of IS*6110* and PGRs vary from strain to strain. The double repetitive element PCR (DRE-PCR), as this method is called, classified 46 clinical isolates into 25 different types *(130)*.

3.4. Spoligotyping

The variability of length and composition of the DR cluster makes it a suitable epidemiologic tool for typing MTB strains that have only a few IS*6110* elements *(121)*. Moreover, the properties of these elements, IS*6110* and the DR cluster, have been combined in a new typing method, the direct variable repeat polymerase chain reaction (DVR-PCR) *(101)*, but this method is too complicated to be performed in a routine clinical laboratory. Recently, Kamberbeek and colleagues *(137)* have extended this methodology by designing a unique PCR to amplify and label all DNA spacer sequences within the DR region. The amplified DNA is subsequently hybridized to short synthetic oligomeric DNA sequences, which are covalently bound to a filter. This method is called "spoligotyping" (from spacer oligotyping) (**Fig. 3**). The spoligotyping technique has been shown to identify outbreaks of tuberculosis and to be suitable for use on clinical samples without culture *(131)*.

4. Summary

The increase in the incidence of TB has led clinical laboratories to improve their existing diagnostic methods. The CDC has recommended the use of certain laboratory methods to facilitate accurate and prompt diagnosis of TB *(132)*. Using a combination of the radiometric BACTEC for primary culture, nucleic acid probes, the BACTEC-NAP test, HPLC, or gas liquid chromatography for species identification and drug susceptibility, the time of reporting has been reduced from months to 2–3 wk in many laboratories. In the US, the proportion of laboratories using these methods increased dramatically along with renewed concern about TB between 1991 and 1994 *(133)*. The fast reporting of TB has helped the physicians in many hospitals to take proper measures to control the disease and treat the patients with appropriate therapeutic regimens.

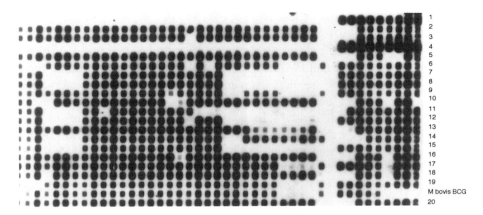

Fig. 3. Hybridization patterns (spoligotypes) of amplified mycobacterial DNA of 20 clinical isolates of MTB and *M. bovis* BCG. 43 spacer sequences interspersed between the DRs were sequenced and bound to a membrane. The PCR amplified DNA products were hybridized to the membrane and detected by ECL detection system (Amersham, England). A black spot represents a positive signal and a blank space means negative signal.

The amplification methods will provide the next generation of tools that can further reduce the time of diagnosis of TB and drug susceptibility testing from weeks to days or even hours. Most of these techniques have been shown to perform well in the research laboratories, but problems of cost and technical reliability remain to be overcome in their transfer to routine clinical laboratories. In the short term, it is likely that they will be reserved for particularly acute clinical problem cases rather than being used in routine TB control. Rapid screening of potential cases of MDR-TB will be a key application. The use of IS*6110*-RFLP fingerprinting along with conventional epidemiology has increased the understanding of the transmission of TB. In some cases, e.g., nosocomial transmission in HIV wards, rapid tests are required to type MTB isolates directly from clinical specimens without the need of culture. Many such methods based on DNA amplification have been developed, and are being evaluated. However, these methods have to be further modified and simplified to become the tests of the future mycobacteriology laboratory. The use of these rapid tests will help in future in the control of the resurgence of tuberculosis. **Table 4** summarizes the various conventional, improved, and rapid molecular tools used in the diagnosis, drug resistance, speciation, and strain typing of MTB.

Finally, the vast majority of TB cases occur in economically-deprived countries. Relatively "high-tech" and expensive approaches based on current DNA amplification techniques do not provide a practical alternative to the traditional

Table 4
Summary of the Methods Used
for Diagnosis, Drug Susceptibility, and Typing of MTB

	Conventional microbiology	Improved technologies	Rapid molecular tools
Diagnosis	Smear (1–2 d)	Bactec (10–14 d)	PCR (all 1–2 d) SDA
	Culture (4–6 wk)	MGIT (5–10 d)	LCR QB replicase
Speciation		Bactec-NAP (1–3 wk)	
	Culture+ Biochemical tests (6–10 wk)	Bactec+Gene Probes (1–3 wk)	PCR+Probes (e.g., 16S rRNA) (1–2 d)
		HPLC (1–3 wk) HPLC-FL (1–3 wk)	
Drug Resistance		Bactec (3–4 wk)	PCR-SSCP (1–2 d)
	Culture (6–10 wk)	Bactec+ Gene Probes (3–4 wk)	PCR + other tests for point mutations 1–2 d)
		MGIT (1–2qk)	Reporter phage (6–8 wk)
Strain		IS6110-RFLP (8–10 wk)	RAPD (1–2 d)
Typing		PFGE (8–10 wk)	Spoligotyping (1–2 d)

d = days, wk = weeks

smear test. Further research will be required to identify improved tools for TB control that can be routinely applied in parts of the world where TB takes its greatest toll on human life.

5. Materials

5.1. Extraction of Genomic DNA of MTB

1. Tris-EDTA buffer (TE buffer): 10 mM Tris-HCl, 1 mM EDTA, pH 8.0.
2. Lysozyme: 10 mg/mL stock.
3. Sodium dodecyl sulfate (SDS): 10% (w/v).
4. Proteinase K: 10 mg/mL.
5. N-acetyl-N, N, N-trimethyl ammonium bromide.
6. Chloroform: isoamyl alcohol (24:1 v/v).
7. Isopropanol.
8. Ethanol.

5.2. Extraction of DNA from Sputum Specimens for PCR

1. *N*-acetyl-L-Cysteine.
2. Sodium hydroxide (NaOH).
3. Sodium citrate (Na citrate).
4. TE buffer (*see* **Section 5.1.** above).
5. Lysozyme: 10 mg/mL stock.
6. Proteinase K: 10 mg/mL.
7. SDS: 10% (w/v).
8. *N*-acetyl-*N*, *N*, *N*-trimethyl ammonium bromide.
9. Chloroform: isoamyl alcohol (24:1 v/v).
10. Ethanol.

6. Methods

6.1. Extraction of Genomic DNA of MTB

1. Scrape the growth of MTB from an LJ slope into 500 μL of TE buffer.
2. Heat the culture at 80°C for 20 min to kill the cells.
3. Centrifuge the cells (3000g for 15 min) and resuspend in 500 μL of TE buffer.
4. Add lysozyme to a final concentration of 1 mg/mL, and incubate the tubes at 37°C for 1 h.
5. Following incubation, add 70 μL of 10% SDS and 6 μL of proteinase K and incubate the mixture at 65°C for 10 min.
6. Add 80 μL of *N*-acetyl-*N,N,N*-trimethyl ammonium bromide. Vortex the tubes briefly and incubate for 10 min at 65°C.
7. Extract DNA by adding an equal volume of chloroform-isoamyl alcohol and vortex the mixture for 10 s. Centrifuge the mixture for 5 min in a micro-centrifuge.
8. Precipitate DNA by adding 0.6 vol of isopropanol at –20°C and centrifuging for 15 min. Wash the pellet once with 70% ethanol and redissolve the air-dried pellet in 20 μL of 0.1X TE buffer.

6.2. Extraction of DNA from Sputum Specimens for PCR

1. Specimens are digested and decontaminated with *N*-acetyl-L-Cysteine (NALC-2%NaOH); add an equal volume of NALC - NaOH (2% NaOH, 1.45% Na Citrate, 0.5% NALC) solution to the sample.
2. Vortex the mixture and allow to stand at room temperature for 15 min for digestion. (If the specimen is too mucoid to liquefy, then some powdered NALC can be added directly to the sample).
3. Concentrate the samples by centrifugation at 3000g for 15 min.
4. The samples can be stored at this stage at –20°C until they are processed for PCR.
5. OR Centrifuge the concentrated sample, and resuspend the pellet in 100 μL of TE buffer.
6. Lyse the cells using lysozyme (1 mg/mL final conc) and incubation at 37°C for 1 h followed by treatment with proteinase K (10 mg/mL) and SDS (10%) at 65°C for 10 min (*see* **Note 4**). The lysate can be stored at –20°C.

7. Thaw the specimens and then place in a heating block at 80°C for 20 min.
8. Centrifuge the lysates for 2 min to pellet debris and use 5–10 μL for the PCR assay.
9. Alternatively DNA can be extracted from the supernatants using phenol: chloroform extraction followed by ethanol precipitation.

7. Notes

1. The following genus specific oligonucleotides (target 16S rRNA) for PCR and digoxigenin labeled genus and species specific oligonucleotides probes are being used:

   Genus specific primers: 5' TGCACACAGGCCACAAGGGA 3'
   5' GAGTTTGATCCTTGGCTCAGGA 3'
   Genus specific probe: 5' TTTCACGAACAACGCGACCAA 3'
   Species specific probes: 5' ACCACAAGACATGCATCCCG 3'(for MTB)
   5' ACCAGAAGACATGCGTCTTG 3'(*M avium*)

   For PCR a 100 μL total reaction volume contains 2.5 U of *Taq* polymerase, 50 m*M* KCl, 10 m*M* Tris-HCl (pH 8.3), 1.5 m*M* MgCl$_2$, 0.01% gelatin, 100 pmol of each of the two primers and 200 μ*M* of each dNTP (dATP, dCTP, dGTP, dTTP). PCR conditions used are 40 cycles of 93°C for 1 min, 60°C for 2 min and 72°C for 6 min.
2. Amplification parameters for amplification of the *rpoB* gene for PCR-SSCP assays. One of the primers is labeled with ^{32}P-dCTP. The 50 μL reaction volume contain 50 m*M* KCl, 10 m*M* Tris-HCl (pH 8.3), 1.5 m*M* MgCl$_2$, 10% glycerol, 200 μ*M* of each dNTP, 0.5 μ*M* of each primer and 1.25 U of *Taq* polymerase. Cycling parameters used are 40 cycles of denaturation at 94°C for 1 min, annealing at 55°C for 1 min and extension at 72°C for 1 min, followed by a final 10-min extension at 72°C.
3. Amplification parameters for amplification of the *rpoB* gene using the Inno-Lipa kit. The primers used are provided with the kit.
   5' GGTCGGCATGTCGCGGATGG 3' biotinylated at 5' end
   5' GCACGTCGCGGACCTCCAGC 3' biotinylated at 5' end
   Amplification conditions used are denaturation at 95°C for 5 min followed by 30 cycles of 95°C for 1 min, 55°C for 1 min, and 72°C for 1 min and a final extension at 72°C for 10 min.
4. While using the commercial kits for PCR assays on clinical samples, the lysis buffer provided with the kit is used to lyse the cells.

References

1. Dolin, P. J., Raviglione, M. C., and Kochi, A. (1994) Global tuberculosis incidence and mortality during 1990–2000. *Bull. World Health Organ.* **72**, 213–220.
2. Barnes, P. F., Bloch, A. B., Davidson, P. T., and Snider, D. E., Jr. (1991) Tuberculosis in patients with human immunodeficiency virus infection. *N. Engl. J. Med.* **324**, 1644–1650.
3. Harries, A. D. (1990) Tuberculosis and human immunodeficiency virus infection in developing countries. *Lancet* **335**, 387–390.
4. Huebner, R. E., Schein, M. F., and Bass, J. B., Jr. (1993) The tuberculin skin test. *Clin. Infect. Dis.* **17**, 968–975.

5. Horsburgh, D. R. (1991) *Mycobacterium avium* complex infection in the acquired immunodeficiency syndrome. *N. Engl. J. Med.* **324,** 1332–1338.

6. Middlebrook, G., Reggiardo, Z., and Tigertt, W. D. (1977) Automatable radiometric detection of growth of *Mycobacterium tuberculosis* in selective media. *Am. Rev. Respir. Dis.* **115,** 1066–1069.

7. Palaci, M., Ueki, Y. M., Soto. D. N., Telles, M. A., da Silva, Curcio, M., and Silva, E. A. M. (1996) Evaluation of Mycobacteria growth indicator tube for recovery and drug susceptibility testing of *Mycobacterium tuberculosis* isolates from respiratory specimens. *J. Clin. Microbiol.* **34,** 762–764.

8. Hanna, B. A., Walters, S. B., Kodsi, S. E., Stitt, D. T., Tierno, P. M., and Tick, L. J. (1994) Detection of *Mycobacterium tuberculosis* directly from patients specimens with the Mycobacteria Growth indicator tube: a new rapid method in "Program and abstracts of the 94th General Meeting of the American Society for Microbiology" (abst. C-112) American Society for Microbiology, D. C., p. 510.

9. Butler, W. R., and Kilburn, J. O. (1988) Identification of major slowly growing pathogenic mycobacteria and *Mycobacterium gordonae* by high performance liquid chromatography of their mycolic acids. *J. Clin. Microbiol.* **26,** 50–53.

10. Jost, K. C., Jr., Dunbar, D. F., Barth, S. S., Headley, V. L., and Elliott, L. B. (1995) Identification of *Mycobacterium tuberculosis* and *M. avium* complex directly from smear positive sputum specimens and BACTEC 12B cultures by high performance liquid chromatography with fluorescence detection and computer driven recognition models. *J. Clin. Microbiol.* **33,** 1270–1277.

11. French, G. L., Chan, C. Y., Poon, D., Cheung, S. W., and Cheng, A. F. (1990) Rapid diagnosis of bacterial meningitis by the detection of a fatty acid marker in CSF with gas chromatography mass spectrometry and selected ion monitoring. *J. Med. Microbiol.* **31,** 21–26.

12. Butler, W. R., Jost, K. C., Jr., and Kilburn, J. O. (1991) Identification of *mycobacteria* by high performance liquid chromatography. *J. Clin. Micrbiol.* **29,** 2468–2472.

13. French, G. L., Teoh, R., Chan, C. Y., Humphries, M. J., Cheung, S. W., and O'Mahony, G. (1987) Diagnosis of tuberculous meningitis by detection of tuberculostearic acid in cerebrospinal fluid. *Lancet* **2,** 117–119.

14. Larsson, L., Mardh, P. A., Odham, G., and Westerdahl, G. (1980) Detection of tuberculostearic acid in biological specimens by means of glass capillary gas-chromatography-electron and chemical ionization mass spectrometry, utilizing selected ion monitoring. *J. Chromaogr.* **163,** 221–224.

15. Drake, T. A., Hindler, J. A., Berlin, O. G. W., and Bruckner, D. A. (1987) Rapid identification of *Mycobacterium avium* complex in culture using DNA probes. *J. Clin. Mirobiol.* **25,** 1442–1445.

16. Lim, S. D., Todd, J., Lopez, J., Ford, E., and Janda, M. (1991) Genotypic identification of pathogenic *Mycobacterium* species by using a non radioactive oligonucleotide probe. *J. Clin. Microbiol.* **29,** 1276–1278.

17. Goto, M., Oka, S., Okuzumi, K., Kimura. S., and Shimada, K. (1991) Evaluation of acridinium ester labelled DNA probes for identification of *Mycobacterium*

tuberculosis and *Mycobacterium intracellulare* complex in culture. *J. Clin. Microbiol.* **29**, 2473–2476.

18. Evans, K. D., Nakasone, A. S., Sutherland, P. A., de la Maza, L. M., and Peterson, E. M. (1992) Identification of *Mycobacterium tuberculosis* and *Mycobacterium avium- M. intracellulare* directly from primary BACTEC cultures by using acridinium ester labeled DNA probes. *J. Clin. Microbiol.* **30**, 2427–2431.

19. Ureda, M. S. (1991) Controlled synthetic oligonucleotide networks for the detection of pathogenic organisms, in *Rapid methods and automation in microbiology and immunology* (Vaheri, A., Tilton, R. C., and Balows, A., eds.), Springer-Verlag, Berlin, pp. 1–5.

20. Liu, J., Robinson, L., Buxton, D., et al. (1994) The use of Q beta signal amplification technology for the rapid detection of *Mycobacterium tuberculosis* complex directly from smear negative sputum specimens in *Abstracts of the 94th general meeting of the American Society for Microbiologists* (Las Vegas), (abst U-98) American Society for Microbiology, Washington, DC.

21. An, Q., Buxton, D., Hendricks, A., Robinson, L., Shah, J., Lu, L., Vera Gracia, M., King, W., and Olive, D. M. (1995) Comparison of amplified Q beta replicase and PCR assays for detection of *Mycobacterium tuberculosis*. *J. Clin. Microbiol.* **33**, 860–867.

22. Eisenach, K. D., Cave, M. D., Bates, J. H., and Crawford, J. T. (1990) Polymerase chain amplification of a repetitive DNA sequence for *Mycobacterium tuberculosis*. *J. Inf. Dis.* **161**, 977–981.

23. Walker, G. T., Nadeau, J. G., Spears, P. A., Schram, J. L., Nycz, C. M., and Shank, D. D. (1994) Multiplex strand displacement amplification (SDA) and detection of DNA sequences from *Mycobacterium tuberculosis* and other *mycobacteria*. *Nucleic Acids Res.* **22**, 2670–2677.

24. Leckie, G., Cao, J. ,Davis, A., Facey, I., Lin, B. C., and Lee, H. (1994) Ligase chain reaction (LCR) DNA amplification for direct detection of *Mycobacterium tuberculosis* in clinical specimens in *Abstracts of the 94th general meeting of the American Society for Microbiologists* (Las Vegas) (abst U-96). American Society for Microbiology, Washington, DC.

25. Anderson, A. B., Thybo, S., Godfrey-Faussett, P., and Stoker, N. G. (1993) Polymerase chain reaction for detection of *Mycobacterium tuberculosis* in sputum. *Eur. J. Clin. Microbiol. Infect. Dis.* **12**, 922–927.

26. Hermans, P. W. M., Schuitema, A. R. J., van Sooloingen, D., Verstynen, C. P. H. J., Bik, E. M., Thole, J. E. R., Kolk, A. H. J., and van Embden, J. D. A. (1990) Specific detection of *Mycobacterium tuberculosis* complex strains by polymerase chain reaction. *J. Clin. Microbiol.* **28**, 1204–1213.

27. Boddinghaus, B., Rogall, T., Flohr, T., Blocker, H., and Bottger, E. C. (1990) Detection and identification of *mycobacteria* by amplification of rRNA. *J. Clin. Microbiol.* **28**, 1751–1759.

28. Brisson-Noel, A., Gicquel, B., Lecossier, D., Levy-Frebault, V., Nassif, X., and Hance, A. J. (1989) Rapid diagnosis of tuberculosis by amplification of mycobacterial DNA in clinical samples. *Lancet* **2**, 1069–1071.

29. Sjobring, U., Meclenburg, M., Anderson, A. B., and Miorner, H. (1990) Polymerase chain reaction for the detection of *Mycobacterium tuberculosis*. *J. Clin. Microbiol.* **28,** 2200–2204.
30. Kaneko, K., Onodera, O., Miyatake, T., and Tsuji, S. (1990) Rapid diagnosis of tuberculous meningitis by polymerase chain reaction (PCR). *Neurology* **40,** 1617, 1618.
31. Cousins, D. V., Wilton, S. D., Francis, B. R., and Gow, B. L. (1992) Use of polymerase chain reaction for rapid diagnosis of tuberculosis. *J. Clin. Microbiol.* **30,** 255–258.
32. Hermans, P. W. M., van Soolingen, D., Dale, J. W., Schuitema, A. R. J., McAdam, R. A., Catty, D., and van Embden, J. D. A. (1990) Insertion element IS986 from *Mycobacterium tuberculosis*: a useful tool for diagnosis and epidemiology of tuberculosis. *J. Clin. Microbiol.* **28,** 2051–2058.
33. De Wit, O., Steyn, L., Shoemaker, S., and Sogin, M. (1990) Direct detection of *Mycobacterium tuberculosis* in clinical specimens by DNA amplification. *J. Clin. Microbiol.* **28,** 2437–2441.
34. Ralphs, N. T., Garrett, S., Morse, R., Cookson, J. B., Andrew, P. W., and Boulnois, G. J. (1991) A DNA primer probe system for the rapid and sensitive detection of *Mycobacterium tuberculosis* complex pathogens. *J. Appl. Bacteriol.* **70,** 221–226.
35. De Wit, D., Wootton, M., Allan, B., and Steyn, L. (1993) Simple method for production of internal control DNA for *Mycobacterium tuberculosis* polymerase chain reaction. *Am. Rev. Respir. Dis.* **144,** 1160–1163.
36. Hermans, P. W. M., van Soolingen, D., Bik, E. M., de Haas, P. E. W., Dale, J. W., and van Embden, J. D. A. (1991) The insertion element IS987 from *Mycobacterium bovis* BCG is located in a hot-spot integration region for insertion elements in *Mycobacterium* complex strains. *Infect. Immun.* **59,** 2695–2705.
37. Thierry, D., Cave, M. D., Eisenach, K. D., Crawford, J. T., Bates, J. H., Gicquel, B., and Guesdon, J. L. (1990) IS*6110,* an IS-like element of *Mycobacterium tuberculosis* complex. *Nucleic Acids Res.* **18,** 188.
38. Kolk, A. H. J., Schuitema, A. R. J., Kuijper, S., van Leeuwen, J., Hermans, P. W. M., van Embden, J. D. A., and Hartskeel, R. A. (1992) Detection of *Mycobacterium tuberculosis* in clinical samples by using polymerase chain reaction and a nonradioactive detection system. *J. Clin. Microbiol.* **30,** 2567–2575.
39. Eisenach, K. D., Sifford, M. D., Cave, M. D., Bates, J. H., and Crawford, J. T. (1991) Detection of *Mycobacterium tuberculosis* in sputum samples using a polymerase chain reaction. *Am. Rev. Respir. Dis.* **144,** 1160–1163.
40. Nolte, F. S., Metchock, B., McGowen, J. E., Jr., Edwards, A., Okwumabna, O., Thurmond, C., Mitchell, P. S., Plikaytis, B., and Shinnick, T. (1993) Direct detection of *Mycobacterium tuberculosis* in sputum by polymerase chain reaction and DNA hybridization. *J. Clin. Microbiol.* **31,** 1777–1782.
41. Shawar, R. M., El-Zaatari, F. A. K., Nataraj, A., and Clarridge, J. E. (1993) Detection of *Mycobacterium tuberculosis* in clinical samples by two step polymerase chain reaction and nonisotopic hybridization methods. *J. Clin. Microbiol.* **31,** 61–65.

42. Clarridge, J. E. R., Shawar, R. M., Shinnick, T. M., and Plikaytis, B. B. (1993) Large scale use of polymerase chain reaction for detection of *Mycobacterium tuberculosis* in a routine mycobacteriology laboratory. *J. Clin. Microbiol.* **31**, 2049–2056.

43. Walker, D. A., Taylor, I. K., Mitchell, D. M., and Shaw, R. (1992) Comparison of polymerase chain reaction amplification of two mycobacterial DNA sequences, IS6110 and the 65kDa antigen gene, in the diagnosis of tuberculosis. *Thorax* **47**, 690–694.

44. Yuen, L. K., Ross, B. C., Jackson, K. M., and Dwyer, B. (1993) Characterization of *Mycobacterium tuberculosis* strains from Vietnamese patients by Southern blot hybridization. *J. Clin. Microbiol.* **31**, 1615–1618.

45. Schulger, N. W., Kinney, D., Harkin T. J., and Rom, W. N. (1994) Clinical utility of the polymerase chain reaction in the diagnosis of infections due to *Mycobacterium tuberculosis*. *Chest* **105**, 1116–1121.

46. Beige, J., Lokies, J., Schaberg, T., Finckh, U., Fischer, M., Mauch, H., Lode, H., Kohler, B., and Rolfs, A. (1995) Clinical evaluation of *Mycobacterium tuberculosis* assay. *J. Clin. Microbiol.* **33**, 90–95.

47. Noordhoek, G. T., Kolk, A. H. J., Bjune, G., Catty, D., Dale, J. W., Fine, P. E., Godfrey Fausset, P., Cho, S. N., Shinnick, T., and Sevenson, S. V. (1994) Sensitivity and specificity of PCR for detection of *Mycobacterium tuberculosis*: a blind comparison study among seven laboratories. *J. Clin. Microbiol.* **32**, 277–284.

48. Savic, B., Sjobring, U., Alugupalli, S., Larrson, L., and Miorner, H. (1992) Evaluation of polymerase chain reaction, tuberculostearic acid analysis, and direct microscopy for the detection of *Mycobacterium tuberculosis* in sputum. *J. Inf. Dis.* **166**, 1177–1180.

49. Plikaytis, B. B., Eisenach, K. D. ,Crawford, J. T., and Shinnick, T. M. (1991) Differentiation of *Mycobacterium tuberculosis* and *Mycobacterium bovis* BCG by a polymerase chain reaction assay. *Mol. Cell. Probes.* **5**, 215–219.

50. Sritharan, V. and Barker, R. (1991) A simple method for diagnosing *M. tuberculosis* infection in clinical samples using PCR. *Mol. Cell. Probes.* **5**, 385–395.

51. Suzuki, Y., Nagata, A., Ono, Y., and Yamada, T. (1988) Complete sequence of the 16S rRNA gene of *Mycobacterium bovis* BCG. *J. Bacteriol.* **170**, 2886–2889.

52. Kolk, A. H. J., Noordhoek, G. T., DeLeeuw, O., Kuijper, S., and van Embden, J. D. A. (1994) *Mycobacterium smegmatis* strain for detection of *Mycobacterium tuberculosis* by PCR used as internal control for inhibition of amplification by PCR used as internal control for inhibition of amplification and for quantification of bacteria. *J. Clin. Microbiol.* **32**, 1354–1356.

53. Kox, L. F. F., Rhientong, D., Medo Miranda, A., Udomsantisuk, N., Ellis, K., van Leeuwen, J., van Heusden, S., Kuijper, S., and Kolk, A. H. J. (1994) A more reliable PCR for detection of *Mycobacterium tuberculosis* in clinical samples. *J. Clin. Microbiol.* **32**, 672–678.

54. Victor, T., du Toit, R., and van Helden, P. D. (1992) Purification of sputum samples through sucrose improves detection of *Mycobacterium tuberculosis* by polymerase chain reaction. *J. Clin. Microbiol.* **30**, 1514–1517.

55. Buck, G. E., O'Hara, L. C., and Summergill, J. T. (1992) Rapid, simple method for treating clinical specimens containing *Mycobacterium tuberculosis* to remove DNA for polymerase chain reaction. *J. Clin. Microbiol.* **30**, 1331–1334.

56. Carpentier, E., Drouillard, B., Dailloux, M., Moingard, D., Vallee, E., Dutilh, B., Maugein, J., Bergogne-Berezin, E., and Carbonnelle, B. (1995) Diagnosis of tuberculosis by amplicor *Mycobacterium tuberculosis* test: a multicenter study. *J. Clin. Microbiol.* **33,** 3106–3110.

57. Dilworth, J. P., Goyal, M., Young, D. B., and Shaw, R. J. (1996) Comparison of polymerase chain reaction for IS6110 and Amplicor in the diagnosis of tuberculosis. *Thorax.* **51,** 320–322.

58. Vlaspolder, F., Singer, P., and Roggeveen, C. (1995) Diagnostic value of an amplification method (Gen-Probe) compared with that of culture for diagnosis of tuberculosis. *J. Clin. Microbiol.* **33,** 2699–2703.

59. Vuorinen, P., Miettinen, A., Vuento, R., and Hallstrom, O. (1995) Direct detection of *Mycobacterium tuberculosis* complex in respiratory specimens by Gen-Probe amplified *Mycobacterium tuberculosis* direct test and Roche Amplicor *Mycobacterium tuberculosis* test. *J. Clin. Microbiol.* **33,** 1856–1859.

60. Goto, K., Okuzumi, K., Sakai, Y., Takewaki, S., Tachikawa, N., Iwamoto, A., Kimura, S., and Shimada, K. (1995) Comparison of amplified *Mycobacterium tuberculosis* direct test (MTD), amplicor mycobacteria kit (Amplicor) and PCR for detection of *Mycobacterium tuberculosis* in clinical specimens. *Kansenshogaku. Zasshi.* **69,** 539–545.

61. Rogall, T., Flohr, T., and Bottger, E. C. (1990) Differentiation of *Mycobacterium* species by direct sequencing of amplified DNA. *J. Gen. Microbiol.* **136,** 1915–1920.

62. Kirschner, P., Rosenau, J., Springer, B., Teschner, K., Feldmann, K., and Bottger, E. C. (1996) Diagnosis of mycobacterial infections by nucleic acid amplification: 18 month prospective study. *J. Clin. Microbiol.* **34,** 304–312.

63. Bloch, A. B., Cauthen, G. M., Onorato, I. M., Dansbury, K. G., Kelly, G. D., Driver, C. R., and Snider, D. E., Jr. (1994) Nationwide survey of drug resistant tuberculosis in the United States. *JAMA.* **271,** 665–671.

64. Friden, T. R., Sterling, T., Pablos-Mendez, A., Kilburn, J. O., Cauthen, G. M., and Dooley, S. W. (1993) The emergence of drug resistant tuberculosis in New York city. *N. Engl. J. Med.* **328,** 521–526.

65. Musser, J. M. (1995) Antimicrobial agent resistance in mycobacteria: molecular genetic insights. *Clin. Microbiol. Rev.* **8,** 496–514.

66. Fischl, M. A., Uttamchandani, R. B., Daikos, G. L., Poblete, R. B., Moreno, J. N., Reyes, R. R., Boota, L. M., Thompson, L. M., Cleary, T. J., and Lai, S. (1992) An outbreak of tuberculosis caused by multiple drug resistant tubercle bacilli among patients with HIV infection. *Ann. Intern. Med.* **117,** 177–183.

67. Honore, N. and Cole, S. T. (1993) Molecular basis of rifampin resistance in *Mycobacterium leprae*. Antimicrob. Agents. *Chemother.* **37,** 414–418.

68. Miller, L. P., Crawford, J. T., and Shinnick, T. M. (1994) The rpoB gene of *Mycobacterium tuberculosis*. Antimicrb. Agents. *Chemother.* **38,** 313–319.

69. Kapur, V., Li, L. L., Iordanescu, S., Hammick, M. R., Wanger, A., Krieswirth, B. N., and Musser, J. M. (1994) Characterization by automated DNA sequencing of mutations in the gene (*rpoB*) encoding the RNA polymerase B subunit in rifampin resistant *Mycobacterium tuberculosis* strains from New York City and Texas. *J. Clin. Microbiol.* **32,** 1095–1098.

70. Telenti, A., Imboden, P., Marchesi, F., Lowrie, D., Cole, S., Colston, M. J., Matter, L., Schopfer, K., and Bodner, T. (1993) Detection of rifampicin resistance mutations in *Mycobacterium tuberculosis*. *Lancet* **341**, 805–811.

71. Middlebrook, G. (1954) Isoniazid-resistance and catalase activity of tubercle bacilli. *Am. Rev. Tuberculosis* **69**, 471–472.

72. Hedgecock, L. and Faucher, I. O. (1957) Relation of pyrogallol-peroxidase activity to isoniazid resistance in *Mycobacterium tuberculosis*. *Am. Rev. Tuberc.* **75**, 670–674.

73. Zhang, Y., Heym, B., Allen, B., Young, D., and Cole, S. (1992) The catalase-peroxidase gene and isoniazid resistance to *Mycobacterium tuberculosis*. *Nature* **358**, 591–593.

74. Heym, B., Zhang, Y., Poulet, S., Young, D., and Cole, S. T. (1993) Characterization of the katG gene encoding a catalase peroxidase required for the Isoniazid susceptibility of *Mycobacterium tuberculosis*. *J. Bacteriol.* **175**, 4255–4259.

75. Zhang, Y. and Young, D. (1993) Molecular mechanisms of isoniazid: a drug at the front line of tuberculosis control. *Trends Microbiol.* **1**, 109–113.

76. Heym, B., Alzari, P. M., Honore, N., and Cole, S. T. (1995) Missense mutations in the catalase peroxidase gene, katG, are associated with isoniazid resistance in *Mycobacterium tuberculosis*. *Mol. Microbiol.* **15**, 235–245.

77. Altamirano, M., Marostenmaki, J., Wong, A., FitzGerald, M., Black,W. A., and Smith, J. A. (1994) Mutations in the catalase-peroxidase gene from isoniazid-resistant *Mycobacterium tuberculosis* isolates. *J. Infect. Dis.* **169**, 1162–1165.

78. Banerjee, A., Dubnau, E., Quemard, A., Balasubramanian, V., Um, K. S., Wilson, T., Collins, D., deLisle, G., and Jacobs, W. R., Jr. (1994) inhA, a gene encoding a target for isoniazid and ethionamide in *Mycobacterium tuberculosis*. *Science* **263**, 227–230.

79. Musser, J. M., Kapur, V., Williams, D. L., Kreisworth, B. N., van Sooloingen, D., and van Embden, J. D. A. (1996) Characteisation of the catalase peroxidase gene (katG) and inhA locus in isoniazid resistant and susceptible strains of *Mycobacterium tuberculosis* by automated DNA sequencing: restricted array of mutations associated with drug resistance. *J. Infect. Dis.* **173**, 196–202.

80. Heym, B., Honore, N., Truffot-Pernot, C., Banerjee, A., Schurra, C., Jacobs, W. R., Jr., van Embden, J. D. A., Grosset, J. H., and Cole, S. T. (1994) Implications of multidrug resistance for the future of short course chemotherapy of tuberculosis: a molecular study. *Lancet* **30**, 277–279.

81. Wilson, T. M. and Collins, D. M. (1996) ahpC, a gene involved in isoniazid resistance of the *Mycobacterium tuberculosis* complex. *Mol. Microbiol.* **19**, 1025–1034.

82. Honore, N. and Cole, S. T. (1994) Streptomycin resistance in mycobacteria. *Antimicrob. Agents. Chemother.* **38**, 238–242.

83. Nair, J., Rouse, D. A., Bai, G. H., and Morris, S. L. (1993) The rpsL gene and streptomycin resistance in single and multiple drug-resistant strains of *Mycobacterium tuberculosis*. *Mol. Microbiol.* **10**, 521–527.

84. Finken, M., Kirschner, P., Meier, A., Wrede, A., and Bottger, E. C. (1993) Molecular basis of streptomycin resistance in *Mycobacterium tuberculosis*: alterations

of the ribosomal protein S12 gene and point mutations within a functional 16S ribosomal RNA pseudoknot. *Mol. Microbiol.* **9,** 1239–1246.

85. Takiff, H. E., Guerrero, C., Phillip, W., et al. (1993) Mutations in the gyrase A gene (gyrA) of *Mycobacterium tuberculosis* confer resistance to fluoroquinolones in "Program and abstracts of the 33rd international conference on Antimicrobial Agents and Chemotherapy" (New Orleans) abstr 1092. American Society for Microbilogy Washington. DC.

86. Takiff, H. E., Cimino, M., Musso, M. C., Weisbrod, T., Martinez, R., Delgado, M. B., Salazar, L., Bloom, B. R., and Jacobs, W. R., Jr. (1996) Eff*lux* pump of the protein antiporter family confers low level fluoroquinolone resistance in *Mycobacterium smegmatis. Proc. Natl. Acad. Sci. USA* **93,** 362–366.

87. Kawa, D. E., Pennell, D. R., Kubista, L. N., and Schell, R. F. (1989) Development of a rapid method for determining the susceptibility of *Mycobacterium tuberculosis* to isoniazid using the Gen-Probe DNA hybridization system. *Antimicrob. Agents Chemother.* **33,** 1000–1005.

88. Jacobs, W. R., Jr., Barletta, R. G., Udani, R., Chan, J., Kalkut, G., Sosne, G., Keiser, T., Sarkis, G. J., Hatfull, G. A., and Bloom, B. R. (1993) Rapid assesment of drug susceptibilities of *Mycobacterium tuberculosis* by means of luciferase reporter phages. *Science* **260,** 819–822.

89. Plikaytis, B. B., Marden, J. L., Crawford, J. T., Woodley, C. L., Butler, W. R., and Shinnick, T. M. (1994) Multiplex PCR assay specific for the multidrug resistant strain W of *Mycobacterium tuberculosis. J. Clin. Microbiol.* **32,** 1542–1546.

90. Nolan, C. M., Elarth, A. M., Barr, H., Saeed, A. M., and Risser, D. R., (1991) An outbreak of tuberculosis in a shelter for homeless men: a description of its evolution and control. *Am. Rev. Respir. Dis.* **143,** 257–261.

91. Reves, R., Blakey, D., Snider, D. E., Jr., and Farer, L. S. (1981) Transmission of multiple drug-resistant tuberculosis: report of a school and community outbreak. *Am. J. Epidemiol.* **113,** 423–435.

92. Snider, D. E., Jones, W. D., and Good, R. C. (1994) The usefulness of phage typing *Mycobacterium tuberculosis* isolates. *Am. Rev. Respir. Dis.* **130,** 1095–1099.

93. Millership, S. E. and Want, S. (1992) Whole cell protein electrophoresis for typing *Mycobacterium tuberculosis. J. Clin. Microbiol.* **30,** 2784–2787.

94. Zainuddin, Z. F. and Dale, J. W. (1990) Does *Mycobacterium tuberculosis* have plasmids? *Tubercle* **71,** 43–49.

95. Hoffner, S. E., Svenson, S. A., Norberg, R., Dias, F., Ghebremichael, S., and Kallenius, G. (1993) Biochemical heterogenity of *Mycobacterium tuberculosis* complex isolates in Guinea-Bissau. *J. Clin. Microbiol.* **31,** 2215–2217.

96. Wasem, C. F., McCarthy, C. M., and Murray, L. W. (1991) Multilocus enzyme electrophoresis analysis of the *Mycobacterium avium* complex and other mycobacteria. *J. Clin. Microbiol.* **29,** 264–271.

97. Imeda, T. (1985) DNA relatedness among selected strains of *Mycobacterium tuberculosis, Mycobacterium bovis* BCG, *Mycobacterium microti* and *Mycobacterium africanum. Int. J. Syst. Bacteriol.* **35,** 147–150.

98. Collins, D. M. and de Lisle, G. W. (1985) DNA restriction endonuclease analysis of *Mycobacterium bovis* and other members of the tuberculosis complex. *J. Clin. Microbiol.* **21**, 562–564.

99. Eisenach, K. D., Crawford, J. T., and Bates, J. H. (1986) Genetic relatedness among strains of the *Mycobacterium tuberculosis* complex. *Am. Rev. Respir. Dis.* **133**, 1065–1068.

100. Small, P. M. and van Embden, J. D. A. (1994) Molecular epidemiology of tuberculosis in "Tuberculosis, pathogenesis, protection and control" (Bloom, B. R., ed.), ASM, Washington, D. C., p. 569–582.

101. Groenen, P. M. A., van Bunchoten, A. E., van Soolingen, D., and van Embden, J. D. A. (1993) Nature of DNA polymorphism in the direct repeat cluster of *Mycobacterium tuberculosis*: application for strain differentiation by a novel method. *Mol. Microbiol.* **105**, 1057–1065.

102. Hermans, P. W., van Soolingen, D., and van Embden, J. D. (1992) Characterization of a major polymorphic tandem repeat in *Mycobacterium kansasii* and *Mycobacterium gordonae*. *J. Bacteriol.* **174**, 4157–4165.

103. Ross, C., Raios, K., Jackson, K., and Dwyer, B. (1992) Molecular cloning of a highly repeated element from *Mycobacterium tuberculosis* and its use as an epidemiological tool. *J. Clin. Microbiol.* **30**, 942–946.

104. Wiid, I. J. F., Werely, C., Beyers, N., Donald, P., and Helden, P. D. (1994) Oligonucleotide (GTG)5 as a marker for strain identification in *Mycobacterium tuberculosis*. *J. Clin. Microbiol.* **32**, 1318–1321.

105. van Embden, J. D., Cave, M. D., Crawford, J. T., Dale, J. W., Eisenach, K. D., Gicquel, B., Herman, P., Martin, C., McAdam, R., Shinnick, T. M., and Small, P. M. (1993) Strain identification of *Mycobacterium tuberculosis* by DNA fingerprinting: recommendations for a standardized methodology. *J. Clin. Microbiol.* **31**, 406–409.

106. Cave, M. D., Eisenach, K. D., McDermott, P. F., Bates, J. H., and Crawford, J. T. (1991) IS6110: conservation of sequence in the *Mycobacterium tuberculosis* complex and its utilization in DNA fingerprinting. *Mol. Cell. Probes.* **5**, 73–80.

107. Otal, I., Martin, C., Vincent-Levy-Frebault, V., Thierry, D., and Gicquel, B. (1991) Restriction fragment length polymorphism analysis using IS*6110* as an epidemiological marker in tuberculosis. *J. Clin. Microbiol.* **29**, 1252–1254.

108. van Soolingen, D., Hermans, P. W. M., de Haas, P. E. W., Soll, D. R., and van Embden, J. D. A. (1991) Occurence and stability of insertion sequences in *Mycobacterium tuberculosis* complex strains: evaluation of an insertion sequence dependent DNA polymorphism as a tool in the epidemiology of tuberculosis. *J. Clin. Microbiol.* **29**, 2578–2586.

109. Godfrey-Faussett, P., Stoker, N. G., Scott, J. A. G., Pasvol. G., Kelly, P., and Clancy, L. (1993) DNA fingerprints of *Mycobacterium tuberculosis* do not change during the development of rifampicin resistance. *Tubercle. Lung. Dis.* **74**, 240–243.

110. Daley, C. L., Small, P. M., Schecter, G. F., Schoolnik, G. K., McAdam, R. A., Jacobs, W. R., Jr., and Hopewell, P. C. (1992) An outbreak of tuberculosis with accelerated progression among persons infected with the human immunodefi-

ciency virus: an analysis using restriction fragment length polymorphism. *N. Engl. J. Med.* **326,** 231–235.

111. Dwyer, B., Jackson, K., Raios, K., Sievers, A., Wilshire, E., and Ross, B. (1993) DNA restriction fragment analysis to define an extended cluster of tuberculosis in homeless men and their associates. *J. Infect. Dis.* **167,** 490–494.

112. Edlin, B. R., Tokars, J. I., Grieco, M. H., Crawford, J. T., Williams, J., Sordillo, E. M., Ong, K. R., Kilburn, J. O., Dooley, S. W., Castro, K. G., Jarvis, W. R., and Holmberg, S. D. (1992) An outbreak of multidrug resistant tuberculosis among hospitalized patients with the acquired immunodeficiency syndrome. *N. Eng. J. Med.* **326,** 1514–1521.

113. Goyal, M., Ormerod, L. P., and Shaw, R. J. (1994) Epidemiology of an outbreak of drug resistant tuberculosis in the UK using restriction fragment length polymorphism. *Clin. Sci.,* **86,** 749–751.

114. Small, P. M., McClenny, N. B., Singh, S. P., Schoolink, G. K., Tompkins, L. S., and Mickelsen, P. A. (1993) Molecular strain typing of *Mycobacterium tuberculosis* to confirm cross-contamination in the mycobacteriology laboratory and modification of procedures to minimise occurence of false positive cultures. *J. Clin. Microbiol.* **32,** 1677–1682.

115. Small, P. M., Shafer, R. W., Hopewell, P. C., Singh, S. P., Murphy, M. I., Desmond, E., Sierra, M. F., and Schoolnick, G. K. (1993) Exogenous reinfection with multidrug resistant *Mycobacterium tuberculosis* in patients with advanced HIV infection. *N. Eng. J. Med.* **328,** 1137–1144.

116. Yang, Z. H., Mtoni, I., Chonde, M., Mwasekaga, M., Fuursted, K., Askgard, D. S., Bennedsen, J., de Haas, P. E. W., Soolingen, D., van Soolingen D., van Embden, J. D. A., and Andersen, A. B. (1995) DNA fingerprinting and phenotyping of *Mycobacterium tuberculosis* isolates from human immunodeficiency virus (HIV)- seropositive and HIV-sernegative patients in Tanzania. *J. Clin. Microbiol.* **33,** 1064–1069.

117. Alland, D., Kalkut, G. E., Moss, A. R., McAdam, R. A., Hahn, J. A., Bosworth, W., Drucker, E., and Bloom, B. R. (1994) Transmission of tuberculosis in New York City: an analysis by DNA fingerprinting and conventional epidemiology. *N. Eng. J. Med.* **330,** 1703–1708.

118. Small, P. M., Hopewell, P. C., Singh, S. P., Antonio, P., Parsonnet, J., Ruston, D. C., Schecter, G. F., Daley, C. L., and Schoolnik, G. K. (1994) The epidemiology of tuberculosis in San Francisco: a population based study using conventional and molecular methods. *N. Engl. J. Med.* **330,** 1703–1709.

119. Genewin, A., Telenti, A., Bernasconi, C., Mordasini, C., Weiss, S., Maurer, A., Reider, H. L., Schopfer, K., and Bodmer, T. (1993) Molecular approach to identify route of transmission of tuber culosis in the community. *Lancet* **342,** 841–844.

120. van Soolingen D., Qian, L., de Haas, P. E. W., Douglas, J. T., Tarore, H., Portaels, F., Qing, H. Z., Enkhasaikan, D., Nymadawa, P., and van Embden, J. D. A. (1995) Predominance of a single genotype of *Mycobacterium tuberculosis* in countries of East Asia. *J. Clin. Microbiol.* **33,** 3234–3238.

121. van Soolingen, D., de Haas, P. E. W., Hermans, P. W. M., Groenen, P. M. A., and van Embden, J. D. A. (1993) Comparison of various repetitive DNA elements as

genetic markers for strain differentiation and epidemiology of *Mycobacterium tuberculosis. J. Clin. Micribiol.* **31,** 1987–1995.

122. Zhang, Y., Mazurek, G. H., Cave, M. D., Eisenach, K. D., Pang, Y., Murphy, D. T., and Wallace, R. J., Jr. (1992) DNA polymorphism in strains of *Mycobacterium tuberculosis* analysed by pulsed field gel electrophoresis: a tool for epidemiology. *J. Clin. Microbiol.* **30,** 1551–1556.

123. Palittapongarnpim, P., Chomyc, S., Fanning, A., and Kunimoto, D. (1993) DNA fingerprinting of *Mycobacterium tuberculosis* isolates by ligation mediated polymerase chain reaction. *Nucleic Acids Res.* **21,** 761–762.

124. Linton, C. J., Jalal, H., Leeming, J. P., and Miller, M. R. (1994) Rapid discrimination of *Mycobacterium tuberculosis* strains by random amplified polymorphic DNA analysis. *J. Clin. Microbiol.* **32,** 2169–2174.

125. Abed, Y., Davin-Regli, A., Bollet, C., and De Micco, P. (1995) Efficient discrimination of *Mycobacterium tuberculosis* strains by 16S - 23S spacer region based random amplified polymorphic DNA analysis. *J. Clin. Microbiol.* **33,** 1418–1420.

126. Haas, W. H., Butler, W. R., Woodley, C. L., and Crawford, J. T. (1993) Mixed linker polymerase chain reaction: a new method for rapid fingerprinting of isolates of the *Mycobacterium tuberculosis* complex. *J. Clin. Microbiol.* **31,** 1293–1298.

127. Ross, B. C. and Dwyer, B. (1993) Rapid, simple method for typing isolates of *Mycobacterium tuberculosis* by using the polymerase chain reaction. *J. Clin. Microbiol.* **31,** 329–334.

128. Pilkaytis, B. B., Crawford, J. T., Woodley, C. L., Butler, W. R., Eisenach, K. D., Cave, M. D., et al. (1993) Rapid amplification based fingerprinting of *Mycobacterium tuberculosis. J. Gen. Microbiol.* **139,** 1537–1542.

129. Goyal, M., Young, D., Zhang, Y., Jenkins, P. A., and Shaw, R. J. (1994) PCR amplification of variable sequence upstream of katG gene to subdivide strains of *Mycobacterium tuberculosis* complex. *J. Clin. Microbiol.* **32,** 3070–3071.

130. Friedman, C. R., Stoeckle, M. Y., Johnson, W. D., Jr., and Riley, L. (1995) Double-repetitive-element PCR method for subtyping *Mycobacterium tuberculosis* clinical isolates. *J. Clin. Microbiol.* **33,** 1383–1384.

131. Goyal, M., van Embden, J. D. A., Young, D. B., and Shaw, R. J. (1995) Evaluation of spoligotyping as an epidemiological test in three different outbreaks of tuberculosis. *Thorax* **50** (Suppl. 2), A35.

132. Tenover, F. C., Crawford, J. T., Huebner, R. E., Geiter, L. J., Horsburgh, C. R., Jr., and Good, R. C. (1993) The resurgence of tuberculosis: is your laboratory ready? *J. Clin. Microbiol.* **31,** 767–770.

133. Bird, B. R., Denniston, M. M., Huebner, R. E., and Good, R. C. (1996) Changing practices in Mycobacteriology: a follow up survey of state and territorial public health laboratories. *J. Clin. Microbiol.* **34,** 554–559.

134. Del Portillo, P., Murillo, A., and Patarroyo, M. E. (1991) Amplification of a species specific DNA fragment of *Mycobacterium tuberculosis* and its possible use in diagnosis. *J. Clin. Microbiol.* **29,** 2163–2168.

135. De Beenhouwer, H., Lhiang, Z., Jannes, G., Mijs, W., Machtelinckx, L., Rossau, R., Traore, H., and Portaels, F. (1995) Rapid detection of rifampicin resistance in

sputum and biopsy specimens from tuberculosis patients by PCR and line probe assay. *Tubercle. Lung Dis.* **76,** 425–430.

136. Goyal, M., Shaw, R. J., Banerjee, D. K., Coker, R. J., Robertson, B. D., and Young, D. B. (1997) Rapid detection of multidrug resistant tuberculosis. *Eur. Respir. J.* **10,** 1120–1124.

137. Kamerbeek, J., Schouls, L., Kolk, A., van Agterveld, M., van Soolingen, D., Kuijper, S., Banschoten, A., Molhuizen, H., Shaw, R., Goyal, M., and van Embden, J. (1997) Simultaneous detection and strain differentiation of Mycobacterium tuberculosis for diagnosis and epiemiology. *J. Clin. Microbiol.* **35,** 907–914.

11

Diagnosis and Epidemiology of Diphtheria

Androulla Efstratiou, Kathryn H. Engler, and Aruni de Zoysa

1. Introduction

Prior to the late 1980s, diphtheria was regarded in many countries as one of "those rare and forgotten" diseases associated with the preimmunization era of the 1940s. Despite the success of many immunization programs, there is still much to be learned about this disease that has made a dramatic "return" in the 1990s, particularly to countries of the former Soviet Union *(1)*. Since 1989, there has been a rapidly expanding epidemic of the disease in these countries, and this has severe implications even for developed countries that have successfully controlled the disease for several decades *(2)*. Diphtheria is also endemic in other countries of the world *(3)*. The increase in international travel, migration from Eastern Europe, and also the emergence of "new strains" of the causative organism, *Corynebacterium diphtheriae,* causing disease have emphasized the importance of both clinical and laboratory awareness. In addition, current immunization programs within each country should be reviewed, particularly for adults, to ensure that population immunity is adequate to prevent the re-emergence of epidemic disease in the Western world.

1.1. History of Diphtheria

Diphtheria is an ancient disease that existed even in the times of Hippocrates, and the first clinical descriptions were made by van Lom in 1560 and Baillou in 1576. However, it was not until 1826 that it was recognized as a clinical entity by Pierre Brettoneau who named the disease after the Greek word "*diphthera,*" meaning "hide or leather," which refers to the consistency of the characteristic membrane produced by the organism in the pharynx or larynx. Loeffler showed that the organism responsible (originally described by Klebs)

From: *Methods in Molecular Medicine, Vol. 15: Molecular Bacteriology: Protocols and Clinical Applications*
Edited by: N. Woodford and A. P. Johnson © Humana Press Inc., Totowa, NJ

was confined to the membrane, and postulated that the organism released a toxin capable of extensive tissue damage.

Roux and Yersin's discovery of the lethal toxin in 1888 was indeed revolutionary in the history of diphtheria. They emphasized the importance of microbiologic diagnosis and clinical management of the patient and their contacts. In 1890, von Behring and Kitasato produced serum antibodies in animals against the toxin and demonstrated that the serum could protect susceptible animals against the disease. In 1901, von Behring won the Nobel prize in medicine for his work on the development of antitoxin. The next major landmark in the history of the disease was the development of toxoid, which occurred in the early 1920s and led to mass immunization in the 1940s and 1950s. The history of diphtheria has demonstrated that mass immunization strategies, microbiologic diagnosis of cases and contacts, effective clinical management, including the isolation of potentially infectious cases, and public health education can all be part of an effective response to both endemic and epidemic disease.

1.2. The Disease and Its Diagnosis

Diphtheria is an infectious disease mainly of the upper respiratory tract classically caused by toxin-producing strains of C. diphtheriae. However, diphtheritic infections of other sites (skin, eyes, ears, gut, urogenital tract, and endocardium) have been described (4). Within the infected area, the toxin diffuses through the mucous membrane and causes necrosis of the mucosal cells resulting in the production of a thick grey pseudomembrane containing fibrin, epithelial cells, bacteria, and polymorphs. The exudate adheres to the fauces, the cervical lymph nodes enlarge, and there is considerable oedema of the neck, a condition classically known as "bullneck." Toxin diffuses into the circulation and binds to various tissues and may cause extensive cardiac and neurological damage in the later stages of the disease. In most cases, the cardiac manifestations occur during the second week of the disease. The more extensive the local lesion and the more delayed the administration of specific treatment, the more frequently myocarditis occurs. The manifestations of neuritis appear later (at 2–6 wk); soft palate paralysis is usually the most common manifestation of diphtheritic neuritis. Case fatality rates have changed little in 50 yr at 5–10%.

The most common form of nonrespiratory diphtheria is cutaneous infection, which is problematic in the Tropics. The lesions are usually chronic and nonhealing skin ulcers, but although the organism may produce toxin, systemic manifestations are uncommon. The ulcer may act as a reservoir for transmission of the respiratory form of the disease, as in the devastating outbreaks that occurred in the United States in the early 1970s (5).

Effective protection against diphtheria is provided by active immunization. However, a full immunization history, and even recent booster immunization does not exclude the diagnosis. Throat cultures, and if possible a portion of the membrane/exudate, should be collected and examined by the microbiology laboratory. The laboratory should always be informed if diphtheria is suspected on clinical and epidemiological grounds. Contacts of the patient and potential carriers should be identified and throat samples should be taken. The Public Health Authorities should be informed, and patients and their contacts should be managed accordingly *(6)*.

Diagnosis is primarily based upon the identification of *C. diphtheriae*, determination of toxigenicity, and epidemiologic characterization of the isolates. In this chapter, the authors describe both conventional and molecular methodologies for the determination of these characteristics. However, it must be emphasized that detection of the gene encoding the toxin does not necessarily imply that the organism is producing the toxin either in vivo or in vitro.

1.3. Treatment

If diphtheria is suspected, then specific treatment with antitoxin and antibiotics should be commenced immediately as microbiologic diagnosis may take 24–48 h or longer to complete. Antitoxin is the main form of treatment; however, antibiotic treatment is necessary to eliminate the organism and reduces the opportunity for further spread. The drugs of choice are usually erythromycin or penicillin. Antibiotics are not a substitute for antitoxin treatment. Guidelines for the management and control of diphtheria are available from the World Health Organization (WHO) *(6)*. The WHO Manual provides recommendations for dosages of antitoxin required to treat the various types of disease.

1.4. Corynebacterium diphtheriae *and Its Toxin*

C. diphtheriae is a Gram-positive pleomorphic rod belonging to the genus *Corynebacterium,* which comprises many species. Most of the species, apart from *C. diphtheriae, C. ulcerans,* and *C. pseudotuberculosis,* are nontoxigenic commensals that may be regarded as opportunistic human pathogens. The potentially toxigenic species acquire this virulence characteristic when infected/lysogenized by the family of β phages or corynephages. The ratio of toxigenic isolates to nontoxigenic isolates appears to have declined whenever mass immunization has been introduced *(7)*. Most *C. diphtheriae* are nontoxigenic in Western Europe *(8)*.

Diphtheria toxin (DT) is a protein with a molecular weight of 58,350 and is released extracellularly into the supernatant during the growth of the organism in vitro. It is synthesized as a single polypeptide chain with two different functional moieties, fragments A and B, which are linked by a disulfide bond. The A chain contains the enzymatic site and the B chain contains the translocation and binding sites of the toxin. After the B chain binds a protein on the host cell surface, the toxin is taken up in an endocytic vesicle. Conformational changes in both the A and B chains, caused by a decreasing pH in the vesicle, permit translocation of the A chain into the host cell cytoplasm. The disulfide bond linking the A and B chains is reduced, and the A chain is released into the cytoplasm. The A chain ADP-ribosylates Elongation Factor2 (EF-2) at a modified histidine residue, thus halting protein synthesis in the host cell. A single A molecule can kill a host cell. The host cell receptor for DT is a protein, histidine binding-elongation factor (HB-EGF) precursor. Heart and nerve cells have more toxin receptors than other cell types; hence, cardiac failure and neurological damage occur in severe forms of the disease. The toxin gene is regulated by an iron-dependent repressor protein (DtxR). The iron-bound form of DtxR binds to the DNA, probably as a dimer, at sites that overlap the -10 region of the promoter and the transcription start site, thus inhibiting transcription *(9)*. Strategies for the treatment and prevention of other potentially fatal diseases have recently focused on the use of DT as it has the potential to be used to kill tumor cells or HIV-infected cells. The strategy is to link the A chain with a molecule that will bind to receptors specific for the targeted cell type. This opens up new approaches to the treatment of cancers, such as melanomas and lymphomas *(10)*.

1.5. Brief Overview of the Epidemiology of Diphtheria, 1940s to the 1990s

During the last 50 yr the diphtheria incidence rate has dramatically declined within Western Europe, the United States, and Canada. However, since 1989, there has been an equally dramatic increase of the disease within countries of the former Soviet Union *(1,2)*. The outbreak spread to the Ukraine in 1991 and to the other 13 Newly Independent States (NIS) in 1993–1994. Since 1989, diphtheria has also re-emerged in other areas of the world *(3,7)* with frightening outcomes. The disease has been designated a "public health emergency" by the World Health Organization (WHO) *(11)*.

The most notable outbreak since the introduction of mass immunization in the 1940s is still ongoing within the former Soviet Union, with more than 50,000 cases being reported from the WHO European Region in 1995 alone, 35,652 cases were from the Russian Federation, 5280 cases from the Ukraine, and the remainder from other countries of the former Soviet Union (particularly Kazakhstan and Tajikistan); there were also a few cases from other countries of Western Europe (**Fig. 1**) *(11)*. Some of these were imported from the former Soviet Union *(12)*.

In the first 6 mo of 1996, reported diphtheria cases in the NIS decreased by 54% to 11,313 cases from the 24,637 cases reported in the first 6 mo of 1995. There is, therefore, evidence of progress toward control of the epidemic in the NIS in 1996. However, WHO still considers the epidemic to be an international

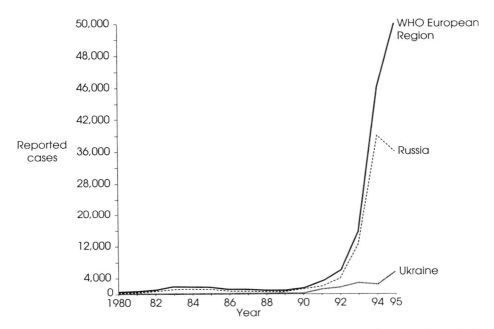

Fig. 1. Reported cases of diphtheria in the WHO European Region, Russia, and the Ukraine 1980–1995. Number of cases reported to WHO Euro, Copenhagen, Denmark.

emergency *(11)*. Small epidemics are also occurring in other regions of the world, for example, in South East Asia, India, Pakistan, Bangladesh, and South America. Therefore, both epidemiologic and microbiologic surveillance is essential in all countries; the microbiological confirmation and surveillance of isolates within a network of designated reference centers is of utmost importance. Molecular microbiologic methodologies are crucial to identify the source of a sporadic isolate if this is not obvious from epidemiologic and clinical data *(3,13)*.

2. Diagnosis of Diphtheria

The main role of the laboratory in the diagnosis of diphtheria is to provide simple, rapid, and reliable methods to assist clinicians in confirming a clinical diagnosis. In advanced cases, a clinical diagnosis of diphtheria normally precedes the microbiologic diagnosis. However, it is often difficult to diagnose diphtheria on clinical grounds as the disease may be confused with other infections, such as severe streptococcal tonsillitis, glandular fever, or Vincent's angina. The laboratory may also aid the clinician by eliminating suspected cases or contacts of diphtheria from further investigation, thus avoiding unnecessary treatment or control measures such as isolation.

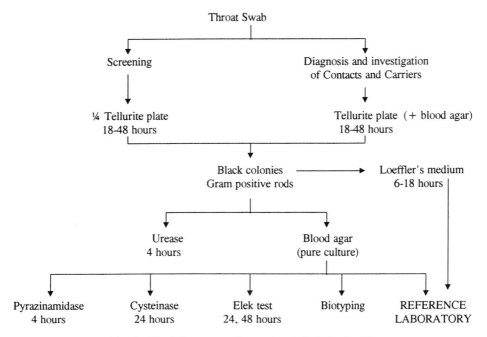

Fig. 2. The laboratory diagnosis of diphtheria *(3)*.

2.1. Conventional Laboratory Diagnosis

Guidelines for the laboratory diagnosis of diphtheria have been published *(14,15)* and are summarized in **Note 1**; a recommended algorithm for laboratory tests is shown in **Fig. 2**. The most important test in the microbiologic diagnosis of diphtheria is the detection of toxigenicity. Currently, the only in vitro phenotypic method readily available to the majority of diagnostic laboratories is the Elek immunoprecipitation test. This test is technically demanding and prone to misinterpretation of nonspecific precipitin lines, particularly in laboratories where the test is not regularly performed. A modified Elek test has been described *(16)* that simplifies the toxigenicity testing of corynebacteria (*see* **Note 2**); a result is available in 16–24 h and nonspecific precipitin lines are not visible.

2.2. Molecular Diagnosis

The polymerase chain reaction (PCR) has been used for the determination of toxigenicity amongst corynebacteria. The majority of assays have focused upon the detection of diphtheria toxin gene (*tox*) sequences that code for the biologically active (fragment A) subunit of the toxin *(17–21)*. However, primers specific for other regions of the gene have also been used *(22)*. The documented primer sequences used by different workers are given in **Table 1**.

Table 1
Primer Sequences Used
for the Detection of Toxigenic Corynebacteria by PCR

	Primer Sequence (5' - 3')	Target	Reference
1	ATCCACTTTTAGTGCGAGAAGGTTCGTCA GAAAACTTTTCTTCGTACCACGGGACTAA	*Tox* A	*17,19,20,21*
2	GTTCGGTGATGGTGCTTCGC CGCCTGACACGATTTCCTGC	*Tox* A	*18*
3	CGGGGATGGTGCTTCGCG CGCGATTGGAAGCGGGGT	Partial sequence of *Tox* A and B	*22*

PCR has a number of advantages compared with conventional phenotypic methods used for the detection of toxigenicity;

1. It is rapid, a result may be obtained within 5–6 h from the selection of colonies to the final result;
2. The interpretation of results is simple;
3. It can be performed directly from primary isolation media (the Elek test requires an additional incubation step on noninhibitory media); and
4. A large number of strains can be simultaneously tested.

A disadvantage with PCR is that some isolates of *C. diphtheriae* have been found that possess toxin genes, but do not express a biologically active protein and are, therefore, for diagnostic purposes, nontoxigenic. Such isolates are relatively rare worldwide; a small number have been isolated in the mid-United States, Canada, and Trinidad *(19,23)*. It is unlikely that these isolates will compromise the use of the PCR assay, except perhaps in those defined locations. Nevertheless, it is advisable to use PCR only as an adjunct to phenotypic tests, such as the Elek test.

Variations within the *tox* and *dtx*R genes may have a significant effect on the choice of primers used in PCR assays for the confirmation of a clinical or microbiologic diagnosis *(24)*. The heterogeneity of *tox* and *dtx*R genes amongst *C. diphtheriae* has been studied using PCR-single-strand conformational polymorphism (PCR-SSCP) *(24)*; the application of this technique is discussed further in **Subheading 3.2.5.**

3. Epidemiologic Investigation

Typing of *C. diphtheriae* strains is important for public health and epidemiologic control as it allows cases to be linked and outbreaks to be traced with greater precision. This information can be used to locate undisclosed sources of infection and for checking that contact tracing has successfully identified all linked cases.

3.1. Traditional Typing Methods

Traditional typing systems have relied on the detection of phenotypic characteristics *(25)*, for example serotyping, biotyping, antibiotic susceptibility patterns, and bacteriophage typing. Epidemiologic typing schemes based on traditional methods, described for *C. diphtheriae,* include biotyping, serotyping, bacteriocin typing, and phage typing.

3.1.1. Biotyping

Biotyping is generally of limited value in epidemiologic investigations because the discriminatory power of the technique is very poor. It has only a limited ability to distinguish distinct strains within a given species. There are four biotypes of *C. diphtheriae,* var. *"gravis,"* var. *"mitis,"* var. *"intermedius,"* and var. *"belfanti" (26)*. The classification of *C. diphtheriae* into different biotypes is based upon their morphological and biochemical properties. The key reactions used are reduction of nitrate, hydrolysis of urea, catalase production, cystinase activity, pyrazinamidase activity, and the fermentation tests for glucose, sucrose, maltose, and starch (**Table 2**) *(15)*. Commercial kits, such as the API CORYNE system (Biomérieux, France) are readily available, easy to interpret, and reliable for the identification of *C. diphtheriae*. In addition, Rosco Diagnostica (Denmark) has developed a range of identification tests, in tablet form, containing chromogenic and modified substrates that detect preformed enzymes that are useful in the rapid identification of *C. diphtheriae (15)*.

3.1.2. Serotyping

Many investigators have developed their own serologic typing systems for *C. diphtheriae (27–30)*. However, because of conflicting serotyping results, poor discriminatory power and the difficulties of preparing standardized antisera, other methods for detecting strain-specific differences have been developed.

3.1.3. Bacteriocin Typing

The discovery of bacteriocins in 1956 by Thibaut and Frédéricq *(31)* opened prospects for their use in the classification of *C. diphtheriae*. Bacteriocins are bacterial protein toxins that are active only on related strains or species. The bacteriocin typing scheme proposed by Meitert in 1972 *(32)* is based on the determination of the susceptibility of test strains against the bacteriocins produced by a set of 20 different defined bacteriocin-producing strains. This method was successfully used by Zamiri et al. *(33)* and Toshach et al. *(34)* to classify *C. diphtheriae* strains, that were untypeable by serotyping. The

Table 2
Biochemical Identification of Corynebacteria

Species	CYS	PYZ	Nitrate	Urea	Fermentation of					Gelatin Liquefaction
					Glucose	Maltose	Sucrose	Starch	Trehalose	
C. diphtheriae										
var *gravis*	+	–	+	–	+	+	–	+	–	–
var *mitis*	+	–	+	–	+	+	–	–	–	–
var *intermedius*	+	–	+	–	+	+	–	–	–	–
var *belfanti*	+	–	–	–	+	+	–	–	–	–
C. ulcerans	+	–	–	+	+	+	–	+	–	–
C. pseudotuberculosis	+	–	–	+	+	+	–	+	+	+
C. pseudodiphtheriticum	–	+	+	+	–	–	–	–	–	–
"C. xerosis"-like	–	+	+	–	+	+	+	–	–	–

CYS, cystinase production on Tinsdale medium.
PYZ, pyrazinamidase activity.

20 bacteriocin-producing strains consisted of 14 *C. diphtheriae,* one *C. ulcerans,* four *C. pseudodiphtheriticum,* and one *"C. xerosis"* strain. In 1973 Gibson and Colman *(35),* using Gillies bacteriocin typing technique *(36),* found that there was a correlation between bacteriocin type and phage type, serotype, and biotype. Bacteriocin typing is complex to perform and the method cannot be standardized.

3.1.4. Phage Typing

The phage typing scheme was first applied systematically to *C. diphtheriae* by Saragea and Maximescu in 1964 *(37).* The original phage typing set consisted of 22 lysogenic phages. The phage types were represented either by patterns of lytic reactions or by isolated reactions. All strains displaying the same lytic pattern or single reactions were considered to belong to the same phage type *(38).* The original phage typing scheme was reproducible, stable, and was the first method to subdivide the three biotypes of *C. diphtheriae* into 21 phage types that correlated with biotype and toxigenicity status. However, the technique had its limitations; it was capable of typing all toxigenic *gravis* strains, but it was not able to type all the nontoxigenic *mitis, gravis,* and *intermedius* strains. This led to the development of an additional phage typing scheme *(38)* and involved the use of an additional set of 33 phages.

The introduction of mass immunization has led to changes in the distribution of biotypes, phage types, and the ratio of toxigenic to nontoxigenic strains. This has, therefore, limited the value of this method. Also, the technique is very labor-intensive and reproducibility is limited.

3.2. Molecular Typing

Limitations of the traditional typing methods have led to the development of a variety of molecular typing schemes for *C. diphtheriae.*

3.2.1. DNA Probes

A variety of DNA probes have been used to study the molecular epidemiology of *C. diphtheriae,* including probes specific for the diphtheria toxin gene (*tox*), insertion sequences, β-phage DNA, and phage attachment sites. In 1983, Pappenheimer and Murphy *(39)* demonstrated the resolving power of molecular methods when they analyzed *C. diphtheriae* isolates using DNA restriction fragment length polymorphisms (RFLPs) with different DNA probes for corynephage-β and different areas of the toxin gene. In 1987, Rappuoli et al. *(40)* used an insertion element of *C. diphtheriae* as a DNA probe to characterize Swedish strains from a diphtheria outbreak in an alcoholic and drug-abusing population and in 1989, Coyle et al. *(41)* studied the molecular epidemiology of the three biotypes of *C. diphtheriae* in the Seattle outbreak by analyzing RFLPs with three DNA probes.

3.2.1.1. DIPHTHERIA TOXIN GENE *(tox)*

This gene has been cloned and sequenced *(42,43)* and is contained in a 1880-bp *Hind*III-*Eco*RI fragment of the phage DNA. A variety of probes specific for the *tox* gene have been described that can be used for determination of the toxigenicity genotype among *C. diphtheriae*.

Groman et al. described three DNA probes that can be used for the detection of the *tox* gene *(23)*. An *Eco*RI-*Xba*I digest of purified phage DNA produces the AB probe: a 2.3-kb fragment contains sequences coding for the A and B fragments of the toxin; the probe is not, however, specific as it contains nontoxin coding sequences at both ends. An *Mbo*I digest of the AB probe produces two fragments of 0.88 kb (A probe) and 1.1 kb (B probe). The A probe is also not specific as it contains 250 bp that are nontoxin encoding. The B probe contains sequences internal to the *tox* gene and is specific. The three probes were used to detect the presence of DNA homologous to the *tox* gene in nontoxigenic isolates of *C. diphtheriae (23)*. A probe for fragment B of the toxin gene was used to study the molecular epidemiology of the Seattle outbreak in 1972–1982 (in conjunction with probes specific for β-phage DNA and phage attachment sites) *(41)*. This probe is 0.98-kb fragment, internal to sequences encoding for fragment B of diphtheria toxin, produced by a *Mbo*I digestion of a pBR322–derived plasmid *(41)*. A probe specific for most of fragment A and the amino-terminal part of fragment B has also been described *(44)* (*see* **Note 3**). This probe is contained in plasmid pDγ2 (ATCC 67011), which is also the source of the insertion sequence probe described in **Subheading 3.2.1.2.**

As with PCR, DNA probes for the toxin gene have identified a number of strains that are positive to the hybridization to the *tox* gene, but which are nontoxigenic by the Elek or in vivo tests *(23,44)*.

3.2.1.2. INSERTION SEQUENCES (IS)

These are transposable DNA elements that can insert themselves into several sites of a bacterial genome. The presence of an IS element in *C. diphtheriae* was first detected in bacteriophage γ *(45)*. The IS element was later shown to be inserted within the *tox* gene and to be responsible for the nontoxigenic phenotype of the γ bacteriophage *(44,45)*. The IS element is contained in plasmid pDγ2 (*see* **Note 4**) and was used as a probe to characterize isolates from the Swedish outbreak of diphtheria *(40,46,47)*. The IS element is present in the same chromosomal location in strains that are closely related, whereas unrelated strains contain different numbers of copies of the IS element integrated into different sites of the chromosome.

3.2.1.3. CORYNEPHAGES AND PHAGE ATTACHMENT SITES

C. diphtheriae is a nonpathogenic bacterium that can become virulent following lysogenization by a corynephage that carries the structural gene encoding for diphtheria toxin. Several phages, both converting and nonconverting have been described. It is now clear that the *tox* gene is part of the phage genome in the converting phages. The most widely studied phage is the β-phage. *C. diphtheriae* may be lysogenized by more than one phage at the same time: multiple lysogeny. Thus recombination between different phages and phage-mediated transfer of the *tox* gene between strains is perhaps important in the evolution of virulence in the organism. It has been shown that a β-like phage that lacks the *tox* gene can acquire it by recombination with a converting phage. Additionally, two artificially mutated phages, each with defective *tox*- genes, but with different mutations may recombine to reconstitute a fully functional gene *(48,49)*. Therefore, naturally occurring *tox* genes constitute a potential reservoir of toxigenicity, as hypothesized by Groman *(23,49)*.

β-phage DNA has been used as a probe to study the molecular epidemiology of the Seattle outbreak *(41)* and in an epidemiologic investigation following the isolation of a toxigenic strain of *C. diphtheriae* from a 10-wk old baby in Manchester (England) *(39)*.

Insertion of the phage occurs via recombination between specific sites on the phage DNA (*att*P) and the bacterial chromosome (*att*B). Most *C. diphtheriae* strains possess two *att*B sites, (*att*B1 and *att*B2), where the corynephages can integrate with similar frequency *(48,50)*. The chromosomal region containing the bacterial attachment sites in *C. diphtheriae* (C7-) has been cloned and partially sequenced, and *att*B1 and *att*B2 are contained in a 3.5-kb *Eco*R1 fragment. This fragment has been cloned into plasmid A364. Sequences homologous to the *att*B region are present in all strains of *Corynebacterium* and the 3.5-kb *Eco*RI fragment of pA364 has been used as a probe for speciation within the genus *(44,49)*. Between one and four hybridization bands may be produced depending on whether the bacterium has one or two *att*B sites, and whether these sites are split into two fragments by the integration of a corynephage. A probe specific for *C. diphtheriae* has also been described; a *Hinc*III digest of pA634 produces a 0.7-kb fragment that contains the region between *att*B1 and *att*B2. This region is present only in *C. diphtheriae* and has been used as a probe specific for this species *(44)*.

3.2.2. Ribotyping

More recently, the use of RFLPs of ribosomal RNA (rRNA) genes has been evaluated to study the molecular epidemiology of *C. diphtheriae* *(51–53)*. RFLPs of rRNA genes are a useful way of discriminating between closely related isolates. rRNA genes are highly conserved and are present as multiple copies in the bacterial genome of all bacteria. The technique relies upon the extraction of chromosomal DNA by the guanidium method, electrophoresis of chromosomal DNA restriction enzyme digests, Southern blotting (transfer of DNA fingerprints to a nylon membrane), followed by hybridization with a nonradioactive (biotinylated) cDNA transcript of RNA (*see* Chapter 2 and **Note 5**). Ribotyping is an effective and discriminatory typing method, and is the main typing tool used to explore the epidemiology of current outbreaks in Europe, particularly in Russia and Eastern Europe. In 1995, De Zoysa et al. *(52)* reported that a predominant clonal group has emerged in Russia and appears to be spreading, not only within Russia, but also to other countries in view of imported cases of diphtheria in Finland, Estonia, and Germany, caused by strains belonging to the predominant clonal group. By the end of 1994, this clonal group had spread to all 15 new independent states of the former Soviet Union *(53)*.

3.2.3. Pulsed-Field Gel Electrophoresis

Another method that is being explored for molecular epidemiologic studies is pulsed-field gel electrophoresis (PFGE; *see* Chapter 3). *C. diphtheriae* strains are digested in agarose blocks using lysozyme, RNase, and proteinase K. The DNA in the agarose blocks is then digested with a rare-cutting enzyme (e.g., *Sfi*I). The fragments are separated by PFGE and the bands visualized by staining the gel in ethidium bromide (EB).

PFGE is highly reproducible and produces well-resolved fragments representing the entire bacterial chromosome in a single gel. Recent data suggest that PFGE is the most discriminating of the current available genotypic methods *(54–56)*. However, PFGE was reported to be less discriminating than ribotyping when applied to isolates of *C. diphtheriae* *(52)*; PFGE was not able to distinguish between the two main ribotype patterns in Russia. PFGE can be performed according to the procedure described by De Zoysa et al. *(52)* (*see* **Note 6**).

3.2.4. PCR Typing

DNA amplification fingerprinting using PCR has become a key procedure in molecular biology. This technique has recently has recently been applied to *C. diphtheriae* and shows promise. In 1994, Mokrousov *(57)* reported arbitrary primed PCR (AP-PCR) typing of *C. diphtheriae*. The technique employs a single, short (typically 10 bp in length) primer, whose nucleotide sequence is not directed at a known genetic locus *(58,59)* (*see* Chapter 6). Such arbitrary primers will amplify one or more unpredictable loci and the PCR reaction will generate a set of fragments; when separated by electrophoresis in agarose gels and stained, each spectrum of products resolves into a banding pattern or fingerprint. Mokrousov *(57)* performed PCR on cell lysates prepared by treatment with lysozyme and sodium hydroxide. A universal primer 5'-GGA TCC AAA ACG GCC AGT-3' was used. AP-PCR typing has potential as it is rapid and simple, but the discriminatory power of the technique and its reproducibility is uncertain.

3.2.5. Single-Strand Conformational Polymorphism (SSCP)

SSCP analysis was first described by Orita et al. in 1989 *(60)*. In this technique double-stranded DNA (i.e., PCR products) is denatured to single-stranded DNA and the products are separated by polyacrylamide gel electrophoresis under nondenaturing conditions. SSCP is capable of detecting both known and unknown single point mutations and polymorphisms in PCR products. This technique has not yet been fully documented for epidemiologic typing of *C. diphtheriae,* but has been used to study variations within the genes encoding for diphtheria toxin (*tox*) and its regulatory element (*dtx*R) from strains of *C. diphtheriae* isolated in Russia and the Ukraine before and during the diphtheria epidemic *(24)*. Twelve sets of overlapping primers were used, eight for the *tox* gene and four for the *dtx*R gene. Significant variation in the number and/or sizes of the amplicons was found within both genes and six different *tox* and 12 different *dtx*R patterns were identified. The majority of epidemic strains from Russia and Ukraine belonged to two *tox* types, designated 3 and 4. Epidemic strains from Ukraine belonged to a *dtx*R type designated 5, and the majority of the epidemic *gravis* strains from Russia belonged to *dtx*R types 2, 5, and 8. A clear association of *tox* and *dtx*R types with strain origin was observed *(24)*.

3.2.6. Multilocus Enzyme Electrophoresis (MLEE)

MLEE has been used in the United States for the characterization of epidemic *C. diphtheriae*. MLEE detects mutations within the highly conserved genes for enzymes that are necessary for cell maintenance. These mutations can affect the electrophoretic mobility of the enzyme and mobility variants of the same enzyme (electromorphs) can be visualized in a gel matrix as bands of different migration rates. Each electromorph of an enzyme is scored as a different allelic form of the gene coding for that enzyme. A profile of the electromorphs defines the electrophoretic type of an individual strain.

Reeves et al. *(61)* used MLEE to subtype clinical isolates of *C. diphtheriae* from the Russian outbreak. They found that MLEE was reproducible and highly discriminatory, but as a typing technique it is labor-intensive and may not be the typing method of choice. MLEE can be performed according to the method described by Reeves et al. *(61)* and Popovic et al. *(53)*.

4. Overview of the Epidemiology of Diphtheria

Based on the recent resurgence of diphtheria within the NIS of the former USSR and outbreaks in other countries (Vietnam, Thailand, Ecuador, Algeria) *(1,2)*, the epidemiology of diphtheria in the vaccination era has significantly changed during the last decade; both adolescents and adults are now likely to be affected during outbreaks *(1)*. For these reasons, serologic, microbiologic, and epidemiologic surveillance measures are essential. Strategies have been formulated by WHO in collaboration with other agencies and are now showing signs of progress in some epidemic areas. At the request of WHO, a European Laboratory Working Group on Diphtheria (ELWGD) was formed in 1993 with participation from more than 14 different countries from Western and Eastern Europe, the United States, Australia, and South East Asia. The aims of the Group are

1. To establish a definitive network of collaborating laboratories, not only in Europe, but globally;
2. To enhance collaboration between all participants;
3. To establish a microbiologic surveillance scheme within Europe; and
4. To establish a database of molecular typing patterns in addition to the harmonization of typing methods for these organisms *(62)*.

4.1. Rationale for Typing

Traditional typing methods have been applied to *C. diphtheriae* with varying success. In the past decade, several groups have applied molecular typing methods to *C. diphtheriae* that have been outlined within this chapter. Molecular typing methods not only confirm the level of differentiation obtained by traditional methods, but allow for further differentiation below the serotype and biotype level. There is no single "best" method; good results are obtained only after detailed epidemiologic studies have defined the true relationships of isolates. The epidemic and nonepidemic isolates from the current outbreaks in the European Region have allowed us to address these issues in collaboration with other centers. Since, the formation of the ELWGD there has been much discussion and evaluation of the various methodologies in order to standardize the procedures used by the various laboratories.

4.2. Microbiologic Surveillance of Diphtheria in Europe and Globally

A microbiologic surveillance has been initiated amongst the ELWGD with close liaison with WHO and allows us to monitor the spread of a particular strain within the European Region. The scheme enables us to provide information on the characteristics of a global collection of *C. diphtheriae*.

Studies thus far have shown that ribotyping seems to be the most useful molecular typing tool for studying the spread of *C. diphtheriae* clones

(52,53,62). To "promote" the method as an internationally recognized technique, the method had to be standardized and agreed among members of the ELWGD, with common restriction enzymes, markers, probes, and gel analysis programs being used by all centers, in addition to common nomenclature and validation of types. This initiative is currently one of the main projects of the ELWGD. From these studies, the authors have confirmed that geographic heterogeneity exists between strains from different regions and that database storage allows the recognition, confirmation, and identification of apparent clones worldwide *(52,53,62)*.

4.3. The Future: Optimism or Complacency?

Predicting future trends in the epidemiology of diphtheria is difficult. Spread of infection to immunized populations following local outbreaks, as described in Scandinavia and North America, showed that diphtheria will always return whenever immunity levels decrease *(41,47)*. This is clearly illustrated in Eastern Europe. With increasing international travel and the emergence of epidemic clones, the existence of diphtheria anywhere in the world poses a threat to the unimmunized and those persons with low levels of immunity. These problems further highlight the importance of microbiologic and epidemiologic surveillance and the use of molecular methodologies. The complete eradication of *C. diphtheriae* as a species will require a new generation of vaccines, aimed not merely at the toxin, but at all the virulence factors that permit the organism to survive in vivo.

5. Notes

1. Laboratory diagnosis. The diagnosis of diphtheria based upon direct microscopy of a smear is unreliable; primary culture onto blood agar and a selective tellurite media such as Hoyle's is recommended. Screening tests recommended for the identification of toxigenic species *(C. diphtheriae, C. ulcerans, and C. pseudotuberculosis)* are the presence of cystinase and the absence of pyrazinamidase activity *(14,15)*. Biotyping of pathogenic strains is performed using simple biochemical tests (**Table 2**); commercial kits such as the API CORYNE system (Biomérieux) or Rosco Diagnostica tablets are also available.

2. Modified Elek test. The modified Elek test *(16)* uses 3 mL of Elek base in a 4.5 cm plate. Test and control strains are inoculated 9 mm from the edge of an antitoxin disc (10 IU/disc). The test provides an accurate result after only 16–24 h incubation, compared to 48 h required by the conventional test. Furthermore, when read at 24 h nonspecific precipitin lines are not present, thus eliminating some of the errors inherent in the conventional Elek test.

3. DNA probes for diphtheria toxin gene. The plasmid pDγ2 (ATCC 67011) contains an insert of 2.5 kb including the *tox* gene of bacteriophage γ from the *Hind*III site to the *Cla*I site in position 1066. This is interrupted at position 176 by a

1450-bp *C. diphtheriae* IS element. Cutting this plasmid with *Bam*HI and *Acc*I, produces five fragments of 3.1, 0.77, 0.55, 0.28, and 0.27 kb. The 3.1-kb fragment contains the vector and the 0.55 and 0.27-kb fragments are internal to the IS element. The 0.77-kb fragment contains nucleotides 297–1066 of the diphtheria toxin gene, a region encoding for most of fragment A and the amino-terminal of fragment B. Following purification and labeling, the 0.77-kb fragment has been used as a probe for toxigenicity by hybridization to Southern blot transfers of *Bam*HI- or *Eco*R1–digests of chromosomal DNA *(44)*.

4. DNA probes for the insertion sequence. The plasmid pDγ2 (ATCC 67011) contains a 2.5-kb insert. It includes the first 1066 bp of the *tox* gene of the bacteriophage γ interrupted in position 176 by the 1450-bp IS element of *C. diphtheriae*. Cutting the plasmid with *Sal*I produces a single 0.6-kb fragment. The fragment should be purified and labeled with biotin and used to probe *Bam*HI-digested chromosomal DNA from *C. diphtheriae* isolates *(44)*.

5. Ribotyping. The modifications made by De Zoysa et al. *(52)* to the DNA extraction method described by Pitcher et al. *(63)* *(see* Chapter 2) are as follows: *C. diphtheriae* isolates are swabbed onto two Columbia blood agar plates and incubated overnight at 37°C. Cells are harvested into 200 μL of lysozyme (50 mg/mL) and incubated at 37°C for 3 h. The cells are treated with 10 μL of proteinase K (10 mg/mL) in lysis buffer (1% sarcosyl in TE) for 1 h at 50°C. The remainder of the extraction procedure is carried out according to Pitcher et al. *(63)*. The DNA yield is determined by measuring the absorbance at λ 260 nm ($1A_{260}$ U double-stranded DNA ~ 50 μg mL^{-1}) *(60)*. 5 μg of DNA are digested with the restriction endonuclease and 2 μg of digest are loaded onto a 1% agarose gel. The digests are electrophoresed in TBE buffer at 27 V for 16 h. The gels are blotted onto a nylon membrane and hybridized with a biotin-labeled cDNA probe derived from the total rRNA of the *C. diphtheriae* type strain NCTC 11397. The rRNA is isolated and labeled using the method described by Pitcher et al. *(64)*. To isolate rRNA from the type strain the cells are grown as follows: 10 mL of Brain Heart Infusion (BHI) broth containing Tween-80 (0.2% v/v) and yeast extract (0.4% v/v) are inoculated with the *C. diphtheriae* type strain (NCTC 11397) and incubated at 37°C for 6–8 h. The cells are then transferred to 200 mL prewarmed BHI broth containing Tween-80 (0.2% v/v) and yeast extract (0.4% v/v) and incubated at 37°C for 24 h.

6. PFGE. The preparation of DNA for PFGE is as follows: A single colony of *C. diphtheriae* is inoculated into 20 mL of BHI broth containing Tween-80 (0.2% v/v) and yeast extract (0.4% v/v) and incubated overnight at 37°C. The culture is centrifuged at 1000*g* for 20 min at 4°C. The cells are washed twice in washing buffer (0.1 *M* Tris, 2.5 *M* NaCl, pH 7.6) and resuspended in 1 mL of washing buffer. 0.5 mL of the suspension is mixed with 0.5 mL of 1.6% low-melting point agarose (Sigma) in a water bath at 50°C and then pipeted into a plug mold and allowed to solidify at 4°C (molds are usually supplied with PFGE equipment; however, specifications of a mold made by in-house facilities has been described by Kaufmann and Pitt) *(65)*. The remainder of the method is per-

formed as described by Kaufmann and Pitt *(65)* for Gram-positive bacteria. Lysostaphin is not used in the lysis buffer and 1 mg/mL lysozyme and 25 µg/mL RNase is freshly added to the lysis buffer. 1 mg/mL proteinase K is freshly added to the proteolysis buffer. The DNA in agarose blocks is digested and electrophoresed as described previously *(52)*.

References

1. Galazka, A. and Robertson, S. E. (1995) Resurgence of diphtheria. *Eur. J. Epidemiol.* **11**, 95–105.
2. Hardy, I. R. B., Dittmann, S., and Sutter, R. W. (1996) Current situation and control strategies for resurgence of diphtheria in newly independent states of the former Soviet Union. *Lancet* **347**, 1739–1744.
3. Efstratiou, A. and George, R. C. (1996) Microbiology and epidemiology of diphtheria. *Rev. Med. Microbiol.* **7**, 31–42.
4. MacGregor, R. R. (1995) *Corynebacterium diphtheriae*, in *Principles and Practice of Infectious Disease* (4th ed.) (Mandell, G. L., Bennet, J. E., and Dolin, R., eds.), Churchill Livingstone, New York, pp. 1865–1872.
5. Harnisch, J. P., Tronca, E., Nolan, C. M., Turck, M., and Holmes, K. K. (1989) Diphtheria among alcoholic urban adults. A decade of experience in Seattle. *Ann. Intern. Med.* **111**, 71–82.
6. Begg, N. (1994) Manual for the management and control of diphtheria in the European Region. Copenhagen: The Expanded Programme on Immunization in the European Region of WHO. ICP/EPI038(B).
7. Galazka, A. and Robertson, S. E. (1995) Diphtheria: changing patterns in the developed world and the industrialized world. *Eur. J. Epidemiol.* **11**, 107–117.
8. Efstratiou, A. George, R. C. and Begg, N. T. (1993) Non-toxigenic *Corynebacterium diphtheriae* in England. *Lancet* **341**, 1592,1593.
9. Pappenheimer, A. M. (1984) The diphtheria bacillus and its toxin: a model system. *J. Hyg. Camb.* **93**, 397–404.
10. Salyers, A. A. and Whitt, D. D. (1994) Diphtheria, in B*acterial Pathogenesis*, (Salyers, A. A. and Whitt, D. D., eds.), American Society for Microbiology, Washington DC., pp 113–121.
11. *World Health Organization Regional Office for Europe. Communicable Disease Report*, number 13, October 1996. Diphtheria update.
12. De Zoysa, A., Efstratiou, A., George, R. C., Vuopio-Varkila, J., Jahkola, M. and Rikushin, Y. (1993) Diphtheria and travel. *Lancet* **342**, 446.
13. Efstratiou, A. (1995) *Corynebacterium diphtheriae*: molecular epidemiology and characterisation studies on epidemic and sporadic isolates. *Microecol. Ther.* **25**, 63–71.
14. Colman, G., Weaver E., and Efstratiou, A. (1992) Screening tests for pathogenic corynebacteria. *J. Clin. Pathol.* **45**, 46–48.
15. Efstratiou, A. and Maple, P. A. C. (1994) Manual for the laboratory diagnosis of diphtheria. Copenhagen: The Expanded Programme on Immunization in the European Region of WHO. ICP/EPI038 (C).

16. Engler, K. H., Glushkevich, T., Mazurova, I. K., George, R. C., and Efstratiou, A. (1997) A modified Elek test for the detection of toxigenic corynebacteria. *J. Clin. Microbiol.* **35**, 495–498.
17. Mikhailovich, V. M., Melnikov, M. G., Mazurova, I. K., Wachsmuth, I. K., Wenger, J. D., Wharton, M., Nakao, H., and Popovic, T. (1995) Application of PCR for detection of toxigenic *Corynebacterium diphtheriae* strains isolated during the Russian diphtheria epidemic, 1990 through 1994. *J. Clin. Microbiol.* **33**, 3061–3063.
18. Lucchini, G. M., Gruner, E., and Altwegg, M. (1992) Rapid detection of diphtheria toxin by the polymerase chain reaction. *Med. Microbiol. Lett.* **1**, 276–283.
19. Pallen, M. J., Hay, A. J., Puckey, L. H., and Efstratiou, A. (1994) Polymerase chain reaction for screening clinical isolates of corynebacteria for the production of diphtheria toxin. *J. Clin. Pathol.* **47**, 353–356.
20. Pallen, M. J. (1991) Rapid screening for toxigenic *Corynebacterium diphtheriae* by the polymerase chain reaction. *J. Clin. Pathol.* **44**, 1025,1026.
21. Aravena-Roman, M., Bowman, R., and O'Neill, G. (1995) Polymerase chain reaction for the detection of toxigenic *Corynebacterium diphtheriae*. *Pathology* **27**, 71–73.
22. Hauser, D., Popoff, M. R., Kiredjian, M., Boquet, P., and Bimet, F. (1993) Polymerase chain reaction assay for diagnosis of potentially toxinogenic *Corynebacterium diphtheriae* strains: correlation with ADP-ribosylation activity assay. *J. Clin. Microbiol.* **31**, 2720–2723.
23. Groman, N., Cianciotto, N., Bjorn, M., and Rabin, M. (1983) Detection and expression of DNA homologous to the *tox* gene in non-toxigenic isolates of *Corynebacterium diphtheriae*. *Infect. Immun.* **42**, 48–56.
24. Nakao, H., Pruckler, J. M., Mazurova, I. K., Narvskaia, O. V., Glushkevich, T., Marijevski, V. F., Kravetz, A. N., Fields, B. S., Wachsmuth I. K., and Popovic, T. (1996) Heterogeneity of diphtheria toxin gene, *tox,* and its regulatory element, *dtx*R, in *Corynebacterium diphtheriae* strains causing epidemic diphtheria in Russia and Ukraine. *J. Clin. Microbiol.* **34**, 1711–1716.
25. Maslow, J. N., Mulligen, M. E., and Arbeit, R. D. (1993) Molecular epidemiology: applications of contemporary techniques to the typing of microorganisms. *Clin. Infect. Dis.* **17**, 153–164.
26. Anderson, J. S., Cooper, K. E., Happold, F. C., and McLeod, J. W. (1933) Incidence and correlation with clinical severity of *gravis, mitis* and *intermediate* types of diphtheria bacillus in a series of 500 cases at Leeds. *J. Pathol Bacteriol.* **36**, 169–182.
27. Ewing, J. O. (1933) The serological grouping of the starch fermenting strains of *C. diphtheriae*. *J. Pathol. Bacteriol.* **37**, 345–351.
28. Hewitt, L. F. (1947) Serological typing of *C. diphtheriae*. *Brit. J. Exp. Pathol.* **28**, 338–346.
29. Robinson, D. T. and Peeney, A. A. P. (1936) Serological types amongst *gravis* strains of *C. diphtheriae* and their distribution. *J. Pathol. Bacteriol.* **43**, 403–415.

30. Wong, S. C. and T'ung, T. (1940) Further studies on type specific protein of *Corynebacterium diphtheriae. Proc. Soc. Exp. Biol. Med.* **43,** 749–754.

31. Thibaut, J. and Fredericq, P. (1956) Actions antibiotiques reciproques chez *C. diphtheriae. C.R. Seances Soc. Biol. Ses. Fil.* **150,** 1513–1514.

32. Meitert, E. and Bica-Popii, V. (1972) Etude des relations antigeniques entre les phages anti-*Corynebacterium hofmanni. Arch. Roum. Pathol. Exp. Microbiol.* **31,** 475, 480.

33. Zamiri, I., McEntegart, M., and Saragea, A. (1972) Diphtheria in Iran. *J. Hyg. Camb.* **70,** 619–625.

34. Toshach, S., Valentine, A., and Sigurdson, S. (1977) Bacteriophage typing of *Corynebacterium diphtheriae. J. Infect. Dis.* **136,** 655–659.

35. Gibson, L. F. and Colman, G. (1973) Diphthericin types, bacteriophage types and serotypes of *Corynebacterium diphtheriae* in Australia. *J. Hyg. Camb.* **71,** 679–689.

36. Gillies, R. R. (1964) Colicine production as an epidemiological marker of *Shigella sonnei. J. Hyg.* **62,** 1.

37. Saragea, A. and Maximescu, P. (1964) Schema provisoire de lysotypie pour *Corynebacterium diphtheriae. Arch. Roum. Pathol. Exp. Microbiol.* **23,** 817–838.

38. Saragea, A., Maximescu, P., and Meitert, E. (1979) *Corynebacterium diphtheriae:* Microbiological methods used in clinical and epidemiological investigations, in *Methods in Microbiology,* (Bergan, T. and Norris, J. R., eds.), Academic, New York, pp. 61–176.

39. Pappenheimer, A. M. and Murphy, J. R. (1983) Studies on the molecular epidemiology of diphtheria. *Lancet* **2,** 923,924.

40. Rappuoli, R., Perugini, M., and Ratti, G. (1987) DNA element of *Corynebacterium diphtheriae* with properties of an insertion sequence and usefulness for epidemiological studies. *J. Bacteriol.* **169,** 308–312.

41. Coyle, M. B., Groman, N. B., Russell, J. Q., Harnisch, J. P., Rabin, M., and Holmes, K. K. (1989) The molecular epidemiology of three biotypes of *Corynebacterium diphtheriae* in the Seattle outbreak, 1972–1982. *J. Infect. Dis.* **159,** 670–679.

42. Greenfield, L., Bjorn, M. J., Horn, G., Fond, D., Buck, G. A., Collier, R. J., and Kaplan, D. A. (1983) Nucleotide sequence of the structural gene for diphtheria toxin carried by corynephage β. *Proc. Natl. Acad. Sci. USA* **80,** 6853–6857.

43. Ratti, G., Rappuoli, R., and Giannini, G. (1983) The complete nucleotide sequence for the gene coding for diphtheria toxin in the corynephage ω (*tox*+) genome. *Nucleic Acids Res.* **11,** 6589.

44. Rappouli, R. and Gross, R. (1990) *Att* sites, *tox* gene and insertion elements as tools for the diagnosis and molecular epidemiology of *Corynebacterium diphtheriae,* in *Gene Probes for Bacteria* (Macario, A. J. L. and Conway de Macario, E., eds.), Academic, San Diego, CA, pp. 205–231.

45. Buck, G. and Groman, N. B. (1981) Genetic elements novel for *Corynebacterium diphtheriae.* Specialised transducing elements and transposons. *J. Bacteriol.* **148,** 143–152.

46. Michel, J. L., Rappuoli, R., Murphy, J. R., and Papenheimer, A. M. Jr. (1982) Restriction endonuclease map of the non-toxigenic corynephage (γ and its relationship to the toxigenic corynephage βc. *J. Virol.* **42,** 510–518.

47. Rappuoli, R., Perugini, M., and Falsen, E. (1988) Molecular epidemiology of the 1984–1986 outbreak of diphtheria in Sweden. *N. Engl. J. Med.* **318**, 12–14.

48. Rappuoli, R., Machel, J. L., and Murphy, J. R. 1983. Integration of corynephages β-tox- ω-tox- and γ-tox- into two attachment sites on the *Corynebacterium diphtheriae* chromosome. *J. Bacteriol.* **153**, 1202–1210.

49. Cianciotto, N., Rappuoli, R., and Groman, N. (1986). Detection of homology to the beta bacteriophage integration site in a wide variety of *Corynebacterium* spp. *J. Bacteriol.* **168**, 103–108.

50. Rappuoli, R. and Ratti, G. (1984) Physical map of the chromosomal region of *C. diphtheriae* containing corynephage attachment sites *att*B1 and *att*B2. *J. Bacteriol.* **158**, 325–330

51. Efstratiou, A., Tiley, S. M., Sangrador, A., Greenacre, E., Cookson, B. D., Chen, S. C. A., Mallon, R., and Gilbert, G. L. (1993) Multiple clones of *Corynebacterium diphtheriae* causing invasive disease. *Clin. Infect. Dis.* **17**, 136.

52. De Zoysa, A., Efstratiou, A., George, R. C., Jahkola, M., Vuopio-Varkila, J., Deshevoi, S., Tseneva, G., and Rikushin, Y. (1995) Molecular epidemiology of *Corynebacterium diphtheriae* from north-western Russia and surrounding countries studied by using ribotyping and pulsed-field gel electrophoresis. *J. Clin. Microbiol.* **33**, 1080–1083.

53. Popovic, T., Kombarova, S. Y., Reeves, M. W., Nakao, H., Mazurova, I. K., Wharton, M., Wachsmuth, I. K., and Wenger, J. D. (1996) Molecular epidemiology of diphtheria in Russia, 1984–1994. *J. Infect. Dis.* **174**, 1064–1072.

54. Arbeit, R. D., Arther, R. D., Dunn, C., Kim, C., Selander, R. K., and Goldstein, R. (1990) Resolution of recent evolutionary divergence among *Escherichia coli* from related lineages; the application of pulsed field electrophoresis to molecular epidemiology. *J. Infect. Dis.* **161**, 230–235.

55. Mulligan, M. E. and Arbeit R. D. (1991) Epidemiologic and clinical utility of typing systems for differentiating among strains of methicillin-resistant *Staphylococcus aureus. Infect. Control Hosp. Epidemiol.* **12**, 20–28.

56. Strulens, M. J., Maes, N., Rost, F., Deplano, A., Jacobs, F., Liesnard, C., Bornstein, N., Grimont, F., Lauwers, S., McIntyre, M. P., and Serruys, E. (1992) Genotypic and Phenotypic methods for the investigation of a nosocomial *Legionella pneumophila* outbreak and efficacy of control measures. *J. Infect. Dis.* **166**, 22–30.

57. Mokrousov, I. V. (1995) Arbitrary PCR typing of *Corynebacterium diphtheriae* strains. Proceedings of the First International Meeting of the European Laboratory Working Group on Diphtheria, London, *UK. PHLS Microbiol. Digest* **12**, 89.

58. Welsh, J. and McClelland, M. (1990) Fingerprinting genomes using PCR with arbitrary primers. *Nucleic Acids Res.* **18**, 7213–7218.

59. Williams, J. G. K., Kubelik, A. R., Livak, K. J., Rafalski, J. A., and Tingey, S. V. (1990) DNA polymorphisms amplified by arbitrary primers are useful as genetic markers. *Nucleic Acids Res.* **18**, 6531–6535.

60. Orita, M., Iwahana, H., Kanazawa, H., Hayashi, K., and Sekiya, T. (1989) Detection of polymorphisms of human DNA by gel electrophoresis as single-strand conformation polymorphisms. *Proc. Natl. Acad. Sci. USA* **86**, 2766–2770.

61. Reeves, M. W. and Wachsmuth, I. K. (1995) Subtyping of clinical *Corynebacterium diphtheriae* by multilocus enzyme electrophoresis and pulsed field gel electrophoresis. Proceedings of the First International Meeting of the European Laboratory Working Group on Diphtheria, *PHLS Microbiology Digest 1995,* **12,** 87,88.
62. Proceedings of the First International Meeting of the European Laboratory Working Group on Diphtheria (1995), (Efstratiou, A., ed.), Public Health Laboratory Service, London.
63. Pitcher, D. G., Saunders, N. A., and Owen, R. J. (1989) Rapid extraction of bacterial genomic DNA with guanidium thiocyanate. *Lett. Appl. Microbiol.* **8,** 151–156.
64. Pitcher, D. G., Owen, R. J., Dyal, P., and Beck, A. (1987) Synthesis of a biotinylated DNA probe to detect ribosomal RNA cistrons in *Providencia stuartii. FEMS Microbiol. Lett.* **48,** 283–287.
65. Kaufmann, M. E. and Pitt, T. L. (1994) Pulsed-field gel electrophoresis of bacterial DNA, in *Methods in Practical Laboratory Bacteriology* (Chart, H., ed.), CRC, pp. 83–92.

12

Diagnosis and Epidemiology
of Infections Caused by *Legionella* spp.

Norman K. Fry and Timothy G. Harrison

1. Introduction

The Legionnaires' Disease (LD) Bacillus was first recognized following a large outbreak of pneumonia at a convention of American military veterans in 1976 *(1)*. In the context of other microbial infections, this is a relatively recent event. Since then, considerable progress has been made in the understanding of legionellae and the epidemiology of the infections they cause. More than 40 species of *Legionella* have been identified and described, and these organisms have been found to be widely distributed in nature. They occur in almost all natural bodies of fresh water *(2)* (although they may be present at only low numbers) and in most man-made aquatic systems *(3)*. It appears that, in their natural environment, legionellae grow in association with other microorganisms and colonize surfaces at the aqueous/solid interface. Here, together with their protozoal hosts, other bacteria, and algae they form a complex consortium of organisms loosely termed "biofilm."

Although legionellae are environmental organisms, approximately half of the named species have been shown to cause infection in man (**Table 1**). Of these species, infections caused by *Legionella pneumophila* are by far the most common, accounting for approx 5% of all pneumonia cases requiring admission to a hospital *(53)*.

2. Detection of Legionellae
in Clinical and Environmental Samples

Legionella infection can vary from a mild respiratory illness to an acute life-threatening pneumonia. The overall mortality rate is about 13%, although this

From: *Methods in Molecular Medicine, Vol. 15: Molecular Bacteriology: Protocols and Clinical Applications*
Edited by: N. Woodford and A. P. Johnson © Humana Press Inc., Totowa, NJ

Table 1
Members of the Family Legionellaceae

Legionella species	Multiple serogroups	Isolated from patients	References
L. pneumophila[a]	16	Y	*4–19*
L. adelaidensis		N	*20*
L. anisa		Y	*21*
L. birminghamensis		Y	*22*
L. bozemanii	2	Y	*23,24*
L. brunensis		N	*25*
L. cherrii		N	*26*
L. cincinnatiensis		Y	*27*
L. dumoffii		N	*23*
L. erythra	2[b]	N	*26,28*
L. fairfieldensis		N	*29*
L. feeleii	2	Y	*30,31*
L. geestiana		N	*32*
L. gormanii		Y	*33*
L. gratiana		N	*34*
L. hackeliae	2	Y	*26,35*
L. israelensis		N	*36*
L. jamestowniensis		N	*26*
L. jordanis		Y	*37*
L. lansingensis		Y	*38*
L. londiniensis		N	*34*
L. longbeachae	2	Y	*39,40*
L. lytica		Y	*41*
L. maceachernii		Y	*26*
L. micdadei		Y	*42*
L. moravica		N	*25*
L. nautarum		N	*34*
L. oakridgensis		N	*43*
L. parisiensis		N	*26*
L. quarteirensis		N	*34*
L. quinlivanii	2	N	*44,45*
L. rubrilucens		N	*26*
L. sainthelensi	2[c]	Y	*46,47*
L. santicrucis		N	*26*
L. shakespearei		N	*48*
L. spiritensis	2	N	*26,49*
L. steigerwaltii		N	*26*
L. tucsonensis		Y	*50*
L. wadsworthii		Y	*51*
L. waltersii		N	*52*
L. worsleiensis		N	*34*

[a.] *L. pneumophila* comprises three subspecies; subsp. *pneumophila*, *fraseri*, and *pascullei (5)*.
[b.] *L. erythra* serogroup 2 is serologically indistinguishable from *L. rubrilucens (28)*.
[c.] *L. sainthelensi* serogroup 2 is serologically indistinguishable from *L. santicrucis (47)*.

can be much higher in certain groups, such as organ transplant patients or others with a compromised immune system. A history of travel is seen in almost 50% of cases and infection is most common in males over 40 yr old *(54)*. In many instances, the clinical picture of legionnaires' disease is sufficiently distinct (when seen in a patient with a suggestive history) to allow suspicion of the diagnosis and for appropriate antibiotic therapy to be initiated. Laboratory tests are, however, essential if a definitive diagnosis is to be established. Hence, detection of the organism, or its components, in samples can help to confirm a clinical diagnosis or to establish a possible source of infection.

2.1. Conventional Methods

2.1.1. Serology

The majority of legionella infections are diagnosed serologically; the indirect immunofluorescent antibody test is the most widely used method *(55)*. The main disadvantage of this approach is that separate antigens are needed for each serogroup being sought. If the assay was to be comprehensive, in excess of 60 antigens would be required. Even with the more modest aim of diagnosing all *L. pneumophila* infections, 16 antigens would be required. In practice only reagents for *L. pneumophila* serogroup 1 have been thoroughly evaluated and even here cross-reactions have been reported *(56)*.

2.1.2. Culture of the Causative Organism

Culture of legionellae from clinical specimens is relatively straightforward, quite sensitive (50–80%), and provides definitive proof of diagnosis. Colonization without infection has not been demonstrated. In addition, isolation of the infecting strain allows epidemiologic typing to be undertaken providing valuable data for the control and prevention of further cases of infection *(57)*. Unfortunately only ~13% of reported cases are confirmed by isolation of the organism *(54)*. This low rate may in part be a reflection of the fact that the diagnosis of legionellosis is often considered only after both the initiation of antibiotic therapy and the failure to identify more common etiologic agents of pneumonia.

Although legionellae are ubiquitous, they are not always isolated from putative environmental sources of infection (e.g., an implicated cooling tower). *Legionella* may be present, but not grown in vitro either because of deficiencies in laboratory media or, in the view of some workers, because the bacteria are in a viable, but nonculturable state *(58,59)*. The use of amoebal enrichment to recover *Legionella* species has been documented *(60)*, and revealed previously undetected sources of infection. The presence in samples of viable legionella, that are nonculturable by standard methods is, thus, an obvious area where molecular methods can play a key role.

2.1.3. Antigen Detection

Despite the advantages of culture, the time taken to obtain results by this method is still measured in days. In contrast, the direct demonstration of legionella antigen in clinical specimens can be achieved within a few hours of specimen collection. Furthermore, a diagnosis may be established by visualization or detection of the organism in tissue even when they are no longer viable, after antibiotic therapy or retrospectively in fixed tissues.

Immunofluorescence using rabbit hyperimmune antisera has been used to diagnose LD since legionellae were first recognized, but the poor sensitivity and specificity of this approach mean that it is not in general use. The issue of specificity has been largely overcome by using reagents made from monoclonal antibodies, the most widely used of these being directed against the *L. pneumophila* major outer membrane protein (MOMP). This reagent reacts with all the serogroups of *L. pneumophila* and so removes the need to examine a specimen with several different antisera. Sensitivity is still considerably lower than that of culture and similarly specific reagents are lacking for most other species of legionella.

The presence of a soluble antigen in the urine of patients with legionellosis has formed the basis for a range of tests, most notably enzyme-linked immunosorbent assays (ELISAs). Legionella antigen may be detectable in the urine very early in the course of the illness, often allowing a diagnosis to be established and treatment optimized shortly after admission to hospital *(61)*. This approach has also proved to be invaluable in the rapid detection of LD outbreaks allowing preventative measures to be implemented as soon as possible. Several kits are now commercially available and are being used increasingly. Unfortunately, these kits are generally only suitable for the detection of *L. pneumophila* serogroup 1 infections.

2.2. Molecular Approaches

The development of molecular methods for the diagnosis of legionella infections has been driven by the need for a rapid, specific assay, capable of detecting all *Legionella* species and serogroups in a single test. Although several promising strategies are developing, none of the techniques discussed here are in widespread routine use (**Tables 2** and **3**).

2.2.1. Nucleic Acid Probes

In 1984, Kohne and colleagues *(82)* described a radiolabeled DNA probe that was complementary to regions of the ribosomal RNAs (rRNAs) of all the twenty-two *Legionella* species known at that time. Subsequently, Gen-Probe (San Diego, CA) modified this to produce a commercial nucleic acid hybridization kit for the detection of legionellae in clinical material. Several evaluation studies have

Table 2
Nucleic Acid Based Assays Using PCR (or *In Situ* Hybridization*) for Detection of *Legionella* spp.

Target	Method of detection	Assessment of specificity	Sample type	Sensitivity	Reference[1] (original derivation)
800 bp fragment of *L.pneumophila* DNA of unknown function	Dot-blot with radiolabelled probe	*L. pneumophila* only (4 other *Legionella* spp. tested)	Seeded water samples	35 cfu[2]	62
5S rDNA	Radiolabelled probe	*Legionella* spp. (9/9 spp. tested) faint bands (on gel) with some non-legionella, but negative by hybridisation	Pure cultures	Single cell	63
mip gene		*L.pneumophila* only		Single cell	
mip gene	Radiolabelled probe	Not assessed	Water (viable vs. culturable)	10 cfu	64, (63)
mip mRNA				10³ cells	
800 bp fragment	Radiolabelled probe	*L.pneumophila* only (22/30)	Pure cultures	50 fg DNA	65, (62)
mip gene		*L.pneumophila* (30/30), *L.micdadei* (1/1), *L.bozemanii* (1/2). Not with any other *Legionella* species (0/32)	68 BAL	25 cfu/ml in seeded BAL	
16S rDNA	EtBr-stained gel[3]	*Legionella* spp. (30/30) *L.pneumophila* (plus 4 other *Legionella* spp.)	Pure cultures	0.01 pg (ca. 3 cells)	66
5S rDNA	EnviroAmp reverse dot-blot	*Legionella* spp. (not assessed)	52 BAL specimens	3X10⁴ (simulated specimen)	67
mip gene		*L.pneumophila* (not assessed)			
800 bp fragment	Dot-blot with radiolabelled probe	*Legionella* spp. (29/29) *Legionella pneumophila* (not assessed)	Water	Single cell	68, (62)
				Single cell	
5S rDNA	EnviroAmp reverse dot-blot	*Legionella* spp. (not assessed)	Water	ca. 10³ cells/ml	69
mip gene		*L.pneumophila* (not assessed)			
16S rDNA	EtBr-stained gel Southern blot with	*Legionella* spp. (7/10 spp. tested)	Bronchial fluid, water, simulated clinical	10 cfu (simulated specimen) 40-200	70

Table 2 (continued)

Target	Method of detection	Assessment of specificity	Sample type	Sensitivity	Reference[1] (original derivation)
5S rDNA	biotinylated probe		specimens	cfu/ml	72, (63)
	Radiolabelled probe	*Legionella* spp. (7/7 spp. tested)	Water	3-4 cfu 1 fg DNA	
mip gene					
mip gene (nested)	EtBr-stained gels	Not assessed	Water	Not assessed	73
mip gene	EtBr-stained gel Southern blot biotinylated probe	*L. pneumophila* (not assessed)	Paired sera 5/5 LD 0/100 controls	1 pg DNA/ml serum	74
5S rDNA	EtBr-stained gel +/- Southern blot with radiolabelled probe	*Legionella* spp. (6/7 spp. tested)	Urine samples (26 from 21 patients plus 30 controls)	1 fg DNA	75
16S rDNA	EIA with DIG-labelled probe	*Legionella* spp. (5/5 spp. tested gave product but only 4/5 detected in EIA)	BAL, 250 specimens	10^2 ml[-1] (simulated specimen)	76
16S rRNA	*In situ* (whole cell) with fluorescent probes, dot-blot with digoxigenin-labelled probe	*L. pneumophila* (14/14) LEG226, LEG705 *Legionella* spp. (17/22) LEG226 *Legionella* spp. (12/22) LEG705	Water, biofilm, infected amoebae	10^2-10^3 cfu/ml seeded tap water - membrane filtration	77
5S rDNA	EtBr-stained gel PCR + REA	Not assessed	Urine, sera	ca. 0.4 µg/ml DNA seeded urine	78
mip gene (nested)	EtBr stained gel chemiluminescent probe	*Legionella* spp. (8/15) *L. pneumophila* (12/13), *L. bozemanii* (1/2), *L. micdadei* (1/2)	Clinical specimens from one patient	<10 cells (estimated from purified DNA [10 fg DNA])	79
mip gene (nested) 5S rDNA	EtBr stained gel + Southern blot with biotinylated probe(s)	Not assessed	Clinical specimens from one patient (8 intratracheal aspirates)	Not assessed (> DFA)	80
23S-5S rDNA spacer region	Reverse dot-blot	*Legionella* spp. (20/20) *L. pneumophila* (22/22), *L. micdadei* (1/1) *L. anisa* (1/1)	Pure cultures	Not assessed	81

[1]Original reference given in parentheses
[2]cfu = colony forming unit
[3]EtBr = ethidium bromide

Table 3
Description of Oligionucleotide Primers and Nucleic Acid Probes Used in the Detection of *Legionella* spp.

Target	Size of amplicon	Primers 5'-3'	Probes 5'-3'	Reference
chromosomal DNA unknown function	800 bp	LEG-1: GTGATGAGGAATCTCGCTG (19) LEG-2: CTGGCTTCTTCCAGCTTCA (19)	LEG-3: GTGCGTTATGGGGTATTGATCACCA (25)	62
5S rDNA	104 bp	L5SL9: ACTATAGCGATTTGGAACCA (20) L5SR93: GCGATGACCTACTTTCGCAT (20)	L5S-1: CTCGAACTCAGAAGTCAAACATTTCCGC GCCAATGATAGTG TGAGGCTTC (50)	63
mip gene	650 bp	LmipL920: GCTACAGACAAGGATAAGTTG (21) LmipR1548: GTTTTGTATGACTTTAATTCA (21)	Lmip-1: TTTGGGGAAGAATTTTAAAATCAAGG CATAGATGTTAATCCGGAAGCAA (50)	64, (63)
mip gene	650 bp	LmipL920: GCTACAGACAAGGATAAGTTG (21) LmipR1548: GTTTTGTATGACTTTAATTCA (21)	Lmip-1: TTTGGGGAAGAATTTTAAAATCAAGG CATAGATGTTAATCCGGAAGCAA (50)	
chromosomal DNA unknown function	800 bp	LEG-1: GTGATGAGGAATCTCGCTG (19) LEG-2: CTGGCTTCTTCCAGCTTCA (19)	LEG-3: GTGCGTTATGGGGTATTGATCACCA (25)	65, (62)
mip gene	600 bp	Lpm-1: GGTGACTGCGGCTGTTATGG (20) Lpm-2: GGCCAATAGGTCCGCCAACG (20)	Lpm-3: CAGCAATGGCTGCAACCGATGCCAC (25)	
16S rDNA	1 kb 700 bp	LG1 451f: CACTTTCAGTGGGGAGGAG (19) LG3 1425r: GACTATCTACTTCTGGTGCA (20) LG1 451f: CACTTTCAGTGGGGAGGAG (19) LP2 1130r: AGTCCCCACCATCACATGCT (20)	LG4 863r: GCG GTC AAC TTA TCG CGT TT (20)	66
5S rDNA mip gene	107 bp	PT 87: GGCGACTATAGCGATTTGGAA (21) PT 161: GGCGACTATAGCGGTTTGGAA (21) PT 163: GCGATGACCTACTTTCGCATGA (22) PT 165: GCGATGACCTACTTTCACATGA (22)		67, EnviroAmp
chromosomal DNA unknown function	800 bp	†LP-1: GTCATGAGGAATCTCGCTG (19) †LP-2: CTGGCTCTTCCAGCTTCA (19) †original designation LEG-1, LEG-2 LEG448-A: GAGGGTTGATAGGTTAAGAGC (21) LEG-854-B: CGGTCAACTTATCGCGTTTGC (21)	LEG-3: GTGCGTTATGGGGTATTGATCACCA (25)	68, (62)
5S rDNA mip gene	107 bp	EnviroAmp		69
16S rDNA	375 bp	LEP1: GTTAAGAGCTGATTAACTG (19) LEP2: TCATATAACCACAACGCTA (21)	375 bp product amplified with LEP1 and LEP2 from *L. pneumophila* sgp1	70
5S rDNA	107 bp	EnviroAmp		71, EnvironAmp, (62)
mip gene 5S rDNA		EnviroAmp		72, (63), EnviroAmp
mip gene				

Table 3 (continued)

Target	Size of amplicon	Primers 5'-3'	Probes 5'-3'	Reference
mip gene (nested)	650 bp	LmipL920: GCTACAGACAAGGATAAGTTG (21) / LmipR1548: GTTTTGTATGACTTTAATTCA (21) / Lmip976: TAAAAATCAAGGCATAGATG (20) / Lmip1427: AGACCTGAGGGAACATAAAT (20)	GGTCGTCTGATTGATTGATGGTACCGTTTTTGA (33)	73, (63), (85)
mip gene		Primer1: GGCCAATAGGTCCGCCAACG (20) / Primer2: GGTGACTGCGGCTGTTATGG (20)	CAGCAATGGCTGCAACCGATGCCAC (25)	74
5S rDNA	108 bp	EnviroAmp		75, EnviroAmp
16S rRNA			LEG226: TCGGACGCAGGCTAATCT (18) / LEG705: CTGGTGTTCCTTCCGATC (18)	77
16S rDNA	366 bp	p1.2: AGGGTTGATAGGTTAAGAGC (20) / cp3.2: CCAACAGCTAGTTGACATCG (20)	CAACCAGTATTATCTGACCG (20)	76
mip gene (nested)	473 bp	TTAGCTACAGACAAGGATAAG (21) / CAATAGGGTCCTACCTGTCTT (21)	TTGATGGCAAAGCGTACTGCT (21)	79, (85)
5S rDNA	185 bp	TTGATGGCAAAGCGTACTGCT (21) / AGCCTATGTCAGTGACAGCTT (21)		78
5S rDNA	104 bp	L5SL9: ACTATAGGCGATTTGGAACCA (20) / L5SR93: GCGATGACCTACTTTCGCAT (20)		
mip gene (nested)	650 bp	LmipL920: GCTACAGACAAGGATAAGTTG (21) / LmipR1548: GTTTTGTATGACTTTAATTCA (21)	Lmip2: AGCCAGGCGTTGTTGTATTGCCAAGTGTT (30)	80, (63)
5S rDNA	489 bp / 104 bp	LmipL997: TAATCCGGAAGCAATGGCTA (20) / LmipR1466: CGGCCAATAGGTCCGCCAAC (20) / L5SL9: ACTATAGGCGATTTGGAACCA (20) / L5SR93: GCGATGACCTACTTTCGCAT (20)	L5S-1b: GAAGTGAAACATTTCCGCGCCAATGATAGT (30)	
23S-5S rDNA internal spacer region		104L: GGCTGATTGTCTTGACCA (18) / 316R: AGGAAGCCTCAACTATCAT (20)	Genus probe: AACCACCTGATACCATCTCGAACTCAGAA (29) / *L. pneumophila* probe: ACGTGAAACGTATCGTGTAAACTCTGACTC (30) / *L. micdadei* probe: ATGTAAATTGCTCAGACAAATGAATACACAGAGTT (34) / *L. anisa* probe: ATGCGAATACAAGATGTAGGTTGGGC (26)	81

[a]Note: A number in parentheses is the length of the nucleotides.

220

been undertaken and the kit has been shown to be reasonably sensitive for the detection of legionellae in clinical specimens when compared with the direct fluorescent antibody (DFA) test, but it is still inferior to culture *(55)*. Furthermore, the specificity of this kit has been questioned *(83)*, and as it incorporates a radioisotope, it has not been widely used. In 1985, Grimont and colleagues described a ^{32}P-labeled DNA probe specific for *L. pneumophila (84)*. The probe was prepared from *Bam*H1-restricted, total genomic *L. pneumophila* DNA from which DNA sequences coding for rRNA had been removed. These authors found that the probe could be made specific for *L. pneumophila*, but was then not very sensitive, detecting a minimum of 10^5 colony forming units (cfu). A similarly disappointing level of sensitivity was obtained by Engleburg and colleagues who used a probe prepared from the plasmid pSMJ31 into which had been cloned the MOMP gene from *L. pneumophila (85)*. The relatively large quantities of *Legionella* nucleic acid required for detection by these probes has limited their use in the diagnosis of infection and in direct detection from the environment.

Another technique that, in the field of microbiology has been primarily used to distinguish phlyogenetic groups, is *in situ* hybridization *(86)* using fluorescent-labeled oligonucleotide probes that can be detected by standard fluorescent microscopy, confocal laser scanning microscopy (CLSM), or captured by charge-coupled device (CCD) cameras. Manz and colleagues used fluorescently labeled probes in conjunction with CSLM to determine the identity of environmental and clinical isolates of suspected legionellae by whole-cell hybridization *(77)*. The results obtained were in good accordance with standard immunofluorescence results. A combination of membrane filtration and whole-cell hybridization allowed the detection of 10^3–10^4 cfu/mL of *L. pneumophila* in tap water. Such techniques, possibly in conjunction with PCR (*in situ* PCR), may prove to be a useful tool in the study of the microbial ecology of this organism in the future.

2.2.2. The Polymerase Chain Reaction

The theoretical sensitivity of PCR (i.e., detection of one copy of the target sequence) has led to its increasing use as an alternative to direct probing, which usually requires a relatively large number of cells. Alternatives to radioisotopic labels now claim equivalent sensitivity without the inherent disadvantages. Thus PCR (in conjunction with nonradioisotopic detection systems) is now seen as the molecular method of choice, promising the advantages of both specificity and sensitivity.

As mentioned in **Subheading 2.1.2.**, exposure to biocides, elevated temperatures *(64)*, or antibiotics may result in the failure to isolate organisms. PCR offers the possibility of detecting organisms even in these circumstances, although the significance of such PCR results must be placed in context. Adequate controls (particularly for false-negatives) preferably containing an internal positive control should always be included.

A wide variety of methods of sample preparation applicable to both clinical and environmental specimens now exist (both commercial and in-house). However, for those seeking to apply molecular techniques to samples other than organisms grown in pure culture, critical comparison of the above methods is still needed.

The PCR-based assays for the diagnosis of *Legionella* infection have focused mainly on three target sites; the macrophage infectivity potentiator gene (*mip*), and the 5S and 16S ribosomal RNA (rRNA) genes. Whereas a variety of formats is becoming available for the detection and confirmation of PCR products (*see* Chapter 5), the majority of amplification products are detected by UV transillumination of ethidium bromide (EB)-stained agarose gels following electrophoresis of the reaction mixtures. Confirmation of the products can be demonstrated by Southern or dot-blot hybridization with labeled oligonucleotides. One reported method for legionellae *(76)* uses an enzyme-linked immunoassay (EIA) format. Here the biotinylated product is captured in streptavidin-coated microtitre plates, and subsequently hybridized with a digoxigenin-labeled oligonucleotide probe, which is detected using an enzyme-linked antibody and the subsequent addition of dye reagents. This method has obvious advantages over gel electrophoresis when large numbers of specimens are to be rapidly processed. However, the technology is largely untested and one disadvantage is that unless the samples are electrophoresed, there is no visual confirmation (by size) of the validity of the result.

Some of the reported methods were originally designed for detection of legionellae in environmental (usually water) samples *(62,64)*, but have been adapted or tested in a clinical context *(65)*. Clinical samples so far investigated by PCR include respiratory tract specimens *(65,67,70,71,76,80)*, sera *(74,78)*, urine *(75,78)*, and body tissues *(71)* (*see* **Table 2**). These assays use so-called *Legionella* genus-specific primers (directed against the rRNA or *mip*-like genes) or *L. pneumophila* species-specific primers (directed against the *mip* mRNA/gene). It should be noted that although *mip*-like sequences occur in species other than *L. pneumophila* species, none of the *mip* PCR assays described can be considered to be genus-specific. Many combine the use of two target genes, rRNA to detect *Legionella* spp. and the *mip* gene to specifically detect *L. pneumophila*. Often single to 10 cell sensitivity is claimed, however, the majority of these assays have only been tested on pure cultures, seeded samples, or small numbers of clinical specimens. Several studies have demonstrated PCR-positive results in culture-negative samples *(70)*. Such results highlight the dilemma arising from using a potentially more sensitive method than the previous "gold standard" method. Independent evaluations of large numbers of clinical samples are needed to determine the ultimate usefulness of these techniques.

2.2.2.1. Mip Gene

The macrophage infectivity potentiator gene encodes a 24-kD surface protein (Mip) that appears to enhance the ability of *L. pneumophila* to infect human macrophages *(87)*. Although all *Legionella* species have Mip-like proteins, until recently, analysis of the *mip* genes has been confined only to *L. pneumophila* and a few *(138)*, other species. This has led to the development of *L. pneumophila*-specific assays *(63)* and assays that detect several *(65)*, or many *(79) Legionella* species. Unfortunately, relatively few clinical specimens have been examined using these methods so their utility is largely unknown. In a retrospective study, Jaulhac and colleagues *(65)* examined 68 frozen bronchioalveolar lavage (BAL) fluid specimens from cases of suspected legionellosis. Eight culture-positive cases were also PCR-positive. Of the 60 culture-negative specimens, seven were positive by PCR of which four came from patients with serologic evidence of legionella infection. The remaining three could not be confirmed by an independent method, but had clinical features of legionellosis.

Perhaps one of the most potentially useful applications of the *mip*-PCR was described by Lindsay and colleagues *(74)*. These workers examined acute and convalescent serum samples from five patients with proven LD and approx 100 patients with no evidence of LD. All sera from the five LD patients yielded the expected 630-bp, band, the identity of which was confirmed by *mip*-specific oligonucleotide probing. No positive PCR results were obtained for specimens from the approx 100 control patients.

2.2.2.2. Ribosomal RNA Genes

The genes that code for rRNAs (including the 5S, 16S, and 23S) are highly conserved. The occurrence of variable regions within the sequences of these genes together with ready access to nucleic acid sequence data, has made them an ideal target for PCR assays of many microbial groups. However, there is a need to evaluate thoroughly (or re-evaluate) such assays with clinical specimens, both for specificity and sensitivity.

In bacteria, the 5S rRNA molecule has approx 120 nucleotides (nt), the 16S approx 1500 nt, and the 23S approx 3000 nt. The 5S rRNA gene has generally been used to discriminate at the level of genus and above, and the 16S rRNA gene at the level of genus and below. Whereas the 23S rRNA gene would allow for an even greater choice in the level of specificity desired, this approach has generally been hindered by lack of sequence information compared to that for 16S rRNA.

1. 5S rRNA. Mahbubani and co-workers *(63)* described the detection of nine *Legionella* species using primers specific for the 5S rRNA gene on pure cultures. Faint DNA bands were seen with some non-*Legionella* species following amplification, however, *Legionella*-specific hybridization was achieved using an internal oligonucleotide probe. Using the same target, Maiwald and coworkers *(72)* applied this system to assess the surveillance of contamination of man-made water systems by legionellae compared with conventional culture. These authors concluded that culture and PCR complemented each other and, although culture-negative (for *Legionella*) samples may be positive by PCR, the converse was also found. Some of these discrepancies may have resulted from the presence of rust and other PCR inhibitors. It is of note that the use of Chelex resin (Biorad) for sample preparation removed this inhibitory effect in most cases. The introduction of Chelex resin for clinical sample preparation has also been described and this is currently in use for a wide variety of PCR-based assays *(67,72)*.

2. 16S rRNA. The 16S rRNA gene has been used as a target for PCR assays for both *Legionella* spp. and *L. pneumophila (66,70,76)*. In one study, *Legionella* genus-specific primers yielded amplification products of the expected size (approx 1 kb) from DNA from all the species of *Legionella* available at that time (30/30), and also the intracellular amoebal pathogen, *Legionella*-like amoebal pathogen-3 (LLAP-3), now *L. lytica (66)*. Subsequent sequence data has revealed that not all of these species have complete homology with the primer sequences. This emphasizes the need for constant re-evaluation of the primers as new species and sequence data from both target and nontarget groups become available. Attempts to obtain an assay specific for *L. pneumophila*, however, resulted in products not only from all three subspecies of *L. pneumophila*, but also four non-*pneumophila* species. Internal probes to confirm the specificity of the amplified products have an obvious role here. As the number of species continues to grow (and increasing sequence heterogeneity is seen), the description of genus-specific and species-specific assays must be regarded with caution. Primers designed to be *Legionella*-specific by other workers have also shown varying specificity (**Table 2**, and **refs. *70,76***).

3. Identification

Identification of legionellae at genus level is straightforward. A bacterium that is a Gram-negative rod, catalase-positive, and grows on complete buffered charcoal yeast extract agar (BCYE), but not on the same medium lacking supplemental *L*-cysteine, can be presumptively identified as a *Legionella* species *(55)*. However, identification to species level is, with the exception of *L. pneumophila*, often very difficult.

3.1. Conventional Methods

3.1.1. Confirmation of Identification at the Genus Level

Legionellae differ from most other Gram-negative bacteria as their cell walls contain large amounts of branched-chain fatty-acids and unusual members of the ubiquinone series which contain 9–15 isoprenyl unit side-chains. Although these characteristics may be relatively easily determined by gas liquid chromatography and mass spectrometry, this method of identification cannot be considered routine.

3.1.2. Identification at the Species Level

Identification of legionellae at the species level is difficult for almost all laboratories *(55)*. However, there is an excellent species-specific monoclonal antibody available to identify *L. pneumophila*, the species that most commonly causes disease. A cheaper, but almost as reliable test for *L. pneumophila* is the ability of this species to hydrolyze hippurate. Strains of a few non-*pneumophila* species are also reported to give a positive result in this test, but in practice these are rarely encountered or give a very weakly positive result.

Identification of other species may be aided by application of a few simple tests, such as autofluorescence under long wavelength UV and the bromocresol-purple spot test. Exposure of legionellae to long wavelength UV differentiates *Legionella* species in to three broad groups (red-, blue/white-, and nonautofluorescent species). The blue-white autofluorescent group comprises eight species and differentiation of these by serologic means is technically very difficult *(55)*. Antigenic analysis of *Legionella* spp. relies on the use of unabsorbed rabbit hyperimmune antisera. Most strains react strongly with one or more of these sera, and by comparison with known strains, a tentative identification can be made. Unfortunately, a significant proportion of strains give multiple reactions and hence confusing results. In these cases, laborious and expensive serum absorption studies or more recently, molecular methods are the only way to obtain a precise identity *(55)*.

3.2. Molecular Methods

3.2.1. Taxonomic Background

The family Legionellaceae and genus *Legionella* were defined in 1979 for a single species *Legionella pneumophila (4)*. The evidence for the recognition of this new species, genus, and family came partly from considerations of the phenotypic features, but mainly from measurements of DNA/DNA pairing which showed low levels of homology between strains of *L. pneumophila* and other Gram-negative bacteria of a similar mol% G+C content (approx 39 mol%).

There followed a succession of phylogenetic studies addressing the taxonomy of the family Legionellaceae *(88–90)*. These in the main concentrated on characterizing the 16S rRNA sequences from *Legionella* species. To date the descriptions of 41 species of *Legionella* have been validly published; all have been established by DNA/DNA and/or 16S rRNA phylogenetic analysis (*see* **Table 1**).

Recently, the 16S rRNA gene sequences of 12 isolates of *Legionella*-like amoebal pathogens have been reported *(91–93)*. In common with typical legionellae, these organisms infect amoebal species, exhibiting the same stages of infective growth, but in contrast they cannot be cultured in vitro. One strain, (LLAP-3), was recovered from a patient with pneumonia by amoebal enrichment, but to date the others have only been recovered from the environment. Phylogenetically all these organisms are clearly legionellae. Ten, including LLAP-3, form a single subgroup, and possibly a single species (*L. lytica*), within the genus. They are most closely related to, but distinct from, the three subspecies of *L. pneumophila*. However, the remaining two, although clearly legionellae have no close evolutionary relationship with any named species and are probably representatives of new species *(93)*.

3.2.2. Identification at the Genus Level

The identification of members of the genus *Legionella* by detection of specific nucleic acid sequences requires the target sequence to be a conserved region of the genome. As mentioned in **Subheading 2.2.2.2.**, rRNA (gene) sequences have been exploited to provide targets for probes of broad specificity and an I^{125}-labeled probe is commercially available (Gen-Probe) for *Legionella* detection and identification. This product appears to be effective for identification, although the results may not always be simple to interpret. One study revealed a number of probe-positive, culture-negative, specimens in which the presence of legionellae could not be confirmed *(83)*. It is possible that these specimens may have contained nonculturable legionellae, such as *L. lytica*, but it is also possible that the sensitivity of the assay was achieved at the cost of specificity.

Use of the PCR represents the best opportunity to overcome such problems. A PCR assay for detection of legionellae in environmental specimens is now commercially available (EnviroAmp, Perkin-Elmer, Norwalk, CT). The kit is based on the use of "genus-specific" primers from within the 5S rRNA gene. However, these primers were based on sequence data only from a few species. As there are now over 40 species and this number is certain to rise, this emphasises the need for continuous re-evaluation of the specificity of such tests.

3.2.3. Identification at the Species Level

As discussed above, measurement of DNA/DNA relative binding remains the "gold-standard" for species identification against which new methods must be judged. However, the method is technically demanding, time-consuming, and results obtained in different laboratories or using alternative methods may not be comparable. In an attempt to overcome some of these problems, Grimont and colleagues developed an alternative molecular approach to identification based on rRNA gene restriction digest patterns *(94)*. A similar method has also been developed that relies upon recognition of the pattern of bands revealed when Southern blotted *Nci*I total restriction fragments are hybridized to a cloned rRNA cistron from *L. pneumophila (95)*. In both methods a pattern of bands is obtained for an unknown strain that can be compared with those of known strains, and the identification established. Each band pattern is unique to a particular species, but one species can have more than one pattern. Consequently, if a novel band pattern is obtained, DNA studies are still needed to determine if they indicate a new species or a new pattern of a known species.

Copping and colleagues have applied PCR of the 16S rDNA followed by restriction enzyme analysis (REA) of the products to allow species-specific identification of the eight species comprising the blue-white autofluorescent group *(96)*. Although requiring multiple digestions, and, hence, being a little cumbersome, this approach might easily be extended to other serologically indistinct groups of species.

It has been proposed that the differentiation of many members of the family Legionellaceae can be achieved by intergenic 16S and 23S rRNA gene sequence length polymorphism analysis *(97)*. Of 38 species examined, seven were indistinguishable (*L. erythra* from *L. rubrilucens*; *L. anisa* and *L. cherrii* from *L. tucsonensis*; *L. quateirensis* from *L. shakespearei*); the number and sizes of the bands varying between the other species. This description was based upon the extrapolation of the pattern from the type strain of each species, but few "wild" strains were examined. However, this method has recently been evaluated using clinical and environmental isolates and computer-aided analysis of the patterns to help standardize results *(98)*. Although some species-specific patterns were identified, intraspecies variation was also evident.

Another recently described method, based on amplification of the 23S-5S rDNA intergenic spacer region (ISR) *(81)*, claims *Legionella* species-specific detection using PCR and reverse dot-blotting. This method involves amplification of the complete ISR together with portions of the 5S and 23S rDNA. A *Legionella* genus probe targets the highly conserved 5S rDNA sequence and *Legionella* species probes target a unique part of the ISR. Individual *Legionella* species-specific probes were designed for *L. pneumophila*, *L. micdadei*, and *L. anisa*, although only one strain of each non-*pneumophila* species was tested (*see* **Table 3**).

4. Epidemiologic Investigations

4.1. Brief Overview of Epidemiology and Rationale for Typing

Legionellae are environmental organisms and no person-to-person transmission has ever been convincingly documented. Consequently, when a case of legionellosis is recognized there is the possibility that others may become infected from the same environmental source if preventative action is not taken.

Furthermore, although outbreaks of LD are uncommon, accounting for only 25% of cases, when they do occur they can be dramatic and severe *(99)*. The primary reason for epidemiologic typing of *Legionella* strains is, therefore, to help identify the environmental source giving rise to cases of legionellosis so that it may be controlled and further cases prevented.

In a few instances, the species or serogroup causing infection is encountered so rarely that isolating the same species or serogroup from an associated environmental site is good evidence that it was the likely source of infection *(100)*. Such situations are, however, rare as although about half of the described species have been implicated in human infection (*see* **Table 1**) the vast majority of cases are caused by *L. pneumophila* and >90% of these are serogroup 1 strains *(55)*. Not only is *L. pneumophila* serogroup 1 the most common cause of legionellosis, but also the serogroup most frequently encountered in the environment. Consequently, isolates of this serogroup must be further differentiated if any convincing epidemiologic link is to be established.

4.2. Conventional Approaches

Perhaps because legionellae were discovered relatively recently, few traditional typing techniques have been developed to discriminate between strains of interest. The notable exception to this is antibody subtyping, which is widely used for the discrimination of *L. pneumophila* serogroup 1 isolates, and even here the reagents used are generally derived from monoclonal antibodies (mAbs).

4.2.1. Monoclonal Antibody Subgrouping

This technique was first applied to type *L. pneumophila* by Plouffe and colleagues in 1983 *(101)*. Since then numerous workers have prepared panels of mAbs and several typing schemes have been described *(102)*. The methods are rapid, simple to undertake, and usually easy to interpret. Typically, the mAbs are used in an indirect immunofluorescent assay, but they have been used in a variety of other formats *(103)*.

To date, the most thoroughly standardized set of mAbs is referred to as the "International panel" This set allows ten clearly defined subgroups to be recognized. Unfortunately, the majority of clinical strains carry an epitope that reacts with the second mAb in the panel. This epitope (referred to as mAb2+) is common to 5 of the 10 subgroups and, hence, the discrimination is very limited where clusters of cases are being sought. Also, there is good evidence that the mAb subgroup of a strain in the environment is not always a stable phenotype *(104,105)*.

4.3. Molecular Methods

The coincidence of the discovery of legionellae and the development of molecular methods for typing has meant that almost all of these techniques have been applied to legionellae as they are developed, but few have been used with any consistency, and so their value is largely unknown. The clear advantage that many of these techniques have over antibody typing is that they are applicable at least across a species (e.g., *L. pneumophila* of any serogroup) and more often across the genus. That being said, there are only a few reports where non-*L. pneumophila* serogroup 1 strains have been studied in any detail *(106–108)*.

4.3.1. Plasmid Analysis

The analysis of plasmid DNA (*see* Chapter 4) was one of the first methods adopted for typing legionellae *(109)*, but for several reasons, has not proved to be very useful. Legionellae appear not to carry a very wide range of plasmids and also, to date, the vast majority of clinical, and all UK outbreak, strains have been plasmidless. The number of plasmids found in one strain usually varies from one to three *(110,111)* and range in size from approx 20–125 mega Daltons. These estimations are often made from gels by comparison with plasmid standards, consequently, there may be significant error in sizing the largest ones. Also, it has been reported that these plasmids may be sensitive to shearing *(112)* and strain variation in susceptibility to cell lysis can yield false-negative results *(113)*. The alkaline lysis method of Kado and Liu (*114*; *see* Chapter 4) has been used by several workers and appears to yield satisfactory results *(105,110,111,115)*. As isolates obtained from a variety of unrelated sources can yield plasmids of similar sizes, further analysis by restriction enzyme analysis, for example, is required to characterize the plasmids definitively *(111)*.

The purpose of these extrachromosomal elements in legionellae is largely unknown, and no virulence markers have been associated with them. The only well-documented function is UV light resistance *(116)*. Whereas no antibiotic resistance has yet been associated with legionella plasmids, the acquisition of transposons conferring such resistance could have severe consequences in the treatment of LD *(117)*.

4.3.2. Isoenzyme Typing

Selander and colleagues *(118)* first used the analysis of electrophoretic mobilities of enzymes from strains of *L. pneumophila* as an indirect method to investigate the genetic structure of populations of this species. They found 62 distinctive electrophoretic types (ETs) among 292 isolates and concluded that *L. pneumophila* was a very heterogeneous species. Given this degree of discrimination, isoenzyme analysis was taken up by several groups of workers as a typing method for use in epidemiologic investigations of outbreaks caused by *L. pneumophila* or *Legionella* spp. *(119,120)*. However, the method has not found wide application because it is technically demanding, very labor intensive, and not easily standardized.

4.3.3. Restriction Enzyme Analysis

One of the most widely used and most effective ways to discriminate between strains of legionella is by comparison of the restriction fragment length polymorphisms (RFLPs) that are generated by digestion of chromosomal DNA by highly specific restriction endonucleases (*see* Chapter 2). A wide range of approaches have been taken to the analysis of these fragments, the simplest being by direct visualization of electrophoretically separated fragments on an agarose gel. Using this approach, van Ketel and colleagues *(121)* analyzed *Eco*RI/*Hin*dIII digests and were able, for example, to distinguish between cases of LD acquired in their hospital, and those imported from hospitals in the surrounding area. The disadvantage of direct visualization is that the resulting REA patterns are often very complex, and so are difficult to interpret and record. Even where modifications are made, such as the use of polyacrylamide and silver staining *(122)* these problems remain. However, several approaches have been successfully developed to overcome these problems.

4.3.3.1. Pulsed-Field Gel Electrophoresis (PFGE)

Restriction endonucleases that only cleave the chromosomal DNA into a few fragments have been used and then pulsed-field gel electrophoresis is applied to separate these large fragments *(123)* (*see* Chapter 3). Although many enzymes have been examined for their suitability *(119,124,125)*, most workers have chosen to use *Sfi*I *(119,126,127)*. The discrimination is good, where only a few strains are being examined, and generally patterns are easily identified and compared. Where large numbers of strains are being studied, or when the analyses extend over a period of months or years, pattern recognition is still a problem. It remains to be seen whether or not computer-aided analysis will enable the discriminatory power of PFGE to be harnessed within the framework of a traditional typing scheme.

4.3.3.2. Ribotyping and RFLP Typing

An alternative to using infrequently cutting restriction endonucleases is to transfer the separated fragments onto membranes (Southern blots) and to probe these with nucleic acid sequences that hybridize to only a few of the transferred fragments. Two types of probe are widely used; homologous or heterologous rRNA *(128,129)* (this approach is called "ribotyping;" *see* Chapter 2) or randomly cloned chromosomal DNA *(130)*. The authors have made extensive use of a RFLP typing method where two randomly cloned chromosomal DNA probes are used to reveal a simple pattern of bands on Southern blots of *Nci*I-digested DNAs *(130)*. The method, which is *L. pneumophila* specific, has been thoroughly evaluated and shown to be highly discriminatory (>100 types), reproducible and stable. Furthermore the patterns obtained are simple enough to be easily identified and allocated to a "type." The technique, in combination with mAb subgrouping, has been successfully applied in the investigation of many outbreaks of legionellosis *(57,99,131)*.

Ribotyping is a more widely reported, but less well-evaluated, technique than the above *(128,129,132,133)*. The usual probe is 16S and 23S rRNA extracted from *Escherichia coli*, although *L. pneumophila* rRNA can also be used. Discrimination is usually poor although by combining the results obtained from four restriction reactions (*Hin*dIII, *Nci*I, *Cla*I, and *Pst*I), Bangsborg and colleagues *(128)* achieved a highly discriminatory method.

4.3.4. PCR-Based Typing Methods

Most recently, studies have focused on PCR-based methodologies with the intention of developing rapid fingerprinting methods that do not require large biomass for DNA extraction (*see* Chapters 6 and 7). A variety of primers have been chosen in these studies. Sandery and colleagues *(134)* prepared a range of random primers from 8–16 nt in length, whereas Gomez-Lus and colleagues *(135)* chose to use a bacteriophage (M13) 21-nt sequence as an arbitrary primer. Other workers have used primers based on the sequence of highly conserved repetitive elements found in most eubacteria *(136,137)*. In most of these studies, reproducibility has been found to be good, but the range and number of strains examined was very limited.

5. Future Goals

It is clear that many of the molecular methods described above offer distinct advantages over conventional approaches for the diagnosis and epidemiologic investigation of infections caused by *Legionella* spp. However, despite a considerable literature, few of these methods are widely, or routinely, in use. This results, in large part, from the rapid rate of technological advance: each new method appearing to offer improved performance over those previously described. It is also because of the lack of any extensive evaluations of most of these methods, hence, it is difficult to assess the relative merits of each approach. This latter point is possibly explained by the relative infrequency with which legionella infections are seen; hence it is difficult to obtain specimens from large series of patients with which to evaluate the techniques. In an attempt to overcome these limitations, multicenter studies are needed, and these are now being initiated. Not only should such studies allow meaningful assessment of the performance of the various assays, but they should also ensure that, where appropriate, procedures are internationally standardized. This is particularly important in, for example, the investigation of outbreaks of travel-associated LD, where simultaneous analysis of samples may be undertaken in laboratories in several different countries *(57)*.

References

1. McDade, J. E., Shepard, C. C., Fraser D. W., Tsai, T. R., Redus, M. A., Dowdle, W. R., and the laboratory investigation team. (1977) Legionnaires' disease: isolation of a bacterium and demonstration of its role in other respiratory disease. *N. Engl. J. Med.* **297,** 1197–1203.

2. Fliermans, C. B., Cherry, W. B., Orrison, L. H., Smith, S. J., Tison, D. L., and Pope, D. H. (1981) Ecological distribution of *Legionella pneumophila. Appl. Environ. Microbiol.* **41,** 9–16.

3. Tobin J. O'H., Swann R. A., and Bartlett C. L. R. (1981) Isolation of *Legionella pneumophila* from water systems: methods and preliminary results. *Br. Med. J.* **282,** 515–517.

4. Brenner, D. J., Steigerwalt, A. G., and McDade, J. E. (1979) Classification of the Legionnaires' disease bacterium: *Legionella pneumophila,* genus novum, species nova of the family Legionellaceae familia nova. *Ann. Intern. Med.* **90,** 656–658.

5. Brenner, D. J., Steigerwalt, A. G., Epple, P., Bibb,. W. F., McKinney, R. M., Starnes, R. W., Colville, J. M., Selander, R. K., Edelstein, P. H., and Moss, C. W., (1988) *Legionella pneumophila* serogroup Lansing 3 isolated from a patient with fatal pneumonia, and descriptions of L. *pneumophila* subsp. *pneumophila* subsp. nov., and *L. pneumophila* subsp. *fraseri* subsp. nov., and L. *pneumophila* subsp. *pascullei* subsp. nov. *J. Clin. Microbiol.* **26,** 1695–1703.

6. McKinney, R. M., Thacker, L., Harris, P. P., Lewallen, K. R., Hebert, G. A., Edelstein, P. H., and Thomason B. M. (1979) Four serogroups of Legionnaires' disease bacteria defined by direct immunofluorescence. *Ann Intern. Med.* **90,** 621–624.

7. Morris, G. K., Patton, C. M., Feeley, J. C., Johnson, S. E., Gorman, G., Martin, W. T., Skaliy, P., Mallison, G. F., Politi, B. D., and Mackel, D. C. (1979) Isolation of the Legionnaires' disease bacterium from environmental samples. *Ann. Intern. Med.* **90,** 664–666.

8. Nagington, J., Wreghitt, T. G., and Smith, D. J. (1979) How many Legionnaires? *Lancet* **ii,** 536–537.

9. England, A. C., McKinney, R. M., Skaliy, P., and Gorman, G. W. (1980) A fifth serogroup of *Legionella pneumophila. Ann. Intern. Med.* **93,** 58–59.

10. McKinney, R. M., Wilkinson, H. W., Sommers, H. M., Fikes, B. J., Sasseville, K. R., Yungbluth, M. M., and Wolf, J. S. (1980). *Legionella pneumophila* serogroup six: isolation from cases of legionellosis, identification by immunofluorescence staining, and immunological response to infection. *J. Clin. Microbiol.* **12,** 395–401.

11. Bibb, W. F., Arnow, P. M., Dellinger, D. L., and Perryman, S. R. (1983) Isolation and characterisation of a seventh serogroup of *Legionella pneumophila. J. Clin. Microbiol.* **17,** 346–348.

12. Bissett, M. L., Lee, J. O., and Lindquist, D. S. (1983) New serogroup of *Legionella pneumophila,* serogroup 8. *J. Clin. Microbiol.* **17,** 887–891.

13. Edelstein, P. H., Bibb, W. F., Gorman, G. W., Thacker, W. L., Brenner, D. J., Wilkinson, H. W., Moss, C. W., Buddington, R. S., Dunn, C. J., Roos, P. J., and Meenhorst, P. L. (1984) *Legionella pneumophila,* serogroup 9: a cause of human pneumonia. *Ann. Intern. Med.* **101,** 196–198.

14. Meenhorst, P. L., Reingold, A. L., Groothuis, D. G., Gorman, G. W., Wilkinson, H. W., McKinney, R. M., Feeley, J. C., Brenner, D. J., and van Furth, R. (1985) Water–related nosocomial pneumonia caused by *Legionella pneumophila* serogroups 1 and 10. *J. Infect. Dis.* **152,** 356–364.

15. Thacker, W. L., Benson, R. F., Wilkinson, H. W., Ampel, N. M., Wing, E. J., Steigerwalt, A. G., and Brenner, D. J. (1986) 11th serogroup of *Legionella pneumophila* isolated from a patient with fatal pneumonia. *J. Clin. Microbiol.* **23,** 1146–1147.

16. Thacker, W. L., Wilkinson, H. W., Benson, R. F., and Brenner, D. J. (1987) *Legionella pneumophila* serogroup 12 isolated from human and environmental sources. *J. Clin. Microbiol.* **25,** 569,570.

17. Lindquist, D. S., Nygaard, G., Thacker, W. L., Benson, R. F., Brenner, D. J., and Wilkinson, H. W. (1988) Thirteenth serogroup of *Legionella pneumophila* isolated from patients with pneumonia. *J. Clin. Microbiol.* **26,** 586–587.

18. Benson, R. F., Thacker, W. L., Wilkinson, H. W., Fallon, R. J., and Brenner, D. J. (1988) *Legionella pneumophila* serogroup 14 isolated from patients with fatal pneumonia. *J. Clin. Microbiol.* **26,** 382.

19. Lück, P. C., Helbig, J. H., Ehret, W., and Ott, M. (1995) Isolation of a *Legionella pneumophila* strain serologically distinguishable from all known serogroups. *Zbl. Bakt.* **282,** 35–39.

20. Benson, R. F., Thacker, W. L., Lanser, J. A., Sangster, N., Mayberry, W. R., and Brenner, D. J. (1991) *Legionella adelaidensis*, a new species isolated from cooling tower water. *J. Clin. Microbiol.* **29,** 1004–1006.

21. Gorman, G. W., Feeley, J. C., Steigerwalt, A., Edelstein, P. H., Moss, C. W., and Brenner, D. J. (1985) *Legionella anisa*: a new species of *Legionella* isolated from potable waters and a cooling tower. *Appl. Environ. Microbiol.* **49,** 305–309.

22. Wilkinson, H. W., Thacker, W. L., Benson, R. F., Polt, S. S., Brookings, E., Mayberry, W. R., Brenner, D. J., Gilley, R. G., and Kirklin, J. K. (1987) *Legionella birminghamensis* sp. nov. isolated from a cardiac transplant recipient. *J. Clin. Microbiol.* **25,** 2120–2122.

23. Brenner, D. J., Steigerwalt, A. G., Gorman, G. W., Weaver, R. E., Feeley, J. C., Cordes, L. G., Wilkinson, H. W., Patton, C., Thomason, B. M., and Sasseville, K. R. L. (1980) *Legionella bozemanii* sp. nov. and *Legionella dumoffii* sp. nov.: classification of two additional species of *Legionella* associated with human pneumonia. *Curr. Microbiol.* **4,** 111–116.

24. Tang, P. W., Toma, S., Moss, C. W., Steigerwalt, A. G., Cooligan, T. G., and Brenner, D. J. (1984) *Legionella bozemanii* serogroup 2: a new etiological agent. *J. Clin. Microbiol.* **19,** 30–33.

25. Wilkinson, H. W., Drasar, V., Thacker, W. L., Benson, R. F., Schindler, J., Potuznikova, B., Mayberry, W. R., and Brenner, D. J. (1988) *Legionella moravica* sp. nov. and *Legionella brunensis* sp. nov. isolated from cooling-tower water. *Ann. Inst. Pasteur.* **139,** 393–402.

26. Brenner, D. J., Steigerwalt, A. G., Gorman, G. W., Wilkinson, H. W., Bibb, W. F., Hackel, M., Tyndall, R. L., Campbell, J., Feeley, J. C., Thacker, W. L., Skaliy, P., Martin, W. T., Brake, B. J., Fields, B. S., McEachern, H. V., and Corcoran, L. K. (1985) Ten new species of *Legionella. Int. J. Syst. Bacteriol.* **35,** 50–59.

27. Thacker, W. L., Benson, R. F., Staneck, J. L., Vincent, S. R., Mayberry, W. R., Brenner, D. J., and Wilkinson, H. W. (1988) *Legionella cincinnatiensis* sp. nov. isolated from a patient with pneumonia. *J. Clin. Microbiol.* **26,** 418–420.

28. Saunders, N. A., Doshi, N., and Harrison, T. G. (1992) A second serogroup of *Legionella erythra* serologically indistinguishable from *Legionella rubrilucens. J. Appl. Bacteriol.* **72,** 262–265.

29. Thacker, W. L., Benson, R. F., Hawes, L., Gidding, H., Dwyer, B., Mayberry, W. R., and Brenner, D. J. (1991) *Legionella fairfieldensis* sp. nov. isolated from cooling tower waters in Australia. *J. Clin. Microbiol.* **29,** 475–478.

30. Herwaldt, L. A., Gorman, G. W., McGrath, T., Toma, S., Brake, B., Hightower, A. W., Jones, J., Reingold, A. L., Boxer, P. A., Tang, P. W., Moss, C. W., Wilkinson, H. W., Brenner, D. J., Steigerwalt, A. G., and Broome, C. V. (1984) A new *Legionella* species, *Legionella feeleii* species nova, causes Pontiac fever in an automobile plant. *Ann. Intern. Med.* **100,** 333–338.

31. Thacker, W. L., Wilkinson, H. W., Plikaytis, B. B., Steigerwalt, A. G., Mayberry, W. R., Moss, C. W., and Brenner, D. J. (1985). Second serogroup of *Legionella feeleii* strains isolated from humans. *J. Clin. Microbiol.* **22,** 1–4.

32. Dennis, P. J., Brenner, D. J., Thacker, W. L., Wait, R., Vesey, G., Steigerwalt, A. G., and Benson, R. F. (1993) Five new *Legionella* species isolated from water. *Int. J. Syst. Bacteriol.* **43,** 329–337.

33. Morris, G. K., Steigerwalt, A., Feeley, J. C., Wong, E. S., Martin, W. T., Patton, C. M., and Brenner, D. J. (1980) *Legionella gormanii* sp. nov. *J. Clin. Microbiol.,* **12,** 718–721.

34. Bornstein, N., Marmet, M., Surgot, M., Nowicki, H., Meugnier, H., Fleurette, J., Ageron, E., Grimont, F., Grimont, P. A. D., Thacker, W. L., Benson, R. F., and Brenner, D. J. (1989) *Legionella gratiana* sp. nov. isolated from French spa water. *Res. Microbiol.* **140,** 541–552.

35. Wilkinson, H. W., Thacker, W. L., Steigerwalt, A. G., Brenner, D. J., Ampel, N. M., and Wing, E. J. (1985) Second serogroup of *Legionella hackeliae* isolated from a patient with pneumonia. *J. Clin. Microbiol.* **22,** 488–489.

36. Bercovier, H., Steigerwalt, A. G., Derhi-Cochin, M., Moss, C. W., Wilkinson, H. W., Benson, R. F., and Brenner D. J. (1986) Isolation of legionellae from oxidation ponds and fishponds in Israel and description of *Legionella israelensis* sp. nov. *Int. J. Syst. Bacteriol.* **36,** 368–371.

37. Cherry, W. B., Gorman, G. W., Orrison, L. H., Moss, C. W., Steigerwalt, A. G., Wilkinson, H. W., Johnson, S. E., McKinney, R. M., and Brenner, D. J. (1982) *Legionella jordanis*: a new species of *Legionella* isolated from water and sewage. *J. Clin. Microbiol.* **15,** 290–297.

38. Thacker, W. L., Dyke, J. W., Benson, R. F., Havlichek, D. H., Robinson-Dunn, B., Stiefel, H., Schneider, W., Moss, C. W., Mayberry, W. R., and Brenner, D. J., (1992) *Legionella lansingensis* sp. nov. isolated from a patient with pneumonia and underlying chronic lymphocytic leukemia. *J. Clin. Microbiol.* **30**, 2398–2401.

39. McKinney, R. M., Porschen, R. K., Edelstein, P. H., Bisset, M. L., Harris, P. P., Bondell, S. P., Steigerwalt, A. G., Weaver, R. E., Ein, M. E., Lindquist, D. S., Kops, R. S. and Brenner, D. J. (1981) *Legionella* longbeachae species nova, another etiologic agent of human pneumonia. *Ann. Intern. Med.* **94**, 739–743.

40. Bibb, W. F., Sorg, R. J., Thomason, B. M., Hicklin, M. D., Steigerwalt, A. G., Brenner, D. J., and Wulf, M. R. (1981) Recognition of a second serogroup of *Legionella longbeachae. J. Clin. Microbiol.* **14**, 674–677.

41. Hookey, J. V., Saunders, N. A., Fry, N. K., Birtles, R. J., and Harrison, T. G. (1996). Phylogeny of Legionellaceae based on small-subunit ribosomal DNA sequences and proposal of *Legionella lytica* comb. nov. for *Legionella*-like amoebal pathogens. *Int. J. Syst. Bacteriol.* **46**, 526–531.

42. Hébert, G. A., Steigerwalt, A. G., and Brenner, D. J. (1980) *Legionella micdadei* species nova: classification of a third species of *Legionella* associated with human pneumonia. *Curr. Microbiol.* **3**, 255–257.

43. Orrison, L. H., Cherry, W. B., Tyndall, R. L., Fliermans, C. B., Gough, S. B., Lambert, M. A., McDougal, L. K., Bibb, W. F., and Brenner, D. J. (1983) *Legionella oakridgensis*: unusual new species isolated from cooling tower water. *Appl. Environ. Microbiol.* **45**, 536–545.

44. Benson, R. F., Thacker, W. L., Waters, R. P., Quinlivan, P. A., Mayberry, W. R., Brenner, D. J., and Wilkinson, H. W. (1989) *Legionella quinlivanii* sp. nov. isolated from water. *J. Curr. Microbiol.* **18**, 195–197.

45. Birtles, R. J., Doshi, N., Saunders, N. A., and Harrison, T. G., (1991) Second serogroup of *Legionella quinlivanii* isolated from two unrelated sources in the United Kingdom. *J. Appl Bacteriol.* **71**, 402–406.

46. Campbell, J., Bibb, W. F., Lambert, M. A., Eng, S., Steigerwalt, A. G., Allard, J., Moss, C. W., and Brenner, D. J. (1984) *Legionella sainthelensi* a new species of Legionelia isolated from water near Mt. St. Helens. *Appl. Environ . Microbiol.* **47**, 369–373.

47. Benson, R. F., Thacker, W. L., Fang, F. C., Kanter, B. Mayberry, W. R., and Brenner, D. J., (1990) Serogroup 2 of *Legionella sainthelensi* isolated from patients with pneumonia. *Res. Microbiol.* **141**, 453–463.

48. Verma, U. K., Brenner, D. J., Thacker, W. L., Benson. R. F., Vesey, G., Kurtz, J. B., Dennis, P. J., Steigerwalt, A. G., Robinson, J. S., and Moss, C. W. (1992) *Legionella shakespearei* sp. nov., isolated from cooling tower water. *Int. J. Syst. Bacteriol.* **42**, 404–407.

49. Harrison, T. G., Saunders, N. A., Doshi, N., Wait, R., and Taylor, A. G. (1988) Serological diversity within the species *Legionella spiritensis. J. Appl. Bacteriol.* **65**, 425–431.

50. Thacker, W L., Benson, R. F., Schifman, R. B., Pugh, E., Steigerwalt, A. G., Mayberry, W. R., Brenner, D. J., and Wilkinson, H. W. (1989) *Legionella tucsonensis* sp. nov. isolated from a renal transplant recipient. *J. Clin. Microbiol.* **27**, 1831–1834.

51. Edelstein, P. H., Brenner, D. J., Moss, C. W., Steigerwalt, A. G., Francis, E. M., and George, W. L. (1982). *Legionella wadsworthii* species nova: a cause of human pneumonia. *Ann. Intern. Med.* **97**, 809–813.

52. Benson, R. F., Thacker, W. L., Daneshvar, M. I., and Brenner, D. J. (1996) *Legionella waltersii* sp. nov. and an unnamed *Legionella* genomospecies isolated from water in Australia. *Int. J. Syst. Bacteriol.* **46**, 631–634.

53. Woodhead, M. A., Macfarlane, J. T., Macrae, A. D., and Pugh, S. F. (1986) The rise and fall of Legionnaires' disease in Nottingham. *J. Infect.* **13**, 293–296.

54. Joseph, C. A., Harrison, T. G., and Watson, J. M. (1993) Legionnaires' disease surveillance: England and Wales, 1992. Comm. Dis. Report. **3**, R124–126.

55. Harrison, T. G., and Taylor, A. G. (1988) *A Laboratory Manual for Legionella.* John Wiley and Sons, Chichester.

56. Marshall, L. E., Boswell, T. C. J., and Kudesia, G. (1994) False positive legionella serology in campylobacter infection: campylobacter serotypes, duration of antibody response and elimination of cross–reactions in the indirect fluorescent antibody test. *Epidemiol. Infect.* **112**, 347–357.

57. Joseph C., Morgan D., Birtles R., Pelaz C., Martín-Bourgón C., Black M., Garcia-Sanchez I., Griffin M., Bornstein N., and Bartlett C. (1996) An international investigation of an outbreak of legionnaires disease among UK and French tourists. *Eur. J. Epidemiol.* **12**, 215–219.

58. Hussong, D., Colwell, R. R., O'Brien, M. O., Weiss, E., Pearson, A. D., Weiner, R. M., and Burge, W. D. (1987). Viable *Legionella pneumophila* not detectable by culture on agar media. *Bio/Technology* **5**, 947–950.

59. Colbourne, J. S., Dennis, P. J., Trew, R. M., Berry, C., and Vesey, G. (1988) *Legionella* and public water supplies. *Wat. Sci. Technol.* **20**, 5–10.

60. Fallon, R. J., and Rowbotham, T. J. (1990) Microbiological investigations into an outbreak of Pontiac Fever due to *Legionella micdadei* associated with use of a whirlpool. *J. Clin. Pathol.* **43**, 479–483.

61. Birtles, R. J., Harrison, T. G., Samuel, D., and Taylor, A. G. (1990) Evaluation of urinary antigen ELISA for diagnosing *Legionella pneumophila* serogroup 1 infection. *J. Clin. Pathol.* **43**, 685–690.

62. Starnbach, M. N., Falkow, S., and Tompkins, L. S. (1989) Species-specific detection of *Legionella pneumophila* in water by DNA amplification and hybridization. *J. Clin. Microbiol.* **27**, 1257–1261.

63. Mahbubani, M. H., Bej, A. K., Miller, R., Haff, L., DiCesare, J., and Atlas, R. M. (1990) Detection of *Legionella* with polymerase chain reaction and gene probe methods. *Mol. Cell. Probes.* **4**, 175–187.

64. Bej, A. K., Mahbubani, M. H., and Atlas, R. M. (1991) Detection of viable *Legionella pneumophila* in water by polymerase chain reaction and gene probe methods. *Appl. Environ. Microbiol.* **57**, 597–600.

65. Jaulhac, B., Nowicki, M., Bornstein, N., Meunier, O., Prevost, G., Piemont, Y., Fleurette, J., and Monteil, H. (1992) Detection of *Legionella* spp. in broncho-alveolar lavage fluids by DNA amplification. *J. Clin. Microbiol.* **30**, 920–924.

66. Fry, N. K. (1992) Analysis of the ribosomal RNA genes of the family Legion-ellaceae for classification and identification. PhD thesis, Council for National Academic Awards, London.

67. Kessler, H. H., Reinthaler, F. F., Pschaid, A., Pierer, K, Kleinhappl, B., Eber, E., and Marth, E. (1993) Rapid detection of *Legionella* species in bronchoalveolar lavage fluids with the EnviroAmp Legionella PCR amplification and detection kit. *J. Clin. Microbiol.* **31**, 3325–3328.

68. Yamamoto, H., Hashimoto, Y., and Ezaki, T. (1993) Comparison of detection methods for *Legionella* species in environmental water by colony isolation, fluo-rescent antibody staining, and polymerase chain reaction. *Microbiol. Immunol.* **37**, 617–622.

69. Palmer, C. J., Tsai, Y.-L., Paszko-Kolva, C., Mayer, C. and Sangermano, L. R. (1993) Detection of *Legionella* species in sewage and ocean water by polymerase chain reaction, direct fluorescent-antibody, and plate culture methods. *Appl. Environ. Microbiol.* **59**, 3618–3624.

70. Lisby, G., and Dessau, R. (1994) Construction of a DNA amplification assay for detection of *Legionella* species in clinical samples. *Eur. J. Clin. Microbiol. Infect. Dis.* **13**, 225–231.

71. Matsiota-Bernard, P., Pitsouni, E., Legakis, N. and Nauciel, C. (1994) Evalua-tion of commercial amplification kit for detection of *Legionella pneumophila* in clinical specimens. *J. Clin. Microbiol.* **32**, 1503–1505.

72. Maiwald, M., Kissel, K., Srimuang, S., von Knebel Doeberitz, M., and Sonntag, H.-G., (1994) Comparison of polymerase chain reaction and conven-tional culture for the detection of legionellas in hospital water samples. *J. Applied Bacteriol.* **76**, 216–225.

73. Catalan, V., Moreno, C., Dasi, M. A., Munoz, C., and Apraiz, D. (1994) Nested polymerase chain reaction for detection of *Legionella pneumophila* in water. *Res. Microbiol.* **145**, 603–610.

74. Lindsay, D., Abraham, W. H., and Fallon, R. J. (1994) Detection of *mip* gene by PCR for diagnosis of Legionnaires' disease. *J. Clin. Microbiol.* **32**, 3068,3069.

75. Maiwald, M., Schill, M., Stockinger, C., Helbig, J. H., Lück, P. C., Witzleb, W., and Sonntag, H.-G. (1995) Detection of *Legionella* DNA in human and guinea pig urine samples by the polymerase chain reaction. *Eur. J. Clin. Microbiol. Infect. Dis.* **14**, 25–33.

76. Jonas, D., Rosenbaum, A., Weyrich., S., and Bhakdi S. (1995) Enzyme-linked immunoassay for detection of PCR-amplified DNA of legionellae in broncho-alveolar fluid. *J. Clin. Microbiol.* **33**, 1247–1252.

77. Manz W., Amann R., Szewzyk R., Szewzyk U., Stenström T-A., Hutzler P., and Schleifer K-H. (1995) In situ identification of Legionellaceae using 16S rRNA-targeted oligonucleotide probes and confocal laser scanning microscopy. *Micro-biology* **141**, 29–39.

78. Murdoch, D. R., Walford, E. J., Jennings, L. C., Light, G. J., Schousboe, M. I., Chereshsky, A. Y., Chambers, S. T., and Town, G. I. (1996) Use of the polymerase chain reaction to detect *Legionella* DNA in urine and serum samples from patients with pneumonia. *Clin. Infect. Dis.* **23,** 475–480.

79. Iwamoto, M., Koga, H., Kohno, S., Kaku, M., and Hara, K. (1995) Detection of *Legionella* species by polymerase chain reaction. *Serodiagn. Immunother. Infect. Dis.* **7,** 99–103.

80. Koide, M., and Saito, A. (1995) Diagnosis of *Legionella pneumophila* infection by polymerase chain reaction. *Clin. Infect. Dis.* **21,** 199–201.

81. Robinson, P. N., Heidrich, B., Tiecke, F., Fehrenbach, F. J., and Rolfs, A., (1996) Species-specific detection of *Legionella* using polymerase chain reaction and reverse dot-blotting. *FEMS Microbiol. Letts.* **140,** 111–119.

82. Kohne, D. E., Steigerwalt, A. G., and Brenner, D. J. (1984) Nucleic acid probe specific for members of the genus *Legionella*, in *Legionella Proceedings of the 2nd International Symposium*, 1984, (Thornsberry, C., Balows, A., Feeley, J. C., and Jakubowski, W., eds.) American Society for Microbiology, Washington DC, pp. 107–108.

83. Laussucq S., Schuster D., Alexander, W. J., Thacker, W. L., Wilkinson, H. W., and Spika, J. S. (1988) False-positive DNA probe test for *Legionella* species associated with a cluster of respiratory illnesses. *J. Clin. Microbiol.* **26,** 1442–1444.

84. Grimont, P. A. D., Grimont, F., Desplaces, N., and Tchen, P. (1985) DNA probe specific for *L. pneumophila. J. Clin. Microbiol.* **21,** 431–437.

85. Engleberg, N. C., Carter, C., Demarsh, P., Drutz, D. J., and Eisenstein, B. I. (1986) A *Legionella*-specific DNA probe detects organisms in lung tissue homogenates from intranasally inoculated mice. *Isr. J. Med. Sci.* **22,** 703–705.

86. Stahl, D. A. and Amann, R. (1991) Development and application of nucleic acid probes, in *Nucleic Acid Techniques in Bacterial Systematics* (Stackebrandt, E., and Goodfellow, M. eds.) John Wiley and Sons, New York, pp. 25–248.

87. Cianciatto, N., Eisenstein, B. I., Mody, C. H., Toews, G. B., and Engleberg, N. C. (1989) A *Legionella pneumophila* gene encoding a species-specific surface protein potentiates initiation of intracellular infection. *Infect. Immun.* **57,** 1255–1262.

88. Ludwig, W., and Stackebrandt, E. (1983) A phylogenetic analysis of *Legionella. Arch. Microbiol.* **135,** 45–50.

89. Woese, C. R., Weisberg, W. G., Hahn, C. M., Paster, B. J., Zablen, L. B., Lewis, B. J., Macke, T. J., Ludwig, W., and Stackebrandt, E. (1985) The phylogeny of purple bacteria: the gamma subdivision. *System. Appl. Microbiol.* **6,** 25–33.

90. Fry, N. K., Warwick, S., Saunders, N. A., and Embley, T. M. (1991) The use of 16S ribosomal RNA analyses to investigate the phylogeny of the family Legionellaceae. *J. Gen. Microbiol.* **137,** 1215–1222.

91. Fry, N. K., Rowbotham, T. J., Saunders, N. A., and Embley, T. M. (1991) Direct amplification and sequencing of the 16S ribosomal DNA of an intracellular *Legionella* species recovered by amoebal enrichment from the sputum of a patient with pneumonia. *FEMS Microbiol. Lett.* **83,** 165–168.

92. Springer, N., Ludwig, W., Drozanski, W., Amann, R., and Schleifer, K. H. (1992) The phylogenetic status of Sarcobium lyticum, an obligate intracellular bacterial parasite of small amoebae. *FEMS Microbiol. Lett.* **96,** 199–202.

93. Birtles, R. J., Rowbotham, T. J., Raoult, D., and Harrison, T. G. (1996) Phylogenetic diversity of intra-amoebal legionellae as revealed by 16S gene sequence comparison. *Microbiology* **142,** 3525–3530.

94. Grimont, F., Lefèvre, M., Ageron, E., and Grimont, P. A. D. (1989) rRNA gene restriction patterns of *Legionella* species: a molecular identification system. *Res. Microbiol.* **140,** 615–626.

95. Saunders, N. A., Harrison, T. G., Kachwalla, N., and Taylor, A. G. (1988) Identification of species of the genus *Legionella* using a cloned rRNA gene from *Legionella pneumophila*. *J. Gen Microbiol.* **134,** 2363–2374.

96. Copping, S. J., Birtles, R. J., Fry, N. K., Doshi, N., Harrison, T. G. Differentiation of the blue-white autofluorescent *Legionella* species by restriction enzyme analysis of PCR amplifed 16S rRNA gene sequences. (Manuscript in preparation).

97. Hookey, J. V., Birtles, R. J., and Saunders, N. A. (1995) Intergenic 16S rRNA gene (rDNA) - 23S rDNA sequence length polymorphisms in members of the family *Legionellaceae*. *J. Clin. Microbiol.* **33,** 2377–2381.

98. Fry, N. K. and Harrison, T. G. (1997) An evaluation of intergenic rRNA gene sequence length polymorphism analysis for the identification of *Legionella* species. *J. Med. Microbiol.*, in press.

99. O'Mahony, M. C., Stanwell-Smith, R. E., Tillet, H. E., Harper, D., Hutchinson, J. G., Farrell, I. D., Hutchinson, D. N., Lee, J. V., Dennis, P. J., Duggal, H. V., Scully, J. A., and Denne, C. (1990) The Stafford outbreak of Legionnaires' disease. *Epidemiol. Infect.* **104,** 361–380.

100. Joly, J. R., Dery, P., Gauvreau, L., Cote, L., and Trepanier, C. (1986) Legionnaires' disease caused by *Legionella dumoffii* in distilled water. *Can. Med. Assoc. J.* **135,** 1274–1277.

101. Plouffe, J. F., Para, M. F., Maher, W. E., Hackman, B., and Webster, L. (1983) Subtypes of *Legionella pneumophila* serogroup 1 associated with different attack rates. *Lancet* **ii,** 649–650.

102. Joly, J. R., McKinney, R. M., Tobin, J. O'H., Bibb, W. F., Watkins, I. D., and Ramsay, D. (1986) Development of a standardized subgrouping scheme for *Legionella pneumophila* serogroup 1 using monoclonal antibodies. *J. Clin. Microbiol.* **23,** 768–771.

103. Bibb, W. F., Arnow, P. M., Thacker, W. L., and McKinney, R. M. (1984) Detection of soluble *Legionella pneumophila* antigens in serum and urine specimens by enzyme-linked immunosorbent assay with monoclonal and polyclonal antibodies. *J. Clin. Microbiol.* **20,** 478–482.

104. Edelstein, P. H., Beer, K. B., and DeBoynton, E. D. (1987) Influence of growth temperature on virulence of *Legionella pneumophila*. *Infect. Immun.* **55,** 2701–2705.

105. Harrison, T. G., Saunders, N. A., Haththotuwa, A., Hallas, G., Birtles, R. J., and Taylor, A. G. (1990) Phenotypic variation amongst genotypically homogeneous *Legionella pneumophila* serogroup 1 isolates: implications for the investigation of outbreaks of Legionnaires' disease. *Epidemiol. Infect.* **104,** 171–180.

106. Harrison, T. G., Saunders, N. A., Haththotuwa, A., Doshi, N., and Taylor, A. G., (1990) Typing of *Legionella pneumophila* serogroups 2-14 strains by analysis of restriction fragment length polymorphisms. *Letts. Applied Microbiol.* **11**, 189–192.

107. Luck, P. C., Helbig, J. H., Hagedorn, H., and Ehret, W. (1995) DNA fingerprinting by pulsed-field gel electrophoresis to investigate a nosocomial pneumonia caused by *Legionella bozemanii* serogroup 1. *Appl. Environ. Microbiol.* **61**, 2759–2761.

108. Lanser, J. A., Adams, M., Doyle, R., Sangster, N., and Steele, T. W. (1990) Genetic relatedness of *Legionella longbeachae* isolates from human and environmental sources in Australia. *Appl. Environ. Microbiol.* **56**, 2784–2790.

109. Brown, A., Vickers, R. M., Elder, E. M., Lema, M., and Garrity, G. M. (1982) Plasmid and surface markers of endemic and epidemic *Legionella pneumophila* strains. *J. Clin. Microbiol.* **16**, 230–235.

110. Stout J. E., Joly J., Para M., Plouffe J., Ciesielski C., Blaser, M. J., and Yu, V. L. (1988). Comparison of molecular methods for subtyping pateints and epidemiologically linked environmental isolates of *Legionella pneumophila*. *J. Inf. Dis.* **157**, 486-495.

111. Nolte, S. N., Conlin, C. A., Roisin, A. J. M., and Redmond, S. R. (1984) Plasmids as epidemiological makers in nosocomial Legionnaires' disease. *J. Infect. Dis.* **149**, 251–256.

112. Mikesell, P., Ezzell, J. W., and Knudson, G. B., (1981) Isolation of plasmids in *Legionella pneumophila* and *Legionella*-like organisms. *J. Infect. Immun.* **31**, 1270–1272.

113. Edelstein, P. H., Nakahama, C., Tobin, J. O'H., Calarco, K, Beer, K., Joly, J. R., and Selander, R. K. (1986) Paleoepidemiological investigation of Legionnaires' disease at Wadsworth Veterans Administration Hospital by using three typing methods for comparison of legionellae from clinical and environmental sources. *J. Clin. Microbiol.* **23**, 1121–1126.

114. Kado C. I. and Liu S-T. (1981). Rapid procedure for detection and isolation of large and small plasmids. *J. Bacteriol.* **145**, 1365–1373.

115. Maher, W. E., Plouffe, J. F., and Para, M. F., (1983) Plasmid profiles of clinical and environmental isolates of *Legionella pneumophila* serogroup 1. *J. Clin. Microbiol.* **18**, 1422–1423.

116. Tully, M., (1991) A plasmid from a virulent strain of *Legionella pneumophila* is conjugative and confers resistance to ultraviolet light. *FEMS. Microbiol. Letts.* **90**, 43–48.

117. Knudson, G. B. and Mikesell, P. A. (1980) A plasmid in *Legionella pneumophila*. *Infect. Immun.* **29**, 1092–1095.

118. Selander, R. K., McKinney, R. M., Whittam, T. S., Bibb, W. F., Brenner, D. J., Nutle, N. S., and Pattison, P. E. (1985) Genetic structure of populations of *Legionella pneumophila*. *J. Bacteriol.* **163**, 1021–1037.

119. Struelens, M. J., Maes, N., Rost, F., Deplano, A., Jacobs, F., Liesnard, C., Bornstein, N., Grimont, F., Lauwers, S., McIntyre, M. P., and Serruys, E. (1992) Genotypic and phenotypic methods for the investigation of a nosocomial *Legionella pneumophila* outbreak and efficacy of control measures. *J. Infect. Dis.* **166**, 22–30.

120. Woods, T. C., McKinney, R. M., Plikaytis, B. D., Steigerwalt, A. G., Bibb, W. F., and Brenner, D. J. (1988) Multilocus enzyme analysis of *Legionella dumoffii. J. Clin. Microbiol.* **26,** 799–803.

121. van Ketel, R. J., ter Schegget, J., and Zanen, H. C. (1984) Molecular epidemiology of *Legionella pneumophila* serogroup 1. *J. Clin. Microbiol.* **20,** 362–364.

122. Haertl, R., and Bandlow, G. (1991) Subtyping of *Legionella pneumophila* serogroup 1 isolates by small-fragment restriction endonuclease analysis. *Eur. J. Clin. Microbiol. Infect. Dis.* **10,** 630–635.

123. Ott, M., Bender, L., Marre, R., and Hacker, J. (1991) Pulsed field electrophoresis of genomic restriction fragments for the detection of nosocomial *Legionella pneumophila* in hospital water supplies. *J. Clin. Microbiol.* **29,** 813–815.

124. Johnson, W. M., Bernard, K., Marrie, T. J., and Tyler, S. D. (1994) Discriminatory genomic fingerprinting of *Legionella pneumophila* by pulsed-field electrophoresis. *J. Clin. Microbiol.* **32,** 2620,2621.

125. Luck, P. C., Birtles, R. J., and Helbig, J. H., (1995) Correlation of MAb subgroups with genotype in closely related *Legionella pneumophila* serogroup 1 strains from a cooling tower. *J. Med. Microbiol.* **43,** 50–54.

126. Schoonmaker, D., Heimberger, T., and Birkhead, G. (1992) Comparison of ribotyping and restriction enzyme analysis using pulsed-field gel electrophoresis for distinguishing *Legionella pneumophila* isolates obtained during a nosocomial outbreak. *J. Clin. Microbiol.* **30,** 1491–1498.

127. Pruckler, J. M., Mermel, L. A., Benson, R. F., Giorgio, C., Cassiday, P. K., Breiman, R. F., Whitney, C. G., and Fields, B. S. (1995) Comparison of *Legionella pneumophila* isolates by arbitrarily primed PCR and pulsed-field gel electrophoresis: analysis from seven epidemic investigations. *J. Clin. Microbiol.* **33,** 2872–2875.

128. Bangsborg, J. M., Gerner-Smidt, P., Colding, H., Fiehn, N-E., Bruun, B., and Hoiby, N. (1995) Restriction fragment length polymorphism of rRNA genes for molecular typing of members of the family Legionellaceae. *J. Clin. Microbiol.* **33,** 402–406.

129. Saunders, N. A., Harrison, T. G., Haththotuwa, A. and Taylor, A. G., (1991) A comparison of probes for restriction fragment length polymorphism (RFLP) typing of *Legionella pneumophila* serogroup 1 strains. *J. Med. Microbiol.* **35,** 152–158.

130. Saunders, N. A., Harrison, T. G., Haththotuwa, A., Kachwalla, N., and Taylor, A. G. (1990) A method for typing strains of *Legionella pneumophila* serogroup 1 by analysis of restriction fragment length polymorphisms. *J. Med. Microbiol.* **31,** 45–55.

131. Watson, J. M., Mitchell, E., Gabbay, J., Maguire, H., Boyle, M., Bruce, J., Tomlinson, M., Lee, J. V., Harrison, T. G., Uttley, A., O'Mahony, M., and Cunningham D. (1994) Piccadilly circus legionnaires' disease outbreak. *J. Pub. Health. Med.* **16,** 341–347.

132. Matsiota-Bernard, P., Thierry, D., Guesdon, J-L., and Nauciel, C. (1994) Molecular epidemiology of *Legionella pneumophila* serogroup 1 by ribotyping with a non-radioactive probe and PCR fingerprinting. *FEMS Immuno. Med. Microbiol.* **9,** 23–28.

133. Mamolen, M., Breiman R. F., Barbaree, J. M., Gunn, R. A., Stone, K. M. Spika, J. S., Dennis, D. T., Mao, S. H., and Vogt, R. L. (1993) Use of multiple molecular subtyping techniques to investigate a Legionnaires' disease outbreak due to identical strains at two tourist lodges. *J. Clin. Microbiol.* **31**, 2584–2588.

134. Sandery, M., Coble, J., and McKersie-Donnolley, S. (1994) Random amplified polymorphic DNA (RAPD) profiling of *Legionella pneumophila*. *Letts. Appl. Microbiol.* **19**, 184–187.

135. Gomez-Lus, P., Fields, B. S., Benson, R. F., Martin, W. T., O'Connor, S. P., and Black, C. M. (1993) Comparison of arbitrarily primed polymerase chain reaction, ribotyping, and monoclonal antibody analysis for subtyping *Legionella pneumophila* serogroup 1. *J. Clin. Microbiol.* **31**, 1940–1942.

136. van Belkum, A., Struelens, M., and Quint, W. (1993) Typing of *Legionella pneumophila* strains by polymerase chain reaction-mediated DNA fingerprinting. *J. Clin. Microbiol.* **31**, 2198–2200.

137. Georghiou, P. R., Doggert, A. M., Kielhofner, M. A., Stout, J. E., Watson, D. A., Lupski, J. R., and Hamill, R. J. (1994) Molecular fingerprinting of *Legionella* species by repetitive element PCR. *J. Clin. Microbiol.* **32**, 2989–2994.

138. Ratcliff, R. M., Donnellan, S. C., Lanser, J. A., Manning, P. A., and Heuzenroeder, M. W. (1997) Interspecies sequence differences in the Mip protien from the genus Legioneall: implications for function and evolutionary relatedness. *Molec. Microbiol.* **25**, 1149–1158.

13

Molecular Methods for *Haemophilus influenzae*

Mark A. Herbert, Derrick Crook, and E. Richard Moxon

1. Introduction

The species *Haemophilus influenzae* belongs to the genus Haemophilus and the family Pasteurellaceae. *H. influenzae* are small, nonmotile, nonspore forming, Gram-negative, pleomorphic rods that range in shape from coccobacilli to long filaments. They require X and V factors (hemin and NAD, respectively) for aerobic growth, and may be facultatively anaerobic *(1)*. Encapsulated *H. influenzae* are classified into six antigenically distinct serotypes (a–f), and have a clonal population structure with two major global subdivisions (I and II) *(2,3)*. Nonencapsulated *H. influenzae* (NCHi) appear to have a nonclonal population composition, but a broader analysis of NCHi in the future may reveal that the population is not so distinct from encapsulated *H. influenzae* *(4)*. Both encapsulated *H. influenzae* and NCHi exhibit wide genetic diversity *(5)*.

Meningitis and epiglottitis are the most life-threatening manifestations of *H. influenzae* type b (Hib) disease and remain prevalent in countries that have not established programs for immunization against Hib. Where national vaccination has been implemented, both invasive disease and carriage have been reduced *(6)*. Capsulate types a and c–f, and NCHi infrequently cause disease, but when they do the spectrum is wide, including meningitis, bacteremia, cellulitis, conjunctivitis, otitis media, chronic bronchitis, pneumonia, osteomyelitis, septic arthritis, obstetric and neonatal infections and, very rarely, epiglottitis *(7,8,9)*. *H. influenzae* biotype aegyptius (*H. aegyptius*) may cause Brazilian Purpuric Fever (BPF), a rare fulminant septicemic illness with a case fatality rate of more than 60%. *H. aegyptius* is conspecific with *H. influenzae*, shares 84–89% DNA homology, and does not merit specific rank *(10)*.

From: *Methods in Molecular Medicine, Vol. 15: Molecular Bacteriology: Protocols and Clinical Applications*
Edited by: N. Woodford and A. P. Johnson © Humana Press Inc., Totowa, NJ

2. Diagnosis

Molecular diagnosis of infectious diseases is in its infancy and, despite distinct theoretical advantages, has many limitations. None of the assays described for *H. influenzae* fulfill the Sackett "guides" *(11)* for the selection of tests with clinical utility; the authors have reviewed the current assays with these "guides" in mind. Two short-falls of molecular diagnostic methods are: first, the assays are often only directed to detecting a single species or genus in circumstances where other genera can cause the same pathology, and second, detection of an organism gives no indication of its susceptibility to antibiotics. Tests aimed at recognizing groups of pathogens causing a single disease *(12)*, such as detection of meningitis caused by *H. influenzae, Neisseria meningitidis,* or *Streptococcus* spp. using a seminested PCR *(13)*, together with tests aimed at identifying antibiotic resistance genes *(14)*, are strategies that may overcome these two intrinsic limitations.

2.1. Diagnostic Probes

Two probes have been used to identify *H. influenzae* in in vitro cultures, a penicillin-binding protein (PBP) gene probe (**Table 1**, probe 1) *(15)* and a chemiluminescent rDNA probe (**Table 1**, probe 2) *(16,17)*. Minor cross-hybridization with other Pasteurellaceae was seen with the PBP probe, except under conditions of high hybridization stringency, and validation has not been undertaken to the point where it could be confidently introduced as a routine test. The chemiluminescent probe identified *H. influenzae* in blood cultures 24–48 h faster than routine methods, with no false-positive or false-negative results in an initial study involving 362 cultures *(16)*; however in a subsequent study, two NCHi were not detected, yielding a sensitivity of 98.4% *(17)*. Though accurate, it has not been demonstrated that early detection assays in clinical practice improve on current techniques.

Two probes have been applied to the detection of *Haemophilus* spp. in body fluids. The PBP gene probe of Malouin (**Table 1**, probe 1) detected *H. influenzae* in sputum, cerebro-spinal fluid (CSF), and blood with 96–97% specificity and 74–100% sensitivity *(18)*, but further validation would be necessary to assess clinical utility. A biotin-labeled whole genome probe (**Table 1**, probe 3) detected several *Haemophilus* spp. in sputum, including non-respiratory pathogens *(19)*, but the probe had only weak specificity, and hybridization to uninfected control sputa was insufficiently examined. In its current form it has not been validated as a standard microbiologic tool.

2.2. DNA Amplification in Diagnosis

PCR is potentially simpler to perform than probe-based assays and has been utilized in the investigation of meningitis *(13,20)*, septicemia *(21)*, exacerba-

Table 1
Probes Applied to the Identification
and Characterization of *Haemophilus influenzae*

Probe	References
Penicillin-binding protein (PBP) gene	*15,18,31*
16S and 23S rRNA gene (rDNA) probes	*16,17,25,46,48,55*
Whole genome probe	*19*
omp gene probes	*24,35,87*
iga gene probes	*32*
pUO38	*2,3,7,40,86*
Bex A gene probe	*20*
Antibiotic resistance genes & plasmids	*48,64*

tions of chronic obstructive airways disease and cystic fibrosis (*see* **Subheading 3.7.**) *(22)*, otitis media *(23,24)*, and in determining ampicillin resistance of *H. influenzae (14)*.

2.2.1. Detection of H. influenzae in Cerebrospinal Fluid (CSF) and Blood

Many *H. influenzae* diagnostic PCR assays have employed two groups of primers developed by van Ketel et al. *(20)* for the detection of *H. influenzae* in CSF (**Table 2**, primer set 1). Group 1 was selected from the *bex A* region of the capsulation locus (*see* **Fig. 1**), and group 2 from the outer membrane protein (OMP) P6 gene, *omp*P6. Forty PCR cycles was exquisitely sensitive and detected 0.01–0.05 pg of *H. influenzae* DNA. In practice, when tested on 200 CSF samples, a sensitivity of 97.5% and a specificity of 95% was observed. When the number of PCR cycles was reduced to 35, the specificity increased to 100%. Internal oligonucleotide probes (**Table 1**, probes 4 and 6) also improved specificity, but did not enhance sensitivity. PCR with the primers of van Ketel have confirmed the presence of Haemophilus spp. in the blood of children with the clinical syndrome BPF; the assay may be useful for detecting cases in remote areas of Brazil, where blood cultures for *H. aegyptius* are frequently negative even in cases with typical fulminant sepsis *(21)*.

In a seminested PCR, designed to detect more than one CSF pathogen, an initial set of primers (ru8 and U_3) directed at 16S rRNA sequences common to the major CSF pathogens, amplified a 1-kb fragment (**Table 2**, primer set 2) *(13)*. Internal primers then amplified rRNA regions unique for *H. influenzae* (primer designated HI), *N. meningitidis*, or *Streptococcus pneumoniae*. When applied to 304 clinical specimens (125 infected by bacteria) the assay had a sensitivity of 94% and a specificity of 96%. Greisen et al. *(25)* employed uni-

Table 2
Primers for the Amplification of *H. influenzae* Genes

Primer set	Primer sequences	Primer name	Gene amplified	Refs
1	CGT TTG TAT GAT GTT GAT CCA GAC T	HI-1	*bexA*	*20,21,23*
	TGT CCA TGT CTT CAA AAT GAT G	HI-2		
	TGA TGA GGT GAT TGC AGT AGG	HI-3		
	or GTG ATT GCA GTA GGG GAT TCG CGC TTT GCA G			
	ACT TTT GGC GGT TAC TCT GT	HI-4	*ompP6*	
	TGT GCC TAA TTT ACC AGC AT	HI-5		
	GCA TAT TTA AAT GCA ACA CCA GCT GCT	HI-6		
2	AAG GAG GTG ATC CAG/A CCG CAG/C G/CTT C	ru8 primer to all eubacteria		*13*
	AAC TC/AC GTG CCA GCA GCC GCG GTA A	U₃ primer to all eubacteria		
	CCT AAG AAG AGC TCA GAG	HI internal primer specific for *H. influenzae*		
3	AGG AGG TGA TCC AAC CGC A	DG74	Universal primers for rRNA genes	*25*
	AAC TGG AGG AAG GTG GGG AT	RW01		
	AAC TGG AGG AAG GTG GGG AC	RDR080		
	GGT TAA GTC CCG CAA CGA GCG C	PL06		
4	CCA GCT TGG TCT CCA TAC TTA AC	HIP61	*ompP6.*	*23*
	TTG AGC AGC ACC ATT CCC TGC	HIP62		
5	TGG GTG CAC GAG TGG GTT AC	TEM (321)	Ampicillin resistance genes	*14*
	TTA TCC GCC TCC ATC CAG TC	TEM (846)		
	ATC AGC CAC ACA AGC CAC CT	ROB (419)		
	GTT TGC GAT TTG GTA TGC GA	ROB (1110)		
6	ATA ACA ACG AAG GGA CTA ACG	O₁	*ompP2*	*35*
	ACC TAC ACC CAC TGA TTT TTC	O₃		
	AGA GTT TGA TCC TGG CTC AG	pAr		
	AAG GAG GTG ATC CAG CCG CA	pH		
7	AAG CAA AAC AAA AAG TTG AGG ATA C	Hinf 1	Derived from the probe of Malouin	*31*
	AAA TTA AAC CTA TTA TTT TGT CGC T	Hinf 7		

246

(**Table 1**, probe 1)

Ref	Sequence	Primer	Target	Note
	CTT TAC CTT TCT TTA AAA AGA TTA C	Hinf 2		
	AAA TCA GAA ATA TTG ATC CAT CAA G	Hinf 3		
	CAG GTG GCC TTT ATA TTG ATT CTA G	Homp 1	*ompP1*	
	AGC GGC TCT ATC TTG TAA TGA CAC A	Homp 3		
	ACA AAT CGG CAT TCA TCC TAA GCA G	Htra 1	Transformation gene cluster	
	GAA ACG TGA ACT CCC CTC TAT ATA A	Htra 3		
8	ATT AGC AAG TAT GCT AGT CTA T	IS1016	Capsulation locus	*80*
	CAA TGA TTC GCG TAA ATA ATG T	bex A		
9	GGC GAT ACA GTG GTT ACT TA	bex B	Capsulation locus	*45*
	GAG CAG CGG CTG ATT AC	ISLOUT		
	GTT ATT ACT TGC GTG ATC GT			
10	CTA CTC ATT GCA GCA TTT GC	a_1	Serotype specific 3-primer sets directed at the capsulation locus	*86*
	GAA TAT GAC CTG ATC TTC TG	a_2		
	AGT GGA CTA TTC CTG TTA CAC	a_3		
	GCG AAA GTG AAC TCT TAT CTC TC	b_1		
	GCT TAC GCT TCT ATC TCG GTG AA	b_2		
	ACC ATG AGA AAG TGT TAG CG	b_3		
	TCT GTG TAG ATG ATG GTT CA	c_1		
	CAG AGG CAA GCT ATT AGT GA	c_2		
	TGG CAG CTG AAA TAT CCT AA	c_3		
	TGA TGA CCG ATA CAA CCT GT	d_1		
	TCC ACT CTT CAA ACC ATT CT	d_2		
	CTC TTC TTA GTG CTG AAT TA	d_3		
	GGT AAC GAA TGT AGT GGT AG	e_1		
	GCT TTA CTG TAT AAG TCT AG	e_2		
	CAG CTA TGA ACA AGA TAA CG	e_3		
	GCT ACT ATC AAG TCC AAA TC	f_1		
	CGC AAT TAT GGA AGA AAG CT	f_2		
	AAT GCT GGA GTA TCT GGT TC	f_3		
11	ACG TAT CTG C	RAPD	Primer A	*54*
12	TAC ATT CGA GGA CCC CTA AGT G	BG2	Arbitrary primers	*22*
	ATG TAA GCT CCT GGG GAT TCA C	ERIC IR		
	AAG TAA GTG ACT GGG GTG AGC G	ERIC 2		

versal 16S rRNA primers (DG74, RW01, RDR080, and PL06) (**Table 2**, primer set 3) to amplify DNA from a broad range of bacteria, and then determined the pathogen's identity with three series of oligonucleotide probes. A universal bacterial, a Gram-positive, and three Gram-negative probes comprised the first series; the second consisted of probes to seven meningitis pathogens, including *H. influenzae*; and the third was directed at bacteria commonly considered contaminants of clinical specimens (**Table 1**, probe 2). A clear limitation of such an approach is the complexity of the procedures, and the Haemophilus spp. probe had an identical sequence to the 16S gene of 53 species within Pasteurellaceae and would therefore cross-hybridize.

2.2.2. PCR in Otitis Media

The fastidious growth characteristics of *H. influenzae* in middle ear effusions often renders it undetectable by conventional culture. PCR has been performed with the primers of van Ketel *(23,24)* or with primers derived from the *omp*P6 gene (**Table 2**, primer set 4) *(24)*. The limit of detection was between 10 pi *(24)* and 10 femtograms *(23)* of DNA. Cross-hybridization with *H. parainfluenzae* occurred, but species could be distinguished by secondary probe hybridization (**Table 1**, probe 4) *(24)*. Identification of *H. influenzae* in middle ear effusions by PCR is likely to have a greater impact on understanding etio-pathogenesis of otitis media, but as a clinical tool, PCR would have limited utility because of the infrequency that diagnostic tympanocentesis is performed.

2.2.3. Ampicillin resistant H. influenzae

PCR can potentially guide therapy in the early stages of bacterial meningitis by detecting resistance genes directly in CSF. Algorithms could be developed, so that if a pathogen were identified, a range of primers would be used to test for the presence of its common resistance genes. One limitation is the number of genes involved. Therefore, with present technology, comprehensive testing would involve multiple time-consuming PCR reactions. Tenover et al. *(14)* described blaTEM and blaROB primers (**Table 2**, primer set 5) that amplified regions of the two most common genes bestowing ampicillin resistance. The presence of blaTEM and blaROB amplicons correlated exactly with antibiotic susceptibility; but as only four ampicillin-resistant *H. influenzae* were examined, the data were insufficient to draw conclusions about clinical utility. However, the approach may have future potential.

3. Epidemiologic Investigation

Phylogenetic relatedness and typing can be addressed with phenotypic and genotypic analyses. The phenotype is subject to alterations in expression, both intrinsic and in response to the environment; genotyping is, therefore, preferable, but no readily accessible technique is available. In the definitive study of the population structure of encapsulated *H. influenzae*, Musser et al. *(3)* analyzed core metabolic enzymes by multilocus enzymes electrophoresis (MLEE). This phenotypic characterization depends on stable traits and correlates with changes in genotype, and provides a measure of genetic distance between strains of *H. influenzae*. Analysis of 2209 isolates demonstrated that encapsulated *H. influenzae* have a clonal population structure with 280 distinct MLEE types spread worldwide, and with two major subdivisions at a genetic distance of 0.66, where potential differences between strains within the species is measured on a scale of 0 representing relatedness, and 1 indicates extreme phylogenetic divergence. The clonal nature of the global population means that within any one geographic locale only a few strains predominate, thus future studies of encapsulated *H. influenzae* will necessitate assays that seek differences, and discriminate between strains that may be highly related. Conversely, NCHi are genetically diverse, with an as yet poorly defined population structure that requires further clarification *(4)*. Typing of NCHi demands assays that emphasize relatedness.

3.1. Restriction Enzyme Analyses

Complete digestion of the genome into fragments of 1–10 kb size is desirable for restriction enzyme analysis (REA; *see* Chapter 2). Enzymes that cut more frequently produce complex banding patterns forming a multitude of small fragments that are difficult to interpret *(26)*. Less frequent cutters produce fragments too large for resolution by usual 0.8% agarose gel electrophoresis, whereas 10–20 bands in the 1–10kb range give optimal clarity *(26)*. Some enzymes, usually those derived from *H. influenzae* (designated with the prefix "*hin*" or "*hae*"), do not restrict homologous DNA well (*see* **Table 3**).

3.1.1. Otitis Media

By comparing the total genomic DNA restriction patterns of strains from middle ear and from nasopharynx, Loos et al. *(27)* provided evidence that the causative bacteria of otitis media in children reach the middle ear via the Eustachian tube. Twelve of 13 pairs of isolates had indistinguishable RFLP, whereas strains compared between children were dissimilar. RFLP was stable and reproducible compared with OMP subtyping. The isolates in one pair were genotypically indistinguishable, but had differing OMP subtypes; the two isolates had possibly undergone altered phenotypic expression between the two microenvironments of the nasopharynx and middle ear. Overall, RFLP offers little advantage to the established technique of OMP subtyping.

Table 3
Restriction Enzymes Employed in *H. influenzae* Studies

Study type	Enzyme sequence	Recognition	References
RFLP of	*Eco*RI	G/AATTC	*27–29*
chromosomal DNA	*Cla*I	AT/CGAT	
	*Pst*I	CTGCA/G	
	*Ssp*I	AAT/AAT	
RFLP of single genes	*Eco*RI		*32,87*
	*Pst*I		
	*Bam*HI	G/GATCC	
RFLP of PCR	*Hae*III	GG/CC	*35,88*
products	*Alu*I	AG/CT	
	*Dra*I	TTT/AAA	
RFLP of plasmids	*Eco*RI		*48,59,61–63,*
	*Pst*I		*65,89*
	*Bam*HI		
	*Ava*I	C/CCGAG	
	*Sac*I	GAGCTC	
	*Hin*dIII	A/AGCTT	
	*Acc*I	GT/AGAC	
	*Mbo*II	GAAGA	
	*Rsa*I	GT/AC	
Chromosomal digests	*Bgl*II	A/GATCT	*2,3,15,18,31,*
for probe analysis	*Bgl*II/*Hin*dIII		*40–42,44,63*
	*Eco*RI		
	*Bgl*II/*Pst*I		
	*Bam*HI/*Bgl*II/*Pst*I		
	*Pst*I		
Pulsed-field	*Sma*I	CCC/GGG	*30,75,76*
gel electrophoresis	*Apa*I	GGGCC/C	
	*Nae*I	GCC/GGC	
	*Eag*I	C/GGCCG	
	*Sac*II	CCGC/GG	
Enzymes used	*Eco*RI		*7,46,48–52,*
in ribotyping	*Bgl*II		*55,66,89*
	*Bam*HI		
	*Eco*RV	GAT/ATC	
Enzymes that do	*Not*I		*49,52,76*
not cut *H. influenzae*	*Sfi*I		
	*Sal*I		
	*Hin*dIII		

RFLP has been applied to clonal analysis, but as only a fraction of the chromosome is appraised, that part containing the restriction enzyme recognition sites, it is a weak technique for determining genetic distance between strains. It is possible that two different clones may have the same profile when cut with a single enzyme, and so RFLP employing several enzymes improves the certainty that two isolates with uniform restriction profiles belong to the same strain or clone. Three-enzyme digests of Hib DNA demonstrated that isolates from four separate sites in an adult were the same strain *(28)*. The majority of 52 Hib isolates from Italy (88.5%), when restricted with three enzymes, had indistinguishable RFLP. Only three patterns were observed in total, and most of the Italian isolates were closely related to Swedish and Dutch strains, which supports the hypothesis that there are a few dominant European Hib clones *(29)*. A limitation of RFLP is that only a proportion of the restricted fragments are analyzable by conventional gel electrophoresis. Pulsed-field gel electrophoresis *(30)* (*see* Chapter 3) separates all DNA fragments, and so determination of strains relatedness could be potentially improved.

3.2. DNA Probes in Epidemiology

Probes have been used for typing *H. influenzae* in outbreak studies. Isolates from a strain collection in Canada probed with the PBP probe of Malouin (**Table 1**, probe 1) *(15,18)* generated 11 typing groups (A-K) and 14 subgroups *(31)*. When applied to clinical specimens however, there was less diversity so that 77% of CSF isolates grouped as types A or B, and 75% of strains causing respiratory tract infections and 86% of isolates from patients with cystic fibrosis grouped as C or E. Similarly, 60 clinical isolates analysed with a probe derived from the well conserved immunoglobulin A1 protease gene (iga) (**Table 1**, probe 5) produced only a few distinct types: three types represented 98% of the strains *(32)*. Analysis with these probes does not discriminate sufficiently between isolates of Hib or NCHi to be of value in strain tracking.

3.3. PCR in Epidemiology

Primers are best directed at genes with conserved portions separated by variable regions, so that they consistently amplify the gene, but the resulting amplicons are of variable length. The *omp*P2 gene *(33,34)* has these qualities and has been used to study the epidemiology of nasopharyngeal NCHi in members of a closed community in the Antarctic (**Table 1**, probe 4; **Table 2**, primer set 6) *(35)*. Restriction of the PCR amplicons with two enzymes resulted in 11 allelic polymorphisms. Isolates with indistinguishable alleles were assumed to be identical, and could be shown to be transferred to new residents with time. Detection of *H. influenzae* was below the limits of the PCR on occasions, as was suggested by the presence, disappearance, and then re-emergence of strains

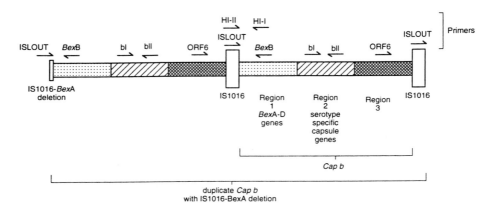

Fig. 1. The duplicate *cap b* locus of *H. influenzae*. IS*1016* and regions 1 and 3 form the chassis, in the center of which are the serotype specific capsule genes. In Hib there are two copies of the chassis, one of which has a partial deletion of IS*1016-bex A*.

in the same individual. Multiplex PCR with primers directed at PBP genes, the ompP1 gene and genes in the transformation cluster (**Table 2**, primer set 7), has been used to distinguish 64 diverse isolates of *H. influenzae* from Canada, but the usefulness of the technique for outbreak studies or strain tracking was not determined *(31)*. Such PCR is less laborious than probe hybridization, but is unlikely to offer any discriminatory advantage.

3.4. Capsular Genotyping

Evaluation of apparent vaccine failure necessitates definitive typing; invasion with Hib in a vaccinee implies vaccine failure, whereas isolation of another capsular type (a, c–f) or NCHi requires careful typing so that type-specific attack rates can be monitored, in part to recognize related strains with an increased invasive propensity. Capsular type is usually determined by serotyping, which is liable to misinterpretation because of cross reactivities, whereas capsular genotyping, with probe pU038 (**Table 1**, probe 6) or by PCR (**Table 2**, primer sets 8, 9, and 10), determines the capsular type unequivocally.

The structure of the *cap b* locus has been well described (*see* **Fig. 1**) and has three regions: Cap region 1 contains *bex A–D* genes necessary for capsular export and in combination with cap region 3 forms the cap chassis, which is flanked by IS1016 forming a compound transposon. Cap region 2 lies in the middle of the chassis and contains serotype-specific genes. The *cap b* locus is composed of multiple (usually two) tandemly repeating cap units, and is unique compared to the other *cap* loci in having a partial deletion of most of one IS1016-bex A region *(36)*. Although a duplicate repeat of *cap* b with a deletion is the ground state in vitro, in vivo as many as 35% of Hib may have undergone *cap* amplification to three to five multimers, and in a gene dose dependent fashion such strains produce up to five times the quantity of capsule *(37)*. Invasive isolates may exhibit a multimeric *cap b* locus, e.g., 15 of 36 isolates from Finland, and 5 of 14 from Washington had more than two copies of *cap (37)*. Passaging in vitro results in a rapid reversion back to two copies. This is consistent with multicopy (≥ 3 copies of *cap*) confering Hib with a survival advantage in vivo. Serotype a strains have a tandem duplication of intact cap loci (with no IS*1016*-bexA deletion), whereas serotypes c and d have one copy. Hib in phylogenetic division II have only one copy, and are less pathogenic than Hib in division I, which generally have the partially deleted duplicated *cap b*.

Probe pUO38 hybridizes to the cap region of *Eco*RI digested chromosomal DNA from the six serotypes *(38–40)*, and 14 banding patterns are produced: a(T), a(N), a(M), b(S), b(G), b(V), b(O1), b(O2), c(1), c(2), d, e, f (C), and f(F). A dendrogram derived from pU038 hybridization is in accord with the phylogenetic divisions determined by MLEE. Recovery of isolates with identical genetic properties from widely disparate geographic regions, at different times over a 40 yr period and with distinct cap RFLP patterns in nonrandom clones of each serotype *(2)*, is consistent with the clonal population structure of Hib as determined by MLEE.

Some NCHi may arise through recombinational loss of one copy of the tandem duplication of *cap b*, and are referred to as b⁻ strains. Probing with pUO38 has clearly defined the b⁻ category, which occurs in vivo and as an in vitro event with a frequency of 0.1 to 0.3% *(41)*. In nature, the majority of NCHi are not the progeny of capsulate ancestors, but belong to a separate diverse population with no cap locus. However, in one study, 20% of NCHi from Finnish 3-yr olds had capsule-specific sequences when probed with pUO38, indicating that a proportion have arisen from encapsulated ancestors *(42)*. Perhaps selection of NCHi derivatives of encapsulated strains is advantageous; capsule may, for instance, interfere with colonization and persistence within the upper respiratory tract. In contrast, probe pUO38 did not hybridize to any of 34 NCHi blood isolates of Pakistani children with pneumonia *(43)*, and only 3 of 24 invasive NCHi in an Oxford prospective survey were b⁻ *(7)*. Hoiseth et al. *(44)* compared NCHi from the nasopharynx with Hib from the blood of patients with invasive Hib disease and found that pU038 hybridized with two of four NCHi; Hib may have caused the disease, but in the interval the nasopharyngeal population may have converted to b⁻.

PCR capsular genotyping (**Table 2**, primer sets 8–10) has been shown to accurately determine the capsular states of *H. influenzae* strains, including b⁻ strains and is easier to perform than probe analysis as there is no need for lengthy DNA extractions, Southern blotting, and hybridization. PCR capsular genotyping recognises four *cap b* patterns, not detected by pUO38 hybridization *(45)*. Two result from simple DNA rearrangements within *cap b*, whereas two are intact tandem repeats, rather than having the IS*1016-bex A* deletion. The significance for virulence of intact tandem repeats in *cap b* is unknown.

3.5. Ribotyping

Ribotyping (*see* Chapter 2) is a genotypic method that avoids the pitfalls of variable expression of phenotypes. Genes encoding rRNA are attractive targets for typing, as they are present at multiple sites on the *H. influenzae* chromosome, and have conserved sequences common to all bacteria. This allows use of universal primers (**Table 2**, primer set 3) and probes (**Table 1**, probe 2), such as *E. coli* 16S and 23S rRNA (commercially available from Bethesda Research Laboratories, Gaithersburg, MD, or Boehringer Mannheim, Germany).

Gene sequences of the rRNA loci are stable and highly conserved even between bacterial species. Sequence variation has been shown to correlate with genetic distance for species other than *H. influenzae*. A phylogenetic dendrogram based on 22 ribotypes of Hib appeared superficially similar to that derived by MLEE in that there were two major divisions *(3)*, but the strains within each division by the two methods were highly discordant; evolutionary divergence based on ribotyping may thus be misleading. Ribotyping of *H. influenzae* is extremely sensitive at recognizing tiny genomic differences and therefore, is suitable for discriminatory, epidemiologic studies *(46)*. However, MLEE is the best reference tool, and is a more precise method than ribotyping if the aim is to discriminate between strains, or to indicate their genetic relatedness.

Nonetheless, ribotyping is a useful technique for differentiating between strains, and more discriminatory than subtyping based on whole cell protein or OMP electrophoretic mobility *(47,48)*. It has been used to develop a subtyping scheme for Hib *(46)*, to subtype invasive *H. influenzae* obtained from adults *(8)*, to differentiate *H. aegyptius* responsible for BPF from strains causing conjunctivitis *(48,49)*, to discriminate between NCHi *(50)*, and to study carriage of NCHi in Aboriginal infants *(51)*. Ribotyping was able to differentiate between 9 of 10 epidemiologically distinct NCHi, but the profiles of two related strains, isolated from the trachea of a septic neonate and from the maternal cervix, were indistinguishable *(50)*. The Hib typing scheme identified 22 subtypes, and when applied to 283 clinical isolates, 14 groups were found, compared with only four by OMP subtyping. Up to 90% of isolates within a locale may be indistinguishable by OMP subtyping (52) and the majority of invasive Hib belong to only a few ribotypes, implying that epidemiologically-distinct strains cannot be distinguished by either method. The laboriousness of rRNA probe hybridization may be overcome by PCR-ribotyping. A new method, long PCR, has improved base-pair fidelity and has made possible the amplification of larger fragments of DNA than previously possible; these could be analyzed by RFLP *(53)* (*see* Chapter 29 for an example).

3.6. RAPD

Randomly amplified polymorphic DNA (RAPD) (*see* Chapter 6) should in theory produce amplicons of the *H. influenzae* genome that are strictly random. Jordans et al. *(54)* chose an RAPD primer previously described for typing *Listeria monocytogenes* (**Table 2**, primer 11). RAPD proved to be a simple technique for NCHi comparisons in outbreaks, with strain discrimination equivalent to OMP typing and similar to ribotyping *(54,55)*. RAPD with six different primers distinguished Hib isolates as effectively as MLEE. The method is likely to be excellent for typing, but has not been assessed for its ability to phylogenetically relate strains.

3.7. Repeat Sequence PCR

Repeated sequences with unknown functions are found in all genomes *(56)*, and in prokaryotes they include those designated repetitive extragenic palindrome (REP) and enterobacterial repetitive intergenic consensus (ERIC) *(56,57)* (*see* Chapter 7). REP are conserved primary sequences with a palindromic structure, located in noncoding regions and distributed throughout the chromosome. These sequences are present in many different species, making them ideal targets for PCR *(57)*. NCHi genomic DNA amplified with ERIC primers (**Table 2**, primer set 12) produced distinct and complex PCR fingerprints of 40 epidemiologically unrelated isolates from patients with chronic obstructive airway disease and cystic fibrosis *(22)*. In the same study, PCR with one ERIC primer in combination with an arbitrary primer, resulted in fewer bands and less heterogeneity. The method is stable and reproducible as evidenced by monomorphic profiles obtained from subcultured colonies, from repeated isolates in the same patient, and from NCHi recovered during a respiratory outbreak. Repeat sequence PCR offers the possibility of a faster and easier high-resolution genotyping technique for epidemiologic studies, with discrimination equivalent to RFLP or OMP subtyping.

3.8. Resistance Genes

Elwell *(58)* suggests that two groups of antibiotic resistance plasmids (R plasmids) exist in Haemophilus spp.: small plasmids specifying β-lactamases, and large plasmids carrying transposable genes determining β-lactam, tetracycline, and chloramphenicol resistance. The evolution of resistance has been studied through plasmid RFLP (i.e., digestion with restriction endonucleases; *see* Chapter 4). Hib isolates from Spain had R-plasmids with only four different RFLP patterns *(59,60)* and epidemiologically unrelated Belgian and Spanish *H. influenzae* had highly comparable R-plasmids by RFLP analysis *(61)*. Plasmid-mediated resistance may thus be the result of the spread of a few ancestral plasmids, or of independent, but identical transposition events into closely related cryptic plasmids.

Anderson et al. *(62)* identified plasmids in *H. influenzae* and Dimopoulou et al. *(63)* recognised the presence of large chromosomally integrated plasmids that could be conjugated into recipient *H. influenzae*. The presence of identical R-plasmids has been useful in relating strains, for instance, in a rehabilitation center outbreak of NCHi respiratory disease *(64)* and of Hib in all household members during a family outbreak *(65)*. Brazilian *H. aegyptius* isolates from conjunctivitis were often found to harbour a 24 Mda plasmid, as did all strains from cases of BPF (**Table 1**, probe 8) *(49,66)*. Eight RFLP patterns were demonstrated for the 24-MDa plasmids, and all strains from cases of BPF had a single type (type 3031) consistent with BPF isolates being clonal.

4. Research

H. influenzae is the first free-living organism the genome of which has been fully sequenced; the magnitude of the event cannot be overstated. The 1.8-Mb genome of *H. influenzae* Rd contains 1743 genes *(67)*, greater than 40% of which, or 736 genes, have no counterparts of known function in genes of other prokaryotes; 1007 genes do have a known function in other organisms, which aids identification of the genes' roles in *H. influenzae* *(68)*. The genetic blue print of the organism will allow us to attribute function, explore genome organization, and understand gene control. Entering the "post-genome era," *H. influenzae* offers exciting possibilities for investigation and control of disease caused by it and related organisms. New virulence determinants can be identified through strategies such as directed and random knockout mutagenesis, and pinpointing repeat sequences that increase susceptibility to copying errors and change the expression of downstream virulence genes. Over expression of "key" pathogenicity genes and purification of protein products may lead to novel antibiotics, vaccines, and industrial enzymes. New methods suggested for determining the interactions of groups of genes are being simultaneously developed for eukaryotes, such as serial analysis of gene expression, or SAGE, and microarray assays that rely on the conversion of mRNA to cDNA, which is then used to probe known genes in libraries *(69)*. Methods such as Signature Tagged Mutagenesis *(70)* have led to the identification of pathogenicity islands in *S. typhimurium*, and are applicable to many prokaryotes, including *H. influenzae*. Animal models are essential for exploring the functions of bacterial genes. The infant rat model *(71,72)* most accurately reflects the pathophysiology of septicemia and meningitis in humans, and the Chinchilla *(73)* represents a model for otitis media.

PFGE in combination with infrequent cutting restriction enzymes allows discrimination of large fragments of DNA *(74)*; it has been used to determine the *H. influenzae* genome size *(75,76)* and may form a universal tool for infection epidemiology. PFGE has not yet been exploited for *H. influenzae*, but for other organisms it can be more discriminatory than either ribotyping or RAPD-PCR *(77)*. Direct sequencing of parts of genomes is also becoming possible, and may be a more accurate method for phylogenetic analysis *(4)*.

5. Future Prospects

H. influenzae other than Hib can produce invasive disease; testimony to this is *H. aegyptius* causing BPF and occasional blood and CSF isolates of the other capsulate types (a, c–f), and of NCHi. Hib in phylogenetic division II are less virulent, and those in division I more invasive *(36)*, with clustering of disease-causing isolates into a few clonal groups *(3)*. These facts suggest that other virulence genes may be in linkage disequilibrium with *cap b*. The serotype b capsule is a major virulence determinant *(78)*, but is not necessary for invasion, as seen by NCHi being able to cause disease. Virulence-enhancing genes may potentially come together in combinations that do not include *cap b*, and "vaccine-escaped" strains could emerge or evolve with the selective pressure of immunization.

An appreciation of the structure of the capsulation locus provides an insight into potential mechanisms of novel strain evolution. Kroll et al. *(36,79)* surmised that the IS*1016-bex* A deletion in the *cap b* chassis is an ancestral mutation that enhances pathogenicity and is a virulence determinant in addition to the specific type b capsule. Capsular type a *H. influenzae* causing meningitis and pneumonia were over-represented in isolates collected in The Gambia *(80)*, and PCR showed that some had serotype a specific genes in the *cap* central region 2 flanked by the *cap b* chassis. Nucleotide sequence divergence between strains is 12% in the chassis, but sequences are identical within region 2, implying that the chassis has undergone divergent evolution, whereas the central region has been conserved, possibly by horizontal transfer *(81)*. *H. influenzae* is highly transformable and, thus, has a remarkable propensity for the uptake of exogenous DNA from other bacteria. It is plausible that Hib acquired its combination of disease-producing genes through horizontal transfer mechanisms, and then relatively recently spread globally, in the same manner as cholera 0139 *(82)*. Multivalent vaccines must be produced to protect against all *H. influenzae*, as well as maintaining surveillance of strains associated with vaccine failures, and remaining vigilant to the possibility of the emergence of "vaccine-escaped" *H. influenzae*.

To assess emergence of new strains, Falla et al. *(39)* probed 215 CSF and blood isolates of *H. influenzae* with pUO38, 72 isolates were from vaccine recipients, and 143 from unvaccinated persons. There was no difference in the *cap b* structure between the two groups and no new strains were seen, but the study was relatively small and undertaken over a short time-span. It provides some reassurance that the likelihood of evolution of a strain of novel enhanced virulence is small.

6. Notes

The authors use the CTAB method for preparation of genomic DNA *(83)*, and the alkaline lysis miniprep (*see* Chapter 4) or CsCl gradient centrifugation methods for isolation of plasmids *(84)*.

References

1. Slack, M. P. (1990) *Haemophilus*, in *Topley and Wilson's: Principles of Bacteriology, Virology and Immunity* (Parker, M. T. and Duerden, B. J., eds.). Edward Arnold, London. pp. 355–382.
2. Musser, J. M., Kroll, J. S., Moxon, E. R., and Selander, R. K. (1988) Clonal population structure of encapsulated *Haemophilus influenzae*. *Infect. Immun.* **56**, 1837–1845.
3. Musser, J. M., Kroll, J. S., Granoff, D. M., Moxon, E. R., Brodeur, B. R., Campos, J., Dabernat, H., Frederiksen, W., Hamel, J., and Hammond, G. (1990) Global genetic structure and molecular epidemiology of encapsulated *Haemophilus influenzae*. *Rev. Infect. Dis.* **12**, 75–111.
4. Pennington, T. H. (1993) *Haemophilus* species and clones. *Rev. Med. Microbiol.* **4**, 50–58.
5. Musser, J. M., Barenkamp, S. J., Granoff, D. M., and Selander, R. K. (1986) Genetic relationships of serologically nontypable and serotype b strains of *Haemophilus influenzae*. *Infect. Immun.* **52**, 183–191.
6. Adams, W. G., Deaver, K. A., Cochi, S. L., Plikaytis, B. D., Zell, E. R., Broome, C. V., and Wenger, J. D. (1996) Decline in childhood *Haemophilus influenzae* type b (Hib) disease in the vaccine era. *JAMA* **269**, 221–226.
7. Falla, T. J., Dobson, S. R., Crook, D. W., Kraak, W. A., Nichols, W. W., Anderson, E. C., Jordens, J. Z., Slack, M. P., Mayon-White, D., and Moxon, E. R. (1993) Population-based study of non-typable *Haemophilus influenzae* invasive disease in children and neonates. *Lancet* **341**, 851–854.
8. Farley, M. M., Stephens, D. S., Brachman, P. S., Jr., Harvey, R. C., Smith, J. D., and Wenger, J. D. (1992) Invasive *Haemophilus influenzae* disease in adults. A prospective, population-based surveillance. *CDC Meningitis Surveillance Group. Ann. Intern. Med.* **116**, 806–812.
9. Moxon, E. R. (1991) *Haemophilus influenzae*, in *Principles and Practice of Infectious Diseases* 5th ed. (Mandell, G. L., Bennett, J. E., and Dolin, R., eds.) Churchill Livingstone, New York, pp. 2039–2044.
10. Casin, I., Grimont, F., and Grimont, P. A. (1986) Deoxyribonucleic acid relatedness between *Haemophilus aegyptius* and *Haemophilus influenzae. Ann. Inst. Pasteur Microbiol.* **137B**, 155–163.
11. Sackett, D. L., Haynes, R. B., Guyatt, G. H., and Tugwell, P. (1991) *Clinical Epidemiology: A Basic Science for Clinical Medicine*. Little Brown, Boston.
12. Tenover, F. C. (1988) Diagnostic deoxyribonucleic acid probes for infectious diseases. *Clin. Microbiol. Rev.* **1**, 82–101.
13. Radstrom, P., Backman, A., Qian, N., Kragsbjerg, P., Pahlson, C., and Olcen, P. (1994) Detection of bacterial DNA in cerebrospinal fluid by an assay for simulta-

neous detection of *Neisseria meningitidis*, *Haemophilus influenzae*, and streptococci using a seminested PCR strategy. *J. Clin. Microbiol.* **32**, 2738–2744.

14. Tenover, F. C., Huang, M. B., Rasheed, J. K., and Persing, D. H. (1994) Development of PCR assays to detect ampicillin resistance genes in cerebrospinal fluid samples containing *Haemophilus influenzae*. *J. Clin. Microbiol.* **3**, 2729–2737.

15. Malouin, F. and Bryan, L. E. (1987) DNA probe technology for detection of *Haemophilus influenzae*. *Mol. Cell Probes* **1**, 221–232.

16. Davis, T. E. and Fuller, D. D. (1991) Direct identification of bacterial isolates in blood cultures by using a DNA probe. *J. Clin. Microbiol.* **29**, 2193–2196.

17. Daly, J. A., Clifton, N. L., Seskin, K. C., and Gooch III, W. M. (1991) Use of rapid, nonradioactive DNA probes in culture confirmation tests to detect *Streptococcus agalactiae*, *Haemophilus influenzae*, and *Enterococcus* spp. from pediatric patients with significant infections. *J. Clin. Microbiol.* **29**, 80–82.

18. Malouin, F., Bryan, L. E., Shewciw, P., Douglas, J., Li, D., van den Elzen, H., and Lapointe, J. R. (1988) DNA probe technology for rapid detection of *Haemophilus influenzae* in clinical specimens. *J. Clin. Microbiol.* **26**, 2132–2138.

19. Terpstra, W. J., Schoone, G. J., ter Schegget, J., van Nierop, J. C., and Griffioen, R. W. (1987) In situ hybridization for the detection of *Haemophilus* in sputum of patients with cystic fibrosis. *Scand. J. Infect. Dis.* **19**, 641–646.

20. van Ketel, R. J., de Wever, B., and van Alphen, L. (1990) Detection of *Haemophilus influenzae* in cerebrospinal fluids by polymerase chain reaction DNA amplification. *J. Med. Microbiol.* **33**, 271–276.

21. Tondella, M. L., Matar, G. M., Perkins, B. A., and Swaminathan, B. (1992) PCR for confirmation of Brazilian purpuric fever. *Lancet* **339**, 936,937.

22. van Belkum, A., Duim, B., Regelink, A., Moller, L., Quint, W., and van Alphen, L. (1994) Genomic DNA fingerprinting of clinical *Haemophilus influenzae* isolates by polymerase chain reaction amplification: comparison with major outer-membrane protein and restriction fragment length polymorphism analysis. *J. Med. Microbiol.* **41**, 63–68.

23. Post, J. C., Preston, R. A., Aul, J. J., Larkins Pettigrew, M., Rydquist White, J., Anderson, K. W., Wadowsky, R. M., Reagan, D. R., Walker, E. S., and Kingsley, L. A. (1995) Molecular analysis of bacterial pathogens in otitis media with effusion. *JAMA* **273**, 1598–1604.

24. Ueyama, T., Kurono, Y., Shirabe, K., Takeshita, M., and Mogi, G. (1995) High incidence of *Haemophilus influenzae* in nasopharyngeal secretions and middle ear effusions as detected by PCR. *J. Clin. Microbiol.* **33**, 1835–1838.

25. Greisen, K., Loeffelholz, M., Purohit, A., and Leong, D. (1994) PCR primers and probes for the 16S rRNA gene of most species of pathogenic bacteria, including bacteria found in cerebrospinal fluid. *J. Clin. Microbiol.* **32**, 335–351.

26. Jordens, J. Z. (1991) Restriction enzyme analysis of chromosomal DNA and its application in epidemiological studies. *J. Hosp. Infect.* **18** Suppl A, 432–437.

27. Loos, B. G., Bernstein, J. M., Dryja, D. M., Murphy, T. F., and Dickinson, D. P. (1989) Determination of the epidemiology and transmission of nontypable *Haemophilus influenzae* in children with otitis media by comparison of total genomic DNA restriction fingerprints. *Infect. Immun.* **57**, 2751–2757.

28. Moeser, J., Costello, P. B., Gillikin, S., and Murphy, T. F. (1991) *Haemophilus influenzae* polyarthritis in an adult: an analysis of serotype b strains. *Rev. Infect. Dis.* **133**, 61–63.

29. Mencarelli, M., Marsili, C., Zanchi, A., Pantini, C., and Cellesi, C. (1993) Genomic DNA fingerprints and phenotypic characteristics of serotype B *Haemophilus influenzae* isolates from Italy. *Eur. J. Epidemiol.* **9**, 353–360.

30. Butler, P. D. and Moxon, E. R. (1990) A physical map of the genome of *Haemophilus influenzae* type b. *J. Gen. Microbiol* **136**, 2333–2342.

31. Cote, S., Sanschagrin, F., Dargis, M., Simard, J. L., Roy, P. H., MacDonald, N. E., Rabin, H. R., Bergeron, M. G., and Malouin, F. (1994) Molecular typing of *Haemophilus influenzae* using a DNA probe and multiplex PCR. *Mol. Cell Probes* **8**, 23–37.

32. Poulson, K., Hjorth, J. P., and Kilian, M. (1988) Limited diversity of the immuno-globulin A1 protease gene (iga) among *Haemophilus influenzae* serotype b strains. *Infect. Immun.* **56**, 987–992.

33. Forbes, K. J., Bruce, K. D., Ball, A., and Pennington, T. H. (1992) Variation in length and sequence of porin (ompP2) alleles of non-capsulate *Haemophilus influenzae*. *Mol. Microbiol.* **6**, 2107–2112.

34. Hansen, E. J., Pelzel, S. E., Orth, K., Moomaw, C. R., Radolf, J. D., and Slaughter, C. A. (1989) Structural and antigenic conservation of the P2 porin protein among strains of *Haemophilus influenzae* type b. *Infect. Immun.* **57**, 3270–3275.

35. Hobson, R. P., Williams, A., Rawal, K., Pennington, T. H., and Forbes, K. J. (1995) Incidence and spread of *Haemophilus influenzae* on an Antarctic base determined using the polymerase chain reaction. *Epidemiol. Infect.* **114**, 93–103.

36. Kroll, J. S., Moxon, E. R., and Loynds, B. M. (1993) An ancestral mutation enhancing the fitness and increasing the virulence of *Haemophilus influenzae* type b. *J. Infect. Dis.* **168**, 172–176.

37. Hoiseth, S. K., Corn, P. G., and Anders, J. (1992) Amplification status of capsule genes in *Haemophilus influenzae* type b clinical isolates. *J. Infect. Dis.* **165**, s114.

38. Kroll, J. S. and Moxon, E. R. (1988) Capsulation and gene copy number at the cap locus of *Haemophilus influenzae* type b. *J. Bacteriol.* **170**, 859–864.

39. Falla, T. J., Crook, D. W., Anderson, E. C., Ward, J. I., Santosham, M., Eskola, J., and Moxon, E. R. (1995) Characterization of capsular genes in *Haemophilus influenzae* isolates from *H. influenzae* type b vaccine recipients. *J. Infect. Dis.* **171**, 1075,1076.

40. Kroll, J. S., Ely, S., and Moxon, E. R. (1991) Capsular typing of *Haemophilus influenzae* with a DNA probe. *Mol. Cell Probes* **5**, 375–379.

41. Hoiseth, S. K., Connelly, C. J., and Moxon, E. R. (1985) Genetics of spontaneous, high- frequency loss of b capsule expression in *Haemophilus influenzae*. *Infect. Immun.* **49**, 389–395.

42. St. Geme, J. W3., Takala, A., Esko, E., and Falkow, S. (1994) Evidence for capsule gene sequences among pharyngeal isolates of nontypeable *Haemophilus influenzae*. *J. Infect. Dis.* **169**, 337–342.

43. Weinberg, G. A., Ghafoor, A., Ishaq, Z., Nomani, N. K., Kabeer, M., Anwar, F., Burney, M. I., Qureski, A. W., Musser, J. M., and Selander, R. K. (1989) Clonal analysis of Hemophilus influenzae isolated from children from Pakistan with lower respiratory tract infections. *J. Infect. Dis.* **160**, 634–643.

44. Hoiseth, S. K. and Gilsdorf, J. R. (1988) The relationship between type b and nontypable *Haemophilus influenzae* isolated from the same patient. *J. Infect. Dis.* **158**, 643–645.

45. Leaves, N. I., Falla, T. J., and Crook, D. W. (1995) The elucidation of novel capsular genotypes of *Haemophilus influenzae* type b with the polymerase chain reaction. *J. Med. Microbiol.* **43**, 120–124.

46. Leaves, N. I. and Jordens, J. Z. (1994) Development of a ribotyping scheme for *Haemophilus influenzae* type b. *Eur. J. Clin. Microbiol. Infect. Dis.* **13**, 1038–1045.

47. Leaves, N. I. (1995) The molecular characterisation of *Haemophilus influenzae*, PhD thesis.

48. Brenner, D. J., Mayer, L. W., Carlone, G. M., Harrison, L. H., Bibb, W. F., Brandileone, Sottnek, F. O., Irino, K., Reeves, M. N., and Swenson, J. M. (1988) Biochemical, genetic, and epidemiologic characterization of *Haemophilus influenzae* biogroup aegyptius (*Haemophilus aegyptius*) strains associated with Brazilian purpuric fever. *J. Clin. Microbiol.* **26**, 1524–1534.

49. Irino, K., Grimont, F., Casin, I., and Grimont, P. A. (1988) rRNA gene restriction patterns of *Haemophilus influenzae* biogroup aegyptius strains associated with Brazilian purpuric fever. *J. Clin. Microbiol.* **26**, 1535–1538.

50. Stull, T. L., LiPuma, J. J., and Edlind, T. D. (1988) A broad-spectrum probe for molecular epidemiology of bacteria: ribosomal RNA. *J. Infect. Dis.* **157**, 280–286.

51. Smith-Vaughan, H. C., Leach, A. J., Shelby-James, T. M., Kemp,K., Kemp, D. J. and Mathews, J. D. (1996) Carriage of multiple ribotypes of non-encapsulated *Haemophilus influenzae* in Aboriginal infants with otitis media. *Epidemiol. Infect.* **116**, 177–183.

52. Bruce, K. D. and Jordens, J. Z. (1991) Characterization of noncapsulate *Haemophilus influenzae* by whole-cell polypeptide profiles, restriction endonuclease analysis, and rRNA gene restriction patterns. *J. Clin. Microbiol.* **29**, 291–296.

53. Barnes, W. M. (1994) PCR amplification of up to 35-kb DNA with high fidelity and high yield from bacteriophage templates. *Proc. Natl. Acad. Sci. USA* **91**, 2216–2220.

54. Jordens, J. Z., Leaves, N. I., Anderson, E. C., and Slack, M. P. (1993) Polymerase chain reaction-based strain characterization of noncapsulate *Haemophilus influenzae*. *J. Clin. Microbiol.* **31**, 2981–2987.

55. Slack, M. P., Crook, D. W., Jordens, J. Z., Anderson, E. C., Falla, T. J., Leaves, N. I., et al. (1992) Molecular and epidemiological aspects of *Haemophilus influenzae* infection. PHLS *Microbiol. Digest* **10**, 122–128.

56. Lupski, J. R. and Weinstock, G. M. (1992) Short, interspersed repetitive DNA sequences in prokaryotic genomes. *J. Bacteriol.* **174**, 4525–4529.

57. Versalovic, J., Koeuth, T., and Lupski, J. R. (1991) Distribution of repetitive DNA sequences in eubacteria and application to fingerprinting of bacterial genomes. *Nucleic Acids Res.* **19**, 6823–6831.

58. Elwell, L. P. (1994) R plasmids and antibiotic resistances, in *Molecular Genetics of Bacterial Pathogenesis* (Miller, V. L., Kaper, J. B., Portnoy, D. A., and Isberg, R. R., eds). American Society for Microbiology, Washington, DC., pp. 17–41.

59. Campos, J., Garcia Tornel, S., Musser, J. M., Selander, R. K., and Smith, A. L. (1987) Molecular Epidemiology of multiply resistant *Haemophilus influenzae* type b in day care centers. *J. Infect. Dis.* **156**, 483–489.

60. Mayer, L. W. (1988) Use of plasmid profiles in epidemiological surveillance of disease outbreaks and in tracing the transmission of antibiotic resistance. *Clin. Microbiol. Rev.* **1**, 228–243.

61. Levy, J., Verhaegen, G., De Mol, P., Couturier, M., Dekegel, D., and Butzler, J. P. (1993) Molecular characterization of resistance plasmids in epidemiologically unrelated strains of multiresistant *Haemophilus influenzae*. *J. Infect. Dis.* **168**, 177–187.

62. Anderson, J. R., Smith, M. D., Kibbler, C. C., Holton, J., and Scott, G. M. (1994) A nosocomial outbreak due to non-encapsulated *Haemophilus influenzae*: analysis of plasmids coding for antibiotic resistance. *J. Hosp. Infect.* **27**, 17–27.

63. Dimopoulou, I. D., Kraak, W. A., Anderson, E. C., Nichols, W. W., Slack, M. P., and Crook, D. W. (1992) Molecular epidemiology of unrelated clusters of multiresistant strains of *Haemophilus influenzae*. *J. Infect. Dis.* **165**, 1069–1075.

64. Sturm, A. W., Mostert, R., Rouing, P. J., van Klingeren, B., and van Alphen, L. (1990) Outbreak of multiresistant non-encapsulated *Haemophilus influenzae* infections in a pulmonary rehabilitation centre. *Lancet* **335**, 214–216.

65. Brightman, C. A., Crook, D. W., Kraak, W. A., Dimopoulou, I. D., Anderson, E. C., Nichols, W. W., and Slack, M. P. (1990) Family outbreak of chloramphenicol-ampicillin resistant *Haemophilus influenzae* type b disease. *Lancet* **335**, 351–352.

66. Swaminathan, B., Mayer, L. W., Bibb, W. F., Ajello, G. W., Irino, K., Birkness, K. A., Garon, C. F., Reeves, M. W., deCunto, Brandileone, M. C., and Sottnek, F. O. (1989) Microbiology of Brazilian Purpuric Fever and diagnostic tests. *J. Clin. Microbiol.* **27**, 605–608.

67. Fleischmann, R. D., Adams, M. D., White, O., Clayton, R. A., Kirkness, E. F., Kerlavage, A. R., et al. (1995) Whole-genome random sequencing and assembly of *Haemophilus influenzae* Rd. *Science* **269**, 496–512.

68. Nowak, R. (1995) Bacterial genome sequence bagged. *Science* **269**, 468–470.

69. Nowak, R. (1995) Entering the postgenome era. *Science* **270**, 368–371.

70. Hensel, M., Shea, J. E., Gleeson, C., Jones, M. D., Dalton, E., and Holden, D. W. (1995) Simultaneous identification of bacterial virulence genes by negative selection. *Science* **269**, 400–403.

71. Moxon, E. R. (1982) Experimental studies of *Haemophilus influenzae* infection in a rat model, in Haemophilus influenzae: *Epidemiology, Immunology, and Prevention of Disease* (Sell, S. H. and Wright, P. F., eds.). Elsevier Biomedical, New York., pp. 57–72.

72. Moxon, E. R., Smith, A. L., Averill, D. R., and Smith, D. H. (1974) *Haemophilus influenzae* meningitis in infant rats after intranasal inoculation. *J. Infect. Dis.* **129**, 154–162.

73. Giebink, G. S. (1982) Experimental otitis media due to *Haemophilus influenzae* in the Chinchilla, in Haemophilus influenzae: *Epidemiology, Immunology, and*

Prevention of Disease (Sell, S. H. and Wright, P. F., eds.). Elsevier Biomedical, New York, pp. 73–86.

74. McClelland, M., Jones, R., Patel, Y., and Nelson, M. (1987) Restriction endonucleasees for pulsed field mapping of bacterial genomes. *Nucleic. Acids. Res.* **15,** 5985–6005.

75. Lee, J. J. and Smith, H. O. (1988) Sizing of the *Haemophilus influenzae* Rd genome by pulsed-field agarose gel electrophoresis. *J. Bacteriol.* **170,** 4402–4405.

76. Kauc, L., Mitchell, M., and Goodgal, S. H. (1989) Size and physical map of the chromosome of *Haemophilus influenzae. J. Bacteriol.* **171,** 2474–2479.

77. Romling, U., Heuer, T., and Tummler, B. (1994) Bacterial genome analysis by pulsed field gel electrophoresis techniques, in *Advances in Electrophoresis* (Chrambach, A., Dunn, M. J., and Radola, B. J., eds). Verlagsgesellschaft mbH., pp. 353–389.

78. Moxon, E. R. and Kroll, J. S. (1988) Type b capsular polysaccharide as a virulence factor of *Haemophilus influenzae. Vaccine* **6,** 113–115.

79. Kroll, J. S., Loynds, B. M., and Moxon, E. R. (1991) The *Haemophilus influenzae* capsulation gene cluster: a compound transposon. *Mol. Microbiol.* **5,** 1549–1560.

80. Kroll, J. S., Moxon, E. R., and Loynds, B. M. (1994) Natural genetic transfer of a putative virulence–enhancing mutation to *Haemophilus influenzae* type a. *J. Infect. Dis.* **169,** 676–679.

81. Kroll, J. S. and Moxon, E. R. (1990) Capsulation in distantly related strains of *Haemophilus influenzae* type b: genetic drift and gene transfer at the capsulation locus. *J. Bacteriol.* **172,** 1374–1379.

82. Karaolis, D. K., Lan, R., and Reeves, P. R. (1994) Molecular evolution of the seventh- pandemic clone of Vibrio cholerae and its relationship to other pandemic and epidemic V. cholerae isolates. *J. Bacteriol.* **176,** 6199–6206.

83. Wilson, K. (1987) Preparation of genomic DNA from bacteria. *Curr. Protocols Molec. Biol.* **4,** 2. 4. 1–2. 4. 5.

84. Sambrook, J., Fritsch, E. F., and Maniatis, T. (1989) *Molecular Cloning: A Laboratory Manual* 2nd ed. Cold Spring Harbor Laboratory Press, Cold Spring Harbor.

85. Falla, T. J., Anderson, E. C., Chappell, M. M., Slack, M. P., and Crook, D. W. (1993) Cross-reaction of spontaneous capsule-deficient *Haemophilus influenzae* type b mutants with type-specific antisera. *Eur. J. Clin. Microbiol. Infect. Dis.* **12,** 147–148.

86. Falla, T. J., Crook, D. W., Brophy, L. N., Maskell, D., Kroll, J. S., and Moxon, E. R. (1994) PCR for capsular typing of *Haemophilus influenzae. J. Clin. Microbiol.* **32,** 2382–2386.

87. Munson, R., Jr., Grass, S., Einhorn, M., Bailey, C., and Newell, C. (1989) Comparative analysis of the structures of the outer membrane protein P1 genes from major clones of *Haemophilus influenzae* type. *Infect. Immun.* **57,** 3300–3305.

88. Smith Vaughan, H. C., Sriprakash, K. S., Mathews, J. D., and Kemp, D. J. (1995) Long PCR-ribotyping of nontypeable *Haemophilus influenzae. J. Clin. Microbiol.* **33,** 1192–1195.

89. Owen, R. J. (1989) Chromosomal DNA fingerprinting—a new method of species and strain identification applicable to microbial pathogens. *J. Med. Microbiol.* **30,** 89–99.

14

The Impact of Molecular Techniques on the Study of Meningococcal Disease

Martin C. J. Maiden

1. Introduction

Neisseria meningitidis, the meningococcus, is normally a harmless commensal bacterium that colonises the naso/oropharynx of humans. This antigenically variable gram-negative diplococcus has the potential, however, to cause rapidly progressing meningitis and fulminant septicemia, either separately or together *(1,2)*. Once present in the bloodstream, meningococci grow rapidly and their highly toxic lipo-oligosaccharides frequently cause extensive tissue damage and severe toxic shock. The progress of the disease is swift, and death often occurs within hours of the onset of symptoms *(3)*. Even in countries where meningococcal infection is relatively rare, it remains a high priority for public health services because of the high mortality rates of fulminant septicemic disease (which can be up to 40% even when intensive supportive therapy is available), the high proportion of sequelae in patients who have recovered (including brain damage and digit or limb loss), and the age groups most susceptible (young children and, to a lesser extent, teenagers) *(4,5)*.

In common with those of other bacterial pathogens, the characterization scheme for the meningococcus (**Fig. 1**) relies on the serologic reactivity of cell surface components *(6)*. The primary level of characterization is the serogroup, which is based on the antigenic properties of the capsular polysaccharide. There are 13 meningococcal serogroups known *(7)*, but only organisms expressing capsules belonging to serogroups A, B, and C commonly cause disease, with serogroup Y and W-135 organisms causing most of the remaining cases *(2)*. Further immunologic characterization of meningococcal isolates is based on "subcapsular" antigens. Serotypes are antigenic variants of the outer membrane porin, PorB, and serosubtypes variants of its relative PorA *(8,9)*. Less com-

From: *Methods in Molecular Medicine, vol. 15: Molecular Bacteriology: Protocols and Clinical Applications*
Edited by: N. Woodford and A. P. Johnson © Humana Press Inc., Totowa, NJ

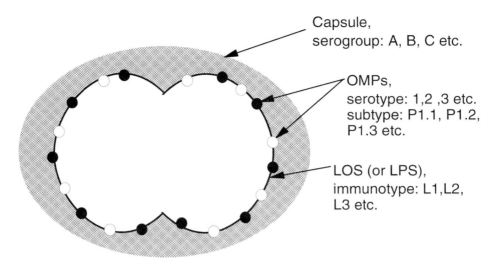

Fig. 1. Serologic typing scheme for *N. meningitidis*. In common with many other pathogenic bacteria, the classification of meningococcal isolates has relied on the identification of
surface components with serologic reagents. Serogroups are defined by the capsular polysaccharide and are designated with capital letters: A, B, C, D, 29E, H, I, J, L, W-135, X, Y, Z.
Further characterization is achieved with subcapsular antigens, the class 2 and 3 outer membrane proteins (OMPs; collectively referred to as PorB) define serotypes, designated with a
number. The larger class 1 OMP *(PorA)* defines serosubtypes, designated by P1. followed by
a number, e.g., P1.7. The PorA protein has two major regions of antigenic variability (VR1
and VR2) and each protein normally has at least two subtypes hence P1.7,16, given in the
order VR1, VR2 separated by a comma. If a designation for the minor variable region (VR3)
is used, (subtypes 6 and 14), this should be given last after a further comma. Lipo-oligosaccharide is also, but less commonly, used in strain characterization. Multiple LOS types are
common within one strain hence L3,7,9 or L1,8,10. The convention is to write the serologic
characteristics in the order given above separated by a colon thus: B:15:P1.7,16:L3,7,9 giving group, serotype, serosubtype(s) and LOS types, respectively.

monly, variation in lipo-oligosaccharide (LOS), the major glycolipid of the
outer leaflet of the meningococcal outer membrane, is used in strain characterization, defining "immunotypes" *(10)*. Originally, immunologic strain characterization was done with polyclonal sera, but monoclonal antibodies are now
used for identifying serotypes, serosubtypes, and immunotypes. However, there
are a number of problems with this serologic scheme *(11)*, including a lack of
comprehensive reagents and poor correlation of serologic characteristics with
the genetic relationships of isolates defined by multilocus enzyme electrophoresis (MLEE) techniques *(12,13)* (*see* **Note 1**).

Meningococcal disease occurs in four distinct epidemiologies *(14,15)*:
endemic, hyperendemic, localized epidemic, and large-scale epidemic/pan-

demic. In western Europe and North America, endemic disease with annual attack rates of 1–5 cases per 100,000 population, is prevalent with occasional hyper-endemic outbreaks (10–15 cases per 100,000, over periods of months or years) or localized disease outbreaks (20–30 cases per 100,000, the outbreaks persisting for a number of weeks). In these geographical areas, serogroup B and C meningococci are the most common cause of infection *(16)*. In contrast, large-scale epidemic/pandemic outbreaks of disease (up to 1000 cases per 100,000) caused by serogroup A meningococci periodically occur in sub-Saharan Africa and China: these represent the most serious manifestation of meningococcal infection and can cause tens of thousands of cases and thousands of deaths within a matter of weeks *(17,18)*.

Currently, there is no effective childhood vaccine that protects against all meningococci, although serogroup A and C polysaccharide-based vaccines can interrupt outbreaks caused by organisms of these serogroups in older children and adults *(19)*. New protein-conjugate A and C polysaccharide vaccines provide the prospect of effective infant vaccines against meningococci expressing capsules of these serogroups *(20)*. Unfortunately, despite intensive research in this area *(21)*, it is unlikely that an infant vaccine capable of eradicating serogroup B disease will be available in the immediate future.

2. Diagnosis

Meningococcal disease is notoriously difficult to diagnose. The initial stages of the syndrome may resemble influenza infection and, in the case of fulminant disease, the early signs of the purpuric rash can be small and easily overlooked or misinterpreted. The rapid progress of the infection means that these difficulties in diagnosis have potentially fatal consequences, as it is important that parenteral antibiotics are administered as soon as possible after onset of the disease. The emphasis on early treatment, preferably before hospitalization *(22)*, and the maintenance of antibiotic therapy on admission to hospital, reduces the likelihood of isolating a meningococcus from clinical samples from approx 50% to less than 5% *(23)*, and has resulted in an increasing proportion of cases that cannot be confirmed by laboratory culture of the organism *(24,25)*. This decrease in laboratory-confirmed cases, and consequent loss of epidemiologic information, is occurring at a time when such data are particularly important because of the testing and introduction of novel vaccines. It has been suggested that for certain patient groups, an improved diagnosis system could increase the number of confirmed cases by 60% *(26)*.

For microbiologic confirmation of a clinical diagnosis, it is necessary to detect meningococci in samples of cerebrospinal fluid (CSF), serum, or whole-blood. When the organism cannot be cultured, this has traditionally been achieved by techniques such as microscopy, coagglutination, or latex agglutination, but these methods are often of low sensitivity and specificity *(27)*. Amplification of bacterial genes from clinical samples by the polymerase chain reaction (PCR) provides alternative methods for detecting bacteria in clinical samples that are rapid and potentially highly sensitive and specific. Meningococcal target genes for which PCR protocols are available, a number of which have already been applied to clinical specimens, are given in **Table 1**. To date, most success has been achieved with CSF samples, but the reluctance of many physicians to take CSF samples *(28)* has led to an increasing interest in using serum or whole-blood samples, which are more readily obtained. However, these specimens are more difficult to use in PCR-based diagnostic tests.

Diagnostic tests that use the PCR require only small quantities of clinical material, are as rapid or faster than culture and the test-specific reagents, the oligonucleotide primers, are inexpensive. Modification of conventional protocols, such as the use of nested PCR (nPCR; *see* Chapter 5), can achieve very high sensitivities, equivalent to one colony forming unit (cfu) per amplification reaction *(29)*. Sensitivity and specificity can be further improved by including hybridization assays with membrane filters (Southern blots) or in microtiter wells (PCR-ELISA; *see* Chapter 5) *(26,30)*. In addition, the amplified genes obtained can be further analyzed, (e.g., by digestion with restriction endonucleases, hybridization with specific probes, or nucleotide sequencing) to provide not only diagnostic, but also epidemiologic data from the same sample.

As it is in principle possible to amplify any bacterial gene by the PCR, careful choice of the target genes for these procedures is important. This choice should be made on the grounds, not only of test specificity and sensitivity, but also of the epidemiologic value of the information that can be obtained by analysis of the amplified genes. Once a gene target has been chosen, extensive evaluation of clinical samples, including double-blinded trials, is necessary before reliance can be placed on a given method for diagnosis. Although double-blinded trials of some methods have been carried out, no extensive comparative analyses of different PCR diagnostic techniques for meningococcal disease have been published to date. The data from a number of separate studies based on different target genes are summarized here.

2.1. 16S Ribosomal RNA (rrn) Gene

Bacteria other than the meningococcus, including *Streptococcus pneumoniae, Haemophilus influenzae*, group B streptococci, and *Listeria monocytogenes* cause meningitis, therefore, the primary requirement of nonculture diagnosis is species identification. All eubacteria have multiple copies of *rrn* genes in their chromosomes (*see* Chapter 2) which comprise conserved and species-specific variable regions of gene sequence *(31,32)*. The method of Olcén et al. *(27,33)*, exploits this by amplification of *rrn* genes from clinical samples derived from

Table 1
PCR Oligonucleotide Primers for the Diagnosis of Meningococcal Disease and Typing of Meningococci

Gene target	Forward primer[a] 5'–3' (name and sequence)		Reverse primer[a] 5'–3' (name and sequence)		Refs.
rrn	u3	AAC TMC GTG CCA GCA GCC GCG GTA	ru8	AAG GAG GTG ATC CAR CCG CAS STT C	*27,33*
	SP	GCT GTG GCT TAA CCA TAG TAG			
	NM	TGT TGG GCA ACC TGA TTG			
	HI	CCT AAG AAG AGC TCA GAG			
	GBS	ACC GGC CTA GAG ATA GGC			
	STREP	GTA CAA CGA GTC GCA AGC			
	LM	GGA GCT AAT CCC ATA AAA CTA		AGA CAT TGG GTW GWR GGK GAR AGT AA	
siaD		AYA TWT TGC ATG TMS CYT TYC CTG A			
porB	27	TTG TAC GGT ACA ATT AAA GCA GGC CGT	28	TTA GAA TTT GTG ACG CAG ACC AAC	*38*
porA	21	CTG TAC GGC GAA ATC AAA GCC GGC GT	22	TTA GAA TTT GTG GCG CAA ACC GAC	*37*
	p1	GCG GCC GTT GCC GAT GTC AGC C	p2	GCG GCA TTA ATT TGA GTG TAG TTG CC	*29*
	p3	CAA AGC CGG CGT GGA AG	p4	GAT CGT AGC TGG TAT TTT CGC C	
dhps	MN1	GGG TCG ACG GTT TCA GAC GGC ATA TAA	NM2	AAG GTA CCG CCC CGT CCT TTT CAG ACG	*46,47*
	NM3	ACG GGC ATT TGC CTG ATG CAC	NM6	CGC CAT CAA TTC GGG CAA ATG	
IS1106	1	ATT ATT CAG ACC GCC GGC AG	2	CCG ATA ATC AGG CAT CCG	*23,30*
			8	TGC CGT CCT GCA ACT GAT GT	*26*

[a]Degenerate bases in the oligonucleotides are indicated as follows: R indicates A or G; Y indicates C or T; K indicates G or T; M indicates A or C; S indicates G or C; and W indicates A or T.

suspected cases of meningitis. A primary PCR (using primers u3 and ru8, **Table 1**) amplifies a 1030-bp product from all eubacterial *rrn* genes. This is followed by one or two secondary amplifications with species-specific primers and ru8 (**Table 1**) that produce differentially sized amplicons that are characteristic for each species. The species-specific products are readily distinguished by separation of the amplicons using agarose gel electrophoresis. The method requires only small samples of CSF (10μL).

The original method, which identified *N. meningitidis* (species-specific primer NM, product size 710 bp, **Table 1**), *H. influenzae* (primer HI, product size 540 bp), and streptococcal species (primer STREP, product size 300 bp), was tested for specificity by analysis of a collection of 133 strains containing 28 bacterial species. The method was then applied to 304 clinical specimens of CSF, including 125 samples from patients with confirmed meningitis, and negative control CSF samples from patients known not to have bacterial meningitis. The sensitivity of the assay for detecting bacteria in clinical specimens was high (ratio of PCR positives to known positives = 117:125 or 0.94; *see* **Note 2**) as was its specificity, (ratio of PCR negative samples to known negative samples = 158:164 or 0.96). The sensitivity for species identifications was also high: the species present in 87 of 98 samples (0.89) were correctly identified. Using agarose gel electrophoresis and ethidium bromide (EB) to detect the PCR products, gave a detection limit of 3×10^2 cfu/mL of CSF *(33)*. It is likely that the detection limit of this approach could be enhanced by the inclusion of a hybridization step, as developed for some other diagnostic PCR systems described below. The assay was recently extended to include primers that identify *S. pneumoniae* (primer SP, product size 937 bp), Group B streptococci (primer GBS, product size 544 bp) and *L. monocytogenes* (primer LM, product size 265 bp). As some of these amplicons have a similar size to others mentioned above, detection of the additional species requires a separate secondary amplification *(27)*.

2.2. Capsular Genes

After confirmation of a meningococcal infection, the most important information for the clinician and public health physician is the serogroup of the causative organism. This information influences the public health measures undertaken (e.g., vaccination, administration of prophylactic antibiotics, and counseling of contacts) *(34)*. A PCR test that distinguishes alleles of *siaD*, the gene encoding a sialyltransferase involved in the synthesis of both serogroup B and serogroup C capsules, has been developed *(34a)*. This procedure is based on a *Taq*1 restriction fragment length polymorphism (RFLP) of the *siaD* gene that appears to distinguish serogroup B and serogroup C isolates. This technique has been adapted to a microtitre well PCR-ELISA hybridization format, using capture probes specific for amplified serogroup B and serogroup C *siaD* alleles (amplified with the primers in **Table 1**).

In studies of clinical specimens, this test was totally specific, in that no false identifications of serogroup were made, and had sensitivities in samples from confirmed cases of 0.81 (17 out of 21) for CSF samples, 0.63 (5 out of 8) for whole blood samples, and 0.30 (3 out of 10) for serum samples. For clinically suspected cases of meningococcal disease, 6 from 14 (0.43) positive results were obtained from CSF samples, 8 from 26 (0.31) serum samples, and 5 from 23 (0.23) whole blood samples. No false-positive results were obtained in this study and the detection limit was estimated at one cell per reaction *(34a)*. However, this technique has been applied only to 12 serogroup B and C isolates and, as one nonpathogenic serogroup H strain gave a positive result with the group C capture probe, further validation of the specificity of the capture probes is required. In addition, expansion of this technique to all five disease-causing serogroups is necessary.

2.3. Serotype and Serosubtype Genes

The *porB* and *porA* genes, which encode the serotype and the serosubtype antigens of the meningococcus, respectively, are probably the best studied of all meningococcal genes and many alleles for both loci have been characterized by nucleotide sequence analysis *(35–40)*. A number of groups have successfully amplified these genes from clinical specimens and work is very active in this area *(29,41)* (Urwin, R. and Fox, A., Manchester Public Health Laboratory, Manchester UK and Bash, M. and Frasch, C., Center for Biologic Evaluation and Research, Bethesda, MD, personal communications).

A nPCR technique has been developed for the detection of the *porA* gene in CSF samples, and combined with nucleotide sequence analysis to provide both diagnostic and epidemiologic information as to the causative organism *(29)*. This work has been extended by a blinded study that included 37 samples from patients with known meningococcal infection, and 49 samples from cases of pneumococcal meningitis, *Haemophilus influenzae* meningitis, suspected borreliosis, and nonbacterial diseases *(41)*. In this study, no false-positives were detected among the 49 negative samples. From the 37 samples from known cases of meningococcal infection, 15 from 16 culture-confirmed cases were positive after nPCR, a sensitivity of 0.94. But detection among clinically defined meningococcal infection was poorer with 15/21 (0.71) of cases giving a PCR product. Of the six negative samples, one gave a positive reaction on dilution (possibly indicating the presence of PCR-inhibitors in this sample) and the remaining five were from patients with septicemia, but no meningitis.

The technique was used to analyze an additional 67 culture-negative samples from patients with suspected meningococcal infection, collected during a phase-three meningococcal vaccine trial in Norway *(42)*. In this set, 10 were positive by nPCR, and from each of these it was possible to sequence the amplified *porA* genes. Four of these samples had the same subtype as the vaccine strain. These data demonstrate that the *porA* nPCR approach provides useful data that contribute to the difficult area of nonculture diagnosis of meningococcal infection and simultaneously provides epidemiologic data *(41)*.

Using appropriate primers (either pair 21 and 22 or pair 27 and 28, **Table 1**, at low annealing stringency) it is possible to amplify both the *porA* and *porB* genes in the same experiment *(38)* and, as the amplified genes have characteristic sizes when examined by agarose gel electrophoresis, this is a potentially useful characteristic for confirming the presence of meningococci. By simultaneously amplifying both genes, the type and subtype of the strain can be determined from the same single experiment used for diagnosis. A number of trials using techniques for the identification of *porB* alleles in amplified CSF samples were being done at the time of writing (Fox, A., personal communication; Frasch, C., personal communication).

2.4. Dihydropterate Synthase (dhps) Gene

Dihydropterate synthase is the target for the sulfonamide drugs and, in the meningococcus, resistance to these drugs has arisen by alterations in the chromosomally encoded *dhps* gene *(43–45)*. This gene has been successfully amplified from at least one clinical specimen (using primers NM1 and NM2, **Table 1**, product size ~ 950 bp) *(46)* and primers that amplify the *dhps* gene from most or all meningococci have been developed (NM1 and NM6; NM3 and NM6) *(47)*. Extensive studies in clinical specimens have not been carried out, but in principle this technique allows the identification of sulfonamide resistance, a commonly determined epidemiologic characteristic, even when the organism is not culturable.

2.5. IS1106

The insertion sequence IS*1106*, which is present in several copies in the meningococcal chromosome *(48)*, has been used by several laboratories as a target for PCR diagnosis, and a number of double-blinded trials of methods based on the IS*1106* sequence have been undertaken using clinical specimens. The original method used agarose gel electrophoresis and EB staining to identify PCR products amplified by primers 1 and 2 (**Table 1**) and detected meningococci in 10 of 11 (0.91) known positives; however, a total of four false-positives were obtained from 32 negative samples (0.13), one of which was not resolved by repeat testing *(23)*. False-positives are a major problem with this target. For example, Davidson et al. *(30)* found 16 out of 34 (0.46) false-positives when CSF samples known to contain bacteria other than the meningococcus were tested using EB-stained gels to identify the PCR products.

The specificity of this test has been improved by the addition of a PCR-ELISA detection step, using oligonucleotide capture probes (*see* Chapter 5) corresponding to internal sequence of the IS*1106* *(26,30)*. Davidson et al. *(30)* used a 23-base oligonucleotide capture probe (TGC CGC CGT TGT TGG AAG GAC TG) that removed false-positive results, but the sensitivity of the PCR-ELISA technique was lower (0.64; 20 positives from a total of 31 known positive CSF samples) than that obtained when Southern hybridization was used to detect the PCR products (0.71; 10 positives from 14 samples). Newcombe et al. *(26)* applied a similar approach, with a 20 base oligonucleotide (GTA CCG ATG CGG AAG GCT AT), to serum and buffy coat samples. These workers also introduced a different amplification primer *(8)* (**Table 1**) and achieved a detection rate of 0.85 (22 positives from 26 known positives) for serum and unity (25 from 25 known positives) for buffy coat preparations. Despite some success with this target, IS*1106* is probably not the best target for nonculture diagnosis (*see* **Note 3**); however, the work on IS*1106* has established the utility of PCR techniques for diagnosis, and the system could be used in addition to one or more of the aforementioned methods for further confirmation of diagnosis.

3. Epidemiologic Investigation

Two approaches are used for the characterization of pathogenic bacteria: The identification of variants of particular cell components or the genes that encode them, and the establishment of genetic relationships among isolates, based on a sample of the whole chromosome. As a consequence of their roles in host-pathogen interactions and potential as components of vaccines, surface antigens or their genes are frequently chosen as typing targets. Characterizing such targets, however, may not give an accurate representation of the genetic relationships among isolates, which are more readily determined by the comparison of a number of genes sampled from diverse locations of the genome. Indeed, for the establishment of genetic relationships among strains, it is preferable to avoid genes likely to be under strong selection, such as those that encode antigens.

Molecular approaches have contributed to the epidemiologic characterization of *N. meningitidis* by defining the biologic bases of traditional typing techniques and by enabling the development of new methods for epidemiologic and genetic characterization of isolates (*see* **Note 4**). Recent work on the serotyping and serosubtyping system for meningococci, which was originally based on antibody reagents that react with variants of the PorB and PorA proteins (**Fig. 1**), illustrates the impact of molecular techniques on the understanding of serologic strain characterization schemes.

Comparisons of the deduced amino acid sequences of many variants of PorA and PorB demonstrated that they are related members of a family of *Neisseria* porins *(37,49–51)* and that their specific serologic reactivities reside in variable surface loops of the porin structure. In the case of PorA, the serosubtyping antigen, most sequence variability resides in the first and fourth surface-exposed loops of the putative porin structure (**Fig. 2**), described as Variable Region 1 (VR1) and VR2: there is a third less variable region corresponding to the fifth putative loop (VR3 or sVR) *(36,37)*. Most of serosubtype-specific monoclonal antibodies react with continuous peptide epitopes that are located within VR1 or VR2. Originally, the designation of serosubtypes was on the basis of antibody reactivity *(52–54)* but, as a consequence of the information obtained by molecular studies, it has been proposed that the peptide sequence of the VRs should be used as the basis for subtype definition *(40,55)*. Under this revised system, members of the same "VR family" share more than 80% peptide sequence identity, and within each family there are minor variants of peptide sequence resulting from amino acid changes, deletions, or insertions. These variants are distinguished by the addition of a letter to the subtype designation (e.g., P1.10a). There are 9 VR1 sequence families and 17 VR2 families at the time of writing. Two subtypes (P1.6 and P1.14) are probably associated with VR3 peptide sequence.

Sequence analyses reveal two reasons for the lack of comprehensive coverage by the current serosubtype monoclonal antibodies (MAbs). In some cases strains express VR families in either VR1, VR2, or both, for which no MAb is available, whereas other nonsubtypable isolates express variants of VRs families that do not react or react poorly with the relevant MAb *(40,55–58)*. To overcome these problems, a dot-blot hybridization method has been developed, which identifies subtype families by hybridization of radiolabeled *porA* genes (amplified with primers 21 and 22; **Table 1**) with panels of cloned DNA encoding meningococcal VRs (available from the author) *(59)*. In a study of 174 isolates that were not characterized by serosubtyping, the DNA dot-blot method identified 149 (0.86) VR1s and 113 (0.65) VR2s. A total of 96 (0.55) strains were characterized at both VR; however, a number of VRs had to be characterized by nucleotide sequence analysis *(40)*.

The serotype antigen, the meningococcal PorB protein, is related to the PorA protein and conforms to a similar structural model, but the antigenic variability of this protein is different *(38,39)*. First, a given meningococcal isolate has one of two mutually exclusive genes at the *porB* locus, encoding either a class 2 or a class 3 PorB protein: There are numerous variants of both class 2 and class 3 PorB proteins. Second, although antigenic diversity among PorB proteins also resides in the putative surface loops of the proposed porin structure, the surface loops of PorB variants are smaller and less diverse. Third, in both class 2 and class 3 PorB proteins, different surface loops vary compared with PorA. Finally, the serotype MAbs do not normally react with continuous peptide

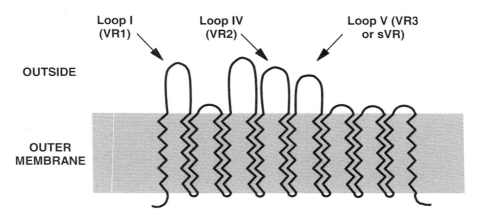

Fig. 2. Molecular basis of serosubtypes in *N. meningitidis*. The proposed ß-barrel secondary structure of the PorA protein in the outer membrane is shown. This model is based on sequence comparisons of meningococcal and other porins and is consistent with the crystal structures of the *Escherichia coli* porins PhoE and OmpF *(110)*. The shaded area represents the outer membrane lipid bilayer and the solid line the folding of the PorA peptide sequence through the membrane. The protein traverses the outer membrane in paired strands of ß-structure, illustrated with zigzags. The first and the last strands are thought to form ß-structure with each other, so that the protein folds to form a cylinder. On the inside of the membrane (in the periplasmic space) there are tight ß-turns between the membrane-traversing strands, but outside there are surface-exposed loops. With the exception of loop III, which is thought to fold within the pore, these surface loops are exposed to the environment. Most variation of peptide sequence among PorA proteins is localized in the largest loops, I (VR1) and IV (VR2), with rather less variation in loop V (VR3 or sVR). Most of the serosubtype antibodies that bind to continuous epitopes bind to peptide sequences present in either VR1 or to VR2. PorB proteins conform to a similar model, except that loops I, IV, and V are smaller and closer to other surface loops in size and serotype-specific monoclonal antibodies do not react with continuous peptides in epitope scanning experiments.

epitopes in epitope scanning experiments (although a continuous human epitope in a class 3 PorB protein has been described after vaccination) *(60)*. DNA-based methods for the identification of serotypes have been developed, that use oligonucleotide probes, defined by nucleotide sequence analyses of genes encoding PorB proteins, to identify differences in surface loops by hybridization *(61)* (Urwin, R. and Fox, A., personal communication). A number of RFLP techniques have been developed for the characterization of meningococci at loci other than *porA* and *porB* *(62–65)*.

Establishing detailed strain relationships by sampling the whole chromo-some is technically and conceptually more complex than characterizing single genes or gene products. The most widely accepted approach is MLEE *(66)*, which is not, strictly speaking, a measure of genetic relationships as is it differ-ences in phenotype that are measured; however, phenotypic differences are used in this method to infer the presence of different alleles that encode vari-ants of each of the proteins included in the analysis (usually metabolic enzymes that can be stained and visualized easily after starch gel electrophoresis). The data from MLEE analyses can be subjected to various statistical procedures, including I_A analysis *(67)*, which tests the degree of clonality within a given species (*see* **Note 5**).

Molecular approaches to the comparison of whole genomes are currently limited to fingerprint analyses, producing patterns from chromosomal DNA, usually after the DNA has been digested with one or more restriction endonu-cleases. In some cases the fingerprints are resolved by pulsed-field gel electro-phoresis (PFGE) (*see* Chapter 3) *(68–71)*, whereas in others conventional electrophoresis is used **(72,73)** (*see* Chapter 2). Southern hybridization with DNA probes has also been used to generate fingerprint patterns *(74)* (*see* Chap-ter 2). Random amplified polymorphic DNAs (or RAPDs), where chromosomal fingerprints are generated by PCR amplification with "random" oligonucle-otide primers *(75)* is an alternative approach (*see* Chapter 6). Whereas these methods are useful in establishing identical strains within collections of iso-lates (*see* **Note 6**), none developed to date is as effective as MLEE in establish-ing genetic relationships among strains, and care has to be taken in applying phylogenetic analyses to fingerprint data (*see* **Note 7**).

4. Research Uses: *Population Biology of the Meningococcus*

Molecular studies, particularly multiple nucleotide sequence comparisons, have had a major influence on the understanding of meningococcal population biology. The population biology of a pathogenic bacterial species is of more than theoretical interest as it has consequences for the interpretation of epide-miologic data and for the design and implementation of novel vaccines.

Reproduction of bacteria by binary fission results in clones: In the absence of mutation each cell receives an exact copy of its mother cell's DNA and is identical to its mother and sister cells. Repeated cell division and accumulation of mutations, when combined with diversity reduction events, such as periodic selection or bottle necking *(76)*, generates a clonal population structure. Clonal populations comprise bacteria that belong to one of a limited number of geneti-cally homogenous lineages, each recognizably derived from an ancestral type. Such populations are characterized by high linkage disequilibrium, or nonran-dom assortment of alleles *(77–79)*. The nonrandom association of alleles means that members of each of these lineages (or clones) will share characteristics, including antigenic types, that distinguish them from other lineages. The practical consequence of this is that the epidemiology of a clonal population, such as pathogenic enteric bacteria, is straightforward as epidemiologic markers are characteristic of lineages *(80)*.

In certain bacteria, such as the meningococcus, this process is disrupted by the transfer of genetic material among lineages (horizontal genetic exchange) *(81)*. The meningococcus is transformable (naturally competent for DNA uptake) and this, together with homologous recombination, provides a mechanism for the transfer of genetic material between cells that do not share a common parent. This results in the reassortment of genetic and serologic markers among clonal lineages and makes following the spread of clones and, hence, epidemiology more difficult than for a strictly clonal species. In some bacterial species, (e.g., *N. gonorrhoeae*) horizontal genetic exchange is so frequent that it largely disrupts the clonal structure of the population *(82,83)*. It is even possible for serologic markers to cross species boundaries using this mechanism *(84)*. Clonal and horizontal spread are illustrated in **Fig. 3** and more complete discussions of these concepts are to be found elsewhere *(15,67,81,85–87)*. For epidemiologic analyses, it is important to distinguish the clonal and horizontal spread of genetic or serologic markers. Where horizontal spread occurs at high frequency, the epidemiologist may well follow the spread of the marker genes rather than individual strains.

The population genetics of *N. meningitidis* is particularly interesting as strains isolated from different epidemiologic situations exhibit diverse population structures *(15)*. At present the reasons for these diverse epidemiologies and population biologies are incompletely understood, but are presumably interrelated. The practical result of this is that the same information (e.g., subtype) has to be interpreted differently in different meningococcal populations. Some examples of the different population structures of epidemiologically diverse meningococci and the effect of these structures on epidemiologic analysis are outlined below.

4.1. Serogroup A Meningococci

Serogroup A meningococci have a clonal population structure, and most isolates of this serogroup belong to one of nine subgroups that can be equated to clonal lineages that share a common ancestor *(88,89)*. Within a given subgroup, PFGE fingerprint patterns are relatively stable during global spread and over time *(68)* and serotype and serosubtype markers are conserved. Thus most serogroup A strains are serotype 4 or 4,21 and a given subgroup is normally uniform for serosubtype *(55,88)*. Interestingly, little sequence diversity is seen in the subtype antigens over periods of up to 50 yr and during epidemic spread

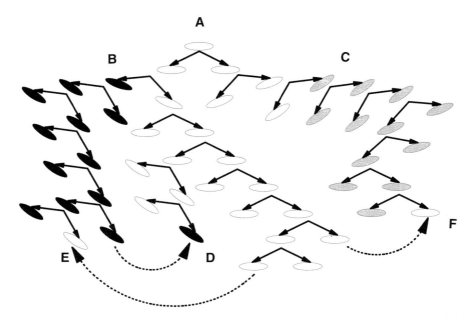

Fig. 3. Clonal (vertical) and nonclonal (horizontal) transfer of genetic material in bacteria. In this illustration an ancestral bacterium (**A**) (shown by a white oval) spreads clonally by binary fission. During this process its descendants acquire mutations giving strains **B** (black) and **C** (grey) which found new lineages. The alleles associated with these strains are passed only from mother to daughter cell, they become characteristic for each lineage and typing reagents that distinguish a gene or gene product characteristic for "black," "white," and "grey" lineages accurately reflect the genetic relationships among the strains. However, if horizontal genetic exchange occurs, illustrated in the figure with broken arrows, the association of descent with particular markers is lost. Hence, strain **E** is descended from B and may be identical to other lineage B strains except for one gene encoding a the typing target, which makes the strain appear to belong to lineage A. Similar arguments hold for strains **D** and **F**.

(55). For these organisms, serogroup, serotype, and serosubtype are helpful in identifying subgroups, although not all subgroups can be distinguished by these criteria alone.

4.2. Serogroup C Meningococci

Many serogroup C isolates are genetically related, belonging to a MLEE defined cluster of electrophoretic types (ETs) referred to as the ET-37 complex *(90,91)*. These organisms cause endemic disease and are occasionallyresponsible for "localized epidemics" within a defined population, for example, a school or military recruit camp. Within a given outbreak of ET-37 meningococci, PFGE fingerprint patterns are conserved; however, in contrast to serogroup A meningococci, the fingerprint patterns of ET-37 related meningococci are diverse among outbreaks (Bygraves, J. A., and Maiden, M. C. J., unpublished observations). The ET-37 organisms are antigenically quite stable, and are normally serogroup C, serotype 2a, or 2b (serotypes defining closely related class 2 PorB proteins), and subtype P1.5,2 (or the related PorA protein, P1.5,10). These organisms are not as antigenically stable as serogroup A, as point mutations have accumulated in the serotype and serosubtype antigens, altering their reactivity with certain monoclonal antibodies *(90)* (Maiden, M. C. J., Malorny, B., Achtman, M., and Feavers, I. M., unpublished observations). Thus, it is not possible to establish the identity of ET-37 complex bacteria from a given outbreak by PFGE fingerprinting, but these bacteria generally have characteristic serologic properties. Unfortunately, these properties are not unique to ET-37 complex meningococci.

4.3. Serogroup B Meningococci Belonging to the ET-5 Complex

The ET-5 complex is a group of related meningococci that have caused hyper-endemic disease in a number of countries since the 1970s *(92,93)*. These organisms generally have conserved PFGE fingerprint patterns, but are variable in their type and subtype antigens. Antigenic change in these organisms has occurred by both gene exchange events and by the accumulation of mutations in VR1 and VR2 *(92–95)*. Thus, serotyping and serosubtyping are poor indicators or relationship among these meningococci, although members of this cluster are identifiable by MLEE and PFGE fingerprint analyses.

4.4. Other Serogroup B Meningococci

The organisms that cause endemic disease in a country such as the UK, where carriage rates are high, are very diverse. These meningococcal populations are probably nonclonal, or consist of numerous short-lived clones. Among endemic disease isolates in the UK, it is rare to see a PFGE fingerprint represented more than once in a sample of such strains unless isolated from the same patient or a close contact *(69,71)*. The serosubtype antigens of these meningococci are also highly diverse with many serosubtype variations and VR family combinations *(40)*. These combinations are not random, perhaps indicating immune selection acting on PorA *(96)*. Identifying trends among such bacteria is difficult and requires more information than is available from simply determining serogroup, serotype, and serosubtype.

5. Future Prospects and Goals

At the time of writing, molecular methods have done much to refine our understanding of meningococci and meningococcal disease, and to identify deficiencies in historic approaches to diagnosis, strain characterization, and epidemiology. As yet, molecular techniques have not provided universally accepted solutions to the problems that they have identified. In particular, none of the currently available molecular techniques for whole-genome analysis are as robust as MLEE and are best used for establishing identity or nonidentity of isolates, rather than detailed genetic characterization. However, there is much room for optimism that solutions based on molecular techniques will be developed in the near future (*see* **Note 8**). The most promising developments of the mid-1990s are bacterial genome sequencing projects *(97)*, which have two implications for epidemiologic characterization of bacteria: Automated sequencing apparatus and software is becoming increasingly available, affordable, and usable; the accessibility of whole genome sequences enables the rational choice of gene targets from any part of the chromosome.

Molecular techniques will produce data of similar or better quality than MLEE when they are used for allele identification, and the best possible means for achieving this is nucleotide sequence analysis (*see* **Note 9**). The sequence analysis approach is becoming feasible with diagnosis and strain characterization increasingly relying on PCR-based approaches. In addition to detailed and accurate typing and diagnostic information, nucleotide sequence data are ideal for basic studies on population biology and evolution. Before nucleotide sequencing is routinely used in this way, further developments are required, for example, nucleotide sequence analysis must become even simpler and cheaper, and appropriate easy to use multiple sequence analysis software developed. There is every prospect that these developments will happen within the next 10 yr and it is realistic to anticipate that routine epidemiologic characterization of bacteria carried out by nucleotide sequence analysis in this time scale (*see* **Note 10**).

6. Notes

1. Given the rapid progress of meningococcal infection and the widespread public concern caused by this alarming disease, clinical and reference laboratories are under great pressure to diagnose meningococcal infection and identify outbreaks of disease quickly and efficiently *(34)*. Although it is not comprehensive, and is in some cases misleading, the serologic typing and subtyping system has the advantage of universal acceptance and is available world-wide *(98)*. The antibodies are currently available from the Standards Processing Division, National Institute for Biological Standards and Control, Blanche Lane, South Mimms, Potters Bar, Hertfordshire, EN6 3QG, UK.

2. Some reports express the sensitivities and specificities of diagnostic tests in terms of a percentage; however, as several of the studies summarized here have many fewer than 100 samples in each group (some as few as 10), in this chapter the author shall report these figures as ratios. In addition, he shall give the actual numbers of samples tested to enable more meaningful comparisons of the various studies.

3. An insertion sequence is a poor choice as a target for a diagnostic approach, as these elements frequently cross species and genus boundaries *(99,100)*. This behavior provides a possible explanation of the false-positive results obtained with the IS*1106*-based diagnostic techniques when EB staining of agarose gels was used as the means of detecting PCR products. The use of this particular target is further complicated by nucleotide sequence rearrangements within the element *(26)*. A further objection is that IS*1106* provides no epidemiologic data.

4. Molecular techniques have provided a wealth of new means of diagnosing bacterial infection and characterizing pathogenic bacteria. The disadvantage of this is that it is easy to develop new, but facile, molecular typing procedures, and many such methods have been published, covering numerous bacterial pathogens. Most of these methods will never be used routinely or standardized among international reference laboratories, and many simply establish identity or nonidentity of isolates. The term YATM (Yet Another Typing Method) has been recently coined to describe such methods *(101)*.

5. MLEE remains the most reliable method for the establishment of the genetic relationship of a given isolate to other meningococci. However, this technique requires isolated strains and is too cumbersome for most laboratories. In addition, the starch gels require careful interpretation and the technique is only really cost effective if large numbers of samples are to be analyzed at any one time. There is a WHO International Reference Laboratory for meningococcal MLEE analysis (Caugant, D. A.,WHO Collaborating Centre for Reference and Research on Meningococci, National Institute for Public Health, 75 Geitmyrsveien, Oslo, N-O462, Norway).

6. Of the many techniques available for identifying whether two organisms are identical or not, PFGE fingerprinting (*see* Chapter 3) is the method of choice when an isolate is available. This technique has the advantages of: being relatively rapid; sampling sites from around the whole chromosome; and not requiring specialized techniques or reagents. It is advisable for most clinical laboratories, and certainly all reference laboratories, to have access to this technology, which can be applied to diverse organisms. There are limitations to the technique: the high-quality chromosomal DNA that is essential for this procedure makes having an isolate a necessity, and when isolates have more than a few band differences, it becomes impossible to draw sensible conclusions on the relatedness of isolates. This is because there are a variety of molecular mechanisms that result in changes in banding patterns, which cannot be distinguished by inspecting band sizes. An alternative, which is faster, is RAPDs (*see* Chapter 6); however, this technique is difficult to control and standardize among, or even within, laboratories and should be applied with care and exhaustive controls.

7. The construction of dendrograms from RFLP or PFGE data has increased in popularity, as commercial software has become available, but there are a number of potential problems with this approach. The mathematical models *(102)* for the construction of such dendrograms assume that restriction fragment changes reflect nucleotide substitution rates, which may not be true, especially in organ-

isms, such as the meningococcus, which frequently recombine. Further difficulties with these approaches include: most RFLPs are generated from a single locus, which may not be representative of the whole chromosome; as PFGE fingerprints get more diverse, it becomes impossible to ensure that bands are identical on the basis of fragment size alone; for both RFLP and PFGE data the restriction fragment sizes are not genuinely independent variables.

8. In cases where isolates are not available, all epidemiologic information must be obtained from meningococcal genes amplified from clinical material for diagnostic purposes. Thus is it sensible to develop new typing techniques that are applicable to these samples and to isolated meningococci, enabling a unified approach to diagnosis, epidemiology, and research. This aim is consistent with the new characterization techniques proposed in **Notes 9** and **10** below. A minimal diagnostic/epidemiologic test should include characterization of at least one of the following: the *rrn* genes (if the species causing the infection is in doubt); the capsular operon (the serogroup of the organism being important for public health measures); and the *porA* and/or *porB* genes (both of which are diagnostic in addition to providing epidemiologic data). Other genes could, of course, be amplified and characterized from clinical samples if more precise strain identification were required.

9. Alleles that have been amplified by the PCR can be characterized by several molecular techniques including: hybridization with specific probes; restriction endonuclease digestion; and nucleotide sequence analysis. Hybridization systems for the serotype *(61)* and the serosubtype antigens *(38)* of the meningococcus have been developed and, although these are more effective than monoclonal antibodies, a set of specific reagents are required and even these systems are not comprehensive *(40)*. Digesting amplified genes with restriction endonucleases and comparing fingerprint patterns generated when the products are separated by electrophoresis is a relatively simple method for comparing amplified genes from many isolates that does not require any specialized reagents *(47,103,104)*. The best of these methods use restriction endonucleases that cut the gene frequently, (e.g., enzymes with 4 bp recognition sites) *(105,106)*. The disadvantage of such techniques is that the fingerprint patterns are not as high resolution as nucleotide sequence, and partial digestion products can result in misleading data *(104,107)*. Nucleotide sequence analysis is the best means of characterizing amplified genes *(40,108)*, or parts of them, but is also the most expensive, time consuming, and technically demanding approach.

To achieve a replacement for MLEE and serotyping, several genes should sequenced, representing diverse chromosomal loci. Although continuity with previous typing methods can be achieved by including genes such as *siaD*, *porA*, *porB*, *dhps*, and perhaps *penA* (involved in ß-lactam resistance *(105)*, a practical, reproducible, and informative alternative to serologic and MLEE analyses can only be achieved with the inclusion of "house keeping" genes not expected to be under selection by antibiotic or immune selection *(109)*.

Note Added in Proof

Since this article was written, the author and colleagues have developed such a method, multilocus sequence typing(MLST), and validated it for meningococci (Maiden et al., submitted).

Acknowledgments

This article was written when the author was on sabbatical leave at the Max-Planck-Intitiut für molekulare Genetik, Berlin, and he is grateful to Mark Achtman and Thomas Trautner for their hospitality and to the Alexander von Humboldt Stiftung for financial support. The author thanks the many colleagues with whom he has interacted over the last 8 yr, particularly Ian Feavers, for his invaluable collaboration, and to Jane Bygraves, Janet Suker, Rachel Urwin, and Alison McKenna for their contributions to the research. Thanks are due to Dennis Jones and Andrew Fox for provision of strains and "front line" experience, Mark Achtman, Jan Poolman, and Wendell Zollinger for the provision of monoclonal antibodies, and to Mark Achtman and Brian Spratt for stimulating discussions on meningococcal population genetics. The author also thanks the following colleagues who kindly provided reprints or prepublication manuscripts, which were used in the preparation of this chapter: Andrew Fox, Carl Frasch, Bjorn-Eric Kristiansen, Jonjoe McFadden, and Per Olcén.

References

1. Cartwright, K. A. V. (ed.) (1995) *Meningococcal Disease* Wiley, Chichester.
2. Peltola, H. (1983) Meningococcal disease: still with us. *Rev. Infect. Dis.* **5,** 71–91.
3. Brandtzaeg, P. (1995) Pathogenesis of meningococcal infections, in *Meningococcal Disease* (Cartwright, K. A. V., ed.), Wiley, Chichester, pp. 71-114.
4. Gold, R. (1987) Clinical aspects of meningococcal disease, in *Evolution of Meningococcal Disease* (Vedros, N. A., ed.), CRC, Boca Raton, FL, pp. 69–97.
5. Steven, N. and Wood, M. (1995) Clinical spectrum of meningococcal disease, in *Meningococcal Disease* (Cartwright, K. A. V., ed.), Wiley, Chichester, pp. 177–205.
6. Branham, S. E. (1953) Serological relationships among meningococci. *Bact. Rev.* **17,** 175–188.
7. Vedros, N. A. (1987) Development of meningococcal serogroups, in *Evolution of Meningococcal Disease* (Vedros, N. A., ed.), CRC, Boca Raton, FL, pp. 33–37.
8. Frasch, C. E. (1987) Development of meningococcal serotyping, in *Evolution of Menigococcal Disease* (Vedros, N. A., ed.), CRC, Boca Raton, FL, pp. 39–55.
9. Frasch, C. E., Zollinger, W. D., and Poolman, J. T. (1985) Serotype antigens of *Neisseria meningitidis* and a proposed scheme for designation of serotypes. *Rev. Infect. Dis.* **7,** 504–510.

10. Scholten, R. J., Kuipers, B., Valkenburg, H. A., Dankert, J., Zollinger, W. D., and Poolman, J. T. (1994) Lipo-oligosaccharide immunotyping of *Neisseria meningitidis* by a whole-cell ELISA with monoclonal antibodies. *J. Med. Microbiol.* **41**, 236–43.

11. Maiden, M. C. J. and Feavers, I. M. (1994) Meningococcal typing. *J. Med. Microbiol.* **40**, 157,158.

12. Caugant, D. A., Mocca, L. F., Frasch, C. E., Frøholm, L. O., Zollinger, W. D., and Selander, R. K. (1987) Genetic structure of *Neisseria meningitidis* populations in relation to serogroup, serotype, and outer membrane protein pattern. *J. Bacteriol.* **169**, 2781–2792.

13. Achtman, M. (1990) Molecular epidemiology of epidemic bacterial meningitis. *Rev. Med. Microbiol.* **1**, 29–38.

14. Schwartz, B., Moore, P. S. and Broome, C. V. (1989) Global epidemiology of meningococcal disease. *Clin. Microbiol. Revs.* **2**, s118–s124.

15. Maiden, M. C. J. and Feavers, I. M. (1995) Population genetics and global epidemiology of the human pathogen *Neisseria meningitidis*, in *Population Genetics of Bacteria* (Baumberg, S., Young, J. P. W., Saunders, J. R., and Wellington, E. M. H., eds.), Cambridge University Press, Cambridge, UK, pp. 269–293.

16. Jones, D. M. (1995) Epidemiology of meningococcal disease in Europe and the USA, in *Meningococcal Disease* (Cartwright, K. A. V., ed.), Wiley, Chichester, pp. 147–157.

17. Moore, P. S. and Broome, C. V. (1994) Cerebrospinal meningitis epidemics. *Sci. Am.* November, 24–31.

18. Achtman, M. (1995) Global epidemiology of meningococcal disease, in *Meningococcal Disease* (Cartwright, K. A. V., ed.), Wiley, Chichester, pp. 159–175.

19. Gotschlich, E. C., Goldschneider, I., and Artenstein, M. S. (1969) Human immunity to the meningococcus IV. Immunogenicity of group A and group C meningococcal polysaccharides. *J. Exp. Med.* **129**, 1367–1384.

20. Twumasi, P. A.,Jr., Kumah, S., Leach, A., O'Dempsey, T. J., Ceesay, S. J., Todd, J., Broome, C. V., Carlone, G. M., Pais, L. B., Holder, P. K., Plikaytis, B.D., and Greenwood, B.M. (1995) A trial of a group A plus group C meningococcal polysaccharide-protein conjugate vaccine in African infants. *J. Infect. Dis.* **171**, 632–638.

21. Frasch, C. E. (1995) Meningococcal vaccines: past, present, and future, in *Meningococcal Disease* (Cartwright, K. A. V., ed.), Wiley, Chichester, pp. 246–283.

22. Cartwright, K. A. V., Reilly, S., White, D. A., and Stuart, J. M. (1992) Early treatment with parenteral antibiotic in meningococcal disease. *Brit. Med. J.* **305**, 147.

23. Ni, H., Knight, A. I., Cartwright, K. A. V., Palmer, W. H., and McFadden, J. (1992) Polymerase chain reaction for diagnosis of meningococcal infection. *Lancet* **340**, 1432–1434.

24. Bohr, V., Rasmussen, N., Hansen, B., Kjersem, H., Jessen, O., Johnsen, N., and Kristensen, H.S. (1983) 875 cases of bacterial meningitis: diagnostic procedures and the impact of preadmission antibiotic: Part III of a three-part series. *J. Infect.* **7**, 193–202.

25. Jones, D.M. and Kaczmarski, E.B. (1994) Meningococcal infections in England and Wales 1994. *Comm. Dis. Rep. Rev.* **2**, R125–R130.

26. Newcombe, J., Cartwright, K. A. V., Palmer, W. H., and McFadden, J. (1996) Polymerase chain reaction of peripheral blood for the diagnosis of meningococcal disease. *J. Clin. Micobiol.* **34,** 1637–1640.

27. Olcén, P., Lantz, P.-G., Backman, A., and Radstrom, P. (1995) Rapid diagnosis of bacterial meningitis by a seminested PCR strategy. *Scand. J. Infect. Dis.* **27,** 537–539.

28. Richards, P. G. and Towu-Aghantse, E. (1986) Dangers of lumbar puncture. *Brit. Med. J.* **292,** 605,606.

29. Saunders, N. B., Zollinger, W. D., and Rao, V. B. (1993) A rapid and sensitive PCR strategy employed for amplification and sequencing of *porA* from a single colony-forming unit of *Neisseria meningitidis. Gene* **137,** 153–162.

30. Davison, E., Borrow, R., Guiver, M., Kaczmarski, E. B., and Fox, A. J. (1996) The adaptation of the IS*1106* PCR to a PCR ELISA format for the diagnosis of meningococcal infection. *Serodiag. Immunother.* **8,** 51–56.

31. Gray, M. W., Sankoff, D., and Cedergren, R. J. (1984) On the evolutionary descent of organisms and organelles: a global phylogeny based on a highly conserved structural core in small subunit ribosomal RNA. *Nucleic Acids. Res.* **12,** 5837–5852.

32. Woese, C. R., Gutell, R., Gupta, R., and Noller, H. F. (1983) Detailed analysis of the higher-order structure of 16S-like ribosomal ribonucleic acids. *Microbiol. Rev.* **47,** 621–669.

33. Radstrom, P., Backman, A., Qian, N., Kragsbjerg, P., Pahlson, C., and Olcén, P. (1994) Detection of bacterial DNA in cerebrospinal fluid by an assay for simultaneous detection of *Neisseria meningitidis, Haemophilus influenzae,* and streptococci using a seminested PCR strategy. *J. Clin. Microbiol.* **32,** 2738–2744.

34. Begg, N. (1995) Outbreak management, in *Meningococcal Disease* (Cartwright, K. A. V., ed.), Wiley, Chichester, pp. 286–305.

34a. Borrow, R., Claus, H., Guiver, M., Smart, L., Jones, D. M., Kaczmarski, E. B., Frosch, M., and Fox, A. J. (1997) Non-culture diagnosis and serogroup determination of meningococcal B and C infection by a sinlyltransferase (*siaD*) PCR ELISA. *Epidemiol. Infect.* **118,** 111–117.

35. Barlow, A. K., Heckels, J. E., and Clarke, I. N. (1989) The class 1 outer membrane protein of *Neisseria meningitidis*: gene sequence and structural and immunological similarities to gonococcal porins. *Mol. Microbiol.* **3,** 131–139.

36. McGuinness, B. T., Barlow, A. K., Clarke, I. N., Farley, J. E., Anilionis, A., Poolman, J. T., and Heckels, J. E. (1990) Deduced amino acid sequences of class 1 protein PorA from three strains of *Neisseria meningitidis. J. Exp. Med.* **171,** 1871–1882.

37. Maiden, M. C. J., Suker, J., McKenna, A. J., Bygraves, J. A., and Feavers, I. M. (1991) Comparison of the class 1 outer membrane proteins of eight serological reference strains of *Neisseria meningitidis. Mol. Microbiol.* **5,** 727–736.

38. Feavers, I. M., Suker, J., McKenna, A. J., Heath, A. B., and Maiden, M. C. J. (1992) Molecular analysis of the serotyping antigens of *Neisseria meningitidis. Infect. Immun.* **60,** 3620–3629.

39. Zapata, G. A., Vann, W. F., Rubinstein, Y., and Frasch, C. E. (1992) Identification of variable region differences in *Neisseria meningitidis* class 3 protein sequences among five group B serotypes. *Mol. Microbiol.* **6,** 3493–3499.

40. Feavers, I. M., Fox, A. J., Gray, S., Jones, D. M., and Maiden, M. C. J. (1996) Antigenic diversity of meningococcal outer membrane protein PorA has implications for epidemiological analysis and vaccine design. *Clin. Diagn. Lab. Immunol.* **3,** 444–450.

41. Caugant, D. A., Høiby, E. A., Frøholm, L. O., and Brandtzaeg, P. (1996) Polymerase chain reaction for case ascertainment of meningococcal meningitis: application to the cerebrospinal fluids collected in the course of the Norwegian serogroup B protection trial. *Scand. J. Infect. Dis.* **28,** 149–153.

42. Bjune, G., Høiby, E. A., Grønnesby, J. K., Arnesen, Ø., Fredriksen, J. H., Halstensen, A., Holten, E., Lindbak, A. K., Nøkleby, H., Rosenqvist, E., Solberg, L. K., Closs, O., Eng, J., Frøholm, L. O., Lystad, A. Bakketeig, L. S., and Hareide, B. (1991) Effect of outer membrane vesicle vaccine against group B meningococcal disease in Norway. *Lancet* **338,** 1093–1096.

43. Fermer, C., Kristiansen, B.-E., Sköld, O., and Swedberg, G. (1995) Sulphonamide resistance in *Neisseria meningitidis* as defined by site-directed mutagenesis could have its origin in other species. *J. Bacteriol.* **177,** 4669–4675.

44. Radstrom, P., Fermer, C., Kristiansen, B.-E., Jenkins, A., Sköld, O., and Swedberg, G. (1992) Transformational exchanges in the dihydropterate synthase gene of *Neisseria meningitidis:* a novel mechanism for the acquisition of sulfonamide resistance. *J. Bacteriol.* **174,** 5961–5968.

45. Kristiansen, B.-E., Radstrom, P., Jenkins, A., Ask, E., Facinelli, B., and Sköld, O. (1990) Cloning and characteriztion of a DNA fragment that confers sulfonamide resistance in a serogroup B, serotype 15 strain of *Neisseria meningitidis. Antimicrob. Agents Chemother.* **34,** 2277–2279.

46. Kristiansen, B.-E., Ask, E., Jenkins, A., Fermer, C., Radstrom, P., and Sköld, O. (1991) Rapid diagnosis of meningococcal meningitis by polymerase chain reaction. *Lancet* **337,** 1568,1569.

47. Kristiansen, B.-E., Fermer, C., Jenkins, A., Ask, E., Swedberg, G., and Sköld, O. (1995) PCR amplicon restriction endonuclease analysis of the chromosomal dhps gene of *Neisseria meningitidis:* a method for studying spread of the disease-causing strain in contacts of patients with meningococcal disease. *J. Clin. Microbiol.* **33,** 1174–1179.

48. Knight, A. I., Ni, H., Cartwright, K. A. V., and McFadden, J. (1992) Identification and characterization of a novel insertion sequence, IS*1106*, downstream of the por gene in B15 *Neisseria meningitidis. Mol. Microbiol.* **6,** 1565–1573

49. Suker, J., Feavers, I. M., and Maiden, M. C. J. (1993) Structural analysis of the variation in the major outer membrane proteins of *Neisseria meningitidis* and related species. *Biochem. Soc. Trans.* **21,** 304–306.

50. Ward, M. J., Lambden, P. R., and Heckels, J. E. (1992) Sequence analysis and relationships between meningococcal class 3 serotype proteins and other porins of pathogenic and non-pathogenic *Neisseria* species. *FEMS Microbiol. Lett.* **94,** 283–290.

51. van der Ley, P., Heckels, J. E., Virji, M., Hoogerhout, P., and Poolman, J. T. (1991) Topology of outer membrane proteins in pathogenic *Neisseria* species. *Infect. Immun.* **59,** 2963–2971.

52. Abdillahi, H. and Poolman, J. T. (1988) *Neisseria meningitidis* group B serosubtyping using monoclonal antibodies in whole-cell ELISA. *Microb. Pathogen.* **4**, 27–32.

53. Abdillahi, H. and Poolman, J. T. (1988) Definition of meningococcal class 1 OMP subtyping antigens by monoclonal antibodies. *FEMS Microbiol. Immunol.* **1**, 139–144.

54. Abdillahi, H. and Poolman, J. T. (1987) Whole-cell ELISA for typing *Neisseria meningitidis* with monoclonal antibodies. *FEMS Microbiol. Lett.* **48**, 367–371.

55. Suker, J., Feavers, I. M., Achtman, M., Morelli, G., Wang, J.-F., and Maiden, M. C. J. (1994) The *porA* gene in serogroup A meningococci: evolutionary stability and mechanism of genetic variation. *Mol. Microbiol.* **12**, 253–265.

56. Suker, J., Feavers, I. M., and Maiden, M. C. J. (1996) Monoclonal antibody recognition of members of the P1.10 variable region family: implications for serological typing and vaccine design. *Microbiology* **142**, 63–69.

57. Rosenqvist, E., Høiby, E. A., Wedege, E., Caugant, D. A., Frøholm, L. O., McGuinness, B. T., Brooks, J. and Lambden, P. R. (1993) A new variant of serosubtype, P1.16 in *Neisseria meningitidis* from Norway associated with increased resistance to bactericidal antibodies induced by a serogroup B outer membrane protein vaccine. *Microb. Pathogen.* **15**, 197–205.

58. Wedege, E., Dalseg, R., Caugant, D. A., Poolman, J. T., and Frøholm, L. O. (1993) Expression of an inaccessible P1.7 subtype epitope on meningococcal class 1 proteins. *J. Med. Microbiol.* **38**, 23–28.

59. Maiden, M. C. J., Bygraves, J. A., McCarvil, J., and Feavers, I. M. (1992) Identification of meningococcal serosubtypes by polymerase chain reaction. *J. Clin. Microbiol.* **30**, 2835–2841.

60. Delvig, A., Wedege, E., Caugant, D. A., Dalseg, R., Kolberg, J., Achtman, M., and Rosenqvist, E. (1995) A linear B-cell epitope on the class 3 outer-membrane protein of *Neisseria meningitidis* recognized after vaccination with the Norwegian group B outer-membrane vesicle vaccine. *Microbiology* **141**, 1593–1600.

61. Bash, M. C., Lesiak, K. B., Banks, S. D., and Frasch, C. E. (1995) Analysis of *Neisseria meningitidis* class 3 outer membrane protein gene variable regions and type identification using genetic techniques. *Infect. Immun.* **63**, 1484–1490.

62. Jordens, J. Z. and Pennington, T. H. (1991) Characterization of *Neisseria meningitidis* isolated by ribosomal RNA gene restriction patterns and restriction endonuclease digestion of chromosomal DNA. *Epidemiol. Infect.* **107**, 253–262.

63. Tondella, M. L., Sacchi, C. T., and Neves, B. C. (1994) Ribotyping as an additional molecular marker for studying *Neisseria meningitidis* serogroup B epidemic strains. *J. Clin. Microbiol.* **32**, 2745–2748.

64. Fox, A. J., Jones, D. M., Gray, S. J., Caugant, D. A., and Saunders, N. A. (1991) An epidemiologically valuable typing method for *Neisseria meningitidis* by analysis of restriction fragment length polymorphisms. *J. Med. Microbiol.* **34**, 265–270.

65. Woods, T. C., Helsel, L. O., Swaminathan, B., Bibb, W. F., Pinner, R. W., Gellin, B. G., Collin, S. F., Waterman, S. H., Reeves, M. W., Brenner, D. J., and Broome, C.V. (1992) Characterization of *Neisseria meningitidis* serogroup C by multilocus enzyme electrophoresis and ribosomal DNA restriction profiles (ribotyping). *J. Clin. Microbiol.* **30**, 132–137.

66. Selander, R. K., Caugant, D. A., Ochman, H., Musser, J. M., Gilmour, M. N., and Whittam, T. S. (1986) Methods of multilocus enzyme electrophoresis for bacterial population genetics and systematics. *Appl. Environ. Microbiol.* **51,** 837–884.

67. Maynard Smith, J., Smith, N. H., O'Rourke, M., and Spratt, B. G. (1993) How clonal are bacteria? *Proc. Natl. Acad. Sci. USA* **90,** 4384–4388.

68. Bygraves, J. A. and Maiden, M. C. J. (1992) Analysis of the clonal relationships between strains of *Neisseria meningitidis* by pulsed field gel electrophoresis. *J. Gen. Microbiol.* **138,** 523–531.

69. Bygraves, J. A. and Maiden, M. C. J. (1991) The resolution of clonal types of *Neisseria meningitidis* by pulsed field gel electrophoresis, in *Neisseriae 1990* (Achtman, M., ed.), Walter de Gruyter, Berlin, Germany, pp. 25–30.

70. Feavers, I. M., Heath, A. B., Bygraves, J. A. and Maiden, M. C. (1992) Role of horizontal genetic exchange, in the antigenic variation of the class 1 outer membrane protein of *Neisseria meningitidis. Mol. Microbiol.* **6,** 489–495.

71. Yakubu, D. E. and Pennington, T. H. (1995) Epidemiological evaluation of *Neisseria meningitidis* serogroup B by pulsed-field gel electrophoresis. *FEMS Immunol. Med. Microbiol.* **10,** 185–189.

72. Bjorvatn, B., Lund, V., Kristiansen, B.-E., Korsnes, L., Spanne, O., and Lindqvist, B. (1984) Applications of restriction endonuclease fingerprinting of chromosomal DNA of *Neisseria meningitidis. J. Clin. Microbiol.* **19,** 763–765.

73. Bjorvatn, B., Hassan King, M., Greenwood, B., Haimanot, R. T., Fekade, D., and Sperber, G. (1992) DNA fingerprinting in the epidemiology of African serogroup A *Neisseria meningitidis. Scand. J Infect. Dis.* **24,** 323–332.

74. Knight, A. I., Cartwright, K. A. V., and McFadden, J. (1992) Phylogenetic and epidemiological analysis of *Neisseria meningitidis* using DNA probes. *Epidemiol. Infect.* **109,** 227–239.

75. Williams, J. G. K., Kubelik, A. R., Livak, K. J., Rafalski, J. A., and Tingey, S. V. (1990) DNA polymorphisms amplified by arbitrary primers are useful as genetic markers. *Nucleic Acids. Res.* **18,** 6531–6535.

76. Achtman, M. (1995) Epidemic spread and antigenic variability of *Neisseria meningitidis. Trend. Microbiol.* **3,** 186–191.

77. Selander, R. K. and Levin, B. R. (1980) Genetic diversity and structure in *Escherichia coli* populations. *Science* **210,** 545–547.

78. Nelson, K., Whittam, T. S., and Selander, R. K. (1991) Nucleotide polymorphism and evolution in the glyceraldehyde-3-phosphate dehydrogenase gene (*gapA*) in natural populations of *Salmonella* and *Escherichia coli. Proc. Natl. Acad. Sci. USA* **88,** 6667–6671.

79. Whittam, T. S., Ochman, H., and Selander, R. K. (1983) Multilocus genetic structure in natural populations of *Escherichia coli. Proc. Natl. Acad. Sci. USA* **80,** 1751–1755.

80. Whittam, T. S. (1995) Genetic population structure and pathogenicity in enteric bacteria, in *Population Genetics of Bacteria* (Baumberg, S., Young, J. P. W., Wellington, E. M. H., and Saunders, J. R., eds.), Cambridge University Press, Cambridge, UK, pp. 217–245.

81. Maiden, M. C. J. (1993) Population genetics of a transformable bacterium: the influence of horizontal genetical exchange on the biology of *Neisseria meningitidis*. *FEMS Microbiol. Lett.* **112,** 243–250.

82. O'Rourke, M. and Stevens, E. (1993) Genetic structure of *Neisseria gonorrhoeae* populations: a non-clonal pathogen. *J. Gen. Microbiol.* **139,** 2603–2611.

83. O'Rourke, M. and Spratt, B. G. (1994) Further evidence for the non-clonal population structure of *Neisseria gonorrhoeae*: extensive genetic diversity within isolates of the same electrophoretic type. *Microbiology* **140,** 1285–1290.

84. Vazquez, J., Berron, S., O'Rourke, M., Carpenter, G., Feil, E., Smith, N. H., and Spratt, B. G. (1995) Interspecies recombination in nature: a meningococcus that has acquired a gonococcal PIB porin. *Mol. Microbiol.* **15,** 1001–1007.

85. Maynard Smith, J. (1995) Do bacteria have population genetics? in *Population Genetics of Bacteria* (Baumberg, S., Young, J. P. W., Wellington, E. M. H., and Saunders, J. R., eds.), Cambridge University Press, Cambridge, UK, pp. 1–12.

86. Maynard Smith, J., Dowson, C. G., and Spratt, B. G. (1991) Localized sex in bacteria. *Nature* **349,** 29–31.

87. Spratt, B. G., Smith, N. H., Zhou, J., O'Rourke, M., and Feil, E. (1995) The population genetics of the pathogenic *Neisseria*, in *Population Genetics of Bacteria* (Baumberg, S., Young, J. P. W., Wellington, E. M. H., and Saunders, J. R., eds.), Cambridge University Press, Cambridge, UK, pp. 143–160.

88. Wang, J.-F., Caugant, D. A., Li, X., Hu, X., Poolman, J. T., Crowe, B. A., and Achtman, M. (1992) Clonal and antigenic analysis of serogroup A *Neisseria meningitidis* with particular reference to epidemiological features of epidemic meningitis in China. *Infect. Immun.* **60,** 5267–5282.

89. Olyhoek, T., Crowe, B. A., and Achtman, M. (1987) Clonal population structure of *Neisseria meningitidis* serogroup A isolated from epidemics and pandemics between 1915 and 1983. *Rev. Infect. Dis.* **9,** 665–682.

90. Wang, J.-F., Caugant, D. A., Morelli, G., Koumaré, B., and Achtman, M. (1993) Antigenic and epidemiological properties of the ET-37 complex of *Neisseria meningitidis*. *J. Infect. Dis.* **167,** 1320–1329.

91. Caugant, D. A., Zollinger, W. D., Mocca, L. F., Frasch, C. E., Whittam, T. S., Frøholm, L. O., and Selander, R. K. (1987) Genetic relationships and clonal population structure of serotype 2 strains of *Neisseria meningitidis*. *Infect. Immun.* **55,** 1503–1513.

92. Caugant, D. A., Frøholm, L. O., Bovre, K., Holten, E., Frasch, C. E., Mocca, L. F., Zollinger, W. D., and Selander, R. K. (1987) Intercontinental spread of *Neisseria meningitidis* clones of the ET-5 complex. *Antonie van Leeuwenhoek J. Microbiol.* **53,** 389–394.

93. Caugant, D. A., Frøholm, L. O., Selander, R. K., and Bovre, K. (1989) Sulfonamide resistance in *Neisseria meningitidis* isolates of clones of the ET-5 complex. *APMIS* **97,** 425–428.

94. McGuinness, B. T., Clarke, I. N., Lambden, P. R., Barlow, A. K., Poolman, J. T., Jones, D. M., and Heckels, J. E. (1991) Point mutation in meningococcal *porA* gene associated with increased endemic disease. *Lancet* **337,** 514–517.

95. Wedege, E., Kolberg, J., Delvig, A., Høiby, E. A., Holten, E., Rosenqvist, E., and Caugant, D. A. (1995) Emergence of a new virulent clone within the electrophoretic type 5 complex of serogroup B meningococci in Norway. *Clin. Diagn. Lab. Immunol.* **2**, 314–321.

96. Gupta, S., Maiden, M. C. J., Feavers, I. M., Nee, S., May, R. M., and Anderson, R. M. (1996) The maintenance of strain structure in populations of recombining infectious agents. *Nat. Med.* **2**, 437–442.

97. Fleischmann, R. D., Adams, M. D., White, O., Clayton, R. A., Kirkness, E. F., Kerlavage, A. R., Bult, C. J., Tomb, J. F., Dougherty, B. A., and Merrick, J. M. (1995) Whole-genome random sequencing and assembly of *Haemophilus influenzae* RD. *Science* **269**, 496–512.

98. Poolman, J. T., Kriz Kuzemenska, P., Ashton, F., Bibb, W., Dankert, J., Demina, A., Frøholm, L. O., Hassan King, M., Jones, D. M., Lind, I., Prakash, K., and Xujing,H. (1995) Serotypes and subtypes of *Neisseria meningitidis*: results of an international study comparing sensitivities and specificities of monoclonal antibodies. *Clin. Diagn. Lab. Immunol.* **2**, 69–72.

99. Kent, L., McHugh, T. D., Billington, O., Dale, J. W., and Gillespie, S. H. (1995) Demonstration of homology between IS*6110* of *Mycobacterium tuberculosis* and DNAs of other *Mycobacterium* spp. *J. Clin. Microbiol.* **33**, 2290–2293.

100. Mulcahy, G. M., Kaminski, Z. C., Albanese, E. A., Sood, R., and Pierce, M. (1996) IS6110-Based PCR method for detection of *Mycobacterium tuberculosis*. *J. Clin. Microbiol.* **34**, 1348–1349.

101. Achtman, M. (1996) A surfeit of YATMs? *J. Clin. Microbiol.* **34**, 1870.

102. Nei, M. and Li, W.-H. (1979) Mathematical model for studying genetic variation in terms of restriction endonucleases. *Proc. Natl. Acad. Sci. USA* **76**, 5269–5273.

103. Kertesz, D. A., Byrne, S. K., and Chow, A. W. (1993) Characterization of *Neisseria meningitidis* by polymerase chain reaction and restriction endonuclease digestion of the *porA* gene. *J. Clin. Microbiol.* **31**, 2594–2598.

104. Peixuan, Z., Xujing, H., and Li, X. (1995) Typing *Neisseria meningitidis* by analysis of restriction fragment length polymorphisms in the gene encoding the class 1 outer membrane protein: application and assessment of epidemics throughout the last 4 decades in China. *J. Clin. Microbiol.* **33**, 458–462.

105. Campos, J., Fuste, M. C., Trujillo, G., Saez Nieto, J., Vazquez, J., Loren, J. G., Vinas, M., and Spratt, B. G. (1992) Genetic diversity of penicillin-resistant *Neisseria meningitidis*. *J. Infect. Dis.* **166**, 173–177.

106. O'Rourke, M., Ison, C. A., Renton, A. M., and Spratt, B. G. (1995) Opa-typing: a high resolution tool for studying the epidemiology of gonorrhoea. *Mol. Microbiol.* **17**, 865–875.

107. Malorny, B., Maiden, M. C. J., and Achtman, M. (1996) The *porA* alleles are identical in subgroup III serogroup A *Neisseria meningitidis* strains isolated in China in the 1960s and 1980s. *J. Clin. Microbiol.* **34**, 1548–1550.

108. Brooks, J. L., Fallon, R. J., and Heckels, J. E. (1995) Sequence variation in class 1 outer membrane protein in *Neisseria meningitidis* isolated from patients with meningococcal infection and close household contacts. *FEMS Microbiol. Lett.* **128**, 145–150.

109. Zhou, J. and Spratt, B. G. (1992) Sequence diversity within the *argF, fbp* and *recA* genes of natural isolates of *Neisseria meningitidis*: interspecies recombination within the *argF* gene. *Mol. Microbiol.* **6,** 2135–2146.

110. Cowan, S. W., Schirmer, T., Rummel, G., Steiert, M., Ghosh, R., Paupit, R. A., Jansonius, J. N., and Rosenbusch, J. P. (1992) Crystal structures explain functional properties of two *E. coli* porins. *Nature* **358,** 727–733.

15

Gonorrhea

Catherine A. Ison

1. Introduction

Gonorrhea is a major sexually transmitted disease (STD) that occurs world-wide. The prevalence has fallen dramatically in most industrialized countries in the last ten years because of effective therapy, contact tracing, and changes in sexual practices since the advent of the Acquired Immunodeficiency Syndrome (AIDS). In contrast, in developing countries the prevalence is high or increasing as a result of both the lack of facilities for diagnosis and of suitable antimicrobial therapy.

The causative agent of gonorrhea is *Neisseria gonorrhoeae*, which is an obligate human pathogen. It colonizes the mucosal epithelium of the lower genital tract with the cervix as the primary site in women, and the urethra in men. Colonization of the rectum primarily occurs in homosexual men, and of the pharynx in men or women who practice oral intercourse. On occasion, complicated infection can occur when the organism ascends to the normally sterile upper genital tract and causes salpingitis or pelvic inflammatory disease in women, and epididymitis or prostatitis in men. Complicated gonococcal infection is more common in women because the primary infection is often asymptomatic and goes undetected and, therefore, untreated. The majority of men present with symptoms and, hence, receive rapid and suitable therapy. In contrast to complicated infection, which can result from inadequate therapy, disseminated gonococcal infection (DGI) is a distinct entity. It occurs when the organism invades into the blood and presents as septicemia, rash, and/or arthritis.

The prevalence of gonorrhea has fallen from 999,937 reported cases in 1975 to 392,848 cases in 1995 in the United States, and from 60,000 cases in 1975 to 12,000 cases in 1995 in England and Wales. The majority of infections present as uncomplicated gonorrhea with a small number of cases of complicated

From: *Methods in Molecular Medicine, vol. 15: Molecular Bacteriology: Protocols and Clinical Applications*
Edited by: N. Woodford and A. P. Johnson © Humana Press Inc., Totowa, NJ

infection. Disseminated gonococcal infection is rare. Although the prevalence of gonococcal infection has declined, the figures mask variation in rates of infection associated with geographic area, sex, ethnicity, and age. A higher prevalence is particularly associated with inner city areas and young age groups in many industrialized countries, and suggests that efforts for intervention should be targeted at communities in these areas.

An additional concern for the treatment of gonorrhea is the continued development and emergence of antibiotic resistance in gonococci. Penicillin was the mainstay of therapy for many years, but because of both plasmid and chromosomally mediated resistance, it is no longer recommended as first-line therapy in many parts of the world. Alternative therapies, such as ciprofloxacin, ceftriaxone, and spectinomycin, are available for treatment in industrialized countries. However, in the developing world, where such antibiotics are not available or are too expensive, inappropriate antibiotics are often used or inadequate dosage is given, which can lead to an increase in resistant strains.

The control of sexually transmitted infections, including gonorrhea, is dependent on rapid and accurate diagnosis, and the administration of an effective antibiotic to break the transmission chain.

2. Diagnosis
2.1. Conventional Methods

Immunologic and molecular techniques have been increasingly used for the diagnosis of sexually transmitted infections, such as *Chlamydia trachomatis*. However, the diagnosis of gonorrhea has remained largely unchanged. Presumptive diagnosis is made by microscopic examination of a urethral or cervical smear for the presence of intracellular Gram-negative cocci, followed by confirmation by growth of the causative organism, *N. gonorrhoeae (1)*. The use of the Gram smear is universal because it is inexpensive, rapid, and can be performed in most STD clinics, often without referral to a microbiology laboratory. The high sensitivity and specificity found with urethral smears from symptomatic men (>95%) enables treatment to be given at their initial clinic presentation. In women the Gram-stained smear is less useful because the sensitivity and specificity are lower (30–40%). However, attempts to develop a replacement test with a higher sensitivity in women have been unsuccessful, and the Gram smear remains the method of choice for the presumptive diagnosis of gonorrhea in most clinics.

Confirmation of the diagnosis of gonorrhea by culture of *N. gonorrhoeae* is considered the "gold standard," and is reputed to have a sensitivity approaching 100%. This will be influenced by the quality of the specimen, provision of a suitable medium, and incubation conditions and the choice of antibiotics for inclusion in the medium to select against normal flora present in the urogenital tract. Whereas culture confirmation takes a minimum of 24–48 h to complete, it has the advantage of a high sensitivity and specificity, and it provides a viable organism for susceptibility testing. There have been a number of attempts to produce antigen detection tests for *N. gonorrhoeae* using immunologic assays, but their lack of sensitivity with specimens from women, together with the high cost and technical expertise required have limited their use *(1)*. In addition, none of the commercially available kits address the problem of testing the susceptibility of the infecting organism.

2.2. Molecular Approaches to Diagnosis

Molecular approaches are attractive because they have the potential for a greater sensitivity and for the detection of *C. trachomatis* in the same specimen or simultaneously. Three techniques have been used for the detection of *N. gonorrhoeae* directly in clinical samples; the use of DNA probes, the polymerase chain reaction (PCR), and the ligase chain reaction (LCR). If one of these methods proves acceptable, the determination of antibiotic susceptibility should also be possible using the same technology.

A prerequisite for any of these molecular techniques is the identification of a DNA sequence specific to *N. gonorrhoeae*. The pathogenic species of *Neisseria* show a high degree of genetic relatedness and therefore, this requirement presents a considerable problem *(2)*. A rapid detection system is also required to give a test that can produce a result in a shorter time than standard methods.

2.2.1. DNA Probes

The initial studies on DNA probes as diagnostic tools used the gonococcal cryptic plasmid, which is rarely found in other species of *Neisseria*, as a probe *(3)*. This 2.6 megadalton (Mda) plasmid is present in the majority of gonococcal strains, but it is not carried by a minority of isolates, particularly those requiring proline, citrulline, and uracil for growth, therefore, limiting the sensitivity of the technique. A more successful approach has been the use of rRNA sequences, and it has been possible to identify sequences specific either for the genus *Neisseria* or for *N. gonorrhoeae* *(4)*. This has been developed using a chemiluminescent assay, which has the added advantage of nonradioactive substrate *(5,6)*.

The commercial kit most widely evaluated is the Gen-Probe PACE-2 assay, which uses a single-stranded DNA probe labeled with an acridinium ester that is complementary to the rRNA of *N. gonorrhoeae*. The sensitivities reported have been between 90–100% *(7–10)* as compared with culture. There has been some concern regarding an appropriate cut-off value to differentiate between positive and negative results, with some workers suggesting that the manufacturers cut-off is too low, resulting in a low specificity or the detection of "false-positives." However, these discrepant specimens, in many instances, also gave a positive result in the probe competition assay (PCA), although confirmation in this manner has been found to be necessary only for samples giving a low signal in the probe assay *(11)*. Results confirmed by two tests suggest these were actually "true" positives, and that isolation of *N. gonorrhoeae* as confirmation of diagnosis may not be as sensitive as was previously thought. An analogous situation may exist to the diagnosis of chlamydial infection, where culture was found to be less sensitive than antigen detection methods. However, when comparisons are made, it must be remembered that culture is entirely dependent on the quality of the specimen and methodology used for isolation.

2.2.2. Polymerase Chain Reaction

In contrast to the diagnosis of chlamydial infection, PCR has not been extensively explored as a method for the diagnosis of gonorrhea. The main reasons for this are the gram stain is a relatively sensitive and inexpensive diagnostic system, and the decrease in the total numbers of cases in the industrialized world has resulted in a reduced market for a diagnostic test for gonorrhea. The development of PCR is dependent, as described above for DNA probe assay, on the choice of suitable primers specific for *N. gonorrhoeae*. Two reports from the same group have described primers based on the *cpp*B gene (*see refs. 12,13* and **Table 1**). This gene is primarily found on the gonococcal cryptic plasmid, but it is also thought that a copy is present on the chromosome *(14)*. There are possible problems with this gene as a target in that it is present in low copy number, and may also be found in a small number of strains of *N. meningitidis* which are occasionally found in the urogenital tract *(15)*. Preliminary studies have shown promising results when used for the detection of *N. gonorrhoeae* alone *(12)*, or in a duplex PCR for detection of both *N. gonorrhoeae* and *C. trachomatis* in clinical samples *(13)*. However, its sensitivity with plasmid-free strains has not been tested. The same primers have been used in a multiplex PCR for the simultaneous detection of *N. gonorrhoeae* and *C. trachomatis* *(16)*. The preliminary evaluation of this test showed promising results with a sensitivity of 92.3% for *N. gonorrhoeae*, and 84.6% for *C. trachomatis*, and a specificity of 100% for both *(16)*. The detection system involves the visualization of PCR products of differing sizes on an agarose gel. If this approach is to be used to screen large numbers of specimens, an alternative detection system that lends itself to automation will be needed.

2.2.3. Ligase Chain Reaction

LCR (*see* Chapter 5) has become an increasingly popular approach for the diagnosis of sexually transmitted diseases. It is attractive for two reasons; it may have a greater specificity than PCR, and an increased sensitivity when

Table 1
Sequence of Oligonucleotides used for Detection of *N. gonorrhoeae*

Source	Sequence (5'–3')	Technique	Refs.
cppB gene	HO1 GCTACGCATACCCGCGTTGC		
	HO3 CGAAGACCTTCGAGCAGACA	PCR	*12,13,16*
opa-2[a]	5' F1-GCCATATGTTGAAACACCGCCC		
	AACCCGATATAATCCGCCCTT-Bio 3'	LCR	*17*
	CGGTGTTTCAACACAATATCGC-F1 3'		
	5' Bio-AAGGGCGGATTATATCGGGTTCC		
opa-3[a]	5' F1-CAACATCAGTGAAAATCTTTTTTTAACC		
	TCAAACCGAATAAGGAGCCGAA-Bio 3'	LCR	*17*
	TTAAAAAAATTTTCACTGATGTTG-F1 3'		
	5' Bio-TTCGGCTCCTTATTCGGTTTGACC		
pilin-2[a]	5' F1-CGGGCGGGGTCGTCCGTTCC		
	TGGAAATAATATATCGATTCTGCG-Bio 3'	LCR	*17*
	AACGGACGACCCCGCCCG-F1 3'		
	5' Bio-TTCGGCTCCTTATTCGGTTTGACC		

[a]Oligonucleotides are shown with their associated fluorescein (F1) or biotin (Bio) haptens.

compared with culture. The development of LCR for gonorrhea has followed its successful use for the detection of *C. trachomatis*. Initial studies have evaluated the use of LCR for gonorrhea alone, but the eventual aim is the detection of *C. trachomatis* and *N. gonorrhoeae* in the same sample. An additional advantage of LCR is that the greater sensitivity may allow the detection of organisms in urine. The use of a specimen that does not require a clinical examination has obvious advantages, particularly for screening in low prevalence populations.

Two sets of primers have been evaluated for use in the LCR based on the multicopy opacity and pilin genes (**Table 1**). Both detected all strains of *N. gonorrhoeae*, but not nonpathogenic *Neisseria* or non-*Neisseria* species when whole cells were tested *(17)*, and they were 100% sensitive and 97.8% specific, as compared with culture, when tested on clinical samples. Subsequent studies have shown that LCR has an apparent enhanced sensitivity when compared with culture, (i.e., culture-negative, LCR-positive), in a similar manner to that found with molecular techniques for the diagnosis of chlamydial infection *(17–20)*. This has led to the use of an expanded "gold standard" for calculating sensitivity, specificity, and predictive values that uses all culture-positive results together with confirmed LCR-positive specimens.

Detection by LCR uses primers directed at the opacity genes, and any results discordant with culture results are confirmed using LCR with primers directed at the pilin genes. This has resulted in sensitivities of 97.3 and 98.5% and speci-

ficities of 99.6 and 99.8% for LCR when used with female and male specimens, respectively *(19)*. Detection of both *C. trachomatis* and *N. gonorrhoeae*
in a single sample has also proved to be highly sensitive when used with
urethral specimens from men or cervical specimens from women *(20)*.
However, there is some discrepancy in the results obtained for the detection of *N. gonorrhoeae* from urine. Smith et al. *(21)* have found the sensitivity
of LCR in urine from women was 94.6%, whereas Buimer et al. *(20)* found the
sensitivity to be only 50%. These workers found a higher sensitivity of 88.9% in
urine from men and suggested that the poor results from women may be a reflection of lower number of organisms present. Whereas screening urine has many
advantages and there is increasing evidence that it can be very useful for screening
populations for chlamydial infection where the number of organisms shed may be
higher, it needs further evaluation as a method of detecting *N. gonorrhoeae*.

3. Antimicrobial Susceptibility

Treatment of *N. gonorrhoeae* is often administered before the susceptibility
of the infecting strain is known. Detection of resistance by methods in routine
use still requires a viable organism *(22)*. Plasmid-mediated resistance to penicillin *(see* below) can be determined using a chromogenic cephalosporin but
detection of plasmid-mediated resistance to tetracycline or chromosomally
mediated resistance to penicillin, ceftriaxone, and ciprofloxacin requires the
use of disk diffusion, breakpoints, or determination of the minimum inhibitory
concentration (MIC). Molecular methods have been used for research purposes,
both for detection in epidemiologic or surveillance studies, and for determination of mechanisms of resistance. Sufficient knowledge is now available to
enable the development of molecular methods for the rapid detection of high-
level resistance directly in the clinical sample.

3.1. Plasmid-Mediated Resistance

Plasmid-mediated resistance to penicillin in *N. gonorrhoeae* is the result of
a TEM-1 type β-lactamase *(see* Chapter 25) that is encoded on a number of
low-molecular weight plasmids *(23)*. The TEM-1 gene is widespread among
members of the Enterobacteriaceae, probably owing to its associations with
transposon Tn2. DNA probes and PCR have been described for the detection
of β-lactamase by gonococci *(24–26)*. DNA probes that detect TEM-1
sequences alone are sensitive, but are not specific for *N. gonorrhoeae (24)*. A
greater degree of specificity is achieved by using oligonucleotides for
sequences in the TEM-1 gene and in the plasmid beyond the Tn2 transposon.
This approach has been described for detection using a DNA probe *(25)* and
PCR *(26)*, and was shown to detect β-lactamase in *N. gonorrhoeae* and in
Haemophilus parainfluenzeae, which carries closely related plasmids. This is

a rapid and sensitive technique, but requires identification of the organism by conventional methods or a species-specific probe.

High-level tetracycline resistance in *N. gonorrhoeae* is the result of the acquisition of the *tetM* determinant (*see* Chapter 31) into the 24.5 Mda conjugative plasmid *(27)*. Detection using PCR initially used primers based on the sequences within the *tetM* determinant of *Ureaplasma urealyticum* and showed 100% sensitivity and specificity in isolates known to be *N. gonorrhoeae (28)*. Two types of *tetM* plasmids have been found in gonococci by DNA sequencing *(29)*, and by PCR using primers, based either on the *U. urealyticum tetM* sequence or on sequences in the neisserial conjugative plasmid *(30)*. It is, therefore, possible to use PCR for detection and for epidemiologic purposes in a single reaction.

3.2. Chromosomally Mediated Resistance

Chromosomal resistance to penicillin in *N. gonorrhoeae* is low-level (MIC, ≥2 µg/mL) and results from the additive effect of mutations at multiple loci, including those designated *penA, mtr,* and *penB (31)*. Therapeutic failure and, hence, resistance to ceftriaxone, a cephalosporin, has not been documented, although reduced susceptibility is known to be associated with the same loci *(32)*. There is no molecular method described to detect chromosomally mediated resistant *N. gonorrhoeae* (CMRNG) and susceptibility to penicillin is currently determined using conventional testing of a viable organism. However, the mutations involved in resistance for *penA (33)* and for *mtr (34)* have now been characterized using molecular methods and, hence, information is available for the development of a DNA-based test.

High-level resistance to ciprofloxacin in *N. gonorrhoeae* has emerged in the last 2 yr and is the result of mutations in the DNA gyrase gene, *gyrA*, and possibly the topoisomerase IV gene, *parC*. Both laboratory mutants *(35)* and clinical isolates *(36,37)* have been shown to have similar changes in the "quinolone resistance determining region" of the *gyrA* gene (*see* Chapter 30). Lower levels of resistance may be associated with mutations in *gyrB* or in cell wall permeability.

4. Epidemiologic Investigation

The epidemiology of gonorrhea has been extensively studied using serologic classification with a panel of twelve monoclonal antibodies (MAbs) to the major outer membrane protein, PI, or Por. The level of discrimination achieved with serotyping alone is low, and in most populations, a small number of serovars predominate. Discrimination has been enhanced by the addition of auxotyping (determination of nutritional requirements) to produce auxotype/serovar (A/S) classes, and this has been used to monitor temporal changes in populations, movement of antibiotic-resistant strains, and for forensic purposes *(38)*. The need for further discrimination has led to the evaluation of a variety of molecular techniques, but currently no single method has been widely accepted. Many of the techniques have been chosen to address a specific problem, such as discrimination between strains of a specific serovar or auxotype, rather than to provide an alternative method for typing all gonococcal strains. All the techniques used have shown that a greater degree of discrimination can be achieved using molecular techniques that are largely independent of serovar or auxotype. This supports the hypothesis that *N. gonorrhoeae* is a panmitic or nonclonal population, as was initially shown by multilocus enzyme electrophoresis *(39–41)*. Extensive recombination is known to occur between isolates of *N. gonorrhoeae*, and it is likely to occur in vivo during mixed infections, resulting in a wide array of genotypes, although occasional clones may occur within this nonclonal population *(42)*. Isolates from linked cases of gonorrhea are likely to show no or only small genetic differences, but unrelated isolates will show considerable variation such that each isolate could appear unique. This questions the use of molecular techniques for classical epidemiologic studies monitoring trends over long time spans as performed with clonal populations, such as *Esherichia coli*. However, these techniques may be more applicable for studying short-term transmission of gonorrhea between sexual contacts in transmission chains or sexual networks,or for forensic purposes that require a highly discriminatory method.

4.1. Plasmid Analysis

N. gonorrhoeae are known to carry a small number of plasmids *(43–48)* as shown in **Table 2**. The plasmid profile differs between strains, but the level of discrimination achievable with this technique (*see* Chapter 4) is low. Knowledge of the plasmid content is most useful for antibiotic-resistant strains and when used in combination with another technique.

4.2. Restriction Endonuclease Fingerprinting

Patterns obtained after digestion of chromosomal DNA with restriction endonucleases, *Hin*dIII *(49–52)* or *Hin*f1 and *Bgl*II *(53)*, followed by agarose gel electrophoresis have been used to discriminate between gonococcal strains. Whereas the method (*see* Chapter 2) has been shown to be discriminatory, the digestion of total DNA produces multiple bands (>50) and minor differences can be difficult to detect. An alternative is to probe digests of total DNA transferred to nitrocellulose or nylon membrane by Southern blotting with a broadspectrum ribosomal RNA (rRNA) probe; this is known as ribotyping or riboprobing *(54–56)* (*see* Chapter 2). Studies using ribotyping have digested DNA with the endonucleases *Hin*cII *(56)*, *Hin*dIII, *Pst*I, *Ava*II, and *Sma*I

Table 2
Plasmids Found in *N. gonorrhoeae*

Molecular weight (MDa)	Function	Reference
2.6	Cryptic	*43*
2.9	Penicillinase	*44*
3.05	Penicillinase	*45*
3.2	Penicillinase	*23*
4.1	Penicillinase	*46*
4.4	Penicillinase	*23*
6.5	Penicillinase	*47*
24.5	Conjugative	*48*
25.2	Conjugative and *tetM*	*27*

(54,55) and have used probes of 16S and 23S rRNA of *E. coli* or a 7.5 kb *Bam*HI-*Pst*I fragment of pKK3535 *(56)*, labeled either with ^{32}P or digoxigenin. This reduces the number of bands to between 10 and 15, and aids the interpretation, but tends to reduce the discrimination unless patterns obtained with more than one enzyme are used.

4.3. Polymerase Chain Reaction

Amplification of gonococcal DNA by PCR for typing purposes has been performed using random primers, RAPD-PCR/AP-PCR *(57–59)* (*see* Chapter 6), and repetitive element sequence-based primers, REP-PCR *(60)* (*see* Chapter 7), as shown in **Table 3**. RAPD-PCR produced between 11 and 15 fragments using the primer OPA-03, and between 15 and 22 fragments with primer OPA-13 as resolved by agarose gel electrophoresis in two studies *(57,58)*. In the larger study of 70 isolates, the discrimination index (0.967 with OPA-03 and 0.978 with OPA-13) was shown to be higher than combinations of auxotype/serovar/plasmid profile *(57)*. O'Rourke and Spratt *(59)* have used AP-PCR to investigate the nonclonal nature of gonococcal isolates. A variety of primers were tested, but the largest number of fragments was obtained using an oligonucleotide Pn-2X-1 corresponding to a sequence within the *Streptococcus pneumoniae* penicillin-binding protein 2X gene. Of the 52 isolates tested belonging to multiple serovars, 12 AP-PCR patterns could be determined. Poh et al. *(60)* have used REP-PCR, which detects polymorphisms in the distances between the repetitive elements on the chromosomes of different isolates, to determine the genetic diversity of isolates belonging to two serovars, IB-2 and IB-6. Six profiles were found, with each group of isolates indicating that the isolates belonging to the two serovars tested in this study were not

Table 3
Sequence of Oligonucleotides used for Molecular Typing of *N. gonorrhoeae*￼

Technique	Sequence[a]	Reference
RAPD/PCR	OPA-03 AGTCAGCCAC	*58*
	OPA-13 CAGCACCCAC	
AP/PCR	Pn-2X-1 AGGACT(TC)TGTTTGGCGTGATAT	*59*
REP/PCR	REPIR-Dt IIINCGNCGNCATCNGGC	*60*
	REP-2-Dt NCGNCTTATCNGGCCTAC	

[a]N = A,C,G, or T; I = inosine.

clonal. These methods were simpler to perform than RFLP of chromosomal DNA and ribotyping, and were shown to be reproducible and discriminatory.

4.4. Pulsed-Field Gel Electrophoresis (PFGE)

PFGE (*see* Chapter 3) has been found to be a useful tool for differentiating patterns generated by digestion of total chromosomal DNA with enzymes that cut infrequently, such as *Spe*I, *Nhe*I, and *Bgl*II *(55,61–65)*. The RFLP patterns produced consist of a small number of high molecular weight fragments that are easier to interpret than restriction endonuclease analysis on conventional agarose or polyacrylamide gel electrophoresis (*see* **Subheading 4.2.**). An analysis of a large collection of strains of differing phenotypes, and from different geographic origin has not been reported. However, studies of individual serovars or auxotypes show this technique to be highly discriminatory.

4.5. Opa-Typing

Diversity in the 11 opa (opacity) genes of *N. gonorrhoeae* has been investigated as a tool for discriminating between linked and unlinked isolates in sexual networks *(66)*. Most typing methods examine variation in conserved or semiconserved genes that are not under pressure from the immune response. However, in order to discriminate isolates from patients within a network, it is necessary to have a method that is of sufficient stability so that isolates that are part of a transmission chain are indistinguishable, but also that is sufficiently discriminatory so that unlinked isolates are each different. *Opa*-typing is performed by amplification of sequences that are conserved between the 11 genes. The PCR product is purified from the remaining primers and digested with a restriction enzyme that recognizes multiple sites; *Taq*1, *Hin*P1, or *Hpa*II. The resulting fragments are end-labeled with ^{32}P, separated by polyacrylamide gel electrophoresis and the patterns visualized by autoradiography.

Retrospective studies have shown that, by using *opa*-typing unlinked isolates appear unique. Isolates from known sexual contacts give RFLP patterns that are indistinguishable from each other, but are also different from all other isolates *(66)*. This high level of discrimination has been achieved by using hypervariable genes that are under immune selection, and has the potential for use in defining sexual networks. There is some evidence that such discrimination may be achievable with PFGE.

4.6. Por Typing or Sequencing

Diversity in the por gene has been determined by amplification by PCR, digestion of the product by restriction enzymes, and separation of the fragments by agarose gel electrophoresis *(67,68)*. Whereas this is a simple technique to perform, the level of discrimination is similar to that obtained with ribotyping. Discrimination can be enhanced by using a combination of different enzymes, but seldom reaches the levels found with PFGE or *opa*-typing. A natural extension of this technique is to determine the DNA sequence of the PCR product *(63,69)*. This has been used only on selected isolates, such as those of similar serovars and isolates from linked patients. This approach is highly discriminatory, but very labor-intensive. It could, however, be a useful adjunct to PFGE or *opa*-typing for identifying isolates from a common source.

5. Conclusion

N. gonorrhoeae is a versatile pathogen that is extremely well-adapted to survival in its human host, and is often used as an example of a mucosal pathogen. Molecular techniques have made a major contribution to our understanding of the mechanisms of colonization, including its ability to exhibit phase and antigenic variation *(70,71)*. In contrast, molecular techniques are not routinely used for diagnosis, determination of susceptibility, or for epidemiologic purposes. However, this chapter has demonstrated that the techniques are applicable in these areas, and that they will be used more extensively in the near future.

References

1. Ison, C. A. (1990) Methods for diagnosing gonorrhoea. *Genitourin. Med.* **66,** 453–459.
2. Muralidhar, B. and Steinman, C. R. (1994) Design and characterization of PCR primers for detection of pathogenic *Neisseriae. Mol. Cell. Probes* **8,** 55–61.
3. Totten, P. A., Homes, K. K., Handsfield, H. H., Knapp, J. S., Perine, P. L., and Falkow, S. (1983) DNA hybridization technique for the detection of *Neisseria gonorrhoeae* in men with urethritis. *J. Infect. Dis.* **148,** 462–471.
4. Rossau, R., Duhamel, M., van Dyck, E., Piot, P., and van Heuverswyn, H. (1990) Evaluation of an rRNA-derived oligonucleotide probe for culture confirmation of *Neisseria gonorrhoeae. J. Clin. Microbiol.* **28,** 944–948.
5. Granato, P. A. and Franz, M. R. (1989) Evaluation of a prototype DNA probe test for the noncultural diagnosis of gonorrhea. *J. Clin. Microbiol.* **27,** 632–635.

6. Panke, E. S., Yang, L. I., Leist, P. A., Magevney, P., Fry, R., and Lee, R. F. (1991) Comparison of Gen-Probe DNA Probe test and culture for the detection of *Neisseria gonorrhoeae* in endocervical specimens. *J. Clin. Microbiol.* **29,** 883–888.

7. Limberger, R. J., Biega, R., Evancoe, A., McCarthy, L., Slivienski, L., and Kirkwood, M. (1992) Evaluation of culture and the Gen-Probe PACE 2 assay for detection of *Neisseria gonorrhoeae* and *Chlamydia trachomatis* in endocervical specimens transported to a state health laboratory. *J. Clin. Microbiol.* **30,** 1162–1166.

8. Hale, Y. M., Melton, M. E., Lewis, J. S., and Willis, D. E. (1993) Evaluation of the PACE 2 *Neisseria gonorrhoeae* assay by three public health laboratories. *J. Clin. Microbiol.* **31,** 451–453.

9. Vlaspolder, F., Mutsaers, J. A. E. M., Blog, F., and Notowicz, A. (1993) Value of a DNA probe assay (Gen-Probe) compared with that of culture for diagnosis of gonococcal infection. *J. Clin. Microbiol.* **31,** 107–110.

10. Schwebke, J. R. and Zajackowski, M. E. (1996) Comparison of DNA probe (Gen-Probe) with culture for the detection of *Neisseria gonorrhoeae* in an urban STD programme. *Genitourin. Med.* **72,** 108–110.

11. Woods, G. L. and Garza, D. M. (1996) Use of Gen-Probe competition assay as a supplement to probes for direct detection of *Chlamydia trachomatis* and *Neisseria gonorrhoeae* in urogenital specimens. *J. Clin. Microbiol.* **34,** 177,178.

12. Ho, B. S. W., Feng, W. G., Wong, B. K. C., and Egglestone, S. I. (1992) Polymerase chain reaction for the detection of *Neisseria gonorrhoeae* in clinical samples. *J. Clin. Pathol.* **45,** 439–442

13. Wong, K. C., Ho, B. S. W., Egglestone, S. I., and Lewis, W. H. P. (1995) Duplex PCR system for simultaneous detection of *Neisseria gonorrhoeae* and *Chlamydia trachomatis* in clinical specimens. *J. Clin. Pathol.* **48,** 101–104.

14. Hagbolm, P., Korch, C., Jonsson, A., and Normark, S. (1986) Intragenic variation by site-specific recombination in the cryptic plasmid of *Neisseria gonorrhoeae*. *J. Bacteriol.* **167,** 231–237.

15. Ison, C. A., Bellinger, C. M., and Walker, J. (1986) Homology of the cryptic plasmid of *Neisseria gonorrhoeae* with plasmids from *Neisseria meningitidis* and *Neisseria lactamica. J. Clin. Pathol.* **39,** 1119–1123.

16. Mahony, J. B., Luinstra, K. E., Tyndall, M., Sellors, J. W., Krepel, J., and Chernesky, M. (1995) Multiplex PCR for detection of *Chlamydia trachomatis* and *Neisseria gonorrhoeae* in genitourinary specimens. *J. Clin. Microbiol.* **33,** 3049–3053.

17. Birkenmeyer, L. and Armstrong, A. S. (1992) Preliminary evaluation of the ligase chain reaction for specific detection of *Neisseria gonorrhoeae. J. Clin. Microbiol.* **30,** 3089–3094.

18. Stary, A., Ching, S-F., Teodorowicz, L., and Lee, H. (1997) Comparison of ligase chain reaction and culture for detection of *Neisseria gonorrhoeae* in genital and extragenital specimens. *J. Clin. Microbiol.* **35,** 239–242.

19. Ching, S., Lee, H., Hook, E. W. III, Jacobs, M. R., and Zenilman, J. (1995) Ligase chain reaction for detection of *Neisseria gonorrhoeae* in urogenital swabs. *J. Clin. Microbiol.* **33,** 3111–1314.

20. Buimer, M., Doornum van, G. J. J., Ching, S., Peerbooms, P. G. H., Plier, P. K., Ram, D., and Lee, H. H. (1996) Detection of *Chlamydia trachomatis* and *Neisseria gonorrhoeae* by ligase chain reaction-based assays with clinical specimens from various sites: implications for diagnosis testing and screening. *J Clin. Microbiol.* **34,** 2395–2400.

21. Smith, K. R., Ching, S., Lee, H., Ohhashi, Y., Hu, H.-Y., Fisher, H. C. III, and Hook, E. W. III. (1995) Evaluation of ligase chain reaction for use with urine for identification of *Neisseria gonorrhoeae* in females attending a sexually transmitted disease clinic. *J Clin. Microbiol.* **33,** 455–457

22. Ison, C. A. (1996) Antimicrobial agents and gonorrhoea: therapeutic choice, resistance and susceptibility testing. *Genitourin. Med.* **72,** 253–257

23. Elwell, L. P., Roberts, M., Mayer, L. W., and Falkow, S. (1977) Plasmid-mediated β-lactamase production in *Neisseria gonorrhoeae. Antimicrob. Agents Chemother.* **11,** 528–533.

24. Huovinen, S., Houvinen, P., and Jacoby, G. A. (1988) Detection of plasmid-mediated β-lactamases with DNA probes. *Antimicrob. Agents Chemother.* **32,** 175–179.

25. Sanchez-Pescador, R., Stempien, M. S., and Urdea, M. S. (1988) Rapid chemiluminescent nucleic acid assays for detection of TEM-1 β-lactamase-mediated penicillin resistance in *Neisseria gonorrhoeae* and other bacteria. *J. Clin. Microbiol.* **26,** 1934–1938

26. Simard, J.-L. and Roy, P. H. (1993) PCR detection of penicillinase-producing *Neisseria gonorrhoeae* in *Diagnostic Molecular Microbiology: Principles and Applications* (Persing, D. H., Smith, T. F., et al., eds.), American Society of Microbiology, Washington, DC, pp. 543–546.

27. Morse, S. A., Johnson, S. R., Biddle, J. W., and Roberts, M. C. (1986) High-level tetracycline resitance in *Neisseria gonorrhoeae* is result of acquisition of streptococcal *tetM* determinant. *Antimicrob. Agents Chemother.* **30,** 664–670

28. Ison, C. A., Tekki, N., and Gill, M. J. (1993) Detection of the *tetM* determinant in *Neisseria gonorrhoeae. Sex Transm. Dis.* **20,** 329–333

29. Gascoyne-Binzi, D. M., Heritage, J., and Hawkey, P. M. (1993) Nucleotide sequences of the tet(M) genes from the American and Dutch type tetracycline resistance plasmid of *Neisseria gonorrhoeae. J. Antimicrob. Chemother.* **32,** 667–676.

30. Xia, M., Pang, Y., and Roberts, M. C. (1995) Detection of two groups of 25.2 MDa *TetM* plasmids by polymerase chain reaction of the downstream region. *Mol. Cell. Probes* **9,** 327–332.

31. Sparling, P. F., Sarubbi, F. A., and Blackman, E. (1975) Inheritance of low-level resistance to penicillin, tetracycline and chloramphenicol in *Neisseria gonorrhoeae. J Bacteriol.* **124,** 740–749

32. Ison, C. A., Bindayna, K. M., Woodford, N., Gill, M. J., and Easmon, C. S. F. (1990) Penicillin and cephalosporin resistance in gonococci. *Genitourin. Med.* **66,** 351–356.

33. Spratt, B. G. (1988) Hybrid penicillin-binding proteins in penicillin resistant strains of *Neisseria gonorrhoeae. Nature* **332,** 173–176.

34. Hagman, K. E., Pan, W., Spratt, B. G., Balthazar, J. T., Judd, R. C., and Shafer, W. M. (1995) Resistance of *Neisseria gonorrhoeae* to antimicrobial hydrophobic agents is modulated by the *mtrCDE* efflux system. *Microbiology* **141,** 611–622.

35. Belland, R. J., Morrison, S. G., Ison, C., and Huang, W. M. (1994) *Neisseria gonorrhoeae* acquires mutations in analogous regions of *gyrA* and *parC* in fluoroquinolone-resistant isolates. *Mol Microbiol.* **14,** 371–380

36. Deguchi, T., Yasuda, M., Asano, M., Tada, K., Iwata, H., Komeda, H., Ezaki, T., Saito, I., and Kawada, Y. (1995) DNA gyrase mutations in quinolone-resistant clinical isolates of *Neisseria gonorrhoeae. Antimcrob. Agents Chemother.* **39,** 561–563.

37. Deguchi, T., Yasuda, M., Nakano, M., Ozeki, S., Ezaki, T., Saito, I., and Kawada, Y. (1996) Quinolone-resistant *Neisseria gonorrhoeae*: correlations of the GyrA subunit of DNA gyrase and the ParC subunit of topoisomerase IV with antimicrobial susceptibility profiles. *Antimicrob. Agents Chemother.* **40,** 1020–1023.

38. Gill, M. J. (1991) Serotyping *Neisseria gonorrhoeae*: a report of the Fourth International Workshop. *Genitourin. Med.* **67,** 53–57.

39. O'Rourke, M. and Stevens, E. (1993) Genetic structure of *Neisseria gonorrhoeae* populations: a non-clonal population. *J. Gen. Microbiol.* **139,** 2603–2611

40. Vazquez, J. A., de la Fuente, L., Berron, S., O'Rourke, M., Smith, N. H., Zhou, J., and Spratt, B. G. (1993) Ecological separation and genetic isolation of *Neisseria gonorrhoeae* and *Neisseria meningitidis. Curr. Biol.* **3,** 567–572.

41. Poh, C. L., Ocampo, J. C., and Loh, G. K. (1992) Genetic relationships among *Neisseria gonorrhoeae* serovars analysed by multilocus enzyme electrophoresis. *Epidemiol. Infect.* **108,** 31–38

42. Gutjahr, T. S., O'Rourke, M., Ison, C. A., and Spratt, B. G. (1997) Arginine, hypoxanthine, uracil-requiring isolates of *Neisseria gonorrhoeae* are a clonal lineage within a non-clonal population. *Microbiology* **143,** 633–640.

43. Korch, C., Hagbolm, P., Ohman, H., Goransson, M., and Normark, S. (1985) Cryptic plasmid of *Neisseria gonorrhoeae*: complete nucleotide sequence and genetic organisation. *J. Bacteriol.* **163,** 430–438

44. Embden, J. D. A. van, Dessens-Kroons, M., and Klingeren, B. van. (1985) A new β-lactamase plasmid in *Neisseria gonorrhoeae. J Antimicrob Chemother.* **15,** 247–250.

45. Yeung, K.-H., Dillon, J. R., Pauze, M., and Wallace, E. (1986) A novel 4.9 kilobase plasmid associated with an outbreak of penicillinase-producing *Neisseria gonorrhoeae. J. Infect. Dis.* **53,** 1162–1165.

46. Gouby, A., Bourg, G., and Raamuz, M. (1986) Previously undescribed 6.6 kilobase R plasmid in penicillinase-producing *Neisseria gonorrhoeae. Antimicrob. Agents Chemother.* **29,** 1095–1097.

47. Brett, M. (1989) A novel gonococcal β-lactamase plasmid. *J. Antimicrob. Chemother.* **23,** 653,654.

48. Sox, T. E., Mohammed, W., Blackman, E., Biswas, G., and Sparling, P. F. (1978) Conjugative plasmids in *Neisseria gonorrhoeae. J. Bacteriol.* **134,** 278–286.

49. Falk, E. S., Bjorvatn, B., Danielsson, D., Kristiansen, B.-E., Melby, K., and Sorensen, B. (1984) Restriction endonuclease fingerprinting of chromosomal DNA of *Neisseria gonorrhoeae. Acta Path. Microbiol. Immunol. Scand. Sect B,* **92,** 271–278.

50. Falk, E. S., Danielsson, D., Bjorvatn, B., Melby, K., Sorensen, B., Kristiansen, B.-E., Lund, S., and Sandstrom, E. (1985) Phenotypic and genotypic characterization of penicillinase-producing strains of *Neisseria gonorrhoeae*. *Acta Path. Microbiol. Immunol. Scand. Sect B,* **93,** 91–97.

51. Falk, E. S., Egglestone, S. I., Digranes, A., Volden, G., and Bjorvatn, B. (1988) Genotypic and phenotypic markers in the differentiation of *Neisseria gonorrhoeae* strains. *APIMS* **96,** 109–116.

52. Dasi, M. A., Nogueira, J. M., Camarena, J. J., Gil, C., Garcia-Verdu, R., Barbera, J. L., and Barbera, J. (1992) Genomic fingerprinting of penicillinase-producing strains of *Neisseria gonorrhoeae* in Valencia, Spain. *Genitourin. Med.* **68,** 170–173.

53. Poh, C. L., Ocampo, J. C., Sng, E. H., and Bygdeman, S. M. (1989) Rapid in-situ generation of DNA restriction endonuclease patterns for *Neisseria gonorrhoeae*. *J. Clin. Microbiol.* **27,** 2784–2788.

54. Ng, L.-K. and Dillon, J. R. (1993) Typing by serovar, antibiogram, plasmid content, riboprobing and isoenzyme typing to determine whether *Neisseria gonorrhoeae* isolates requiring proline, citrulline and uracil for growth are clonal. *J Clin. Microbiol.* **31,** 1555–1561.

55. Li, H. and Dillon, J. R. (1995) Utility of ribotyping restriction endonuclease analysis and pulsed-field gel electrophoresis to discriminate between isolates of *Neisseria gonorrhoeae* of serovar IA-2 which require arginine, hypoxanthine or uracil for growth. *J. Med. Microbiol.* **43,** 208–215

56. Poh, C. L., Khng, H. P., Lim, C. K., and Loh, G. K. (1992) Molecular typing of *Neisseria gonorrhoeae* by restriction fragment length polymorphisms. *Genitourin. Med.* **68,** 106–110.

57. Camarena, J. J., Nogueira, J. M., Dasi, M. A., Moreno, F., Garcia, R., Ledesma, E., Llorca, J., and Hernandez, J. (1995) DNA amplification fingerprinting for subtyping *Neisseria gonorrhoeae* strains. *Sex Transm. Dis.* **22,** 128–136

58. Dasi, M. A., Camarena, J. J., Ledesma, E., Garcia, R., Moreno, F., and Nogueira, J. M. (1993) Random amplification of polymorphic DNA of penicillinase-producing *Neisseria gonorrhoeae* strains. *Genitourin. Med.* **69,** 404,405.

59. O'Rourke, M. and Spratt, B. G. (1994) Further evidence for the non-clonal population structure of *Neisseria gonorrhoeae*: extensive genetic diversity within isolates of the same electrophoretic type. *Microbiology* **140,** 1285–1290.

60. Poh, C. L., Ramachandran, V., and Tapsall, J. W. (1996) Genetic diversity of *Neisseria gonorrhoeae* IB-2 and IB-6 isolates revealed by whole-cell repetitive element sequence-based PCR. *J Clin. Microbiol.* **34,** 292–295.

61. Ng, L.-K., Carballo, M., and Dillon, J. R. (1995) Differentiation of *Neisseria gonorrhoeae* isolates requiring proline citrulline and uracil by plasmid content, serotyping and pulse-field gel electrophoresis. *J Clin. Microbiol.* **33,** 1039–1041.

62. Poh, C. L., Loh, G. K., and Tapsall, J. W. (1995) Resolution of cloncal subgroups among *Neisseria gonorrhoeae* IB-2 and IB-6 serovars by pulsed-field gel electrophoresis. *Genitourin. Med.* **71,** 145–149.

63. Poh, C. L., Lau, Q. C., and Chow, V. T. K. (1995) Differentiation of *Neisseria gonorrhoeae* IB-3 and IB-7 serovars by direct sequencing of protein IB gene and pulsed-field gel electrophoresis. *J. Med. Microbiol.* **43,** 201–207.

64. Xia, M., Whittington, W. L., Holmes, K. K., Plummer, F. A., and Roberts, M. C. (1995) Pulsed-field gel electrophoresis for genomic analysis of *Neisseria gonorrhoeae. J. Infect. Dis.* **171,** 455–458.

65. Xia, M., Roberts, M. C., Whittington, W. L., Holmes, K. K., Knapp, J. S., Dillon, J. R., and Wi, T. (1996) *Neisseria gonorrhoeae* with decreased susceptibility to ciprofloxacin: pulsed-field gel electrophoresis typing of strains from North America, Hawaii, and the Phillipines. *Antimicrob. Agents Chemother.* **40,** 2349,2440.

66. O'Rourke, M., Ison, C. A., Renton, A. M., and Spratt, B. G. (1995) Opa-typing: a high resolution tool for studying the epidemiology of gonorrhoea. *Mol. Microbiol.* **17,** 865–875.

67. Lau, Q. C., Chow, V. T. K., and Poh, C. L. (1995) Differentiation of *Neisseria gonorrhoeae* strains by polymerase chain reaction and restriction length poylmorphism of outer membrane protein IB genes. *Genitourin. Med.* **71,** 363–366.

68. Ison, C. A., O'Rourke, M., Anwar, N., Renton, A. R., and Spratt, B. G. (1995) Molecular typing of *Neisseria gonorrhoeae* as a method of predicting isolates from sexual contacts. Abstracts of Eleventh meeting of the International Society for STD Research. No. 29, p. 37.

69. Cooke, S. J., Paz, de la H., Poh, C. L., Ison, C. A., and Heckels, J. E. (1997) Variation within serovars of *Neisseria gonorrhoeae* detected by structural analysis of outer membrane protein PIB and by molecular typing techniques. *Microbiology* **143,** 1415–1422.

70. Meyer, T. F., Pohlner, J., and Putten, van J. P.M. (1994) Biology of the pathogenic *Neisseriae,* in *Current Topics in Microbiology and Immunology* (Dangl, J. L., ed.), vol. 192, Springer-Verlag, Berlin, Heidelberg, pp. 283-317.

71. Robertson, B. D. and Meyer, T. F. (1992) Genetic variation in pathogenic bacteria. *Trend. Genet.* **8,** 422–427.

16

Chancroid

Stephen A. Morse, David L. Trees, and Patricia A. Totten

1. Introduction

Chancroid is a genital ulcerative disease (GUD). These diseases are common throughout the world and include syphilis, genital herpes, chancroid, lymphogranuloma venereum, and donovanosis. Chancroid is particularly common in Africa, Asia, and Latin America where its incidence may exceed that of syphilis as a cause of genital ulceration *(1,2)*. However, chancroid is considered an uncommon sexually transmitted infection in the United States.

The causative agent of chancroid is the fastidious gram-negative bacillus, *Haemophilus ducreyi*. The taxonomic position of *H. ducreyi* as a *Haemophilus* species has been questioned for a number of years. *H. ducreyi* was originally placed in the genus *Haemophilus* because of its requirement for hemin (X-factor) and because it has a G + C content that is within the accepted range for *Haemophilus* species. However, DNA hybridization data indicated that *H. ducreyi* was unrelated to true haemophili, such as *H. influenzae (3)*. Sequencing of the 16S ribosomal RNA (rRNA) of the type strain CIP542 *(4)* and two additional strains *(5)* confirmed that *H. ducreyi* belonged to the family Pasteurellaceae, as do other haemophili. However, owing to the divergence in the 16S rRNA between *H. ducreyi* and other *Haemophilus* species, *H. ducreyi* was assigned to cluster 4, whereas the true haemophili were assigned to cluster 1 *(5)*. There has been renewed interest in this pathogen and in chancroid owing to the association between genital ulcers and HIV infection, and the recognition of the disease as a cause of GUD in the US.

2. Diagnosis
2.1. DNA Probes

Parsons et al. *(6,7)* were the first investigators to report the development of DNA probes for *H. ducreyi*. These investigators used high-titered polyclonal

From: *Methods in Molecular Medicine, Vol. 15: Molecular Bacteriology: Protocols and Clinical Applications*
Edited by: N. Woodford and A. P. Johnson © Humana Press Inc., Totowa, NJ

antiserum against formalin-killed *H. ducreyi* strain ATCC 33922 to screen an *Eco*RI *H. ducreyi* genomic library in lambda gt11. Three DNA inserts coding for proteins that were recognized by the *H. ducreyi* antiserum were selected and subcloned into a pUC13 plasmid vector. The ^{32}P-labeled probes, designated pLP1, pLP4, and pLP8, reacted strongly with 16 strains of *H. ducreyi*; no reactivity was observed with *Treponema pallidum, Neisseria gonorrhoeae*, and Herpes simplex virus (HSV) DNA. The probes easily detected 10^4 colony forming units (cfu) of *H. ducreyi*; weaker reactions were observed with 4.9×10^3 cfu, and negative reactions were obtained with 1.4×10^3 cfu. The probes also detected the presence of *H. ducreyi* DNA in rabbit lesion exudates; in addition, three of four positive reactions were obtained from specimens that were negative for *H. ducreyi* by culture. pLP8 also reacted weakly with relatively high numbers of *Haemophilus* spp. and *Pasteurella* spp. ($3–6 \times 10^7$ cfu), although negative reactions were observed at lower cell concentrations ($10^5–10^6$ cfu). Subsequent sequencing of the insert in pLP8 *(8)* revealed that it encoded *H. ducreyi* homologs of *groE* and *groEL*. These are highly conserved genes and may explain the hybridization observed with *Haemophilus* spp. and *Pasteurella* spp.

Two additional *H. ducreyi* probes were developed by Chui et al. *(9)*, based on published sequences of the 16S rRNA gene of *H. ducreyi (4,5)*. Although these probes were specific for *H. ducreyi*, neither reacted with all strains of *H. ducreyi* that were tested. Because of concerns about the sensitivity and specificity of DNA probes for detection of *H. ducreyi* in specimens, several investigators have developed detection techniques based on the polymerase chain reaction (PCR).

2.2. Polymerase Chain Reaction

PCR (*see* Chapter 5) appears to offer a more sensitive and specific approach for the diagnosis of chancroid. Three PCR assays for chancroid have been developed. Chui et al. *(9)* used broad primers based on eubacterial 16S rRNA gene sequences to amplify a 303 bp sequence from members of the Pasteurellaceae and Enterobacteriaceae. Using two *H. ducreyi*-specific probes internal to this sequence, they obtained 100% sensitivity with 51 strains from six continents that were isolated over a 15-yr period. The clinical utility of PCR was compared with that of culture, using 100 clinical specimens from men with genital ulcers consistent with a clinical diagnosis of chancroid. Swab specimens were transported to the laboratory in phosphate-buffered saline containing chenodeoxycholate (1 mg/mL). After extraction, the DNA was amplified by means of a PCR protocol that involved 25 cycles of amplification, followed by use of Southern blot hybridization to detect the PCR product. A sensitivity of 83% and a specificity of 67% relative to culture were obtained. The sensitivity could be increased to 98% after three rounds of 25 cycles; however, the specificity compared with that of culture decreased to 51%.

Johnson et al. *(10)* developed a PCR assay using a pair of primers selected from sequences of an anonymous fragment of DNA cloned from *H. ducreyi*. The 1100 bp PCR product was detected on Southern transfers with a ^{32}P-labeled probe consisting of the entire cloned sequence. The specificity of the PCR assay was determined using 118 isolates of *H. ducreyi* and 25 isolates belonging to related genera or organisms found in genital ulcers. No amplification was observed for any bacterium other than *H. ducreyi*, with the exception of a single strain of *H. parainfluenzae*, which gave inefficient amplification of a 500 bp fragment that did not hybridize with the 1100 bp fragment amplified from *H. ducreyi* DNA. The utility of this assay was assessed by using specimens from 217 consecutive patients who sought treatment for genital ulcers at a sexually transmitted diseases clinic. HSV cultures, darkfield examination, and syphilis serology were performed on each patient. Among a subgroup of 183 men whose genital ulcer contained a single etiologic agent (as determined by standard laboratory tests), PCR for *H. ducreyi* had a sensitivity of 65% and a specificity of 52% relative to culture. Further studies revealed that the presence of inorganic phosphate in the collection media inhibited the activity of *Taq* polymerase *(10,11)*. In a second series of 96 men with genital ulcers, Johnson et al. *(11)* used the sample collection and preparation protocol described by Chui et al. *(9)* and obtained a PCR sensitivity of 100% and a specificity of 84% when compared with culture. The decreased specificity likely reflected the poor sensitivity of culture.

Orle et al. *(12)* reported on the development of a commercial multiplex PCR assay that permits the simultaneous amplification of DNA targets from *H. ducreyi, T. pallidum*, and HSV types-1 and -2. This assay was evaluated on genital ulcer specimens from 101 consecutive patients. With respect to chancroid, 24 of 25 culture-positive specimens were positive by PCR and an additional 11 culture-negative specimens were PCR-positive. A confirmatory PCR assay utilizing a different target gene suggested that these were not false-positive results. When commercially available, this PCR assay should improve the diagnosis of chancroid and other genital ulcer diseases.

A comparison of target genes and primer sequences for each of the three procedures are presented in **Table 1**. The procedure of Johnson et al. *(10,11)* is described in detail (*see* **Subheading 6.1.**) since it is the one with which the authors have the most experience.

3. Epidemiologic Investigation

The epidemiology of chancroid is poorly understood because of the lack of typing methods that permit differentiation between strains of *H. ducreyi*. Strains of *H. ducreyi* have been phenotypically characterized by methods such as outer membrane protein profiles *(13,14)*, indirect immunofluorescence *(15)*,

Table 1
Comparison of Target Sites and Primer Sequences for Various
***H. ducreyi* PCR Procedures**

Target gene	Primers (5'-3')	Reference
groEL	ATGGTACAGGTTTAGATGATGCCTTAGATG	*8*
	AACTACGCGTGCTTTAATTTGTGCTTCATC	
16S rRNA gene	AGGTGCTGCATGGCCTGTC	*9*
	CTAGCGATTCCGACTTCA	
unknown	CCCCGACACTTTTACACGCGCT	*10,11*
	GCCAGCCAGTGACGCCGATGCC	
16S rRNA gene	CAAGTCGAACGGTAGGCACGAAG	*12*
	TTCTGTGACTAACGTCAATCAATTTTG	

enyzme profiles *(16)*, and lectin typing *(17,18)*. However, these methods are generally unsatisfactory as they lack the power to differentiate isolates into many types. Ribotyping, which is based on restriction fragment length polymorphisms of rRNA, genes that are highly conserved and are present in multiple copies on the genome, has been used to investigate the molecular epidemiology of genetically diverse bacteria *(19,20)* (*see* Chapter 2). Sarafian et al. (21) and later Brown and Ison (22) determined the potential of ribotyping for differentiating among strains of *H. ducreyi*. The choice of restriction enzyme was found to be important as incubation of *H. ducreyi* DNA with *Eco*R1, *Cla*I, *Hin*fI, *Hae*III, *Sau*3A, *Hpa*II, *Pst*I, *Sma*I, *Bam*HI, *Kpn*I, *Not*I, *Dra*I, *Sal*I, or *Bgl*I either resulted in no digestion, incomplete digestion, or in the production of fragments that were too small to be effectively separated by the electrophoresis system used *(21,22)*. Several restriction enzymes gave banding patterns after hybridization with a rRNA gene probe that were suitable for typing. These were *Hin*dIII, *Hin*cII, *Ava*II, *Bst*EII, and *Bgl*II. Sarafian et al. *(21)* examined 44 *H. ducreyi* strains and observed eight distinct patterns (ribotypes) using *Hin*dIII, and four distinct patterns with *Hin*cII digested DNA. Analysis of 86 additional strains increased the number of *Hin*dIII ribotypes to 12 *(23)*. Combining results obtained with two different enzymes can provide additional discrimination. Brown and Ison *(22)* found that combining *Bst*EII and *Bgl*II ribotypes gave the highest index of discrimination; however, Sarafian et al. *(21)* found that the combination of *Hin*dIII and *Hin*cII provided no additional resolution. Flood et al. *(24)* used ribotyping and plasmid analysis to characterize strains isolated during an outbreak of chancroid in San Francisco, and found that this outbreak was caused by multiple strains, and that some strains were acquired as the result of sexual exposure during travel away from San Francisco.

4. Research Uses

4.1. Pathogenesis of Chancroid

An understanding of the pathogenesis of chancroidal disease is imperative so that rational intervention strategies can be devised. However, little is known about the interaction of the causative organism, *H. ducreyi*, with its host. Several potential virulence factors of *H. ducreyi* have been identified, including a unique lipo-oligosaccharide (LOS) *(25–27)*, cytotoxic and cytolytic activities *(28,29,40,42)*, pili *(30)*, a hemoglobin-binding protein *(31,32,46)*, and the ability to attach to and invade into epithelial cells and possibly fibroblasts *(33,34)*.

One way to examine the pathogenic potential of a particular virulence factor is to make isogenic mutants, differing only in expression of the virulence gene in question, and to analyze their virulence in tissue culture or animal models. Several genetic systems have been developed to manipulate *H. ducreyi* DNA. DNA is readily isolated from *H. ducreyi*, cloned into plasmids, and expressed using *Escherichia coli* promoters. Conversely DNA can be introduced back into *H. ducreyi* by cloning *H. ducreyi* DNA into the shuttle vector pLS88 *(35)* and transforming this DNA into *H. ducreyi* using electroporation *(36)*. Random mutagenesis of *H. ducreyi* genes has been accomplished using transposons Tn*916* and Tn*1545*-Δ3, allowing the identification of several genes, including a LOS biosynthesis gene and the hemolysin genes *(29,37,38)*. Finally, mutation of specific genes in *H. ducreyi* has been achieved by insertion of an antibiotic resistance cassette into the *H. ducreyi* target gene cloned into an *E. coli* plasmid that will not replicate in *H. ducreyi*, followed by electroporation of this plasmid into *H. ducreyi* with selection for the antibiotic resistance of the insertional cassette *(36)*. These techniques will be described in detail in **Subheadings 6.4.–6.6.**

Various cell lines have been used to evaluate the virulence attributes of various *H. ducreyi* strains. Primary cultures of human foreskin epithelial cells (keratinocytes) and fibroblasts have been used to show attachment of *H. ducreyi* to these cell types. Invasion into epithelial cells has been demonstrated *(29)*, but is controversial for fibroblasts *(33,39)*. The interaction of *H. ducreyi* with these two cell types are particularly relevant because they represent the cells actually affected in chancroidal ulcers. Different toxic activities of *H. ducreyi* have been defined based on the target cell type affected. A cytolytic activity was identified based on its ability to lyse red blood cells, and was thus termed a hemolysin *(29,40)*. A cell-associated cytotoxic activity for fibroblasts has been described *(28,41)* as has an extracellular cytotoxic activity directed toward HEp-2 and HeLa cells, but not fibroblasts *(42)*. Thus, the cellular location of the toxin and the target range affected indicate that the latter two toxins are separate activities. Using isogenic strains created by the techniques described in this chapter, the authors and others were able to demonstrate that the contact-dependent cytotoxin is dependent upon expression of the hemolysin *(43,43a)*.

Animal models for chancroidal disease have also been developed. Purcell et al. *(44)* showed that rabbits housed at 15–17°C develop lesions when subcutaneously inoculated with *H. ducreyi* into their shaved backs. Similarly, the ears of swine (which presumably have a lower temperature than the core temperature of swine) will develop lesions after inoculation with *H. ducreyi* *(44a)*. Humans, inoculated on their forearms with suspensions of *H. ducreyi* will elicit lesions, as will male *Macaca nemestina* primates inoculated on their foreskins *(44b,44c)*. All models have their unique advantages and disadvantages for evaluating the pathogenesis of chancroid, the contribution of different virulence factors of *H. ducreyi* to disease, and the development of immunity to infection.

4.2. Cloning and Expression of H. ducreyi DNA in E. coli

Gene libraries of *H. ducreyi* DNA cloned into plasmids in *E. coli* have been used to isolate specific genes, such as the *H. ducreyi* hemolysin *(29)*. Thus, the authors have prepared *Sau*3A libraries of *H. ducreyi* DNA in pUC19 using standard techniques, allowing the insertion of 8–10-kb fragments of *H. ducreyi* DNA *(29)*. These clones were then used to screen for hemolytic activity encoded by the *H. ducreyi* hemolysin genes. Many other *E. coli* plasmid vectors and cloning systems have been used to express *H. ducreyi* genes in *E. coli* *(32,37,45,46)* Once the appropriate *H. ducreyi* clones have been isolated in *E. coli*, standard *E. coli* techniques for expression, subcloning and mutagenesis can be performed.

4.3. Introduction of DNA into H. ducreyi by Electroporation

A major breakthrough in the study of *H. ducreyi* genes was the demonstration that DNA could be introduced into this organism by electroporation *(36)*. Thus, electroporation has been used to mutagenize *H. ducreyi* either by transposons Tn*916* or Tn*1545*-Δ3 or by marker exchange with cloned mutated genes *(25,32–34)*. Finally, electroporation can be used to introduce cloned *H. ducreyi* genes back into *H. ducreyi* using the shuttle vector pLS88, allowing complementation of mutated genes and expression of defined DNA segments in trans. The authors have successfully used the techniques outlined in this chapter (*see* **Subheadings 6.4.–6.6.**) to electroporate DNA into *H. ducreyi* strain 35000.

5. Materials

5.1. PCR

1. Dacron-tipped swabs.
2. Specimen collection buffer: 50 m*M* Tris-HCl, 1.5 *M* NaCl, 0.1 % w/v sodium chenodeoxycholate, pH 7.6.

3. Extraction buffer. 0.1 *M* Tris-HCl, 0.1 *M* NaCl, 5 m*M* EDTA, 1% w/v sodium dodecyl sulfate (SDS), 0.1% w/v sodium chenodeoxycholate.
4. Proteinase K solution. 1 mg/mL proteinase K in 25 m*M* Tris-HCl, 5 m*M* CaCl$_2$, pH 7.5.
5. Buffer-saturated phenol (Gibco BRL, Gaithesburg, MD).
6. Chloroform:isoamyl alcohol, 24:1.
7. 3 *M* sodium acetate, pH 9.0.
8. 100% ethanol.
9. PCR Buffer (10X stock). 100 m*M* Tris-HCl, 500 m*M* KCl, 15 m*M* MgCl$_2$, 0.1% w/v gelatin.
10. dNTP solutions. 10 m*M* solutions of each nucleotide prepared in 5 m*M* HEPES buffer, pH 7.4.
11. Oligonucleotide primers for PCR:
    SJ1A: 5'CCCCGACACTTTTACACGCGCT3'
    SJ2A: 5'GCCAGCCAGTGACGCCGATGCC3'.
12. *Taq* Polymerase (Perkin-Elmer, Foster City, CA).
13. Nitrocellulose or Nylon membranes.
14. (^{32}P) dCTP.
15. Hybridization solution. 0.5% w/v nonfat dry milk, 1% w/v SDS, 0.02% w/v Ficoll 400, 5 m*M* EDTA, prepared in 5X SSC.
16. 20X SSC. 3 *M* NaCl, 0.3 *M* sodium citrate, 4 m*M* Tris-HCl, pH 7.6.
17. HCV Monitor lysis buffer (Roche Pharmaceuticals, Basel, Switzerland).
18. GlasPac DNA glass-binding compound (National Scientific Supply Company, San Rafael, CA).
19. Wash buffer. 50% ethanol, 10 m*M* Tris-HCl pH 7.5, 100 m*M* NaCl.
20. TE buffer. 10 m*M* Tris-HCl pH 8.0, 0.1 m*M* EDTA.

5.2. Ribotyping

1. Incubator set to 33–35°C.
2. Horizontal Gel Electrophoresis System (20 × 20 cm). BRL Model H4 (Bethesda Research Laboratories, Gaithersburg, MD) or a similar model.
3. Plastic shield to protect body from radiation.
4. Geiger-Muller counter to monitor for possible contamination or spills.
5. GC Agar base (Becton Dickinson, Cockeysville, MD).
6. IsoVitaleX (Becton Dickinson).
7. Fetal bovine serum.
8. Hemoglobin.
9. 20X SSC: 175.3 g NaCl and 88.2 g sodium citrate per liter, pH 7.0. Store at room temperature.
10. TES buffer: 10 m*M* Tris-HCl, pH 8.0, 1 m*M* EDTA, and 100 m*M* NaCl. Autoclave at 120°C for 30 min. Store at room temperature.
11. 10% SDS.
12. Ribonuclease A (RNase A) (Boehringer Mannheim, Indianapolis, IN). Dissolve at a concentration of 10 mg/mL in 10 m*M* Tris-HCl, pH 7.5, 15 m*M* NaCl. Heat to 100°C for 15 min. Allow to cool slowly to room temperature. Dilute to 2 mg/mL in 10 m*M* Tris-HCl, pH 7.5, 15 m*M* NaCl. Dispense into aliquots and store at −20°C.

13. Proteinase K (Boehringer Mannheim), 20 mg/mL in distilled water. Dispense into aliquots and store at –20°C.
14. Buffer-saturated phenol (various suppliers).
15. 5 M sodium perchlorate.
16. Chloroform.
17. Isopropanol.
18. 70% Ethanol stored at –20°C.
19. Restriction endonucleases (*Hind*III, *Hinc*II) are supplied in concentrated form by the manufacturers (e.g, U.S. Biochemical, Cleveland, OH; BRL; Amersham International, Amersham, UK). Store at –20°C.
20. Incubation buffers (10X) are supplied by the manufacturer. Store at –20°C.
21. Sterile distilled water.
22. TE buffer: 10 mM Tris-HCl, 1 mM EDTA (pH 8.0). Autoclave at 120°C for 30 min. Store at room temperature.
23. Tris-borate (TBE) buffer for electrophoresis. 89 mM Tris-borate and 2 mM EDTA Dilute stock solution 1:10 with distilled water to make working solution.
24. 1% agarose (electrophoresis grade, ultra pure, BRL) in working solution of TBE buffer.
25. Loading dye. 0.25% bromophenol blue, 0.25% xylene cyanol, 25% Ficoll (type 400) in distilled water. Solution can be stored at room temperature.
26. Ethidium bromide (EB) stock solution. 10 mg/mL. Add 0.1 g ethidium bromide to 10 mL distilled water. Mix well and store in the dark at 4°C. Always wear gloves when handling EB powder or solution.
27. Molecular size markers. A 1-kb DNA ladder (BRL) can be used to provide molecular size markers. The size of the markers can be determined using a fluorescent ruler placed next to the gel when it is photographed; the size of the bands can be determined relative to the fluorescent ruler.
28. Transfer membrane. Nylon membranes are resistant to tearing and can be stripped and rehybridized several times. The authors use 20 × 20 cm MagnaGraph nylon transfer membranes, 0.45 u (Micron Separations, Inc., Westboro, MA).
29. 0.25 M HCl. prepare by diluting 83.3 mL of concentrated HCl with distilled water to 1 L.
30. Denaturation solution. 0.5 M NaOH, 1.5 M NaCl. Store at room temperature.
31. Neutralization solution. 1.5 M NaCl, 0.5 M Tris-HCl, pH 7.0. Store at room temperature.
32. Whatman 3MM filter paper.
33. Saran wrap.
34. DNA 5'-end labeling kit (BRL or equivalent source) containing T4-polynucleotide kinsase, buffer, and sterile deionized water.
35. 16S and 23S rRNA from *E. coli* (Boehringer Mannheim), supplied in a solution containing 2 μg/mL.
36. (γ-^{32}P)ATP, 10 pM (40 μCi).
37. 0.1 M Tris-HCl buffer, pH 9.5.
38. 3 M Sodium acetate, pH 5.5.
39. 100% ethanol, stored at –20°C.
40. Tris-EDTA-SDS buffer. 10 mM Tris-HCl, pH 8.0, 5 mM EDTA, 0.1% SDS.

41. Sephadex G-50 (medium grade). Add 5 g Sephadex G-50 to 50 mL of Tris-EDTA-SDS buffer in a 100 mL screw-capped bottle. Boil for 3 h with occasional stirring. The gel should always remain immersed in the buffer. Cool and store at 4°C until used.
42. Blue disposable plastic pipet tips (7 cm in length and 1 cm in diameter at the large end).
43. Sterile siliconized glass wool.
44. Denhardt's solution (100X): 10 g of Ficoll 400, 10 g of polyvinyl-pyrrolidine, 10 g of bovine serum albumin (Pentax Fraction V) and distilled water to a final volume of 500 mL. Filter through a disposable Nalgene filter. Dispense into 25 mL aliquots and store at –20°C.
45. 1 *M* Tris-HCl, pH 7.4
46. Formamide (deionized). Mix 50 mL of formamide and 5 g of mixed-bed ion exchange resin (AG 501- X8, 20–50 mesh; Bio-Rad, Richmond, CA). Stir for 30 min at room temperature. Filter twice through Whatman No. 1 filter paper. Dispense into 1 mL aliquots and store at –20°C.
47. Prehybridization buffer. 18 mL of 20X SSC, 6 mL of 100X Denhardt's solution, 36 mL of distilled water. This volume of buffer is sufficient for two blots.
48. Hybridization buffer. 10 mL of 20X SSC, 5 mL of 100X Denhardt's solution, 2 mL of 1 *M* Tris-HCl, pH 7.4, 1 mL of 10% SDS, 32 mL of formamide, 50 mL of distilled water. This amount of buffer is sufficient for two blots.
49. Washing solution I. 2X SSC, 0.1% SDS.
50. Washing solution II. 0.25X SSC, 0.1% SDS.

5.3. Electroporation

1. Electroporator (e.g., Bio-Rad Gene Pulser II, Hercules, CA).
2. Incubator set to 35°C.
3. *H. ducreyi* strain 35000.
4. CM (charcoal medium) plates (*see* **Subheading 6.3.1.**).
5. CM plates with antibiotics (*see* text).
6. 100X GGC. Dissolve 10 g of glucose, 1 g of glutamine, and 2.6 g of cysteine in enough water to total 100 mL. Filter sterilize, and freeze at –20°C in 10 mL aliquots (*see* **Note 1**).
7. 2X GCP. Mix 15 g of proteose peptone #3 (Difco), 1 g of starch (soluble potato, Difco), 4 g of dibasic potassium phosphate, 1 g of monobasic potassium phosphate, 5 g of NaCl, and 500 mL of distilled water. Autoclave and store at room temperature until use (*see* **Note 2**).
8. Hd broth *(47)*. 50 mL of 2X GCP, 38 mL of sterile water, 10 mL of fetal bovine serum (e.g.k Biologos, Inc, Naperville, IL), 1 mL of 100X GGC, 1 mL of catalase (10 mg/mL, Sigma #C-6665), filter sterilized. Incubate overnight at 35°C to check for sterility.
9. Sterile cold 10% glycerol.
10. Electroporation cuvets, 1 mm gap (e.g., BTX Electroporator cuvets PlusTM, BTX, San Diego, CA).

6. Methods

6.1. PCR

6.1.1. Specimen Collection

1. Collect the specimen from an ulcer with a moistened dacron-tipped swab.
2. Twirl or agitate the swab in 1 mL of collection buffer. Express the fluid from the swab by pressing it against the side of the vial.
3. Heat the collection buffer containing the specimen to 95°C for 10 min in a heat block. After the specimens have been heated and cooled, they may be frozen and stored at –70°C until analyzed by PCR.

6.1.2. Extraction of Nucleic Acids

1. Transfer 200 μL of the heat-treated specimen in collection buffer to a 1.5 mL microfuge tube. Add 180 μL of extraction buffer and 20 μL of proteinase K solution. Incubate the specimen at 37°C for 2 h.
2. Extract the aqueous specimen once with an equal volume of buffer-saturated phenol and then with an equal volume of chloroform:isoamyl alcohol.
3. Transfer the aqueous phase to a clean tube and precipitate the nucleic acids by adding 0.12 vol of 3 *M* sodium acetate and 3 vol of cold ethanol.
4. Collect the nucleic acids by centrifugation and reconstitute in 100 μL of distilled water.

6.1.3. Polymerase Chain Reaction (see **Note 3**)

1. Into an appropriate-sized microcentrifuge tube or PCR tube (tube used will depend on thermocycler used), add water (40 μL), 10X PCR buffer (10 μL), dNTP solutions (4 μL each), primers SJ1A and SJ2A (5 μL [100 pM] each), prepared specimen (22 μL), *Taq* polymerase (2 μL); total volume =100 μL.
2. Positive and negative controls for specimen preparation and amplification are included with each run to control for sensitivity and contamination. For specimen preparation, the positive control consists of 10^4 *H. ducreyi* cells suspended in collection buffer and the negative control consists of collection buffer only. Each of these is processed along with the specimens. For amplification, the positive control consists of the DNA equivalent of 10^4 *H. ducreyi* cells and the negative control contains no DNA.
3. Amplification is performed for 40 cycles under the following conditions: 1.5 min at 95°C, 2 min at 67°C, and 4 min at 72°C.

6.1.4. Detection of PCR Products (see **Note 4**)

After amplification, the reaction products (amplicons) are analysed using a dot blot.

1. Amplicons are alkali-denatured and neutralized as described by Southern **(48)** (*see* Chapter 2) and immediately applied to Nytran membranes (Schlecher and Schuell, Keene, NH).

2. Dry and bake the membranes at 80°C for 2 h.
3. Incubate membranes in hybridization buffer at 68.5°C for 1–2 h.
4. Hybridizations are carried out in sealed bags in the same mixture at 68.5°C for 16–18 h, with DNA probes (*see* **Note 5**), labeled with (^{32}P) dCTP by the random primer method *(49)* as recommended by the manufacturer (Gibco/BRL).
5. Wash membranes twice for 10 min in 5X SSC with 0.5% SDS at 68.5°C and three times for 15 min in 2X SSC with 0.5% SDS at 68.5°C.
6. The extent of hybridization is determined by autoradiography on XAR5 film (Kodak, Buffalo, NY).

6.1.5. Preparation of Other Clinical Specimens for PCR

6.1.5.1. ULCER SPECIMENS (*SEE* SUBHEADING 6.1.1.)

6.1.5.2. CULTURE (AS DESCRIBED BY CHUI ET AL. *[9]*).

1. Bacterial colonies are scraped from plates (described above) with a swab and washed twice with 12 mM Tris pH 7.6.
2. The cell suspension is adjusted to an optical density of 0.12 at 540 nm in the same Tris buffer.
3. One milliliter of the standardized suspension is pelleted and resuspended in 100 µL of Tris buffer.
4. Crude DNA is obtained by subjecting the cell suspension to three cycles of boiling and freezing. Ten microliters of each sample is used for PCR analysis.

6.1.5.3. URINE SPECIMENS (DAPTED FROM STACY-PHILLIPS ET AL. *[50]*).

1. Warm HCV Monitor lysis buffer in a 37°C water bath and mix by inversion to dissolve any precipitate.
2. Thaw urine at room temperature.
3. Using a plugged tip, add 1 mL of urine to a microcentrifuge tube and centrifuge for 5 min at 14,000 rpm.
4. Remove supernatant and add 0.6 mL of HCV Monitor lysis buffer, vortex, and incubate at 65°C for 10 min.
5. Add 25 µL of GlasPac matrix to bind the DNA, vortex and incubate at room temperature for 15 min.
6. Centrifuge 1 min at 14,000 rpm, remove and discard supernatant.
7. Resuspend GlasPac matrix in 1 mL of wash buffer, vortex, and centrifuge for 1 min at 14,000 rpm and remove supernatant.
8. Repeat wash step twice more.
9. After last wash step, remove as much supernatant as possible and dry the pellet for 15 min at room temperature with top of tube open.
10. Add 225 µL of TE buffer, vortex to resuspend pellet and incubate for 5 min at 50°C.
11. Centrifuge for 2 min at 14,000 rpm and transfer 200 µL of sample to a clean microcentrifuge tube, being careful not to disturb the GlasPac pellet.
12. The specimen is now ready for PCR analysis.

6.2. Ribotyping

6.2.1. Growth of H. ducreyi

1. *H. ducreyi* is a fastidious microorganism and requires an enriched medium for growth. We have found that chocolate agar (GC medium base containing 1% [v/v] IsoVitaleX, 1% [w/v] hemoglobin, and 10% [v/v] fetal bovine serum) is best for the growth of cells used to prepare DNA for ribotyping. Heart infusion agar base (Becton Dickinson) supplemented with 1% IsoVitaleX and 5% fetal bovine serum or Mueller Hinton agar with chocolatized horse blood (50 mL/L), 1% IsoVitaleX and 5% fetal bovine serum can also be used (*see* **Note 6**).
2. Incubate the inoculated plates in a candle jar or with 5% CO_2 at 33–35°C for 48 h.

6.2.2. Isolation of Genomic DNA

In order to obtain readable and reproducible electrophoretic patterns after cleavage by restriction endonucleases, the extracted DNA must be of high molecular weight and free from inhibitors that might interfere with endonuclease activity. Several methods have been used to obtain DNA from *H. ducreyi (22,23)*. We prefer using a modification of the method described by Brenner et al. *(51)*.

1. Remove cells from agar plates using a swab and suspend in 1.5 mL of 1X SSC. A 0.5 mL aliquot of this suspension is placed in an Eppendorf microcentrifuge tube and the cells sedimented by centrifugation at 15,600*g* for 1 min. (*see* **Note 7**).
2. Suspend the cells in 0.5 mL of TES, sediment by centrifugation, and resuspend in 0.5 mL of TES.
3. Add 30 µL of 10% SDS and 15 µL of RNase A and incubate the tubes at 37°C for 30 min.
4. Add 3 µL of proteinase K and continue incubation at 37°C for an additional 30 min.
5. Add sodium perchlorate (5 *M*) to the lysate in a ratio of 1:5; extract the DNA twice with phenol and twice with chloroform (*see* **Note 8**).
6. Precipitate the DNA with 0.6 vol of isopropanol, wash five times in ice-cold 70% ethanol, and dry before suspending in distilled water.
7. Dilute an aliquot of the DNA sample 1:20 in distilled water and measure the absorbance at 260 and 280 nm. The absorbance ratio A260/280 should be about 1.8–1.9 and A260 should be >1.0 (*see* **Note 9**).
8. The DNA concentration of the diluted sample can be estimated by its absorbance at 260 nm (a solution containing 50 µg/mL of double-stranded DNA has an absorbance of 1 at 260 nm *(52)*.
9. The DNA is digested with restriction endonucleases (*see* **Note 10**), electrophoresed through agarose gels and then transferred to nylon membrane *(48,53)* using methods described elsewhere (*see* Chapter 2).

6.2.6. Radioactive Labeling of the rRNA Probe

The method used to radiolabel the 16S and 23S rRNA probe is essentially that described by Grimont and Grimont *(54)*.

1. Pipet 25 μL of rRNA solution into a sterile microfuge tube. Add 2.5 μL of 3 *M* sodium acetate solution and 60 μL of cold ethanol. Mix by vortexing and store at −70–80°C for 30 min (or overnight at −20°C).
2. Centrifuge for 10 min in a refrigerated microcentrifuge to pellet the rRNA. Remove the supernatant taking care not to disturb the pellet.
3. Dry the pellet in a vacuum dessicator for 5 min. Release the vacuum slowly to avoid dislodging the dried rRNA.
4. Resuspend the rRNA in 25 μL of 0.1 *M* Tris-HCl buffer and heat at 90°C for 5 min then quickly cool in ice.
5. Behind a plastic bench shield add the following to the resuspended rRNA (25 μL): 8 μL of 5X exchange buffer, 3 μL of (γ-^{32}P)ATP, 2 μL of T4 polynucleotide kinase and 2 μL of sterile distilled water. Spin down in a microcentrifuge for 10 s. Incubate at 37°C for 30 min.
6. Prepare a chromatography column by plugging the small end of a blue pipet tip with a small amount of siliconized glass wool. Hold it inserted in a tube of smaller diameter. Fill the column with swollen Sephadex G-50 suspension to 5 mm from the top. It is essential that the Sephadex is constantly covered with buffer throughout. Wash the column with 5 mL of Tris-EDTA-SDS buffer.
7. The separation of the labeled rRNA from the unincorporated (γ-^{32}P)ATP by gel filtration is performed behind a plexiglass shield. Place the Sephadex column on an Eppendorf tube (1.5 mL capacity); add the 40 μL of radioactive rRNA sample on top of the column; rinse the Eppendorf tube that contained the radioactive rRNA with 60 μL of TE buffer and transfer this to the column. Immediately add 500 μL of TE buffer to the column. Collect all the eluate (~ 600 μL) in the Eppendorf tube. The eulate should contain radioactive rRNA without free nucleotides.
8. Estimate the activity of the labeled rRNA by spotting 1 μL onto a nitrocellulose filter, dry at 37°C for 30 min and place in a scintillation vial with 5 mL of scintillation fluid. Measure the radioactivity in a liquid scintillation counter (*see* **Note 11**).

6.2.7. Hybridization

Hybridization of denatured DNA fragments firmly bound to the nylon membrane requires three distinct steps: prehybridization step, hybridization step, and a washing step. The following method has been adapted from different sources *(52,57)* and is suitable for two 20 × 20 cm membranes.

1. Insert both membranes into heat-sealable plastic bags, add 20 mL of prehybridization buffer, seal, and incubate on a rocker platform for 30 min to 1 h at room temperature.
2. Open each bag at one corner and remove the prehybridization buffer. Add 20 mL of hybridization buffer and reseal the bag. Incubate on a rocker platform for 1–2 h at room temperature.
3. Open the bag again. Add the labeled probe (15–20 × 10^6 cpm), mix well, and reseal the bag. Place the bag in a water-containing plastic box in a 37°C shaking water bath. Incubate overnight while shaking.

4. Open the bag at one corner and remove the hybridization buffer. Add 50 mL of 2X SSC, 0.1% SDS solution, mix, and empty.
5. Remove the membranes with forceps and place in a plastic box containing 200 mL of 2X SSC, 0.1% SDS. Incubate at 37°C with shaking for 15 min. Repeat this procedure with 200 mL of fresh buffer.
6. Replace buffer with 0.25X SSC, 0.1% SDS and wash twice as above.
7. Dry the nylon membrane between two paper towels. Then wrap the membrane in Saran wrap and place the wrapped membrane in a film cassette containing an intensifying screen. In a dark room, insert film (Kodak X-Omat; Eastman Kodak, Rochester, N.Y.), close the cassette, and place in a freezer at –80°C. The exposure time may vary from 1–3 d (depending on the radioactivity bound to the membrane).
8. Open the cassette and develop the film in a dark room. If the film is over- or under-exposed, insert new film into the cassette and expose again for a different time keeping in mind the half-life of the ^{32}P-labeled probe.

6.2.8. Interpretation

Figure 1 demonstrates the typical patterns observed after digestion of *H. ducreyi* DNA with *Hinc*II (**Fig. 1A**) and *Hind*III (**Fig. 1B**). A set of standard strains for ribotyping (*Hind*III and *Hinc*II) are given in **Table 2**.

6.3. Electroporation

6.3.1. Preparation of CM Plates (47)

1. To prepare sufficient agar for 25 plates; add to a glass container 18 g of GC agar base (Difco),1 g of activated charcoal, and 250 mL of distilled water.
2. In another glass container, add 5 g of hemoglobin and 250 mL of distilled water
3. Mix the media in both flasks with a stirring bar until they dissolve (about 20 min), then autoclave and place in a 55°C water bath to cool (about 20 min).
4. Pour the contents of the flask containing the hemoglobin solution into the flask with the GC agar base. Aseptically add 5 mL of 100X GGC.
5. Pour 20 mL of media into each plate, then leave the plates at room temperature overnight to solidify. The next morning, package the plates, inverted, in plastic bags and place in a refrigerator until use. Incubate two plates overnight at 35°C to check for sterility of each batch (*see* **Note 12**).

6.3.2. Preparation of DNA for Electroporation

The DNA used for electroporation must be free of salts or the solution will arc when subjected to electroporation. Thus, DNA prepared by CsCl centrifugation is preferred. However, the authors have successfully transformed plasmid DNA prepared by the alkaline lysis miniprep technique *(53)* after dialysis.

6.3.2.1. PREPARATION OF DNA ISOLATED BY CsCl GRADIENTS

1. Plasmid DNA is isolated by CsCl gradient centrifugation by the standard alkaline lysis technique *(53)*. After the plasmid band is removed from the gradient, the ethidium is extracted with CsCl-saturated n-butanol.

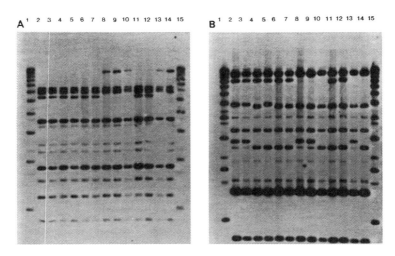

Fig. 1. Southern blots of *H. ducreyi* DNA, digested with *Hinc*II **(A)** or *Hin*dIII **(B)**, separated by agarose gel electrophoresis, and hybridized with ^{32}P-labeled 16S and 23S *E. coli* rRNA. Lanes: 1 and 15, 1-kb DNA ladder labeled with ^{32}P by nick translation; (A) *Hinc*II ribotypes: 1 (lanes 2–7, 11,12); 2 (lanes 8–10, 13, 14). (B) *Hin*dIII ribotypes: 1 (lanes 2 and 3); 2 (lanes 4,6,7,11,12); 3 (lane 5); 4 (lanes 8,9,13); 5 (lane 14).

Table 2
Standard Ribotyping Strains of *H. ducreyi*[a]

Isolate	Geographic source	Ribotype *Hin*dIII	*Hinc*II
ATCC 33940 (CIP542)	Hanoi, Viet Nam	1	1
V-1159	Seattle, WA	2	
C148	Kenya, Africa	3	
HD-181	Atlanta, GA	4	2
HD-179	Orange County, CA	5	
HD-355	Kenya, Africa	6	4
HD-239	Orange County, CA	7	
HD-342	Kenya, Africa	8	3
HD-142	San Francisco, CA	9	
HD-192	San Francisco, CA	10	
HD-302	Thailand	11	
HD-303	Thailand		5
HD-305	Thailand	12	

[a]Standard ribotyping strains may be obtained from CDC.

2. Transfer 100 μL of the DNA solution (in CsCl) to a microcentrifuge tube, add 200 μL of distilled water, mix, then add 900 μL of ethanol and stand at room temperature for 10 min.
3. Centrifuge for 10 min to pellet the DNA, remove the supernatant, then add 70% ethanol.
4. Centrifuge for 5 min to repellet the DNA, pour off the supernatant, then remove the last amount of liquid with a micropipetor.
5. Spin the tube again (10 s pulse) to collect any residual ethanol. Remove the resulting liquid with a micropipetor.
6. Suspend the DNA in 20 μL of distilled water. This DNA should be very concentrated (~1 mg/mL) and is now ready for electroporation.

6.3.2.2. SMALL-SCALE PREPARATION OF DNA BY ALKALINE LYSIS *(53)* (*SEE* CHAPTER 4)

DNA prepared by a miniprep isolation technique is not pure and the solution may arc when pulsed by the electroporator. However, the salts can be easily and rapidly removed from the plasmid preparation by dialysis using the following method:

1. Fill a plastic Petri dish with 20 mL of water.
2. Place 24 mm vs 0.25 mm filter (Millipore, Bedford MA) on top of the water with forceps.
3. Place 10–20 μL of plasmid DNA on a small area on top of the floating filter and allow it to dialyze for at least 10 min.
4. This DNA is now ready for electroporation.

6.3.3. Electroporation of DNA into H. ducreyi

1. Subculture a portion of a frozen stock culture of *H. ducreyi* strain 35000 onto a CM plate and allowed to incubate in a candle jar at 35°C for 2 d.
2. Subculture the growth from this plate to two fresh CM plates and incubate overnight.
3. Late in the following day, inoculate this growth onto four fresh CM plates (less than a week old) and, using a swab, cover the whole plate. These plates are then incubated for 15–18 h.
4. The next morning, remove the growth with a sterile cotton swab, suspend in 4 mL of sterile 10% glycerol, and transfer to four microcentrifuge tubes.
5. Centrifuge the bacterial suspension for 1 min to pellet the bacteria. Remove the supernatant, add 500 μL of fresh 10% glycerol, and suspend the bacteria by repeated pipetting with a 1-mL pipetman.
6. Repeat **step 5** three times to wash the bacteria in 10% glycerol. In the second wash, the bacterial suspension is combined into two tubes, and in third wash, combined into one tube. After the last centrifugation, remove all of the supernatant from the pellet with a micropipetor and add 10% glycerol at a volume approximately equal to the cell pellet (~ 200 μL). Suspend the bacterial pellet in the resulting small volume by repeated pipeting with a 1-mL pipetman and vortexing (*see* **Note 13**).

7. Turn on the Bio-Rad Gene Pulser II and set to 1.8 V, 200 Ohms, and 25 μF capacitance (*see* **Note 14**).
8. Prepare sterile tubes containing 1 mL of Hd broth.
9. Transfer 30 μL of the bacterial suspension to a sterile microfuge tube and mix with 2–10 μL of the plasmid DNA by pipeting up and down with the micropipetor. Transfer this mixture quickly to a 0.1-cm gap electroporator cuvet, place in the cuvette holder, and pulse (*see* **Note 15**).
10. As quickly as possible, transfer the contents of the cuvet to the prepared tube containing Hd broth. Use the sterile pipetor provided with the BTX cuvet to rinse out the cuvette with a portion of the Hd broth in this tube.
11. Incubate the tubes without shaking at 35°C for 6 h (*see* **Note 16**).
12. Transfer the contents of the tubes to a microcentrifuge tube, centrifuge for 1 min to pellet the bacteria, pour off the supernatant, and suspend the pellet in the remaining liquid.
13. The bacterial suspension is spread on the selection plate using a flamed bent glass rod. CM plates used for selection contain either 20 μg/mL kanamycin, 10 μg/mL tetracycline, or 2 μg/mL chloramphenicol, depending upon antibiotic resistance cassette used for selection. Incubate the plates for 2–3 d.
14. Colonies appearing on the plates should be transferred to a fresh selection plate and tested for the appropriate phenotype.

6.4. Mutation of H. ducreyi Genes by Transpositional Mutagenesis

Two systems of transpositional mutagensis have been used for *H. ducreyi* strain 35000: Tn*916* and Tn*1545*-Δ3 *(36,41,42)*. In the authors' experiments Tn*1545*-Δ3 yielded at least 100-fold more transposon mutants per electroporation experiment than Tn*916* *(42)*. Thus, transposon libraries containing a collection of transposon mutants in strain 35,000, can be easily prepared using Tn*1545*-Δ3. The authors have used both transposon systems to generate derivatives of *H. ducreyi* strain 35000 altered in expression of the *H. ducreyi* hemolysin (**Fig. 2**).

1. Prepare plates of CM media containing kanamycin (for use with transposon Tn*1545*-Δ3, which encodes kanamycin resistance) ; just before pouring the 500 mL of CM medium into petri dishes (*see* **Subheading 6.3.1.**), add 0.4 mL of kanamycin (25 mg/mL). The final concentration of kanamycin in the plates is 20μg/mL. The stock solution of kanamycin is prepared by mixing 250 mg of kanamycin in 10 mL of water. It is not filter sterilized and is stored at –20°C.
2. Prepare plates of CM media containing tetracycline (for use with transposon Tn*916*, which encodes tetracycline resistance) ; just before pouring the 500 mL of CM medium into Petri dishes (*see* **Subheading 6.3.1.**), add 0.25 mL of tetracycline (20 mg/mL) The final concentration of tetracycline in the plates is 10 μg/mL) The stock solution of tetracycline is prepared by mixing 200 mg of tetracycline (Sigma) in 10 mL of 50% ethanol. This solution is not filter sterilized and is stored at –20°C.

A

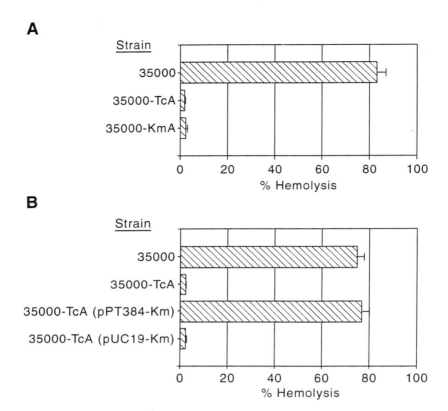

Fig. 2. Identification of the hemolysin gene of *H. ducreyi* strain 35000 by transposon mutagenesis and complementation. **(A)** Hemolytic phenotype of *H. ducreyi* 35000 and transposon mutants by the liquid hemolysis assay. Strains 35000-TcA and 35000-KmA are nonhemolytic transposon mutants generated by transposon mutagenesis with Tn*916* and Tn*1545*-Δ3, respectively. **(B)** Complementation of the hemolytic defect in strain 35000-TcA by the cloned *H. ducreyi* hemolysin gene, as determined by the liquid hemolysis assay. Strain 35000, but not 35000-TcA is hemolytic in this assay. The hemolytic defect in 35000-TcA is restored by the cloned hemolysin gene (pPT384-KmA), but not vector alone (pUC19-KmA).

3. DNA from pAM120 (containing Tn*916*) or pMGC20.1 (containing Tn*1545*-Δ3, *see* **Note 17**) is prepared as above (*see* **Subheading 6.3.2.1.**). This DNA should be at a concentration of >1 mg/mL.

4. The *H. ducreyi* strain to be mutated is electroporated with plasmid pAM120 or pMCG20.1 (*see* **Subheading 6.3.3.**). Transpositional mutants are isolated on CM plates containing 10 μg/mL tetracycline (for selection of tranposition of Tn*916* from pAM120) or 20 μg/mL kanamycin (for selection of Tn*1545*-Δ3 transposition from pMCG20.1).

5. These plates are incubated for 48–72 h, then the resulting colonies are screened for loss of the appropriate phenotype.

6.5. Directed Mutagenesis of H. ducreyi Genes

As described above, genes in *H. ducreyi* can be mutated using transposons Tn*916* and Tn*1545*-Δ3. However, once an *H. ducreyi* gene is cloned into *E. coli*, a more direct method of mutagenesis can be employed. Thus, the target *H. ducreyi* gene is cloned into plasmids in *E. coli* that do not replicate in *H. ducreyi*, (e.g., pUC18) and a portion of the gene is deleted and replaced by a chloramphenicol resistance cassette *(36)* using standard cloning techniques. The resulting plasmid is then used for electroporation into *H. ducreyi*, mutants are selected on plates containing choloramphenicol, then screened for loss of the appropriate phenotype.

1. Prepare 500 mL of media for CM plates and, just before pouring the medium into the Petri dishes, add 0.29 mL of chloramphenicol (35 mg/mL). The stock solution of chloramphenicol is prepared by dissolving 200 mg in 10 mL of methanol. It is not filter sterilized and is stored at −20°C. The final concentration of chloramphenicol in the agar medium is 2 µg/mL (*see* **Note 18**).
2. The recombinant plasmid containing the chloramphenicol cassette inserted in the target *H. ducreyi* gene is prepared as described above (*see* **Subheading 6.3.2.1.**). This DNA should be at a concentration of >1 mg/mL (*see* **Note 19**).
3. The purified plasmid is electroporated into *H. ducreyi* as previously described (*see* **Subheading 6.3.3.** and **Note 20**).
4. The electroporation mixtures are plated on CM plates containing 2 µg/mL chloramphenicol, then screened for loss of the appropriate phenotype.
5. Strains with the appropriate phenotype are further screened by Southern blot to determine whether they have the chloramphenicol cassette inserted in the appropriate place in the chromosome. The latter step is particularly important because not all chloramphenicol resistant transformants will have be the appropriate phenotype.

6.6. Complementation of H. ducreyi Genes using pLS88

Plasmid pLS88 is a suitable shuttle vector for transferring genes on plasmids into *H. ducreyi* because this plasmid will replicate in both *H. ducreyi* and *E. coli*. Thus, genes can be cloned into this plasmid in *E. coli*, then used to complement transposon mutants in *H. ducreyi*. The authors have used this vector to complement the hemolysin defect in our transposon mutants.

1. Prepare CM plates containing kanamycin (20 µg/ml, *see* **Subheading 6.4.**)
2. The *H. ducreyi* gene used to complement a mutation in *H. ducreyi* is cloned into shuttle vector pLS88, using restriction sites that do not inactivate the kanamycin gene on this plasmid.
3. As the number of transformants obtained after electroporation with pLS88 derivatives is so high, plasmid DNA prepared by the miniprep method, rather

than DNA isolated on CsCl gradients, can be used. This plasmid DNA is subjected to electroporation with the appropriate strain followed by selection on CM plates containing 20 μg/mL kanamycin.

7. Notes

1. IsoVitaleX™ (BBL), or XV factor supplement (PML microbiologicals, Tualatin, OR) can be substituted for GGC in CM plates.
2. This medium may have a slight sediment after storage. Be sure to mix before use.
3. As with any PCR procedure, when performing *H. ducreyi* PCR it is important to take extreme care in specimen and amplicon handling in order to avoid contamination. Separate rooms and/or biological hoods should be used for specimen preparation, PCR reaction set-up and end-product analysis. In between each preparation and analysis, work areas should be thoroughly cleaned with bleach (10%) to destroy any DNA present that could lead to contamination of specimens or PCR reactions.
4. An alternative detection method is to perform agarose gel electrophoresis on the amplicons to detect the 1.1-kb product. This is followed by Southern hybridization *(48)*, using the probe described below, for confirmation.
5. The probe is generated by PCR with the primers *PrSJ5A* and *PrSJ8A* using DNA from the *H. ducreyi* type strain CIP542 (the melting temperature of the primers is 62°C). After synthesis by PCR, the probe is purified by agarose gel electrophoresis. The subsequent product lacks homology with the original primers as it represents an internal portion of the target gene. An amount equivalent to 10^7 counts/mL is used for hybridization.
6. Freshly made agar plates (less than 1-wk old) are optimal for the growth of *H. ducreyi*.
7. Growth from two plates should be used for those strains that grow well, although the growth from four plates should be used for those that do not grow well.
8. Do not use buffer-saturated phenol more than 1-mo old.
9. Ratios below 1.7 indicate significant protein contamination, whereas ratios above 1.9 indicate the presence of RNA.
10. Inefficient DNA cleavage can be caused by using nonoptimal conditions or by the presence of contaminants, such as SDS, phenol, chloroform, ethanol, and salts, all of which may inhibit endonuclease activity. Contaminants can be removed by dialysis of the DNA sample against TE buffer.
11. Alternative methods of labeling probes for ribotyping include labeling rRNA using reverse transcriptase *(55)* (*see* Chapter 2) and labeling cloned rRNA genes by primer extension *(56)*.
12. The growth of *H. ducreyi* appears more luxuriant and less cohesive on fresh plates than on plates more than 1-wk old, presumably because the fresh plates are more moist. Thus, for growth of *H. ducreyi* for electroporation, the authors like to use plates that are only a few days old. The authors have found it convenient to store 500-mL aliquots of CM in glass bottles at room temperature. As media is needed, these bottles are melted in a microwave oven, cooled by immersion in a 55°C

water bath, mixed with 5 mL of 100X GGC, and poured into plates as needed.

13. This final suspension is difficult to pipet because of the high density of bacteria, but a smooth, thick suspension with no clumps, and a density of approx 1–5 × 10^{10} bacteria/mL can be achieved after repeated pipeting.

14. A Cel-Porator Electroporation System (GIBCO-BRL, Gaithersburg, MD) and a field strength 16.2 over a 0.15 cm distance has also been used for electroporation *(36)* as has a Bio-Rad Gene Pulser System set to 2.5 V, 200 Ohms, and 25 µF capacitance *(29)*.

15. If the bacterial/DNA suspension contains too much salt, it will arc when pulsed and the bacteria will not be transformed. This problem can be solved by either adding less DNA, dialyzing the DNA, or washing the bacteria a few more times.

16. This incubation allows the bacteria to express the newly acquired antibiotic resistance gene before plating on the selection medium. Other investigators have used chocolate agar plates incubated for 6 h for growth of *H. ducreyi* for expression. The authors have found the two techniques comparable.

17. Plasmid pMS1, containing Tn*1545*-Δ3 has also been used for transposon mutagenesis of *H. ducreyi (38)*. This plasmid, unlike pMGC20.1, has an antibiotic marker (chloramphenicol) in the plasmid outside of the transposon. Thus, transposon mutants can be screened for acquisition of the transposon (kanamycin resistance), but not the plasmid carrier (chloramphenicol resistance).

18. The concentration of chloramphenicol in the selection plates is critical. The authors have found that plates containing either 1 or 2 µg/mL chloramphenicol is satisfactory, although they get a slight background growth on plates containing 1 µg/mL chloramphenicol. If plates containing >5 µg/mL chloramphenicol are used for selection, the transformation efficiency is markedly reduced.

19. It is also important to use a very concentrated solution of plasmid DNA for these experiments (> 1 µg per electroporation experiment).

20. Hansen et al. *(36)* have reported that the efficiency of mutagenesis is increased when linear DNA is used. Thus, plasmid DNA is cut with a restriction enzyme that cuts within the vector sequence, and concentrated solutions of this DNA is used for electroporation.

References

1. Ortiz-Zepeda, C., Hernandez-Perez, E., and Marroquin-Burgos, R. (1994) Gross and microscopic features in chancroid: a study in 200 new culture-proven cases in San Salvador. *Sex. Transm. Dis.* **21,** 112–117.

2. Piot, P. and Plummer, F. A. (1990) Genital ulcer adenopaty syndrome, in *Sexually Transmitted Diseases,* 2nd ed. (Holmes, K. K., Mardh, P.-A., Sparling, P. F., Weisner, P. J., Cates, Jr., W., Lemon, S. M., and Stamm, W. E., eds.), McGraw-Hill, New York, pp. 711–716.

3. Casin, I., Grimont, F., Grimont, P. A. D., and Sanson-Le Pors, M.-J. (1985) Lack of deoxyribonucleic acid relatedness between *Haemophilus ducreyi* and other *Haemophilus* species. *Int. J. Syst. Bacteriol.* **35,** 23–25.

4. Rossau, R., Duhamel, M., Jannes, G., Decourt, J. L., and Van Heuverswyn, H. (1991) The development of specific rRNA-derived oligonucleotide probes for *Haemophilus ducreyi*, the causative agent of chancroid. *J. Gen. Microbiol.* **137,** 277–285.

5. Dewhirst, F. E., Paster, B. J., Olsen, I., and Fraser, G. J. (1992) Phylogeny of 54 representative strains of species in the family Pasteurellaceae as determined by comparison of 16S rRNA sequences. *J. Bacteriol.* **174,** 2002–2013.

6. Parsons, L. M., Shayegani, M., Waring, A. L., and Bopp, L. H. (1989) DNA probes for the identification of *Haemophilus ducreyi*. *J. Clin. Microbiol.* **27,** 1441–1445.

7. Parsons, L. M., Shayegani, M., Waring, A. L., and Bopp, L. H. (1990) Construction of DNA probes for the identification of *Haemophilus ducreyi* in *Gene Probes for Bacteria* (Macario, A. J., and Conway de Macario, E., eds.), Academic, New York, pp. 69–94.

8. Parsons, L. M., Waring, A. L., and Shayegani, M. (1992) Molecular analysis of the *Haemophilus ducreyi groE* heat shock operon. *Infect. Immun.* **60,** 4111–4118.

9. Chui, L., Albritton, W., Paster, B., Maclean, I., and Marusyk, R. (1993) Development of the polymerase chain reaction for diagnosis of chancroid. *J. Clin. Microbiol.* **31,** 659–664.

10. Johnson, S. R., Martin, D. H., Cammarata, C., and Morse, S. A. (1994) Development of a polymerase chain reaction assay for the detection of *Haemophilus ducreyi*. *Sex. Transm. Dis.* **21,** 13–23.

11. Johnson, S. R., Martin, D. H., Cammarata, C., and Morse, S. A. (1995) Diagnosis of chancroid by a polymerase chain reaction assay: alterations in sample preparation increase sensitivity. *J. Clin. Microbiol.* **33,** 1036–1038.

12. Orle, K. A., Gates, C. A., Martin, D. H., Body, B. A., and Weiss, J. B. (1996) Simultaneous PCR detection of *Haemophilus ducreyi, Treponema pallidum*, and Herpes Simplex Viruses Types-1 and -2 from genital ulcers. *J. Clin. Microbiol.* **34,** 49–54.

13. Odumeru, J. A., Ronald, A. R., and Albritton, W. L. (1983) Characteriztion of cell proteins of *Haemophilus ducreyi* by polyacrylamide gel electrophoresis. *J. Infect. Dis.* **148,** 710–714.

14. Taylor, D. N., Escheverria, P., Hanchalay, S., Pitarangsi, C., Slootmans, L., and Piot, P. (1985) Antimicrobial suscpetibility and characterization of outer membrane proteins of *Haemophilus ducreyi* isolated in Thailand. *J. Clin. Microbiol.* **21,** 442–444.

15. Slootmans, L., Vanden Berghe, D. A., and Piot, P. (1985) Typing *Haemophilus ducreyi* by indirect immunofluorescence assay. *Genitourin. Med.* **61,** 123–126.

16. Van Dyck, E. and Piot, P. (1987) Enzyme profiles of *Haemophilus ducreyi* strains isolated on different continents. *Eur. J. Clin. Microbiol.* **6,** 40–43.

17. Korting, H. C., Abeck, D., Johnson, A. P., Ballard, R. C., Taylor-Robinson, D., and Braun-Falco, O. (1988) Lectin typing of *Haemophilus ducreyi*. *Eur. J. Clin. Microbiol. Infect. Dis.* **7,** 676–680.

18. Schalla, W. O. and Morse, S. A. (1994) Epidemiological applications of lectins to agents of sexually transmitted diseases in *Lectin-Microorganism Interactions* (Doyle, R. J. and Slifkin, M., eds.), Marcel Dekker, New York, pp. 111–142.

19. Irino, K., Grimont,F., Casin, I., Grimont, P. A. D., and The Brazilian Purpuric Fever Study Group. (1988) rRNA gene restriction patterns of *Haemophilus influenzae* biogroup aegyptius associated with Brazilian purpuric fever. *J. Clin. Microbiol.* **26**, 1535–1538.

20. Yogev, D., Halachmi, D., Kenny, G. E., and Razin, S. (1988) Distinction of species and strains of mycoplasmas (mollicutes) by genomic DNA fingerprints with an rRNA gene probe. *J. Clin. Microbiol.* **26**, 1198–1201.

21. Sarafian, S. K., Woods, T. C., Knapp, J. S., Swaminathan, B., and Morse, S. A. (1991) Molecular characterization of *Haemophilus ducreyi* by ribosomal DNA fingerprinting. *J. Clin. Microbiol.* **29**, 1949–1954.

22. Brown, T. J. and Ison, C. A. (1993) Non-radioactive ribotyping of *Haemophilus ducreyi* using a digoxigenin labelled cDNA probe. *Epidemiol. Infect.* **110**, 289–295.

23. Ballard, R. and Morse, S. A. (1996) Chancroid in *Atlas of Sexually Transmitted Diseases and AIDS,* 2nd ed. (Morse, S. A., Moreland, A., and Holmes, K. K., eds.), Mosby-Wolfe, London, pp. 47–63.

24. Flood, J. M., Sarafian, S. K., Bolan, G. A., Lammel, C., Engelman, J., Greenblatt,R. M., Brooks, G. F., Back,A., and Morse, S. A. (1992) Multistrain outbreak of chancroid in San Francisco, 1989-1991. *J. Infect. Dis.* **167**, 1106–1111.

25. Melaugh, W., Phillips, N. J., Campagnari, A. A., Karalus, R., and Gibson, B. W. (1992) Partial characterization of the major lipooligosaccharide from a strain of *Haemophilus ducreyi*, the causative agent of chancroid, a genital ulcer disease. *J. Biol. Chem.* **267**, 13,434–13,439.

26. Odumeru, J. A., Wiseman, G. M., and Ronald, A. R. (1985) Role of lipopolysaccharide and complement in susceptibility of *Haemophilus ducreyi* to human serum. *Infect. Immun.* **50**, 495–499.

27. Odumeru, J. A., Wiseman, G. M., and Ronald, A. R. (1987) Relationship between lipopolysaccharide composition and virulence of *Haemophilus ducreyi*. *J. Med. Microbiol.* **23**, 155–162.

28. Alfa, M. J. (1992) Cytopathic effect of *Haemophilus ducreyi* for human foreskin cell culture. *J. Med. Microbiol.* **37**, 43–50.

29. Totten, P. A., Norn, D. V., and Stamm, W. E. (1995) Characterization of the hemolytic activity of *Haemophilus ducreyi. Inf. Immun.* **63**, 4409–4416.

30. Brentjens, R. J., Ketterer, M., Apicella, M. A., and Spinola, S. M. (1996) Fine tangled pili expressed by *Haemophilus ducreyi* are a novel class of pili. *J. Bacteriol.* **178**, 808–816.

31. Elkins, C. (1995) Identification and purification of a conserved heme-regulated hemoglobin-binding outer membrane protein from *Haemophilus ducreyi. Infect. Immun.* **63**, 1241–1245.

32. Elkins, C., Chen, C., and Thomas, C. E. . (1995) Characterization of the HgbA locus encoding a hemoglobin receptor from *Haemophilus ducreyi. Infect. Immun.* 2194–2200.

33. Lammel, C. J., Dekker, N. P., Palefsky, J., and Brooks, G. F. (1993) In-vitro model of *Haemophilus ducreyi* adherence to and entry into eukaryotic cells of genital origin. *J. Infect. Dis.* **167**, 642–650.

34. Totten, P. A., Lara, J. C., Norn, D. V., and Stamm, W. E. (1994) *Haemophilus ducreyi* attaches to and invades cultured human foreskin epithelial cells in vitro. *Infect. Immun.* **62,** 5632–5640.

35. Willson, P. J., Albritton, W. L., Slaney, L., and Setlow, J. K. (1989) Characterization of a multiple antibiotic resistance plasmid from *Haemophilus ducreyi. Antimicrob. Agents Chemother.* **33,** 1627–1630.

36. Hansen, E. J., Latimer, J. L., Thomas, S. E., Helminen, M., Albritton, W. L., and Radolf, J. D. (1992) Use of electroporation to construct isogenic mutants of *Haemophilus ducreyi. J. Bacteriol.* **174,** 5442–5449.

37. Palmer, K. L. and Munson, J. R. S. (1995) Cloning and characterization of the genes encoding the haemolysin of *Haemophilus ducreyi. Mol. Microbiol.* **18,** 821–830.

38. Stevens, M. K., Cope, L. D., Radolf, J. D., and Hansen, E. J. (1995) A system for generalized mutagenesis of *Haemophilus ducreyi. Infect. Immun.* **63,** 2976–2982.

39. Alfa, M. J., Degagne, P., and Hollyer, T. (1993) *Haemophilus ducreyi* adheres to but does not invade cultured human foreskin cells. *Infect. Immun.* **61,** 1735–742.

40. Palmer, K. L., Grass, S., and Munson, R. S., Jr. (1994) Identification of a hemolytic activity elaborated by *Haemophilus ducreyi. Infect. Immun.* **62,** 3041–3043.

41. Hollyer, T. T., DeGagne, P. A., and Alfa, M. J. (1994) Characterization of the cytopathic effect of *Haemophilus ducreyi. Sex. Trans. Dis.* **21,** 247–257.

42. Purven, M. and Lagergard, T. (1992) *Haemophilus ducreyi,* a cytotoxin-producing bacterium. *Infect. Immun.* **60,** 1156–1162.

43. Alfa, M. J., DeGagne, P., and Totten, P. A. (1996) *Haemophilus ducreyi* hemolysin acts as a contact-cytotoxin and damages human foreskin fibroblasts in cell culture. *Infect. Immun.* **43,** 2349–2352.

43a. Palmer, K. L., Goldman, W. E., and Munson, R. S. Jr. (1996) An isogenic haemolysin-deficient mutant of *Haemophilus ducreyi* lacks the ability to produce cytopathic effects on human foreskin fibroblasts. *Mol. Microbiol.* **21,** 13–19.

44. Purcell, B. K.,. Richardson, J. A., Radolf, J. D., and Hansen, E. J. (1991) A temperature-dependent rabbit model for production of dermal lesions by *Haemophilus ducreyi. J. Infect. Dis.* **164,** 359–367.

44a. Hobbs, M. M., San Mateo, L. R., Orndorff, P. E., Almond, G., and Kawula, T. H. (1995) Swine model of *Haemophilus ducreyi* infection. *Infect. Immun.* **63,** 3094–3100.

44b. Spinola, S. M., Wild, L. M., Apicella, M. A., Gaspari, A. A., and Campagnari, A. A. (1994) Experimental human infection with *Haemophilus ducreyi. J. Infect. Dis.* **169,** 1146–1150.

44c. Totten, P. A., Morton, W. R., Knitter, G. H., Clark, A. M., Kiviat, N. B., and Stamm, W. E. (1994) A primate model for chancroid. *J. Infect. Dis.* **169,** 1284–1290.

45. Spinola, S. M., Griffiths, G. E., Shanks, K. L., and Blake, M. S. (1993) The major outer membrane protein of *Haemophilus ducreyi* is a member of the OmpA family of proteins. *Infect. Immun.* **61,** 1346–1351.

46. Stevens, M. K., Porcella, S., Klesney-Tait, J., Lumbley, S., Thomas, S. E., Norgard, M. V., Radolf, J. D., and Hansen, E. J. (1996) A hemoglobin-binding outer membrane protein is involved in virulence expression by *Haemophilus ducreyi* in an animal model. *Infect. Immun.* **64,** 1724–1735.

47. Totten, P. A. and Stamm, W. E. (1994) Clear broth and plate media for culture of *Haemophilus ducreyi. J. Clin. Microbiol.* **32,** 2019–2023.
48. Southern, E. M. (1975) Detection of specific sequences among DNA fragments separated by gel electrophoresis. *J. Mol. Biol.* **98,** 503–517.
49. Feinberg, A. P. and Vogelstein, B. (1983) A technique for labeling DNA restriction endonuclease fragments to high specific activity. *Anal. Biochem.* **132,** 6–13.
50. Stacy-Phillips, S., Mecca, J. J., and Weiss, J. B. (1995) Multiplex PCR assay and simple preparation method for stool specimens detect enterotoxigenic *Escherichia coli* DNA during course of infection. *J. Clin. Microbiol.* **33,** 1054–1059.
51. Brenner, D. J., McWhorter, A. C., Leete Knutson, J. K., and Steigerwalt, A. G. (1982) *Escherichia vulneris*: a new species of Enterobacteriaceae associated with human wounds. *J. Clin. Microbiol.* **15,** 1133–1140.
52. Sambrook, J., Fritch, E. F., and Maniatis, T. (1989) *Molecular Cloning. A Laboratory Manual,* 2nd ed, Cold Spring Harbor Laboratory, New York.
53. Ausubel, F. M., et al. (ed). (1995) *Current Protocols in Molecualr Biology,* Wiley, New York.
54. Grimont, F., and Grimont, P. A. D. (1991) DNA fingerprinting, in *Nucleic Acid Techniques in Bacterial Systematics* (Stackebrandt, E. and Goodfellow, M., eds.), Wiley, Chichester, England, pp. 247–279.
55. Picard-Pasquier, N., Quaqued, M., Picard, B., Goullet, P., and Krishnamoorthy, R. (1989) A simple sensitive method of analyzing bacterial ribosomal DNA polymorphism. *Electrophoresis* **10,** 186–189.
56. Brosius, J., Ullrich, A., Raker, M. A., Gray, A., Dull, T. J., Gutell, R. R., and Noller, H. F. (1981) Construction and fine mapping of recombinant plasmids containing the *rrnB* ribosomal RNA operon of *Escherichia coli. Plasmid* **6,** 112–118.
57. Altwegg, M., Altwegg-Bissig, R., Demarta, A., Peduzzi, R., Reeves, M. W., and Swaminathan, B. (1988) Comparison of four typing methods for *Aeromonas* species. *J. Diarrhoeal. Dis. Res.* **6,** 88–94.

17

Mycoplasma and Ureaplasma Infections

Claire B. Gilroy and David Taylor-Robinson

1. Introduction

Organisms of the class Mollicutes (meaning soft-skin) have regressively evolved, by genome reduction, from Gram-positive bacterial ancestors, namely certain clostridia *(1)*. The taxonomy of the class Mollicutes, containing four orders, five families, and eight genera, is shown in **Table 1** *(2)* The term "mollicute" is sometimes trivially used to describe any organism in the class. The term "mycoplasma" might be used best to describe any member of the genus *Mycoplasma*, but is also used, as in this chapter, in a trivial way to refer to any organism in the class.

Mycoplasmas are the smallest and simplest prokaryotes capable of self-replication; information is provided by a genome that may be as small as 580 kb and is estimated to code for less than 500 genes *(3)*. Some of the key differences between mycoplasmas and eubacteria are outlined in **Table 2** *(4)*. Keeping the number of structural elements, metabolic pathways, and components of protein synthesis to an essential minimum, places mycoplasmas closest to the concept of "minimum cells" *(5)*. They have adopted a parasitic mode of life, securing from the host the many nutrients that they cannot synthesize and are fastidious in their growth requirements as a result of the small genome. Growth rates in broth media vary from 1 h for some ureaplasmas to 6 h for *Mycoplasma pneumoniae*. Studies of mycoplasmas, perhaps more than those of any other group of organisms, have benefited in recent years from the input of molecular biology and genetics. The need for a molecular approach has been particularly felt as classic genetics could not be applied to most mycoplasmas because of the difficulty in cultivation, as well as the use of the UGA codon to encode tryptophan instead of the (almost) universal STOP signal. Conjugation and/or cell fusion techniques are emerging, as are vectors suitable for mycoplasmal

From: *Methods in Molecular Medicine, vol. 15: Molecular Bacteriology: Protocols and Clinical Applications*
Edited by: N. Woodford and A. P. Johnson © Humana Press Inc., Totowa, NJ

Table 1
Taxonomy and Properties of Mycoplasmas (class Mollicutes; Adapted from ref. 2)

Classification	Current no. of recognized species	Genome size (kbp)	Mol% G + C content	Distinctive properties	Habitat
Order I: Mycoplasmatales					
Family I: Mycoplasmataceae					
Genus I: *Mycoplasma*	100	580–1300	23–41	Optimum growth at 37°C	Humans, animals, plants, insects
Genus II: *Ureaplasma*	6	730–1160	27–30	Urease positive	Humans, animals
Order II: Entomoplasmatales					
Family I: Entomoplasmataceae					
Genus I: *Entomoplasma*	5	790–1140	27–29	Optimum growth at 30°C	Plants, insects
Genus II: *Mesoplasma*	12	870–1100	27–30		Plants, insects
Family II: Spiroplasmataceae					
Genus I: *Spiroplasma*	17	940–2240	25–31	Helical filaments	Arthropods (including insects), plants
Order III: Acholeplasmatales					
Family I: Acholeplasmataceae					
Genus I: *Acholeplasma*	13	About 1600	27–36		Animals, plants, insects
Order IV: Anaeroplasmatales					
Family I: Anaeroplasmataceae					
Genus I: *Anaeroplasma*	4	About 1600	29–33	Oxygen-sensitive obligate anaerobes	Bovine-ovine rumen
Genus II: *Asteroleplasma*	1	About 1600	40	Oxygen-sensitive obligate anaerobes	Bovine-ovine rumen
Uncultivated, unclassified MLOs		500–1185	23–29	Uncultivated as yet	Plants, insects

Table 2
Properties Distinguishing Mycoplasmas from Eubacteria (Adapted from ref. 4)

Property	Mycoplasmas	Eubacteria
Cell wall	Absent	Present
Peptidoglycan	Absent	Present
Genome size	500–1700	>1500
Mol% G + C content of genome	23–41	25–75
No. of detectable cell proteins	About 400	1000
No. of rRNA operons	1–2	1–10
5S rRNA length (nucleotides)	104–113	>114
No. of *tuf* genes	1	1 or 2
RNA polymerase	Rifampin resistant	Rifampin sensitive
UGA codon usage	Trytophan codon in *Mycoplasma, Spiroplasma,* and *Ureaplasma* spp. but not in *Acholeplasma* spp.	Stop codon
DNA polymerase complex	One enzyme in *Mycoplasma* and *Ureaplasma* spp., three in *Acholeplasma* and *Spiroplasma* spp.	Three enzymes

gene expression. Tools used for the investigation of nucleic acids, genes, and proteins have been applied not only to the detection and characterization of mycoplasmas, but also to their taxonomic and phylogenetic properties *(6)*.

Species of *Mycoplasma* that have been isolated to date from humans are listed in **Table 3** *(7)*. *M. buccale, M. faucium, M. lipophilum, M. orale, M. salivarium,* and possibly *M. fermentans* are considered to be representative of the normal human oropharyngeal flora *(8)*; *M. salivarium* resides primarily in the gingival crevices. Mycoplasmas are also common inhabitants of the urogenital tract. *M. hominis, M. fermentans,* and *Ureaplasma urealyticum* have all been isolated from the urogenital tract of healthy individuals, and, therefore, might be considered normal flora. However, some of these species are found both in healthy individuals and in association with disease which complicates attempts to clarify their etiologic role in disease. Factors to be considered in assessing their pathogenic potential include site(s) of colonization in the genital tract, the number of organisms present, and perhaps strain differences.

As many mycoplasmas are inhabitants of the mucosal membranes of the respiratory, urogenital, or gastrointestinal tracts, direct host-to-host transmission of organisms occurs through oral-to-oral, genital-to-genital, or oral-to-genital contact. Certain mycoplasmas that are part of the normal flora of the oropharynx and lower genital tract are most likely acquired by oral-to-oral and

Table 3
Primary Sites of Colonization, Pathogenicity, and Metabolism of Mycoplasmas of Human Origin (Adapted and Updated from ref. 7)

Species	Primary site of colonization		Pathogenicity	Metabolism of:	
	Oropharynx	Genitourinary tract		Glucose	Arginine
M. buccale	+	-	-	-	+
M. faucium	+	-	-	-	+
M. fermentans	+	?+	?	+	+
M. genitalium	-	+	+	+	-
M. hominis	-	+	+	-	+
M. lipophilum	+	-	-	-	+
M. orale	+	-	-	+	+
M. penetrans	-	+	?	+	+
M. pneumoniae	+	-	+	+	-
M. pirum	?	?+	?	+	+
M. primatum	-	+	-	-	+
M. salivarium	+	-	-	-	+
M. spermatophilum	-	+	-	-	+
U. urealyticum[a]	-	+	+	-	-

[a]Metabolizes urea.

by genital-to-genital contact, respectively. However, changes in sexual practices have resulted in apparent alterations in the host tissue location of mycoplasmas, so that some commonly found in the oropharynx can occur in the ano/urogenital tract and vice versa. Sexually transmitted human diseases for which a mycoplasmal involvement has been reasonably assured are mentioned below, together with other diseases that are not considered to be sexually transmitted, but that are caused by mycoplasmas usually residing in the urogenital tract.

2. Diagnosis

Diagnostic techniques for mycoplasmas have undergone a major revolution since earlier published methodologies over a decade ago *(9,10)*. The use of improved media and the application of new molecular techniques for the detection and identification of such organisms, has seen a significant expansion in the range of hosts colonized or infected with these organisms and in the number of newly characterized *Mycoplasma* species.

The problem in establishing a role for mycoplasmas in human sexually transmitted diseases has always been complicated largely by the existence of the commensal mycoplasmal flora in the urogenital tract, and the approaches to defining such a role are outlined later. The technical procedures required to determine whether a mycoplasmal infection has occurred, are presented briefly, and are mainly those concerned with detection of the organisms. However, before any type of detection can begin, samples must be taken and considerable care is required in their collection and transport to the laboratory, particularly if culture is to be attempted (*see* **Notes 1** and **2**).

2.1. Detection of Organisms

Historically, culture was the technique most widely used for the detection of mycoplasmas, but the advent first of DNA probes and, more recently, the polymerase chain reaction (PCR) has seen considerable advances.

2.1.1. Culture

Many of the current culture media formulations for mycoplasmas are entirely based, or with only minor alterations, on the medium originally described by Derrick Edward *(11)*. However, in the late 1960s and early 1970s, this formulation was reported to be inadequate in meeting the growth requirements of the newly discovered ureaplasmas. Several major modifications in the Edward medium resulted in successful cultivation of these organisms *(12,13)*. Also in the early 1970s, spiroplasmas were discovered in various plant diseases and insect hosts, and more intense efforts were required to develop culture media for other more fastidious spiroplasmas from arthropod hosts. These efforts culminated in the development of the medium formulation designated SP-4. The reported application of this medium for primary isolation and maintenance of a variety of new and fastidious mycoplasmas of human origin since 1979, such as *M. genitalium (14)* and *M. penetrans (15)* has affirmed its overall value in meeting the nutritional needs of many different mycoplasmas. SP-4 also enhances primary isolation of other mycoplasmas of human origin, particularly *M. fermentans (16)* and *M. pneumoniae (17)*. However, although medium formulations are important, the quality of the components may be even more so. The development of a successful medium is through trial and error, and components need to be pretested for their abilities to support growth. Quality control with a fastidious isolate is important.

Inoculation of specimens into liquid medium, which is then serially diluted, followed by subculture to liquid or agar media, provides the most sensitive method for the isolation of most mycoplasmas. Jensen et al. *(18)* recently reported an alternative method in which they passaged 11 *M. genitalium*-PCR positive (primers from **ref.** *19*) urethral nongonococcal urethritis (NGU) specimens in Vero cell cultures that were monitored with a PCR assay. Vero cell-passaged material was inoculated into acellular medium when the PCR was strongly positive. This technique resulted in the isolation of four strains of *M. genitalium*, and may prove to be applicable for the isolation of other fastidious mycoplasmas.

2.1.2. DNA Probes

The first DNA probes applied to the diagnosis of mycoplasmal infections consisted of ribosomal RNA (rRNA) genes of *M. capricolum* cloned into the *Escherichia coli* plasmid pBR325. The recombinant plasmid was named pMC5 *(20)* and has become one of the most popular DNA probes in mycoplasmal studies. The highly conserved rRNA genes were effective in detecting and identifying mycoplasmas contaminating cell cultures by Southern blot hybridization with labeled pMC5 as a probe *(21)*.

To provide more specific probes, synthetic oligonucleotides 15–40 nucleotides in length complementary to variable species-specific regions of mycoplasmal 16S rRNA genes can be applied. Although the majority of rRNA probes are for detection of contamination of tissue cultures, a number of probes for the detection of human mycoplasmas are given in **Table 4** *(22–25)*. Another class of DNA probes consists of chromosomal segments specific for a certain mycoplasmal species. These segments are derived from a genomic library of the specific mycoplasma. Species-specific probes derived from genomic libraries have been developed for *M. pneumoniae* and *M. genitalium (26)*. Dotblot hybridization with these probes, labeled either radioactively or with a variety of nonradioactive molecules (e.g., digoxigenin or biotin), has enabled 10^4–10^5 colony-forming units (cfus) to be detected, a level of sensitivity that often is insufficient for use in a clinical laboratory *(27,28)*.

Table 4
Features of rRNA Oligonucleotide Probes for the Detection of Mycoplasmas Isolated from Humans

Designation	Region[a]	Specificity	Oligonucleotide sequence	Ref.
MP20	V6	*M. pneumoniae*	5' CTCTAGCCATTACCTGCTAA 3'	*22*
MP30	V3	*M. pneumoniae*[b]	5' CCACCTGTCACTCGGTTAACCTCCATTATG 3'	*22*
Mcc1	S5b-U6	*M. hominis group*[c]	5' GCCCCACTCGTAAGAGGCATG 3'	*23*
Mcc2	U6	*M. pneumoniae group*	5' GCACGTTTGCAGCCCTAGACA 3'	*23*

[a]Target region of the 16S rRNA according to the nomenclature of Gray et al. *(24)*.
[b]Cross-hybridizes with *M. genitalium*.
[c]Groups defined by Weisburg et al. *(25)*.

2.1.3. PCR

The introduction of PCR technology in the late 1980s superseded all the previously developed DNA probes and kits. PCR tests are several orders of magnitude more sensitive than those based on direct hybridization with a DNA probe. The high sensitivity of the PCR is of value when the number of organisms in a clinical specimen is small, or when there are other difficulties in culturing. PCR is fast; a single DNA sequence can be copied over a billion times within 3 h. The first reports of the application of PCR assays to the diagnosis of mycoplasmal infections appeared in 1989 *(29,30)*. Since then the number of reports on the PCR as a tool in mycoplasmal diagnosis has been rising at an exponential rate. In particular, use of PCR has increased immensely the detection of certain mycoplasmas, *M. genitalium* being a good example. This mycoplasma was isolated in 1980 from urethral specimens from 2 of 13 men with NGU *(14)*. Until very recently, however, despite the isolation of five strains of *M. genitalium* from extragenital sites together with *M. pneumoniae*, no other isolates from the urogenital tract had been reported despite several attempts at recovery *(31,32)*. Nevertheless, the undoubted existence of *M. genitalium* in the urogenital tract was shown by the use of PCR technology and, as mentioned above, Jensen et al. *(18)* have recently developed an alternative method for the isolation of this very fastidious organism. The detection of other mycoplasmas, including *U. urealyticum*, has also benefited from the use of PCR assays rather than culture. Although *U. urealyticum* is not as fastidious as *M. genitalium* and often can be isolated with ease, it can be very difficult, if not impossible, to culture from certain specimens; amniotic fluids, endotracheal aspirates of newborns *(33)*, and synovial fluid from a septic arthritis patient have been shown to be positive only by PCR, despite numerous attempts at culture. The PCR assays outlined in **Table 5** *(19,29,30,33–45)* are those for mycoplasmal species that have been found in humans. None of the technical aspects appears to be unique to mycoplasmal PCRs. For detailed descriptions of the methodology and evaluation of PCRs reference is made to Rolfs et al. *(46)*.

2.2. Sample Collection and Preparation

Similar considerations apply to specimens from any anatomical site. Well-taken throat swabs are at least as good as sputum samples for detecting *M. pneumoniae* and other respiratory mycoplasmas because of the adherence of the organisms to the respiratory mucosa. The same applies to the recovery of mycoplasmas from the male urogenital tract where urethral swabbing is the usual approach, although a urine specimen may be preferred because of its noninvasiveness.

Specimen preparation for PCR testing is an important parameter that should be optimized. Since mycoplasmas do not have a cell wall, boiling the sample after concentration of the organisms by centrifugation should be sufficient to make their DNA accessible. This may be adequate for testing cell culture samples, although treatment of the sample with detergents and proteinase K is preferable *(47)*. For urine specimens, Wang and Lo *(34)* recommend treatment of the urine sediment with proteinase K followed by heating to 95°C for 10 min. However, some samples contain undefined inhibitors of the PCR reaction that reduce the efficiency of amplification so that DNA extraction has to be employed *(48)*. This is a serious drawback that may hamper the adoption of PCR for general use in routine clinical laboratories. PCR inhibitors may be ruled out by the use of internal controls, of which the β-globin gene or HLA genes are good examples. The internal control constructed by Ursi et al. *(49)*, by inserting a fragment of foreign plasmid DNA within a *M. pneumoniae* sequence to be amplified, was effective in the detection of false-negative PCR results. In many cases, the effect of PCR inhibitors in tested specimens may be abolished, or reduced, simply by diluting the specimens *(33)*.

3. Epidemiologic Investigation

Although epidemiologic investigation is often the long-term aim of studies involving genital mycoplasmas, most of the recent reports have focused on their etiologic role, rather than their epidemiology. The molecular and other ways in which a genital mycoplasmal infection may be detected have been outlined. It is, however, important for the reader to understand what is required before a mycoplasma can be incriminated as a cause of disease, and which genital-tract diseases truly have a mycoplasmal etiology. The criteria that ideally should be fulfilled before ascribing a mycoplasma as the cause of disease are set out below and are an expansion of Koch's postulates.

1. The isolation rate from patients with the disease is significantly greater than from subjects without the disease.

Table 5
Published PCRs for Mycoplasmas Isolated from Humans

Mycoplasma	Target DNA	Primers	Probe	Reference
M. fermentans	Insertion-sequence like element	RW004 GGACTATTGTCTAAACAATTTCCC RW005 GTTATTCGATTTCTAAATCGCCT	RW006 GCTGTGGCCATTCTCTTCTTCTACGTT	(34)
		RW003 TTCTCCTGTAGTTTGATTTTGCC RW004 same as above RW006 same as probe RW006	RW005 same as primer RW005	(35)
	16S rRNA gene	RNAF1 CAGTCGATAATTTCAAATACTC RNAF2 GGTACCGTCAAAACAAAAT	RNAF3 ATGAAGATTACGGAAAAGAGCNTTTCTTCGCT GGA	(36)
M. genitalium	MgPa gene[a]	Mg1 TGTCTATGACCAGTATGTAC Mg2 CTGCTTTGGTCAAGACATCA Mg3 GTAATTAGTTACTCAGTAGA	probe made using primers Mg4 ATCAAACCCTGCTTGTAATG Mg5 ACTTGTTCCTATAGTAGTGAT	(37)
		MG1 AGTTGATGAAACCTTAACCCCTTGG MG2 CCGTTGAGGGGTTTTCCATTTTTGC	MG3 GACCATCAAGGTATTTCTCAACAGC	(19)
		MG1 same as (19) MG2 same as (19) Mg1a GGTTAACTTACCACTGGCCTTTGATC Mg2 same as (37) Mg3 same as (37)	Same probe as (19) same probe as (37)	(38)
		G3A GCTTTAAACCTGGTAACCAGATTGACT G3B GAGCGTTAGAGATCCCGTTCTGTTA	3A3B CCTTTGATTGTAACTGTT	(39)
M. hominis	16S rRNA gene	MYCHOMP ATACATGCATGTCGAAGCGAG MYCHOMN CATCTTTTAGTGCGCCTTAC	MYCHOMS CGCATGGAACCGCATGGTTCCGTTG	(40)

343

Table 5 (continued)

M. penetrans	16S rRNA gene	MYCPENTP CATGCAAGTCGGACGAAGCA MYCPENTN AGCATTTCCTCTTCTTACAA	MYCPENETS[b] CATGAGAAAATGTTTAAAGTTCGTTTG	(40)
M. pirum	16S rRNA gene	MYCPIRP TACATGCAAGTCGATCGGAT MYCPIRN CATCCTATAGCGGTCCAAAC	MYCPIRS1 CAAATGTACTATCGCATGAGAAACATTT	(41)
M. pneumoniae	P1 adhesin gene[c]	CAATGCCATCAACCCGCGCTTAACC CGTGGTTTGTTGACTGCCACTGCCG	np[d]	(30)
M. pneumoniae	tuf gene[e]	Mpn38 TACTCGTTACGACCAAATCGATAAG Mpn39 GTTCAACTGTAATCGAGGTATTG	Mpn46 TCCACGTGAGCGGAGTTAA	(42)
	cloned fragment	GAAGCTTATGTACAGGTTGG ATTACCATCCTTGTTGTAAGG	CGTAAGCTATCAGCTACATGGAGG	(29)
U. urealyticum	urease gene	U5 CAATCTGCTCGTGAAGTATTAC U4 ACGACGTCCATAAGCAACT	np	(33)
	urease gene	14b CCAGGAAAAGTAGTACCAGGAGC C72b CTCCTAATCTAACGCTATCACC	np	(43)

[a]Gene for the attachment protein designated MgPa by Hu et al. (44).
[b]Published sequence is incorrect. The correct sequence is given in the table.
[c]Gene for the attachment protein designated P1 by Hu et al. (45).
[d]np = no probe sequence was given in the reference.
[e]tuf gene encoded the elongation factor protein Tu.

2. More organisms of the particular mycoplasmal species are recovered from patients with the disease than from subjects without the disease.
3. An antibody response to the mycoplasma occurs in patients with the disease and occurs significantly more often in them than it does in subjects without the disease.
4. There is a clinical response to an antibiotic to which the mycoplasma is susceptible in vitro and the response is accompanied by elimination of the mycoplasma.
5. An antibiotic that inhibits the mycoplasma, but not other putative causal agents (a differential antibiotic) produces beneficial clinical effect.
6. The mycoplasma, when introduced into an animal species, produces disease similar to that from which it was isolated, that is susceptible to antibiotic therapy and is associated with an antibody response.
7. The mycoplasma, when given experimentally to human volunteers, produces a disease that is similar to that occurring naturally, and the induced disease is associated with an antibody response to the mycoplasma and with susceptibility to antibiotic therapy.
8. A specific antibody that is naturally induced or through immunization protects against development of the disease.

It is clear, however, that it may be difficult or, indeed, impossible to fulfill all the criteria for some diseases. This is particularly so for diseases that involve the upper genital tract. In some instances, however, a sufficient number of the criteria may be met to strongly suggest etiologic involvement.

Many conditions have been associated with infection by mycoplasmas, but using the criteria mentioned, few can be considered to have a mycoplasmal cause. However, *M. genitalium* is considered to be a cause of NGU in men by several groups *(2,48,50,51)*. *U. urealyticum* has also been attributed as a cause of epididymitis *(52)* and *M. hominis* as a cause of pelvic inflammatory disease in women *(53)*. Further details of these associations are shown in **Table 6** *(48,50–55)* and in the references mentioned therein.

There are several conditions that are not in the usual sense considered to be sexually transmitted, but are caused sometimes by mycoplasmas that by virtue of their dominant urogenital tract colonization, are sexually transmitted. Several examples are outlined in **Table 7** *(56–60)*.

4. Research Uses

The availability of molecular methods not only enhances diagnosis, but also other investigations, some of which are mentioned briefly.

4.1. Genome Sequencing

In the authors' opinion, one of the most important events in the field of mycoplasmology (if not bacteriology) was the sequencing of the entire genome of *M. genitalium* by Fraser et al. *(61)* This constitutes the first major step in the complete deciphering and molecular characterization of the machinery of a

Table 6
Sexually Transmitted Diseases Attributed to Mycoplasmas

Disease	Mycoplasma	Strength of the assoc.	Refs.
NGU	*M. genitalium*	+++	*48,50,51,54*
	U. urealyticum	++	*48*
Epididymitis	*U. urealyticum*	+	*52*
Pelvic inflammatory disease	*M. hominis*	+	*53*
Sexually acquired reactive arthritis	*U. urealyticum*	++	*55*

++++ = overwhelming; +++ = strong; ++ = moderate; + = weak.

Table 7
Other Diseases Attributed to Sexually Transmitted Mycoplasmas

Disease	Mycoplasma	Strength of assoc.[a]	Refs.
Pyelonephritis	*M. hominis*	++	*56*
Post-partum and post-abortion fever	*M. hominis*	+++	*57*
	U. urealyticum	++	
Arthritis in hypo-gammaglo-bulinaemia and in patients on immunosuppression	*M. hominis*	++++	*58,59*
	U. urealyticum	++++	
Pneumonia, chronic lung disease in very low birth weight infants	*U. urealyticum*	++	*60*

[a]As for **Table 6**.

living cell, and should help in relating structure to function. Also, a survey of the genes and their organization permits the description of a minimal set of genes required for survival.

4.2. Cultivation

As discussed above, most mycoplasmas are nutritionally fastidious, and continued improvements in culture media are likely to aid in the discovery of new species. Furthermore, cultivation is important because it is obvious that detection by molecular techniques does not provide viable organisms that may be required for antibiotic susceptibility testing and other investigations. The novel approach to improving in vitro cultivation by inoculating mycoplasmal media with cocultured eukaryotic cells that were monitored by PCR *(18)*, used for *M. genitalium*, followed the successful coculturing of previously unculturable spiroplasmas on insect cell lines by Hackett et al. *(62)*. Great care must be taken to ensure the eukaryotic cells are free of contaminating mycoplasmas initially. This is best achieved by using a generic PCR assay; the main features of two commercially available kits are shown in **Table 8**.

4.3. Pathogenicity

Most mycoplasmas tenaciously adhere to the epithelial linings of the respiratory or urogenital tract, and may be considered to be surface pathogens. With the increasing incidence of immunocompromised patients, evidence is accumulating for invasion of tissues and the intracellular location of some mycoplasmas, notably *M. fermentans, M. penetrans*, and *M. genitalium (63–65)*. The adhesion of mycoplasmas to host cells is a prerequisite for colonization by the organism and for infection. Some of the research to establish virulence factors in mycoplasmas has involved the use of molecular methods to examine adhesins, the best defined of which are those of *M. pneumoniae* and *M. genitalium (66,67)*.

4.4. Animal Models

The greater sensitivity of a PCR for *M. genitalium* compared with culture was first established by examining sequential vaginal specimens taken from experimentally infected mice *(37)*. Mice and hamsters infected in this way and examined by the PCR assay have also been used in immunologic and antibiotic studies. The use of the PCR as a diagnostic tool should allow animal experimentation with other mycoplasmas that are difficult to culture.

4.5. Mycoplasmal Genetics

Genetic studies on mycoplasmas have been hampered by the nonavailability of suitable selectable markers and gene transfer systems *(68)*. The lack of a cell wall would be expected to facilitate the introduction of foreign DNA into the cells. However, natural transfer by spontaneous mating or by cell fusion only occurs at low efficiencies *(69,70)*. Increased transformation and transfection efficiencies have been achieved in the presence of polyethylene glycol (PEG) and even better efficiencies using electroporation *(71,72)*. Major efforts have been directed at developing vectors for cloning and expressing exogenous DNA in mycoplasmas. Transfection experiments with the replicative form (RF) of *Spiroplasma citri* virus SpV1 carrying an insert of the adhesin P1 gene of *M. pneumoniae* resulted in the expression of this part of the adhesin in the spiroplasma, as *S. citri* reads UGA as tryptophan *(73)*. Attempts at producing

Table 8
Comparison of Two Commercially Available, PCR-Based Kits for the Detection of Mycoplasmas in Tissue Cultures

Supplier	Stratagene	Boehringer
Product name	Mycoplasma PCR Primer Set	Mycoplasma PCR ELISA[a]
Catalogue number	302007	1 663 925
No. of tests per kit	50	96
Mycoplasmas detected	*M. orale, M. hyorhinis, M. fermentans, M. arginini,* and *Acholeplasma laidlawi*	*M. orale, M. hyorhinis, M. fermentans, M. arginini, M. salivarium, M. gallisepticum, M. hominis, M. bovis, M. californicum, M. bovoculi, M.* Pg50 *bovis* group, *M. bovigenitalium, M. hyopneumoniae, A. laidlawi,* and *U. urealyticum*
Detection methodology	PCR amplification followed by visualization of products on agarose gel stained with ethidium bromide.	PCR amplification followed by a photometric ELISA
Sensitivity	10^3–10^4 cfu/mL	10^3 cfu/mL
Time to complete test	4–5 h	6 h

[a]May cross-react with *Clostridium* species.

shuttle vectors capable of replication both in *S. citri* and *E. coli* have been reported *(74)*. The vectors were constructed by insertion of the *S. citri* origin of replication, *oriC*, and the *tetM* determinant into ColE1-derived plasmids and could be shuttled from *E. coli* to *S. citri* and back to *E. coli*. This approach could be used in the future for studying genital-tract mycoplasmas.

5. Future Prospects and Goals

The use of the PCR has certainly enhanced studies involving mycoplasmas and has been essential for the study of *M. genitalium*. Future advances in the use of PCR technology, for example a quantitative PCR, would be extremely useful in etiological and epidemiological studies. Detection of PCR products is another area that is evolving to meet the needs of clinical laboratories. By using nonradioactive methods, coupled with automation and the microtitreplate format, very large numbers of samples could be processed in a relatively short period of time.

The ligase chain reaction (LCR), which is being used very successfully in the detection of HIV and, more recently, *Chlamydia trachomatis (24)*, might be used for the detection of various mycoplasmas once the sets of probes have been defined and optimized.

6. Notes

1. Sample collection. Swabs and other specimens should be taken without contact with antiseptics, analgesics, or lubricants, since some of these substances inhibit the growth of mycoplasmas. Swabs should be immediately expressed in mycoplasmal transport medium (or sometimes 2SP, which is a sucrose-phosphate medium for the transport of chlamydial samples) and should not be allowed to dry. Taking the medium to the location of the patient is a sound policy. The swab should not be broken off into the medium as the swab stick may contain possible inhibitors. It should be agitated in the medium, expressed against the side of the container (usually a 1.5–2 mL microcentrifuge tube), and then discarded. It is important to prevent any specimen from drying, so all samples should be inoculated into transport media as soon as possible, particularly when the volume of specimen is small. Blood samples should also be processed soon after collection, as the recovery of genital mycoplasmas declines appreciably after 1 h. An appropriate volume to collect is 2 mL of heparinized blood in 18 mL of mycoplasma growth medium or blood culture medium.
2. Sample storage. Once mycoplasmal medium has been inoculated it should be chilled to 4°C and transported to the laboratory as soon as possible, preferably within 24 h. If transportation is not possible for a few days, the inoculated medium should be frozen to –70°C or frozen in liquid nitrogen, and transported frozen. Frozen specimens may be stored, if necessary, for long periods at –70°C *(75)* or in liquid nitrogen. When the specimens are to be examined, they should be rapidly thawed in a 37°C waterbath. Specimens that have never been frozen are preferable, because there is always some loss of viability through freezing.

References

1. Woese, C. R., Maniloff, J., and Zablen, L. B. (1980) Phylogenetic analysis of the mycoplasmas. *Proc. Natl. Acad. Sci. USA* **77,** 494–498.
2. Lackey, P. C., Ennis, D. M., Cassell, G. H., Whitley, R. J., and Hook, E. W. (1995) The etiology of non-gonococcal urethritis. *Clin. Infect. Dis.* **21,** 759 (Abstract).
3. Muto, A. (1987) The genome structure of *Mycoplasma capricolum. Isr. J. Med. Sci.* **23,** 334–341.
4. Razin, S. (1992) Mycoplasma taxonomy and ecology. in *Mycoplasmas: Molecular Biology and Pathogenesis* (Maniloff, J., McElhaney, R. N., Finch, L. R., and Baseman, J. B., eds.), Amer. Soc. Microbiol. Washington D.C, pp. 3–22.
5. Morowitz, H. J. (1984) The completeness of molecular biology. *Isr. J. Med. Sci.* **20,** 750–753.
6. Bove, J. M. (1993) Molecular features of mollicutes. *Clin. Infect. Dis.* **17,** S10–S31.

7. Krause, D. C. and Taylor-Robinson, D. (1992) Mycoplasmas which infect humans, in *Mycoplasmas: Molecular Biology and Pathogenesis* (Maniloff, J., McElhaney, R. N., Finch, L. R., and Baseman, J. B., eds.), Amer. Soc. Microbiol., Washington, DC, pp. 417–444.

8. Tully, J. G. (1993) Current status of the mollicute flora of humans. *Clin. Infect. Dis.* **17**, S2–9.

9. Razin, S. and Tully, J. G. (eds.) (1983) *Methods in Mycoplasmology*, vol. 1, *Mycoplasma Characterization*, Academic, New York.

10. Razin, S. and Tully, J. G. (eds.) (1983) *Methods in Mycoplasmology*, vol. 2, *Diagnostic Microplasmology*, Academic, New York.

11. Edward, D. G. (1947) A selective medium for pleuropneumonia-like organisms. *J. Gen. Microbiol.* **1**, 238–243.

12. Freundt, E. A. (1983) Culture media for classic mycoplasmas, in *Methods in Mycoplasmology*, Academic, New York, pp. 127–135.

13. Shepard, M. C. (1983) Culture media for ureaplasmas, in *Methods in Mycoplasmology* (Razin, S. and Tully, J. G., eds.), Academic, New York, pp. 137–146.

14. Tully, J. G., Taylor-Robinson, D., Cole, R. M., and Rose, D. L. (1981) A newly discovered mycoplasma in the human urogenital tract. *Lancet* **i**, 1288–1291.

15. Lo, S., Hayes, M. M., Wang, R. Y.-H., Pierce, P. F., Kotani, H., and Shih, J.-W. (1991) Newly discovered mycoplasma isolated from patients with HIV. *Lancet* **338**, 1415–1418.

16. Dawson, M. S., Hayes, M. M., Wang, R. Y.-H., Armstrong, D., Budzko, D. B., Kundsin, R. B., and Lo, S. (1993) Detection and isolation of *Mycoplasma fermentans* from urine of HIV positive patients with AIDS. *Arch. Pathol. Lab. Med.* **117**, 511–514.

17. Tully, J. G., Rose, D. L., Whitcomb, R. F., and Wenzel, R. P. (1979) Enhanced isolation of *Mycoplasma pneumoniae* from throat washings with a newly modified culture medium. *J. Infect. Dis.* **139**, 478–482.

18. Jensen, J. S., Hansen, H. T., and Lind, K. (1996) Isolation of *Mycoplasma genitalium* strains from the male urethra. *J. Clin. Microbiol.* **34**, 286–291.

19. Jensen, J. S., Uldum, S. A., Sondergard Andersen, J., Vuust, J., and Lind, K. (1991) Polymerase chain reaction for detection of *Mycoplasma genitalium* in clinical samples. *J. Clin. Microbiol.* **29**, 46–50.

20. Amikam, D., Razin, S., and Glaser, G. (1982) Ribosomal RNA genes in *Mycoplasma. Nucleic Acids Res.* **10**, 4215–4222.

21. Razin, S., Amikam, D., and Glaser, G. (1984) Mycoplasmal ribosomal RNA genes and their use as probes for detection and identification of Mollicutes. *Isr. J. Med. Sci.* **20**, 758–761.

22. Gobel, U., Maas, R., Haun, G., Vinga Martins, C., and Stanbridge, E. J. (1987) Synthetic oligonucleotide probes complementary to rRNA for group-and species-specific detection of mycoplasmas. *Isr. J. Med. Sci.* **23**, 742–746.

23. Mattsson, J. G. and Johansson, K. E. (1993) Oligonucleotide probes complementary to 16S rRNA for rapid detection of mycoplasma contamination in cell cultures. *FEMS Microbiol. Lett.* **107**, 139–144.

24. Gray, M. W., Sankoff, D., and Cedergren, R. J. (1984) On the evolutionary descent of organisms and organelles: A global phylogeny based on a highly conserved structural core in small subunit ribosomal RNA. *Nucleic Acids Res.* **12,** 5837–5852.

25. Weisburg, W. G., Tully, J. G., Rose, D. L., Petzel, J. P., Oyaizu, H., Yang, D., Mandelco, L., Sechrest, J., Lawrence, T. G., and Van Etten, J. (1989) A phylogenetic analysis of the mycoplasmas: basis for their classification. *J. Bacteriol.* **171,** 6455–6467.

26. Hyman, H. C., Yogev, D., and Razin, S. (1987) DNA probes for detection and identification of *Mycoplasma pneumoniae* and *Mycoplasma genitalium. J. Clin. Microbiol.* **25,** 726–728.

27. Razin, S., Hyman, H. C., Nur, I. and Yogev, D. (1987) DNA probes for detection and identification of mycoplasmas (Mollicutes). *Isr. J. Med. Sci.* **23,** 735–741.

28. Marmion, B. P., Williamson, J., Worsick, D. A., Kok, T.-W., and Harris, R. J. (1993) Experience with newer techniques for the laboratory detection of *Mycoplasma pneumoniae* infection: Adelaide, 1978–1992. *Clin. Infect. Dis.* **17 (Suppl. 1),** S90–99.

29. Bernet, C., Garret, M., de Barbeyrac, B., Bébéar, C., and Bonnet, F. (1989) Detection of *Mycoplasma pneumoniae* by using the polymerase chain reaction. *J. Clin. Microbiol.* **27,** 2492–2496.

30. Jensen, J. S., Sondergard Andersen, J., Uldum, S. A., and Lind, K. (1989) Detection of *Mycoplasma pneumoniae* in simulated clinical samples by polymerase chain reaction. *Acta Path. Microbiol. Immunol. Scand.* **97,** 1046–1048.

31. Samra, Z., Borin, M., Bukowsky, Y., Lipshitz, Y., and Sompolinsky, D. (1988) Non-occurence of *Mycoplasma genitalium* in clinical specimens. *Eur. J. Clin. Microbiol. Infect. Dis.* **7,** 49–51.

32. Taylor-Robinson, D., Furr, P. M., and Hanna, N. F. (1985) Microbiological and serological study of non-gonococcal urethritis with special reference to *Mycoplasma genitalium. Genitourin. Med.* **61,** 319–324.

33. Blanchard, A., Hentschel, J., Duffy, L., Baldus, K., and Cassell, G. H. (1993) Detection of *Ureaplasma urealyticum* by polymerase chain reaction in the urogenital tract of adults, amniotic fluid and in the respiratory tract of newborns. *Clin. Infect. Dis.* **17 (Suppl. 1),** S148–S153.

34. Wang, R. Y.-H. and Lo, S. (1993) PCR detection of *Mycoplasma fermentans* infection in blood and urine, in *Diagnostic Molecular Microbiology, Principles and Applications.* (Persing,D. H., Smith, T. F., Tenover, F. C., and White A. J., eds.), Amer. Soc. Microbiol. Washington, DC, pp. 511–516.

35. Katseni, V. L., Gilroy, C. B., Ryait, B. K., Ariyoshi, K., Bieniasz, P. D., Weber, J. N., and Taylor-Robinson, D. (1993) *Mycoplasma fermentans* in individuals seropositive and seronegative for HIV-1 [see comments]. *Lancet* **341,** 271–273.

36. Blanchard, A., Hamrick, W., Duffy, L., Baldus, K., and Cassell, G. H. (1993) Use of the polymerase chain reaction for detection of *Mycoplasma fermentans* and *Mycoplasma genitalium* in the urogenital tract and amniotic fluid. *Clin. Infect. Dis.* **17 (Suppl. 1),** S272–S279.

37. Palmer, H. M., Gilroy, C. B., Furr, P. M., and Taylor-Robinson, D. (1991) Development and evaluation of the polymerase chain reaction to detect *Mycoplasma genitalium. FEMS Microbiol. Lett.* **61,** 199–203.

38. Deguchi, T., Gilroy, C. B., and Taylor-Robinson, D. (1995) Comparison of two PCR-based assays for detecting *Mycoplasma genitalium* in clinical samples. *Eur. J. Clin. Microbiol. Infect. Dis.* **14,** 629–631.

39. Cadieux, N., Lebel, P., and Brousseau, R. (1993) Use of a triplex polymerase chain reaction for the detection and differentiation of *Mycoplasma pneumoniae* and *Mycoplasma genitalium* in the presence of human DNA. *J. Gen. Microbiol.* **139,** 2431–2437.

40. Grau, O., Kovacic, R., Griffais, R., Launay, V., and Montagnier, L. (1994) Development of PCR-based assays for the detection of two human mollicute species, *Mycoplasma penetrans* and *M. hominis*. *Mol. Cell Probes.* **8,** 139–148.

41. Grau, O., Kovacic, R., Griffais, R., and Montagnier, L. (1993) Development of a selective and sensitive polymerase chain reaction assay for the detection of *Mycoplasma pirum*. *FEMS Microbiol. Lett.* **106,** 327–333.

42. Luneberg, E., Jensen, J. S., and Frosch, M. (1993) Detection of *Mycoplasma pneumoniae* by polymerase chain reaction and non-radioactive hybridisation in microtitre plates. *J. Clin. Microbiol.* **31,** 1088–1094.

43. Teng, K., Li, M., Yu, W., Li, H., Shen, D., and Liu, D. (1994) Comparison of PCR with culture for detection of *Ureaplasma urealyticum* in clinical samples from patients with urogenital infections. *J. Clin. Microbiol.* **32,** 2232–2234.

44. Hu, P. C., Schaper, U., Collier, A. M., Clyde, W. A., Horikawa, M., Huang, Y. S., and Barile, M. F. (1987) A *Mycoplasma genitalium* protein resembling *Mycoplasma pneumoniae* attachment protein. *Infect. Immun.* **55,** 1126–1131.

45. Hu, P. C., Cole, R. M., Huang, Y. S., Graham, J. A., Gardner, D. E., Collier, A. M., and Clyde, W. A. (1982) *Mycoplasma pneumoniae* infection: role of a surface protein in the attachment organelle. *Science* **216,** 313–315.

46. Rolfs, A., Schuller, I., Finckh, U., and Weber-Rolfs, I. (1992) *PCR: Clinical Diagnostics and Research*. Springer-Verlag, Berlin.

47. Spaepen, M., Angulo, A. F., Marynen, P., and Cassiman, J.-J. (1992) Detection of bacterial and mycoplasma contamination in cell cultures by polymerase chain reaction. *FEMS Microbiol. Lett.* **99,** 89–94.

48. Horner, P. J., Gilroy, C. B., Thomas, B. J., Naidoo, R. O., and Taylor-Robinson, D. (1993) Association of *Mycoplasma genitalium* with acute non-gonococcal urethritis. *Lancet* **342,** 582–585.

49. Ursi, J.-P., Ursi, D., Leven, M., and Pattyn, S. R. (1992) Utility of an internal control for the polymerase chain reaction: application to detection of *Mycoplasma pneumoniae* in clinical samples. *Acta Pathol. Microbiol. Immunol. Scand.* **100,** 635–639.

50. Jensen, J. S., Orsum, R., Dohn, B., Uldum, S., Worm, A. M., and Lind, K. (1993) *Mycoplasma genitalium*: a cause of male urethritis? *Genitourin. Med.* **69,** 265–269.

51. Janier, M., Lassau, F., Casin, I., Grillot, P., Scieux, C., Zavaro, A., Chastang, C., Bianchi, A., and Morel, P. (1995) Male urethritis with and without discharge: A clinical and microbiological study. *Sex. Trans. Dis.* **22,** 244–252.

52. Jalil, N., Doble, A., Gilchrist, C., and Taylor-Robinson, D. (1988) Infection of the epididymis by *Ureaplasma urealyticum*. *Genitourin. Med.* **64,** 367–368.

53. Miettinen, A., Paavonen, J., Jansson, E., and Leinikki, P. (1983) Enzyme immunoassay for serum antibody to *Mycoplasma hominis* in women with acute pelvic inflammatory disease. *Sex. Trans. Dis.* **10 (Suppl.),** 289–293.

54. Deguchi, T., Komeda, H., Yasuda, M., Tada, K., Iwata, H., Asano, M., Ezaki, T., and Kawada, Y. (1995) *Mycoplasma genitalium* in non-gonococcal urethritis [letter]. *Int. J. STD. AIDS* **6,** 144,145.

55. Munday, P. E. (1985), Persistent and recurrent non-gonococcal urethritis, in *Clinical Problems in Sexually Transmitted Diseases* (Taylor-Robinson, D., ed.), Martinus Nijhoff, Dordrecht, pp. 15–35.

56. Thomsen, A. C. (1978) Occurrence of mycoplasmas in the urinary tracts of patients with acute pyelonephritis. *J. Clin. Microbiol.* **8,** 84–88.

57. Eschenbach, D. A. (1993) *Ureaplasma urealyticum* and premature birth. *Clin. Infect. Dis.* **17 (Suppl.),** 100–106.

58. Furr, P. M., Taylor-Robinson, D., and Webster, A. D. (1994) Mycoplasmas and ureaplasmas in patients with hypogammaglobulinaemia and their role in arthritis: microbiological observations over twenty years. *Ann. Rheum. Dis.* **53,** 183–187.

59. Taylor-Robinson, D., Furr, P. M., and Webster, A. D. (1986) *Ureaplasma urealyticum* in the immunocompromised host. *Pediatr. Infect. Dis.* **5,** S236–238.

60. Wang, E. E. L., Cassell, G. H., Sanchez, M., Regan, J. A., Payne, N. R., and Liu, P. P. (1993) *Ureaplasma urealyticum* and chronic lung disease of prematurity: critical appraisal of the literature of causation. *Clin. Infect. Dis.* **17(Suppl. 1),** 112–116.

61. Fraser, C. M., Gocayne, J. D., White, O., Adams, M. D., Clayton, R. A., Fleischmann, R. D., Bult, C. J., Kerlavage, A. R., Sutton, G., Kell, J. M., Fritchman, J. L., Weidman, J. F., Small, K. V., Sandusky, M., Furhrmann, J., Nguyen, D., Utterback, T. R., Saudek, D. M., Phillips, C. A., Merrick, J. M., Tomb, J.-F., Dougherty, B. A., Bott, K. F., Hu, P., Lucier, T. S., Peterson, S. N., Smith, H. O., Hutchison, C. A., and Venter, J. C. (1995) The minimal gene complement of *Mycoplasma genitalium. Science* **270,** 397–403.

62. Hackett, K. J., Lynn, D. E., Williamsom, D. L., Ginsberg, A. S., and Whitcomb, R. F. (1986) Cultivation of the Drosophila sex-ratio spiroplasma. *Science* **232,** 1253–1255.

63. Stadtlander, C. T. K.-H., Watson, H. L., Simecka, J. W., and Cassell, G. H. (1993) Cytopathogenicity of *Mycoplasma fermentans* (including strain incognitus). *Clin. Infect. Dis.* **17(Suppl. 1),** S289–S301.

64. Lo, S., Hayes, M., Kotani, H., Pierce, P., Wear, D., Newton, P. I. I., Tully, J., and Shih, J. (1993) Adhesion onto and invasion into mammalian cells by *Mycoplasma penetrans*: A newly isolated mycoplasma from patients with AIDS. *Mod. Pathol.* **6,** 276–280.

65. Jensen, J., Blom, J., and Lind, K. (1994) Intracellular location of *Mycoplasma genitalium* in cultured Vero cells as demonstrated by electron microscopy. *Int. J. Exp. Pathol.* **75,** 91–98.

66. Razin, S. and Jacobs, E. (1992) Mycoplasma adhesion. *J. Gen. Microbiol.* **138,** 407–422.

67. Baseman, J. B. (1993) The cytadhesins of *Mycoplasma pneumoniae* and *M. genitalium*, in *Mycoplasma Cell Membranes* (Rottem, S. and Kahane, I., eds.), Plenum, NY, pp. 243–259.
68. Dybvig, K. (1990) Mycoplasmal genetics. *Ann. Rev. Microbiol.* **44,** 81–104.
69. Barroso, G. and Labarere, J. (1988) Chromosomal gene transfer in *Spiroplasma citri. Science* **241,** 959–961.
70. Roberts, M. C. and Kenny, G. E. (1987) Conjugal transfer of transposon Tn*916* from S*treptococcus faecalis* to *Mycoplasma hominis. J. Bacteriol.* **169,** 3836–3839.
71. King, K. W. and Dybvig, K. (1991) Plasmid transformation of *Mycoplasma mycoides* subsp. *mycoides* is promoted by high concentrations of polyethyleneglycol. *Plasmid* **26,** 108–115.
72. Hedreyda, C. T., Lee, K. K., and Krause, D. C. (1993) Transformation of *Mycoplasma pneumoniae* with Tn4001 by electroporation. *Plasmid* **30,** 170–175.
73. Marais, A., Bové, J. M., Dallo, S. F., Baseman, J. B., and Renaudin, J. (1993) Expression in *Spiroplasma citri* of an epitope carried on the G fragment of the cytadhesin P1 gene of *Mycoplasma pneumoniae. J. Bacteriol.* **175,** 2783–2787.
74. Ye, F., Renaudin, J., Bové, J. M., and Laigret, F. (1994) Cloning and sequencing of the replication origin (*oriC*) of the *Spiroplasma citri* chromosome and construction of autonomously replicating artificial plasmids. *Curr. Microbiol.* **29,** 23–29.
75. Furr, P. M. and Taylor-Robinson, D. (1990) Long-term viability of stored mycoplasmas and ureaplasmas. *J. Med. Microbiol.* **31,** 203–206.

18

Application of Molecular Methods to the Study of Infections Caused by *Salmonella* spp.

E. John Threlfall, Mike D. Hampton, and Anne M. Ridley

1. Introduction

Disease caused by any member of the genus *Salmonella* is termed salmonellosis. The type of disease and its symptoms are generally related to the infecting species and reflect the invasiveness and virulence of the organism. For example, enteric fevers are systemic diseases usually resulting from infection with *Salmonella typhi*, *S. paratyphi* A, B, or C. Salmonellosis is caused by more than 2200 different salmonella serotypes, which can be classified into three groups according to their adaptation to human and animal hosts. One group of serotypes can be regarded as those as organisms that cause enteric fever only in humans and higher primates. Members of this group, which includes *S. typhi*, *S. paratyphi* A, B, and C are restricted to humans and higher primates and are not found in food animals. A second group causes diseases in specific animals (e.g., *S. dublin*—cattle, *S. pullorum*—poultry, *S. choleraesuis*—pigs). However, when some members of this group cause infections in humans the disease is frequently invasive and can be life-threatening (e.g., *S. cholerae-suis*, *S. dublin*). The third group, which includes the great majority of the remaining 2000+ serotypes, typically causes mild-to-moderate enteritis in humans, which is often self-limiting, but which can be severe in the young, the elderly, and in patients with other underlying complications. This group includes the four serotypes most common in humans in England and Wales at the present time: *S. enteritidis*, *S. typhimurium*, *S. virchow*, and *S. hadar*. The great majority of serotypes of this third group are zoonotic in origin and have as their reservoirs animals used for food, particularly cattle, poultry, and pigs.

In England and Wales, salmonellosis accounts for about 30,000 reported infections per annum and salmonellas are second only to campylobacters as

From: *Methods in Molecular Medicine, Vol. 15: Molecular Bacteriology: Protocols and Clinical Applications*
Edited by: N. Woodford and A. P. Johnson © Humana Press Inc., Totowa, NJ

the most common cause of gastrointestinal disease at the present time *(1)*. As stated above, the most common serotypes are the zoonotic serotypes *S. enteritidis*, *S. typhimurium*, *S. virchow*, and *S. hadar*, which account for about 85% of infections in humans. *S. typhimurium* is also an important cause of disease in food animals, particularly cattle, and outbreaks caused by multidrug-resistant clones have been responsible for severe economic losses in herds over the last 20 yr *(2)*. *S. enteritidis* became the predominant serotype in humans in the late 1980s and outbreaks related to poultry and poultry products, particularly shell eggs, have been reported in many countries throughout the world *(3)*. *S. virchow* and *S. hadar* are both poultry-related serotypes, and although the disease caused by these serotypes in their food animal host may not of necessity be severe, they both cause salmonellosis in humans which, particularly for *S. virchow*, may develop into an invasive illness *(4)*.

To combat salmonellosis, laboratory-based surveillance of both human and animal infections is a necessary prerequisite of any intervention strategy. For this purpose, detailed and reproducible methods of strain identification and subdivision of strains within the most common serotypes are essential. In the PHLS Laboratory of Enteric Pathogens (LEP), which is the national center for salmonellas from humans in England and Wales, the policy has been to use phenotypic methods—serotyping—for the primary identification of strains, and phage typing and antibiogram analysis for the subdivision of serotypes of clinical or epidemiologic importance. These phenotypic methods are supplemented when appropriate by a range of molecular methods based on analysis of plasmid and chromosomal DNA (genotyping). This hierarchical approach has now been used in numerous outbreak investigations, and DNA-based methods are being increasingly used for subdivision both within serotype and phage type *(5)*.

2. Phenotypic Methods of Strain Identification

The genus *Salmonella* is defined as two taxons, namely *Salmonella enterica* and *S. bongori*, of which the primary taxon is *S. enterica*. Both taxons belong to the enterobacteriaceae and are Gram-negative, facultatively anaerobic, nonsporing motile rods. Using a range of biochemical reactions *(6,7)*, *S. enterica* is divided into six subspecies (**Tables 1** and **2**) of which the primary subspecies is Group I—*S. enterica* subsp. *enterica* Within subspecies, strains are classified into serotypes on the basis of diversity within the lipopolysaccharide (O) and flagellar protein (H) antigens *(8)* and, as stated above, over 2200 such serotypes are now recognized *(9)*. Although the great majority of these are capable of causing disease in humans, in practice only a relatively small number are of epidemiologic importance.

The primary phenotypic method for the subdivision of serotypes of clinical epidemiologic importance is phage typing and internationally-accepted phage

Table 1
Typical Biochemical Reactions for Subspecies Differentiation

Test or substrate	Subspecies						
	I	II	IIIa	IIIb	IV	V	VI
Dulcitol	+	+	−	−	−	+	d
Lactose	−	−	−	+	−	−	d
ONPG	−	−	+	+	−	+	d
Salicin	−	−	−	−	+	−	−
Sorbitol	+	+	+	+	+	+	−
Malonate	−	+	+	+	−	−	−
Mucate	+	+	+	d	−	+	+
Gelatine	−	+	+	+	+	−	+
Growth in KCN	−	−	−	−	+	+	−

d = different reactions.

Table 2
Salmonella **Subspecies**

I	*S. enterica* subsp. *Enterica*
II	*S. enterica* subsp. *Salmae*
IIIa	*S. enterica* subsp. *Arizonae*
IIIb	*S. enterica* subsp. *Diarizonae*
IV	*S. enterica* subsp. *Houtenae*
VI	*S. enterica* subsp: *Indica*
V	*S. bongori*

typing schemes have been developed for *S. typhi*, *S. paratyphi* A and B, and for the epidemiologically-important zoonotic serotypes *S. enteritidis*, *S. typhimurium*, *S. virchow*, *S. hadar*, *S. pullorum*, and *S. agona* (for a review, *see* **ref. 10**).

Antibiogram analysis can prove an invaluable additional phenotypic method for subdivision both within serotype and phage type, and in the LEP all salmonellas are screened for resistance to a range of antimicrobial agents, including ampicillin, chloramphenicol, gentamicin, kanamycin, streptomycin, sulphonamides, tetracyclines, trimethoprim, furazolidone, nalidixic acid, and ciprofloxacin *(11)*. The data provide a framework for observing changes in the incidence of drug resistance and for investigating contributory factors. For example, antibiogram analysis has been of critical importance in investigating the spread of a multiresistant strain of *S. typhimurium* definitive phage type (DT) 104, epidemic in humans and bovine animals in England and Wales since 1991 *(12–14)*, and to the increasing incidence of resistance to ciprofloxacin in zoonotic salmonellas *(15)*.

3. DNA-Based Methods

3.1. Analysis of Plasmids

Molecular typing methods based on the characterization of plasmid DNA which have been used for the differentiation of *Salmonella* include plasmid profile typing, plasmid fingerprinting, and the identification of plasmid-mediated virulence genes.

3.1.1. Plasmid Profile Typing

This method is based on the numbers and molecular weights (MWs) of carried plasmids following the extraction of partially-purified plasmid DNA and agarose gel electrophoresis (AGE; *see* Chapter 4, this volume). It has been used in support of epidemiologic investigations for differentiation both within serotype and phage type, but is restricted to those serotypes that possess plasmids and is of limited use in serotypes in which the majority of isolates contain only one plasmid. Plasmid profile typing has been particularly useful for discrimination within certain phage types of *S. typhimurium* (16,17), but has been of only limited value in *S. enteritidis*, a serotype in which the majority of isolates carry a single plasmid of 38 megadaltons (MDa) (18). Furthermore, the interpretation of results in strains with plasmids coding for drug resistance may be difficult, as such plasmids are often not an integral part of the genotype of the strain and can be lost if antibiotic selective pressure is withdrawn.

3.1.2. Plasmid Fingerprinting

Further discrimination may be achieved by cleaving plasmid DNA with restriction endonucleases. The resultant plasmid "fingerprint" may be used to discriminate between plasmids of similar MWs, or alternatively to determine the degree of molecular relatedness between plasmids of different MWs. As with plasmid profile typing, plasmid fingerprinting is limited to plasmid-carrying strains. Furthermore, carriage of multiple plasmids can make the results extremely difficult to interpret. Methods for the extraction and fingerprinting of plasmid DNA from salmonellas have been described (19,20), as well as elsewhere in this volume (*see* Chapter 4).

3.1.3. Identification of Plasmid-Mediated Virulence Genes

Certain serotypes possess plasmids regarded as "serotype-specific" (SSP) (21). Although SSPs from different serotypes are fundamentally unrelated, they all possess a highly conserved area that comprises the *salmonella plasmid viru-lence (spv)* genes responsible for the virulence for BALB/c mice (22,23). In general, such SSPs are nonconjugative and do not carry genes coding for drug resistance. The presence or absence of SSPs can be a useful epidemiologic marker and a large number of strains can be rapidly screened using a DNA probe specific for the spv genes. Further discrimination may be achieved by hybridization analysis following restriction endonuclease fingerprinting of SSP DNA *(24)*. For this purpose, a 3.5 kb pair region of the *spv* operon, which is delineated by the recognition site of the restriction enzyme *Hind*III, may be used. This 3.5 kb region carrying the *spvCDorfE* genes has been cloned from the *S. dublin* SSP and can be labeled and used as a probe by standard methods. Alternatively, oligoprobes may be constructed from the published sequences of the *spv* genes and used as probes. Details of the relevant gene sequences can be obtained from publications by Norel et al. *(25,26)* and Gulig et al. *(27,28)*.

3.2. Methods Based on the Analysis of Chromosomal DNA

The methods described above are applicable only to plasmid-carrying strains. More recent methods have sought to identify regions of heterogeneity within the bacterial chromosome. Such methods identify differences in patterns following the digestion of genomic DNA with restriction endonucleases, identify areas of heterogeneity using DNA probes that hybridize with complementary sequences in the target molecule following Southern blotting and DNA hybridization (restriction fragment length polymorphism [RFLP] typing), or utilize the polymerase chain reaction (PCR) to amplify areas of DNA between known or arbitrarily-generated sequences.

Two methods of genomic restriction enzyme fingerprinting have been used for the differentiation of salmonellas, namely total DNA restriction pattern analysis and pulsed-field gel electrophoresis (PFGE). Details of both these methods have been published previously *(29–31)*; a detailed description of the methodology for PFGE is provided elsewhere in this volume (*see* Chapter 3). Of these two methods of chromosomal fingerprinting, that which is now most commonly used for salmonellas is PFGE. Methods utilizing DNA hybridization technology to detect variation in single or multicopy gene sequences across the bacterial genome and which have been used for the subtyping of *Salmonella* include ribotyping; random cloned chromosomal sequence (RCCS) typing; and insertion sequence (IS) *200* fingerprinting.

3.2.1. PFGE

PFGE is a generic variation of DNA restriction pattern analysis that can separate linear DNA fragments of a magnitude ranging from in excess of 6 Mb pairs to less than 10 kb with excellent resolution. This is achieved by alternating the direction of the electrical current between two field orientations for a predetermined time period (i.e., the field is pulsed). For PFGE, total genomic DNA is digested with "rare-cutting" restriction enzymes, cleaving the DNA into between 5 and 30 fragments that can be separated on a single gel to provide a macrorestriction fingerprint of the whole bacterial genome. These PFGE-generated genomic fingerprints are highly reproducible and, by varying the pulse time, it is possible to increase resolution between fragments within predetermined size ranges. This method is becoming increasingly used for the fingerprinting of *Salmonella*, and has been used for subdivision both within serotype *(32–34)* and phage type *(35,36)*. An example of the use of PFGE to subdivide isolates of *S. montevideo* is shown in **Fig. 1.** As the method is reproducible, it has been proposed that PFGE could provide the basis for a definitive scheme for the genotypic subtyping of salmonellas at both a national and international level *(34)*.

3.2.2. Ribotyping

Ribotyping (*see* Chapter 2, this volume) is a form of RFLP analysis using DNA probes derived from ribosomal ribonucleic acid (rRNA) itself, or an rRNA operon (*rrn*), to detect variation in the copy number and location of 16S and/or 23S rRNA loci in restriction endonuclease-cleaved chromosomal DNA. There are normally seven *rrn* operons in most bacterial species, the RNA products of which (23S, 16S, 5S) are characterized by their sedimentation coefficient when centrifuged. Parts of the rRNA genes are very highly conserved and a single probe derived from these sequences will hybridize to homologous sequences in DNA from many bacterial species. Initially used as a taxonomic tool, DNA fragments carrying rRNA genes were detected by hybridization with an isotopically-labeled *Escherichia coli* 16 + 23S rRNA probe *(37)*. More recently, plasmid pKK3535, which contains a rRNA gene operon, has been extensively used for ribotyping *(38)*. With the advent of gene sequencing and PCR-based methodology, more specific probes have been constructed and, in particular, an internal 550 bp probe from the published sequence of the 16S *E. coli rrnB* gene has been used to provide a molecular fingerprint based on the distribution of conserved 16S *rrn* genes across the genome for several salmonella serotypes *(39–42)*.

Details of the methods used for ribotyping vary from laboratory to laborotory, but comprehensive accounts of the methods used for the determination of rRNA gene restriction patterns, and the development of bacterial species-specific rRNA probes using PCR have been published elsewhere *(43)* and in this volume (*see* Chapter 2).

3.2.3. Random Cloned Chromosomal Sequence (RCCS) Typing

Observations of heterogeneity in chromosomal restriction enzyme-generated fragment patterns can be enhanced by using randomly-cloned sequences of chromosomal DNA as single-stranded gene probes. This method of typing, which was originally developed by Tompkins and coworkers in the USA *(44)*,

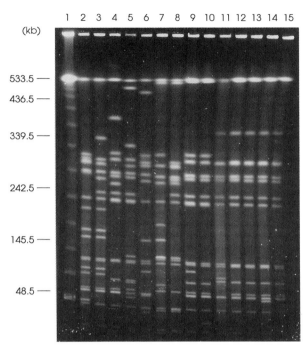

Fig. 1. *Xba*I-generated pulsed field profiles of *Salmonella montevideo*. Lanes contain: 1, Phage λ DNA (48.5 kb) concatamers (Sigma); 2–15, isolates of *S. montevideo* from humans in England and Wales, 1995–1996.

provided a method for the differentiation of *S. typhimurium*, but was of limited use for *S. dublin* and *S. enteritidis*. As yet, RCCS typing has not been fully evaluated for discrimination within phage type.

3.2.4. Insertion Sequence (IS200) Fingerprinting

Insertion sequences (ISs) are a class of mobile genetic elements found in bacteria and which contain only those genes necessary for their own transposition. They may be present in several genetic locations both in plasmids and on the chromosome. A comprehensive review of IS elements in *Salmonella* and *Mycobacterium* has recently been published *(45)*. For salmonellas, the *Salmonella*-specific 708 bp insertion sequence IS*200*, first described by Lam and Roth in 1983 *(46)*, has been shown to be distributed on conserved loci on the chromosome of many salmonella serotypes with copy numbers ranging from 1 to 25 *(47)*. In some serotypes (e.g., *S. brandenburg* *[48]* and *S. infantis* *[49]*), identification of the number and distribution of IS*200* elements in the genome has provided a method of discrimination suitable for epidemiologic investigations. In *S. typhi*, 15 clonal lineages were identified in 49 phage types, with two lineages predominating *(50)* and preliminary studies have suggested that IS*200* fingerprinting may be as discriminatory as phage typing for *S. typhimurium (51)*.

The main advantages of the RFLP-based methods described above are that they can provide fingerprints that are easy to interpret and reproduce and, for certain gene sequences, may provide information about strain phylogeny. Furthermore, the advent of gene sequencing and PCR has provided rapid methods of constructing highly specific gene probes based on the published sequences of, for example, the *E. coli rrnB* gene or IS*200*. The main disadvantages are that both *rrn* gene fingerprinting and IS*200* analysis may be nondiscriminatory. Furthermore, in the case of IS*200* fingerprinting, it must be remembered that not all serotypes or phage types possess such elements.

3.3. PCR-Based Methods

PCR-based methods that have been used for the fingerprinting of DNA in salmonella outbreaks include random amplified polymorphic DNA (RAPD) analysis, which has also been called arbitrarily-primed PCR (AP-PCR), and fingerprinting based on the presence of repetitive extragenic palindromic sequences (REPs), or enterobacterial repetitive intergenic consensus sequences (ERICs) in the bacterial genome.

3.3.1. RAPD or AP-PCR Typing

RAPD/AP-PCR typing (*see* Chapter 6, this volume) is based on the amplification of DNA fragments of unknown sequence following low stringency annealing of short (10–20 bp) oligonucleotide primers. The resulting fragments are separated by conventional AGE to produce a characteristic fingerprint. For salmonellas, RAPD/AP-PCR has been used for the differentiation of *S. enteritidis (52–54)* and *S. typhimurium (54)*. An example of the use of RAPD for the differentiation of *S. gold-coast* in an outbreak situation in the UK is shown in **Fig. 2.** RAPD/AP-PCR fingerprinting has the advantage of rapidity but as yet the reproducibility is variable.

3.3.2. REP-PCR and ERIC-PCR

Several families of short, repetitive DNA sequences have been identified in the bacterial genome, of which the best characterized are repetitive extragenic palindromic sequences (REPs), which are 38 bp long *(55)*, and enterobacterial repetitive intergenic consensus sequences (ERICs), which are 22 bp long *(56)*. Primers based on REP and ERIC sequences can be used in the amplification of *Salmonella* DNA. The bands obtained after agarose gel electrophoresis of the PCR products (REP-PCR, ERIC-PCR) represent amplification of DNA between adjacent repetitive sequences (*see* Chapter 7, this volume).

1 2 3 4 5 6 7 8 9 10 11 12 13 14 15 16 17 18 19

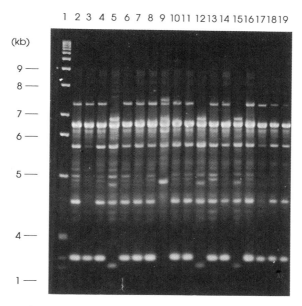

Fig. 2. RAPD profiles in *Salmonella gold-coast*. Lanes contain: 1, A 1-kb ladder (Gibco); 2–20, RAPD fingerprints of isolates of *S. gold-coast* from humans and food samples (primer sequence: ATC TGG CAG C *[58]*).

For salmonellas, the use of these two methods for strain differentiation is currently being evaluated in several laboratories. However, in a recent comparative study of RAPD and ERIC-PCR for the discrimination of *S. typhimurium* and *S. enteritidis*, it was concluded that the potential of RAPD for strain discrimination was considerably greater than that offered by ERIC-PCR *(57)*.

4. Discussion

DNA-based molecular typing and fingerprinting methods are used in the LEP for the discrimination of salmonellas within known phenotypic types (e.g., serotypes, phage types, antibiograms) and when other typing methods are inadequate or not available. Methods such as those described above are used both individually and in combination (e.g., plasmid profile, ribotyping, IS*200* fingerprinting and PFGE). However, it is essential that before use for subtyping or in an outbreak situation, the methods have been epidemiologically validated.

One of the main advantages of the use of DNA fingerprinting for the subtyping of *Salmonella* is that the degree of discrimination can often be much greater than that achieved by phenotypic methods, and can reveal subtle differences in DNA structure that cannot be shown phenotypically. However, there are certain inherent limitations in methodology that should be taken into account in the interpretation of results and their application to outbreak investigations. For example, different enzymes can give different results and may extend subdivision within defined PFGE types. It may be very difficult to interpret the significance of small differences in PFGE-generated molecular fingerprints as may result, for example, from the presence of high molecular weight plasmid DNA. Similarly, it can be difficult to evaluate the importance of small differences in band position, particularly in relation to migration through the gel, and also to decide what significance to place on differences in gel profile (i.e., how many "differences" are necessary in a fingerprint before two strains can be said to be epidemiologically-distinct). Furthermore, for some of the "new" methods of fingerprinting, such as PFGE, to be standardized and upgraded from "in house" "fingerprinting" to a general purpose typing scheme with defined reference strains, it is necessary to have access to a large and continually evolving data bank. It is also essential to have an agreed definition at an international level as to what constitutes a PFGE "fingerprint," and what may be regarded as a variation within a fingerprint.

For molecular typing schemes for *Salmonella*, or indeed any other bacterial species, to be recognized internationally it is essential to have agreed standardization of type nomenclature, methods, conditions and interpretation criteria, and even reference strains for designated molecular "types." For meaningful comparison of data, strict controls are essential. This is particularly important for PCR-based methods, in which there are inherent problems with reproducibility. As a consequence of this, although PCR-based methods may be used for "in house" typing, in general they are not as yet suitable for interlaboratory studies. However, standardization in the design of thermocyclers, coupled with improved protocols should result in methods such as RAPD/AP-PCR being increasingly used in salmonella epidemiology.

Despite these problems, for salmonellas there is little doubt that the need to produce highly reproducible molecular data rapidly coupled with increasing automation will result in the development of international databanks that can be accessed by many laboratories. Furthermore, in the not too distant future molecular methods will become the norm for the subtyping and fingerprinting of *Salmonella* for epidemiology.

References

1. Wall, P. G., de Louvois, J., Gilbert, R. J., and Rowe, B (1996) Food poisoning: notifications, laboratory reports, and outbreaks—where do the statistics come from and what do they mean? *CDR Rev.* **6,** R93–R100
2. Wray, C. and Davies, R. H. (1996) A veterinary view of salmonella in food animals. *PHLS Microbiol. Dig.* **13,** 44–48.
3. Rodrigue, D. C., Tauxe, R. V., and Rowe, B. (1990) International increase in *Salmonella enteritidis*: a new pandemic? *Epidemiol. Infect.* **105,** 21–27.

4. Threlfall, E. J., Hall, M. L. M., and Rowe, B. (1992) Salmonella bacteraemia in England and Wales. *J. Clin. Pathol.* **45**, 34–36.
5. Threlfall, E. J., Powell, N. G., and Rowe, B. (1994) Differentiation of salmonellas by molecular methods. *PHLS Microbiol. Dig.* **11**, 199–202.
6. Cowan, S. T. and Steel, L. J. (1974) in *Manual for the Identification of Medical Bacteria,* 2nd ed. Cambridge University Press, Cambridge, UK.
7. Le Minor, L. (1988) Typing of *Salmonella* species. *Eur. J. Clin. Microbiol. Infect. Dis.* **7**, 214–218.
8. Kauffmann F. (1972) *Serological Diagnosis of* Salmonella *Species.* Munksgaard, Copenhagen.
9. Rowe, B. and Hall, M. L. M. (1989) Kauffmann–White Scheme 1989. London. Central Public Health Laboratory.
10. Threlfall, E. J. and Frost, J. A. (1990) The identification, typing and fingerprinting of *Salmonella*: laboratory aspects and epidemiological application. *J. Appl. Bacteriol.* **68**, 5–16.
11. Frost, J. A. (1994) Testing for resistance to antibacterial drugs, in *Methods in Practical Laboratory Bacteriology,* (Chart, H. ed.), CRC, New York. pp. 73–82.
12. Threlfall, E. J., Frost, J. A., Ward, L. R., and Rowe, B. (1994) Epidemic in cattle of *S typhimurium* DT 104 with chromosomally-integrated multiple drug resistance. *Vet. Rec.* **134**, 577.
13. Wall, P. G., Morgan, D., Lamden, K., Ryan, M., Griffin, M., Threlfall, E. J., Ward, L. R., and Rowe, B. (1994) A case-control study of infection with an epidemic strain of multiresistant Salmonella typhimurium DT 104 in England and Wales. *Comm. Dis. Rep.* **4**, R130-135.
14. Evans, S. J. and Davies, R. H. (1996) Case control study of multiple-resistant *Salmonella typhimurium* DT104 infection of cattle in Great Britain. *Vet. Rec.* **139**, 557–558.
15. Frost, J. A., Kelleher, A., and Rowe, B. (1996) Increasing ciprofloxacin resistance in salmonellas in England and Wales, 1991–1994. *J. Antimicrob. Chemother.* **37**, 85–91.
16. Threlfall, E. J., Frost, J. A., Ward, L. R., and Rowe, B. (1990) Plasmid profile typing can be used to subdivide phage type 49 of *Salmonella typhimurium* in outbreak investigations. *Epidemiol. Infect.* **104**, 243–251.
17. Wray, C., Mclaren, I., Parkinson, N. M., and Beedell Y. (1987) Differentiation of Salmonella typhimurium DT204c by plasmid profile and biotyping *Vet. Rec.* **121**, 514–516.
18. Threlfall, E. J., Hampton, M. D., Chart, H., and Rowe B. (1994) Use of plasmid profile typing for surveillance of *Salmonella enteritidis* phage type 4 from humans, poultry and eggs. *Epidemiol. Infect.* **112**, 25–32.
19. Woodford, N., Johnson, A. P., and Threlfall, E. J. (1994) Extraction and fingerprinting of bacterial plasmids, in *Methods in Practical Laboratory Bacteriology.* (Chart, H. ed.), CRC, Boca Raton, FL, pp. 93–105.
20. Threlfall, E. J. and Woodford, N. (1995) Plasmid profile typing and plasmid fingerprinting, in *Methods in Microbiology Volume 46: Diagnostic Bacteriology Protocols,* (Howard, J. J. and Whitcombe, D. eds.), Humana, Totowa, NJ, pp. 225–236.

21. Helmuth, R., Stephan, R., Bunge, C., Hoog, B., Steinbeck, A., and Bulling, E. (1985) Epidemiology of virulence-associated plasmids and outer membrane protein patterns within seven common *Salmonella* serovars. *Infect. Immun.* **48**, 175–182.

22. Williamson, C. M., Pullinger, G. D., and Lax A. J. (1988) Identification of an essential virulence region on *Salmonella* plasmids. *Microbial Pathol.* **5**, 469–473.

23. Lax, A. J., Pullinger, G. D., Spink, J. M., Qureshi, F., Wood, M. W., and Jones, P. W. (1993) Plasmid genes involved in virulence in *Salmonella*, in *Biology of Salmonella*, (Cabello, F., Hormaeche, C., Mastroeni, P., and Bonina, L., eds.), Plenum, New York, pp. 181–190.

24. Hampton, M. D., Threlfall, E. J., Frost, J. A., Ward, L. R., and Rowe, B. (1995) *Salmonella typhimurium* DT 193: differentiation of an epidemic phage type by antibiogram, plasmid profile, plasmid fingerprint and *s*almonella *p*lasmid virulence (*spv*) gene probe. *J. Appl. Bacteriol.* **78**, 402–408.

25. Norel, F., Pisano, M. -R., Nicoli, J., and Popoff, M. Y. (1989a) Nucleotide sequence of the plasmid-borne virulence gene *mkfA* encoding a 28kDa polypeptide from *Salmonella typhimurium*. *Res. Microbiol.* **140**, 263–265.

26. Norel, F., Pisano, M. -R., Nicoli, J., and Popoff, M. Y. (1989b) Nucleotide sequence of the plasmid-borne virulence gene *mkfB* from *Salmonella typhimurium*. *Res. Microbiol.* **140**, 263–265.

27. Gulig, P. A. and Chiodo, V. A. (1990) Genetic and DNA sequence analysis of the *Salmonella typhimurium* virulence plasmid gene encoding the 28,000-molecular weight protein. *Infect. Immun.* **58**, 2561–2568.

28. Gulig, P. A., Caldwell, A. L., and Chiodo, V. A. (1990) Identification, genetic analysis and DNA sequence of a 7.8 kb virulence region of the *Salmonella typhimurium* virulence plasmid. *Mol. Microbiol.* **6**, 1395–1441.

29. Giovannetti, L. and Ventura, S. (1995) Application of total DNA restriction pattern analysis to identification and differentiation of bacterial strains, in *Methods in Microbiology, vol. 46: Diagnostic Bacteriology Protocols,* (Howard, J. J. and Whitcombe, D., eds.), Humana, Towata, NJ, pp. 165–179.

30. Hillier, A. J. and Davidson, B. E. (1995) Pulsed field gel electrophoresis, in *Methods in Microbiology, vol. 46: Diagnostic Bacteriology Protocols,* (Howard, J. J. and Whitcombe, D., eds.), Humana, Towata, NJ, pp. 149–164.

31. Kaufmann, M. E. and Pitt, T. L. (1994) Pulsed-field gel electrophoresis, in *Methods in Practical Laboratory Bacteriology,* (Chart, H. ed.), CRC, Boca Raton, FL, pp. 123–138.

32. Thong, K. -L., Cheong, Y. -M., Puthucheary, S. Koh, C. -L., and Pang T. (1994) Epidemiologic analysis of sporadic Salmonella typhi isolates and those from outbreaks by pulsed-field gel electrophoresis. *J. Clin. Microbiol.* **32**, 1135–1141.

33. Threlfall, E. J., Hampton, M. D., Ward, L. R., and Rowe, B. (1996) Application of pulsed-field gel electrophesis to an international outbreak of *Salmonella agona*. *Emerging Infect. Dis.* 130–132.

34. Punia, P., Ridley, A. M., Hampton, M. D., Ward, L. R. Rowe, B., and Threlfall, E. J. (1997) Pulsed-field electrophoretic fingerprinting of *Salmonella indiana* and its epidemiological applicability. *J. Appl. Microbiol.,* in press.

35. Powell, N. G., Threlfall, E. J., Chart, H., and Rowe, B. (1994) Subdivision of *Salmonella enteritidis* PT 4 by pulsed-field gel electrophoresis: potential for epidemiological surveillance. *FEMS Microbiol. Lett.* **119**, 193–198.

36. Powell, N. G., Threlfall, E. J., Chart, H., Schofield, S. L., and Rowe, B. (1995) Correlation of change in phage type with pulsed-field profile and 16S *rrn* profile in *Salmonella enteritidis* phage types 4, 7 and 9a. *Epidemiol. Infect.* **114**, 403–411.

37. Grimont, F. and Grimont, P. A. D. (1986) Ribosomal ribonucleic acid gene restriction patterns as potential taxonomic tools. *Ann. Inst. Pasteur* **137B**, 165–175.

38. Martinetti, G. and Altwegg, M. (1990) rRNA gene restriction patterns and plasmid analysis as a tool for typing *Salmonella enteritidis*. *Res. Microbiol.* **141**, 1151–1162.

39. Altwegg, M., Hickman-Brenner, F. W., and Farmer, J. J., III. (1989) Ribosomal RNA gene restriction patterns provide increased sensitivity for typing *Salmonella typhi* strains. *J. Inf. Dis.* **160**, 145–149.

40. Stanley, J., Chowdry-Baquar, N., and Threlfall, E. J. (1993) Genotypes and phylogenetic relationships of *Salmonella typhimurium* are defined by molecular fingerprinting of IS*200* and 16S *rrn* loci. *J. Gen. Microbiol.* **139**, 1133–1140.

41. Gruner, E., Martinetti, G., Lucchini, G., Hoop, R. K., and Altwegg, M. (1994) Molecular epidemiology of Salmonella enteritidis. *Eur. J. Epidemiol.* **10**, 85–89.

42. Powell, N. G., Threlfall, E. J., Chart, H., Schofield, S. L., and Rowe, B. (1995) Correlation of change in phage type with pulsed field profile and 16S rrn profile in *Salmonella enteritidis* phage types 4, 7 and 9a. *Epidemiol. Infect.* **114**, 403–411.

43. Grimont, F., and Grimont, P. (1995) Determination of rRNA gene restriction patterns, in *Methods in Microbiology Volume 46: Diagnostic Bacteriology Protocols*, (Howard, J. J. and Whitcombe, D., eds.), Humana, Totowa, NJ, pp. 181–200.

44. Tompkins, L. S., Troup, N., Labaigne-Roussel, A., Cohen, M. L. (1986) Cloned, random chromosomal sequences as probes to identify *Salmonella* species. *J. Infect. Dis.* **152**, 156–162.

45. Stanley, J. and Saunders, N. (1996) DNA insertion sequences and the molecular epidemiology of *Salmonella* and *Mycobacterium*. *J. Med. Microbiol.* **45**, 236–251.

46. Lam, S. and Roth, J. R. (1983) IS*200*: a *Salmonella*-specific insertion sequence. *Cell* **34**, 951–960.

47. Gibert, I., Barbe, J., and Casadesus, J. (1990) Distribution of insertion sequence IS200 in *Salmonella* and *Shigella*. *J. Gen. Microbiol.* **36**, 2555–2560.

48. Baquar, N., Burnens, A., and Stanley, J. (1994) Comparative evaluation of molecular typing of strains from a national epidemic due to *Salmonella brandenburg* by rRNA gene and IS*200* probes and pulsed-field gel electrophoresis. *J. Clin. Microbiol.* **32**, 1876–1880.

49. Pelkonen, S., Romppanen, E. -L., Siitonen, A., and Pelkonen, J. (1994) Differentiation of *Salmonella* serovar *infantis* from human and animal sources by fingerprinting IS*200* and 16S *rrn* loci. *J. Clin. Microbiol.* **32**, 2128–2133.

50. Threlfall, E. J., Torre, E., Ward, L. R., Dávalos-Pérez, A., Rowe, B., and Gibert, I. (1994) Insertion sequence IS*200* fingerprinting of *Salmonella typhi*: an assessment of epidemiological applicability. *Epidemiol. Infect.* **112**, 253–261.

51. Stanley, J. S., Chowdry-Baquar, N., and Threlfall, E. J. (1993) Genotypes and phylogenetic relationships of Salmonella typhimurium are defined by molecular fingerprinting of IS*200* and 16S *rrn* loci. *J Gen Microbiol* **139,** 1133–1140.

52. Fadl, A. A., Nguyen, A. V., and Khan, M. I (1995) Analysis of *Salmonella enteritidis* isolates by arbitrarily primed PCR. *J. Clin. Microbiol.* **33,** 987–989.

53. Lin, A. W., Usera, M. A., Barrett, T. J., and Goldsby, R. A. (1996) Application of random amplified polymorphic DNA analysis to differentiate strains of *Salmonella enteritidis*. *J. Clin. Microbiol.* **34,** 870–876.

54. Hilton, A. C., Banks, J. G., and Penn, C. W. (1996) Random amplification of polymorphic DNA (RAPD) of *Salmonella*: strain differentiation and characterisation of amplified sequences. *J. Appl. Bacteriol.* **81,** 575–584.

55. Stern, M. J., Ames, G. L. F., Smith, N. H., Robinson, E. C., and Higgins, C. F. (1984) Short, interspersed repetitive DNA sequences in prokaryotic genomes. *J. Bacteriol.* **174,** 4525–4529.

56. Versalovic, J., Koeuth, T., and Lupski, J. R. (1991) Distribution of repetitive DNA sequences in eubacteria and application to fingerprinting of bacterial genomes. *Nucleic Acids. Res.* **19,** 6823–6831

57. Millemann,Y., Lesages-Descauses, M. -C., Lafont, J. -P., and Chaslus-Dancla, E. (1996) Comparison of random amplified polymorphic DNA analysis and enterobacterial repetitive intergenic consensus-PCR for epidemiological studies of *Salmonella*. *FEMS Immunol. Med. Microbiol.* **14,** 129–134.

58. Kantama, L., Jayanetra, P., and Bangtrakulnonth, A. (1995) Epidemiological study of Salmonella enteritidis outbreak in Thailand by random amplified polymorphic DNA (RAPD) technique. Southeast *Asian J. Trop. Med. Publ. Health* **26,** 49–51.

19

Cholera

Timothy J. Barrett and Daniel N. Cameron

1. Introduction

1.1. Historical Background

Few diseases invoke public fear as readily as cholera. In its most severe state, cholera can cause death from hypotensive shock within 12 h of the first symptom. Cholera typically occurs in epidemics, spreading rapidly within the community, especially if hygeinic conditions are poor. Fortunately, effective water treatment has limited the spread of cholera in most of the developed world, and the treatment of cholera by oral rehydration has dramatically reduced the mortality rate.

Whereas there are accounts of disease resembling cholera dating from ancient times, there is considerable debate among medical historians as to whether cholera existed outside of Asia before the early 1800s *(1)*. The first of seven major cholera pandemics (widespread epidemics) began in 1817, when cholera spread from its endemic focus in India *(2)*. The exact beginning and end of each pandemic is open to interpretation, but it is generally held that the sixth pandemic ended by 1923, and the world outside of Asia was cholera-free by 1950. Strains from the fifth and sixth pandemic are known to have been *Vibrio cholerae* O1 of the classical biotype, and it is assumed (but not known) that the previous four pandemics were also caused by classical strains *(1)*.

In 1961, the El Tor biotype of *V. cholerae* O1 spread from Indonesia to other countries in Asia, beginning the seventh pandemic. From Asia, cholera soon spread to the Middle East, Africa, southern Europe, the Pacific, and eventually to South America *(3)*. When cholera reappeared in South America after a nearly 100-yr absence, it quickly spread from a few coastal towns in Peru to affect every country in Central and South America, except Uruguay, within 2 yr. The seventh pandemic continues today, with cholera still occurring in

From: *Methods in Molecular Medicine, Vol. 15: Molecular Bacteriology: Protocols and Clinical Applications*
Edited by: N. Woodford and A. P. Johnson © Humana Press Inc., Totowa, NJ

most of these areas. Prior to 1961, the El Tor biotype was known to cause isolated cases of cholera-like illness, but it had not caused any outbreaks, and was not thought to have epidemic potential *(4)*. It is unknown whether changes in the organism or the environment may have allowed the spread of this "new" form of cholera.

A similar story began in 1992. Prior to this year, only toxigenic *V. cholerae* strains of serogroup O1 had caused epidemic cholera. Although there were 137 other "O" antigenic (lipopolysaccharide) types, identification of a *V. cholerae* isolate as a "non-O1" serotype eliminated concern that the isolate represented a potential new epidemic. In late 1992, however, epidemic disease indistinguishable from cholera, but not because of *V. cholerae* O1, appeared in India and Bangladesh *(5)*. The organism proved to be a new serotype of *V. cholerae*, and was given the antigenic designation O139. This discovery opens the possibility that additional serogroups of *V. cholerae* may have the potential to cause epidemic cholera, and in fact may have done so in the years before typing sera was available. Previous cholera pandemics suggest that this strain may spread from Asia, beginning an eighth pandemic *(6)*, but such has not occurred as of this writing.

1.2. Microbiology of V. cholerae

The discovery of the cholera bacillus is generally attributed to Robert Koch, who described the organism in 1883 and demonstrated that this comma-shaped organism was the cause of epidemic cholera. Decades later, the work of Pacini was recognized as predating that of Koch by nearly 30 yr, and the name he originally gave to the organism, *Vibrio cholera*, was reflected in the new systematic name *Vibrio cholerae (1)*. DNA-DNA hybridization studies show that *V. cholerae* is a single, well-defined species, easily classifiable by phenotypic characterization *(7)*. Because of the ease with which *V. cholerae* can be distinguished from other *Vibrio* species by standard biochemical tests, molecular methods are seldom applied to the identification of this species. Within the species, however, the pathogenic potential of individual isolates may greatly vary.

As described above, only isolates of serogroups O1 and O139 have been shown to possess epidemic potential. Within serogroup O1, some isolates do not produce cholera enterotoxin (CT), and do not appear to be pathogenic on the basis of volunteer studies *(8)*. Occasionally, however, CT-negative strains are isolated from persons with diarrhea, so their pathogenic potential cannot be entirely dismissed *(9)*. Isolates of *V. cholerae* belonging to serogroups other than O1 and O139 typically do not produce CT, and are not associated with epidemic potential (**Table 1**). These organisms are common inhabitants of estuarine environments, are sometimes isolated from wounds or other extraintestinal sites, and may even cause some cases of gastroenteritis through

Table 1
Characteristics of *V. cholerae* Serogroups

Serogroup	Serologic subtypes (no.)	Biotypes (no.)	CT production typical	Epidemic potential
O1	3	2	yes	yes
O2-O138	0	1	no	no
O139	0	1	yes	yes

unknown mechanisms *(10)*. In summary, the ability to produce CT and the presence of the O1 or O139 antigen are the most important factors in determining the public health significance of an isolate of *V. cholerae*. Molecular methods for the detection of the CT gene, and investigations into the molecular basis of serotypes O1 and O139 will be discussed in more detail below.

2. Diagnosis

2.1. Serotyping of V. cholerae Isolates

The major antigen used in the typing of *V. cholerae* isolates by serologic reactivity is the O antigen. Although *V. cholerae* does have a flagellar antigen, its value in serotyping is limited because of the presence of shared flageller ("H") epitopes among all *Vibrio* species *(11)*. When applied to *V. cholerae*, the term "serotype" usually refers to different antigenic forms of the O1 antigen, rather than a combination of O and H antigens. There are three antigenic forms of the O1 serogroup: Inaba, Ogawa, and Hikojima. Inaba strains produce A and C antigens, Ogawa strains produce A and B, and Hikojima strains produce A, B, and C. Specific Inaba and Ogawa antisera are made by absorption with the other serotype. Hikojima strains are extremely rare and unstable. Some authors prefer not to use this designation, and report the serotype as Inaba or Ogawa, depending on which reaction is stronger *(12)*. Strains of *V. cholerae* O1 have been shown to shift between the Inaba and Ogawa serotypes, both in nature (during the course of an outbreak) and in the laboratory *(13)*. The mechanisms by which these changes occur, and the implications for replacing traditional serotyping with molecular methods are discussed below.

2.2. Mechanism of Antigenic Variation in V. cholerae O1 and O139

The subdividing of *V. cholerae* O1 isolates by serotype is hindered not only by the limited number of serotypes, but also by the phenomenon of antigenic shift. The ability to alter the antigenic determinants recognized by a host's immune system is highly advantageous to the organism in evading host immune responses. In fact, the immune response of the host may be necessary to provide the selective pressure to switch serotypes *(14)*. Serotype conversion occurs at a rate of 10^{-5} for Ogawa to Inaba and much less for Inaba to Ogawa *(15)*.

The gene complex responsible for O-antigen biosynthesis (designated *rfb*) was initially localized to a 19 kb *Sst*I fragment, containing 21 open reading frames *(16)*. Comparison of DNA sequences in this region between an Ogawa and an Inaba isolate indicated high sequence conservation. Only three differences were found, including a single base change in an intergenic region, a small deletion in *rfbR*, and a single base deletion in *rfbT (17)*. Further analysis of additional strains showed that only alterations in the *rfbT* gene were necessary for serotype conversion.

Inaba strains are essentially *rfbT* mutants of Ogawa strains, resulting from the absence of a specific epitope on the LPS molecule caused by the truncation of the RfbT protein. Any mutation resulting in a defective RfbT protein will switch an Ogawa strain to the Inaba serotype, but the reverse is only possible if the specific *rfbT* mutation in that strain is corrected *(18)*. The latter event was observed during the epidemic in Latin America in which only Inaba strains were initially isolated, but Ogawa strains shown to be indistinguishable by other subtyping methods were later isolated (Cameron, unpublished). The potential for replacing serotyping of *V. cholerae* O1 by molecular methods is discussed below.

With the emergence of *V. cholerae* O139 in 1992, a non-O1 *V. cholerae* strain was credited for the first time as the cause of epidemic cholera *(5)*. This new serogroup of *V. cholerae* differs from serogroup O1 isolates by the absence of most *rfb* genes. Probes specific to each gene in the operon indicated that only *rfbQ*, *rfbR*, and *rfbS* remain *(19)*. Specific serotypes within *V. cholerae* O139 have not been recognized.

2.3. Detection of the Cholera Toxin Gene

Though other virulence factors have been recognized, the potential for producing CT remains the most important factor in assessing the significance of *V. cholerae* isolates. As discussed above, nearly all diarrhea-associated isolates of *V. cholerae* are CT-producing, and only CT-producing isolates have caused epidemic cholera *(20)*. Cholera toxin is composed of two polypeptides, forming the A and B subunits. The B subunit is made up of five identical elements and binds to GM1, an epithelial cell surface receptor. The A subunit is responsible for the adenylate cyclase activation of small intestine epithelial cells *(21)*. Two methods, DNA hybridization with gene probes and the polymerase chain reaction (PCR), have been successfully used to detect the toxin gene.

Probes to *eltAB* encoding the heat-labile enterotoxin (LT) of *E. coli* were first utilized to identify toxin-producing strains of *V. cholerae* due to its high homology to *ctxAB* which encodes CT (**Table 2**) *(22)*. Probes to *ctxAB* carried

Table 2
Probes used to Detect Cholera Toxin or El Tor Hemolysin Genes in *Vibrio cholerae* Isolates

Target	Probe	Size (bp)	Ref.
eltA	*Hinc*II fragment of pEWD299	1275	*22*
eltB	*Eco*RI-*Hind*II fragment of pEWD299	590	*22*
ctxAB	*Eco*RI-*Bgl*II fragment of pCVD002	4000	*23*
ctxA	5'-GCA AGA GGA ACT CAG ACG GG-3'	20	*24*
ctxA	PCR fragment (*see* **Table 2** for primers)	564	*37*
hlyA	5'-CGG CAT TCA TCT GAA TGA T-3'	19	*33*

on an *Eco*RI-*Bgl*II fragment were soon developed, using radioactive ^{32}P as a label *(23)*. Cloning and purifying DNA fragments proved time-consuming and tedious. Oligonucleotides to the *ctxA* subunit provided equivalent results without the need to prepare fragment DNA *(24)*. Oligonucleotides are stable, commercially available, and tend to hybridize more rapidly than fragment probes. A limited number of reporter molecules, and the possibility of a gene mutation at the hybridization site that could render a false negative result, are the major disadvantages to their use.

Radiolabeled probes were used to detect the CT gene by hybridization to colony blots and Southern blots. Colony blots permit the detection of many toxin-producing isolates simultaneously and are advantageous for processing large quantities of isolates. Southern hybridization of *ctx* showed that some isolates of *V. cholerae* have multiple copies of the *ctx* gene. Classical isolates, and most US Gulf Coast isolates, have two copies of *ctx*, but El Tor isolates have only one copy *(25)*.

DNA probing was limited by the need to use radioactive materials for detection, since the sensitivity of the technique was much higher than that obtained with nonradioactive markers at the time these methods were developed. However, the development of nonradioactive labels such as alkaline phosphatase *(26)* and digoxigenin *(27)* has made labeling of both fragments and oligonucleotides routine, eliminating this technical disadvantage. Although difficulty in labeling oligonucleotides with digoxigenin has been observed (Cameron, unpublished), this problem has been solved with the use of a digoxigenin-labeled PCR-generated fragment probe *(28)*.

PCR utilizes a DNA polymerase to amplify a specific DNA sequence (amplicon), usually from a target gene, defined by two short, specific oligonucleotides (primers) that flank the region to be amplified. The amplicon can then be visualized by agarose gel electrophoresis. The *ctxA* locus has provided a stable region to amplify and thus identify the presence of cholera toxin *(29)*.

Several PCR primers and protocols have been developed (**Table 3**), all of which use standard PCR strategies (*see* Chapter 5).

PCR has many advantages over other detection techniques. It is often faster, specific, and more sensitive. Unlike DNA probes, PCR can directly detect the presence of *ctx* sequences in environmental or clinical samples such as food *(30)* or stool *(31)*. However, some diffficulty in amplification may occur because of substances in food or stool that can inhibit DNA polymerase and hamper PCR results (*see* **Notes 1–3**).

2.4. Biotyping of V. cholerae O1

The separation of *V. cholerae* O1 strains into classical and El Tor biotypes has been traditionally based on phenotypic characterization including hemolysis of sheep red blood cells, bacteriophage susceptibility, Voges-Proskauer reaction, polymyxin B susceptibility, and hemagglutination of chicken red blood cells. Early isolates of the El Tor biotype were strongly hemolytic, but most recent isolates are not, greatly diminishing the value of this test *(32)*. More recently, molecular methods have been developed to distinguish between biotypes. Even though El Tor strains are no longer hemolytic, they still carry a hemolysin gene that is not present in classical strains. An oligonucleotide probe (**Table 2**) has been developed to detect the presence of this gene, and appears to be biotype specific *(33)*. An alternative approach uses DNA sequence differences in the gene encoding the toxin coregulated pilus (*tcpA*). In a multiplex polymerase chain reaction assay, classical and El Tor strains give different size amplicons (**Table 3**) *(34)*.

3. Epidemiologic Investigation

The differentiation between different strains of the same species has become a vital part of epidemiologic investigations. Epidemiologic studies of cholera generally fall into one of two categories: outbreak investigations and surveillance. The questions raised in these two types of investigations are somewhat different, and different methods may be more appropriate for each category. During outbreak investigations, the critical question is whether a particular isolate (human, food, or environmental) is the "outbreak strain." Rapid methods for screening large numbers of isolates may be more appropriate for this type of investigation. More sensitive methods can then be applied to a smaller subset of isolates. A "yes or no" answer is sufficient; the genetic relatedness of that isolate to others is not particularly important. Since subtyping patterns are compared with a single outbreak pattern, visual interpretation of gels is sufficient.

In surveillance studies, it is necessary to identify the source (or likely source) of the isolate. In the United States, for example, it would be critical to public health investigations to distinguish between the *V. cholerae* O1 strain endemic

Table 3
Sequences of PCR Primers Used in Analysis of *V. cholerae* Isolates

5'-3' Sequence[a]	Target	No. bases	Size (bp)	Ref.
F: CTC AGA CGG GAT TTG TTA GGC ACG	*ctx*	24	302	*29*
R: TCT ATC TCT GTA GCC CCT ATT ACG		24		
F: CGG GCA GAT TCT AGA CCT CCT G	*ctx*	22	564	*28*
R: CGA TGA TCT TGG AGC ATT CCC AC		23		
F: GTG GGA ATG CTC CAA GAT CAT CG	*ctx*	23	431	*31*
R: ATT GCG GCA ATC GCA TGA GGC GT		23		
F: TGA AAT AAA GCA GTC AGG TG	*ctx*	20	779	*30*
R: GTG ATT CTG CAC ACA AAT CAG		21		
F:CAC GAT AAG AAA ACC GGT CAA GAG AGC	*tcpA*,	24	617	*34*
R:ACC AAA TGC AAC GCC GAA TGG	classical	24		
F:GAA GAA GTT TGT AAA AAG AAG AAC AC	*tcpA*,	26	471	*34*
R:GAA GGA CTT TTT ACG TG	El Tor	17		

[a]F, forward primer; R, reverse primer.

to the Gulf Coast, the Latin American strain, or strains from Asia or Africa. Such studies require the highest level of strain discrimination, and the ability to accurately compare subtyping patterns with a library of previously seen patterns (*see* **Subheading 5.2.**). The usefulness of each of the most commonly used methods for epidemiologic investigations of cholera is discussed below.

These techniques are often used in outbreak investigations to support epidemiologic data. Most commonly, these techniques involve the subtyping or grouping of isolates based on protein or DNA similarities. Molecular techniques are particularly useful in characterizing individual isolates, since phenotypic traits are often insufficient to discriminate between various strains of *V. cholerae (35)*.

3.1. Multilocus Enzyme Electrophoresis

Multilocus enzyme electrophoresis (MEE) measures the mobility of different allelic forms of enzymes (isozymes) involved in the standard metabolic processes of bacterial cells. Crude cellular protein extracts are electrophoresed through a starch gel and detected with a specific precipitating substrate. Each electrophoretic mobility is considered to indicate a different isozyme. Strains of *V. cholerae* have been characterized by a panel of 16 enzymes *(36)*. In order to be detectable, differences between isozymes must be significant enough to cause a structural change that affects the mobility of the enzyme *(35)*.

The discriminatory power of MEE for epidemiologic investigations of *V. cholerae* O1 outbreaks is limited. Studies of classical isolates showed minimal diversity, with the majority of isolates in one electrophoretic type or ET

(35). Nontoxigenic isolates are much more diverse, differing at multiple loci *(37)*. Toxigenic El Tor strains were separated into four major ETs: Australia (ET1), US Gulf Coast (ET2), seventh pandemic (ET3), and Latin America (ET4). Toxigenic strains differed in mobility of leucine aminopeptidase and diaphorase 1 *(36)*. The majority of isolates from the current epidemic in Latin America differ from ET3 isolates only by leucine aminopeptidase. However, a few isolates were recently detected that possessed the same ET as other seventh pandemic strains *(38)*. It is unknown whether these isolates are potential precursors to the Latin American epidemic or revertant mutations of the peptidase N gene. Isolates of *V. cholerae* O139 are also ET3 *(39)*.

3.2. Ribotyping (see *Note 4*)

Ribotyping (*see* Chapter 2) of *V. cholerae* isolates uses a labeled *E. coli* probe specific for genes encoding rRNA to hybridize with Southern blots of restriction fragments generated by digestion with *Bgl*I *(40)*. High levels of strain discrimination are possible with this technique. Ribotyping adds a greater level of discrimination to analysis of restriction fragment length polymorphisms (RFLPs) by minimizing the total number of fragments observed. It has the advantage over other probes in that it is not specific for a particular organism and, therefore, requires no special DNA preparation to generate the same probe for different organisms.

Ribotyping of classical isolates produced several highly related ribotype patterns. Koblavi et al. identified four closely related riboytpes that were distinct from El Tor isolates *(40)*. Analysis of isolates from a re-emergence of classical cholera in the 1980s in Bangladesh demonstrated that these isolates were identical to the sixth pandemic *(41)*. Supporting this observation, a ribotyping scheme placed classical isolates within the same group, dividing them into seven subgroups *(42)*. This same scheme separated El Tor isolates into several groups. Isolates from endemic sources were separated from other El Tor strains. Isolates from the US Gulf Coast belonged to a single ribotype, whereas isolates from Australia were separated into two similar ribotypes different from the U.S. Gulf Coast strain. Several additional ribotypes were associated with Asia and Africa. The Latin American strain possessed a ribotype found among isolates from Africa and Asia.

3.3. Pulsed-Field Gel Electrophoresis (see *Note 5*)

Pulsed-field gel electrophoresis (PFGE; *see* Chapter 3) separates DNA fragments of very high molecular weight by imbedding DNA in agarose plugs to reduce mechanical shearing and using restriction enzymes that infrequently cut DNA (usually six or eight base cutters). DNA fragments are separated by electrophoresing at an angle (usually 120°), alternating the direction of the current, and ramping the time of the switching to maximize fragment separation.

PFGE using *Not*I to restrict genomic DNA identified multiple patterns within ETs and ribotypes, though occasionally MEE and ribotyping subdivided isolates with the same PFGE pattern *(43)*. PFGE of *V. cholerae* O139 isolates identified four very similar patterns, grouping them as a single strain *(39)*. PFGE is reproducible, stable over time, and relatively rapid in comparison with MEE and methods requiring DNA hybridization. PFGE is the most discriminating of the molecular techniques used in subtyping *V. cholerae*, but has the disadvantage of generating a virtually unlimited number of different patterns. As with all RFLP analysis, the significance of variability between patterns is usually unknown. Lack of fragments or additional fragments may be caused by the acquisition or loss of plasmids or phage. Thus, minor differences in PFGE patterns may have no epidemiologic importance *(43)*.

3.4. DNA Sequencing of ctx

DNA sequencing of *ctxB* depends on limited heterogeneity within the gene. Results from sequencing may have significant implications for development of vaccines, use of diagnostic tests for CT antibody, and design of PCR primers. Automation and standardization have increased reproducibility and improved quality of laboratory results. An advanced laboratory with a robot and automated sequencer could sequence 400–800 bases from 10–20 isolates in 12 h *(44)*. Sequencing of a variety of isolates showed two differences in DNA sequence resulting in separation of three different "genotypes." Interestingly, classical and US Gulf Coast strains had identical *ctxB* sequences. Isolates endemic to Australia had a unique genotype, whereas those from the seventh pandemic and Latin America were of the same genotype *(44)*. Although providing interesting information about the possible evolutionary relationship of these strains, the value of this method for molecular epidemiology is limited.

3.5. Plasmid Profile Analysis

Plasmid analysis (*see* Chapter 4) was originally used to characterize classical *V. cholerae* isolates from El Tor strains. Classical strains typically have two plasmids, 21 and 3 megadaltons in size *(45)*. Plasmids carried by El Tor isolates are usually associated with antimicrobial resistance *(46)*. Since plasmids are extrachromosomal DNA elememts, they may not be maintained within particular isolates without external pressure. Plasmid analysis of *V. cholerae* is most useful for epidemiologic studies when a strain of interest carries an unusual plasmid.

3.6. Southern Hybridization with Vibriophages

The presence of vibriophage VcA-1 and VcA-3 has been used to identify classical and US Gulf Coast El Tor strains, respectively *(47)*. Hybridization of radiolabeled phage DNA to *Hin*dIII genomic digests separated US toxigenic El

Tor isolates from foreign toxigenic El Tor isolates *(24)*. In addition, some subtyping of US Gulf Coast isolates was possible. However, differences in banding patterns were minor. Loss of phage DNA caused by continued laboratory handling and differences in phage insertion sites *(48)* could alter results.

4. Current Research

4.1. Pathogenesis

The pathogenesis of cholera has been the subject of intensive investigations since at least 1959 when S. N. De demonstrated the enterotoxigenicity of culture-free supernatants of *V. cholerae* O1 *(49)*. CT still appears to be by far the most important toxin in causing diarrhea, but other toxins such as the zonula occludens toxin *(50)* and accessory cholera enterotoxin *(51)* also contribute to fluid loss. Since excellent recent reviews of these and other virulence factors of *V. cholerae* O1 are available elsewhere *(52)* no attempt will be made here to assess their importance in pathogenesis.

Several authors *(53,54)* have used molecular probes or PCR to survey isolates of *V. cholerae* for the presence of these three toxin genes. All three genes were found in virtually every isolate in which at least one of the genes was present, and they were found only in strains in which the gene for toxin coregulated pili (TCP) was also present *(55)*. These findings are not surprising in light of the recent publication by Waldor and Mekalanos showing that the three toxin genes, along with six other genes or open reading frames, are carried by a filamentous bacteriophage that uses TCP as a receptor *(56)*. This discovery is of enormous importance in understanding the origin(s) of epidemic cholera, as discussed below. It also raises serious concerns about using nontoxigenic *V. cholerae* O1 as a live vaccine, since the vaccine strain could potentially acquire the complete array of toxin genes through bacteriophage infection.

Although it would certainly be folly to assume that the pathogenesis of cholera is completely understood, many researchers are now focusing on understanding the ecology of *V. cholerae* and the origins of virulence, rather than on the mechanisms of pathogenesis themselves.

4.2. The Origin(s) of Epidemic Cholera

Molecular approaches to understanding the evolution of *V. cholerae* may help solve the great enigma of the origin of epidemic cholera. *V. cholerae* is part of the normal bacterial flora of estuarine environments *(57)*. In areas away from human fecal contamination, *V. cholerae* isolates, even those of serogroup O1, are nearly always CT-negative *(52)*. This observation led to the theory that the epidemic form of *V. cholerae* O1 evolved from a nonepidemic ancestor at some ancient time, and that environmental isolates of toxigenic *V. cholerae* O1 today are always of human origin (i.e., represent fecal contamination). Recent evidence suggests that strains with epidemic potential may have evolved more recently, and that other pathogenic serotypes (such as O139) may emerge in the future.

Phylogenetic analysis of *V. cholerae* based on DNA sequencing of the *asd* locus suggests that the sixth pandemic, seventh pandemic, and US Gulf Coast isolates are separate clones that evolved independently from different lineages of environmental non-O1 *V. cholerae* strains *(58)*. Nearly identical *asd* sequences were found in different O groups, and O1 isolates were found in different lineages, suggesting horizontal transfer of O antigen within the species. These data suggest that epidemic cholera has evolved more than once from nonepidemic environmental isolates. On the other hand, current evidence suggests that *V. cholerae* O139 evolved from epidemic *V. cholerae* O1 rather than environmental strains *(19,59)*. Further studies using DNA sequencing and other types of phylogenetic analysis are needed to support these hypotheses.

5. Future Goals
5.1. Improved Diagnostic Tests

Detection of the cholera toxin gene by DNA probes or PCR has been a highly reliable method for identifying toxin-producing strains of *V. cholerae*. The greatest improvements in diagnostic methods for detecting the cholera toxin gene are likely to come from general improvements in molecular methodology rather than new probes or PCR. Better methods for extracting DNA from stool or food samples are required to make optimal use of the power of PCR. Such methods may well come from research on organisms other than *V. cholerae.*

The second critical element in assessing isolates of *V. cholerae* O1 is serotype. When it was first reported that conversion between the Inaba and Ogawa serotypes resulted from a single base change in the *rfbT* gene, it seemed possible that serotyping might be replaced by molecular methods targeting that mutation. Unfortunately, sequencing of the *rfbT* gene from several isolates associated with the Latin American epidemic revealed multiple mutations resulting in a truncated RfbT protein (Cameron, unpublished data). Although it would be possible to design a multiplex PCR to detect each of the known mutations, it is likely that there are also many unknown mutations. It thus seems unlikely that traditional serotyping will be replaced by molecular methods anytime soon.

5.2. Standardization of Typing Methods

Efforts to standardize methods of molecular typing are critical to extending the usefulness of these methods in epidemiologic investigations. Unless two laboratories are using the same methods, comparison of typing results are meaningless. It is not uncommon, for example, for two isolates of *V. cholerae* O1 to have the same ribotype using one enzyme, and different ribotypes using a second enzyme. An international committee is currently attempting to standardize typing methods for *Listeria monocytogenes* *(60)*, and similar efforts for *V. cholerae* are clearly warranted. In addition to laboratory methods, typing nomenclature must also be standardized. The term "*V. cholerae* O1 Inaba" conveys the same meaning to everyone in the cholera field, but "ribotype 5" does not. The development of gel analysis software (*see* Chapter 3) should greatly facilitate these efforts. Laboratories can now electronically exchange image files and make direct, computer-assisted comparisons with their own data.

5.3. DNA Sequencing for Molecular Epidemiology

Ultimately, DNA sequencing is likely to replace other methods for studying the molecular epidemiology of *V. cholerae*. Automated equipment and the sequencing of PCR-generated templates has greatly increased the speed and ease with which DNA sequencing can be done. As with other organisms to which this technique can be applied, the key to the successful use of DNA sequencing in molecular epidemiology is the selection of target genes. Some genes, such a *pepN* *(61)* or *ctxB* *(44)* are relatively stable and provide only the broadest grouping of strains. Others genes, such as *asd*, are more variable and more useful for fine strain discrimination *(58)*.

6. Notes

1. The primary complications arising from molecular analysis of *V. cholerae* isolates are related to the purity of the DNA preparation. In nearly every molecular technique discussed, DNA preps are subject to rapid degradation, most likely caused by the presence of some type of DNase. Each technique employs varying strategies to compensate for this degradation.

2. PCR uses a small amount of template DNA cells to avoid false-positive results. Enough cells can be obtained by touching a loop to a single colony *(27)*. The cells are suspended in 200 µL of sterile deionized water and lyzed by boiling for 10 min for *pepN* or 20 min for *ctx*. Extended boiling reduces the amount of final PCR product, presumably by destroying the DNA. Only 2–3 µL of boiled cells are necessary in the PCR reaction. PCR of *ctx* fragments is less dependent on cell concentration than PCR for *pepN* *(61)*.

3. To control for contamination, PCR preparation should preferably be conducted in a separate room or area other than PCR amplification and electrophoresis. Proper positive and negative controls should be processed with each PCR test group. Negative controls include samples with water instead of DNA, as well as isolates known to be negative for the particular amplified product (e.g., nontoxigenic strains for PCR of *ctx*). Nonspecific hybridization of primers can also be troublesome. Bands of an inappropriate size should be interpreted as negative *(27)*. Faint bands of the correct size should be retested, possibly with an increase in the annealing temperature to 60°C *(28)*, and verified with hybridization to a gene probe.

4. Preparation of DNA for ribotyping must be performed by phenol:chloroform extractions. Guanidium extractions have been used successfully with other organisms (*see* Chapter 2), but produced limited results when attempted on isolates of *V. cholerae*. Although results from a guanidium preparation were obtainable, the DNA was often degraded within a week, making subsequent testing on the same sample preparation impossible *(42)*.

5. PFGE uses an additional lysis step to prevent potential contamination with DNases *(43)*. Plugs are lysed for one hour at 37°C in lysis buffer (1 *M* NaCl, 10 m*M* Tris-HCl, pH 8.0, 100 m*M* EDTA, 0.5% sarkosyl, 0.2% sodium deoxycholate, 1 mg/mL lysozyme). After this preliminary lysis step, plugs are subjected to overnight treatment in ESP buffer (1% SDS, 500 m*M* EDTA, pH 8.0, 1 mg/mL proteinase K) at 50°C. Proteinase K is effective from 50–65°C and the high concentration of EDTA prevents the activity of DNases, but does not inhibit proteinase K. Several washes (at least six for 30 min each) in TE (10 m*M* Tris-HCl, pH 8.0, 1 m*M* EDTA) are necessary to remove the ESP solution.

References

1. Politzer, R. (Ed.) (1959) *Cholera*, World Health Organization, Geneva.
2. Barua, D. (1992) History of cholera, in *Cholera* (Barua, D. and Greenough, W. B., III, eds.), Plenum, NY, pp.1–36.
3. Tauxe, R. V. and Blake, P. A. (1992) Epidemic cholera in Latin America. *JAMA* **267**, 1388–1390.
4. Blake, P. A. (1994): Historical perspectives on pandemic cholera, in Vibrio cholerae *and Cholera*, (Wachsmuth, I. K., Blake, P. A., and Olsvik, O., eds.), ASM Press, Washington, DC, pp. 293–296.
5. Cholera working group (1993) Large epidemic of cholera-like disease in Bangladesh caused by *Vibrio cholerae* O139 synonym Bengal. *Lancet* **342**, 430,431.
6. Swerdlow, D. L. and Ries, A. A. (1993) *Vibrio cholerae* non-O1 - the eighth pandemic? *Lancet* **342**, 382,383.
7. Baumann, P. A., Furniss, A. L., and Lee, J. V. (1984) Genus I. Vibrio, in *Bergey's Manual of Systematic Bacteriology*, vol. 1, (Krieg, N. R. and Holt, J. G., eds.), Williams & Wilkens, Baltimore, pp. 518–533.
8. Levine, M. M., Black, R. E., Clements, M. L., Cisneros, L., Saah, A., Nalin, D. A., Gill, D. M., Craig, J. P., Young, C. R., and Ristaino, P. (1982) The pathogenicity of nonenterotoxigenic *Vibrio cholerae* serogroup O1 biotype El Tor isolated from sewage water in Brazil. *J. Infect. Dis.* **145**, 296–299.
9. Morris, J. G., Jr., Picardi, J. L., Lieb, S., Lee, J. V., Roberts, A., Hood, M., Gunn, R. A., Blake, P. A. (1984) Isolation of nontoxigenic *Vibrio cholerae* O Group 1 from a patient with severe gastrointestinal disease. *J. Clin. Microbiol.* **19**, 296,297.
10. Morris, J. G., Jr. (1990) Non-O group 1 *Vibrio cholerae*: a look at the epidemiology of an occasional pathogen. *Epidemiol. Rev.* **12**, 179–191.
11. Shinoda, S., Kariyama, R., Ogawa, M., Takeda, Y., and Miwatani, T. (1976) Flagellar antigens of various species of the genus *Vibrio* and related genera. *Int. J. Syst. Bacteriol.* **26**, 97–101.

12. Kelly, M. T., Hickman-Brenner, F. W., and Farmer, J. J., III (1991) Vibrio, in *Manual of Clinical Microbiology,* 5th ed. (Balows, A., Hausler, W. J., Jr, Herrman, K. O., Isenberg, H. D., and Shadomy, H. J., eds.), American Society for Microbiolgy, Washington, DC, pp. 384–395.

13. Sack, R. B. and Miller, C. E. (1969) Progressive changes of vibrio serotypes in germ-free mice infected with *Vibrio cholerae. J. Bacteriol.* **99,** 688–695.

14. Svennerholm, A. M., Jonson, G., and Holmgren, J. (1994). Immunity to *Vibrio cholerae* infection, in Vibrio cholerae *and Cholera: Molecular to Global Perspectives,* (Wachsmuth, I. K., Blake, P. A., and Olsvik, O., eds.), American Society for Microbiology, Washington, DC, pp. 257–271.

15. Bhaskaran, K. (1960) Recombination of characters between mutant stocks of *Vibrio cholerae* strain 162. *J. Gen. Microbiol.* **23,** 47–54.

16. Ward, H. M., Morelli, G., Kamke, M., Morona, R., Yeadon, J., Hackett, J. A., and Manning, P. A. (1987) A physical map of the chromosomal region determining O-antigen biosynthesis in *Vibrio cholerae* O1. *Gene* **55,** 197–204.

17. Stroeher, U. H., Karageorgos, L. E., Morona, R., and Manning, P. A. (1992) Serotype conversion in *Vibrio cholerae* O1. *Proc. Natl. Acad. Sci. USA* **89,** 2566–2570.

18. Manning, P. A., Stroeher, U. H., and Morona, R. (1994) Molecular basis for O-antigen biosynthesis in *Vibrio cholerae* O1: Ogawa-Inaba switching, in Vibrio cholerae *and Cholera: Molecular to Global Perspectives,* (Wachsmuth, I. K., Blake, P. A., and Olsvik, O., eds.), American Society for Microbiology, Washington, DC, pp. 77–94.

19. Stroeher, U. H., Jedani, K. E., Dredge, B. K., Morona, R., Brown, M. H., Karageorgos, L. E., Albert, M. J., and Manning, P. A. (1995) Genetic rearrangements in the *rfb* region of *Vibrio cholerae* O1 and O139. *Proc. Natl. Acad. Sci. USA* **92,** 10,374–10,378.

20. Finkelstein, R. A. (1988) Cholera, the cholera enterotoxins, and the cholera enterotoxin- related enterotoxin family, in *Immunological and Molecular Genetic Analysis of Bacterial Pathogens* (Owen, P. and Foster, T. J., eds.), Elsevier Science Publisher, Amsterdam, pp. 85–102.

21. Spangler, B. D. (1992) Structure and function of cholera toxin and related *Escherichia coli* heat-labile enterotoxin. *Microbiol. Rev.* **56,** 622–647.

22. Mosely, S. L. and Falkow, S. (1980) Nucleotide sequence homology between the heat-labile enterotoxin of *Escherichia coli* and *Vibrio cholerae* deoxyribonucleic acid. *J. Bacteriol.* **144,** 444–446.

23. Lockman, H. A. and Kaper, J. B. (1983) Nucleotide sequence analysis of the A2 and B subunits oif *Vibrio cholerae* enterotoxin. *J. Biol. Chem.* **258,** 13,722–13,726.

24. Almeida, R. J., Cameron, D. N., and Wachsmuth, I. K. (1992) Vibriophage VcA-3 as an epidemic strain marker for the U. S. Gulf Coast *V. cholerae* O1 clone. *J. Clin. Microbiol.,* **30,** 300–304.

25. Mekalanos, J. J. (1983) Duplication and amplification of toxin genes in *Vibrio cholerae. Cell,* **35,** 253–263.

26. Wright, A. C., Guo, Y., Johnson, J., Nataro, J. P., and Morris, J. G., Jr. (1992) Development and testing of a nonradioactive DNA oligonucleotide probe that is specific for *Vibrio cholerae* cholera toxin. *J. Clin. Microbiol.* **30,** 2302–2306.

27. Centers for Disease Control and Prevention (1994) Laboratory methods for the diagnosis of *Vibrio cholerae*. U. S. Department of Health and Human Services, Public Health Division, Washington, DC.

28. Fields, P. I., Popovic, T., Wachsmuth, K., and Olsvik, O. (1992) Use of polymerase chain reaction for detection of toxigenic *Vibrio cholerae* O1 strains from the Latin American cholera epidemic. *J. Clin. Microbiol.* **30,** 2118–2121.

29. Shirai, H., Nishibuchi, M., Ramamurthy, T., Bhattacharya, S. K., Pal, S. C., and Takeda, Y. (1991) Polymerase chain reaction for detection of cholera enterotoxin operon of *Vibrio cholerae*. *J. Clin. Microbiol.* **29,** 2517–2521.

30. Koch, W. H., Payne, W. L., Wentz, B. A., and Cebula, T. A. (1993) Rapid polymerase chain reaction method for detection of *Vibrio cholerae* in foods. *App. Environ. Microbiol.* **59,** 556–560.

31. Varela, P., Pollevick, G. D., Rivas, M., Chinen, I., Binsztein, N., Frasch, A., and Ugalde, R. A. (1994) Direct detection of *Vibrio cholerae* in stool samples. *J. Clin. Microbiol.* **32,** 1246–1248.

32. Barrett, T. J. and Blake, P. A. (1981) Epidemiologic usefulness of changes in hemolytic activity of *Vibrio cholerae* biotype El Tor during the seventh pandemic. *J. Clin. Microbiol.* **13,** 126–129.

33. Alm, R. A. and Manning, P. A. (1990) Biotype-specific probe for *Vibrio cholerae* serogroup O1. *J. Clin. Microbiol.* **28,** 823–824.

34. Keasler, S. P. and Hall, R. H . (1993) Detecting and biotyping *Vibrio cholerae* O1 with multiplex chain reaction. *Lancet* **341,** 1661.

35. Wachsmuth, K., Olsvik, O., Evins, G. M., and Popovic, T. (1994) Molecular epidemiology of cholera, in Vibrio cholerae *and Cholera: Molecular to Global Perspectives* (Wachsmuth, I. K., Blake, P. A., and Olsvik, O., eds.), American Society for Microbiology, Washington, DC, pp. 357–370.

36. Wachsmuth, I. K., Evins, G. M., Fields, P. I., Olsvik, O., Bopp, C. A., Wells, J. G., Carrillo, C., and Blake, P. A. (1993) The molecular epidemiology of cholera in Latin America. *J. Infect. Dis.* **167,** 621–626.

37. Chen, F., Evins, G. M., Cook, W. L., Almeida, R., Hargrett-Bean, N., and Wachsmuth, K. (1991) Genetic diversity among toxigenic and nontoxigenic *Vibrio cholerae* O1 isolated from the Western Hemisphere. *Epidemiol. Infect.* **107,** 225–233.

38. Evins, G. M., Cameron, D. N., Wells, J. G., Greene, K. D., Popovic, T., Giono-Cerezo, S., Wachsmuth, I. K., and Tauxe, R. V. (1995) The emerging diversity of the electrophoretic types of *Vibrio cholerae* in the Western Hemisphere. *J. Infect. Dis.* **172,** 173–179.

39. Popovic, T., Fields, P. I., Olsvik, O., Wells, J. G., Evins, G. M., Cameron, D. N., Farmer, J. J., III, Bopp, CA., Wachsmuth, K., Sack, R. B., Albert, M. J., Nair, G. B., Shimada, T., and Feeley, J. C. (1995) Molecular subtyping of toxigenic *Vibrio cholerae* O139 causing epidemic cholera in India and Bangladesh, 1992–1993. *J. Infect. Dis.* **171,** 122–127.

40. Koblavi, S., Grimont, F., and Grimont, P. A. D. (1990) Clonal diversity of *Vibrio cholerae* O1 evidenced by rRNA restriction patterns. *Res. Microbiol.* **141,** 645–657.

41. Faruque, S. M., Alim, A. R. M. A., Rahman, M. M., Siddique, A. K., Sack, R. B., and Albert, M. J. (1993) Clonal relationships among classical *Vibrio cholerae* O1 strains isolated between 1961 and 1992 in Bangladesh. *J. Clin. Microbiol.* **31**, 2513–2516.

42. Popovic, T., Bopp, C. A., Olsvik, O., and Wachsmuth, K. (1993) Epidemiologic application of a standardized ribotype scheme for *Vibrio cholerae* O1. *J. Clin. Microbiol.* **31**, 2474–2482.

43. Cameron, D. N., Khambaty, F. M., Wachsmuth, I. K., Tauxe, R. V., and Barrett, T. J. (1994) Molecular characterization of *Vibrio cholerae* O1 strains by pulsed-field gel electrophoresis. *J. Clin. Microbiol.* **32**, 1685–1690.

44. Olsvik, O., Wahlberg, J., Petterson, B., Uhlen, M., Popovic T., Wachsmuth, I. K., and Fields, P. I. (1993) Use of automated sequencing of PCR-generated amplicons to identify three types of cholera toxin subunit B in *Vibrio cholerae* O1 strains. *J. Clin. Microbiol.* **31**, 22–25.

45. Cook, W. L., Wachsmuth, K., Johnson, S. R., Birkness, K. A., and Samadi, A. R. (1984) Persistence of plasmids, cholera toxin genes and prophage DNA in classical *V. cholerae* O1. *Infect. Immun.* **45**, 222–226.

46. Weber, J. T., Mintz, E. D., Canizares, R., Semiglia, A., Gomez, I., Sempertegui, R., Davila, A., Greene, K. D., Puhr, N. D., Cameron, D. N., Tenover, F. C., Barrett, T. J., Bean, N. H., Ivey, C., Tauxe, R. V., and Blake, P. A. (1994) Epidemic cholera in Ecuador: multidrug-resistance and transmission by water and seafood. *Epidemiol. Infect.* **112**, 1–11.

47. Gerdes, J. C. and Romig, W. (1975) Complete and defective bacteriophages of classical *Vibrio cholerae*: relationship to the kappa type bacteriophage. *J. Virol.* **15**, 1231–1238.

48. Goldberg, S. and Murphy, J. R. (1983) Molecular epidemiological studies of the United States Gulf Coast V*ibrio cholerae* strains: integration site of mutator vibriophage VcA-3. *Infect. Immun.*, **42**, 224–230.

49. De, S. N. (1959) Enterotoxicity of bacteria-free culture filtrate of *Vibrio cholerae*. *Nature (London)* **183**, 1533,1534.

50. Fasano, A., Baudry, B., Pumplin, D. W., Wasserman, S. S., Tall, B. D., Ketley, J. M., and Kaper, J. B. (1991) *Vibrio cholerae* produces a second enterotoxin, which affects intestinal tight junctions. *Proc. Natl. Acad. Sci. USA* **88**, 5242–5246.

51. Trucksis, M., Galen, J. E., Michalski, J., Fasano, A., and Kaper, J. B. (1993) Accessory cholera enterotoxin (Ace), the third toxin of a *Vibrio cholerae* virulence cassette. *Proc. Natl. Acad. Sci. USA* **90**, 5267–5271.

52. Kaper, J. B., Morris, J. G.,Jr., and Levine, M. M. (1995) Cholera. *Clin. Microbiol. Rev.* **8**, 48–86.

53. Echeverria, P., Hoge, C. W., Bodhidatta, L., Serichantalergs, O., Dalsgaard, A., Eampokalap, B., Perrault, J., Pazzaglia, G., O'Hanley, P., and English, C. (1995) Molecular characterization of *Vibrio cholerae* O139 isolates from Asia. *Am. J. Trop. Med. Hyg.,* **52**, 124–127.

54. Kurazono, H., Pal, A., Bag, P. K., Nair, G. B., Karasawa, T., Mihara, T., and Takeda, Y. (1995) Distribution of genes encoding cholera toxin, zonula occludens

toxin, accessory cholera toxin, and El Tor hemolysin in *Vibrio cholerae* of diverse origins. *Microbial Pathogenesis.* **18,** 231–235.

55. Taylor, R. K., Miller, V. I., Furlong, D. B., and Mekalanos, J. J. (1987) Use of *phoA* gene fusions to identify a pilus colonization factor cooridinately regulated with cholera toxin. *Proc. Natl. Acad. Sci. USA* **84,** 2833–2837.
56. Waldor, M. K. and Mekalanos, J. J. (1996) Lysogenic conversion by a filamentous phage encoding cholera toxin. *Science,* **272,** 1910–1914.
57. Colwell, R. R. and Spira, W. M. (1992) The ecology of *Vibrio cholerae,* in Cholera (Barua, D. and Greenough, W. B., III, eds.), Plenum Medical Book Company, NY, pp. 107–127.
58. Karaolis, D. K. R., Lan, R., and Reeves, P. R. (1995) The sixth and seventh cholera pandemics are due to independent clones separately derived from environmental, nontoxigenic, non-O1 *Vibrio cholerae. J. Bacteriol.* **177,** 3191–3198.
59. Pajni, S., Sharma, C., Bhasin, N., Ghosh, A., Ramamurthy, T., Nair, G. B., Ramajayam, S., Das, B., Kar, S., Roychowdhury, S., and Ghosh, R. K. (1995) Studies on the genesis of *Vibrio cholerae* O139: identification of probable progenitor strains. *J. Med. Microbiol.* **42,** 20–25.
60. Bille, J. (1994) Report on the WHO-sponsored international collaborative study of subtyping methods for *Listeria monocytogenes.* Seventh International Congress of Bacteriology and Applied Microbiology, International Union of Microbiological Societies Congress, Prague, Czech Republic, July 3–8, 1994.
61. Cameron, D. N. and Fields, P. I. (1995) Use of polymerase chain reaction to differentiate between multilocus enzyme electrophoretic types of *Vibrio cholerae* O1. Abstacts of the 95th Annual Meeting of the American Society for Microbiology.

20

Diagnosis and Investigation
of Diarrheagenic *Escherichia coli*

James P. Nataro and Juan Martinez

1. Introduction

Although most *Escherichia coli* are harmless commensals of the human intestine, certain specific, highly-adapted *E. coli* strains are capable of causing urinary tract, systemic or enteric/diarrheagenic infection. Diarrheagenic *E. coli* are divided into six distinct categories, or pathotypes, each with a distinct pathogenic scheme (**Table 1**). Combined, diarrheagenic *E. coli* have emerged as perhaps the most important enteric pathogens of man. In the developing world, the *E. coli* categories account for more cases of gastroenteritis among infants than any other cause *(1)*. In addition, *E. coli* are also the most common cause of traveller's diarrhea, which afflicts more than one million travellers to the developing world annually *(1)*. Enterohemorrhagic *E. coli* (EHEC) are the cause of hemolytic uremic syndrome (HUS), which has become a major foodborne threat in many parts of the developed world *(2)*.

Molecular diagnostic techniques have proven to be very useful tools in the detection of diarrheagenic *E. coli*, largely because such organisms are difficult to distinguish from the nonpathogenic *E. coli* of the intestinal microflora. Over the past several years, a large number of molecular diagnostic techniques have been introduced to facilitate the detection of diarrheagenic *E. coli* in human stools and environmental samples. The number of molecular tests presents the investigator or clinician with a myriad of options from which to choose the best detection method. Criteria to be considered when choosing a test include:

1. What genetic target should be detected?
2. What is the most sensitive and specific means of detecting this gene or target?

From: *Methods in Molecular Medicine, Vol. 15: Molecular Bacteriology: Protocols and Clinical Applications*
Edited by: N. Woodford and A. P. Johnson © Humana Press Inc., Totowa, NJ

Table 1
Categories of Diarrheagenic *E. coli*

Category	Toxins	Invasion	Virulence plasmid	Adhesin	Clinical syndrome
ETEC	LT, ST	—	Many	CFA/I, CFA/II, CFA/IV, others	Watery diarrhea
EPEC	—	+	60 MDa	Bundle-forming pilus	Watery diarrhea of infants
EHEC	SLT-1, SLT-2	—	60 MDa[a]	Intimin, Fimbriae[a]	Hemorrhagic colitis, HUS
EAEC	EAST1[a]	?	65 MDa[a]	AAF/I, AAF/II	Watery, persistent diarrhea
EIEC	EIET[a]	+++	140 MDa	Ipa's(?)	Watery diarrhea, dysentery
DAEC	?	?	?	F1845[a]	Watery diarrhea

[a]Role in pathogenesis unproven

3. Given that different methods have various sensitivities and specificities, which test is most practical for my laboratory in terms of cost, man-hours, speed, reliability, and so on?

This chapter describes the most promising molecular methods available for the detection of diarrheagenic *E. coli* and addresses these issues in the context of clinical diagnosis, epidemiology, and pathogenesis research applications.

2. Tests Available for the Detection of Diarrheagenic *E. coli*
2.1. Enterotoxigenic E. coli *(ETEC)*

ETEC cause watery diarrhea that can range from mild self-limiting disease to severe purging. The organism is a major cause of weanling diarrhea in the developing world and the major cause of traveller's diarrhea *(3–5)*. ETEC elicit diarrhea by colonization of the small bowel mucosa, followed by elaboration of the heat-labile (LT) and/or heat-stable (ST) enterotoxins. Colonization of the mucosal surface is usually effected by fimbriae known as colonization factor antigens (CFAs) *(6)*. There are a very large number of different CFAs among human ETEC (and a similarly large number among animal ETEC), although a relatively small number are overrepresented among clinical isolates *(6)*.

ETEC were among the first pathogenic microorganisms for which molecular diagnostic techniques were successfully developed. As early as 1982, Moseley et al. introduced DNA probes for the detection of the LT- and ST-encoding genes in stool and environmental samples *(7)* (**Table 2**). Since that time, several advances have been made in ETEC detection, but genetic tech-

Table 2
Polynucleotides Currently Used
by the Authors for Detection of Diarrheagenic *E. coli*

Category	Probe designation or plasmid source	Probe fragment description	Ref.
ETEC - LT[a]	pCVD403	1.2 kb *Hinc*II	7
EPEC - EAF	pJPN16	1.0 kb *Bam*HI-*Sal*I	26
EPEC - EAE	pCVD434	1.0 kb *Sal*I-*Stu*I	32
EHEC - SLT-I	pJN37-19	1.1 kb *Bam*HI	82
EHEC - SLT-II	pNN110-18	0.84 kb *Sma*I-*Pst*I	82
EHEC - Plasmid	pCVD419	3.4 kb *Hind*III	41
EAEC	pCVD432	0.7 kb *Eco*RI-*Pst*I	59
EIEC	pPS2.5	2.5 kb *Hind*III	68
DAEC	pSLM852	0.39 kb *Pst*I	75

[a]The authors do not use an ST polynucleotide fragment probe (*see* **Note 1**).

niques continue to attract the most attention and use. It should be stressed that there is no perfect test for ETEC. Detection of colonization factors is impractical because of the great number of such factors and their heterogeneity; detection of LT and ST toxins defines an ETEC isolate, yet some such isolates will express colonization factors specific for animals and, thus, lack human pathogenicity.

The LT gene itself serves as a good polynucleotide probe (**Table 2**), providing good sensitivity and specificity when labeled with radioisotopes *(7,8)* or with enzymatic, nonisotopic detection systems *(9)*. Several different protocols have been published in which nonisotopic labeling methods have proven useful for LT detection *(10–15)*; the authors now use a highly reliable alkaline phosphatase-based detection system for use in detecting polynucleotide probe colony blot hybridization (*see* **Note 1**).

The ST polynucleotide probe initially yielded poor sensitivity and specificity, presumably because of the small size of the gene (JPN and Kaper, J., unpublished observations). For this reason, oligonucleotide probes have been developed that are substantially more sensitive and specific for ST detection *(12)* (**Table 3** lists authors' recommendations for nucleotide sequences of oligonucleotides used for probing and PCR of diarrheagenic *E. coli*). An LT oligonucleotide has also been developed (12), but this reagent has few advantages over the enzymatically detected LT fragment probe. Recently, a trivalent oligonucleotide probe has been proposed that may be of use in detecting genes for LT, ST, and the EHEC shiga-like toxin genes (*see* **Subheading 2.3.**); this probe showed promise in an early report *(11)*.

Table 3
Nucleotide Sequences of PCR Oligonucleotide Primers and Oligonucleotide Probes Recommended by the Authors for Detection of Diarrheagenic *E. coli*

Category	Factor	PCR oligos (ref)	Oligonucleotide Probe (ref)
ETEC	STI	1) TTAATAGCACCCGGTACAAGCAGG 2) CTTGACTCTTCAAAAGAGAAAATTAC (19)	(STH)GCTGTGAATTGTGTTGTAATCC (STP)GCTGTGAACTTTGTTGTAATCC (80)
	LT	1) GGCGACAGATTATACCGTGC 2) CCGAATTCTGTTATATATGTC (12)	GCGAGAGGAACACAAACCGG (12)
EPEC	EAE	*see footnote below*	
	EAF	1) CAGGGTAAAAGAAAGATGATAA 2) TATGGGGACCATGTATTATCA (35)	TATGGGGACCATGTATTATCA (28)
	BFP	1) AATGGTGCTTGCGCTTGCTGC 2) GCCGCTTTATCCAACCTGGTA (36)	GCTACGGTGTTAATATCTCTGGCG (JPN, unpublished)
EHEC	EAE	1) CAGGTCGCGTCGTGTCTGCTAAA 2) TCAGCGTGGTTGGATCAACCT (47) (O157:H7-specific)	ACTGAAAGCAAGCGGTGGTG (81)
	SLTI	1) TTTACGATAGACTTCTCGAC 2) CACATATAAATTATTTCGCTC (48) (SLT-I AND II)	GATGATCTCAGTGGGCGTTC (42)
	SLTII	ABOVE	TCTGAAACTGCTCCTGTGTA (42)
	PLASMID	1) ACGATGTGGTTATTCTGGA 2) CTTCACGTCACCATACATAT (48)	CCGTATCTTATAATAAGACGGATGTTGG (48)
EAEC	PLASMID	1) CTGGGCGAAAGACTGTATCAT 2) CAATGTATAGAAATCCGCTGTT (62)	NONE
EIEC	*ial*	1) CTGGATGGTATGGTGAGG 2) GGAGGCCAACAATTATTTCC (67)	CCATCTATTAGAATACCTGTG (69)

Primer pairs are arbitrarily labeled 1) and 2); each primer is written 5′–3′. References for each reagent are listed at the end of the appropriate sequence. See text for abbreviations and discussion. No primers specific for EPEC EAE have been published.

Although it is generally preferable to identify *E. coli* as members of diarrheagenic categories after isolation, it is technically possible to perform DNA hybridization directly on stool samples that have been fixed to paper blots *(16)*. For ETEC, this may be particularly valuable because of the high number of ETEC typically shed in the stools of infected individuals *(13)*.

Several PCR assays for ETEC have proven to be quite sensitive and specific *(12,13,17–20)*, both when used directly on clinical samples as well as on isolated bacterial colonies. A useful adaptation of PCR is the "multiplex" PCR assay *(13,18)*, in which several PCR primers are combined, with the aim of detecting one of several different diarrheagenic *E. coli* pathotypes in a single reaction. In multiplex PCR, a second step of non- isotopic probe hybridization is generally required to differentiate the PCR products.

2.2. Enteropathogenic E. coli (EPEC)

EPEC have emerged as common causes of watery diarrhea among infants in the developing world; yet documented diarrheal outbreaks among infants in the developed world may still occur *(1,21)*. The pathogenesis of EPEC features a characteristic attaching and effacing lesion on intestinal epithelial cells *(22)*; the genes responsible for attaching and effacing are encoded on the chromosome as a closely linked "pathogenicity island" *(23)*. Intimate attachment itself is conferred by the protein intimin, the product of the *eae*A gene. In addition, EPEC typically possess a plasmid that promotes adherence to the intestinal mucosa via a type 4 fimbrial antigen (the Bundle-Forming Pilus or BFP) *(24)*.

Potential targets for molecular diagnosis of EPEC, therefore, include both plasmid and chromosomal genes. The chromosomal locus most often detected is the *eae*A gene. Two plasmid-borne loci include the *bfp*A gene, which is the gene encoding the pilin subunit of the BFP, and the empirically-derived EPEC Adherence Factor ("EAF") probe, which is located adjacent to the BFP-encoding gene cluster.

The EAF probe, the first molecular diagnostic reagent for EPEC, is a one kilobase (kb) *Bam*HI-*Sal*I DNA fragment (**Table 2**) that is adjacent to a HEp-2 adherence-inactivating insertion in the 60 megadalton (MDa) plasmid of strain E2348/69 *(25,26)*. In most studies, the sensitivity and specificity of the EAF probe in detecting the localized pattern of HEp-2 adherence has proven to be greater than 95% *(26)*. However, it is known now that the EAF probe is merely a marker for the highly conserved EAF plasmid *(27)*, and that other plasmid fragments could prove to be at least as powerful as the EAF probe fragment in detecting EPEC.

Jerse et al. also described an oligonucleotide probe consisting of 21 bases derived from one end of the EAF fragment *(28)* (**Table 3**). The sensitivity and specificity of the EAF oligonucleotide probe are similar to those of the EAF polynucleotide probe. However, the EAF oligonucleotide hybridization proto-col is faster than that of the EAF fragment probe, and the oligonucleotide is able to differentiate positives and negatives much more clearly than the fragment probe.

Recent studies have identified specific genes that encode the BFP itself, which is now known to be the molecular structure conferring localized adherence upon EPEC strains *(29,30)*. In a study published by Giron et al. *(31)*, an 850-bp DNA fragment that comprises the *bfp*A gene was used as a DNA probe and compared with results generated by the original EAF probe. These investigators found that although there was general agreement between the two probes, the BFP probe was slightly more sensitive than the EAF probe. Moreover, these data suggest that the BFP fimbrial subunit is a highly conserved protein among EPEC strains.

Jerse et al. found that an *eae*A fragment hybridized with all of 89 EPEC strains on colony blot *(32)* (**Table 2**). However, it is known now that the attaching and effacing locus is not found exclusively among EPEC, but is also present in EHEC, some *Hafnia* strains, some *Citrobacter freundii* strains, and may even be found in nonpathogenic *E. coli* isolates *(33,34)*. Therefore, detection of *eae*A alone is not as sensitive or specific for EPEC as is the use of the BFP or EAF gene probes. In developing countries where EPEC are prevalent, however, the positive predictive value of *eae*A detection may be very high.

Two studies have been published describing the development of PCR for EPEC. Franke et al. *(35)* chose two oligonucleotide primers from the EAF probe sequence (which does not encode a fimbrial gene) (**Table 3**). Using PCR, these investigators found positive reactions with all of 151 EAF probe-positive EPEC and none of 277 EAF probe-negative strains. Gunzberg et al. *(36)* have also developed an EPEC PCR, using instead a *bfp*A gene target (**Table 3**). These investigators detected 10 out of 10 localized adherence (LA)-positive EPEC and none of 95 other bacteria that were LA-negative. One would expect that PCR for the BFP will yield similar results to those obtained using BFP polynucleotide probe hybridization.

2.3. Enterohemorrhagic E. coli (EHEC)

EHEC typically cause an afebrile bloody colitis and, in about 10% of patients, this infection can be followed by HUS *(2)*. Like EPEC, EHEC elicit an attaching and effacing lesion of the intestinal mucosa (this time of the colon), a phenotype that requires a functional *eae*A chromosomal gene *(37)*. The elaboration of an oligomeric shiga-like toxin (SLT-I or SLT-II) contributes to the hemorrhagic colitis and to the development of the systemic sequelae of disease *(38)*. EHEC adhere to the mucosa by virtue of a plasmid-borne fimbrial antigen that is distinct from the BFP *(39)*; the plasmid itself is conserved among

EHEC strains as is the EAF plasmid among EPEC *(39)*. Thus, EHEC generally hybridize with the *eae*A probe, but not with BFP or EAF probes *(31)*. In addition, EHEC will hybridize with genetic sequences from the *slt1* and/or *slt2* genes *(40)*, and with a fragment derived empirically from the EHEC plasmid *(41)* (**Table 2**).

Molecular approaches are assuming greater importance in the detection of EHEC strains. Currently, the most common approach to identification of EHEC in the United States is the cultivation of *E. coli* isolates on MacConkey medium containing sorbitol, since inability to ferment sorbitol is characteristic of EHEC O157:H7. Colonies unable to ferment sorbitol are then serologically identified as O157:H7. This approach is useful for identification of O157:H7 strains; however, an increasing number of EHEC isolates are of serotypes other than O157:H7 and are sorbitol-fermenters *(42)*.

DNA polynucleotide probes have been extensively used for the diagnosis of EHEC. The first fragment probe to achieve wide use was derived empirically from the EHEC plasmid *(41)* (**Table 2**); recently, sequence analysis has shown that this fragment encodes the gene for a hemolysin protein *(43)*. Although this probe has aided the epidemiologic study of the organism, the experience of the authors is that a substantial proportion of EHEC isolates are probe-negative, rendering it insufficiently sensitive for general surveillance or diagnosis.

EHEC may express the genes encoding SLT-I and/or SLT-II, which share approx 60% nucleotide identity *(38,44)*. Brown and coworkers selected oligonucleotides from the A and B subunit genes of SLT-I, SLT-II, and an animal variant known as SLTII-V *(44)*. These workers showed that at the relatively low hybridization temperature of 45°C, the oligonucleotide homologous to the A subunits hybridized with SLT-I, SLT-II, and SLTII-V-producing strains. When the temperature was raised to 53°C, each hybridized only with strains producing the specific toxin variant. Against a collection of EHEC from various parts of the world, the A-I oligonucleotide yielded a sensitivity of 97%, and a specificity of 100% in identifying any SLT-producing *E. coli* (at 45°C). SLT-I- and II-specific oligonucleotides have also been tested, yielding excellent sensitivity and specificity *(42)* (**Table 3**).

PCR has been developed for the detection of *slt* and *eae*A genes *(19,45–48)*. These tests are extremely sensitive in the detection of EHEC and may prove especially useful when diagnosing EHEC infection late in the course, when HUS has already supervened.

None of the three molecular targets used for EHEC diagnosis will produce satisfactory results when detected alone. Neither the *eae*A nor the *slt* genes are specific for EHEC isolates; the EHEC plasmid probe on the other hand, although specific, is not sufficiently sensitive for use as a single tool. Therefore, any approach to EHEC detection will require the use of two or more of the tests. The combination of *eae*A and *slt* detection has proven to provide adequate sensitivity and specificity in the authors' hands and others *(48)* and detects EHEC isolates regardless of their serotype and biochemical properties.

2.4. Enteroaggregative E. coli (EAEC)

EAEC are pathogens associated with persistent diarrhea in the developing world and have been implicated recently in the developed world as causes of both outbreaks and sporadic diarrhea among AIDS patients *(49–52)*. EAEC infection is manifested clinically by the presence of watery mucoid diarrhea, but blood may be found in a significant proportion of patients' stools. At least two enterotoxins have been described in EAEC *(53–55)*, as well as a number of adherence factors *(56,57)*. EAEC infection is diagnosed by the isolation of *E. coli* from the stools of patients and the demonstration of the aggregative adherence (AA) pattern in the HEp-2 assay. Analysis of small bowel aspirates has not increased yield *(58)*. The presence of the typical "stacked brick" pattern of adherence in the HEp-2 assay remains the gold standard for the identification of EAEC.

The large plasmids present in most EAEC strains share a high degree of DNA homology *(59,60)*. Baudry et al. *(59)* selected a 1.0-kb *Sau*3a fragment from the plasmid of strain 17-2 that hybridized with a fragment of similar size from the 65-MDa plasmid of strain 042 (**Table 2**). This DNA probe was hybridized against a collection of EAEC strains from various parts of the world. In the initial report, 56 (89%) of the 63 EAEC strains (by HEp-2 assay) were positive with the EAEC probe. Of 376 strains representing normal flora and other diarrheagenic categories, only two hybridized with the probe. Subsequent experience with the EAEC probe has revealed that the correlation of probe-positivity with AA varies by location. In some studies, the correlation achieves the 89% sensitivity reported by Baudry et al. *(61)*, whereas in other studies the sensitivity may be substantially lower *(58)*. The epidemiologic significance of probe-positive vs probe-negative strains is as yet undetermined.

PCR has also been developed using oligonucleotide primers derived from the probe sequence, analysis of which suggests a single cryptic open reading frame *(62)* (**Table 3**). Sensitivity and specificity of the EAEC PCR is similar to that of the AA probe.

2.5. Enteroinvasive E. coli (EIEC)

EIEC are biochemically, genetically, and pathogenetically closely related to *Shigella* spp. *(63)*. Both characteristically cause an invasive inflammatory colitis, but either may also elicit a watery diarrhea syndrome indistinguishable from that caused by other *E. coli* pathogens. The pathogenesis of disease caused by EIEC and *Shigella* involves cellular invasion and spread, and requires specific chromosomal and plasmid-borne virulence genes *(64)*. The molecular diagnosis of *Shigella* and EIEC relies on detection of these virulence-related genes; however, since the virulence genes of EIEC and Shigella are highly conserved, techniques currently available are unable to distinguish between the two. Thus, identification of an EIEC strain as such first requires that *E. coli* be isolated, then further identified as those containing EIEC/*Shigella*-specific loci.

Two polynucleotide probes have been described for detection of EIEC and *Shigella*. Probe pMR17 is a 17-kb *Eco*RI fragment derived from the invasiveness plasmid (pInv) of a *Shigella flexneri* serotype 5 strain *(65,66)*. *ial* is a 2.5-kb *Hind*III fragment isolated from the pInv of an EIEC strain *(67,68)* (**Table 2**). Both of these probes have been shown to be virtually 100% sensitive and specific for EIEC strains that have retained their virulence *(65)*. It should be noted, however, that EIEC strains may lose all or part of the pInv plasmid on in vitro passage or storage, and, therefore, strains should be hybridized with the probe(s) as soon as possible after they are shed in the feces.

A 21-base oligonucleotide derived from *ial* has been tested as a probe for the detection of *Shigella* and EIEC; this probe proved to be identical in sensitivity and specificity with the polynucleotide probe *(69)* (**Table 3**). When labeled with alkaline phosphatase, this probe was capable of detecting as few as 1000 colony forming units (cfu) of *Shigella flexneri* mixed with human stool, after a 12-h incubation at 37°C.

Frankel et al. derived PCR primers from the *ial* locus that were sensitive and specific for the detection of EIEC and *Shigella* *(69,70)* (**Table 3**). This PCR system was able to detect as few as 10 cfu of *S. flexneri* in stool without enrichment of the sample. The PCR product was verified using the *ial* alkaline phosphatase probe described above. The *ial* PCR is also useful in a multiplex PCR system to simultaneously identify EIEC with other *E. coli* categories *(70)*.

2.6. Diffusely Adherent E. coli (DAEC)

DAEC cause a watery diarrhea syndrome in adults and children outside of infancy *(9,71)*. The pathogenesis of DAEC diarrhea is not as yet elucidated, but several virulence-related characteristics have been identified. Most DAEC strains express a surface fimbria designated F1845 that may be encoded either by the chromosome or a plasmid *(72)*. DAEC strains have been shown to induce characteristic elongated projections from the surface of epithelial cells in culture *(73)*; it is not known whether either of these phenotypes is associated with pathogenicity.

DAEC are defined by the presence of the DA pattern in the HEp-2 adherence assay *(74)*. A DAEC DNA probe is available, which consists of a 390-bp polynucleotide fragment (**Table 2**) derived from the *daa*C gene *(75)*; *daa*C encodes a molecular usher necessary for expression of the F1845 fimbriae. Approximately 75% of DAEC strains from around the world are positive with the F1845 gene probe *(61)*. Because of the genetic relatedness of F1845 with other members of the Dr family of adhesins, false positive reactions with the DA probe may occur, albeit with unknown frequency. No PCR assay has yet been described to identify DAEC.

3. Clinical Diagnosis of Diarrheagenic *E. coli* Disease

The need to diagnose diarrheagenic *E. coli* disease in a particular patient carries with it certain implications. First, one must be able to detect the organism within, or after isolation from, a clinical sample. In most cases, the microbiologist will want to isolate the organism in pure culture for confirmation and, perhaps, antibiotic susceptibility testing. Tests for clinical diagnostic purposes should have great sensitivity and specificity and be available on demand; cost and ability to batch many specimens are lower priorities. For these reasons, PCR assays are often preferred over DNA probe analysis, although use of nonradioactive nucleic acid probes may be satisfactory for this purpose as well. In most cases, testing five lactose-positive colonies isolated on MacConkey or Eosin-methylene blue agar is adequate, although the sensitivity will be modestly improved by testing more colonies, especially in the asymptomatic patient. If EIEC are sought, several lactose-negative colonies should be chosen as well.

A common problem when performing PCR on stools is the presence of nonspecific inhibitory substances in the stool itself. Whereas methods have been proposed to circumvent this problem, chelation of nucleic acid directly onto a glass matrix bead has been shown to produce the greatest sensitivity in head-to-head comparison *(13)* (*see* **Notes 2** and **3**).

Although PCR is useful as a clinical test, there is the potential problem of product carryover from one reaction to the next, and also the possibility of amplifying nucleic acids that are well below the level of clinical or epidemiologic significance, either because the organism is a rare constituent of the gut microflora, or because the organisms are no longer viable. For this reason, the authors suggest that if PCR is utilized, every effort should be made to isolate the implicated strains so that confirmation of their presence is possible.

The isolation of ETEC, EHEC, EPEC, or EIEC from a stool generally implies that this organism is responsible for the patient's symptoms. However, whether all isolates of EAEC and DAEC are pathogenic is not clear, and so a critical question in the management of patients from whom EAEC or DAEC are isolated is whether or not the isolate is in fact the etiologic agent in question. The authors currently accept an EAEC or DAEC as a likely cause of diarrhea in one of the following circumstances:

1. When the patient presents in the course of a documented outbreak,
2. When the patient's isolate can be shown to belong to one of the common serotypes associated with disease (EAEC only) *(76)*,
3. When the patient exhibits persistent diarrhea and stools repeatedly yield EAEC or DAEC as the predominant flora in the absence of another enteric pathogen, and/or
4. When persistent symptoms resolve after eradication of a persistently present EAEC or DAEC strain.

4. Epidemiologic Investigation

The epidemiology of diarrheagenic *E. coli* has been greatly facilitated by the development of molecular diagnostic methods *(77)*. Such techniques not only enable the detection of diarrheagenic *E. coli* in the stools of symptomatic or asymptomatic individuals, but also allow the identification of these pathogens in food, environmental, or household samples. DNA probes have proven to be useful in this regard in that they are able to detect low levels of the organism in a background of other pathogens, but PCR has proven even more powerful. Often, PCR is even more sensitive on environmental samples than in stools, because of the presence of nonspecific inhibitors of PCR in the latter *(13)*.

One important consideration when performing a DNA probe study is to decide how many pure colonies should be selected from each stool specimen to provide optimal detection of diarrheagenic *E. coli*. To generate a preliminary answer to this question, the authors analyzed a study of Chilean infants in which EPEC were found to be associated with diarrheal disease *(9)*. Within this cohort there were 178 episodes of EPEC diarrhea, but EPEC were found in only 11 samples from diarrhea-negative controls. The chance of any single colony being EAF probe-positive from a diarrheal stool (that was already known to generate at least one EAF-positive colony) was 62%. This implies that in the majority of EPEC-infected individuals, EPEC became the predominant *E. coli* isolate within the stool. In controls, the probability of being EAF probe-positive for any one particular colony was 50%. Thus, looking at three colonies from each stool specimen would yield a sensitivity in detecting an EAF-positive specimen of about 95%; adding a fourth colony would increase the sensitivity to 98%. In the controls, three colonies produced a sensitivity of 88%, whereas adding another colony increased the yield to 94%.

5. Research Uses

Molecular diagnostic techniques have also been useful in pathogenesis studies of diarrheagenic *E. coli*. For example, the finding that only EAF-positive

EPEC were associated with diarrheal disease helped to implicate this plasmid and its adherence factor as virulence properties of EPEC *(26)*. This observation was confirmed in volunteer studies *(78)*. Use of DNA probes for ETEC has demonstrated that ST-producing ETEC are more reproducibly and more strongly associated with diarrhea than are LT-producing ETEC *(9)*. A similar observation can be made for SLT, as SLT-II predominates over SLT-I in cases of HUS *(9)*. It should be noted, however, that such observations merely offer hypotheses that need to be confirmed by more rigorous molecular approaches.

The authors utilize molecular diagnostic approaches in the performance of vaccine trials both on the inpatient volunteer ward as well as in the field. The use of molecular probes can greatly simplify surveillance for a given pathogen in a vaccinated community and, perhaps more importantly, such approaches can serve to differentiate attenuated vaccine constructs from wild-type pathogens. Such an approach has been successfully used in field trials of attenuated cholera vaccines (Nataro, J. P. and Kaper, J. B., unpublished data).

6. Future Prospects and Goals

Molecular diagnostic techniques will become increasingly used in the detection of infectious agents of human disease. The diarrheagenic *E. coli* represent one group of organisms in which this approach is particularly fruitful. Future directions of this research will focus on the identification of critical virulence factors for improved detection of new pathogens such as EAEC and DAEC. In addition, one can expect continued improvements in the methodology of nonisotopic nucleic acid-based systems, especially in the development of rapid, sensitive, and highly reliable detection methods that can be readily implemented in clinical laboratories, hopefully in both the developed as well as developing worlds.

7. Notes

1. The authors have developed and used a nonisotopic detection system for polynucleotide probes, based upon commercially available reagents, which produces highly reliable results in the testing of *E. coli* by colony blot. This method works well for LT, EAF, AA, DA, SLT (I and II), *eae*A, EHEC, EIEC DNA fragment probes, all using identical reaction and washing conditions. Such an approach allows us to simultaneously test many colonies in parallel reactions. The authors' method has the advantage of avoiding the use of radioisotopes and of producing probes that are stable at −20°C for many months. The authors use a different approach for detection of ST, utilizing an oligonucleotide conjugated to alkaline phosphatase. Although this is not optimal in that it requires performance of the ST probing separately and using different temperatures and times from the other probes, the authors have not yet developed oligonucleotide probes for all the other categories that have similar melting temperatures and other performance variables.

The polynucleotide probe method they use entails transfer of bacterial growth to Whatman filter paper, lysis *in situ*, and removal of debris using Proteinase K, followed by hybridization and detection. Fragment probes are labeled by using nick translation incorporation of biotin-dATP, followed by addition of a streptavidin-alkaline phosphatase conjugate and detection of the enzyme using nitroblue-tetrazolium and bromo-chloro-indolyl-phosphate (*see* Chapter 2). The nick translation and detection reactions are both performed using commercially-available kits (Bio-Prime DNA Labeling System, and BluGENE Detection System; both from Gibco/BRL Life Technologies, Gaithersberg, MD). The method is essentially as described by Gicquelais et al. *(79)* with the following modifications:

a. We lyse colonial growth, after transfer to Whatman filter paper, by microwave radiation at the highest setting in a standard microwave oven. Since ovens vary, they recommend that the lysis solution (1.5 mL per paper filter) is allowed to boil for 10 s while in the microwave, but that the solution should not be allowed to completely boil away. This may require optimization for any particular microwave oven. When using steam (as described in the original reference *[79]*) instead of a microwave, the filters can be placed in the steamer as soon as the lysis solution (0.5 N NaOH, 1.5 M NaCl) is added; if using the microwave, it is recommended that the filters are allowed to incubate with shaking in the solution for 10 min before heating.

b The authors perform all their hybridizations using Whatman 541 paper filters, in glass Petri dishes and use 1.5 mL of each solution per paper filter (multiple filters can be stacked). The Petri dishes should be washed in Alconox detergent and rinsed in 70% ethanol prior to use to destroy any contaminating phosphatases.

c. After lysis, filters are treated with Proteinase K to remove bacterial debris *(79)*. This should be done at 37–50°C with gentle agitation.

d. The authors have found that after labeling fragment probes with the BioPrime system, 2.5 µL of probe per paper blot produces sufficient signal intensity.

2. The presence of inhibitors of the polymerase chain reaction in human stool has been the subject of several investigations in pursuit of optimal pathogen detection strategies. The identity of these inhibitors is still not known, although it is presumed that food by-products or the presence of anaerobic bacteria account for most of these inhibitory effects. Most commonly, stool samples are simply diluted until the inhibitors are no longer present in sufficient concentration *(12)*. The other popular method involves extraction of nucleic acids from the test material, either with or without digestion of RNA *(67)*. Chelation of nucleic acids by binding to a glass bead, or immunomagnetic separation of the target pathogen from other microorganisms have produced better results in separating the target from background inhibitors (13) (*see* **Note 3**).

3. The glass matrix nucleic acid capture method as described by Stacy-Phipps et al. *(13)* is performed as follows: Approximately 100 mg of stool is suspended in 1.5 mL Dulbecco's phosphate-buffered saline (PBS) in a microcentrifuge tube and centrifuged at 2800g for 1 min to separate debris. The supernatant is transferred to a

new tube and the tube centrifuged at 16,000g for 5 min. The supernatant of this spin is discarded and the pellet is suspended in 0.6 mL GuSCN binding/lysis buffer (5.3 M guanidine thiocyanate, 10 mM dithiothreitol, 1% Tween-20, 300 mM sodium acetate, 50 mM sodium citrate, pH 7.0), and the mixture is incubated at 65°C for 10 min. Fifty microliters of resuspended glass matrix (GlasPac; National Scientific Supply Company, San Rafael, CA) is added to the solution, the mixture is centrifuged at 16,000g for 1 min and the supernatant is discarded. The matrix is resuspended in 1 mL of wash buffer (50% ethanol, 10mM Tris-HCl, pH 7.5, 100 mM NaCl). Following centrifugation at 16,000g for 1 min, the wash step is repeated twice. The matrix pellet is dried at room temperature for 5 min. Bound DNA is eluted by incubation in 100 μL of 10 mM Tris-HCl, pH 8.0, 0.1 mM EDTA at 50°C for 5 min with periodic mixing. Following centrifugation at 16,000g for 2 min, the eluate is transferred to a new tube and used for PCR.

References

1. Levine, M. M. (1987) *Escherichia coli* that cause diarrhea: enterotoxigenic, enteropathogenic, enteroinvasive, enterohemorrhagic, and enteroadherent. *J. Infect. Dis.* **155**, 377–389.
2. Pickering, L. K., Obrig, T. G., and Stapleton, F. B. (1994) Hemolytic-uremic syndrome and enterohemorrhagic *Escherichia coli*. *Pediatr. Infect. Dis. J.* **13**, 459–476.
3. Albert, M. J., Faruque, S. M., Faruque, A. S. G., Neogi, P. K. B., Ansaruzzaman, M., Bhuiyan, N. A., Alam, K., and Akbar, M. S. (1995) Controlled study of *Escherichia coli* diarrheal infections in Bangladeshi children. *J. Clin. Microbiol.* **33**, 973–977.
4. Levine, M. M. (1985) *Escherichia coli* infections. *N. Engl. J. Med.* **313**, 445–447.
5. Echeverria, P., Taylor, D. N., Donohue-Rolfe, A., Supawat, K., Ratchtrachenchai, O., Kaper, J., and Keusch, G. T. (1987) HeLa cell adherence and cytotoxin production by enteropathogenic *Escherichia coli* isolated from infants with diarrhea in Thailand. *J. Clin. Microbiol.* **25**, 1519–1523.
6. Cassels, F. J. and Wolf, M. K. (1995) Colonization factors of diarrheagenic *E. coli* and their intestinal receptors. *J. Indust. Microbiol.* **15**, 214–226.
7. Moseley, S. L., Echeverria, P., Seriwatana, J., Tirapat, C., Chaicumpa, W., Sakuldaipeara, T., and Falkow, S. (1982) Identification of enterotoxigenic *Escherichia coli* by colony hybridization using three enterotoxin gene probes. *J. Infect. Dis.* **145**, 863–869.
8. Lanata, C. F., Kaper, J. B., Baldini, M. M., Black, R. E., and Levine, M. M. (1985) Sensitivity and specificity of DNA probes with the stool blot technique for detection of *Escherichia coli* enterotoxins. *J. Infect. Dis.* **152**, 1087–1090.
9. Levine, M. M., Ferreccio, C., Prado, V., Cayazzo, M., Abrego, P., Martinez, J., Maggi, L., Baldini, M. M., Martin, W., Maneval, D., Kay, B., Guers, L., Lior, H., Wasserman, S. S., and Nataro, J. P. (1993) Epidemiologic studies of *Escherichia coli* diarrheal infections in a low socioeconomic level peri-urban community in Santiago, Chile. *Am. J. Epidemiol.* **138**, 849–869.

10. Rademaker, C. M. A., Martinez-Martinez, L., Perea, E. J., Jansze, M., Fluit, A. C., Glerum, J. H., and Verhoef, J. (1993) Detection of enterovirulent *Escherichia coli* associated with diarrhoea in Seville, Southern Spain, with non-radioactive DNA probes. *J. Med. Microbiol.* **38**, 87–89.

11. Begaud, E. and Germani, Y. (1992) Detection of enterotoxigenic *Escherichia coli* in faecal specimens by acetylaminofluorene-labelled DNA probes. *Res. Microbiol.* **143**, 315–325.

12. Schultsz, C., Pool, G. J., Van Ketel, R., De Wever, B., Speelman, P., and Dankert, J. (1994) Detection of enterotoxigenic *Escherichia coli* in stool samples by using nonradioactively labeled oligonucleotide DNA probes and PCR. *J. Clin. Microbiol.* **32**, 2393–2397.

13. Stacy-Phipps, S., Mecca, J. J., and Weiss, J. B. (1995) Multiplex PCR assay and simple preparation method for stool specimens detect enterotoxigenic *Escherichia coli* DNA during course of infection. *J. Clin. Microbiol.* **33**, 1054–1059.

14. Abdul, A. A., Faruque, S. M., Ahmad, Q. S., Hossain, K. M., Mahalanabis, D., and Albert, M. J. (1994) Evaluation of a non-radioactive chemiluminescent method for using oligonucleotide and polynucleotide probes to identify enterotoxigenic *Escherichia coli*. *J. Diar. Dis. Res.* **12**, 113–116.

15. Jablonski, E., Moomaw, E. W., Tullis, R. H., and Ruth, J. L. (1987) Prepration of oligodeoxynucleotide-alkaline phosphatase conugates and their use as hyrbidization probes. *Nucleic Acids Res.* **14**, 6115–6128.

16. Sommerfelt, H., Bhan, M. K., Srivastava, R., and Bhatnagar, S. (1993) Evaluation of DNA-DNA hybridization for the direct detection of enterotoxigenic *Escherichia coli* in stool blots. *Scand. J. Infect. Dis.* **25**, 457–463.

17. Tornieporth, N. G., John, J., Salgado, K., De Jesus, P., Latham, E., Melo, M. C. N., Gunzburg, S. T., and Riley, L. W. (1995) Differentiation of pathogenic *Escherichia coli* strains in Brazilian children by PCR. *J. Clin. Microbiol.* **33**, 1371–1374.

18. Lang, A. L., Tsai, Y.-L., Mayer, C. L., Patton, K. C., and Palmer, C. J. (1994) Multiplex PCR for detection of the heat-labile toxin gene and Shiga-like toxin I and II genes in *Escherichia coli* isolated from natural waters. *Appl. Environ. Microbiol.* **60**, 3145–3149.

19. Olsvik, O. and Strockbine, N. A. (1993) PCR detection of heat-stable, heat-labile, and Shiga-like toxin genes in *Escherichia coli*, in *Diagnostic Molecular Microbiology: Principles and Applications,* (Persing, D. H., Smith, T. F., Tenover, F. C., and White, T. J., eds.), Rochester, NY, Mayo Foundation, pp. 271–276.

20. du Toit, R., Victor, T. C., and Van Helden, P. D. (1993) Empirical evaluation of conditions influencing the polymerase chain reaction: enterotoxigenic *Escherichia coli* as a test case. *Eur. J. Clin. Chem.* **31**, 225–231.

21. Robins-Browne, R. M. (1987) Traditional enteropathogenic *Escherichia coli* of infantile diarrhea. *Rev. Infect. Dis.* **9**, 28–53.

22. Donnenberg, M. S. and Kaper, J. B. (1992) Enteropathogenic *Escherichia coli*. *Infect. Immun.* **60**, 3953–3961.

23. McDaniel, T. K., Jarvis, K. G., Donnenberg, M. S., and Kaper, J. B. (1995) A genetic locus of enterocyte effacement conserved among diverse enterobacterial pathogens. *Proc. Natl. Acad. Sci. USA* **92,** 1664–1668.

24. Giron, J. A., Ho, A. S. Y., and Schoolnik, G. K. (1991) An inducible bundle-forming pilus of enteropathogenic *Escherichia coli. Science* **254,** 710–713.

25. Baldini, M. M., Nataro, J. P., and Kaper, J. B. (1986) Localization of a determinant for HEp-2 adherence by enteropathogenic *Escherichia coli. Infect. Immun.* **52,** 334–336.

26. Nataro, J. P., Baldini, M. M., Kaper, J. B., Black, R. E., Bravo, N., and Levine, M. M. (1985) Detection of an adherence factor of enteropathogenic *Escherichia coli* with a DNA probe. *J. Infect. Dis.* **152,** 560–565.

27. Nataro, J. P., Maher, K. O., Mackie, P., and Kaper, J. B. (1987) Characterization of plasmids encoding the adherence factor of enteropathogenic *Escherichia coli. Infect. Immun.* **55,** 2370–2377.

28. Jerse, A. E., Martin, W. C., Galen, J. E., and Kaper, J. B. (1990) Oligonucleotide probe for detection of the enteropathogenic *Escherichia coli* (EPEC) adherence factor of localized adherent EPEC. *J. Clin. Microbiol.* **28,** 2842–2844.

29. Donnenberg, M. S., Giron, J. A., Nataro, J. P., and Kaper, J. B. (1992) A plasmid-encoded type IV fimbrial gene of enteropathogenic *Escherichia coli* associated with localized adherence. *Mol. Microbiol.* **6,** 3427–3437.

30. Zhang, H.-Z., Lory, S., and Donnenberg, M. S. (1994) A plasmid-encoded prepilin peptidase gene from enteropathogenic *Escherichia coli. J. Bacteriol.* **176,** 6885–6891.

31. Giron, J. A., Donnenberg, M. S., Martin, W. C., Jarvis, K. G., and Kaper, J. B. (1993) Distribution of the bundle-forming pilus structural gene (*bfp*A) among enteropathogenic *Escherichia coli. J. Infect. Dis.* **168,** 1037–1041.

32. Jerse, A. E., Gicquelais, K., and Kaper, J. B. (1991) Plasmid and chromosomal elements involved in the pathogenesis of attaching and effacing *Escherichia coli. Infect. Immun.* **59,** 3869–3875.

33. Schmidt, H., Rüssmann, H., and Karch, H. (1993) Virulence determinants in nontoxinogenic *Escherichia coli* O157 strains that cause infantile diarrhea. *Infect. Immun.* **61,** 4894–4898.

34. Schmidt, H., Russman, H., Schwarzkopf, A., Aleksic, S., Heesemen, J., and Karch, H. (1994) Prevalence of attaching and effacing *Escherichia coli* in stool samples from patients and controls. *Int. J. Med. Microbiol. Virol. Parasitol. Infect. Dis.* **281,** 201–213.

35. Franke, J., Franke, S., Schmidt, H., Schwarzkopf, A., Wieler, L. H., Baljer, G., Beutin, L., and Karch, H. (1994) Nucleotide sequence analysis of enteropathogenic *Escherichia coli* (EPEC) adherence factor probe and development of PCR for rapid detection of EPEC harboring virulence plasmids. *J. Clin. Microbiol.* **32,** 2460–2463.

36. Gunzberg, S. T., Tornieporth, N. G., and Riley, L. W. (1995) Identification of enteropathogenic *Escherichia coli* by PCR-based detection of the bundle-forming pilus gene. *J. Clin. Microbiol.* **33,** 1375–1377.

37. Donnenberg, M. S., Tzipori, S., McKee, M. L., O'Brien, A. D., Alroy, J., and Kaper, J. B. (1993) The role of the *eae* gene of enterohemorrhagic *Escherichia coli* in intimate attachment in vitro and in a porcine model. *J. Clin. Invest.* **92,** 1418–1424.

38. O'Brien, A. D., Tesh, V. L., Donohue-Rolfe, A., Jackson, M. P., Olsnes, S., Sandvig, K., Lindberg, A. A., and Keusch, G. T. (1992) Shiga toxin: Biochemistry, genetics, mode of action, and role in pathogenesis. *Curr. Top. Microbiol. Immunol.* **180,** 65–94.

39. Karch, H., Heesemann, J., Laufs, R., O'Brien, A. D., Tacket, C. O., and Levine, M. M. (1987) A plasmid of Enterohemorrhagic *Escherichia coli* O157:H7 is required for expression of a new fimbrial antigen and for adhesion to epithelial cells. *Infect. Immun.* **55,** 455–461.

40. Karch, H. and Meyer, T. (1989) Evaluation of oligonucleotide probes for identification of Shiga-like-toxin-producing *Escherichia coli. J. Clin. Microbiol.* **27,** 1180–1186.

41. Levine, M. M., Xu, J., Kaper, J. B., Lior, H., Prado, V., Tall, B., Nataro, J., Karch, H., and Wachsmuth, K. (1987) A DNA probe to identify Enterohemorrhagic *Escherichia coli* of O157:H7 and other serotypes that cause hemorrhagic colitis and hemolytic uremic syndrome. *J. Infect. Dis.* **156,** 175–182.

42. Gunzer, F., Bohm, H., Russman, H., Bitzan, M., Aleksic, S., and Karch, H. (1992) Molecular detection of sorbitol-fermenting *Escherichia coli* O157 in patients with hemolytic-uremic syndrome. *J. Clin. Microbiol.* **30,** 1807–1810.

43. Schmidt, H., Karch, H., and Beutin, L. (1994) The large-sized plasmids of enterohemorrhagic *Escherichia coli* O157 strains encode hemolysins which are presumably members of the *E. coli*-hemolysin family. *FEMS Microbiol. Lett.* **117,** 189–196.

44. Brown, J. E., Sethabutr, O., Jackson, M. P., Lolekha, S., and Echeverria, P. (1989) Hybridization of *Escherichia coli* producing Shiga-like toxin I, Shiga-like toxin II, and a variant of Shiga-like toxin II with synthetic oligonucleotide probes. *Infect. Immun.* **57,** 2811–2814.

45. Karch, H., Janetzki-Mittman, C., Aleksic, S., and Datz, M. (1996) Isolation of enterohemorrhagic *Escherichia coli* O157 strains from patients with hemolytic-uremic syndrome by using immunomagnetic separation, DNA-based methods, and direct culture. *J. Clin. Microbiol.* **34,** 516–519.

46. Brian, M. J., Frosolono, M., Murray, B., Miranda, A., Lopez, E. L., Gomez, H. F., and Cleary, T. G. (1992) Polymerase chain reaction for diagnosis of enterohemorrhagic *Escherichia coli* infection and hemolytic uremic syndrome. *J. Clin. Microbiol.* **30,** 1801–1806.

47. Gannon, V. P. J., Rashed, M., King, R. K., and Thomas, E. J. G. (1993) Detection and characterization of the *eae* gene of shiga-like toxin-producing *Escherichia coli* using polymerase chain reaction. *J. Clin. Microbiol.* **31,** 1268–1274.

48. Fratamico, P. M., Sackitey, S. K., Wiedmann, M., and Deng, M. Y. (1995) Detection of *Escherichia coli* O157:H7 by multiplex PCR. *J. Clin. Microbiol.* **33,** 2188–2191.

49. Savarino, S. J. (1993) Enteroadherent *Escherichia coli*: A heterogeneous group of *E. coli* implicated as diarrhoeal pathogens. *Trans. Roy. Soc. Trop. Med. Hyg.* **87 Suppl. 3,** 49–53.

50. Smith, H. R., Scotland, S. M., Willshaw, G. A., Rowe, B., Cravioto, A., and Eslava, C. (1994) Isolates of *Escherichia coli* O44:H18 of diverse origin are enteroaggregative. *J. Infect. Dis.* **170,** 1610–1613.

51. Wanke, C. A., Schorling, J. B., Barrett, L. J., Desouza, M. A., and Guerrant, R. L. (1991) Potential role of adherence traits of *Escherichia coli* in persistent diarrhea in an urban Brazilian slum. *Pediatr. Infect. Dis. J.* **10**, 746–751.

52. Cravioto, A., Tello, A., Navarro, A., Ruiz, J., Villafan, H., Uribe, F., and Eslava, C. (1991) Association of *Escherichia coli* HEp-2 adherence patterns with type and duration of diarrhoea. *Lancet* **337**, 262–264.

53. Eslava, C., Villaseca, J., Morales, R., Navarro, A., and Cravioto, A. (1993) Identification of a protein with toxigenic activity produced by enteroaggregative *Escherichia coli*. *Abstr. Gen. Meet. Amer. Soc. Microbiol.* **B105**, 44. (Abstract).

54. Savarino, S. J., Fasano, A., Watson, J., Martin, B. M., Levine, M. M., Guandalini, S., and Guerry, P. (1993) Enteroaggregative *Escherichia coli* heat-stable enterotoxin 1 represents another subfamily of *E. coli* heat-stable toxin. *Proc. Natl. Acad. Sci. USA* **90**, 3093–3097.

55. Savarino, S. J., Fasano, A., Robertson, D. C., and Levine, M. M. (1991) Enteroaggregative *Escherichia coli* elaborate a heat-stable enterotoxin demonstrable in an in vitro rabbit intestinal model. *J. Clin. Invest.* **87**, 1450–1455.

56. Nataro, J. P., Yikang, D., Cookson, S., Cravioto, A., Savarino, S. J., Guers, L. D., Levine, M. M., and Tacket, C. O. (1995) Heterogeneity of enteroaggregative *Escherichia coli* virulence demonstrated in volunteers. *J. Infect. Dis.* **171**, 465–468.

57. Nataro, J. P., Deng, Y., Maneval, D. R., German, A. L., Martin, W. C., and Levine, M. M. (1992) Aggregative adherence fimbriae I of enteroaggregative *Escherichia coli* mediate adherence to HEp-2 cells and hemagglutination of human erythrocytes. *Infect. Immun.* **60**, 2297–2304.

58. Fang, G. D., Lima, A. M., Martin, C. C., Barrett, L. J., Nataro, J. P., and Guerrant, R. L. (1992) Aggregative HEp-2 cell adherent *Escherichia coli* and *Cryptosporidium*: important pathogens in hospitalized children with persistent diarrhea in Northeast Brazil. *Abstr. 32nd Intersci. Conf. Antimicrob. Agents Chemother. Abstr.* **686**, (Abstract).

59. Baudry, B., Savarino, S. J., Vial, P., Kaper, J. B., and Levine, M. M. (1990) A sensitive and specific DNA probe to identify enteroaggregative *Escherichia coli*, a recently discovered diarrheal pathogen. *J. Infect. Dis.* **161**, 1249–1251.

60. Vial, P. A., Robins Browne, R., Lior, H., Prado, V., Kaper, J. B., Nataro, J. P., Maneval, D., Elsayed, A., and Levine, M. M. (1988) Characterization of enteroadherent-aggregative *Escherichia coli*, a putative agent of diarrheal disease. *J. Infect. Dis.* **158**, 70–79.

61. Kang, G., Mathan, M. M., and Mathan, V. I. (1995) Evaulation of a simplified HEp-2 cell adherence assay for *Escherichia coli* isolated from south Indian children with acute diarrhea and controls. *J. Clin. Microbiol.* **33**, 2204–2205.

62. Schmidt, H., Knop, C., Franke, S., Aleksic, S., Heeseman, J., and Karch, H. (1995) Development of PCR for screening of enteroaggregative *Escherichia coli*. *J. Clin. Microbiol.* **33**, 701–705.

63. Acheson, D. W. K. and Keusch, G. T. (1995) Shigella and enteroinvasive *Escherichia coli*, in *Infections of the Gastrointestinal Tract*, (Blaser, M. J., Smith, P. D., Ravdin, J. I., Greenberg, H. B., and Guerrant, R. L., eds.), New York, Raven, pp. 763–784.

64. Goldberg, M. B. and Sansonetti, P. J. (1993) Shigella subversion of the cellular cytoskeleton: a strategy for epithelial colonization. *Infect. Immun.* **61**, 4941–4946.

65. Gomes, T. A. T., Toledo, R. F., Trabulsi, L. R., Wood, P. K., and Morris, J. G. (1987) DNA probes for identification of enteroinvasive *Escherichia coli*. *J. Clin. Microbiol.* **25**, 2025–2027.

66. Wood, P. K., Morris, J. G., Small, P. L., Sethabutr, O., Toledo, M. R., Trabulsi, L., and Kaper, J. B. (1986) Comparison of DNA probes and the Sereny test for identification of invasive Shigella and *Escherichia coli* strains. *J. Clin. Microbiol.* **24**, 498–500.

67. Schoolnik, G. K. (1993) PCR detection of Shigella species and enteroinvasive *Escherichia coli*, in *Diagnostic Molecular Microbiology: Principles and Applications*, (Persing, D. H., Smith, T. F., Tenover, F. C., and White, T. J., eds.), Rochester, NY, Mayo Foundation, pp. 277–281.

68. Small, P. L. and Falkow, S. (1986) Development of a DNA probe for the virulence plasmid of Shigella spp. and enteroinvasive *Escherichia coli*, in *Microbiology-1986,* (Lieve, L., Bonventre, P. F., Morello, J. A., Silver, S. D., and Wu, W. C., eds.), Washington, DC, American Society for Micrrobiology, pp. 121–124.

69. Frankel, G., Riley, L., Giron, J. A., Valmassoi, J., Friedman, A., Strockbine, N., Falkow, S., and Schoolnik, G. A. (1990) Detection of *Shigella* in feces using DNA amplification. *J. Infect. Dis.* **161**, 1252–1256.

70. Frankel, G., Giron, J. A., Vlamossoi, J., and Schoolnik, G. (1989) Multi-gene amplification: simultaneous detection of three virulence genes in diarrheal stool. *Mol. Microbiol.* **3**, 1729–1734.

71. Gunzburg, S. T., Chang, B. J., Elliott, S. J., Burke, V., and Gracey, M. (1993) Diffuse and enteroaggregative patterns of adherence of enteric *Escherichia coli* isolated from aboriginal children from the Kimberley region of Western Australia. *J. Infect. Dis.* **167**, 755–758.

72. Bilge, S. S., Apostol, J. M., Jr., Aldape, M. A., and Moseley, S. L. (1993) mRNA processing independent of RNase III and RNase E in the expression of the F1845 fimbrial adhesin of *Escherichia coli*. *Proc. Natl. Acad. Sci. USA* **90**, 1455–1459.

73. Yamamoto, T., Kaneko, M., Changchawalit, S., Serichantalergs, O., Ijuin, S., and Echeverria, P. (1994) Actin accumulation associated with clustered and localized adherence in *Escherichia coli* isolated from patients with diarrhea. *Infect. Immun.* **62**, 2917–2929.

74. Nataro, J. P., Kaper, J. B., Robins Browne, R., Prado, V., Vial, P., and Levine, M. M. (1987) Patterns of adherence of diarrheagenic *Escherichia coli* to HEp-2 cells. *Pediatr. Infect. Dis. J.* **6**, 829–831.

75. Bilge, S. S., Apostol, J. M., Jr., Fullner, K. J., and Moseley, S. L. (1993) Transcriptional organization of the F1845 fimbrial adhesin determinant of *Escherichia coli*. *Mol. Microbiol.* **7**, 993–1006.

76. Nataro, J. P. (1995) Enteroaggregative and diffusely adherent *Escherichia coli*, in *Infections of the Gastrointestinal Tract,* (Blaser, M. J., Smith, P. D., Ravdin, J. I., Greenberg, H. B., and Guerrant, R. L., eds.), New York, Raven, pp. 727–737.

77. Taylor, D. N. and Echeverria, P. (1993) Diarrhoeal disease: current concepts and future chellenges. Molecular biological approaches to the epidemiology of diarrhoeal diseases in developing countries (Review). *Trans. Roy. Soc. Trop. Med. Hyg.* **87,** 3–5.
78. Levine, M. M., Nataro, J. P., Karch, H., Baldini, M. M., Kaper, J. B., Black, R. E., Clements, M. L., and O'Brien, A. D. (1985) The diarrheal response of humans to some classic serotypes of enteropathogenic *Escherichia coli* is dependent on a plasmid encoding an enteroadhesiveness factor. *J. Infect. Dis.* **152,** 550–559.
79. Gicquelais, K. G., Baldini, M. M., Martinez, J., Maggi, L., Martin, W. C., Prado, V., Kaper, J. B., and Levine, M. M. (1990) Practical and economical method for using biotinylated DNA probes with bacterial colony blots to identify diarrhea-causing Escherichia-Coli. *J. Clin. Microbiol.* **28,** 2485–2490.
80. Murray, B. E., Mathewson, J. J., DuPont, H. L., and Hill, W. E. (1987) Utility of oligodeoxyribonucleotide probes for detecting enterotoxigenic *Escherichia coli*. *J. Infect. Dis.* **155,** 809–811.
81. Willshaw, G. A., Scotland, S. M., Smith, H. R., Cheasty, T., Thomas, A., and Rowe, B. (1994) Hybridization of strains of *Escherichia coli* O157 with probes derived from the *eae*A gene of enteropathogenic E. coli and the *eae*A homolog from a Vero cytotoxin-producing strain of E. coli O157. *J. Clin. Microbiol.* **32,** 897–902.
82. Newland, J. W. and Neill, R. J. (1988) DNA probes for Shiga-like toxins I and II and for toxin-converting bacteriophages. *J. Clin. Microbiol.* **26,** 1292–1297.

21

Campylobacter Infections

Species Identification and Typing

Janet R. Gibson and Robert J. Owen

1. Introduction

Campylobacters are the most frequently identified cause of acute bacterial diarrhea in humans in England and in many other developed countries. Although *C. jejuni* and *C. coli* are numerically the most important species in cases of campylobacter enteritis, there is a growing awareness that some of the other 13 species of *Campylobacter* may also be linked with human disease. However, precise identification of these organisms is rarely carried out in clinical laboratories, and the significance of most species or subtypes in causing human disease is unknown.

The epidemiology of human enteric infections caused by campylobacters is not entirely clear. Most cases in England are apparently sporadic with a seasonal peak occurring in late spring, but the sources of the infections remain largely undetermined. Large outbreaks occur infrequently and are usually associated with the consumption of contaminated water or unpasteurized milk. Although feces from infected individuals can contain 10^6–10^9 campylobacters per gram *(1)*, the organisms are thought to be rarely transmitted from person to person, despite a reported low infective dose *(2)*.

The cost implications of campylobacteriosis in health care are substantial, as are the costs of controlling contamination in water and in the food industry, especially as contaminated poultry and raw milk have been identified as vehicles of infection *(3)*.

In this chapter, the authors shall briefly review the current phenotypic identification and typing methods for campylobacters, and shall describe the main genotyping methods available for epidemiologic studies of *C. jejuni*.

From: *Methods in Molecular Medicine, Vol. 15: Molecular Bacteriology: Protocols and Clinical Applications*
Edited by: N. Woodford and A. P. Johnson © Humana Press Inc., Totowa, NJ

2. Speciation

Campylobacters are relatively inert in most conventional biochemical tests, and there are no convenient methods available that readily identify all species. They are microaerophilic, grow at 37°C, and are oxidase-positive. *C. jejuni* is distinguished from other species by the ability to hydrolyze hippurate. However, truly hippurate-negative isolates of *C. jejuni* have been reported *(4)*, and their identity has been confirmed using genetic methods. Definitive identification of the less common species relies on time-consuming genetic analysis such as 16S ribosomal RNA (rRNA) gene sequencing, or DNA-DNA hybridization. During the 1980s, considerable effort was directed toward developing DNA probes for the identification and detection of campylobacters *(5)*, but the methodology is relatively complex and has not been widely applied.

More recently, species-specific PCR-based assays have been described *(6–9)*, mostly based on the rRNA gene sequences. For instance, Eyers et al. *(6)* and Bastyns et al. *(7)* developed PCR primers targeted at 23S rRNA gene sequences that are specific for a wide range of campylobacters important in both human and veterinary medicine. Restriction fragment length polymorphisms of PCR products obtained from amplification of the 16S rRNA gene have also been used to discriminate between species of *Campylobacter (8)*. However, these assays have yet to be fully evaluated and further rapid PCR-based methods are needed to identify the full range of *Campylobacter* species.

3. Epidemiologic Investigation

3.1. Conventional Typing Methods

A diverse range of conventional phenotyping methods, differing widely in discriminatory power, have been applied to the typing of *C. jejuni* over the past 20 yr *(10)*. Biotyping, serotyping, and phage typing have been the most widely used, and are reviewed below.

3.1.1. Lior Biotyping Scheme

The extended biotyping scheme of Lior, based on hippurate hydrolysis, rapid H_2S production, and DNA hydrolysis with toluidine blue-DNA agar, is a simple scheme for separating strains of the most commonly encountered "thermophilic" species *(11)*. It provides low discrimination, however, with only four *C. jejuni*, two *C. coli*, and two *C. lari* biotypes. Strains of *C. jejuni* are usually either biotype I or II. The scheme does not reliably distinguish between *C. coli* and *C. lari*, and is not applicable to other species with potential clinical importance, such as *C. upsaliensis, C. fetus*, and *C. hyointestinalis*.

3.1.2. Resistotyping

The Preston biotyping scheme uses 11 resistotyping tests and four basic identification tests to provide a numerical code *(12)*. To date, the scheme has not been widely used and there have been no reported evaluations. A disk system of biotyping, based on six of the Preston resistotyping tests (tetracycline 18 µg; nalidixic acid 30 µg; metronidazole 5 µg; 2,3,5-triphenyltetrazolium chloride 1000 µg; 5-fluorouracil 80 µg; and sodium arsenite 30 µg), has been commercially developed (Mast Diagnostics, Bootle, UK) and may facilitate rapid biotyping in the future. Although the commercial disk scheme is simple to perform and appears discriminatory, a high proportion of strains (up to 44%) within the two most commonly occurring Penner serotypes (HS1 and HS4 complex) were sensitive to all agents *(13)*.

3.1.3. Serotyping

Two serotyping systems, one based on heat-stable (HS) soluble antigens (Penner scheme) *(14)*, and the other on thermolabile (HL) antigens (Lior scheme) *(15)*, have gained international recognition for typing *C. jejuni* and *C. coli*. There is currently no central or commercial supply of antisera, and so the standardization and comparison of results from different laboratories using in-house antisera is problematic. Even so, the HS scheme is recognized to provide a useful basis for initial typing of isolates from disparate sources.

3.1.4. Phage Typing

Two phage typing systems for *C. jejuni* and *C. coli*, known as the Lior *(16)* and the Preston *(17)* schemes, have been developed and the bacteriophages and propagating strains deposited in national culture collections (ATCC and NCTC). Neither scheme has been widely used, and consequently, data on typability and discriminatory power of the schemes are limited. The associations between phage type and other phenotypic properties, such as serotype or biotype, are currently unclear.

3.2. Genotypic Typing Methods

3.2.1. Genomic DNA Digests

3.2.1.1. RESTRICTION ENZYME ANALYSIS

Analysis of restriction endonuclease digest patterns produced by high frequency cutting enzymes is referred to as chromosomal restriction-enzyme analysis (REA), bacterial restriction-endonuclease digest analysis (BRENDA), or genomic DNA digestion (GDD) (*see* Chapter 2). Such analyses were first applied to strains within various species of the genus *Campylobacter* because of the limitations of biochemical and other phenotypic typing methods. The species examined by REA include *C. jejuni, C. coli, C. lari, C. fetus* subsp. *fetus*, and *C. upsaliensis (18–21)*. The DNA digest patterns with commonly used restriction endonucleases, such as *Eco*RI, *Hae*III, *Hin*dIII, *Bst*II, and *Xho*I, appear to be highly stable and provide a sensitive means of differentiating individual strains. Densitometric scanning and numerical analysis has facilitated, to some extent, the comparison of patterns and has enabled relationships between some *Campylobacter* species to be determined *(19)*. In general, however, the complex DNA patterns are difficult to analyze and interpret, and although occasionally used on limited sets of strains *(18–21)*, they are unsuitable for general typing purposes.

3.2.1.2. PULSED-FIELD GEL ELECTROPHORESIS (PFGE)

An alternative method that simplifies DNA restriction digests involves the use of enzymes that cut the chromosomal DNA into a few large fragments, which are then separated by PFGE *(see* Chapter 3). In 1991, Yan and colleagues *(22)* showed that PFGE analysis of *Sma*I-restricted genomic DNA provided a reliable means of differentiating *C. jejuni* from *C. coli*, and was useful for intraspecific strain differentiation. The importance of PFGE as a new subtyping method was established in the author's subsequent systematic investigations of *Sma*I PFGE profiles of DNA from *C. jejuni* and *C. coli* representing various Penner heat-stable serogroups *(23–25)*. Considerable diversity was observed (**Fig. 1**) with differences evident between strains of the same serotype, as well as between those of different serotypes. The authors' data highlight the value of PFGE profiling as a subtyping method because it groups as well as distinguishes between strains within each species *(23–25)*. Strains that yield similar *Sma*I profiles may be further subtyped by analysis with an alternative endonuclease such as *Kpn*I *(see* **Notes 1** and **2**). One potential problem of the technique is that DNAse-positive strains of *C. jejuni* degrade chromosomal DNA during standard preparation procedures before PFGE analysis. However, this problem can be avoided if the DNAse is inactivated by formaldehyde fixation of the bacterial cells before preparation of DNA for PFGE *(26)* *(see* **Note 3**).

Recent studies on campylobacters show that PFGE offers high typability and discrimination as well as reproducibility *(23–25)*. Although the PFGE patterns are readily amenable to scanning and storage on computer for the compilation of pattern databases *(see* Chapter 3), such analysis is time-consuming and as such is unsuitable for the routine screening of large numbers of strains. Therefore, PFGE should be considered as a supplementary technique for studying selected sets of isolates (e.g., suspected outbreak clusters). It is necessary, however, first to obtain a database of the patterns relevant to the primary screening divisions (e.g., each HS serotype) in order to interpret the PFGE

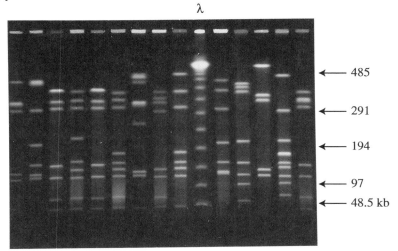

λ

← 485

← 291

← 194

← 97

← 48.5 kb

Fig. 1. Examples of *Sma*I genomic DNA digests of *C. jejuni* separated by PFGE. The gel (1% agarose) was subjected to electrophoresis for 22 h at 200 V and 14°C with ramped pulse times from 10 s to 35 s. The lanes contain digests from separate strains with fragments ranging in size from approx 40–500 kb. The tenth lane contains size markers comprising successively larger concatemers (48.5 kb) of lambda DNA.

profiles. A lack of pattern variation in strains of a particular serotype may reflect the clonal nature of that serotype, rather than indicate epidemiologic relatedness. For example, the authors' studies have shown that sporadic strains of HS11 and HS27 show a low degree of genetic variability, whereas strains of the HS1 and HS4 complex comprise many PFGE types.

3.2.2. Ribotyping

Ribotyping is a technique in which labeled probes are used to detect rRNA genes in restriction fragments on a Southern blot (*see* Chapter 2). The resultant pattern of bands is referred to as the ribopattern. Several different rRNA gene probes have been used to study campylobacters over the past few years. These include:

1. A cDNA of 16S + 23S rRNA of *Escherichia coli*, the so-called universal probe, has been successfully used for ribotyping *C. jejuni* and *C. coli*, and several other *Campylobacter* spp. *(27)*. Most analyses were based on *Hae*III digests, which provided profiles of six to eight bands, although Patton et al. *(28)* suggested that *Pvu*II and *Pst*I digests also provided a sensitive ribotyping system. *E. coli* rRNA is a commercially available reagent (*see* Chapter 2).
2. A cDNA of 16S + 23S rRNA from *C. jejuni* NCTC 11168 gave similar results to the above universal probe *(29)*.

3. A 16S rRNA-specific intragenic probe (1500 bp) generated by PCR amplification from DNA of *C. jejuni* NCTC 11168 (*see* **Note 4**) has been used extensively in the authors' recent work *(23–25,30)* to detect the three copies of the 16S rRNA gene, which appear to be a feature of thermophilic campylobacters. When using an enzyme such as *Hae*III, which does not cut within the 16S rRNA gene, only three bands are detected. It is necessary, therefore, to use at least two endonucleases to create enough pattern variation to give adequate discrimination for general typing purposes. The authors' systematic studies of *C. jejuni* and *C. coli* have used a combination of *Hae*III and *Pst*I profiles, neither of which have cutting sites within the *Campylobacter* 16S rRNA gene. Such ribopatterns are highly reproducible, and their simplicity facilitates rapid analysis of results. To date, the authors have identified approx 52 *Hae*III and 29 *Pst*I ribopatterns, or about 80 combined ribotypes among several hundred strains of *C. jejuni*. These form the basis of a recently proposed 16S ribotyping scheme that covers all Penner HS serotypes of *C. jejuni (30)*.

Associations between ribotype and HS serotype are complex and some serotypes (e.g., HS1) have a predominant and characteristic ribotype, whereas strains of other serotypes (e.g., HS19) comprise several ribotypes. Also, certain ribotypes are a feature of a number of different serotypes, (e.g., HS1, HS2, and HS63 all share a predominant *Pst*I:*Hae*III combined ribotype pattern).

The main disadvantage of ribotyping is technical complexity and, in the absence of expensive automated equipment, it involves several time-consuming steps: DNA extraction, electrophoresis, blotting, hybridization, and detection of bands (*see* Chapter 2). However, it does offer 100% typability and provides a useful way of checking culture purity, because mixed cultures are indicated if the 16S ribopattern comprises more than three distinct bands.

3.2.3. PCR-Based Techniques

Several PCR-based DNA fingerprinting techniques have been successfully applied in the epidemiologic investigation of campylobacter infections *(31–40)*. These offer the advantages of simplicity, speed, and relative low cost for consumables. However, the results have yet to be systematically compared with those of other more established typing methods.

Amplification of *C. jejuni* DNA can be achieved from whole cells, present on agar plates or on storage beads, without the need to isolate DNA. Light suspensions of bacteria are prepared in distilled water for use as template DNA. DNAse-positive strains, suspended in distilled water, require immediate boiling for 10 min prior to use in the PCR mix.

3.2.3.1. Fla-Typing

DNA profiles are obtained by restriction digestion of PCR-amplified flagellin gene sequences. Primers that may be used to amplify the *flaA* gene are given in **Note 5**. The discrimination provided by the technique depends on the enzyme used; low discrimination with *Hin*fI *(31)*, but greater discrimination with *Dde*I *(32)*, *Alu*I *(33)*, or a combination of *Pst*I and *Eco*RI *(34)*. The authors' studies indicate that *Dde*I *fla*-typing is overall less discriminatory than PFGE analysis, although strains within certain *Sma*I PFGE types can be further differentiated on the basis of their *fla* profiles. The method appears to have potential because of its relative simplicity, apparent reproducibility, and the possibility of good discrimination with the appropriate enzyme. However, there have been no attempts to standardize methods between laboratories, and consistent associations with either Penner or Lior serotypes have not emerged *(31,34,35)*. Also, the long-term stability of the *fla* genes in comparison with that of the entire genome is unknown.

3.2.3.2. RAPD PROFILING

Randomly amplified polymorphic DNA (RAPD) sequences are generated using a single short (usually 10-mer) oligonucleotide primer chosen without regard to sequences in the target genome (*see* Chapter 6). The divergence of primer site locations within the genome results in the generation of different sized fragments (i.e., distinct band profiles) for differing strains. Although RAPD analysis has been applied to campylobacters (*see* **Note 6**), in particular to outbreak investigations *(36–38)*, problems remain with the reproducibility aspects of the technique. Profiles are complex and vary according to the quality of the template DNA, the reaction mix components (particularly Mg^{2+} ion concentration), the source of *Taq* polymerase, and the thermocycler employed.

3.2.3.3. PCR BASED ON REPETITIVE SEQUENCES

Primers based on repetitive, extragenic palindromic (REP) elements, and enterobacterial repetitive intergenic consensus (ERIC) sequences (*see* Chapter 7), together with an arbitrary primer, have been used to subtype strains of *C. jejuni* and *C. upsaliensis* associated with outbreaks *(39,40)*. Iriate and Owen *(40)* concluded that, although the fingerprints for *C. jejuni* were highly discriminatory, they lacked serotype specificity and did not consistently match "types" defined by other molecular methods. However, the profiles were reproducible, provided that defined technical protocols were adhered to precisely. The technique had demonstrable potential for the rapid initial investigation of outbreak strains.

3.2.4. Plasmid Profiling

Plasmid profiling (*see* Chapter 4) has been used only to a limited extent for typing *C. jejuni* because most strains give low (<30%) plasmid recovery rates *(41)*. Higher plasmid contents have been found in *C. coli*, especially in porcine isolates, and the method may have more potential for typing *C. upsaliensis* with at least 60% of strains carrying one or more plasmids *(42,43)*. Overall, the method appears to offer low discrimination for typing campylobacters, but it may be useful as a strain fingerprinting feature if all other properties are identical *(38)*.

4. Future Goals

Some key aspects that need to be considered in the future development of *Campylobacter* speciation and subtyping are:

1. Further research into methods that allow rapid speciation, particularly of hippurate-negative species.
2. Standardization of genetic methods as far as possible so that results from different laboratories can be compared with confidence.
3. Continued development and evaluation of PCR-based procedures, such as *fla*-profiling. Extensive data collection is required in order to define types within such schemes and to assess their value as secondary typing methods.

As yet, no single typing method for *C. jejuni* has been identified that allows both high discrimination and high throughput of strains. A combination of screening and subtyping methods (e.g., serotyping and PFGE analysis) is essential to provide the level of discrimination necessary for epidemiologic surveillance purposes.

5. Notes

1. PFGE. The DNA present in a 3 × 6 × 1 mm block is digested with 20 U of *Sma*I for 6 h at 25°C. *Kpn*I digests also require 20 U of enzyme for 6 h (37°C), although better results are obtained if the enzyme solution is renewed after 3 h.
2. The PFGE running conditions used in the authors' laboratory for the DNA digests are: *Sma*I, 200 V for 22 h with ramped pulse times from 10–35 s; *Kpn*I, 200 V for 40 h with ramped pulse times from 2–40 s.
3. To inactivate endogenous DNAse activity in *Campylobacter* spp. prior to PFGE block preparation, 900 µL of cell suspension equivalent to a MacFarlands 5 standard is added to 100 µL of formaldehyde solution (37–40%, BDH) in a 1.5 mL microcentrifuge tube. After incubation for 1 h at room temperature, the cells are washed three times in 1 mL of saline, and finally resuspended in approx 700 µL of saline. The volume of saline used for cell resuspension is adjusted according to the estimated cell loss that occurred during the washing procedure. The solution may require repeated pipeting in order to ensure even redistribution of the bacteria. In some instances, the cells adhere to the walls of the tube, and initially do not spin down. These are harvested either by removing the supernatant and recentrifuging the tube, or by washing the sides of the tube with saline. Once washed, the cell suspension is incorporated into agarose and lysed following the usual procedures *(26)* (*see* Chapter 3).

4. The primers used to produce the 1500-bp 16S rDNA probe correspond to nucleotides 7–27 and 1492–1510 of the *E. coli* 16S *rrnB* sequence. These are:
Forward: 5'-AAGAGTTTGATACCTGGCTCAG-3'
Reverse: 5'-GGTTACCTTGTTACGACTT-3'
The amplification cycle consists of an initial denaturation of target DNA at 94°C
for 5 min followed by 28 cycles of denaturation at 94°C for 1 min, primer annealing at 52°C for 1 min, and extension at 72°C for 2 min. The product is analyzed
by electrophoresis in 0.7% agarose, purified, and labeled by random-priming.

5. Primers that may be used to amplify the *fla*A gene:
Forward[*]: 5'-ATGGGATTTCGTATTAAC-3'
Reverse[†]: 5'-GCACC(CT)TTAAG(AT)GT(AG)GTTACACCTGC-3'
[*]Primer pg50, described by Oyofo et al. *(44)*. This is located on the plus strand
beginning at the N terminus.
[†]Primer RAA19, described by Alm et al. *(34)*. This binds to both flagellin genes
274 bp from the 3' end (sequence from *C. coli*).
Amplification of DNA involves an initial denaturation step at 94°C for 2 min,
followed by 25 cycles of 94°C for 15 s (denaturation), 45°C for 15 s (annealing),
and 72°C for 30 s (extension). Thirty cycles are used when amplification is performed directly on cells in distilled water.

6. The primer OPA-11 (Operon Technology Res. Inc., Alameda, CA) with sequence
5'-CAATCGCCGT-3' was selected after preliminary testing of a number of different
primers *(37)* as it gave a reproducible and easy-to-read pattern of bands. A reaction
volume of 100 μL was made up by addition of 2.5 U of *Taq* polymerase (BCL),
10 mM Tris-HCL, pH 8.3 at 25°C; 50 mM KCl; 2.0 mM MgCl$_2$, 200 μM of each
deoxynucleotide triphosphate, 0.3 μM of the primer, 10 μL of the cell suspension
(OD$_{260}$ = 0.15), and sterile distilled water. This solution was overlaid with 100 μL of
paraffin oil and cycled through the following temperature profile: one cycle of 94°C
for 1 min, and 45 cycles of 94°C for 1 min, 36°C for 1 min, and 72°C for 2 min.

References

1. Blaser, M. J., Hardesty, H. L., Powers, B., and Wang, W. L. L. (1980) Survival of
Campylobacter fetus subsp. *jejuni* in biological milieus. *J. Clin. Microbiol.* **11,**
309–313.
2. Robinson, D. A. (1981) Infective dose of *Campylobacter jejuni* in milk. *B.M.J.*
282, 1584.
3. Healing, T. D., Greenwood, M. H., and Pearson, A. D. (1992) *Campylobacter* and
enteritis. *Rev. Med. Microbiol.* **3,** 159–167.
4. Totten, P. A., Patton, C. M., Tenover, F. C., Barrett, T. J., Stamm, W. E.,
Steigerwalt, A. G., Lin, J. Y., Holmes, K. K., and Brenner, D. J. (1987) Prevalence and characterization of hippurate-negative *Campylobacter jejuni* in King
Country, Washington. *J. Clin. Microbiol.* **25,** 1747–1752.
5. Owen, R. J. (1995) Recent developments in the molecular systematics of
campylobacters and arcobacters in *Molecular Approaches to Food Safety,* (Eklund, M.,
Richard, J. L., and Mise, K., eds.), Alaken, Fort Collins, CO, pp. 19–38.

6. Eyers, M., Chapelle, S., Van Camp, G., Goossens, H., and De Wachter, R. (1993) Discrimination among thermophilic *Campylobacter* species by polymerase chain reaction amplification of 23S rRNA gene fragments. *J. Clin. Microbiol.* **31,** 3340–3343.

7. Bastyns, K., Chapelle, S., Vandamme, P., Goossens, H., and De Wachter, R. (1994) Species-specific detection of campylobacters important in veterinary medicine by PCR amplification of 23S rDNA areas. *System. Appl. Microbiol.* **17,** 563–568.

8. Cardarelli-Leite, P., Blom, K., Patton, C. M., Nicholson, M. A., Steigerwalt, A. G., Hunter, S. B., Brenner, D. J., Barrett, T. J., and Swaminathan, B. (1996) Rapid identification of *Campylobacter* species by restriction fragment length polymorphism analysis of a PCR-amplified fragment of the gene encoding for 16S rRNA. *J. Clin. Microbiol.* **34,** 62–67.

9. Giesendorf, B. A. J., Van Belkum, A., Koeken, A., Stegeman, H., Henkens, M. H. C., van der Plas, J., Gossens, H., Niesters, H. G. M., and Quint, W. G. V. (1993) Development of species-specific DNA probes for *Campylobacter jejuni, Campylobacter coli,* and *Campylobacter lari* by polymerase chain reaction fingerprinting. *J. Clin. Microbiol.* **31,** 1541–1546.

10. Owen, R. J. and Gibson, J. R. (1995) Update on epidemiological typing of *Campylobacter. PHLS Microbiol. Digest* **12,** 2–6.

11. Lior, H. (1984) New extended biotyping scheme for *Campylobacter jejuni, Campylobacter coli,* and "*Campylobacter laridis.*" *J. Clin. Microbiol.* **20,** 636–640.

12. Bolton, F. J., Wareing, D. R. A., Skirrow, M. B., and Hutchinson, D. N. (1992) Identification and biotyping of campylobacters in *Identication Methods in Applied and Environmental Microbiology,* (Board, R. G. ,Jones, D., and Skinner, F. A., eds.), SAB Technical Series 29. Academic Press, London, pp. 151–161.

13. Owen, R. J., Lorenz, E., and Gibson, J. (1997) Application of the MAST resistotyping scheme to *Campylobacter jejuni* and *Campylobacter coli. J. Med. Microbiol.* **46,** 34–38.

14. Penner, J. L. and Hennessy, J. N. (1980) Passive hemagglutination technique for serotyping *Campylobacter fetus* subsp. *jejuni* on the basis of soluble heat-stable antigens. *J. Clin. Microbiol.,* **12,** 732–737.

15. Lior, H., Woodward, D. L., Edgar, J. A., Laroche, L. J., and Gill, P. (1982) Serotyping of *Campylobacter jejuni* by slide agglutination based on heat-labile antigenic factors. *J. Clin. Microbiol.* **15,** 761–768.

16. Khakhria, R. and Lior, H. (1992) Extended phage-typing scheme for *Campylobacter jejuni* and *Campylobacter coli. Epidem. Infect.* **108,** 403–414.

17. Salama, S. M., Bolton, F. J., and Hutchinson, D. N. (1990) Application of a new phage typing scheme to campylobacters isolated during outbreaks. *J. Epidemiol. Infect.* **104,** 405–411.

18. Bradbury, W. C., Pearson, A. D., Marko, M. A., Congi, R. V., and Penner, J. L. (1984) Investigation of a *Campylobacter jejuni* outbreak of serotyping and chromosomal restriction endonuclease analysis. *J. Clin. Microbiol.* **19,** 342–346.

19. Owen, R. J., Costas, M., and Dawson C. (1989) Application of different chromosomal DNA fingerprints to specific and subspecific identification of *Campylobacter* isolates. *J. Clin. Microbiol.* **27**, 2338–2343.

20. Owen, R. J. and Hernandez, J. (1990) Genotypic variation in *"Campylobacter upsaliensis"* from blood and faeces of patients in different countries. *FEMS Microbiol. Lett.* **72**, 5–10.

21. Lind, L., Sjogren, E., Melby, K., and Kaijser, B. (1996) DNA fingerprinting and serotyping of *Campylobacter jejuni* isolates from epidemic outbreaks. *J. Clin. Microbiol.* **34**, 892–896.

22. Yan, W., Chang, W., and Taylor, D. E. (1991) Pulsed-field gel electrophoresis of *Campylobacter jejuni* and *Campylobacter coli* genomic DNA and its epidemiological application. *J. Infect. Dis.* **163**, 1068–1072.

23. Owen, R. J., Sutherland, K., Fitzgerald, C., Gibson, J., Borman, P., and Stanley, J. (1995) Molecular subtyping scheme for serotypes HS1 and HS4 of *Campylobacter jejuni*. *J. Clin. Pathol.* **33**, 872–877.

24. Stanley, J. D., Linton, D., Sutherland, K., Jones, C., and Owen, R. J. (1995) High-resolution genotyping of *Campylobacter coli* identifies clones of epidemiologic and evolutionary significance. *J. Infect. Dis.* **172**, 1130–1134.

25. Gibson, J. R., Fitzgerald, C., and Owen, R. J. (1995) Comparison of PFGE, ribotyping and phage typing in the epidemiological analysis of *Campylobacter jejuni* serotype HS2 infections. *Epidem. Infect.* **115**, 215–225.

26. Gibson, J. R., Sutherland, K., and Owen, R. J. (1994) Inhibition of DNase activity in PFGE analysis of DNA from *Campylobacter jejuni*. *Lett. Appl. Microbiol.* **19**, 357–358.

27. Fayos, A., Owen, R. J., Desai, M., and Hernandez, J. (1992) Ribosomal RNA gene restriction fragment diversity amongst Lior biotypes and Penner serotypes of *Campylobacter jejuni* and *Campylobacter coli*. *FEMS. Microbiol. Lett.* **95**, 87–94.

28. Patton, C. M., Wachsmuth, I. K, Evins, G. M., Kiehlbauch, J. A., Plikaytis, B. D., Troup, N., Tompkins, L., and Lior, H. (1991) Evaluation of 10 methods to distinguish epidemic associated *Campylobacter* strains. *J. Clin. Microbiol.* **29**, 680–688.

29. Owen, R. J., Desai, M., and Garcia, S. (1993) Molecular typing of thermotolerant species of *Campylobacter* with ribosomal RNA gene patterns. *Res. Microbiol.* **144**, 709–720.

30. Fitzgerald, C., Owen, R. J., and Stanley, J. (1996) Comprehensive ribotyping scheme for heat-stable serotypes of *Campylobacter jejuni*. *J. Clin. Microbiol.* **34**, 265–269.

31. Owen, R. J., Fitzgerald, C., Sutherland, K., and Borman, P. (1994) Flagellin gene polymorphism analysis of *Campylobacter jejuni* infecting man and other hosts and comparison with biotyping and somatic antigen serotyping. *Epidemiol. Infec.* **113**, 221–234.

32. Nachamkin, I., Bohachick, K., and Patton, C. M. (1993) Flagellin gene typing of *Campylobacter jejuni* by restriction fragment length polymorphism analysis. *J. Clin. Microbiol.* **31**, 1531–1536.

33. Birkenhead, D., Hawkey, B. M., Heritage, J., Gascoyne-Binzi, D. M., and Kite, P. (1993) PCR for the detection and typing of campylobacters. *Lett Appl. Microbiol.* **17**, 235–237.

34. Alm, R. A., Guerry, P., and Trust, T. J. (1993) Distribution and polymorphism of the flagellin genes from isolates of *Campylobacter coli* and *Campylobacter jejuni. J. Bacteriol.* **175**, 3051–3057.

35. Burnens, A. P., Wagner, J., Lior, H., Nicolet, J., and Frey, J. (1995) Restriction fragment length polymorphisms among the flagellar genes of the Lior heat-labile serogroup reference strains and field strains of *Campylobacter jejuni* and *C. coli. Epidemiol. Infect.* **114**, 423–431.

36. Mazurier, S., Van de Guessen, A., Heuvelman, K., and Wernars, K. (1992) RAPD analysis of *Campylobacter* isolates; fingerprinting without the need to purify DNA. *Lett. Appl. Microbiol.* **14**, 260–262.

37. Owen, R. J. and Hernandez, J. (1993) Ribotyping and arbitrary-primer PCR fingerprinting of campylobacters, in *New Techniques in Food and Beverage Microbiology,* (Kroll, R. G., Gilmour, A., and Sussman, M., eds.), SAB Tech Ser 31, Blackwell, Oxford, pp. 265–285.

38. Fayos, A., Owen, R. J., Hernandez, J., Jones, C., and Lastovica, A. (1993) Molecular subtyping by genome and plasmid analysis of *Campylobacter jejuni* serogroups 01 and 02 (Penner) from sporadic and outbreak cases of human diarrhoea. *Epidemiol. Infect.* **111**, 415–427.

39. Giesendorf, B. A. J., Goossens, H., Niesters, H. G. M., van Belkum, A., Koeken, A., Enditz, H. P., Stegeman, H., and Quint, W. G. V. (1994) Polymerase chain reaction-mediated DNA fingerprinting for epidemiological studies on *Campylobacter* spp. *J. Med. Microbiol.* **40**, 141–147.

40. Iriarte, M. and Owen, R. J. (1996) Repetitive and arbitary primer DNA sequences in PCR-mediated fingerprinting of outbreak and sporadic isolates of *Campylobacter jejuni. FEMS Immunol. Med. Microbiol.* **648**, 17–22.

41. Lind, L., Sjogren, E., Welinder-Olsson, C., and Kaijser, B. (1989) Plasmids and serogroups in *Campylobacter jejuni. APMIS* **97**, 1097–1102.

42. Owen, R. J. and Hernandez, J. (1990) Occurrence of plasmids in "*Campylobacter upsaliensis*" (catalase negative or weak group) from geographically diverse patients with gastroenteritis or bacteraemia. *Eur. J. Epidemiol.* **6**, 111–117.

43. Stanley, J., Jones, C., Burnens, A., and Owen, R. J. (1994) Distinct genotypes of human and canine isolates of *Campylobacter upsaliensis* determined by 16S rRNA gene typing and plasmid profiling. *J. Clin. Microbiol.* **32**, 1788–1794.

44. Oyofo, B. A., Thornton, S. A., Burr, D. H., Trust, T. J., Pavloskis, O. R., and Guerry, P. (1992) Specific detection of *Campylobacter jejuni* and *Campylobacter coli* by using polymerase chain reaction. *J. Clin. Microbiol.* **30**, 2613–2619.

22

Detection and Typing of *Helicobacter pylori*

Robert J. Owen and Janet R. Gibson

1. Introduction

Helicobacter pylori is a curved microaerobic bacterium that was first isolated from human antral gastric biopsy material in 1982 by Marshall and colleagues in Perth, Western Australia *(1)*. Since then, enormous interest has developed in the micro-organism that now appears to be one of the most common human bacterial pathogens, estimated to be infecting at least one third of the world population *(2)*. Although most individuals appear to be asymptomatic, *H. pylori* is implicated as a key risk factor in a number of gastrointestinal diseases, including duodenal and gastric ulceration and gastric cancer *(3)*. Most clinical and basic research aspects have been extensively reviewed *(4)*, but some key features of the general epidemiology of *H. pylori* infections are as follows:

1. Humans are almost the exclusive host.
2. Gastric mucosa is the principal site of colonization.
3. Distribution is worldwide.
4. Seroprevalence in developed countries is 40–50%.
5. Seroprevalence in developing countries is 60–90%.
6. Prevalence of infection increases with age.
7. Spread is believed to be mainly by person-to-person contact.

This chapter will review present methods for detection and typing of *H. pylori*.

2. Diagnosis
2.1. Conventional Methods

The routine diagnosis of *H. pylori* infection relies on a number of invasive and noninvasive detection methods (**Table 1**). Such methods have been

From: *Methods in Molecular Medicine, Vol. 15: Molecular Bacteriology: Protocols and Clinical Applications*
Edited by: N. Woodford and A. P. Johnson © Humana Press Inc., Totowa, NJ

Table 1
Methods Commonly Used
for Routine Clinical Laboratory Detection of *H. pylori*

Invasive methods	Noninvasive methods
Culture from biopsy	Serologic tests (IgG ELISA kits)
Histology of biopsy	Urea breath test (C^{13} and C^{14})
Biopsy rapid urease test	Culture from feces
Culture from gastric juice	

described in a number of excellent articles *(3,5)*, but are continually undergoing development. Histology of biopsy specimens is the most widely used invasive method and is usually combined with culture of the specimen. Serologic tests are most commonly used for noninvasive detection, and a wide selection of ELISA kits are commercially available. Most of these methods offer high sensitivity and specificity *(6)*.

2.2. Molecular Methods

Various molecular methods have been developed as alternative approaches for the detection of *H. pylori* in clinical material, and these may be resorted to when conventional tests give negative results. The two main approaches available are as follows:

2.2.1. DNA-DNA Hybridization

The first attempt at molecular detection of *H. pylori* was by the use of *in situ* hybridization. The application involved the use of a biotinylated genomic DNA probe to detect organisms in wax-embedded biopsies that were sequential specimens from a single patient *(7)*. An oligonucleotide probe also was developed as a means of detecting *H. pylori* in gastric tissue, and a good correlation with other conventional methods (urease test, staining, and culture) was reported *(8)*. These and other probes *(9–11)* developed for *H. pylori* are listed in **Table 2**. Although these studies demonstrated the feasibility of using probes for detection of *H. pylori*, the approach has rarely been used only in practice.

2.2.2. PCR Assays

PCR-based assays for detecting *H. pylori* have become the most widely investigated molecular detection method, and assays based on various target gene sequences have been developed and applied to a range of clinical specimens. Most work has been done on detecting *H. pylori* in gastric biopsies, and occasionally gastric aspirates (**Table 3**). Generally, the evaluations *(10,12–18)* showed good agreement between the PCR assays and other diagnostic meth-

Table 2
DNA Probes for Detection and Identification of *H. pylori*

Probe	Test format	Ref.
Genomic DNA	Dot blot	*9*
Genomic DNA	*In situ*	7
16S r RNA Oligo-probe 22 bp	Dot blot	*8*
Genomic DNA	Dot blot	*8*
Cloned fragment 1.9 kb	Southern blot	*10*
Cloned fragment 1.2kb	Southern blot	*11*

Table 3
Detection of *H. pylori* in Gastric Biopsies and Gastric Aspirates by PCR

Target	Specimen[a]	Country[b]	No. of individuals tested	Ref.
cryptic	B, GA	US	33,37	*10*
ureA	B, EB	UK	23,5	*12* (see **Note 1**)
rRNA	B	US	15	*13*
surface antigen	B	SW	27	*14*
ureA	B	NL	66	*15* (see **Note 1**)
rRNA	B, GA	UK	23	*16*
ureA	B	TW	17	*17* (see **Note 1**)
rRNA	EB	US	95	*18*

[a]B,biopsy; GA, gastric aspirate; EB, embedded biopsy.
[b]TW, Taiwan; NL, Netherland; SW, Sweden.

ods, which included culture and/or histology (>97%) and serology (100%), as well as the rapid urease test and urea breath test. PCR assays have also been increasingly applied as a means of detecting *H. pylori* in dental plaque, saliva, and feces (**Table 4**), although the results from such studies are more difficult to assess *(14,16,19–25)*. Detection levels varied considerably from study to study and as culture was generally negative for samples, it was difficult to determine specificity and sensitivity precisely, even with the inclusion of appropriate controls.

2.3. Comments

PCR assays appear to have the potential for the greatest sensitivity (1–1000 cfu) and specificity (96%), although in practice that may not be achievable because of inhibitors in sample extracts, particularly from dental plaque and feces. PCR assays may be valuable as an additional laboratory diagnostic test for *H. pylori*

Table 4
Detection of *H. pylori* in the Mouth and Feces by PCR

Target	Specimen[a]	Country	No. of individuals	Ref.
Surface antigen	SAL	SW	19	*14*
ureA	DP	UK	45	*19* (*see* **Note 1**)
rRNA	DP	USA	18	*20*
Surface antigen	DP	SW	19	*21*
rRNA	SAL	UK	13	*16* (*see* **Note 1**)
ureA	DP	UK	6	*22* (*see* **Note 1**)
rRNA	FE	UK	seeded[b]	*23*
rRNA	FE	UK	42	*24*
rRNA	FE	NL	24	*25*

[a]DP, dental plaque; SAL, saliva; FE, feces.
[b]1 pg detected 1.9×10^7 to 7.4×10^4 cells/mL.

detection when there has been a delay in culturing biopsies, or for checking culture-negative specimens from seropositive patients. They also may be useful for checking contamination of endoscopes and other equipment. Staining, rapid urease test, and serology all provided similar levels of sensitivity, and were marginally more sensitive than culture. The oligonucleotide and genomic probes were the least sensitive methods with levels of detection of 5×10^3 cfu and 5×10^4 cfu, respectively.

3. Epidemiologic Investigation

A range of different typing and fingerprinting techniques have been applied to *H. pylori* (**Table 5**) with varying degrees of success. The main methods are reviewed as follows:

3.1. Conventional Typing Methods

3.1.1. Biotyping

H. pylori is inactive in biochemical tests most commonly used to characterize medically-important bacteria, and is generally homogeneous with respect to enzymic profiles *(26)*. However, there are some differences in preformed enzymes between strains, and these form the basis of the API Zym (Bio Merieux, Basingstoke, Hants) biotyping system, which uses five enzymes (acid phosphatase, leucine arylamidase, naphthol phosphohydrolase, C4, and C8 esterases) to distinguish four biotypes. In a study of 126 strains of *H. pylori* from ten countries, 85% belonged to just one biotype (biotype 2) *(27)*, indicating that biotyping provided a low degree of discrimination.

Table 5
Phenotyping and Genotyping Methods for *H. pylori*

Phenotyping	Genotyping
Biotyping	Electrophoretic profiles of genomic DNA digests
Serotyping	—frequent-cutting enzymes
Coagglutination (protein A)	—low-frequency-cutting enzymes with PFGE
Immunoblotting	Ribotyping
Total protein profiling	—16 + 23S rRNA genes
Outer membrane profiling	—16S rRNA gene
Lectin agglutination	Urease gene hybridization patterns
Multilocus enzyme	—urease A, C, D genes
electrophoresis (MLEE)	PCR-based profiling
	—RAPD and REPs
	—urease gene RFLPs
	—flagellin gene RFLPs
	Plasmid profiling

3.1.2. Serotyping

Serotyping based on either flagellar *(28)* or heat-stable lipopolysaccharide *(29)* antigens appears to have potential for discriminating between isolates of *H. pylori*. However, neither method has been systematically developed, so serotyping is not yet an available option for the clinical or reference laboratory.

3.2. Molecular Typing Methods

Analysis of *H. pylori* DNA by a number of different techniques has revealed considerable diversity between isolates, with those from different patients rarely having the same genetic fingerprint. So far, such results have not provided evidence either of clonality among *H. pylori* strains, or of divisions within the species that may be considered to represent types.

3.2.1. Restriction Endonuclease Digest Patterns

Restriction endonuclease analysis (*see* Chapter 2) has been widely applied to *H. pylori* to compare strains from different individuals, including family sets, from different gastric sites in the same patient, as well as to compare those obtained before and after treatment *(26)*. However, the endonucleases most commonly used for such studies, *Hae*III and *Hin*dIII, produce complex, multibanded patterns. Strains from different individuals generally produce unique profiles, and it has proved difficult to interpret the differences between such patterns. Low-frequency-cutting enzymes, such as *Not*I and *Nru*I have been used in pulsed-field gel electrophoretic (PFGE) analysis *(30)* (*see* Chapter 3), but are not yet widely applied to epidemiologic studies. Although such profiles were highly discriminatory, the authors found that DNA from about 25% of strains may not be cut by a particular restriction endonuclease (unpublished observations).

3.2.2. Southern Blot Analysis

Southern blot hybridization analysis allows simplification of the banding patterns obtained following digestion with frequent-cutting endonucleases. A number of different genes have been targeted for detection, including the 16S and 23S ribosomal RNA (rRNA) genes, urease genes, and *H. pylori*-specific genes. The major findings from such studies are as follows:

3.2.2.1. RIBOTYPING (*SEE* CHAPTER 2)

Hybridization probes based on either the 16S rRNA gene sequence alone *(31)*, or both the 16S and 23S rRNA gene sequences *(32)*, have been used to detect the two copies of the genes present in *H. pylori*. The complexity of the banding pattern depends on the restriction endonuclease used and on which probe is employed. Ribopatterns obtained with either *Hae*III or *Hin*dIII digests probed with 16S + 23S rRNA sequences are highly discriminatory, with distinctive profiles of three to five bands, ranging in size from 1 kb to10 kb. The patterns are reproducible and may be stored on computer for future comparative analysis *(32)*.

Ribotyping has been used to fingerprint isolates from individuals in the UK as well as from various other parts of the world and to monitor pre- and posttreatment strains of *H. pylori (26)*. The results showed that approx 60% of UK patients are each colonized by a unique strain with a uniform genotype, but up to 40% of individuals contained strains with more than one DNA subtype (one or two band differences). Although isolates from unrelated individuals are genotypically distinct, similarities have been observed in strains from some family members *(33)*. Ribotyping has proved to be suitable for studies of genomic variation in defined populations, such as sequential patient isolates, but has not led to the delineation of distinct types. The technique is laborintensive and is, therefore, not suitable for the examination of large numbers of strains.

3.2.2.2. UREASE GENE PROBES

H. pylori produces a large quantity of urease that hydrolyzes urea to form ammonia and carbon dioxide, and enables the organism to survive the acidic conditions of the stomach. At least eight genes are involved in urease production and, of these, the *ureA* and *ureB* gene polypeptides correspond to the two structural subunits of the enzyme. *Hin*dIII-generated restriction fragment length polymorphisms (RFLPs) containing the *ureA* and *ureC* and *ureD* genes were detected using PCR-derived probes *(34)*. The 411-bp *ureA* probe (*see* **Note 1**) hybridized to one fragment whereas the larger 1.7-kb *ureC* and *ureD* probe (*see* **Note 2**) highlighted between two and four bands. Analysis of the combined *ureA* and *ureC* and *ureD* profiles subdivided 64 strains into 24 patterns, 11 of which contained more than one strain. A key finding of the study was the greater homogeneity observed among strains from patients with disease symptoms (ulcers and gastritis) compared with strains from asymptomatic individuals.

3.2.2.3. PROBES TO H. PYLORI-SPECIFIC FRAGMENTS

Chromosomal DNA fragments, cloned randomly from the *H. pylori* type strain (NCTC 11637) have been used to probe RFLPs on Southern blots *(35)*. Although these detected undefined genetic loci, considerable diversity in hybridization patterns was observed, and the method proved to be highly discriminatory.

3.2.3. PCR-based techniques

The advent of PCR provided a means for the rapid fingerprinting of isolates of *H. pylori*. DNA fragment patterns have been obtained either directly using arbitrary primers, or primers to repetitive sequences, or indirectly through endonuclease digestion of PCR product (PCR-RFLP analysis).

3.2.3.1. ARBITRARILY PRIMED PCR

This fingerprinting technique, AP-PCR (also known as RAPD, randomly amplified polymorphic DNA), (*see* Chapter 6), uses a single short (usually a 10 mer) primer to amplify a number of fragments in the bacterial genome, and has been used to distinguish between clinical isolates of *H. pylori (36)*.

3.2.3.2. PCR BASED ON REPETITIVE EXTRAGENIC SEQUENCES

Repetitive extragenic pallindromic (REP) sequences have been identified in many bacterial genomes (*see* Chapter 7). Primers based on REP sequences have been used to amplify short spacer fragments falling between the repeat motifs in *H. pylori (37)*. Cluster analysis of the DNA fingerprints generated by these primers suggested that strains grouped according to gastric pathology (e.g., peptic ulcer or gastritis). However, the method has not been evaluated for typing purposes.

3.2.3.3. PCR-RFLP ANALYSIS

This technique involves PCR amplification of all or part of a selected gene or cluster of genes using specific primers. The product, usually 1.5–2.5 kb is then digested with a restriction endonuclease and the resultant fragments are separated by gel electrophoresis to produce a banding pattern. Both the gene and the restriction endonuclease need to be carefully chosen to achieve the required degree of interstrain discrimination. *H. pylori* genes subjected to such analysis have included the urease, flagellin, and 48-kDa stress protein genes. Interstrain variability within the *ureA* and *ureB* structural genes (*see* **Note 3**), as well as of within the *ureC* and *ureD* genes *(26,38–40)* was established. Double enzyme digests of *ureA* with *Hae*III and *Alu*I further enhanced discrimination *(40)*. Urease gene-based RFLP analysis compared favorably with ribotyping in the level of discrimination, and had the advantage of speed and simplicity. In contrast, PCR-RFLP analysis of the flagellin (*flaA* and *flaB*) genes (*see* **Note 4**) showed less interstrain variability, but the level of discrimination depended on the enzyme used (e.g., *Alu*I > *Mbo*I > *Hind*III > *Msp*I) *(41,42)*. *Alu*I and *Sau*3A digests of PCR product from the 48-kDa stress protein have also been used to fingerprint strains of *H. pylori* *(40)*.

4. Future Prospects and Objectives

Unlike many other causative agents of communicable disease, no baseline information exists on strain types of *H. pylori* because of the lack of general epidemiologic typing schemes. Progress on the development of such schemes has been slow, as recognized outbreaks of infection do not occur, and it is difficult to obtain representative numbers of strains from gastric biopsy material for detailed epidemiologic surveillance. Current knowledge of the epidemiology of the organism is predominantly based on patient serologic data. There is a need for improved typing systems in order to carry out transmission studies and to investigate possible reservoirs of infection. Detailed typing is also essential to develop an understanding of strain population distribution and evolution, as well as for monitoring strains pre- and posttreatment. In particular, typing of *H. pylori* should be aimed at establishing the specific types associated with different gastric pathologies. In summary, future research priorities should include:

1. Development of improved and rapid phenotypic and genotypic systems for strain typing with the construction of databases of internationally recognized types.
2. Studies on the identification of pathogenic genotypes and their incidence, as well as their associations with different gastric pathologies.
3. Development of improved conventional bacteriological media formulations and molecular methods for the detection and isolation of *H. pylori* from clinical samples (feces, dental plaque, gastric juice, and saliva) as well as for testing environmental samples.
4. Studies on modes and routes of person-to-person transmission with special reference to families, children in playgroups and schools, and other institutionalized populations.

5. Notes

1. A 411-bp internal fragment of the *ureA* gene corresponding to nucleotides [nt] 304 to 714, is amplified by PCR using the primers (5'-GCCAATGG-TAAATTAGTT-3' and 5'-CTCCTTAATTGTTTTTAC-3') described by Clayton and colleagues *(12)*. Genomic DNA of *H. pylori* NCTC 11637 (50 ng) is amplified in a reaction mixture containing 200 µ*M* deoxynucleoside triphosphates (dNTPs), 0.2 µ*M* of each primer and 2.5 U of Taq DNA polymerase. The reaction involves an initial heating step at 95°C for 5 min, followed by 30 cycles of denaturation at 94°C for 30 s, primer annealing at 48°C for 1 min and primer extension at 72°C for 1.5 min. The final step is extension for 5 min at 72°C. The products are analyzed by electrophoresis in 1% agarose, purified, and labeled by random-priming for use as probes.

2. The 1.7-kb probe spanning the *ureC* and *ureD* genes corresponds to nt 521 to 2240 and is amplified using the primers 5'-TGGGACTGATGGCGTGAGGG-3' and 5'-ATCATGACATCAGCGAAGTTAAAAATGG-3' *(34)*. Genomic DNA of *H. pylori* NCTC 11637 is amplified in the reaction mixture described in **Note 1**. The reaction involves an initial heating step at 95°C for 5 min followed by 20 cycles of denaturation at 95°C for 1 min, annealing at 60°C for 1 min and extension at 72°C for 2 min with a final extension step at 72°C for 5 min.

3. The 2.4-kb fragment containing the *ureA* and *ureB* genes is amplified using the forward primer 5'-AGGAGAATGAGATGA-3' and the reverse primer 5'-ACTTTATTGGCTGGT-3' of Foxall et al. *(43)*. Amplification involves 30 cycles of 94°C for 1 min (denaturation), 40°C for 1 min (annealing), and 72°C for 1 min (extension).

4. Oligonucleotide primers used for the amplification of an internal 1.5-kb *fla*A fragment are based on the *H. pylori* strain 898-1 *fla*A sequence *(44)*. The primers, which are located near the 5' and 3' ends of the *fla*A gene have the sequence 5'-ATGGCTTTTCAGGTCAATAC-3' (forward position 139) and 5'-CCTTAAGATAT-TTTGTTGAACG-3' (reverse position 1659). The amplification cycle consists of an initial denaturation of target DNA at 95°C for 5 min followed by 25 cycles of denaturation at 94°C for 1 min, primer annealing at 60°C for 1 min, and extension at 72°C for 2 min with a final extension step for 5 min at 72°C.

References

1. Marshall, B. and Warren, J. R. (1983) Unidentified curved bacilli on gastric epithelium in active chronic gastritis. *Lancet* **1,** 1273–1275.
2. Taylor, D. N. and Blaser, M. J. (1991) The epidemiology of *Helicobacter pylori* infection in Helicobacter pylori, in *Peptic Ulceration and Gastritis,* (Marshall, B. J., McCallum, R. W., and Gurrent, R. L., eds.), Blackwell Scientific Publications, Oxford, pp. 46–54.
3. Calam, J. (1995) *Clinicians' Guide to* Helicobacter pylori. Chapman and Hall Medical, London.

4. Calam, J. (1995) *Helicobacter pylori* in *Baillière's Clinical Gastroenterology International Practise and Research,* vol 9. Baillière Tindall, London.

5. Mégraud, F. (1996) Diagnosis of *Helicobacter pylori* in *Helicobacter pylori: Techniques for Clinical Diagnosis and Basic Research,* (Lee, A. and Mégraud, F., eds.), W. S. Saunders, London, pp. 507–518.

6. Feldman, R. A., Deeks, J. J., Evans, S. J. W., and the *Helicobacter pylori* Serology Study Group. (1995) Multi-laboratory comparison of eight commercially available *Helicobacter pylori* serology kits. *Eur. J. Clin. Microbiol. Infect. Dis.* **14,** 428–433.

7. Van den Berg, F. M., Zijlmans, H., Langenberg, W., Rauws E., and Schipper M. (1989) Detection of *Campylobacter pylori* in stomach tissue by DNA in situ hybridisation. *J. Clin. Pathol.* **42,** 995–1000.

8. Morotomi, M., Hoshina, S., Green. P., Neu, H. C., LoGerfo, P., Wantanabe, I., and Weinstein, I. B. (1989) Oligonucleotide probe for detection and identification of *Campylobacter pylori. J. Clin. Microbiol.* **27,** 2652–2655.

9. Wetherall, B. L., McDonald, P. J., and Johnson, A. M. (1988) Detection of *Campylobacter pylori* DNA by hybridisation with non-radioactive probes in comparison with a ^{32}P-labelled probe. *J. Med. Microbiol.,* **26,** 257–263.

10. Valentine, J. L., Arthur, R. R., Mobley, H. L. T., and Dick, J. D. (1991) Detection of *Helicobacter pylori* by using the polymerase chain reaction. *J. Clin. Microbiol.* **29,** 689–695.

11. Desai, M., Linton, D., Owen, R. J., Cameron, H, and Stanley, J. (1993) Genetic diversity of *Helicobacter pylori* indexed with respect to clinical symptomology, using a 16S rRNA and a species-specific DNA probe. *J. Appl. Bacteriol.,* **75,** 574–582.

12. Clayton, C. L., Kleanthous, H., Coates, P. J., Morgan, D. D., Tabaqchali, S. (1992) Sensitive detection of *Helicobacter pylori* by using polymerase chain reaction. *J. Clin. Microbiol.,* **30,** 192–200.

13. Engstrand, L., Nguyen, A., Graham, D. Y., E-Zaatari, F. A. K. (1992) Reverse transcription and polymerase chain reaction amplification of rRNA for detection of *Helicobacter* species. *J. Clin. Microbiol.,* **30,** 2295–2301.

14. Hammar, M., Tyszkiewicz, T., Wadstrom, T., and O'Toole, P. W. (1992) Rapid detection of *Helicobacter pylori* in gastric biopsy material by polymerase chain reaction. *J. Clin. Microbiol.* **30,** 54–58.

15. Van Zwet, A. A., Thijs, J. C., Kooistra-Smid, A. M. D., Schirm, J., and Snijder, J. A. M. (1993) Sensitivity of culture compared with that of polymerase chain reaction for detection of *Helicobacter pylori* from antral biopsy samples. *J. Clin. Microbiol.* **31,** 1918–1920.

16. Mapstone, N. P., Lynch, D. A. F., Lewis, F. A., Axon, A. T. R., Tompkins, D. S., Dixon, M. F., and Quirke, P. (1993) PCR identification of *Helicobacter pylori* DNA in the mouths and stomachs of patients with gastritis using PCR. *J. Clin. Microbiol.* **46,** 540–543.

17. Wang, J. T., Lin, J. T., Sheu, J. C., Yang, J. C., Chen, D. S., and Wang, T. H. (1993) Detection of *Helicobacter pylori* in gastric biopsy tissue by polymerase chain reaction. *Eur. J. Clin. Microbiol. Infect. Dis.* **12,** 367–371.

18. Weiss, J., Mecca, J., Da Silva, E., and Gassner, D. (1994) Comparison of PCR and other diagnostic techniques for detection of *Helicobacter pylori* infection in dyspeptic patients. *J. Clin. Microbiol.* **32,** 1663–1668.
19. Banatvala, N., Lopez, C. R., Owen, R. J., Abdi, Y., Davies, G., Hardie, J., and Feldman, R. (1993) *Helicobacter pylori* in dental plaque. *Lancet* **341,** 380.
20. Nguyen, A. H., Engstrand, L., Genta, R. M., Graham, D. Y., and El-Zaatari, F. A. K. (1993) Detection of *Helicobacter pylori* in dental plaque by reverse transcription-polymerase chain reaction. *J. Clin. Microbiol.* **31,** 783–787.
21. Olsson, K., Wadstrom, T., and Tyszkieweiz. (1993) *H. pylori* in dental plaques. *Lancet* **341,** 956–957.
22. Owen, R. J., Hurtardo, A., Banatvala, N., Feldman, R. A., and Hardie, J. (1994) Direct DNA fingerprinting of *Helicobacter pylori* in dental plaque by PCR amplification and restriction analysis of urease A gene sequence. *Serodiagn. Immun. Ther. Infect. Dis.* **6,** 196–202.
23. Ho S-A., Hoyle, J. A., Lewis, F. A. *et al.* (1991) Direct polymerase chain reaction test for detection of *Helicobacter pylori* in humans and animals. *J. Clin. Microbiol.* **29,** 2543–2549.
24. Mapstone, N. P., Lynch, D. A. F, Lewis, F. A., Axon, A. T. R., Tompkins, D. S., Dixon, M. F., and Quirke, P. (1993) PCR identification of *Helicobacter pylori* in faeces from gastritis patients. *Lancet* **341,** 447.
25. Van Zwet, A. A., Thijs, J. C., Kooistra-Smid, A. M. D., Schirm, J., and Snijder, J. A. M. (1994) Use of PCR with feces for detection of *Helicobacter pylori* infections in patients. *J. Clin. Microbiol.* **32,** 1346–1348.
26. Owen, R. J. (1995) Bacteriology of *Helicobacter pylori* in *Baillière's Clinical Gastroenterology International Practise and Research,* vol 9, (Calam, J., ed.), Baillière Tindall, London, pp. 415–446.
27. Owen, R. J., Desai, M. (1990) Preformed enzyme profiling of *Helicobacter pylori* and *Helicobacter mustelae* from human and animal sources. *Lett. Appl. Microbiol.* **11,** 103–105.
28. Lior, H. (1991) Serological characterisation of *H. pylori* : a provisional serotyping scheme. *Ital. J. Gastroenterol.* **23, (Suppl. 2),** 42.
29. Mills, S. D., Kurjanczyk, L. A., and Penner, J. L. (1992) Antigenicity of *Helicobacter pylori* lipopolysaccharides. *J. Clin. Microbiol.* **30,** 3275–3180.
30. Taylor, D. E., Eaton, M., Chang, N., and Salama, S. M. (1992) Construction of *Helicobacter pylori* genomic map and demonstration of diversity at the genome level. *J. Bacteriol.* **174,** 6800–6806.
31. Linton, D., Moreno, M., Owen, R. J., and Stanley, J. (1992) 16S rrn gene copy number in *Helicobacter pylori* and its application to molecular typing. *J. Appl. Bacteriol.* **73,** 501–506.
32. Owen, R. J., Hunton, C., Bickley, J., Moreno, M., and Linton, D. (1992) Ribosomal RNA gene restriction patterns of *H. pylori*: analysis and appraisal of *Hae*III digests as a molecular typing system. *Epidemiol. Infect.* **108,** 35–47.

33. Bamford, K. B., Bickley, J., Collins, J. S. A., Johnson, B. T., Potts, S., Boston, V., Owen, R. J., and Sloan, J. M. (1993) *Helicobacter pylori*: comparison of DNA fingerprints provides evidence for intrafamilial infection. *Gut* **34**, 1348–1350.

34. Desai, M., Linton, D., Owen, R. J., and Stanley, J. (1994) Molecular typing of *Helicobacter pylori* isolates from asymptomatic, ulcer and gastritis patients by urease gene polymorphism. *Epidemiol. Infect.* **112**, 151–160.

35. Stanley, J., Moreno. M. J., Jones, C., and Owen, R. J. (1992) Molecular typing of *Helicobacter pylori* by chromosomal and plasmid DNA organisation. *Mol. Cell. Probes* **6**, 305–312.

36. Akopyanz, N., Bukanov, N. O., Wadstrom, T. U., Kresovich, S., and Berg, D. E. (1992) DNA diversity among clinical isolates of *Helicobacter pylori* detected by PCR-based RAPD fingerprinting. *Nucleic Acids Res.* **20**, 5137–5142.

37. Go, M. F., Tran, L., Chan, K. Y., Versalovic, J., Koeuth, T., Graham, D. Y., and Lupiski, J. R. (1993) Cluster analysis of REP-PCR-DNA fingerprints of *H. pylori* genomes yields evidence for disease-specific clustering: implications for peptic ulcer disease pathogenesis. *Acta Gastro-enterologica Belgica* **56**, 55.

38. Romero Lopez, C., Owen, R. J., and Desai, M. (1993) Differentiation between isolates of *Helicobacter pylori* by PCR-RFLP analysis of urease A and B genes, and comparison with ribosomal RNA gene patterns. *FEMS Microbiol. Lett.* **110**, 37–44.

39. Hurtado, A. and Owen, R. J. (1993) Urease gene polymorphisms in *Helicobacter pylori* from family members. *Med. Microbiol. Lett.* **2**, 386–393.

40. Clayton, C. L., Kleanthous, H., Morgan, D. D., Puckey, L., and Tabaqchali, S. (1993) Rapid fingerprinting of *Helicobacter pylori* by polymerase chain reaction and restriction fragment length polymorphism analysis. *J. Clin. Microbiol.* **31**, 1420–1425.

41. Hurtado, A., Owen, R. J., and Desai, M. (1994) Flagellin gene profiling of *Helicobacter pylori* infecting symptomatic and asymptomatic individuals. *Res. Microbiol.* **145**, 585–594.

42. Forbes, K. J., Fang, Z., and Pennington, T. H. (1995) Allelic variation in the *Helicobacter pylori* flagellin genes *fla*A and *fla*B: its consequences for strain typing schemes and population structure. *Epidemiol. Infect.* **114**, 257–266.

43. Foxall, P. A., Hu, L-T., and Mobley, H. L. T. (1992) Use of polymerase chain reaction-amplified *Helicobacter pylori* urease structural genes for differentiation of isolates. *J. Clin. Microbiol.* **30**, 739–741.

44. Leying, H., Suerbaum, S., Geis, G., and Haas, R. (1992) Cloning and characterisation of a *Helicobacter pylori* flagellin gene. *Mol. Microbiol.* **6**, 2863–2874.

23

Nosocomial Infections Caused by Staphylococci

Ariane Deplano and Marc J. Struelens

1. Introduction

Staphylococcus aureus and coagulase-negative species of staphylococci (CNS), particularly *S. epidermidis*, are among the most frequently isolated bacteria from patients with nosocomial infection. Conventional methods used in the clinical microbiology laboratory for identification, susceptibility testing, and epidemiologic typing of staphylococci have several limitations. Identification to species level is often limited to *S. aureus*, based on ability to produce coagulase, thermostable nuclease, and clumping factor. However, several newly-described species of staphylococci may express some of these characters, whereas atypical strains of *S. aureus* do not, leading to misidentification *(1,2)*. Furthermore, among the 17 species of staphylococci currently recognized to be indigenous to humans, the potential of *S. lugdunensis* to cause serious infections is being increasingly recognized *(1,2)*. Therefore, further speciation of CNS is clinically relevant. Distinction of true CNS infection from specimen contamination by skin flora can be addressed by demonstrating the repeated isolation of the same clone from multiple specimens, e.g., blood cultures *(1)*. Currently available, rapid biochemical test systems enable identification of staphylococci to species level with 60–90% accuracy *(1)*. Antimicrobial susceptibility pattern, the only routinely available strain marker, is not reliable for clonal delineation because of phenotypic variation among clonally derived isolates *(1)*.

Selection of effective therapy for staphylococcal infection, and avoidance of excess use of glycopeptides, depend on the timely detection of methicillin-resistant strains. The main mechanism of β-lactam resistance in all staphylococcal species involves expression of an additional low-affinity penicillin-binding protein (PBP 2a), encoded by the *mec*A gene *(3,4)*. Conventional susceptibility tests lack sensitivity and specificity for distinguishing between low-expression

From: *Methods in Molecular Medicine, Vol. 15: Molecular Bacteriology: Protocols and Clinical Applications*
Edited by: N. Woodford and A. P. Johnson © Humana Press Inc., Totowa, NJ

class methicillin-resistant staphylococci and high-expression β-lactamase producers *(4)*.

Epidemiologic typing of staphylococci causing outbreaks of nosocomial infections, notably epidemic strains of methicillin-resistant *S. aureus* (MRSA), is required in many hospitals to guide infection control measures. Phage-typing, once the mainstay method for *S. aureus* typing, suffers from many drawbacks, including the need for strain referral to a specialized center and limited discrimination, typability, and reproducibility *(5)*.

Genomic diversity at species and subspecies level among staphylococcal populations, as well as conserved genetic determinants of antimicrobial resistance, have made possible the development of genotypic methods with superior performance over phenotypic methods for speciation, clonal delineation, and detection of methicillin resistance. The purpose of this chapter is to review these applications and to assess their current impact on the management and investigation of staphylococcal infections.

2. Diagnosis

2.1. Aims and Performance Criteria

Rapid and accurate identification of staphylococci to the species level, particularly *S. aureus*, and, where appropriate, other non*aureus* human staphylococci, is of diagnostic importance. Detection of methicillin resistance in a separate or in a combined assay, is equally relevant. For both purposes, sensitivity, specificity, turnaround time, and applicability to the routine clinical microbiology work flow, either as a confirmatory or first line test, will be considered for evaluation of performance. Although most of these molecular diagnostic assays were developed for culture characterization, their potential application to direct detection in clinical specimens will be addressed.

2.2. Species Identification and Detection of Methicillin Resistance

2.2.1. Ribotyping

Table 1 describes Southern blot hybridization techniques used for taxonomic study of staphylococci by analysis of ribosomal RNA (rRNA) gene restriction polymorphism, or ribotyping *(6–11; see* Chapter 2). Either *E. coli* 16S-23S rRNA or a *Bacillus subtilis* rDNA probe can be used. The selection and number of restriction endonucleases used for cleavage of chromosomal DNA is critical for discrimination. Currently, the enzymes *Hind*III, *Eco*RI, and *Cla*I appear optimal for species delineation. The use of combined *Hind*III/ *Eco*RI analysis enabled accurate identification of 27 of 28 staphylococcal species (with the exception of *S. hyicus* strains from diverse animal sources) *(7)* and the recognition of the new species *S. pasteuri (11)*. Hybridization with

Table 1
Ribotyping Techniques Designed for Species Identification of Staphylococci

No. of target species	Enzyme(s)	Probe	Label/detection system	Refs.
7	*Hind*III, *Eco*RI	16 + 23S rRNA from *E. coli*	[32]P/autoradiography	**6**
15	*Ava*I, *Bam*HI, *Cla*I, *Eco*RI, *Hae*III, *Pst*I, *Pvu*II, *Sac*I	16 + 23S rRNA from *E. coli* (Pharmacia)	[32]P/autoradiography	**7**
13	*Hind*III, *Eco*RI	pBA2 (2.3 kb	[32]P/autoradiography	**8,9**
31		*B. subtilis* DNA encoding 16S rRNA cloned in pBR322)		
8	*Hind*III, *Eco*RI	pBA2	[32]P/autoradiography; AAF[a]/colorimetry	**10**
27	*Eco*RI	pBA2	[32]P/autoradiography; AAF/colorimetry	**11**

[a]Acetyl-aminofluorene.

nonradioactively labeled probes (e.g., labeled with acetyl-aminofluorene) provides better resolution of hybridizing bands and more accurate pattern matching than the use of radioactively labeled probes *(10)*. Strains belonging to different species present distinct hybridization patterns (HP), but heterogeneity of patterns may be observed among strains within a given species. More diverse patterns are found in *S. aureus* and *S. epidermidis* than in more ecologically confined species, such as *S. capitis*. Patterns consisting of 4–10 bands common to all strains within the same species can be used as reference "core ribotypes" for visual matching of patterns with isolates to be identified *(6,8)*. Alternatively, computer-assisted clustering of Dice coefficients of similarity (*see* Chapter 3) between all ribotypes can assist with identification: for a given enzyme, the intrataxon similarity varied between 62 and 95%, whereas intertaxa similarity ranged between 17 and 67% *(7)*. By using two enzymes, clustering to a single taxon with more than 65% similarity was found for all staphylococcal strains of human origin *(7)*. Ribotyping performs accurately for the identification of atypical staphylococcal strains that are difficult to classify by conventional tests and provides rapid screening of new taxa *(7,10,11)*.

Ribotyping is thus a powerful identification method for staphylococcal cultures. It is useful in reference laboratories for the identification of atypical strains. However, the manual method is laborious, technically complex and too lengthy for application in the routine clinical microbiology laboratory. Recently, a fully automated system has become commercially-available (RiboPrinter, Qualicon, Wilmington, DE) that performs the entire procedure, from an isolated colony on an agar plate to a print-out of a normalized ribotype that has been matched against a database, within 8 h. However, despite these advances, ribotyping is not applicable to direct detection in clinical samples because of the large amount of high-quality DNA required, and the difficulty of recognizing patterns obtained from mixed bacterial populations.

2.2.2. DNA Probes for Species Identification of Staphylococci

Several DNA probes have been developed and clinically validated for identification of *S. aureus* and for detection of the *mec*A gene in staphylococcal cultures (**Table 2**). For identification of *S. aureus*, DNA probes complementary to species-conserved sequences found in 16S rRNA and in the thermostable nuclease gene *nuc* have been constructed and used in nonradioactive solution-phase or dot/colony blot hybridization formats, respectively (**Table 2**). Both showed 100% accuracy *(12–14)*. The commercially available AccuProbe[R] provides results in 1 h and can be directly applied on positive blood cultures *(13)*, but it requires the use of a special luminometer *(12)*. Reagent and labor costs were about threefold greater than conventional tests in one study *(13)*. The plasmid pIP1608 can be applied as a DNA probe for direct detection and enumeration of *S. aureus* in mixed cultures by colony blot hybridization with primary isolation plates *(14)*.

Very recently, Goh et al. reported preliminary results of a more universal strategy for DNA probe-mediated species identification of staphylococci *(15)*. By using a set of consensus primers based on highly conserved regions within the single-copy 60 kilodalton (kDa) heat shock protein (HSP60) gene, they amplified from genomic DNA of six reference staphylococcal species (**Table 2**) a 600 bp PCR product that showed 100% accuracy in species identification of clinical isolates by dot blot hybridization. Further studies of HSP60 genes as universal targets for microbial identification appear warranted.

2.2.3. DNA Probes for Detection of the Methicillin Resistance Determinant

Several dot-blot hybridization assays using plasmids containing internal *mec* gene fragments encoding the PBP 2a protein were evaluated for detection of methicillin-resistant *S. aureus* and CNS isolates (**Table 2**). A commercially developed liquid-phase assay, using two oligonucleotides for separation and capture/detection, respectively, was more recently reported *(22)*. These DNA probes accurately detect fully resistant strains (oxacillin MIC >8 µg/mL), but show occasional discrepancies with the various phenotypic methods (MIC or oxacillin agar spread plate) used for characterization of borderline-resistant

Table 2
DNA Hybridization Techniques Designed for Species Identification of Staphylococci and for Methicillin Resistance Detection

Target gene	Probe(s)	Hybridization format	Label/detection system	Sensitivity S. aureus	Sensitivity CNS	Specificity S. aureus	Specificity CNS	Ref.
A. Species identification:								
S. aureus 16S rRNA	AccuProbe S. aureus (Gen Probe)	Liquid-phase	Acridinium ester/ chemiluminescence	100	NA[a]	100	NA	*12*
S. aureus 16S rRNA	Accuprobe S. aureus (Gen Probe)	Liquid-phase	Acridinium ester/ chemiluminescence	100	NA	100	NA	*13*
S. aureus thermonuclease (nuc) gene	pIP1608 (186 bp intragenic nuc fragment cloned in pUC18)	Colony/dot blot	^{32}P/autoradiography; Fluorescein/ chemiluminescence	100	NA	100	NA	*14*
HSP60 gene (S. aureus, 5 CNS species[b])	600 bp PCR product with HSP60 consensus primers	Dot blot	Digoxigenin/ chemiluminescence	100	100	100	100	*15*
B. Methicillin-resistance detection:								
mecA	pSA1(528 bp intragenic mec A fragment cloned in pBluescript sK)	Dot blot	Digoxigenin/colorimetry	76	100	100	100	*16*
mecA	pSA12	Dot blot	Digoxigenin/colorimetry	NA	90	NA	100	*17*
mecA	1100 bp intragenic mecA fragment cloned in pUC18	Dot blot	^{32}P/autoradiography; digoxigenin/ colorimetry	100	99	100	95	*18*
mecA	1157 bp internal mecA fragment cloned in pTZ 219	Dot blot	^{32}P/autoradiography	100	ND[c]	100	ND	*19*

(continued)

Table 2 (continued)

Target gene	Probe(s)	Hybridization format	Label/detection system	Sensitivity S. aureus	Sensitivity CNS	Specificity S. aureus	Specificity CNS	Ref.
mecA	25 bp oligonucleotide intragenic to *mecA*	Colony blot	Alkaline phosphatase/ colorimetry	99	100	95	89	*20*
mecA	1.3 kb intragenic *mecA* fragment pBluescript sK	Dot blot	[32]P/autoradiography	100	NA	100	NA	*21*
mecA	- oligo 1 - oligo 2 (Ciba-Corning Diagnosis Corp.)	Liquid phase DNA capture assay	Paramagnetic particle/ magnetic separation Acridinium ester/ chemiluminescence	100	100	100	99	*22*

[a]Not applicable.
[b]*S. epidermidis, S. haemolyticus, S. schleiferi, S. saprophyticus,* and *S. lugdunensis.*
[c]Not done.

strains. This may relate to the presence of other resistance mechanisms in some *S. aureus* strains, including modified-affinity PBPs and hyperproduction of β-lactamase *(22)*, which remain of ill-defined clinical relevance. In some low-level methicillin-resistant strains of *S. haemolyticus* with false-negative hybridization results, a positive signal was obtained after culture in the presence of methicillin, suggesting high-frequency deletion of the *mec* gene (17). As a rule, the specificity of *mec* probes was higher (typically 100%) with *S. aureus* than with CNS strains, in which *mec* occasionally appears cryptic (**Table 2**). Target DNA does not need to be high quality or quantitated for dot-blot hybridization assays *(18)*. However, it is important to use multiple negative and positive control samples in each blot to ensure uniformity of exposure to the labeled probe and to avoid false-negative results *(18)*. The time needed to detect hybridization signals is related to the amount and type of probe label and, with some systems, use of prolonged exposure or unsufficiently stringent conditions can generate false-positive results.

Although quite accurate, most of these DNA probe assays are too labor-intensive to be easily applied as first line or even confirmatory susceptibility tests in a diagnostic laboratory.

2.2.4. PCR Techniques for Species Identification of Staphylococci

PCR amplification of specific gene sequences offers a sensitive, rapid, and relatively simple molecular diagnostic tool applicable in clinical microbiology to both characterization of pure cultures and direct microbial detection in clinical specimens. Template DNA of adequate quality for PCR can be obtained by a rapid colony lysis procedure, based on short treatment with lysostaphin and proteinase K followed by boiling. Similarly, simple lysis techniques have been used with broth cultures or blood cultures. **Table 3** summarizes PCR methods designed for species identification and detection of the *mec* determinant in staphylococcal strains, whereas **Table 4** recapitulates multiplex-PCR assays for combined species identification and *mec* detection *(23–26)*.

Two strategies of PCR-mediated identification of staphylococci have been successfully explored. The first relies on detection of species-conserved genes that are present in all strains of *S. aureus*, such as the *nuc* and *fem*A genes *(23–25)*. The second approach is based in species-specific length polymorphism of amplified spacer regions lying between transfer RNA (tRNA) genes *(26)* or 16S-23S rRNA genes *(27)* by use of outwardly directed sets of consensus primers.

The *fem*A detection PCR has been developed in a format amenable to colorimetric microtiter plate reading by using the enzymatic detection (ED-PCR) principle (*see* Chapter 5). In this assay, DNA is amplified using a pair of primers labeled with biotin and dinitrophenol, respectively. Amplified products are captured into streptavidin-coated microtiter wells and detected with an enzyme-linked antidinitrophenol antibody *(24,31)*.

Table 3
PCR Techniques for Identification of Staphylococci and Detection of Methicillin Resistance

Author (Reference)	Target gene (species)	Template DNA	Detection of PCR products	Size of amplified DNA product	Primers (sequence 5'-3')	Sensitivity		Specificity	
						S. aureus	CNS	S. aureus	CNS
A. Species identification:									
Unal (23)	femA (S. aureus)	Colony lysate	Agarose electrophoresis	990 bp (position: 766-1755)	Primer5: CGAGGTCATTGCAGCTT GCTTAC Primer6: CTAGACCAGCATCTTCAGC	100	NA*	100	NA
Kizaki (24)	femA (S. aureus)	Colony lysate	Biotin-streptavidin solid phase capture + dinitrophenol-antibody enzyme linked colorimetry (ED-PCR)	467 bp	Primer 3: ACAGTCAAAGAGTTTGGTGC Primer 4: TCTAACACTGAGTGATAACG	100	NA	100	NA
Brakstad (25)	nuc (S. aureus)	Broth lysate; Simulated blood culture; Swab clinical samples	Agarose electrophoresis + hybridization with 33 bp nuc probe	270 bp (position: 48-328)	1: GCGATTGATGGTGATACGGTT 2: AGCCAAGCCTTGACGAACTAAAGC	100	NA	100	NA
Welsh (26)	Inter-tRNA gene spacer regions (S. aureus, CNS species)	Purified genomic DNA	Polyacrylamide electrophoresis	Fingerprint 60-650 bp DNA fragments	T5A: AGTCCGGTGCTCTAACCAACTG T5B: AATGCTCTACCAACTGAACT T3A: GGGGGTTCGAATTCCCGCCGG CCCCA (used alone or in pairs)	100	100	100	100
Jensen (27)	Inter-16S-23S rRNA gene spacer regions (S. aureus, CNS species)	Purified genomic DNA	Polyacrylamide electrophoresis	Fingerprint 400-700 bp DNA fragments	G1: GAAGTCGTAACAAGC L1: CAAGGCATCCACCGT	100	100	100	100
Maes (28)	Inter-tRNA gene spacer regions (S. aureus, CNS species)	Colony lysate	Metaphor™ agarose blectrophoresis	Fingerprint 60-600 bp DNA fragments	T5A: AGTCCGGTGCTCTAACCAACTG T3B: AGGTCGCGGGTTCGAATCC	100	100	100	100

438

B. Methicillin resistance detection:

Reference	Target (species)	Sample	Detection method	Product size (position)	Primers				
Murakami (28, 29)	mecA (S aureus)	Colony lysate	Agarose electrophoresis + hybridization with (17 bp) mecA probe	533 bp (position 1282-1814)	RSM2647: AAAATCGATGGTAAAGGTT GGC RSM2648: AGTTCTGCAGTACCGGATTT GC	98	96	97	90
Tokue (30)	mecA (S aureus)	DNA extract	Agarose electrophoresis	1338 bp (position 478-1816)	MR1: GTGGAATTGGCCAATACAGG MR2: TGAGTTCGTCAGTACCGGAT	100	ND[b]	86	ND
Ubukata (31)	mecA (S. aureus, 7CNS species)	Colony lysate, blood cultures	Biotin-streptavidin solid phase capture + dinitrophenol-antibody enzyme linked colorimetry (ED-PCR)	150 bp (position 181-330)	MRS1: GAAATGACTGAACGTCCGAT MRS2: GCGATCAATGTTACCGTAGT	100	87-100	98-100	60-100
Unal (23)	mecA (S aureus, 5 CNS species)	Colony lysate	Agarose electrophoresis	1107 bp (position 466-1573)	Primer1: GACCGAAACAATGTGGAATT GGCC Primer2: CACCTTGTCCGTAACCTGAA TCAGC	NA	NA	NA	NA
Unal (32)	mecA (S aureus)	Colony lysate	Agarose electrophoresis	1817 bp (position:37-1854)	primer1: GTTGTAGTTGTCGGGTTTGG primer2: CCACCCAATTTGTCTGCCAG TTTCTCC	100	ND	100	ND
Predari (17)	mecA (S aureus)	Colony lysate	Agarose electrophoresis	528 bp (position: 516-1044)	1: dGGGATCATAGCGTCATTATTC 2: dAACGATTGTGACACGATAGCC	ND	100	ND	100
Zambardi (33)	mecA (S aureus) gyrA	Colony lysate	Agarose electrophoresis	533 bp (position 1282-1598) 280 bp (position codon 44-138)	RSM2647: AAAATCGATGGTAAAGGTT GGC RSM2648: AGTTCTGCAGTACCGGAT TTGC P1: AGTACATCGTCGTATACTATATGG P2: ATCACGTAACGTTCAAGTGTG	88	ND	100	ND
Geha (34)	mecA (S aureus, 3 CNS species) 16S rDNA	Broth lysate	Agarose electrophoresis	310 bp (position: 318-627) 479 bp gene	-mecA1: GTAGAAATGACTGAACGTCC GATAA mecA2: CCAATTCCACATTGTTTCGGT CTAA X: GGAATTCAAA(T/G, T:))GAATTGAC GGGGGC Y:CGGGATCCCAGGCCCGGGAACGTATTCAC	100	93	98	100

[a]Not analyzed.
[b]Not done.

Table 4
Multiplex PCR Techniques for Combined Identification and Detection of Methicillin Resistance

Author (Reference)	Target gene	Template DNA	Detection of PCR products	Size of amplified DNA product	Primers (sequence 5'-3')	Sensitivity		Specificity	
						S.aureus	CNS	S.aureus	CNS
Brakstad (35)	mecA	Broth lysate	Agarose electrophoresis	533 bp (position 1282-1598)	RSM2847: AAATCGATGGTAAAGGTT GGC; RSM2848::AGTTCTGCAGTACCGGATTT GC	100	93	100	95
	nuc			267 bp (position 49-327)	P1: GCGATTGATGGTGATACGGTT; P2: AGCCAAGCCTTGACGAACTAAAGC	100	NA*	100	NA
Kizaki (24)	mecA	Colony lysate	Agarose electrophoresis	640 bp	1: AGTTGTAGTTGTCGGGTTTG; 2/ AGTGGAACGAAGGTATCATC				
	femA			467 bp	3/ ACAGCTAAAGAGTTTGGTGC; 4/ TCTAACACTGAGTGATAACG	100	NA	100	NA
Vannuffel (36)	mecA	Colony lysate	Agarose electrophoresis	310 bp (position: 885-1194)	M1: TGGCTATCGTGTCACAATCG; M2: CTGGAACTTGTTGAGCAGAG	100	100	100	98
	femA			686 bp (position: 217-902)	F1: CTTACTTACTGGCTGTACCTG; F2: ATGTCGCTTGTTATGTGC	100	NA	100	NA
	IS431			444 bp (position: 32-476)	C1: AGGATGTTATCACTGTAGCC; C2: GATGATCAATGACAGTCAGG				

PCR analysis of intergenic tRNA or rRNA spacer polymorphism requires accurate size determination of small amplified DNA products, which is often achieved by using polyacrylamide gel separation *(26,27)*. Initial reports showed that strains of five staphylococcal species could be differentiated by these methods *(26,27)*. Further evaluation of the tRNA spacer polymorphism analysis for identification of a large collection of clinical isolates of 17 staphylococcal species showed this method have an accuracy of more than 95% *(28)*. The routine use of these assays is hindered by the complexity of pattern analysis and by the limited accuracy and precision of size determination using gel electrophoresis, which depends on inclusion of multiple size markers and computer-assisted normalization of DNA patterns (*see* Chapter 3).

3. Epidemiologic Investigation
3.1. Aims and Performance Criteria

Molecular typing is used to differentiate epidemiologically related and unrelated isolates on the basis of genomic characteristics. Typing techniques serve to confirm or delineate epidemics, to identify the potential source and pattern of transmission and to test the efficacy of control measures. Accurate typing systems should fulfill several criteria of performance, including typability, reproducibility, stability, discriminatory power, epidemiologic concordance, and typing system concordance *(37–40)*. Typability is the proportion of isolates that can be analysed in the typing system and assigned to a type, and should ideally include all isolates. Reproducibility refers to the ability of the typing system to assign the same type when a strain is repeatedly tested. Stability is the biological feature of clonally derived isolates to express constant characters over time and generations. The stability of a typing method is acceptable if the system recognizes clonal relatedness and does not lead to misclassification of subclonal variants as being epidemiologically unrelated. Discriminatory power estimates the probability that epidemiologically-unrelated isolates are classified in distinct types and can be calculated following Simpson's Index of Diversity (D) *(41)*. In practice, a typing system, or combination of systems, displaying D values ≥0.95 is acceptable *(37–39)*. Epidemiologic concordance is the faculty of a typing method to classify all well-documented epidemiologically-related isolates into the same type *(40)*. Evaluation of typing system concordance is interesting to evaluate if isolates initially considered to be epidemiologically related are concordantly found related by a number of independent systems or, if not, should be re-examined for possible misclassification *(37–39)*.

3.2. Plasmid Analysis

Plasmid analysis (*see* Chapter 4) was the earliest molecular typing method applied to staphylococci, based on determination of number and size of plasmid DNA. It is a simple and low-cost typing method useful for the investigation of epidemic or endemic infections caused by staphylococcal species that harbor multiple plasmids, such as *S. epidermidis*, which can carry as many as 12 plasmids *(42–55)*. The plasmid diversity in staphylococci varies by species (**Table 5**). For some species, plasmid typing shows poor typability and low discrimination; the latter is typical of species, such as *S. lugdunensis*, in which a few highly-conserved plasmid are common to most strains *(57)*.

Archer et al. *(43)* showed plasmid diversity among *S. epidermidis* skin isolates compared with the identical plasmid profile exhibited by most isolates of *S. epidermidis* from cases of prosthetic-valve endocarditis. However, plasmid instability occurs in vivo during chronic colonization or infection by a single clone or along the course of a prolonged outbreak, as reported for both *S. aureus* and *S. epidermidis (45,50,60,61)*. Therefore, some variation in plasmid size or restriction pattern does not necessarily indicate strain difference, especially if that variation can be explained by a single genetic event, such as acquisition or loss of a plasmid *(51,61,64)*.

Artefacts can complicate interpretation of pattern of intact plasmid because different molecular forms (supercoiled, circular, or linear) of a plasmid migrate differentially in agarose gel *(51)* (*see* Chapter 4). DNA extraction and separation techniques can influence plasmid yield and conformation-dependent banding patterns. To minimize artefacts, it has been recommended that plasmid DNA should be purified in non-denaturing conditions and Tris-borate-EDTA used as electrophoresis buffer *(43,50,57)*.

Restriction endonuclease analysis of plasmid DNA (REAP) is used to establish plasmid relatedness between strains in a more reproducible and discriminating manner than can be achieved by patterns of intact plasmids (**Table 5**). REAP has been shown to be useful, especially in combination with analysis of chromosomal polymorphism, for the epidemiologic investigation of MRSA outbreaks *(58,62–64)* However, typability of *S. aureus* isolates varies widely between studies (44–100%). Several authors estimate relatedness of REAP patterns by computation of Dice coefficients (*see* Chapter 3). Epidemiologically related strains are defined based on REAP profiles differing by less than three DNA fragments, corresponding to a Dice coefficient ≥85% *(51,58,61)*. More heterogeneous plasmid profiles were found among MSSA strains than MRSA strains. These findings support the view that MRSA arose from a limited number of clones. Plasmid profile generally correlates well with antibiotic resistance pattern, in accordance with the acquisition of plasmid-borne resistance determinants that can be demonstrated by hybridization studies using probes for cognate genes and transposons *(45,55,58)*.

Table 5
Performance of Plasmid Profile and Restriction Endonuclease Analysis of Plasmid (REAP) for Typing of Staphylococci

Species	Typeability (%)	Reproducibility (%)	Stability (%)	Discrimination[a] (%)	Refs.
A. Plasmid profile:					
S. epidermidis	81–100	90–100	47–92	48–100	*42–55*
S. hominis	33	NT[b]	NT	100	*48,49*
S. haemolyticus	71–79	NT	NT	NA[c]	*56,83*
S. lugdunensis	66	NT	NT	55	*57*
S. aureus	44–100	94–96	94	48–89	*58–66*
S. schleiferi	19	NT	NT	87[d]	*67*
B. REAP:					
S. aureus	44–100	100	88	89–92	*59–62, 64–67*
S. epidermidis	100	100	100	NA	*51*
S. schleiferi	19	NT	NT	93[d]	*68*

[a]Estimated by D index *(41)* based on epidemiologically unrelated and typeable isolates.
[b]Not tested.
[c]Insufficient data available for computing D index.
[d]Estimated on the six typeable strains.

3.3. Genomic DNA Restriction Analysis

Conventional restriction endonuclease analysis (REA) of genomic staphylococcal DNA (*see* Chapter 2) using frequently-cleaving enzymes (including *Eco*RI, *Hind*III, *Cla*I, and *Bgl*II) generates between 50 and 100 small fragments to produce complex DNA banding patterns by agarose gel electrophoresis. As summarized in **Table 6**, REA shows good typeability, stability, and reproducibility in several studies. It has been successfully applied to epidemiologic investigations of infections with *S. aureus, S. haemolyticus,* or *S. epidermidis (52,69–75),* species for which it exhibits a discriminatory power (D) ranging between 81–100%. In contrast, REA had poor discrimination for typing strains of *S. lugdunensis (57)* and *S. schleiferi (68).* A majority of epidemiologically unrelated *S. lugdunensis* strains displayed conserved REA patterns *(57).*

Several authors used correlation coefficients, like Dice *(70),* to estimate relatedness between REA profiles. This was performed using computer-assisted laser densitometry, usually in a restricted size range in which DNA fragments showed good resolution.

Two approaches overcome the complexity and improve the resolution of genomic REA patterns: Southern blot transfer of restriction fragments and

Table 6
Performance of Total Chromosomal DNA Restriction Enzyme Analysis
(REA) for Typing of Staphylococci

Species	No. of strains	Enzyme	Typeability (%)	Reprod. (%)	Discrimination[a] (%)	Ref.
S. aureus	40	*Eco*RI	100	100	81	*69*
S. aureus	20	*Bgl*II, *Sal*I	100	NT[b]	83	*70*
S. aureus	92	*Bgl*II	100	100	97	*71*
S. aureus[c]	6	*Eco*RI	100	100	34	*72*
S. aureus	24	*Bgl*II	100	NT	NA[d]	*73*
S. epidermidis	36	*Eco*RI	100	100	NA	*74*
S. epidermidis	45	*Bcl*I	100	NT	100	*52*
S. haemolyticus	31	*Eco*RI, *Sal*I	100	NT	NA	*56*
S. haemolyticus	12	*Bam*HI, *Cla*I, *Hin*dIII	100	NT	NA	*75*
S. schleiferi	31	*Eco*RI, *Cla*I *Hin*dIII, *Pst*I *Pvu*II	100	NT	0	*68*
S. lugdunensis	30	*Eco*RI	100	NT	35	*57*
		*Pst*I	100	NT	40	
		*Pvu*II	100	NT	13	
S. hominis	2	*Eco*RI	100	100	100	*72*
S. hyicus	10	*Eco*RI	100	100	38	*72*
S. chromogenes	12	*Eco*RI	100	100	71	*72*
S. simulans	6	*Eco*RI	100	100	0	*72*
S. xylosus	3	*Eco*RI	100	100	67	*72*

[a]Estimated by D index *(41)* based on epidemiologically unrelated and typeable isolates.
[b]Not tested.
[c]Staphylococcal strains from veterinary origin.
[d]Insufficient data available for computing D index.

hybridization to labeled DNA probes, and use of infrequently cleaving enzymes that generate less than 30 large genomic fragments, followed by separation by pulsed field gel electrophoresis (PFGE), a technique also called macro-restriction analysis (*see* Chapter 3).

3.4. Genome RFLP Analysis by Southern Hybridization

Restriction DNA fragments transferred to a solid membrane and hybridized with a labeled DNA probe show hybridization patterns that are easy to analyze and compare. Differences in patterns reflect rearrangements or mutations that alter the distribution of restriction sites within the genome or manifest variation in the chromosomal location of target sequences. The discriminatory power of the RFLP method increases when the target sequence is present in multiple copies located in various regions of the genome.

Numerous DNA and RNA probes have been described for typing of staphylococci, based on the following genomic elements:

1. 16S and 23S rRNA genes, present in approximately nine copies (or *rrn* alleles) in staphylococcal genomes (ribotyping);
2. Genes present in single copy such as *mec*A or the accessory gene regulator (*agr* locus);
3. Mobile genetic elements, including insertion sequences (IS*256*, IS*257*, IS*431*, IS*1181*) and transposons (Tn*554*, Tn*4001*).

3.4.1. Ribotyping

Probes used for ribotyping (*see* Chapter 2) include 16S and 23S rRNA from *E. coli* or *S. aureus*, or plasmids pBA2 and pKK3585 containing cloned fragments encoding 16S rRNA from *B. subtilis* and the entire *rrn*B operon of *E. coli*, respectively. This technique shows complete typeability and excellent reproducibility and stability for all staphylococcal species investigated. The use of several different enzymes, in parallel or in combination, is required to optimize the discriminatory power of ribotyping (**Table 7**). There is no consensus on which combination of enzymes to apply nor on criteria for interpreting ribotype pattern differences. For example, some authors consider ribotypes that vary by a single band to indicate different strains *(8, 77, 78)*, whereas others consider these to represent subtypes *(76)*. In general, ribotyping is less discriminating than other methods indexing chromosomal polymorphism for *S. aureus* typing *(76)*. Although the manual method is laborious and takes up to a week to perform, an automated system is commercially available that produces results in 8 h.

3.4.2. RFLP Analysis Using Single Gene Polymorphism

Discrimination with this type of probe has only given moderate or poor discrimination (**Table 8**). Therefore, multiple probes must be used concurrently or in association with other DNA probes (e.g., for IS elements). Thus, this typing method, although very reproducible, can only be considered as a complementary method. In addition, it is also a time-consuming technique, like all Southern hybridization procedures.

3.4.3. RFLP Analysis Using Repetitive Elements

Insertion sequence IS*256* is associated with aminoglycoside resistance genes and occurs in mutiple locations in the genomes of staphylococci. It appears to be one of the most discriminating RFLP probes for typing methicillin and gentamicin resistant strains of *S. aureus* and *S. epidermidis* (**Table 8**). However,

Table 7
Performance of Ribotyping[a] for Staphylococcal Strain Delineation

Organism	No. of strains	Enzyme	Typeability (%)	Reprod. (%)	Discrimination[b] (%)	Ref.
S. aureus	25	*Hin*dIII, *Eco*RI	100	100	87	**8**
S. aureus	59	*Hin*dIII, *Cla*I	100	100	72–82	**76**
S. aureus	44	*Hin*dIII, *Eco*RI, comb.	100	100	89–95	**77**
S. aureus	50	*Eco*RI	100	NT[c]	91	**78**
S. aureus	102	*Hin*dIII, *Eco*RI	100	100	48–71	**79**
MRSA	61	*Cla*I	100	NT	NA[d]	**80**
MRSA	81	*Eco*RI, *Hin*dIII, *Cla*I	100	100	50–64	**81**
MRSA	26	*Eco*RI, *Hin*dIII, *Cla*I	100	NT	50–67	**82**
S. epidermidis	45	*Bcl*I	100	100	NA	**58**
S. epidermidis	53	*Hin*dIII, *Eco*RI, comb.	100	NT	NA	**83**
S. epidermidis	51	*Eco*RI	100	100	76	**84**
S. epidermidis	86	*Eco*RI, *Hin*dIII, comb.	100	NT	69–91	**85**
S. schleiferi	31	*Eco*RI, *Id*II, *Cla*I, comb.	100	NT	74–86	**68**
S. haemolyticus	45	*Cla*I	100	NT	29	**86**

[a]*See* **Table 1** for previously described plasmids and RNA probes.
[b]Estimated by D index based on epidemiologically-unrelated and typeable isolates.
[c]Not tested.
[d]Insufficient data available for computing D index.

most other strains are not typable with this probe. Other insertion sequences and transposons used for genomic RFLP analysis also showed incomplete typability and detected only limited diversity. Likewise, the bacteriophage M13 protein III tandem repeat sequence, which can be used to reveal polymorphism

Table 8
Performance of RFLP Typing of Staphylococci by Southern Hybridization with Various DNA Probes

Organism	No. of strains	Enzyme	DNA target	Typeability (%)	Reprod. (%)	No. of types/ No. of strains[a]	Ref.
S. aureus	59	ClaI	mecA	63	100	2/37	76
		ClaI	Tn554	63	100	2/37	
		ClaI	agr	100	100	5/59	
		ClaI	aacA-aphD	49	75	2/29	
		ClaI	IS257/431	60	100	9/59	
S. aureus	102	EcoRI	aacA-aphD	61	NT[b]	15/59	79
		EcoRI	aacA-aphD + part of IS256	61	100	42/59	
		HindIII	aacA-aphD + part of IS256	61	100	35/59	
S. aureus	46	HaeIII	cna	61	NT	3/28	87
		HaeIII	fnbA- fnbB	100	NT	11/46	
		HindIII	hlb	100	NT	11/46	
S. aureus	52	HindIII	IS1181	79	NT	16/41	88
MRSA	61	ClaI	ClaI	100	NT	1/61	80
		ClaI	ClaI	100	NT	2/61	
MRSA	42	ClaI	mecA	100	NT	3/74	89
MRSA	74	ClaI	Tn554	100	NT	1/42	
MRSA	79	ClaI	mecA	100	NT	2/79	90
		ClaI	Tn554	89	NT	4/70	
MRSA	189	ClaI	mecA	100	NT	7/189	91
		ClaI	Tn554	98	NT	8/185	
MRSA	52	EcoRI	M13 tandem repeat	100	NT	10/52	92
MRSA	17	EcoRI	IS256	100	100	14/17	93
MRSA	43	ClaI	mecA	100	NT	3/43	94
		ClaI	Tn554	95	NT	3/42	

(continued)

Table 8 (continued)

Organism	No. of strains	Enzyme	DNA target	Typeability (%)	Reprod. (%)	No. of types/ No. of strains[a]	Ref.
S. epidermidis	18	*Bcl*I	random genomic	100	NT	8/18	52
			S. epidermidis DNA				
S. epidermidis	48	*Eco*RI	IS256	100	NT	35/48	55
S. epidermidis	53	*Eco*RI	*aacA-aphD*	64	NT	17/53	83
		*Hind*III	*aacA-aphD*	64	NT	2/53	
		*Eco*RI	*aacA-aphD* + part of IS256	64	NT	23/53	
		*Hind*III	*aacA-aphD* + part of IS256	64	NT	19/53	
S. epidermidis	15	*Hind*III	Tn554	42	NT	2/15	95
		*Cla*I	*mecA*	83	NT	2/15	
			pcrA	100	NT	3/15	
			gyrA	100	NT	3/15	
			agr	100	NT	3/15	
S. haemolyticus	12	*Hind*III	*mecA*	100	NT	2/12	75
			Tn554	100	NT	3/12	
			β lactamase	100	NT	3/12	
			aacA-aphD	100	NT	3/12	
S. haemolyticus	45	*Cla*I	IS431	100	NT	24/45	86

[a]Number of types among typeable isolates; insufficient data available for computation of D index in most studies.
[b]Not tested.

in eukaryotic and in bacterial genomes, was found only moderately discriminating for typing MRSA strains *(92)*. Again, these techniques are not very efficient because they take several days to perform and are labor-intensive.

3.5. Genome Macrorestriction Analysis Resolved by Pulsed-Field Gel Electrophoresis (PFGE)

Genome macrorestriction analysis (*see* Chapter 3) has been shown to provide very high discrimination between strains of a variety of bacterial and yeast pathogens, leading several authors to recommend it as a typing method of choice *(37–39)*. It has been widely applied to *S. aureus*, especially MRSA strains, for which a discrimination ranging between 87 and 100% has been reported (**Table 9**). Some reference centers now use PFGE as a replacement for phage typing as their main tool for *S. aureus* outbreak investigation *(5)*. The performance of PFGE for typing strains of other staphylococcal species has been less extensively evaluated. Although a variety of electrophoretic systems and pulsing protocols have been employed, the contour-clamped homogeneous electric field (CHEF) and field-inversion gel electrophoresis (FIGE) formats appear to be the most popular (**Table 9**). The latter two systems showed similar levels of resolution in comparative studies *(76,102)*. However, FIGE requires less costly equipment.

Different investigators have used a variety of criteria for interpreting minor and major variations in PFGE patterns obtained with a single or several rare cutters. Whereas some authors have used single fragment variation to distinguish types *(97)*, others have categorized isolates displaying up to three *(71)* or six *(96)* fragment differences between PFGE patterns as subclonal variants, found to be clustered by quantitative pattern similarity analysis and construction of dendrograms *(71,96)*. Recently, working parties in Europe and United States have recommended the latter, more flexible approach based on the number of likely genetic events separating genomic variants *(38,40)*. Studies of in vivo genomic stability of *S. aureus* strains recovered over months or years in individual patients or during extensive outbreaks support this view *(62,71,100)*.

Despite the relative complexity of interpretation of results, PFGE appears an excellent typing method for *S. aureus* and *S. epidermidis*. It is, nonetheless, a method that requires specialized equipment, a substantial amount of expertise to obtain reproducible results, and is relatively expensive and time-intensive. Rapid DNA preparation protocols are available that considerably reduce the turnaround time of PFGE typing *(112)*. Intercenter reproducibility is a desirable goal for surveillance studies of MRSA, but this still requires further standardization of the method *(102)*. Commercially prepared reagent kits and PFGE systems are currently evaluated in comparison with in house protocols for MRSA typing.

Table 9
Performance of Macrorestriction Analysis Using PFGE for Typing of Staphylococci

Organism	No. of strains	Enzyme	PFGE system[a]	Typeability (%)	Reprod. (%)	No. of types/ No. of isolates[b]	Discrimination[c] (%)	Ref.
MRSA	39	*Sma*I	CHEF	100	100	13/39	87	62
MRSA	25	*Sma*I	CHEF	100	100	10/25	NA[d]	67
MRSA	61	*Sma*I	CHEF	100	NT[e]	5/31	NA	80
MRSA	239	*Sma*I	TAFE	100	100	26/239	NA	81
MRSA	85	*Sma*I	CHEF	100	NT	21/85	NA	89
MRSA	43	*Sma*I	CHEF	100	NT	16/43	NA	94
MRSA	58	*Sma*I, *Bss*HII	FIGE	100	100	19/58	NA	96
MRSA	111	*Sma*I, *Not*I	CHEF	100	NT	31/111	93	97
MRSA	29	*Sma*I	CHEF	100	NT	10/29	NA	98
MRSA	30	*Sma*I	CHEF	100	NT	10/30	90	99
MRSA	28	*Sma*I	CHEF	100	NT	16/28	NA	100
MRSA	50	*Sma*I	CHEF	100	100	10/50	NA	101
MRSA	12	*Sma*I	CHEF, FIGE	100	0[f]	12/12	100	102
MSSA	3	*Sma*I	FIGE	100	100	3/3	NA	103
MRSA	175	*Sma*I	TAFE	100	100	9/175	NA	104
S. aureus	300	*Sma*I	CHEF	100	100	98/300	NA	5
S. aureus	78	*Sma*I	CHEF	100	NT	10/78	NA	65

Species		Enzyme	Method					Ref
S. aureus	92	SstII	CHEF	100	100	33/92	98	71
S. aureus	59	SmaI	CHEF, FIGE	100	67	25/59	90–93	76
S. aureus	82	SmaI	CHEF	100	NT	42/82	99	106
S. aureus	69	SmaI	CHEF	100	100	36/69	89	105
S. aureus	70	SmaI	CHEF	100	NT	40/70	NA	107
S. aureus	47	SmaI, ApaI	CHEF	100	NT	27/47	NA	108
S. epidermidis	5	SmaI	FIGE	100	100	3/5	NA	103
S. aureus	95	SmaI, SstII	CHEF	100	NT	13/55	NA	109
S. epidermidis	86	SmaI	RAGE	100	100	21/86	85	110
S. aureus	51	SmaI	CHEF	100	NT	45/51	NA	111
S. epidermidis	12	SmaI	FIGE	100	NT	6/12	NA	112
S. haemolyticus	45	SmaI	CHEF	100	NT	24/45	NA	86

[a]PFGE systems are abbreviated as follows: CHEF, contour-clamped homogeneous electric fields; TAFE, transverse alternating field electrophoresis; FIGE, field-inversion gel electrophoresis; RAGE, rotating alternating gel electrophoresis.

[b]Number of types among typeable isolates.

[c]Estimated by D index (41) based on epidemiologically unrelated and typeable isolates.

[d]Insufficient data available for computing D index.

[e]Not tested.

[f]Interlaboratory reproducibility between three different laboratories: no laboratory obtained identical patterns.

3.6. PCR-Based Methods of Genome Fingerprinting

PCR-based methods described to date for the genomic typing of strains of staphylococci are based on the following strategies:

1. Analysis of single gene allelic polymorphism, based on specific amplification (by using convergent primers at high stringency) of a known, hypervariable element and by determination of its size, RFLP, single-strand conformational polymorphism (SSCP) or nucleotide sequence;
2. Analysis of interrepeat element spacer length polymorphism by using divergent primers complementary to multiple copy elements at high stringency;
3. Use of the universal approach of randomly amplified polymorphic DNA (RAPD) analysis, or arbitrarily primed PCR (AP-PCR), both of which are based on low stringency amplification of arrays of spacer DNA elements lying between inversely repeated motifs of unknown sequence in the target genome, by using single primers of arbitrary sequence (*see* Chapter 6; **Table 10**).

Single gene polymorphism analysis of suitably chosen hypervariable regions (e.g., the coagulase gene 81-bp tandem repeat sequences) showed moderate to good discrimination between *S. aureus* strains in initial studies, especially when high resolution techniques, such as partial sequence determination, have been used (**Table 10**). However, the coagulase gene PCR-RFLP procedure exhibited lower discrimination than several other genomic typing systems in a large comparative study *(76)*. It is nevertheless a rapid, simple, and highly reproducible method.

Repetitive element PCR (rep-PCR) (*see* Chapter 7) typing was initially developed based on repeat motifs of unclear function described in Gram-negative bacteria, like the enterobacterial repetitive intergenic consensus (ERIC) sequences *(37,38)* and later extended to other repeat motifs found in staphylococci, including mycoplasmal repeat elements, rDNA and insertion sequences (**Table 10**). Preliminary results suggest both good reproducibility and high discrimination with these techniques, which await further evaluation in comparison with other high-resolution genotyping methods.

Arbitrarily-primed PCR analysis showed variable, but often high, discriminatory power for typing *S. aureus* strains (**Table 10**). Good correlation with other genotyping systems was found *(123–126)*. Discrimination can be optimized by increasing the number of primers used separately or in combination and by empirically selecting those yielding maximal polymorphism *(82,99,123,125)*. A study of intercenter reproducibility of AP-PCR typing of a well-defined collection of *S. aureus* strains indicated wide variability in the number and composition of DNA types obtained in the different centers, but overall agreement in correct clustering by pattern similarity of the majority of isolates according to epidemiologic origin *(126)*. These findings suggest that

Table 10

Performance of PCR-Based Methods for Typing Staphylocci

Organism (Reference)	Target DNA	Amplicon analysis	Reproducibility	No. of types/ No. of isolates	Discrimination (%)
A. Single gene polymorphism:					
S. aureus (76)	coagulase gene	RFLP AluI	NT	7/ 59	32
S. aureus (100)	coagulase gene	Single fragment size	NT[a]	5/ 39	NA[b]
		RFLP AluI	NT	9/ 39	NA
S. aureus (113)	coagulase gene	Single fragment size	NT	18/ 69	87
		RFLP AluI	NT	18/ 69	87
S. aureus (114)	coagulase gene	Single fragment size	NT	1/ 50	NA
		RFLP AluI	NT	1/ 50	NA
	X region of protein A (spa)	RFLP HinfI	NT	1/ 50	NA
S. aureus (115)	coagulase gene	Single fragment size	NT	6/ 30	76
		RFLP AluI	NT	22/ 30	97
		Sequencing	100	12/ 12	100
MRSA (116)	X region of protein A (spa)	RFLP RsaI	NT	9/ 47	68
MRSA (117)	X region of protein A (spa)	RFLP RsaI	100	16/ 54	77
MRSA (118)	mecA + mec- associated HVR[d] (dru)	Single fragment size, SSCP, Sequencing	NT	5/ 61	NA
MR S. epidermidis (118)	mecA + mec- associated HVR (dru)	Single fragment size, SSCP	NT	3/ 15	NA
MR S. haemolyticus (118)	mecA + mec- associated HVR (dru)	Single fragment size, SSCP	NT	2/ 11	NA
B. Repetitive element- spacer length polymorphism (Rep- PCR):					
MRSA (119)	RepMP3[e] spacer	Multiple fragment patterns	100	8/ 170	NA
S. aureus (120)	16S- 23S rRNA spacer	Multiple fragment patterns, sequencing	99	35/ 322	NA
S. aureus (121)	Tn916- 16S rRNA spacer	Multiple fragment patterns	100	15/ 24	91
MRSA (122)	IS256 spacer	Multiple fragment patterns	100	29/ 90	97

Table 10 (continued)

C. Arbitrarily- primed PCR (AP- PCR) fingerprinting:

MRSA (123)	Combination of ERIC, mecA and arbitrary sequence spacer	Multiple fragment patterns	NT	4 to 23/ 44	NA
MRSA (124)	Arbitrary sequence spacer	Multiple fragment patterns	NT	12 to 25/ 26	93 to 98
S. aureus (82)	Combination of ERIC, mecA and arbitrary sequence spacer	Multiple fragment patterns	100	9 to 13/ 26	81 to 96
S. aureus (125)	Combination of ERIC, mecA and arbitrary sequence spacer	Multiple fragment patterns	NT	2 to 15/ 48	NA
S. aureus (126)	Combination of ERIC and arbitrary sequence spacer	Multiple fragment patterns	0'	6 to 29/ 59	NA
MR CNS (127)	Arbitrary sequence spacer	Multiple fragment patterns	NT	11/ 18	100

AP-PCR can be used reliably for local outbreak investigations, but still lacks standardization for large-scale surveillance.

In summary, rep- and AP-PCR analysis are attractive typing systems because of their technical simplicity, relative low-cost in capital investment and reagents, and short turnaround time. Current limitations of these techniques include the difficulty to obtain reproducible results, especially for low stringency protocols, and the lack of general rules for the interpretation of results, particularly for scoring variation in DNA products amplified at low-yield. Enhanced resolution and standardization of analysis of PCR fragments can be obtained by incorporating fluorophore-labeled oligonucleotide primers in the reaction, using polyacrylamide gel electrophoresis and performing computer-assisted amplimer pattern analysis by automated laser scanning systems *(119)*.

3.7. Current Methods of Choice for Typing Staphylococci

The comparative evaluation of staphylococcal typing methods is progressing *(76,81,82,99,124–126)*, but firm conclusions cannot yet be drawn as to the most efficient, partly because new methods are continuously developed and existing methods are being refined. Moreover, only a few studies *(76)* have sufficiently used large and well-defined collections of related and unrelated isolates of a given species to validly define the performance of the methods *(38)*. At present, no ideal system has emerged and a combination of two or three molecular methods, preferably combined with reliable phenotypic data, such as the antimicrobial resistance profile, is to be recommended for the epidemiologic assessment of clonal relatedness among a set of isolates *(76)*. PFGE analysis is one of the most discriminating typing systems available, but this method requires careful interpretation of minor variations to determine clonal relatedness *(38,40,96,101)*. PCR-based fingerprinting methods provide efficient genotyping of staphylococci, but the limited reproducibility of low stringency AP-PCR typing *(126)* suggests it is most useful for initial screening of clonal relatedness in outbreak situations where a limited number of isolates can be processed in a single assay. For longitudinal studies, fine tuning of PCR-based fingerprinting methods is in progress. High stringency, interrepeat amplimer length polymorphism analysis appears promising for this purpose (**Table 10**). The combination of RFLP typing using several DNA probes or of PCR-RFLP analysis of several polymophic loci offers another reliable, yet more labor-intensive, approach to define and follow the clonal dissemination of MRSA strains over extended periods and geographic areas *(79,80,89–91,94, 99)*.

4. Research Uses

In this section, selected methods that are useful for the genotypic detection of antibiotic resistance determinants, other than methicillin resistance (*see* above), will be summarized. Genetic methods used for the mapping and functional analysis of these genes and for the analysis of virulence genes of staphylococci are beyond the scope of this review.

4.1. PCR Detection of Genes Conferring Resistance to Aminoglycosides and to Macrolides, Lincosamides, and Streptogramins

Aminoglycoside resistance (*see* Chapter 28) is usually mediated in staphylococci by aminoglycoside-modifying enzymes, or combination thereof, including the bifunctional enzyme AAC(6')-APH(2"), APH(3')-III, and ANT(4,4"). These enzymes are encoded by well-conserved genes that are located on transferable plasmids or on the chromosome, and can be mobilized by transposons. A synthetic DNA probe has been designed to detect and map the location of the bifunctional enzyme gene *(128)*. A multiplex PCR assay was shown to correlate completely with biochemical assays for detection of all three enzymes in MRSA strains *(129)* (**Table 11**).

Bacterial resistance to macrolides is usually determined by production of Erm methlytransferases that methylate an adenine in 23S rRNA, thereby reducing target affinity and producing cross-resistance to macrolides, lincosamides, and the B class of streptogramins (MLS$_B$ phenotype) (*see* Chapter 31). Although there exists a large diversity of *erm* genes encoding these rRNA methylases, degenerate oligonucleotide primers have been designed for detection of *erm* genes in gram-positive cocci *(130)* (**Table 11**). Further identification requires hybridization with specific intragenic DNA probes or sequence determination of the amplified product.

4.2. PCR Detection of Mutations in Topoisomerase Genes Conferring Resistance to Fluoroquinolones

The most common mechanisms of resistance to fluoroquinolones in *S. aureus* clinical isolates or in in vitro-selected mutants, are alterations of the target proteins, DNA gyrase, and topoisomerase IV *(131)*. All described point mutations conferring resistance map within narrow regions called the quinolone resistance determining regions *(131)* (*see* Chapter 30). PCR amplification of these regions can be used to detect these mutations by RFLP analysis, SSCP analysis or, ideally, sequence determination of the amplified product *(132,133)*.

5. Future Prospects and Goals

Molecular methods applicable to the diagnosis of staphylococcal infections are promising, especially multiple-target amplification assays aimed at detection of genes indicative of the species and of antimicrobial resistance of immediate therapeutic relevance, such as the *mec* determinant. Current efforts should

Table 11
PCR Detection of Genes Conferring Resistance to Aminoglycosides and to Macrolides-Lincosamides-Streptogramins B (MLS)

Antimicrobial resistance	Resistance determinant	Gene	Primer sequence (5' to 3')	Ref.
Aminoglycosides	AAC (6') - APH (2")	*aacA-aphD*	1 : CCA AGA GCA ATA AGG GCA TACC 2 : CAC ACT ATC ATA ACC ACT ACC G	*129*
	APH (3')-III	*aphA3*	1 : CTG ATC GAA AAA TAC CGC TGC 2 : TCA TAC TCT TCC GAG CAA AGG	
	ANT (4',4")	*aadC*	1 : CTG CTA AAT CGG TAG AAGC 2 : CAG ACC AAT CAA CAT GGC ACC	
MLS	Erm methyltransferases	*erm*	1 : GAA/G ATI GGI III GGI AAA/G GGI CA 2 : AAC/T TGA/G TTC/T TTI GTA/G AA	*130*

457

aim at further simplification of nucleic acid amplification and detection proto-cols. Ultimately, a greater level of automation would be desirable to integrate these techniques into the routine clinical microbiology workflow. Additional clinical studies are needed to better define the clinical impact on diagnostic decisions and management of patients with staphylococcal infections, includ-ing infection control procedures. Both clinical efficacy in therapeutic and prevention strategies based on results of these tests, as well as their cost-effectiveness and cost-benefit ratio are to be evaluated. These studies should define where and when molecular assays, performed on clinical specimens or primary cultures, may be indicated either as confirmation or first line tests. More work is also necessary to determine the procedures for quality assurance of these assays and proficiency testing of laboratories performing them.

Molecular techniques have already established their superior usefulness as compared with phenotypic assays for the epidemiologic investigation of noso-comial staphylococcal infections. Current efforts are directed at both further optimizing and standardizing several genotyping systems, and at performing comprehensive critical comparisons of their performance, including those sys-tems for which commercial versions are available. These studies are most effective when performed by multiple centers. Computer-assisted analysis of molecular typing data is another topic of current efforts to achieve longitudinal epidemiologic typing. When successfully standardized, common software may enable the electronic exchange of data and database sharing between workers interested in epidemiologic tracking of epidemic clones. An unresolved debate concerns to what extent these genotyping assays can and should optimally be used, either in hospital clinical microbiology laboratories or in reference labo-ratories. It is likely that a combination of the two levels of testing is warranted to ensure both timely local monitoring and interventions, and a more global surveillance of major problems such as MRSA epidemics. Which techniques are most appropriate at each level remains to be defined. It is likely that, as new technology emerges and become refined, molecular typing will further enhance our understanding of the epidemiology of staphylococcal infections, which, in turn, should contribute to more effective prevention strategies.

References

1. Kloos, W. E. and Bannerman, T. L. (1994) Update on clinical significance of coagulase-negative Staphylococci. *Clin. Microbiol. Rev.* **7,** 117–140.
2. Vandenesch, F., Eykyn, S. J., and Etienne, J. (1995) Infections caused by newly-described species of coagulase-negative staphylococci. *Rev. Med. Microbiol.* **6,** 94–100.
3. Hartman, B. J. and Tomasz, A. (1984) Low-affinity penicillin-binding protein associated with β-lactamase resistance in *Staphylococcus aureus. J. Bacteriol.* **158,** 513–516.

4. Chambers, H. F. (1988) Methicillin-resistant staphylococci. *Clin. Microbiol. Rev.* **1,** 173–186.

5. Bannerman, T. L., Hancock, G. A., Tenover, F. C., and Miller, J. M. (1995) Pulsed-field gel electrophoresis as a replacement for bacteriophage typing of *Staphylococcus aureus. J. Clin. Microbiol.* **33,** 551–555.

6. Thomson-Carter, F. M., Carter, P. E., and Pennington, T. H. (1989) Differentiation of staphylococcal species and strains by ribosomal RNA gene restriction patterns. *J. Gen. Microbiol.* **135,** 2093–2097.

7. Bialkowska-Hobrzanska, H., Harry, V., Jaskot, D., and Hammerberg, O. (1990) Typing of coagulase-negative staphylococci by southern hybridization of chromosomal DNA fingerprints using a ribosomal RNA probe. *Eur. J. Clin. Microbiol. Infect. Dis.* **9,** 588–594.

8. De Buyser M. L., Morvan A., Grimont F., and El Solh N. (1989) Characterization of *Staphylococcus* species by ribosomal RNA gene restriction patterns. *J. Gen. Microbiol.* **135,** 989–999.

9. De Buyser, M. L., Morvan, A., Aubert, S., Dilasser, F., and El Solh, N. (1992) Evaluation of a ribosomal RNA gene probe for the identification of species and subspecies within the genus *Staphylococcus. J. Gen. Microbiol.* **138,** 889–899.

10. Chesneau, O., Aubert, S., Morvan, A., Guesdon, J. L., and El Solh, N. (1992) Usefulness of the ID32 Staph System and a method based on rRNA gene restriction site polymorphism analysis for species and subspecies identification of staphylococcal clinical isolates. *J. Clin. Microbiol.* **30,** 2346–2352.

11. Chesneau, O., Morvan, A., Grimont, F., Labischinski, H., and El Solh, N. (1993) *Staphylococcus pasteuri* sp. nov., isolated from human, animal, and food specimens. *Int. J. Syst. Bacteriol.* **43,** 237–244.

12. Freney, J., Meugnier, H., Bes, M., and Fleurette, J. (1993) Identification of *Staphylococcus aureus* using a DNA probe: Accuprobe[R]. *Ann. Biol. Clin.* **51,** 637–639.

13. Davis, T. E. and Fuller, D. D. (1991) Direct identification of bacterial isolates in blood cultures by using a DNA probe. *J. Clin. Microbiol.* **29,** 2193–2196.

14. Chesneau, O., Allignet, J., and El Solh, N. (1993) Thermonuclease gene as a target nucleotide sequence for specific recognition of *Staphylococcus aureus. Mol. Cell. Probes* **7,** 301–310.

15. Goh, S. H., Potter, S., Wood, J. O., Hemmingsen, S. M., Reynolds, R. P., and Chow, A. W. (1996) HSP60 gene sequences as universal targets for microbial species identification: studies with coagulase-negative staphylococci. *J. Clin. Microbiol.* **34,** 818–823.

16. Ligozzi, M., Rossolini, G. M., Tonin, E. A., and Fontana, R. (1991) Nonradioactive DNA probe for detection of gene for methicillin resistance in *Staphylococcus aureus. Antimicrob. Agents Chemother.* **35,** 575–578.

17. Predari, S. C., Ligozzi, M., and Fontana, R. (1991) Genotypic identification of methicillin-resistant coagulase-negative staphylococci by polymerase chain reaction. *Antimicrob. Agents Chemother.* **35,** 2568–2573.

18. Archer, G. L. and Pennell, E. (1990) Detection of methicillin resistance in staphylococci by using a DNA probe. *Antimicrob. Agents Chemother.* **34,** 1720–1724.

19. de Lancastre, H., Sa Figueiredo, A., Urban, C., Rahal, J., and Tomasz, A. (1991) Multiple mechanisms of methicillin resistance and improved methods for detection in clinical isolates of *Staphylococcus aureus. Antimicrobial. Agents Chemother.* **35,** 632–639.

20. Shimaoka, M., Yoh, M., Segawa, A., Takarada, Y., Yamamoto, K., and Honda, T. (1994) Development of enzyme-labeled oligonucleotide probe for detection of mecA gene in methicillin-resistant *Staphylococcus aureus. J. Clin. Microbiol.* **32,** 1866–1869.

21. Richard, P., Meyran, M., Carpentier, E., Thabaut, A., and Drugeon, H. B. (1994) Comparison of phenotypic methods and DNA hybridization for detection of methicillin-resistant *Staphylococcus aureus. J. Clin. Microbiol.* **32,** 613–617.

22. Kolbert, C. P., Connolly, J. E., Lee, M. J., and Persing, D. H. (1995) Detection of the staphylococcal *mecA* gene by chemiluminescent DNA hybridization. *J. Clin. Microbiol.* **33,** 2179–2182.

23. Unal, S., Hoskins, J., Flokowitsch, J. E., Wu, C. Y. E., Preston, D. A., and Skatrud, P. L. (1992) Detection of methicillin-resistant Staphylococci by using the polymerase chain reaction. *J. Clin. Microbiol.* **30,** 1685–1691.

24. Kizaki, M., Kobayashi, Y., and Ikeda, Y. (1994) Rapid and sensitive detection of the *fem*A gene in staphylococci by enzymatic detection of polymerase chain reaction (ED-PCR): comparison with standard PCR analysis. *J. Hosp. Infect.* **28,** 287–295.

25. Brakstad, O. D., Aasbakk, K., and Maeland, J. A. (1992) Detection of Staphylococcus aureus by Polymerase Chain Reaction amplification of the *nuc* gene. *J. Clin. Microbiol.* **30,** 1654–1660.

26. Welsh, J. and McClelland, M. (1992) PCR-amplified length polymorphisms in tRNA intergenic spacers for categorizing staphylococci. *Mol. Microbiol.* **6,** 1673–1680.

27. Jensen, M. A., Webster, J. A., and Straus, N. (1993) Rapid identification of bacteria on the basis of Polymerase Chain Reaction-amplified ribosomal DNA spacer polymorphisms. *Appl. Environ. Microbiol.* **59,** 945–952.

28. Maes, N., De Gheldere, Y., De Ryck, R., Vancechoutte, M., Meugnier, M., Etienne, J., and Struelens, M. J. (1997) Rapid and accurate identification of *Staphylococcus* species by tRNA intergenic sporer length polymorphism analysis. *J. Clin. Microbiol.* **35,** 2477–2481.

29. Murakami, K., Minamide, W., Wada, K., Nakamura, E., Teraoka, H., and Watanabe, S. (1991) Identification of methicillin-resistant strains of staphylococci by Polymerase Chain Reaction. *J. Clin. Microbiol.* **29,** 2240–2244.

30. Tokue, Y., Shoji, S., Satoh, K., Watanabe, A., and Motomiya, M. (1992). Comparison of a Polymerase Chain Reaction assay and a conventional microbiologic method for detection of methicillin-resistant *Staphylococcus aureus. Antimicrob. Agents Chemother.* **36,** 6–9.

31. Ubukata, K., Nakagami, S., Nitta, A., Yamane, A., Kawakami, S., Sugiura, M., and Konno, M. (1992) Rapid detection of the *mecA* gene in methicillin-resistant staphylococci by enzymatic detection of Polymerase Chain Reaction products. *J. Clin. Microbiol.* **30,** 1728–1733.

32. Unal, S., Werner, K., DeGirolami, P., Barsanti, F., and Eliopoulos, G. (1994) Comparison of tests for detection of methicillin-resistant *Staphylococcus aureus* in a clinical microbiology laboratory. *Antimicrobial Agents Chemother.* **38**, 345–347.

33. Zambardi, G., Reverdy, M. E., Bland, S., Bes, M., Freney, J., and Fleurette, J. (1994) Laboratory diagnosis of oxacillin resistance in *Staphylococcus aureus* by a multiplex Polymerase Chain Reaction assay. *Diagn. Microbiol. Infect. Dis.* **19**, 25–31.

34. Geha, D. J., Uhl, J. R., Gustaferro, C. A., and Persing, D. H. (1994) Multiplex PCR for identification of methicillin-resistant staphylococci in the clinical laboratory. *J. Clin. Microbiol.* **32**, 1768–1772.

35. Brakstad, O. G., Maeland, J. A., and Tveten, Y. (1993) Multiplex polymerase chain reaction for detection of genes for *Staphylococcus aureus* thermonuclease and methicillin resistance and correlation with oxacillin resistance. *APMIS* **101**, 681–688.

36. Vannuffel, P., Gigi, J., Ezzedine, H., Vandercam, B., Delmee, M., Wauters, G., and Gala, J. L. (1995) Specific detection of methicillin-resistant *Staphylococcus* species by multiplex PCR. *J. Clin. Microbiol.* **33**, 2864–2867.

37. Arbeit, R. D. (1995) Laboratory procedures for the epidemiologic analysis of microorganisms, in *Manual of Clinical Microbiology,* (Murray, P. R., Barton, E. J., Pfeller, M. A., Tenover, F. C., and Yolken, R. H., eds.), American Society for Microbiology, Washington, DC, pp. 190–208.

38. Struelens, M. J. and the European Strudy Group on Epidemiological Markers (1996) Consensus guidelines for appropriate use and evaluation of microbial epidemiologic typing systems. *Clin. Microbial. Infect.* **2**, 2–11.

39. Struelens, M. J. (1996) Laboratory methods in the investigation of outbreaks of hospital-acquired infection, in *Surveillance of Nosocomial Infections,* (Emmerson, A. M. and Ayliffe, G. A. J., eds.), *Baillière's Clin.Infect.Dis.* **3**, 267–288.

40. Tenover, F. C., Arbeit, R. D., Goering, R. V., Mickelsen, P. A., Murray, B. E., Persing, D. H., and Swaminathan, B. (1995) Interpreting chromosomal DNA restriction patterns produced by pulsed-field gel electrophoresis: criteria for bacterial strain typing. *J. Clin. Microbiol.* **33**, 2233–2239.

41. Hunter, P. R. (1990) Reproducibility and indices of discriminatory power of microbial typing methods. *J. Clin. Microbiol.* **28**, 1903–1905.

42. Parisi, J. T. and Hecht, D. W. (1980) Plasmid profiles in epidemiologic studies of infections by *Staphylococcus epidermidis. J. Infect. Dis.* **141**, 637–643.

43. Archer, G. L., Vishniavsky, N., and Stiver, H. G. (1982) Plasmid pattern analysis of *Staphylococcus epidermidis* isolates from patients with prosthetic valve endocarditis. *Infect. Immun.* **35**, 627–632.

44. Archer, G. L., Karchmer, A. W., Vishniavsky, N, and Johnston, L. (1984) Plasmid-pattern analysis for the differentiation of infecting from noninfecting *Staphylococcus epidermidis. J. Infect. Dis.* **149**, 913–920.

45. Mickelsen, P. A., Plorde, J. J., Gordon, K. P., Hargiss, C., McClure, J., Fritz, D., Schoenknecht, F. D., Condie, F., Tenover, F. C., and Tompkins, L. S. (1985) Instability of antibiotic resistance in a strain of *Staphylococcus epidermidis* isolated from an outbreak of prosthetic valve endocarditis. *J. Infect. Dis.* **152**, 50–58.

46. Valentine, C. R., Yandle, S. H., Marsik, F. J., Ebright, J. R, and Dawson, M. S. (1988) Evaluation of the variety of plasmid profiles in *S epidermidis* isolates from hospital patients and staff. *Infect. Control Hosp. Epidemiol.* **9,** 441–446

47. Etienne, J., Brun, Y., El Solh, N., Delorme, V., Mouren, C., Bes, M., and Fleurette, J. (1988) Characterization of clinically significant isolates of *Staphylococcus epidermidis* from patients with endocarditis. *J. Clin. Microbiol.* **26,** 613–617.

48. Ludlam, H. A., Noble, W. C., Marples, R. R., and Phillips I. (1989) The evaluation of a typing scheme for coagulase-negative staphylococci suitable for epidemiological studies. *J. Med. Microbiol.* **30,** 161–165.

49. Thore, M., Kühn, I., Löfdahl, S., and Burman, L. G. (1990) Drug-resistant coagulase-negative skin staphylococci. *Epidemiol. Infect.* **105,** 95–105.

50. Etienne, J., Bes, F., Renaud,M., Brun, Y., Greenland, T. B., Freney, J., and Fleurette, J. (1990) Instability of characteristics among coagulase-negative staphylococci causing endocarditis. *J. Med. Microbiol.* **32,** 115–122.

51. Hartstein, A. I., Morthland, V. H., and Rashad, A. L. (1991) Reproducibility of *Staphylococcus epidermidis* plasmid profiles. *Diagn. Microbiol. Infect. Dis.* **14,** 275–280.

52. Wilton, J., Jung, K., Vedin, I., Aronsson, B., and Flock, J. I. (1992) Comparative evaluation of a new molecular method for typing *Staphylococcus epidermidis. Eur. J. Clin. Microbiol. Infect. Dis.* **11,** 515–521.

53. Dryden, M. S., Talsania, H. G., Martin, S., Cunningham, M., Richardson, J. F., Cookson, B., Marples, R. R., and Phillips, I. (1992) Evaluation of methods for typing coagulase-negative staphylococci. *J. Med. Microbiol.* **37,** 109–117.

54. Herwaldt, L. A., Hollis, R. J., Boyken, L. D., and Pfaller, M. A. (1992) Molecular epidemiology of coagulase-negative staphylococci isolated from immunocompromised patients. *Infect. Control and Hosp. Epidemiol.* **13,** 86–92.

55 Loncle, V., Casetta, A., Buu-Hoi, A., and El Solh, N. (1993) Analysis of pristinamycin-resistant *Staphylococcus epidermidis* isolates responsible for an outbreak in a Parisian hospital. *Antimicrob. Agents Chemother.* **37,** 2159–2165.

56. Renaud, F., Etienne, J., Bertrand, A., Brun, Y., Greenland, T. B., Freney, J., and Fleurette, J. (1991) Molecular epidemiology of *Staphylococcus haemolyticus* strains isolated in an Albanian hospital. *J. Clin. Microbiol.* **29,** 1493–1497.

57. Etienne, J., Poitevin-Later, F., Renaud, F., and Fleurette, J. (1990) Plasmid profiles and genomic DNA restriction endonuclease patterns of 30 independent *Staphylococcus lugdunensis* strains. *FEMS Microbiol. Lett.* **67,** 93–98

58. Bigelow, N., Ng, L.-K., Robson, H. G., and Dillon, J. R. (1989) Strategies for molecular characterisation of methicillin- and gentamicin-resistant *staphylococcus aureus* in a Canadian nosocomial outbreak. *J. Med. Microbiol.* **30,** 51–58.

59. Coia, J. E., Hussain-Noor, I., and Platt, D. J. (1988) Plasmid profiles and restriction enzyme fragmentation patterns of plasmids of methicillin-sensitive and methicillin-resistant isolates of *Staphylococcus aureus* from hospital and the community. *J. Med. Microbiol.* **27,** 271–276.

60. Hartstein, A. I., Morthland, V. H., Eng, S., Archer, G. L., Schoenknecht, F. D., and Rashad, A. L. (1989) Restriction enzyme analysis of plasmid DNA and bacteriophage typing of paired *Staphylococcus aureus* blood culture isolates. *J. Clin. Microbiol.* **27,** 1874–1879.

61. Coia, J. E., Thomson-Carter, T., Baird, D., and Platt, D. J. (1990) Characterisation of methicillin-resistant *Staphylococcus aureus* by biotyping, immunoblotting and restriction enzyme fragmentation patterns. *J. Med. Microbiol.* **31,** 125–132.

62. Hartstein, A. I., Phelps, C. L., Kwok R. Y. Y., and Mulligan, M. E. (1995) In vivo stability and discriminatory power of methicillin-resistant *Staphylococcus aureus* typing by restriction endonuclease analysis of plasmid DNA compared with those of other molecular methods. *J. Clin. Microbiol.* **33,** 2022–2026.

63. Udo, E. E., Pearman, J. W., and Grubb, W. B. (1993) Genetic analysis of community isolates of methicillin-resistant *Staphylococcus aureus* in Western Australia. *J. Hosp. Infect.* **25,** 97–108

64. Trilla, A., Nettleman, M. D., Hollis, R. J., Fredrickson, M., Wenzel, R. P., and Pfaller, A. (1993) Restriction endonuclease analysis of plasmid DNA from methicillin-resistant *Staphylococcus aureus*: clinical application over a three-year period. *Infect. Control Hosp. Epidemiol.* **14,** 29–35.

65. Sabria-Leal, M., Morthland, V. H., Pedro-Boter, M. L., Sopena, N., Gimenez-Perez, M., Branchini, M. L. M., and Pfaller, M. A. (1994) Molecular epidemiology for local outbreaks of methicillin resistant *Staphylococcus aureus* (MRSA). *Eur. J. Epidemiol.* **10,** 325–330.

66. Fang, F. C., McClelland, M., Guiney, D. G., Jackson, M. M., Hartstein, A. I., Morthland, V. H., Davis, C. E., McPherson, D. C., and Welsh J. (1993) Value of molecular epidemiologic analysis in a nosocomial methicillin-resistant *Staphylococcus aureus* outbreak. *JAMA* **270,** 1323–1328.

67. Yuk-Fong, Liu P., Shi, ZY., Lau, YJ., Hu, BS., Shyr, JM., Tsai, WH., Lin, YH., and Tseng, CY. (1996) Use of restriction endonuclease analysis of plasmids and pulsed-field gel electrophoresis to investigate outbreaks of methicillin-resistant *Staphylococcus aureus* infection. *Clin. Infect. Dis.* **22,** 86–90.

68. Grattard, F., Etienne, J., Pozzetto, B., Tardy, F., Gaudin, O. G., and Fleurette, J. (1993) Characterization of unrelated strains of *Staphylococcus schleiferi* by using ribosomal DNA fingerprinting, DNA restriction patterns, and plasmid profiles. *J. Clin. Microbiol.* **31,** 812–818.

69. Burnie, J. P., Matthews, R. C., Lee, W., and Murdoch, D. (1989) A comparison of immunoblot and DNA restriction patterns in characterising methicillin-resistant isolates of *Staphylococcus aureus*. *J. Med. Microbiol.* **29,** 255–261.

70. Hall, L. M. C., Jordens, J. Z., and Wang, F. (1989) Methicillin-resistant *Staphylococcus aureus* from China characterized by digestion of total DNA with restriction enzymes. *Epidemiol. Infect.* **103,** 183–192.

71. Struelens, M. J., Deplano, A., Godard, C., Maes, N., and Serruys, E. (1992) Epidemiologic typing and delineation of genetic relatedness of methicillin-resistant *Staphylococcus aureus* by macrorestriction analysis of genomic DNA by using pulsed-field gel electrophoresis. *J. Clin. Microbiol.* **30,** 2599–2605.

72. Matthews, K. R., Jayarao, B. M., and Oliver, S. P. (1992) Restriction endonuclease fingerprinting of genomic DNA of *Staphylococcus* species of bovine origin. *Epidemiol. Infect.* **109,** 59–68.

73. Jordens, J. Z., and Hall, L. M. C. (1988) Characterisation of methicillin-resistant *Staphylococcus aureus* isolates by restriction endonuclease digestion of chromosomal DNA. *J. Med. Microbiol.* **27,** 117–123.

74. Renaud, F., Freney, J., Etienne, J., Bes, M., Brun, Y., Barsotti, O., André, S., and Fleurette, J. (1988) Restriction endonuclease analysis of *Staphylococcus epidermidis* DNA may be a useful epidemiological marker. *J. Clin. Microbiol.* **26,** 1729–1734.

75. Low, D. E., Schmidt, B. K., Kirpalani, H. M., Moodie, R., Kreiswirth, B., Matlow, A., and Ford-Jones, E. L. (1992) An endemic strain of *Staphylococcus haemolyticus* colonizing and causing bacteremia in neonatal intensive care unit patients. *Pediatrics* **89,** 696–700.

76. Tenover, F. C., Arbeit, R., Archer, G., Biddle, J., Byrne, S., Goering, R., Hancock, G., et al. (1994) Comparison of traditional and molecular methods of typing isolates of *Staphylococcus aureus. J. Clin. Microbiol.* **32,** 407–415.

77. Preheim, L., Pitcher, D., Owen, R., and Cookson, B. (1991) Typing of methicillin resistant and susceptible S*taphylococcus aureus* strains by ribosomal RNA gene restriction patterns using a biotinylated probe. *Eur. J. Clin. Microbiol. Infect. Dis.* **10,** 428–436.

78. Blumberg, H. M., Rimland, D., Kiehlbauch, J. A., Terry, P. M., and Wachsmuth, I. K. (1992) Epidemiologic typing of *Staphylococcus aureus* by DNA restriction fragment length polymorphisms of rRNA genes: elucidation of the clonal nature of a group of bacteriophage-nontypeable, ciprofloxacin-resistant, methicillin-susceptible *S. aureus* isolates. *J. Clin. Microbiol.* **30,** 362–369.

79. Monzon-Moreno, C., Aubert, S., Morvan, S., and El Solh, N. (1991) Usefulness of three probes in typing isolates of methicillin-resistant *Staphylococcus aureus* (MRSA). *J. Med. Microbiol.* **35,** 80–88.

80. de Lencastre, H., Couto, I., Santos, I., Melo-Cristino, J., Torres-Pereira, A., and Tomasz, A. (1994) Methicillin-resistant *Staphylococcus aureus* disease in a Portuguese hospital: characterization of clonal types by a combination of DNA typing methods. *Eur. J. Clin. Infect. Dis.* **13,** 64–73.

81. Prevost, G., Jaulhac, B., and Piemont, Y. (1992) DNA fingerprinting by pulsed-field gel electrophoresis is more effective than ribotyping in distinguishing among methicillin-resistant *Staphylococcus aureus* isolates. *J. Clin. Microbiol.* **30,** 967–973.

82. van Belkum, A., Bax, R., and Prevost, G. (1994) Comparison of four genotyping assays for epidemiological study of methicillin-resistant *Staphylococcus aureus. Eur. J. Clin. Microbiol. Infect. Dis.* **13,** 420–424.

83. Walcher-Salesse, S., Monzon-Moreno, C., Aubert, S., and El Solh N. (1992) An epidemiological assessment of coagulase-negative staphylococci from an intensive care unit. *J. Med. Microbiol.* **36,** 321–331.

84. Crichton, P. B., Anderson, L. A., Philips, G., Davey, P. G., and Rowley, D. I. (1995) Subspecies discrimination of staphylococci from revision arthroplasties by ribotyping. *J. Hosp. Infect.* **30,** 139–147.

85. Izard, N. C., Hächler, H., Grehn, M., and Kayser, F. H. (1992) Ribotyping of coagulase-negative staphylococci with special emphasis on intraspecific typing of *Staphylococcus epidermidis. J. Clin. Microbiol.* **30,** 817–823.

86. Degener, J. E., Heck, M. E. O. C., van Leeuwen, W. J., Heemskerk, C., Crielaard, A., Joosten, P., and Caesar, P. (1994) Nosocomial infection by *Staphylococcus haemolyticus* and typing methods for epidemiological study. *J. Clin. Microbiol.* **32,** 2260–2265.

87. Smeltzer, M. S., Pratt, F. L., Gillaspy, F. G., and Young, L. A. (1996) Genomic fingerprinting for epidemiological differentiation of *Staphylococcus aureus* clinical isolates. *J. Clin. Microbiol.* **34,** 1364–1372.

88. Derbise, A., Dyke, K. G. H., and El Solh, N. (1994) Isolation and characterization of IS*1181*, an insertion sequence from *Staphylococcus aureus. Plasmid* **31,** 251–264.

89. Teixeira, L. A., Resende, C. A., Ormonde, L. R., Rosenbaum, R., Figueiredo, A. M. S., de Lencastre, H., and Tomasz, A. (1995) Geographic spread of epidemic multiresistant *Staphylococcus aureus* clone in Brazil. *J. Clin. Microbiol.* **33,** 2400–2404.

90. Kreiswirth, B. N., Lutwick, M., Chapnick, E. K., Gradon, J. D., Lutwick, L. I., Sepkowitz, D. V., Eisner, W., and Levi, M. H. (1995) Tracing the spread of methicillin-resistant *Staphylococcus aureus* by southern blot hybridization using gene-specific probes of mec and Tn*554. Microbial Drug Res.* **1,** 307–313.

91. Angeles Dominguez, M., de Lencastre, H., Linares, J., and Tomasz, A. (1994) Spread and maintenance of a dominant methicillin-resistant *Staphylococcus aureus* (MRSA) clone during an outbreak of MRSA disease in a Spanish hospital. *J. Clin. Microbiol.* **32,** 2081–2087.

92. Wei, M. Q., Groth, D. M., Mendis, A. H. W., Sampson, J., Wetherall, J. D., and Grubb, W. B. (1992) Typing of methicillin-resistant *Staphylococcus aureus* with an M13 repeat probe. *J. Hosp. Infect.* **20,** 233–245.

93. Wei, M. Q., Udo, E. E., and Grubb, W. B. (1992) Typing of methicillin-resistant *Staphylococcus aureus* with IS*256. FEMS Microbiol. Lett.* **99,** 175–180.

94. Santos Sanches, I., Ramirez, M., Troni, H., Abecassis, M., Padua, M., Tomasz, A., and de Lencastre, H. (1995) Evidence for the geographic spread of a methicillin-resistant *Staphylococcus aureus* clone between Portugal and Spain. *J. Clin. Microbiol.* **33,** 1243–1246.

95. Nesin, M., Projan, S. J., Kreiswirth, B., Bolt, Y., and Novick, R. P. (1995) Molecular epidemiology of *Staphylococcus epidermidis* blood isolates from neonatal intensive care unit patients. *J. Hosp. Infect.* **31,** 111–121.

96. El-Adhami, W., Roberts, L., Vickery, A., Inglis, B., Gibbs, A., and Stewart, P. R. (1991) Epidemiological analysis of a methicillin-resistant *Staphylococcus aureus* outbreak using restriction fragment length polymorphisms of genomic DNA. *J. Gen. Microbiol.* **137,** 2713–2720.

97. Ichiyama, S., Ohta, M., Shimokata, K., Kato, N., and Takeuchi, J. (1991) Genomic DNA fingerprinting by pulsed-field gel electrophoresis as an epidemiological marker for study of nosocomial infections caused by methicillin-resistant *Staphylococcus aureus. J. Clin. Microbiol.* **29,** 2690–2695.

98. Wei, M. Q., Wang, FU, and Grubb, W. B. (1992) Use of contour-clamped homogeneous electric field (CHEF) electrophoresis to type methicillin-resistant *Staphylococcus aureus*. *J. Med. Microbiol.* **36**, 172–176.

99. Kluytmans, J., van Leeuwen, W., Goessens, W., Hollis, R., Messer, S., Herwaldt, L., Bruining, H., Heck, M., Rost, J., van Leeuwen, N., van Belkum, A., and Verbrugh, H. (1995) Food-initiated outbreak of methicillin-resistant *Staphylococcus aureus* analyzed by pheno- and genotyping. *J. Clin. Microbiol.* **33**, 1121–1128.

100. Trzcinski, K., Hryniewicz, W., Claus, H., and Witte, W. (1994) Characterization of two different clusters of clonally related methicillin-resistant *Staphylococcus aureus* strains by conventional and molecular typing. *J. Hosp. Infect.* **28**, 113–126.

101. Jorgensen, M., Givney, R., Pegler, M., Vickery, A., and Funnell, G. (1996) Typing multidrug-resistant *Staphylococcus aureus*: conflicting epidemiological data produced by genotypic and phenotypic methods clarified by phylogenetic analysis. *J. Clin. Microbiol.* **34**, 398–403.

102. Cookson, B. D., Aparicio, P., Deplano, A., Struelens, M., Goering, R., and Marples, R. (1996) Inter-centre comparison of pulsed-field gel electrophoresis for the typing of methicillin-resistant *Staphylococcus aureus*. *J. Med. Microbiol.* **44**, 179–184.

103. Goering, R. V. and Duensing, T. D. (1990) Rapid field inversion gel electrophoresis in combination with an rRNA gene probe in the epidemiological evaluation of staphylococci. *J. Clin. Microbiol.* **28**, 426–429.

104. Prévost, G., Pottecher, B., Dahlet, M., Bientz, M., Mantz, J. M., and Piémont, Y. (1991) Pulsed field gel electrophoresis as a new epidemiological tool for monitoring methicillin-resistant *Staphylococcus aureus* in an intensive care unit. *J. Hosp. Infect.* **17**, 255–269.

105. Schlichting, C., Branger, C., Fournier, J. M., Witte, W., Boutonnier, A., Wolz, C., Goullet, P., and Döring, G. (1993) Typing of *Staphylococcus aureus* by pulsed-field gel electrophoresis, zymotyping, capsular typing, and phage typing: resolution of clonal relationships. *J. Clin. Microbiol.* **31**, 227–232.

106. Nada, T., Ichiyama, S., Osada, Y., Ohta, M., Shimokata, K., Kato, N., and Nakashima, N. (1996) Comparison of DNA fingerprinting by PFGE and PCR-RFLP of the coagulase gene to distinguish MRSA isolates. *J. Hosp. Infect.* **32**, 305–317.

107. Couto, I., Melo-Cristino, J., Fernandes, M. L., Garcia, T., Serrano, N., Saldago, M. J., Torres-Pereira, A., Santos Sanches, I., and de Lencastre, H. (1995) Unusually large number of methicillin-resistant *Staphylococcus aureus* clones in a Portuguese hospital. *J. Clin. Microbiol.* **33**, 2032–2035.

108. Carles-Nurit, M. J., Christophle, B., Broche, S., Gouby, A., Bouziges, N., and Ramuz, M. (1992) DNA polymorphisms in methicillin-susceptible and methicillin-resistant strains of *Staphylococcus aureus*. *J. Clin. Microbiol.* **30**, 2092–2096.

109. Pantucek, R., Götz, F., Doskar, J., and Rosypal, S. (1996) Genomic variability of *Staphylococcus aureus* and the other coagulase-positive *Staphylococcus* species estimated by macrorestriction analysis using pulsed-field gel electrophoresis. *Int. J. Syst. Bacteriol.* **46**, 216–222.

110. Huebner, J., Pier, G. B., Maslow, J. N., Muller, E., Shiro, H., Parent, M., Kropec, A., Arbeit, R. D., and Goldmann, D. A. (1994) Endemic nosocomial transmission of *Staphylococcus epidermidis* bacteremia isolates in a neonatal intensive care unit over 10 years. *J. Infect. Dis.* **169,** 526–531.

111. Witte, W., Cuny, Ch., and Claus, H. (1993) Clonal relatedness of *Staphylococcus aureus* strains from infections in humans as deduced from genomic DNA fragment patterns. *Med. Microbiol. Lett.* **2,** 72–79.

112. Goering, R. V. and Winters, M. A. (1992) Rapid method for epidemiological evaluation of Gram-positive cocci by field inversion gel electrophoresis. *J. Clin. Microbiol.* **30,** 577–580.

113. Goh, S. H., Byrne, S. K., Zhang, J. L., and Chow, A. W. (1992) Molecular typing of *Staphylococcus aureus* on the basis of coagulase gene polymorphisms. *J. Clin. Microbiol.* **30,** 1642–1645.

114. Schwarzkopf, A., Karch, H., Schmidt, H., Lenz, W., and Heesemann, J. (1993) Phenotypical and genotypical characterization of epidemic clumping factor-negative, oxacillin-resistant *Staphylococcus aureus. J. Clin. Microbiol.* **31,** 2281–2285.

115. Schwarzkopf, A. and Karch, H. (1994) Genetic variation in *Staphylococcus aureus* coagulase genes: potential and limits for use as epidemiological marker. *J. Clin. Microbiol.* **32,** 2407–2412.

116. Frénay, H. M. E., Theelen, J. P. G., Schouls, L. M., Vandenbroucke-Grauls, M. J. E., Verhoef, J., Van Leeuwen, W. J., and Mooi, F. R. (1994) Discrimination of epidemic and nonepidemic methicillin-resistant *Staphylococcus aureus* strains on the basis of protein A gene polymorphism. *J. Clin. Microbiol.* **32,** 846,847.

117. Frénay, H. M. E., Bunschoten, A. E., Schouls, L. M., van Leeuwen, W. J., Vandenbroucke-Grauls, C. M. J. E., Verhoef, J., and Mooi, F. R. (1996) Molecular typing of methicillin-resistant *Staphylococcus aureus* on the basis of protein A gene polymorphism. *Eur. J. Clin. Microbiol. Infect. Dis.* **15,** 60–64.

118. Nishi, JI., Miyanohara, H., Nakajima, T., Kitajima, I., Yoshinaga, M., Maruyama, I., and Miyata, K. (1995) Molecular typing of the methicillin resistance determinant (*mec*) of clinical strains of *Staphylococcus* based on *mec* hypervariable region length polymorphisms. *J. Lab. Clin. Med.* **126,** 29–35.

119. Del Vecchio, V. G., Petroziello, J. M., Gress, M. J., McCleskey, F. K., Melcher, G. P., Crouch, H. K., and Lupski, J. R. (1995) Molecular genotyping of methicillin-resistant *Staphylococcus aureus* via fluorophore-enhanced repetitive-sequence PCR. *J. Clin. Microbiol.* **33,** 2141–2144.

120. Gürtler, V. and Barrie, H. D. (1995) Typing of *Staphylococcus aureus* strains by PCR-amplification of variable-length 16S-23S rDNA spacer regions: characterization of spacer sequences. *Microbiology* **141,** 1255–1265.

121. Cuny, C. and Witte, W. (1996) Typing of Staphylococcus aureus by PCR for DNA sequences flanked by transposon Tn*916* target region and ribosomal binding site. *J. Clin. Microbiol.* **34,** 1502-1505.

122. Deplano, A., Vanechoute, M., Verschrogen, G., and Struelens, N. J. (1997) Typing of *Staphylococcus aureus* and *Staphylococcus epidermis* strains by PCR analysis of inter-IS*256* sporer length polymorphism. *J. Cin. Microbiol.* **35,** 2580–2587.

123. van Belkum, A., Bax, R., Peerbooms, P., Goessens, W. H. F., van Leeuwen, N., and Quint, W. G. V. (1993) Comparison of phage typing and DNA fingerprinting by polymerase chain reaction for discrimination of methicillin-resistant *Staphylococcus aureus* strains. *J. Clin. Microbiol.* **31,** 798–803.

124. Saulnier, P., Bourneix, C., Prevost, G., and Andremont, A. (1993) Random amplified polymorphic DNA assay is less discriminant than pulsed-field gel electrophoresis for typing strains of methicillin-resistant *Staphylococcus aureus. J. Clin. Microbiol.* **31,** 982–985.

125. Struelens, M. J., Bax, R., Deplano, A., Quint, W. G. V., and van Belkum, A. (1993) Concordant clonal delineation of methicillin-resistant *Staphylococcus aureus* by macrorestriction analysis and polymerase chain reaction genomic fingerprinting. *J. Clin. Microbiol.* **31,** 1964–1970.

126. van Belkum, A., Kluytmans, J., van Leeuwen, W., Bax, R., Quint, W., Peters, E., Fluit, A. D., Vandenbroucke-Grauls, C., van den Brule, A., Koeleman, H., Melchers, W., Meis, J., Elaichouni, A., Vaneechoutte, M., Moonens, F., Maes, N., Struelens, M., Tenover, F., and Verbrugh, H. (1995) Multicenter evaluation of arbitrarily primed PCR for typing of *Staphylococcus aureus* strains. *J. Clin. Microbiol.* **33,** 1537–1547.

127. Bingen, E., Barc, MC., Brahimi, N., Vilmer, E., and Beaufils, F. (1995) Randomly amplified polymorphic DNA analysis provides rapid differentiation of methicillin-resistant coagulase-negative *staphylococcus* bacteremia isolates in a pediatric hospital. *J. Clin. Microbiol.* **33,** 1657–1659.

128. Jordens, J. Z. and Hall, L. M. C. (1989) Chromosomally-encoded gentamicin resistance in "epidemic" methicillin-resistant *Staphylococcus aureus*: detection with a synthetic oligonucleotide probe. *J. Antimicrob. Chemother.* **23,** 327–334.

129. Vanhoof, R., Godard, C., Content, J., Nyssen, H. J., Hannecart-Pokorni, E., and the Belgian Study Group of Hospital Infections (GDEPIH/GOSPIZ) (1994) Detection by polymerase chain reaction of genes encoding aminoglycoside-modifying enzymes in methicillin-resistant *Staphylococcus aureus* isolates of epidemic phage types. *J. Med. Microbiol.* **41,** 282–290.

130. Arthur, M., Molinas, C., and Mabilat, C. (1993) PCR detection of *erm* erythromycin resistance genes by using degenerate oligonucleotide primers. in: *Diagnostic Molecular Microbiology: Principles and Applications,* (Persing, D. H., ed.), American Society for Microbiology, Washington, DC.

131. Ferrero, L., Cameron, B., and Crouzet, J. (1995) Analysis of gyrA mutations in stepwise-selected ciprofloxacin-resistant mutants of *Staphylococcus aureus. Antimicrob. Agents Chemother.* **39,** 1554–1558.

132. Tokue, Y., Sugano, K., Saito, D., Noda, T., Ohkura, H., Shimosato, Y., and Sekiya, T. (1994) Detection of novel mutations in the *gyr*A gene of *Staphylococcus aureus* by nonradioisotopic single-strand conformation polymorphism analysis and direct DNA sequencing. *Antimicrob. Agents Chemother.* **38,** 428–431.

133. Goswitz, J. J., Willard, K. E., Fasching, C. E., and Peterson, L. R. (1992) Detection of gyrA gene mutations associated with ciprofloxacin resistance in methicillin-resistant *Staphylococcus aureus*: analysis by polymerase chain reaction and automated direct DNA sequencing. *Antimicrob. Agents Chemother.* **36,** 1166–1169.

24

Application of Molecular Techniques to the Study of Nosocomial Infections Caused by Enterococci

Teresa M. Coque, Prema Seetulsingh, Kavindra V. Singh, and Barbara E. Murray

1. Introduction

Enterococci are components of the normal bowel flora of humans and other animals, and have traditionally been considered to be of relatively low virulence in healthy individuals. However, they are increasingly important nosocomial pathogens and have been cited as the leading organism isolated from hospital-acquired infections, and the third leading cause of nosocomial bacteremia in the United States in a recent National Nosocomial Infection Surveillance (NNIS) system report of the Centers for Disease Control (1). The increase in enterococcal infections has been associated with the emergence of resistance to multiple antibiotics, in particular resistance to ß-lactams, high-level aminoglycoside resistance, and resistance to glycopeptides. Concern that antibiotic resistance will continue to spread and will increasingly render conventional antimicrobial chemotherapy inadequate for serious enterococcal infections has stimulated interest in methods to improve the diagnosis and epidemiologic investigation of infections caused by enterococci.

The aim of this chapter is to review the modern molecular methods that have been used over the past decade in the identification of enterococci at the genus and species level, in the detection of resistance to antibiotics and in the assessment of the degree of relatedness of isolates. The authors will also briefly mention some experimental approaches that may become useful in the future for understanding the pathogenesis of infections caused by enterococci.

From: *Methods in Molecular Medicine, Vol. 15: Molecular Bacteriology: Protocols and Clinical Applications*
Edited by: N. Woodford and A. P. Johnson © Humana Press Inc., Totowa, NJ

2. Identification

The identification of enterococci traditionally has been based on biochemical characteristics and reaction with Lancefield serological group D antisera. However, accurate identification is sometimes difficult since some species can exhibit unusual fermentation behavior *(2–5)*, and others can only be differentiated by one or two biochemical tests *(6)*. The use of molecular techniques (nucleic acid hybridization and DNA amplification) and protein analysis (penicillin-binding proteins [PBPs] and whole-cell protein profiling) may provide a valuable alternative to serologic and biochemical characterization for identification of enterococci.

2.1. Nucleic Acid Techniques

Nucleic acid techniques have the advantage over phenotypic identification systems of not being affected by variations in gene expression. Although preparation of genomic DNA from gram-positive organisms has been cumbersome in the past, simple protocols for small-scale preparation of DNA amenable for hybridization and/or the polymerase chain reaction (PCR) have now been successfully applied to enterococci *(7–14)* (*see* **Notes 1–3**). For colony hybridization, preparation of colony lysates, and denaturation of enterococcal genomic DNA can be performed by modified standard protocols used for other gram-positive organisms *(15)*.

2.1.1. Specific Probes for the Genus

One commercially available probe for identification of *Enterococcus* spp. is a DNA oligomer complementary to enterococcal ribosomal RNA (rRNA) sequences (AccuProbe Enterococcus, Gen Probe, San Diego, CA) *(6,16,17)*. Identification using the AccuProbe has been compared with standard biochemical characterization for 313 enterococcal isolates of different species, 253 isolates of 12 streptococcal species, and 164 isolates representing 13 other bacterial genera, demonstrating a sensitivity and specificity of 100% *(16)*. This probe was able to detect all species of enterococci, with the exception of *E. cecorum*, *E. columbae*, and *E. saccharolyticus*, species that have less genetic and phenotypic similarity with other enterococcal species *(6)*. The probe can be applied to purified colonies as well as directly to pellets from blood culture broths *(16,17)*. Although it constitutes a rapid (40 min) and discriminatory procedure for characterization of enterococci, it has the disadvantages of cost, the need for special equipment, and the inability to identify these organisms to the species level.

2.1.2. Identification to the Species Level

Different types of probes have been used for identification to the species level. Oligonucleotide DNA probes complementary to 23S rRNA sequences specific for *E. faecalis* or *E. faecium* have been applied to isolates from milk samples using colony lysis *(18)*, and by whole cell *in situ* hybridization with fluorescent-labeled probes *(19)*. Specificity of these probes was evaluated against a very limited number of enterococcal isolates ($n = 9$, one strain per species), 22 lactococci, and six isolates representing five other bacterial genera.

Biotin-labeled probes generated from genomic DNA of ATCC strains of *E. faecalis*, *E. faecium*, *E. gallinarum*, *E. raffinosus*, and *E. casseliflavus* have been used for the characterization of more than 100 enterococcal isolates previously identified by standard procedures and recovered from different geographical areas *(5)*. These probes hybridized strongly to *Eco*RI-digested genomic DNA of the corresponding species and weakly to DNA from different species except for five isolates; four of these five were biochemically identified as *E. faecium* and one as *E. gallinarum*, but their DNA hybridized to *E. gallinarum* and *E. casseliflavus* probes, respectively. In the same work, analysis of *Sma*I digested genomic DNA patterns by pulsed-field gel electrophoresis (PFGE) *(see* Chapter 3) was used together with hybridization to the above probes. Although patterns can vary widely within the same species, the size range of *Sma*I restriction fragments was relatively conserved in each enterococcal species and was helpful in classifying the above strains as nonmotile *E. gallinarum* and nonpigmented *E. casseliflavus*, respectively.

Probes specific for *E. faecium*, *E. gallinarum*, and *E. casseliflavus* have also been developed from studies of the intrinsic antibiotic resistance of these species *(20–22)*. The chromosomal gene *aac(6′)-Ii* of *E. faecium* confers moderate levels of resistance to tobramycin, kanamycin, sisomicin and netilmicin and was found in all of 58 *E. faecium* isolates studied, but not in 73 isolates of 13 other enterococcal species *(20,23)* (**Table 1**). The genes *vanC-1* and *vanC-2* confer low-level resistance to vancomycin *(21,22)* and have been found in *E. gallinarum, and E. casseliflavus/flavescens*, respectively, but not in the other enterococcal species tested, indicating that they are species-specific *(21,22,24)*. These genes have been detected by both hybridization *(21,22,25)* and PCR *(24–27)* (**Table 1**). Similarly, the genes coding for the D-alanine-D-alanine (D-Ala:D-Ala) ligases *(ddl)* in *E. faecalis* and *E. faecium* ($ddl_{E.\ faecalis}$, $ddl_{E.\ faecium}$) were found in all *E. faecalis* and *E. faecium*, respectively, which were studied, but not in the remaining enterococcal species, indicating they are specific for those species *(24,26)* (**Table 1**).

The presence of other genes in different enterococcal species has been examined in the authors' laboratory. These include *efaA* (GeneBank accession number U3756) that shows homology to streptococcal adhesins *(28)*, and two other genes (one with homology to the *E. faecalis* autolysin and one with homology to p54 protein of *E. faecium*, of unknown function) detected in the authors' laboratory by immunoscreening *Escherichia coli* DNA libraries of

Table 1
Oligonucleotides Used for Generation of Probes by PCR

Gene	Sequence (5' to 3')	5' Position	Genebank accession number (ref.)	Ref. for the use of the primers	Other primers used
vanA	+GCAAGTCAGGTGAAGATGG	675	X56895	*104*	*11,24,25,73,125*
	−ACCTCGCCAACAACTAACGC	1049	*(115)*		
vanB	+ACCCGTCTTGTTGAAGCCGGCAC	168	L6138	*104*	*24,25*
	−CAAAAAAGATCAACACGCGCAAGCCC	529	*(116)*		
vanB2	+ATTGTCTGGATCCCCTATG	30	L15304	*62*	
	−GCAAGCCCTCTGCATCAAG	540	*(117)*		
vanC	+GAAGACAACAGGAAGACCGC	126	M75132	*25*	
	−ATCGCATCACAAGCACCAATC	921	*(118)*		
vanC1	+GGTATCAAGGAAACCTC	246	M75132	*24*	
	−CTTCCGCCATCATAGCT	1067	*(118)*		
canC2, vanC3	+CTCCTACGATTCTCTTG	455	L29638, L29639	*24*	
	−CGAGCAAGACCTTTAAG	855	*(22)*		
ddlE. faecalis	+ATCAAGTACAGTTAGTCTT	98	U00457	*24,26*	
	−ACGATTCAAAGCTAACTG	1038			
ddlE. faecium	+GCAAGGCTTCTTAGAGA	NA	NA	*24,26*	
	−CATCGTGTAAGCTAACTTC	NA			
aac (6')-Ii	+GCGGTAGCAGCGGTAGACCAAG	307	L12710	*23*	
	−GCATTTGGTAAGACACCTACG	630	*(20)*		
aph(3')-IIIa	+CCGCTGCGTAAAAGATACG	585	V01547	*95*	*124*
	−CTCCAATCAGGCTTGATCC	1271	*(119)*		
aac(6')-aph(2")	+GAGATTGGTTGTTCTGAAATG	163	M13771	(Personal observation)	
	−GTCTGGACTTGACTCACTTC	1817	*(120)*		
aac(6')-aph(2")	+GATGATGATTTTCCTTTGATG	346	M13771	(Personal observation)	

472

Gene	Primer sequence			
gyrA	−CTACCATTTTCGATAAAATTCCTG	1687	(115)	23,30
	+CGGGATGAACGAATTGGGTGTGA	NA	NA	
	−AATTTACTCATACGTGCTTCGG	NA	(30)	
asa1	+GATTCTTCGATTGTGTTGTAAAACG	4016	X17214	107
	−GGTGCCACAATCAAATTAGG	4372	(121)	
cylA	+ACTCGGGATTGATAGGC	6655	L37110	106
	−GCTGCTAAAGCTGCGCTT	7326	(122,123)	
efaA	+CGTTAGCTGCTTGCGGGAATC	185	U3756	107
	−CCATACTACGTTTATCGACAC	898	(28)	(Personal observation)

+, sense primer; −, antisense primer

NA = Not available

Gene which code for ( ): ^ddl^E. faecalis and ^ddl^E. faecium (D-Ala:D-Ala) ligases (ddl) in E. faecalis and E. faecium, respectively): aac(6')-Ii ('-N-aminoglycoside acetyltransferase; aph(3')-IIIa (3'-5" aminoglycoside phosphotransferase); aac(6')-aph(2") (6'-aminoglycoside acetyltransferase 2'-aminoglycoside phosphotranferase); gyrA from E. faecalis (subunit A of DNA gyrase); asa1 (aggregation substance encoded by sex-pheromone plasmid pAD1); cylA (component A of the cytolysin operon of E. faecalis) enterococcal DNA with sera from patients with enterococcal endocarditis *(29)*. All three genes hybridized to DNA from 144 clinical isolates of *E. faecalis*, but not to DNA from 60 isolates of seven other enterococcal species by colony hybridization using high stringency conditions (personal observation). Similarly, *gyrA* from *E. faecalis*, that codes for the A subunit of DNA gyrase *(30)*, hybridized to DNA from 19 *E. faecalis*, but not to DNA from 42 isolates of six other enterococcal species by colony hybridization *(23)*. These data suggest that these probes are specific for *E. faecalis*. Probes for *gyrA* and *efaA* were generated from *E. faecalis* by PCR using the primers shown in **Table 1**, whereas the two other probes were intragenic fragments cloned in the authors' laboratory *(29)*. Also, a probe containing a 2.7-kb *Dra*I-*Eco*RI fragment of the muramidase gene *(31)* was found in two *E. hirae*, but not in 10 *E. faecalis*, eight *E. faecium*, two *E. gallinarum*, two *E. casseliflavus*, two *E. solitarius*, one *E. raffinosus*, and one *E. mundtii* by colony lysis hybridization indicating its possible application for *E. hirae* identification.

2.1.3. Advantages and Drawbacks of Nucleic Acid Techniques

Molecular methods constitute discriminatory and sensitive procedures for identification of enterococcal strains, including isolates with ambiguous biochemical characteristics *(5,21,24,27)*. Whereas most of the methodology described in **Subheading 2.1.2.** is currently restricted to research laboratories, the possibility of performing one test instead of the many required for standard biochemical identification makes the future of this technology promising. An advantage of PCR is that it allows a test to be performed in a shorter period of time than hybridization, but a more limited number of samples can be studied at a time. Hybridization allows processing of many samples per assay, which makes this procedure useful for large-scale studies.

2.2. Protein Analysis

Analysis of PBPs has been applied to the characterization of enterococcal species and compared with identification by standard methods *(3,32–34)*. Species-specific patterns of at least five PBPs have been observed in 11 enterococcal species *(3,33,34)*. Williamson et al. examined 20 *E. faecalis*, 25 *E. faecium*, 10 *E. durans*, 11 *E. gallinarum*, six *E. casseliflavus*, five *E. avium*, four *E. hirae*, one *E. malodoratus*, and one *E. mundtii*, finding a single PBP pattern per species, except for *E. faecium* which showed two similar profiles (with or without PBP5*) *(3)*. In this study, three strains (one *E. gallinarum*, one *E. casseliflavus*, and one *E. avium*) could not be classified according to their PBP profile. Isolates of *E. raffinosus* (*n* = 31), *E. solitarius* (*n* = 1), and *E. pseudoavium* (*n* = 1) were analyzed in other studies *(33,34)*. Although these strains presented a species-specific general pattern, some extra PBPs were reported for *E. raffinosus* by Grayson et al. *(33)*.

Whole-cell protein profiling has also demonstrated utility for characterization of *Enterococcus* spp. *(2,35)*. Different strains of the same species can show some differences in these profiles, but general species-specific patterns showing 30–35 bands with major differences in the 40–60 kDa and 20–30 kDa regions, were observed in 54 isolates of 16 species including CDC reference strains and isolates recovered from human infections, animal, or environmental sources *(35)*. This technique was also able to identify *E. faecium* strains with atypical biochemical characteristics *(2)*.

Protein analysis has been shown to be useful for differentiating species with similar characteristics such as *E. faecium*, *E. durans*, or *E. hirae (3,32)* or between *E. raffinosus* and *E. avium* strains *(33)*. However, these methods are time-consuming, require special equipment, and need further standardization of reference banding patterns, especially for some species for which only a few strains have been evaluated.

3. Epidemiologic Investigation

3.1. Problems of Conventional Methods of Typing

Conventional phenotypic typing methods such as biotyping, antibiograms, serotyping, phage typing, and bacteriocin typing have been applied to enterococci in the past *(36–38)*. Biotypes, antibiograms, and serotypes often show few differences within an enterococcal species, and are generally felt to be insufficiently discriminatory for most epidemiologic investigations. Bacteriophage typing and bacteriocin typing have been used with some success, but special reagents are required and a large number of tests may need to be performed *(38)*. Multilocus enzyme electrophoresis (MLEE) and biochemical fingerprinting have also been used to type enterococci. They have been found to be reproducible and moderately discriminatory, and may be convenient methods to use if the reagents and equipment are already in use in a laboratory *(39–41)*.

3.2. Nucleic Acid Analysis

3.2.1. Plasmid Analysis

Whole plasmid analysis has been used in the investigation of outbreaks caused by *E. faecalis* with *(42)* and without high-level of resistance to gentamicin (HLRG) *(43)*, ampicillin-resistant *E. raffinosus (44)*, and vancomycin-resistant *E. faecium (25)*. Restriction enzyme analysis (REA) of plasmids has been used to generate plasmid digestion profiles from ampicillin-resistant enterococci (*E. faecium*, *E. raffinosus*) from single hospitals *(45)*, ampicillin-resistant non-ß-lactamase-producing *E. faecium* isolates from diverse geographic areas *(46)*, HLRG *E. faecium* isolates from different continents *(47)*, HLRG *E. faecalis (48–51)*, ß-lactamase-producing, aminoglycoside-resistant

E. faecalis (52), and multidrug-resistant (ampicillin-resistant, HLRG, and van-comycin-resistant) *E. faecium* in a university-affiliated hospital *(53)*. Although most authors have used *Eco*RI for REA of the plasmids *(45,49,51–54)*, combination of two or more enzymes may improve plasmid differentiation *(46,47,55)*.

Differences found using whole plasmid analysis and REA of plasmids have led some investigators to suggest that both techniques should be used in combination for typing purposes *(46)*. In some studies, plasmid analysis has been used in combination with PFGE *(25,45,46,50,53,54)* or ribotyping *(47)*. The combination of plasmid analysis and PFGE may be helpful since some investigators have reported disagreement between these techniques *(25,42,46,54)*.

One disadvantage of plasmid analysis is that its usefulness is limited to plasmid-containing strains. Another disadvantage is that plasmids can be gained, lost, or changed in vivo, resulting in changes in the profile of epidemiologically related isolates. Also, this method may present technical problems such as difficulty in extracting plasmids (*see* **Note 4**; and Chapter 4). Despite these limitations, this method can be useful for typing and for the investigation of outbreaks caused by enterococci if methods based on analysis of chromosomal DNA are not available.

3.2.2. Analysis of Chromosomal DNA

3.2.2.1. Conventional Electrophoresis

Some of the drawbacks associated with plasmid analysis can be overcome by examining chromosomal DNA. One method consists of digesting genomic DNA with restriction endonucleases that have multiple recognition sites within the genome, and then analyzing the patterns generated by the fragments (usually <20 kb in size) by conventional electrophoresis (*see* Chapter 2). This method, referred to as RFLP (Restriction Fragment Length Polymorphism), or BRENDA (Bacterial Restriction Endonuclease Digestion Analysis) has been applied to enterococci isolated from hospitalized patients *(56,57)* and community based individuals *(58)* (**Table 2**). In the study by Hall et al. *(57)*, RFLP using *Sst*I was performed to type nosocomial isolates of *E. faecalis* and *E. faecium* from different sources, blood culture isolates (mostly *E. faecalis* and from different patients), and multiple isolates repeatedly recovered from the same patients (mostly *E. faecalis*), demonstrating nosocomial spread of some strains. Bingen et al. used *Hin*dIII and *Pvu*II for epidemiologic investigation of 16 vancomycin-resistant *E. faecium* in a children's hospital showing evidence of genetic unrelatedness among strains *(56)*. Finally, DNA banding patterns obtained by chromosomal digestion with *Sal*I were used for characterization of 180 strains (121 *E. faecium*, 37 *E. durans*, 21 *E. faecalis*,

Table 2
Common Restriction Endonucleases used for Analysis for Enterococci

Enzyme	Method	Species studied	No. of fragments generated[a]	Reference
*Apa*I	PFGE	*E. faecium*	17–27	*46,53,59*
*Eag*I	PFGE	*E. faecalis*	NA	*63*
		E. faecium	NA	
*Not*I	PFGE	*E. faecalis*	NA	*48*
			11–16	(Personal observation)
*Sma*I	PFGE	*E. faecalis*	12–20	*5,48,66*
*Sma*I	PFGE	*E. faecium*	10–20	*5,46,62,65*
*Sma*I	PFGE	*E. casseliflavus*	≥15	*5, 64*
*Sma*I	PFGE	*E. gallinarum*	~10–15	*5*
*Sma*I	PFGE	*E. raffinosus*	~21–25	*5*
*Hind*III	PFLP	*E. faecium*	Too many to count	*56*
*Pvu*II	RFLP	*E. faecium*	Too many to count	*56*
*Sal*I	RFLP	*E. faecalis*	10–20	*58,72*
		E. faecium	(in the 1.6–5.0 kb	
		E. durans	range)	
*Sst*I	RFLP	*E. faecalis*	8–19	*57*
		E. faecium	11–16	
			(in the 1.6–8.0 kb	
			range)	

[a]NA, not available.

and 1 *E. casseliflavus*) recovered from an out-patient clinic (*n* = 15) and from the geographically isolated British Antarctic Survey Base (*n* = 160) *(58)*. Isolates from the two communities did not share common RFLP patterns, but identical profiles were observed in enterococci within each community group. Diversity was more frequently observed among *E. faecium* (24 RFLP patterns) than among *E. faecalis* or *E. durans* (10 and 4 RFLP patterns, respectively) *(58)*. RFLP was felt to be more discriminatory than ribotyping in the three mentioned studies *(56–58)*.

Conventional electrophoresis is simple, easy to perform, does not require specialized apparatus and, hence, is less expensive than PFGE. Results are also available relatively quickly: DNA can be extracted from isolates in 1 d and results can be obtained on the second day *(57)*. However, the results reported usually involve analysis of only 8–20 bands in the 1.6–8 kb range *(57,58)*; considering that enterococcal genomes have an estimated size of ~ 2500–3000 kb *(59)*, a large part of the genome is left unresolved. Moreover, typing based on the use of small fragments may be assessing plasmid DNA or insertions in plasmid or chromosomal DNA.

3.2.2.2. PULSED-FIELD GEL ELECTROPHORESIS

PFGE was first applied to enterococci by our group *(60)* and has now been used by a number of other laboratories. Various enzymes have been used to generate restriction endonuclease digestion patterns of enterococci, but the most frequently used enzyme is *Sma*I (**Table 2**) *(5,45,46,48,50,53,59–66)*. Although the CHEF (contour clamped homogeneous electric fields) BioRad DRII is a commonly used apparatus, other variations, (e.g., field inversion gel electrophoresis [FIGE] *[67,68]*), have also been used. Though the two techniques were considered to yield comparable results in the study by Green et al., one difference was that better band separation was noted in the 50–200-kb range with FIGE, whereas with PFGE, better resolution of bands larger than 250 kb range was obtained *(67)*. However, by altering the settings for CHEF, one can resolve fragments in the lower size range as well *(65)*.

PFGE has been established as both discriminatory and reproducible for sub-species strain differentiation (*see* Chapter 3). However, a few caveats are worth mentioning. First, analysis by PFGE requires specialized apparatus and is fairly labor- and time-intensive (for technical details, *see* **Notes 5–11**). As with any electrophoretic technique, caution should be exercised when interpreting patterns. Single band differences may simply result from genetic switching *(69)* rather than real differences between strains, and even multiple changes could result from the acquisition of a single large transposon. For example, transferable chromosomal elements containing the *vanB* glycopeptide resistance gene cluster have been as large as 250 kb, or about one tenth the size of the genome *(70)*. As with chromosomal fingerprints obtained by conventional electrophoresis (previous section), the possibility of additional bands being caused by the presence of extrachromosomal plasmids should also be considered (*see* **Note 12**) *(71)*.

PFGE has been compared with MLEE for typing *E. faecalis* isolates recovered over a 20 yr period from intercontinental sources *(39)*, and for typing vancomycin-resistant *E. faecium* isolated from animal foodstuffs and fecal samples from community-based individuals *(41)*. In the study by Tomayko et al., clonal groups defined by PFGE were also recognized by MLEE, confirming the previously reported clonal spread of ß-lactamase-producing *E. faecalis* to six hospitals in five states *(39,66)*, However, MLEE recognized broader clonal groups whereas PFGE was able to discriminate isolates within some MLEE clones. PFGE may be preferred when recent spread and limited genetic diversity of isolates are suspected, whereas MLEE may identify ancestral relationships among isolates *(39)*,

PFGE has been applied to study the relatedness of ß-lactamase-producing, HLRG *E. faecalis* isolates collected over a 7 yr period from the same hospital *(71)*. The results showed that the chromosomal digestion patterns of some isolates varied more than others, but all were felt to be a single strain. The identity of some of the PFGE patterns from isolates recovered over a 4 yr period emphasizes that identical patterns do not prove recent contact between patients. In such instances, plasmid analysis might be considered to try to assess a more recent relationship.

3.2.3. Ribotyping

The use of rRNA-based probes (*see* Chapter 2) is another technique that has been applied to the investigation of the molecular epidemiology of enterococci *(47,56–58,62,72–75)*. Enzymes that have been used to generate DNA fingerprints include *Eco*RI *(58,72,74,75)*, *Hin*dIII *(56,58,62,74,75)*, *Bsc*I *(57)*, *Bam*HI *(47,73)*, *Pvu*II *(56,62,75)*, and *Pst*I *(75)*. Although hybridization patterns obtained have been found to be reproducible and relatively simple to interpret, ribotyping is a time-consuming and relatively complex methodology. In a number of studies, PFGE *(62,74)* or conventional electrophoresis *(56–58)* appeared to be superior for strain differentiation.

3.2.4. Insertion Sequences Analysis: IS6770 Typing

The insertion sequence IS*6770* appears to be predominantly an enterococcal element and it has been found in varying copy number in genomic DNA from >90% of enterococci tested *(76)*. Patterns generated by hybridization of *Eco*RI- or *Hin*dIII-digested genomic DNA with an internal IS*6770* probe have been used to type more than 100 enterococcal isolates including vancomycin-resistant *E. faecium* from outbreaks from different locations and clinical isolates of *E. faecalis*. Results were in agreetment with those obtained by PFGE and FIGE *(76)*.

3.2.5. PCR Fingerprinting

PCR-based fingerprinting procedures have been described for molecular typing of various different bacterial species *(77)*. RAPD analysis (using Random Amplified Polymorphic DNA fingerprint patterns; *see* Chapter 6) has been used for epidemiologic characterization of vancomycin-resistant *E. faecium* *(78)* and the authors have applied REP-PCR (*see* Chapter 7) to *E. faecalis* (unpublished results). This method has shown good correlation with the results of PFGE, although some discrepancies were observed. PCR-ribotyping (using primers to amplify rRNA or tRNA sequences) has also been applied to enterococci *(9,13,79)*. However, only a few isolates were included in these studies, and the results were not compared with those of well-established typing methods. The use of PCR fingerprinting is currently restricted to research laboratories with experience and proper equipment.

3.2.6. Analysis of Resistance Elements

Detection of antibiotic resistance at the gene level has been used to study the distribution and spread of specific resistance genes *(25,80–82)*. This approach also overcomes some of the limitations of the phenotypic methods such as poor detection of low-level resistance to glycopeptides and prediction of lack of synergism of aminoglycosides with ß-lactams in specific cases. These molecular techniques can also be useful when isolates have an atypical phenotype (e.g., isolates with the *vanB* genotype that are resistant to both vancomycin and teicoplanin) *(25,83–86)*. However, the possibility of detecting genes not expressed (which can give a false-positive result) *(42)* and the emergence of new genes not recognized by these probes are potential limitations of this approach *(87)*.

3.2.6.1. AMINOGLYCOSIDE RESISTANCE (*SEE* CHAPTER 28)

Detection of high-level resistance to aminoglycosides (HLRA) in enterococci is clinically relevant because it predicts lack of synergy between the aminoglycoside in question and cell-wall active agents, whose use in combination is the treatment of choice for endocarditis *(36)*. HLRA in enterococci can be caused by modifying enzymes and/or, for streptomycin, by mutation in a ribosomal gene(s) *(36,88)*. The main limitation of the phenotypic approach is that resistance to synergism can also occur without HLRA *(82,89–92)*. The rate of modification of amikacin and netilmicin by certain modifying enzymes may not be sufficient to confer HLR to these aminoglycosides, although the bactericidal and synergistic effect is abolished *(88)*. Similarly, the chromosomally encoded AAC(6')-Ii specific for *E. faecium* strains does not confer HLR to tobramycin, kanamycin, or sisomicin, although it confers resistance to synergism *(20,93)*. Although still rare, some enterococci with moderate resistance to gentamicin (MICs = 256–500 µg/mL) or streptomycin (MICs 256–1024 µg/mL) may also be resistant to the bactericidal effect of penicillin plus these aminoglycosides *(91,94)*.

Hybridization to probes for genes encoding aminoglycoside modifying enzymes demonstrated 100% agreement with the deduced enzyme content for AAC(6')-APH(2"), APH(3'), ANT(4'), and 87% agreement for ANT(6') in one study in which more than 550 enterococcal clinical isolates were studied *(82)*. These genes have been detected by hybridization *(20,42,80–82,95)* or PCR *(80,96)* (**Table 1**). The use of multiple primers simultaneously permits the detection of different genes *(80)*, which is a very frequent event in enterococci *(97)*.

3.2.6.2. GLYCOPEPTIDE RESISTANCE (*SEE* CHAPTER 29)

Detection of glycopeptide resistance is not without some difficulties since automated methods and disk diffusion may fail to detect strains with low-level resistance to vancomycin *(98–100)*. Different phenotypes have been defined (VanA, VanB and VanC) based on the susceptibility to vancomycin and teicoplanin *(101)* and the genes involved in generating these phenotypes (*vanA*, *vanB*, and *vanC*) can be detected by hybridization to specific probes

(25,73,83,86) or by PCR *(24,25)* (**Table 1**). Multiplex PCR can be useful to detect different genes *(24)*. Both PCR and restriction mapping combined with hybridization to probes generated from within the *van* cluster have been used to determine the similarity of the *vanA* gene cluster among vancomycin-resistant enterococci from different locations *(61,102,103)*.

Different probes for *vanA* and *vanB* gene have been used in different studies *(25,62,73,101,104)*. In one study, cross hybridization was reported for some VanB isolates that hybridized weakly to a *vanA* probe *(25,101)*, probably because of the degree of similarity between these two genes.

4. Research Uses
4.1. Detection of Possible Virulence Markers

As a result of the alarming increase in antibiotic resistance, the treatment of some enterococcal infections has become very difficult and, in some instances, there is no available therapy. Thus, the study of possible virulence properties may be particularly relevant as a way to search for new options to treat or prevent infections by these organisms.

Hemolysin, gelatinase, and aggregation substance have been suggested as possible virulence factors in enterococci *(105)*. Using published gene sequences to generate probes by PCR, a large number of enterococcal clinical and fecal isolates have been examined in different studies *(106,107)* (**Table 1**). Although these traits are more common in nosocomial than community isolates of *E. faecalis*, their absence in more than 40% of endocarditis isolates suggests that other factors should also be important *(106,107)*. Moreover, none of these genes was found in *E. faecium*.

Immunologic studies on the antigenic composition of *E. faecalis* isolates causing endocarditis have led to the recognition of common antigens in these strains (37 kDa, 40 kDa, and 70 kDa) *(108–110)*. To further characterize such antigens, immunoscreening of *E. coli* DNA libraries of enterococcal DNA with serum from patients with endocarditis has been performed, and has led to the characterization of antigens expressed in vivo *(28,29)*. DNA sequence analysis of the genes encoding those antigens has revealed similarities with adhesins from some oral streptococci (*efaA*) *(28)*, an autolysin of *E. faecalis* and the p54 protein of *E. faecium*, among others *(29)*.

4.2. Analysis of Mutants

A methodology that generates targeted insertion mutants in *E. faecalis* from DNA manipulated in *E. coli* has recently been used to create auxotrophic mutants of *E. faecalis* based on purine and pyrimidine biosynthesis genes *(111)*. Inactivation of potential virulence genes by similar methods could be used to generate *E. faecalis* mutants that could be tested in vitro and in animal models.

5. Notes

1. Genomic DNA extraction. InstaGene™ matrix (Bio-Rad, Hercules, CA) following the manufacturer's instructions has been used in the authors' laboratory for the past few years to extract genomic DNA from enterococci and other gram-positive or gram-negative organisms for PCR. The vancomycin resistance genes and some of the other genes discussed in the chapter have been amplified from various sources of genomic DNA extracted by this method. The only drawback of this method is that the authors cannot quantitate the DNA, and it appears that if too much bacterial growth is used, there is an inhibitory effect on PCR amplification.

2. Optimization of PCR reactions. In the authors' experience, InVitrogen's PCR optimizer kit (InVitrogen, San Diego, CA) provides a good range of buffers to optimize the PCR amplification conditions. This kit not only has given consistent results, but has also saved time by minimizing the need for repetition of PCR reactions.

3. Confirmation of PCR products. When dealing with a new PCR product, the authors' usually verify that the correct fragment has been generated by sequencing of the fragment or by hybridization to an intragenic probe. They have occasionally found that a fragment of the correct size, when submitted for sequencing, was unrelated to the sequence desired; this is particularly a problem when degenerate primers are used. If the restriction sites of the amplified products are known, they can also be utilized to verify the validity of the product *(11)*.

4. Extraction of plasmid DNA from enterococci for epidemiologic investigations has been performed using modified mini-lysis methods *(47,49,112,113)*. The authors' have tried various techniques for *E. faecalis*, including cesium chloride gradients *(114)*, and although good restriction endonuclease digestion products can be generated, they often get poor results and smeared bands when they have analyzed undigested plasmids. Currently, the authors' usually use the method described in the study by Woodford et al. *(47)*, to extract plasmid DNA from enterococci (*see* Chapter 4). Extraction of plasmids from *E. faecium* by this method has shown consistency, and the DNA obtained generates good patterns of digested and undigested plasmids on agarose gels. In the authors' experience, this method has not been as consistent when used for *E. faecalis*; the addition of mutanolysin has helped with some strains, but not others. The authors' prefer to reprecipitate the plasmid DNA in order to remove degraded proteins or detergents from the samples. A general rule of thumb is that shearing of large plasmid molecules can be avoided by using large-bore pipet tips. In addition, vortexing should be avoided before removal of chromosomal DNA, as this minimizes the chances of carrying over any sheared chromosomal DNA into the final plasmid samples.

5. Culture medium. Brain-heart-infusion broth is their medium of choice for the growth of enterococci and the authors' have used this medium for growth when analyzing DNA from some other gram-positive organisms, such as group B streptococci (GBS) and staphylococci. Antibiotic supplementation of the medium is optional.

6. Lysis solution. The lysis solution previously published (6 mM Tris-HCl pH = 7.6, 1 M NaCl, 100 mM EDTA pH = 7.5, 0.5% Brij 58, 0.2% deoxycholate, 0.5% sodium lauroyl sarcosine, 20 µg RNase (DNase free) per mL, 1 mg lysozyme per mL)

(60) has worked well in our hands not only with enterococci, but also with GBS, *Aeromonas* spp., *Klebsiella* spp. *E. coli, Shigella* spp., and *Acinetobacter* without modification. For *Staphylococcus aureus*, 5–30 U/mL (and for *S. epidermidis*, 40–45 U/mL) of lysostaphin aids cell lysis.

7. Storage of DNA plugs. Washes of DNA plugs with 10 m*M* Tris-HCl - 1 m*M* EDTA (TE buffer), pH 8.0 at the end of plug preparation have been sufficient to remove proteinase K present in the plugs. The authors' do not use phenyl methyl sulfonyl fluoride to inactivate proteinase K. They save DNA plugs in TE buffer at 4°C, and in their hands they remain in good condition for 6 mo to more than 2 yr.

8. Digestion of DNA embedded in plugs. The authors always rinse the plugs stored in TE buffer for 4–5 min with autoclaved distilled water before placing them in a reaction mixture for restriction digestion as this helps to minimize the effects of residual detergents or other inhibitors present. Also, when restriction enzymes that require incubation temperatures above 50°C are used, the reaction mix should be overlaid with two drops of sterile mineral oil in order to prevent fluid loss from the reaction mixture.

9. Pulse-ramping time for enterococci. *Sma*I-digested DNA with a pulse ramping time of 5–35 s for *E. faecalis* and 2–28 s for *E. faecium* (since *Sma*I digestion of chromosomal DNA of *E. faecium* generally yields smaller fragments) usually gives adequate resolution of DNA fragments present. Pulse ramping times can be increased or reduced when focusing on a specific size range of DNA fragments for better resolution *(65)*.

10. CHEF gels and hybridization. The authors always use 1.6% SeaKem Gold agarose GTG grade (FMC Bioproducts, Rockland, ME) for CHEF gels when we plan to transfer the DNA to filters. Ultraviolet nicking of DNA fragments for 120 s followed by 20 min denaturation in 0.4 *M* sodium hydroxide (NaOH) and 24–36 h transfer from the back side of the gel using 0.4 *M* NaOH as the transfer buffer gives efficient transfer of DNA fragments (<6kb–>1000 kb) onto nylon membranes. They usually use ~4 × 10^7 counts per minute when hybridizing the CHEF gels. In the authors' experience, the use of radiolabeled oligonucleotide probes has not been successful as no positive hybridization signals could be detected; this is likely because of the small amount of radioactivity that is incorporated into these oligonucleotide sequences plus the small amount of target DNA present on filters.

11. Sterility during all the steps and general safety precautions as described in the literature must be followed.

12. Localizing plasmids on CHEF gels. The authors have identified the presence of some plasmid bands among the genomic DNA bands by extracting the plasmid separately, mixing the whole plasmids with Incert agarose, casting them in the plug molds and then processing in the same way as genomic DNA samples. Specific DNA probes or the whole plasmid can be used to localize these bands. As expected, undigested plasmids (of ~ 60 kb) migrated differently from the linearized derivative and, for unknown reasons, tend to produce smeary bands when longer pulse ramping times are used. The authors' have not analyzed plasmids of other sizes in pulsed-field gels.

References

1. Schaberg, D. R., Culver, D. H., and Gaynes, R. P. (1991) Major trends in the microbial etiology of nosocomial infection. *Am. J. Med.* **91,** 72S–75S.
2. Teixeira, L. M., Facklam, R. R., Steigerwalt, A. G., Pigott, N. E., Merquior, V. L. C., and Brenner, D. J. (1995) Correlation between phenotypical characteristics and DNA relatedness within Enterococcus faecium strains. *J. Clin. Microbiol.* **33,** 1520–1523.
3. Williamson, R., Gutmann, L., Horaud, T., Delbos, F., and Acar, J. (1986) Use of penicillin-binding proteins for the identification of enterococci. *J. Gen. Microbiol.* **132,** 1929–1937.
4. Vincent, S., Minkler, P., Bincziewski, B., Etter, L., and Shlaes, D. M. (1992) Vancomycin resistance in *Enterococcus gallinarum. Antimicrob. Agents Chemother.* **36,** 1392–1399.
5. Donabedian, S., Chow, J. W., Shlaes, D. M., Green, M., and Zervos, M. J. (1995) DNA hybridization and contour-clamped homogeneous electric field electrophoresis for identification of enterococci to the species level. *J. Clin. Microbiol.* **33,** 141–145.
6. Facklam, R. R., and Sahm, D. A. (1995). Enterococcus, in *Manual of Clinical Microbiology, sixth ed.* (Murray, P. R., Baron, E. J., Pfaller, M. A., Tenover, F. C., and Yolken, R. H., eds.), American Society of Microbiology, Washington, DC, pp. 308–314.
7. Pitcher, D. G., Saunders, N. A., and Owen, R. J. (1989) Rapid extraction of bacterial genomic DNA with guanidium thiocyanate. *Lett. App. Microbiol.* **8,** 151–156.
8. Wilson, K. (1994) Preparation of genomic DNA from bacteria. 2.4.1.-2.4.2. Greene, Brooklyn, NY.
9. McClelland, M., Petersen, C., and Welsh, J. (1992) Length polymorphisms in tRNA intergenic spacers detected by using the polymerase chain reaction can distinguish streptococcal strains and species. *J. Clin. Microbiol.* **30,** 1499–1504.
10. Hynes, W. L., Ferretti, J. J., Gilmore, M. S., and Segarra, R. E. (1992) PCR amplification of streptococcal DNA using crude cell lysates. *FEMS Microbiol. Lett.* **94,** 139–142.
11. Klare, I., Heier, H., Claus, H., Reissbrodt, R., and Witte, W. (1995) vanA-mediated high-level glycopeptide resistance in *Enterococcus faecium* from animal husbandry. *FEMS Microbiol. Lett.* **125,** 165–172.
12. Dutka-Malen, S., Leclercq, R., Coutant, V., Duval, J., and Courvalin, P. (1990) Phenotypic and genotypic heterogeneity of glycopeptide resistance determinants in gram-positive bacteria. *Antimicrob. Agents Chemother.* **34,** 1875–1879.
13. Cocconcelli, P. S., Porro, D., Galandini, S., and Senini, L. (1995) Development of RAPD protocol for typing of strains of lactic bacteria and enterococci. *Lett. App. Microbiol.* **21,** 376–379.
14. Tomayko, J. F., Zscheck, K. K., Singh, K. V., and Murray, B. E. (1996) Comparison of the ß-lactamase gene cluster in clonally distinct strains of *Enterococcus faecalis. Antimicrob. Agents Chemother.* **40,** 1170–1174.
15. Sambrook, J., Fritsch, E. F., and Maniatis, T. (1989). *Molecular cloning. A Laboratory Manual, 2nd Ed* CSHL.

16. Daly, J. A., Clifton, N. L., Seskin, K. C., and Manford Gooch, W. (1991) Use of rapid, nonradioactive DNA probes in culture confirmation tests to detect *Streptococcus agalactiae, Haemophilus influenzae,* and *Enterococcus* spp. from pediatric patients with significant infections. *J. Clin. Microbiol.* **29,** 80–82.

17. Davis, T. E., and Fuller, D. D. (1991) Direct identification of bacterial isolates in blood cultures by using a DNA probe. *J. Clin. Microbiol.* **29,** 2193–2196.

18. Betzl, D., Ludwig, W., and Scheleifer, K. H. (1990) Identification of lactococci and enterococci by colony hybridization with 23S rRNA-targeted oligonucleotides probes. *Appl. Environ. Microbiol.* **56,** 2927–2929.

19. Beimfohr, C., Krause, A., Amann, R., Ludwig, W., and Schleifer, K. H. (1993) In situ identification of lactococci, enterococci and streptococci. *Sys. Appl. Microbiol.* **16,** 450–456.

20. Costa, Y., Galimandd, M., Leclercq, R., Duval, J., and Courvalin, P. (1993) Characterization of the chromosomal aac(6')-Ii gene specific for *Enterococcus faecium. Antimicrob. Agents Chemother.* **37,** 1896–1903.

21. Leclercq, R., Dutka-Malen, S., Duval, J., and Courvalin, P. (1992) Vancomycin resistance gene vanC is specific to *Enterococcus gallinarum. Antimicrob. Agents Chemother.* **36,** 2005–2008.

22. Navarro, F., and Courvalin, P. (1994) Analysis of genes encoding D-alanine-D-alanine ligase-related enzymes in *Enterococcus casseliflavus* and *Enterococcus flavescens. Antimicrob. Agents Chemother.* **38,** 1788–1793.

23. Coque, M. T., and Murray, B. E. (1995) Identification of *Enterococcus faecalis* strains by DNA hybridization and pulsed-field electrophoresis gel. *J. Clin. Microbiol.* **33,** 3368–3369.

24. Dutka-Malen, S., Evers, S., and Courvalin, P. (1995) Detection of glycopeptide resistance genotypes and identification to the species level of clinically relevant enterococci by PCR. *J. Clin. Microbiol.* **33,** 24–27.

25. Clark, N. C., Cooksey, R. C., Hill, B. C., Swenson, J. M., and Tenover, F. C. (1993) Characterization of glycopeptide-resistant enterococci from U.S. hospitals. *Antimicrob. Agents Chemother.* **37,** 2311–2317.

26. Dutka Malen, S., Evers, S., and Courvalin, P. (1995) Detection of glycopeptide resistance genotypes and identification to the species level of clinically relevant enterococci by PCR (Published erratum). *J. Clin. Microbiol.* **33,** 1434.

27. Cartwright, C. P., Stock, F., Fahle, G. A., and Gill, V. J. (1995) Comparison of pigment production and motility tests with PCR for reliable identification of intrinsically vancomycin-resistant enterococci. *J. Clin. Microbiol.* **33,** 1931–1933.

28. Lowe, A. M., Lambert, P. A., and Smith, A. W. (1995) Cloning of an *Enterococcus faecalis* endocarditis antigen: homology with adhesin from some oral streptococci. *Infect. Immun.* **63,** 703–706.

29. Xu, Y., Jiang, L., Murray, B. E., and Weinstock, G. M. (1996) *Enterococcus faecalis* antigens in human infections. in press.

30. Korten, V., Huang, W. M., and Murray, B. E. (1994) Analysis by PCR and direct DNA sequencing of *gyrA* mutations associated with fluoroquinolone resistance in *Enterococcus faecalis. Antimicrob. Agents Chemother.* **38,** 2091–2094.

31. Chu, C., Kariyama, R., Daneo-Moore, L., and Shockman, G. (1992) Cloning and sequence analysis of the muramidase-2 gene from *Enterococcus hirae. J. Bacteriol.* **174,** 1619–1625.

32. Fontana, R., Cerini, R., Longoni, P., Grossato, A., and Canepari, P. (1983) Identification of a streptococcal penicillin-binding protein that reacts slowly with penicillin. *J. Bacteriol.* **156,** 1343–1350.

33. Grayson, M. L., Eliopoulos, G. M., Wennersten, C. B., Ruoff, K. L., Klimm, K., Sapico, F. L., Bayer, A. S., and Moellering, R. C. (1991) Comparison of *Enterococcus raffinosus* with *Enterococcus avium* on the basis of penicillin susceptibility, penicillin-binding protein analysis, and high-level aminoglycoside resistance. *Antimicrob. Agents Chemother.* **35,** 1408–1412.

34. Collins, M. D., Facklam, R. R., Farrow, J. A. E., and Williamsom, R. (1989) *Enterococcus raffinosus* ssp. nov., *Enterococcus solitarius* ssp. nov. and *Enterococcus pseudoavium* ssp. nov. *FEMS Microbiol. Lett.* **57,** 283–288.

35. Merquior, V. L. C., Peralta, J. M., Faklam, R. R., and Teixeira, L. M. (1994) Analysis of whole-cell protein profiles as a tool for characterization of *Enterococcus species. Curr. Microbiol.* **28,** 149–153.

36. Murray, B. E. (1990) The life and times of the enterococcus. *Clin. Microbiol. Rev.* **3,** 46–65.

37. Coudron, P. E., Mayhall, C. G., Facklam, R. R., Spadors, A. C., Lamb, V. A., Lybrand, M. R., and Dalton, H. P. (1984) *Streptococcus faecium* outbreak in a neonatal intensive care unit. *J. Clin. Microbiol.* **20,** 1044–1048.

38. Kuhnen, E., Richter, F., Richter, K., and Andries, L. (1988) Establishment of a typing system for group D streptococci. *Zentralbl Bakteriol Parasitenkd Infektionskr Hyg Abt 1 Orig Reihe A.* **267,** 322–330.

39. Tomayko, J. F., and Murray, B. E. (1995) Analysis of *Enterococcus faecalis* isolates from intercontinental sources by multilocus enzyme electrophoresis and pulsed-field gel electrophoresis. *J. Clin. Microbiol.* **33,** 2903–2907.

40. Kuhn, I., Burman, L. G., Hoeggman, S., Tullus, K., and Murray, B. E. (1995) Biochemical fingerprinting compared with ribotyping and pulsed-field gel electrophoresis of DNA for epidemilogical typing of enterococci. *J. Clin. Microbiol.* **33,** 2812–2817.

41. Klare, I., Heier, H., Claus, H., Bohme, G., Marin, S., Seltmann, G., Harenbeck, R., Antanassova, V., and Witte, W. (1995) *Enterococcus faecium* strains with *vanA* mediated high level resistance isolated from animal foodstuffs and fecal samples of humans in the community. *Microbial. Drug Res.* **1,** 265–272.

42. Huycke, M. M., Spiegel, C. A., and Gilmore, M. S. (1991) Bacteremia caused by hemolytic, high-level gentamicin-resistant *Enterococcus faecalis. Antimicrob. Agents Chemother.* **35,** 1626–1634.

43. Luginbuhl, L. M., Rotbart, H. A., Facklam, R. R., Roe, M. H., and Elliot, J. A. (1987) Neonatal enterococcal sepsis: case-control study and description of an outbreak. *Pediatr. Infect. Dis.* **6,** 1022–1030.

44. Chirurgi, V. A., Oster, S. E., Goldberg, A. A., Zervos, M. J., and McCabe, R. E. (1991) Ampicillin-resistant *Enterococcus raffinosus* in an acute-care hospital: case-control study and antimicrobial susceptibilities. *J. Clin. Microbiol.* **29,** 2663–2665.

45. Boyce, J. M., Opal, S. M., Potter-Bynoe, G., LaForge, R. G., Zervos, M. J., Furtado, G., Victor, G., and Medeiros, A. A. (1992) Emergence and nosocomial transmission of ampicillin-resistant enterococci. *Antimicrob. Agents Chemother.* **36,** 1032–1039.

46. Donabedian, S. M., Chow, J. W., Boyce, J. M., McCabe, R. E., Markowitz, S. M., Coudron, P. E., Kuritza, A., Pierson, C. L., and Zervos, M. J. (1992) Molecular typing of ampicillin resistant, non-ß-lactamase producing *Enterococcus faecium* from diverse geographical areas. *J. Clin. Microbiol.* **30,** 2757–2761.

47. Woodford, N., Morrison, D., Cookson, B., and George, R. C. (1993) Comparison of high-level gentamicin-resistant *Enterococcus faecium* isolates from different continents. *Antimicrob. Agents Chemother.* **37,** 681–684.

48. Antalek, M. D., Mylotte, J. M., Lesse, A. J., and Sellick, J. A. (1995) Clinical and molecular epidemiology of *Enterococcus faecalis* bacteremia, with special reference to strains with high level resistance to gentamicin. *CID* **20,** 103–109.

49. Zervos, M. J., Kauffman, C. A., Therasse, P. M., Bergman, A. G., Mikesell, T. S., and Schaberg, D. R. (1987) Nosocomial infection by gentamicin-resistant *Streptococcus faecalis*: an epidemiologic study. *Ann. Intern. Med.* **106,** 687–691.

50. Thal, L. A., Chow, J. W., Patterson, J. E., Perri, M. B., Donabedian, S., Clewell, D. B., and Zervos, M. J. (1993) Molecular characterization of highly resistant *Enterococcus faecalis* isolates lacking high level streptomycin resistance. *Antimicrob. Agents Chemother.* **37,** 134–137.

51. Patterson, J. E., Masecar, B. L., Kauffman, C. A., Schaberg, D. R., Hierholzer, W. J., Jr., and Zervos, M. J. (1988) Gentamicin resistance plasmids of enterococci from diverse geographic areas are heterogeneous. *J. Infect. Dis.* **158,** 212–216.

52. Rhinehart, E., Smith, N. E., Wennersten, C., Gorss, E., Freeman, J., Eliopoulos, G. M., Moellering, R. C., Jr., and Goldmann, D. A. (1990) Rapid dissemination of β-lactamase-producing, aminoglycoside-resistant *Enterococcus faecalis* among patients and staff on an infant-toddler surgical ward. *N. Engl. J. Med.* **323,** 1814–1818.

53. Boyce, J. M., Opal, S. M., Chow, J. W., Zervos, M. J., Potter-Bynoe, G., Sherman, C. B., Romulo, M. C., Fortna, S., and Medeiros, A. A. (1994) Outbreak of multidrug resistant *Enterococcus faecium* with transferable *vanB* class vancomycin resistance. *J. Clin. Microbiol.* **32,** 1148–1153.

54. Murray, B. E., Lopardo, H. A., Rubeglio, E. A., Frosolono, M., and Singh, K. V. (1992) Intrahospital spread of a single gentamicin-resistant, β-lactamase-producing strain of *Enterococcus faecalis* in Argentina. *Antimicrob. Agents Chemother.* **36,** 230–232.

55. Patterson, J. E., Wanger, A., Zscheck, K. K., Zervos, M. J., and Murray, B. E. (1990) Molecular epidemiology of β-lactamase-producing enterococci. *Antimicrob. Agents Chemother.* **34,** 302–305.

56. Bingen, E. H., Denamur, E., Lambert-Zechovsky, N. Y., and Elion, J. (1991) Evidence for the genetic unrelatedness of nosocomial vancomycin-resistant *Enterococcus faecium* strains in a pediatric hospital. *J. Clin. Microbiol.* **29,** 1888–1892.

57. Hall, L. M., Duke, B., Guiney, M., and Williams, R. (1992) Typing of *Enterococcus* species by DNA restriction fragment analysis. *J. Clin. Microbiol.* **30**, 915–919.

58. Lacoux, P. A., Jordens, J. Z., Fenton, C. M., Guiney, M., and Pennington, T. H. (1992) Characterization of enterococcal isolates by restriction enzyme analysis of genomic DNA. *Epidemiol. Infect.* **109**, 69–80.

59. Singh, K. V., and Murray, B. E. (1994) Revised estimates of enterococcal chromosomal sizes. *DNA Cell Biol.* **13**, 1145,1146.

60. Murray, B. E., Singh, K. V., Heath, J. D., Sharma, B. R., and Weinstock, G. M. (1990) Comparison of genomic DNAs of different enterococcal isolates using restriction endonucleases with infrequent recognition sites. *J. Clin. Microbiol.* **28**, 2059–2063.

61. Van der Auwera, P., Pensart, N., Korten, V., Murray, B. E., and Leclercq, R. (1996) Incidence of oral glycopeptides on the fecal flora of human volunteers: selection of highly glycopeptide resistant enterococci. *J. Infect. Dis.* **173**, 1129–1136.

62. Plessis, P., Lamy, T., Donnio, P. Y., Autuly, F., Grulois, I., Le Prisé, P. Y., and Avril, J. L. (1996) Epidemiologic analysis of glycopeptide-resistant *Enterococcus* strains in neutropenic patients receiving prolonged vancomycin administration. *Eur. J. Clin. Microbiol. Infect. Dis.* **14**, 959–963.

63. Christie, C., Hammond, J., Reising, S., and Patterson, J. E. (1994) Clinical and molecular epidemiology of enterococcal bacteremia in a pediatric teaching hospital. *J. Pediatrics.* **125**, 392–399.

64. Morris, T., Brecher, S. M., Fitzsimmons, D., Durbin, A., Arbeit, R. D., and Maslow, J. N. (1995) A pseudoepidemic due to laboratory contamination deciphered by molecular analysis. *Infect. Con. Hosp. Epidemiol.* **16**, 82–87.

65. Miranda, A. G., Singh, K. V., and Murray, B. E. (1991) DNA fingerprinting of *Enterococcus faecium* by pulsed-field gel electrophoresis may be a useful epidemiologic tool. *J. Clin. Microbiol.* **29**, 2752–2757.

66. Murray, B. E., Singh, K. V., Markowitz, S. M., Lopardo, H. A., Patterson, J. E., Zervos, M. J., Rubeglio, E., Eliopoulos, G. M., Rice, L. B., Goldstein, F. W., Jenkins, S. G., Caputo, G. M., Nasnass, R., Moore, L. S., Wong, E. S., and Weinstock, G. (1991) Evidence for clonal spread of a single strain of β-lactamase-producing *Enterococcus faecalis* to six hospitals in five states. *J. Infect. Dis.* **163**, 780–785.

67. Green, M., Barbadora, K., Donabedian, S., and Zervos, M. J. (1995) Comparison of field inversion gel electrophoresis with contour clamped homogeneous electric field electrophoresis as a typing method for *Enterococcus faecium*. *J. Clin. Microbiol.* **33**, 1554–1557.

68. Boyle, J. F., Soumakis, S. A., Rendo, A., Herrington, J. A., Gianarkis, D. G., Thurberg, B. E., and Painter, B. G. (1993) Epidemiologic analysis and genotypic characterization of a nosocomial outbreak of vancomycin-resistant enterococci. *J. Clin. Microbiol.* **31**, 1280–1285.

69. Tenover, F. C., Arbeit, R. D., Goering, R. V., Mickelsen, P. A., Murray, B. E., Persing, D. H., and Suaminathan, B. (1995) Interpreting chromosomal DNA restriction patterns produced by pulsed field gel electrophoresis: criteria for bacterial strain typing. *J. Clin. Microbiol.* **33**, 2233–2239.

70. Quintiliani, R., Jr., and Courvalin, P. (1994) Conjugal transfer of the vancomycin resistance determinant *vanB* between enterococci involves the movement of large genetic elements from chromosome to chromosome. *FEMS Microbiol. Lett.* **119,** 359–364.

71. Seetulsingh, P. S., Tomayko, J. F., Coudron, P. E., Markowitz, S. M., Skinner, C., Singh, K. V., and Murray, B. E. (1996) Chromosomal DNA restriction endonuclease digestion patterns of beta-lactamase producing *Enterococcus faecalis* isolated from a single hospital over a 7-year period. *J. Clin. Microbiol.* **34,** 1892–1896.

72. Jordens, J. Z., Bates, J., and Griffiths, D. T. (1994) Faecal carriage and nosocomial spread of vancomycin-resistant *Enterococcus faecium. J. Antimicrob. Chemother.* **34,** 515–528.

73. Woodford, N., Morrison, D., Johnson, A. P., Briant, V., George, R. C., and Cookson, B. (1993) Application of DNA probes for rRNA and *vanA* genes to investigation of a nosocomial cluster of vancomycin-resistant enterococci. *J. Clin. Microbiol.* **31,** 653–658.

74. Gordillo, M. E., Singh, K. V., and Murray, B. E. (1993) Comparison of ribotyping and pulsed-field gel electrophoresis for subspecies differentiation of *Enterococcus faecalis. J. Clin. Microbiol.* **31,** 1570–1574.

75. Bates, J., Jordens, J. Z., and Griffiths, D. T. (1994) Farm animals as a putative reservoir for vancomycin-resistant enterococcal infection in man. *J. Antimicrob. Chemother.* **34,** 507–516.

76. Thorisdottir, A. S., Carias, L. L., Marshall, S. H., Green, M., Zervos, M. J., and Giorgio, C. (1994) IS6770, an insertion-like sequence useful for determining the clonal relationship of clinical enterococcal isolates. *J. Infect. Dis.* **170,** 1539–1548.

77. van Belkum, A. (1994) DNA fingerprinting of medically important microorganisms by use of PCR. *Clin. Microbiol. Rev.* **7,** 174–184.

78. Barbier, N., Saulnier, P., Chachaty, E., Dumontier, S., and Andremont, A. (1996) Random amplified polymorphic DNA typing versus pulsed-field gel electrophoresis for epidemiological typing of vancomycin resistant enterococci. *J. Clin. Microbiol.* **34,** 1096–1099.

79. Kostman, J. R., Alden, M. B., Mair, M., Edlind, T. D., LiPuma, J. J., and Stull, T. L. (1995) A universal approach to bacterial molecular epidemiology by polymerase chain reaction ribotyping. *J. Infect. Dis.* **171,** 204–208.

80. van Asselt, G. J., Vliegenthart, J. S., Petit, P. L. C., van de Klundert, J. A. M., and Mouton, R. P. (1992) High-level aminoglycoside resistance among enterococci and group A streptococci. *J. Antimicrob. Chemother.* **30,** 651–659.

81. Weems, J. J., Lowrance, J. H., Baddour, L. M., and Simpson, W. A. (1989) Molecular epidemiology of nosocomial, multiply aminoglycoside resistant *Enterococcus faecalis. J. Antimicrob. Chemother.* **24,** 121–130.

82. Ounissi, H., Derlot, E., Carlier, C., and Courvalin, P. (1990) Gene homogeneity for aminoglycoside-modifying enzymes in gram-positive cocci. *Antimicrob. Agents Chemother.* **34,** 2164–2168.

83. Quintiliani, R., Jr., Evers, S., and Courvalin, P. (1993) The *vanB* gene confers various levels of self-transferable resistance to vancomycin in enterococci. *J. Infect. Dis.* **167**, 1220–1223.

84. Hayden, M. K., Trenholme, G. M., Schultz, J. E., and Sahm, D. F. (1993) In vivo development of teicoplanin resistance in a VanB *Enterococcus faecium* isolate. *J. Infect. Dis.* **167**, 1224–1227.

85. Green, M., Binczewski, B., Pasculle, A. W., Edmund, M., Barbadora, K., Kusne, S., and Shlaes, D. (1993) Constitutively vancomycin-resistant *Enterococcus faecium* resistant to synergistic ß-lactam combinations. *Antimicrob. Agents Chemother.* **37**, 1238–1242.

86. Swenson, J. M., Clark, N. C., Ferraro, M. J., Sahm, D. F., Doern, G., Pfaller, M. A., Reller, L. B., Weinstein, M. P., Zabransky, R. J., and Tenover, F. C. (1994) Development of a standardized screening method for detection of vancomycin-resistant enterococci. *J. Clin. Microbiol.* **32**, 1700–1704.

87. Sahm, D. F., and Gilmore, M. S. (1994) Transferability and genetic relatedness of high level gentamicin resistance among enterococci. *Antimicrob. Agents Chemother.* **38**, 1194–1196.

88. Leclercq, R., Dutka-Malen, S., Brisson-Noel, A., molinas, C., Derlot, E., Arthur, M., Duval, J., and Courvalin, P. (1992) Resistance of enterococci to aminoglycosides and glycopeptides. *CID* **15**, 495–501.

89. Leclercq, R., Bismuth, R., and Duval, J. (1992) New high content disks for determination of high level aminoglycoside resistance in clinical isolates of *Enterococcus faecalis. Eur. J. Clin. Microbiol. Infect. Dis.* **11**, 356–360.

90. Sahm, D. F., Boonlayangoor, S., Iwen, P. C., Baade, J. L., and Woods, G. L. (1991) Factors influencing determination of high level aminoglycoside resistance in *Enterococcus faecalis. J. Clin. Microbiol.* **29**, 1934–1939.

91. Bantar, C. E., Micucci, M., Fernandez Canigia, L., Smayevski, J., and Bianchini, H. M. (1993) Synergy characterization for *Enterococcus faecalis* strains displaying moderately high level gentamicin and streptomycin resistance. *J. Clin. Microbiol.* **31**, 1921–1923.

92. Moellering, R. C., Jr., Murray, B. E., Schoenbaum, S. C., Adler, J., and Wennersten, C. B. (1980) A novel mechanism of resistance to penicillin-gentamicin synergism in *Streptococcus faecalis. J. Infect. Dis.* **141**, 81–86.

93. Wennersten, C. B., and Moellering, R. C., Jr. (1980). Mechanism of resistance to penicillin-aminoglycoside synergism in Streptococcus faecium, in *Current Chemotherapy and Infectious Diseases* Vol. 1, (Nelson, J. and Grassi, C., eds.), American Society for Microbiology, Washington, DC, pp. 710–712.

94. Thal, L. A., Chow, J. W., Mahayni, R., Bonilla, H., Perri, M. B., Donabedian, S. A., Silverman, J., Taber, S., and Zervos, M. J. (1995) Characterization of antimicrobial resistance in enterococci of animal origin. *Antimicrob. Agents Chemother.* **39**, 2112–2115.

95. Coque, T. M., Arduino, R. C., and Murray, B. E. (1995) High-level resistance to aminoglycosides: Comparison of community and nosocomial fecal isolates of enterococci. *Clin. Infect. Dis.* **20**, 1048–1051.

96. Kaufhold, A., Podbielski, A., Horaud, T., and Ferrieri, P. (1992) Identical genes confer high level of resitance to gentamicin upon *Enterococcus faecalis*, *Enterococcus faecium*, and *Streptococcus agalactiae*. *Antimicrob. Agents Chemother.* **36,** 1215–1218.

97. Patterson, J. E., and Zervos, J. M. (1990) High-level gentamicin in *Enterococcus*: microbiology, genetic basis, and epidemiology. *Rev. Infect. Dis.* **12,** 644–652.

98. Tenover, F. C., Swenson, J. M., O'Hara, C. M., and Stocker, S. A. (1995) Ability of commercial and reference antimicrobial susceptibility testing methods to detect vancomycin resistance in enterococci. *J. Clin. Microbiol.* **33,** 1524–1527.

99. Tenover, F. C., Tokars, J., Swenson, J., Paul, S., Spitalny, K., and Jarvis, W. (1993) Ability of clinical laboratories to detect antimicrobial agent resistant enterococci. *J. Clin. Microbiol.* **31,** 1695–1699.

100. Snell, J. J. S., Brown, D. F. J., Perry, S. F., and George, R. (1993) Antimicrobial susceptibility testing of enterococci: results of a survey conducted by the United Kingdom National External Quality Assesment Scheme for Microbiology. *J. Antimicrob. Chemother.* **32,** 401–411.

101. Woodford, N., Johnson, A. P., Morrison, D., and Speller, D. C. E. (1995) Current perspectives on glycopeptide resistance. *Clin. Microbiol. Rev.* **8,** 585–615.

102. Handwerger, S., Skoble, J., Discotto, L. F., and Pucci, M. J. (1995) Heterogeneity of the *vanA* gene cluster in clinical isolates of enterococci from Northeastern United States. *Antimicrob. Agents Chemother.* **39,** 362–368.

103. Aarestrup, F. M., Ahrens, P., Madsen, M., Pallesen, L. V., Poulsen, R. L., and Westh, H. (1996) Glycopeptide susceptibility among Danish *Enterococcus faecium* and *Enterococcus faecalis* isolates of animal and human origin and PCR identification of the genes within the VanA cluster. *Antimicrob. Agents Chemother.* **40,** 1938–1940.

104. Coque, T. M., Tomayko, J. F., Ricke, S. C., Okhyusen, P. C., and Murray, B. E. (1996) Vancomycin Resistant Enterococci from Nosocomial, Community and Animal Sources in the United States. submitted.

105. Jett, B. D., Huycke, M. M., and Gilmore, M. S. (1994) Virulence in enterococci. *Clin. Microbiol. Rev.* **7,** 462–478.

106. Coque, T. M., Patterson, J. E., Steckelberg, J. M., and Murray, B. E. (1995) Incidence of hemolysin, gelatinase and aggregative substance among enterococci isolated from patients wtih endocarditis and other infections and from feces of hospitalized and community-based persons. *J. Infect. Dis.* **171,** 1223–1229.

107. Huycke, M. M., and Gilmore, M. S. (1995) Frequency of aggregation substance and cytolysin genes among enterococcal endocarditis isolates. *Plasmid* **34,** 152–156.

108. Aitchison, E. J., Lambert, P. A., Smith, E. G., and Farrell, I. D. (1987) Serodiagnosis of *Streptococcus faecalis* endocarditis by immunoblotting of surface protein antigens. *J. Clin. Microbiol.* **25,** 211–215.

109. Lambert, P. A., Shorrock, P. J., Aitchison, E. J., Domingue, P. A. G., Power, M. E., and Costerton, J. W. (1990) Effect of in vivo growth conditions upon expression of surface protein antigens in *Enterococcus faecalis*. *FEMS Microbiol. Immun.* **64,** 51–54.

110. Shorrock, P. J., Lambert, P. A., Aitchison, E. J., Smith, E. G., Farrell, I. D., and Gutschik, E. (1990) Serological response in *Enterococcus faecalis* endocarditis determined by enzyme-linked immunosorbent assay. *J. Clin. Microbiol.* **28**, 195–200.

111. Li, X., Weinstock, G. M., and Murray, B. E. (1995) Generation of auxotrophs mutants in *Enterococcus faecalis. J. Bacteriol.* **177**, 6866–6873.

112. Anderson, D. G., and McKay, L. L. (1983) Simple and rapid method for isolating large plasmid DNA from lactic streptococci. *Appl. Environ. Microbiol.* **46**, .

113. Weaver, K. E., and Clewell, D. B. (1988) Regulation of the pAD1 sex pheromone response in *Enterococcus faecalis*: construction and characterization of *lacZ* transcriptional fusions in a key control region of the plasmid. *J. Bacteriol.* **170**, 4343–4352.

114. Wanger, A. R., and Murray, B. E. (1990) Comparison of enterococcal and staphylococcal beta-lactamase plasmids. *J. Infect. Dis.* **161**, 54–58.

115. Dutka-Malen, S., Molinas, C., Arthur, M., and Courvalin, P. (1990) The VANA glycopeptide resistance protein is related to D-alanyl-D-alanine ligase cell wall biosynthesis enzymes. *Mol. Gen. Genet.* **224**, 364–372.

116. Evers, S., Sahm, D. F., and Courvalin, P. (1993) The *vanB* gene of vancomycin-resistant Enterococcus faecalis V583 is structurally related to genes encoding D-Ala:D-Ala ligases and glycopeptide-resistance proteins VanA and VanC. *Gene* **124**, 143–144.

117. Gold, H. S., Ünal, S., Cercenado, E., Thauvin-Eliopoulos, C., Eliopoulos, G. M., Wennersten, C. B., and Moellering, R. C., Jr. (1993) A gene conferring resistance to vancomycin but not teicoplanin in isolates of *Enterococcus faecalis* and *Enterococcus faecium* demonstrates homology with *vanB, vanA* and *vanC* genes of enterococci. *Antimicrob. Agents Chemother.* **37**, 1604–1609.

118. Dutka-Malen, S., Molinas, C., Arthur, M., and Courvalin, P. (1992) Sequence of the vanC gene of *Enterococcus gallinarum* BM4174 encoding a D-alanine:D-alanine ligase-related protein necessary for vancomycin resistance. *Gene* **112**, 53–58.

119. Trieu-Cuot, P., and Courvalin, P. (1983) Nucleotide sequence of the *Streptococcus faecalis* plasmid gene encoding the 3'5"-aminoglycoside phosphotransferase type III. *Gene* **23**, 331–341.

120. Ferretti, J. J., Gilmore, K. S., and Courvalin, P. (1986) Nucleotide sequence analysis of the gene specifying the bifunctional 6'-aminoglycoside acetyltransferase 2"-aminolgycoside phosphotransferase enzyme in Streptococcus faecalis and identification and cloning of gene regions specifying the two activities. *J. Bacteriol.* **167**, 631–638.

121. Galli, D., and Wirth, R. (1991) Comparative analysis of *Enterococcus faecalis* sex pheromone plasmids identifies a single homologous DNA region which codes for aggregation substance. *J. Bacteriol.* **173**, 3029–3033.

122. Segarra, R. A., Booth, M. C., Morales, D. A., Huycke, M. M., and Gilmore, M. S. (1991) Molecular characterization of the *Enterococcus faecalis* cytolysin activator. *Infect. Immun.* **59**, 1239–1246.

123. Gilmore, M. S., Segarra, R. A., Booth, M. C., Bogie, C. P., Hall, L. R., and Clewell, D. B. (1994) Genetic structure of the *Enterococcus faecalis* plasmid pAD1-encoded cytolytic toxin system and its relationship to lantibiotic determinnts. *J. Bacteriol.* **176**, 7335–7344.

124. van der Klundert, J. A. M., and Vlegenthart, J. S. (1993). PCR detection of genes coding for aminoglycoside-modifying enzymes, in *Diagnostic Molecular Microbiology. Principles and Applications* (Persing, D. H., Smith, T. F., Tenover, F. C., and Whiote, T. J., eds.), American Society for Microbiology, Washington, DC, pp. 547–552.
125. Torres, C., Reguera, J. A., Sanmartin, M. J., Perez-Diaz, J. C., and Baquero, F. (1994) *vanA*-mediated vancomycin-resistant *Enterococcus* spp. in sewage. *J. Antimicrob. Chemother.* **33,** 553–561.

25

Molecular Approaches for the Detection and Identification of β-Lactamases

David J. Payne and Christopher J. Thomson

1. Introduction

β-lactamases confer resistance to β-lactam antibiotics, which are the most widely used family of antibiotics. It is, therefore, essential that one can identify the production of β-lactamases by clinical isolates and have effective ways of distinguishing the different enzymes. This is necessary for epidemiologic surveys, predicting future resistance trends, and to ensure that patients receive the appropriate β-lactam or alternative therapy.

In 1985, approx 20 different plasmid β-lactamases had been characterized *(1)*. These β-lactamases were effectively distinguished by isoelectric focusing (IEF) by virtue of their different isoelectric points (pI) (*see* Chapter 27). Today, approx 200 different β-lactamases have been characterized, and it is likely that more novel enzymes will continue to be identified. The majority of β-lactamases are serine active site enzymes, although a variety of metallo active site enzymes are now being identified *(2)*. Therefore, IEF is no longer sufficient to allow different β-lactamases to be identified reliably. In fact, currently 47 β-lactamases focus between pI 5 and pI 6, and 36 enzymes focus between pI 8 and pI 9. In addition, several important TEM β-lactamases have the same pI and are, hence, indistinguishable by IEF. Although, molecular DNA probing techniques have become very effective for identifying different β-lactamases, in some cases this approach must still be combined with kinetic and IEF data to identify a β-lactamase. This chapter reviews the use of molecular techniques to detect and distinguish β-lactamases, whereas the following chapter gives a detailed guide to performing IEF and kinetic studies on these enzymes.

From: *Methods in Molecular Medicine, Vol. 15: Molecular Bacteriology: Protocols and Clinical Applications*
Edited by: N. Woodford and A. P. Johnson © Humana Press Inc., Totowa, NJ

2. Diagnosis

The first indication of the presence of β-lactamases is usually the observation of increased levels of resistance in clinical isolates as detected by routine sensitivity testing. In some species, for example, *Haemophilus influenzae* or *Neisseria gonorrhoeae*, the presence of a β-lactamase in itself is enough information to make a judgment on treatment options, and in these cases often a simple nitrocefin hydrolysis test is performed to confirm the presence or absence of β-lactamase production *(3)*. In other cases, most notably the Enterobacteriaceae, it is not the presence of a β-lactamase that is important, but the type of enzyme and, in particular, whether the enzyme has an extended-spectrum of activity or increased resistance to β-lactamase inhibitors. It has become apparent, however, that the emergence of some of the initial mutations in the TEM and SHV β-lactamases that give rise to an extended-spectrum or inhibitor resistance may not be detected by the use of standard clinical breakpoints or sensitivity testing procedures *(4,5)*. To overcome this shortfall, several sensitivity testing techniques have been devised to highlight the occurrence of these enzymes.

To date, all the TEM- and SHV-derived β-lactamases that confer resistance to third-generation cephalosporins (referred to as the extended-spectrum TEM and SHV β-lactamases) remain susceptible to β-lactamase inhibitors such as clavulanic acid *(6)*. This property can be used to detect these enzymes in clinical isolates by double-disk testing using an amoxicillin/clavulanic acid disk, and a disk containing a third generation cephalosporin. If the increased resistance to the cephalosporin is a result of extended spectrum β-lactamase activity, then the clavulanic acid will inhibit this and a distinctive dumbell-shaped zone of inhibition will be produced *(7)*. Such results can be difficult to interpret and, more recently, an attempt has been made to standardize the procedure with the incorporation of clavulanic acid into cephalosporin-containing E-test strips *(8)*.

A different set of mutations in the TEM gene confers no resistance to later generation cephalosporins, but increases resistance to β-lactamase inhibitors. The initial detection of these TEM derived inhibitor-resistant β-lactamases is more problematic. As yet, no satisfactory detection system has been devised, although it appears that this may be facilitated by sensitivity testing inhibitor combinations with a low fixed concentration of inhibitor rather than testing at a fixed ratio of β-lactam:β-lactamase inhibitor *(9)*.

3. Epidemiologic Uses

3.1. Detection of β-Lactamase Genes by Hybridization

The use of DNA probes to detect β-lactamase genes is well-established, and the general procedures involved are outlined in **Fig. 1** *(10)*. Studies

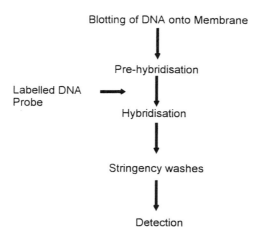

Fig. 1. General procedure for the detection of β-lactamase genes by DNA probing.

with β-lactamase DNA probes have played a major role in defining the epidemiology of β-lactamases in clinical isolates. In any analysis of DNA by hybridization, the DNA to be tested has to be immobilized onto a suitable membrane. The procedure chosen will, to some extent, be determined by the information required. For example, the location of a β-lactamase gene on a plasmid can be ascertained by restriction digestion of the plasmid followed by Southern blotting of the gel *(11)*. However, if only a positive or negative result is required, then the DNA to be tested can be applied directly to the membrane by dot blotting *(12)*. Once the DNA has been immobilized then hybridization can be tested against the probe DNA.

3.1.1. Generation of DNA for Use as a Gene Probe

The first step in making a gene probe is generating a fragment of DNA to label. Previously, β-lactamase DNA probes were usually made from restriction digests of the β-lactamase gene (**Tables 1** and **2**). Occasionally, DNA probes containing regions upstream or downstream of the β-lactamase gene have been used and have led to low discrimination and false positive identifications. Consequently, the use of intragenic gene probes is strongly recommended. The quickest way to manufacture a gene probe is by polymerase chain reaction (PCR) of an intragenic region of the target gene and a number of PCR primers for a variety of β-lactamase genes are detailed in **Tables 1** and **2**. An alternative to a PCR generated probe is the use of an oligoprobe (an oligonucleotide). In this instance the DNA can be generated by a DNA synthesizer.

Table 1
Examples of TEM- and SHV-Derived Probes and PCR Primers (in Most Cases only Examples of Intragenic Gene Probes are Shown)

β-lactamase	Size	Description of probe or PCR primers	Type of label or utility of PCR product	Plasmid	Cross hybridization (in addition to self hybridization)	Refs.
TEM-1	504	Obtained by PCR. Primers: TAATTGTTGCCGGGAAGCTA TCGGGCCGCGCGTAGGCATGAT	Radioactive	pBR322	TEM-2-7,20,22	13,14
TEM-1	656	*Hin*fI and *Taq*I, intragenic	Radioactive	pBR322	TEM-2	15
TEM-1	15	ACTTCTAGTCAACCC	Radioactive	N/A	None reported[+]	15
TEM-2	15	ACTTCTATTCAACCC	Radioactive	N/A	None reported[+]	15
TEM-1	11	AACGCGAGGAC	Radioactive	N/A	TEM-26 and other TEMs	16
TEM-1	850	*Nde*I-*Eco*RII	Biotin	pUC19	None reported[+]	17
TEM-1	424	*Bgl*I and *Hinc*II, intragenic	Radioactive/ Digoxygenin	pMON60, pBR322	TLE-1, TEM-2, YOU-1, YOU-2	18,19, 20,21
TEM-1	526	Obtained by PCR. Primers: TGGGTGCACGAGTGGGTTAC TTATCCGCCTCCATCCAGTC	Diagnostic PCR	n/a	TEM-4/ 9 and other TEMs	22
SHV-1	780	*Puv*II-*Puv*II	Radioactive	pMON38	SHV-2	21
SHV-1	467	NotI-PstI Intragenic	Biotin	pCLL3411	SHV-7	17
SHV-3	467	NotI-PstI, intragenic	Radioactive	pHUC37	SHV-6	23
SHV-3	626	Obtained by PCR. Primers: AGCAGGGCGACAATCCCGCG AATGGCAATCAGCGCTTCCC	Radioactive	pUD18	SHV-1-5, OHIO-1	13
SHV-1-5	475	Obtained by PCR. Primers: TCAGCGAAAAACACCTTG TCCCGCAGATAAATCACCA	PCR-SSCP	n/a	n/a	24
OHIO-1	223	*Pst*I-*Hae*II, intragenic	Radioactive	pSK04	SHV-1, (LEN-1)	25

( ) = weak hybridization, [+]No cross hybridization reported, but probe would probably hybridize with other TEM genes.

Examples of Other β-Lactamase Probes and PCR Primers (in Most Cases only Examples of Intragenic Gene Probes are Shown)

β-lactamase	Size	Description of probe or PCR primers	Type of probe or utility of PCR product	Plasmid	Cross hybridization (in addition to self hybridization)	Refs.
OXA-1	15	Oligoprobe: CCAAAGACGTGGATG	Radioactive	N/A	None reported	26
OXA-1a	315	BglII-BglII, intragenic	Radioactive	pMON300 PMON301	OXA-4	18,21
OXA-2a	510	HincII-HincII, intragenic	Radioactive	pMON21	(OXA-3)	21
PSE-1	587	Obtained by PCR: Primers: GGGGCTTGATGCTCACTCCA	Radioactive	RPL11	PSE-4, CARB-2-4	13
PSE-1 a	1300	BamHI and BglII part of the PSE-1 gene)	Radioactive	pMON810 pMON811	PSE-4, CARB-3	18,21
PSE-2 a	460	RsaI-XbaI, intragenic	Radioactive	pMON234	OXA-6 (OXA-5)	21
PSE-4 a	180	BglII-XbaI	Radioactive	pMON209	PSE-1 CARB-3	21
ROB-1	250	Sau3A, intragenic	Radioactive	pMON401	None reported	18
ROB-1	240	DraI-DraI, intragenic	Radioactive	pMON401	None reported	27
ROB-1	692	Obtained by PCR. Primers: ATCAGCCACACAAGCCACCT GTTTGCGATTTGGTATGCGA	Diagnostic PCR	N/A	None reported	22
IMP-1*	450	HindIII-HindIII, intragenic	Radioactive	pHIP29	None reported	28
CfiA*	726	Obtained by PCR. Primers: TCCATGTCTTTTCCCTGTCGCAGTT GCACTTCAAAGCCATAGCCCGAA	Radioactive	pJST241	None reported	29
Ent. cloacae Amp C	530	NruI-SphI	Digoxigenin	pEC1E	None reported	30
E. coli Amp C	685	Pst1-XhoI Intragenic	Digoxigenin	pNU6	None reported	20
P. mirabilis carbenicillinase	2.6Kb	HindIII-HindIII	Radioactive	pPM79P1	None reported	31

() = weak hybridization; (a) = Probe used in study to compare IEF with DNA probing for the detection of β-lactamases in 122 clinical isolates, *metallo-β-lactamases.

3.1.2. Labeling of β-Lactamase Gene Probes and Detection of Hybridization

Many different methods for labeling β-lactamase probes have been developed and compared. Traditionally, hybridization of DNA probes was detected by radioactive labeling. However, the reliance on radioactivity precluded the widespread use of probing technology in general laboratories, and alternatives were sought. For example, Zwadyk et al. *(32)* developed a biotinylated probe for TEM-1, but illustrated that this was 100 times less sensitive than the ^{32}P-labeled probe. Probes labeled with digoxigenin have also been developed and compared with IEF and biotin-labeled probes. In this work it was shown that the digoxigenin-labeled probe was as sensitive as IEF, but the biotin-labeled probe had poor specificity *(33)*. The development of chemiluminescent detection systems has enabled the use of fluorescein-labeled DNA probes that offer similar sensitivity to radioactive probes without the practical complications of radioactivity. An outline of the protocol for making such probes is presented.

3.1.3 Fluorescein-Labeled β-Lactamase Probes

Fluorescein-labeled β-lactamase probes (ECL™, Amersham Life Science, Amersham Place, Little Chalfont, Buckinghamshire, England) are detected with the enzyme horseradish peroxidase (HRP) that catalyzes the oxidation of luminol. The horseradish peroxidase is attached to an antifluorescein antibody that binds to the fluorescein-labelled probe (**Fig. 2**). The oxidized luminol then decays to the ground state via a light emitting pathway that can be detected on photographic film. The use of oligonucleotide probes labeled with fluorescein has been successfully used for detecting and differentiating related genes encoding trimethoprim resistance *(34)*. Protocols for these steps are described in detail in the manufacturer's guidelines (Amersham Life Science).

3.2. Detection of β-Lactamase Genes by PCR

PCR has been extensively used in the detection and analysis of resistance genes in bacterial isolates *(35,36)* and in the detection of specific micro-organisms directly in clinical and environmental specimens *(37–39)*. However, it is only relatively recently that the possibility of employing PCR in the direct detection of resistance genes in clinical specimens has been investigated. PCR has been used to detect rifampicin resistance in *Mycobacterium tuberculosis* directly in sputum samples *(40,41)* and, more recently, for detecting ampicillin resistance genes in cerebrospinal fluid (CSF) samples containing *H. influenzae (22)*. In the latter case, PCR primers were used that were specific for the bla_{TEM} and bla_{ROB} genes. **Tables 1** and **2** illustrate these primers and the size of the PCR product obtained. Correlation was obtained between the result of minimum inhibitory concentration (MIC) testing, β-lactamase production as determined by nitrocefin and PCR testing.

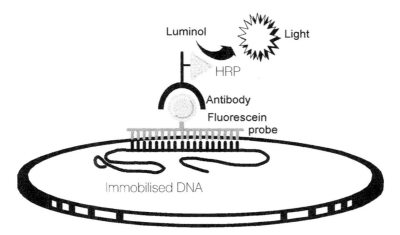

Fig. 2. Use of fluorescein-labeled β-lactamase gene probes.

3.2.1. PCR Approaches for Studying the Heterogeneity of β-Lactamase Genes in a Species

Jones et al. *(42)* used PCR to study the heterogeneity of the chromosomal *ampC* gene in 91 clinical isolates of Citrobacter spp. Primers were chosen based on the *ampC* genes from *C. freundii* OS60, *C. freundii* I113, and *C. diversus* NF85 to give PCR products of 870, 870, and 620 bp, respectively. Sixty-five of the isolates gave a PCR product with one of the sets of primers, three of these PCR products were larger than expected. Twenty-six isolates gave no product, but 21 of these had *ampC* genes, which hybridized with one of the PCR products. Such a study illustrates how PCR can potentially identify novel β-lactamase genes and also rapidly determine the heterogeneity of a β-lactamase gene within a particular species. This study highlights the potential of PCR for typing at a molecular level.

3.3. Characterization of TEM β-Lactamases

This section explains how PCR and hybridization studies can be utilized to study a particular series of β-lactamases using the TEM β-lactamases as an illustration. The gene encoding TEM-1 β-lactamase is the most widespread plasmid-mediated β-lactamase gene found in clinical isolates. This enzyme is capable of conferring resistance to penicillins and some of the early cephalosporins. Since 1982 *(43)* there have been reports of TEM enzymes that differ from TEM-1 by a variety of one to four amino acid changes. These amino acid changes reconfigure the active site of TEM-1, enabling it to hydrolyze and confer resistance to penicillins and first, second, and third generation cephalosporins. Other mutations in the TEM gene can confer resistance to β-lactamase inhibitors *(5,44)* although, to date, no TEM β-lactamase has been characterized that confers both an extended-spectrum and inhibitor-resistant phenotype simultaneously. Therefore, it is necessary to identify point mutations in the TEM enzymes as they will provide information on the expected resistance profile of the producing organism.

3.3.1. PCR for the Sequencing of TEM-Related Enzymes

Amplification of TEM β-lactamases for sequence analysis (**Subheading 4.1.**) can be achieved by using primers from the conserved regions at either end of the TEM structural gene. The sequences of the primers *(45)* are:

5'-GAAGACGAAAGGGCCTCGTG -3'   Forward
5'-GGTCTGACAGTTACCAATGC -3'   Reverse

The PCR reaction is set up according to the schedule in **Table 3**.

3.3.2. Oligotyping to Identify Point Mutations in TEM β-Lactamases

Detailed characterization of the variant enzymes is difficult. An intragenic TEM probe will identify an extended-spectrum TEM, but will not distinguish it from TEM-1. Isoelectric focusing is inappropriate as these enzymes all focus in a very narrow pI range (**Subheading 1.**). Oligotyping can be used to examine point mutations in a particular β-lactamase gene. The use of this technique to distinguish the plethora of TEM β-lactamases is reviewed below.

Oligotyping provides the best and most time-effective way of distinguishing these enzymes. This methodology utilizes a set of oligonucleotides that are specific for each of the point mutations. The techniques used to label and detect hybridization are the same as those listed in **Subheading 3.1.** This technique was first devised by Ouellette et al. *(15)* where oligonucleotides were devised to distinguish TEM-1 from TEM-2, which differ by one amino acid. Mabilat and Courvalin *(46)* developed the method further using 12 radioactive oligoprobes (17 mers) to probe five different loci, and this was used to distinguish 14 different extended-spectrum TEM genes, including seven novel TEMs, in 265 clinical isolates (**Table 4**). Henquell et al. *(44)* have used oligotyping to characterise changes in the TEM gene associated with resistance to β-lactamase inhibitors among 107 inhibitor-resistant TEM enzymes. Fifteen oligonucleotides were used to study three different loci in the TEM gene, which facilitated the identification of 10 different types of inhibitor-resistant TEM β-lactamases, IRT-1 to 10 (**Table 5**).

It is important to note that IEF and biochemical data can be used in conjunction with the oligotyping data to confirm observations. In addition, as oligotyping probes only a limited number of amino acid changes, other changes in the sequence may also exist, so care is required in interpretation of the data.

Table 3
PCR Amplification Procedure

Segment	Temperature	Time	Ramp rate	Repeats	Function
1	96°C	30 s	48°C/min		DNA denaturation
	50°C	30 s	10°C/min	X 1	Primer annealing
	72°C	90 s	30°C/min		Primer extension
2	96°C	15 s	48°C/min		DNA denaturation
	50°C	30 s	10°C/min	X 20-24	Primer annealing
	72°C	90 s	30°C/min		Primer extension
3	72°C	5 min	—	X 1	Final extension

Table 4
Characterization of Extended-Spectrum TEM β-Lactamases Using Oligo-Probes to Detect Point Mutations *(46)*

Amino acid altered[a]	Sequence of oligo probe[b]	Enzymes with particular amino acid mutation
Gln 39	ACCCAACT**G**ATCTTCAG	TEM-1, 4, 5, 6, 9, 10, 12, 15, 17, 19
Lys 39	ACCCAACT**T**ATCTTCAG	TEM-2, 3, 7, 8, 11, 13, 14, 16, 18
Glu 104	TGAGTACT**C**AACCAAGT[c]	TEM-1, 2, 5, 7, 10, 11, 12, 13, 19
Lys 104	TGAGTACT**T**AACCAAGT[c]	TEM-3, 4, 6, 8, 9, 14, 15, 16, 17, 18
Arg 164	TTCCCAAC**G**ATCAAGGC[c]	TEM-1, 2, 3, 4, 13, 14, 15, 17, 18, 19
Ser 164	TTCCCAAC**T**ATCAAGGC[c]	TEM-5, 7, 8, 9, 10, 12
His 164	GTTCCCAA**T**GATCAAGG	TEM-6, 11, 16
Gly 238	ACGCTCAC**C**GGCTCCAG[c]	TEM-1, 2, 5, 6, 7, 9, 10, 12, 13, 16, 17, 18,
Ser 238	ACGCTCAC**T**GGCTCCAG[c]	TEM-3, 4, 8, 14, 15, 19
Thr261	TCCCCGTC**G**TGTAGATA	TEM-1, 2, 3, 5, 6, 7, 8, 10,11,12, 15, 16,17,18, 19
Met261	TCCCCGTC**A**TGTAGATA	TEM-4, 9, 13, 14

[a]Ambler numbering *(47)*.
[b]The underlined nucleotide indicates the point mutation.
[c]Biotinylated versions of these probes have also been used *(48)*.

If an ambiguity remains after oligotyping and biochemical studies, complete sequencing of the gene may be required (*see* **Subheading 4.**). Oligotyping using biotinylated probes has also been successfully developed *(48)*. Oligotyping has been used to characterize TEM-variant enzymes but this procedure could also be developed to characterize the SHV-variant enzymes, and any other resistance genes that differ by only point mutations.

Table 5
Characterization of Inhibitor Resistant TEM β-Lactamases Using Oligo-Probes to Detect Point Mutations *(44)*

Amino acid altered[a]	Sequence of oligo probe[b]	Enzymes with particular amino acid mutation
Met69	TTTCCAATGATGAGCACT	TEM-1
Leu69	TTTCCAATG**C**TGAGCACT	TEM-33 (IRT-5)[c], 35 (IRT-4) , 39 (IRT-10)
Leu69	TTTCCAATG**T**TGAGCACT	TEM-35 (IRT-4)
Val69	TTTCCAATG**G**TGAGCACT	TEM-34 (IRT- 6) , 38 (IRT-9)
Ile69	TTTCCAATGAT**T**AGCACT	TEM-32 (IRT-3)
Ile69	TTTCCAATGAT**A**AGCACT	TEM-32 (IRT- 3) , 37 (IRT-8)
Ile69	TTTCCAATGAT**C**AGCACT	No hydridisation detected
Arg244	CGTGGGTCTCGCGGTATC	TEM-1
Cys244	CGTGGGTCT**T**GCGGTATC	TEM-31 (IRT-1)
Ser 244	CGTGGGTCT**A**GCGGTATC	TEM-30 (IRT-2)
Asn276	ATGAACGAAATAGACAG	TEM-1
Asp276	ATGAACGA**G**ATAGACAG	TEM-35 (IRT-4), 36 (IRT-7), 37 (IRT-8), 39 (IRT-10)

[a]Ambler numbering *(47)*.
[b]The underlined nucleotide indicates the point mutation.
[c]IRT, inhibitor resistant TEM.

An alternative approach for identifying and characterizing β-lactamases that differ by point mutations is PCR single conformational polymorphism (PCR-SSCP). This method is based on the fact that a single nucleic acid change will affect the electrophoretic mobility of a short sequence of DNA and this technique has been used to distinguish SHV-1 to -5 (**Table 1**) *(24)*. PCR-SSCP may also characterise mutations in the TEM-derived β-lactamases. However, on many occasions β-lactamase genes may have nucleic acid changes that do not result in amino acid changes (so-called silent mutations). Therefore, in such a scenario, PCR-SSCP could indicate that two β-lactamases differ although they may have identical amino acid sequences and confer an identical profile of resistance.

4. Research Issues

4.1. Direct Sequence Analysis of β-Lactamases

An intragenic gene probe may indicate the type of β-lactamase produced by a clinical isolate (e.g., TEM- or SHV-derived) and oligoprobing enables known point mutations to be identified. However, the definitive confirmation and characterization of such an enzyme necessitates establishing the DNA nucleotide

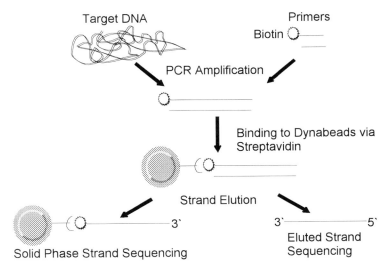

Fig. 3. DNA sequencing of PCR product.

sequence of the gene encoding it. Fortunately, this can now be quickly performed without the need for time-consuming cloning procedures. The suggested procedure circumvents the need for cloning by amplifying the resistance gene by PCR and then directly sequencing the PCR product by employing the DYNAL® magnetic bead separation system. The overall strategy is outlined in **Fig. 3**. The system is based on the principle that one of the primers employed to PCR the resistance gene is labeled with biotin: This allows the two DNA strands of the PCR product to be separated with the aid of magnetic beads that bind to the biotin-labeled strand of the PCR product *(49,50)*. Once separated, the sequencing of the single stranded β-lactamase DNA can be performed. This procedure can be applied to any gene provided suitable primers are available to permit amplification by PCR.

5. Future Prospects and Goals

The application of PCR technology to β-lactamases has an array of potential benefits in both the clinical setting and β-lactamase research. The ability of PCR to detect β-lactamase genes directly in clinical specimens (**Subheading 3.2.**) creates many new opportunities to assist clinical therapy in the early stages of life-threatening diseases. The challenge here is to design approaches or kits that would enable easy access to the technology for the clinical microbiology laboratory.

The obvious advantages of PCR for β-lactamase research is that it expedites the generation of gene probes and facilitates the sequencing of β-lactamase

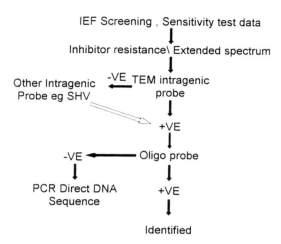

Fig. 4. Scheme for identifying TEM or SHV β-lactamases.

genes. PCR approaches have clearly superseded the traditional approaches of generating probes from restriction fragments. PCR also has other applications, **Subheading 3.2.1.** describes the use of PCR to study the heterogeneity of β-lactamase genes in Citrobacter spp. This approach should be applied to other β-lactamase genes and may illustrate a diversity of β-lactamases produced by a species which was previously assumed to possess only one type of β-lactamase.

Over the last 5 yr the use of DNA probing approaches to detect β-lactamases have been complicated by the plethora of enzymes that differ by only a few amino acids. For example, DNA from a clinical isolate that hybridizes with a TEM-1 probe indicates the enzyme could be any of over 30 different enzymes, each of which is capable of providing a very different spectrum of resistance compared to TEM-1. Secondary probing approaches, such as oligoprobing, have been successfully developed to overcome this difficulty, and in the future these types of approaches may be more reliable. Despite all these applications, there will undoubtedly be instances where detailed biochemical evaluation, and ultimately sequencing of the whole gene, are required for unequivocal identification. **Figure 4** suggests a scheme for identifying TEM or SHV β-lactamases.

Gene probes will play a very important role in defining the future clinical epidemiology of metallo-β-lactamases (**Table 2**), identification of these genes by probing will not be complicated by single amino acid changes. These enzymes confer resistance to most commonly used β-lactams and are produced by common pathogens, consequently, it is important to gauge their clinical significance.

6. Notes

1. Identification of cryptic genes. Although a positive hybridization indicates the presence of a particular β-lactamase gene, it provides no indication of whether the gene is being expressed and conferring resistance. This introduces the concept of covert or silent genes. As the use of genetic approaches to detect resistance genes is a relatively new concept, the incidence of silent genes or remnant genetic structures is not known, but they have been observed a few times. Jouvenot et al. *(51)* observed positive hybridization with a TEM-1 probe with 16 clinical isolates that did not appear to produce a TEM β-lactamase. This may have been caused by the TEM-1 probe containing 200 bp of extraneous DNA outside the TEM-1 gene. Ouellette et al. *(26)* also observed that five strains of *Pseudomonas aeruginosa*, which did not produce TEM-1, hybridized with a TEM-1 oligonucleotide probe (a 15 mer). However, DNA from these strains did not hybridize with a 656-bp TEM-1 probe. Therefore, it is probable that there was some fortuitous hybridization between the oligoprobe and homologous sequences in the strains of *P. aeruginosa*. These studies illustrate the importance of determining whether examples of silent genes may be an artefact of the probe used.

2. The *CfiA* cryptic metallo-β-lactamase gene. Probing studies by Podglagen et al. *(29)* illustrated that carbapenem-sensitive clinical isolates of *Bacteroides fragilis* carried a cryptic *CfiA* metallo-β-lactamase gene that became expressed when the isolates were exposed to imipenem. The isolates consequently became resistant to carbapenems and almost all other clinically important β-lactam antibiotics. Activation of this gene was a consequence of the insertion of a novel 1.3 kb insertion sequence immediately upstream of the *CfiA* gene *(52)*. This implies that the *CfiA* gene may be more widespread than previously appreciated, and the selective pressure applied by the predicted increase in carbapenem consumption may result in a significant increase in the extent of *CfiA* expression. In the clinical setting, such false-positive results caused by the detection of a cryptic gene may not be serious since the likely implication for the patient would be a change in antibiotic. In fact, although the gene may be silent in vitro, it may become activated in vivo, in which case the change to an alternative antibiotic would be advantageous.

3. Immunological approaches. Hybridization and PCR approaches are targeted at the DNA that encodes the β-lactamase. However, there are a few examples of where antibody approaches have been used to identify strains producing β-lactamases. These are reviewed below. A monoclonal antibody (MAbs) for a chromosomally encoded *P. aeruginosa* β-lactamase *(53)* has been shown to have no cross-reactivity with 23 different β-lactamases from seven different Bush groups, thus indicating the specificity of the MAbs *(54)*. In addition, antisera to TEM-1 have been shown to cross-react with TEM-2 and TLE-1, and to a lesser extent with SHV-1, but no reaction was observed with HMS-1 or the OXA or PSE enzymes. Antisera to OXA-1 react with OXA-4 and vice versa.

4. A study by Morin et al. *(55)*, however, illustrated that although 16 MAbs to TEM-1 cross-reacted with TEM-1, TLE-1 and, to a lesser extent, with SHV-1, they also

exhibited a reaction with OXA-1,3,6, and 7 and AER-1. This illustrates that diverse β-lactamases may have quite similar epitopes. Therefore, in certain circumstances, immunological approaches for the identification of β-lactamases can be less effective than DNA techniques. In addition, preparation of the antisera or MAb is much more time consuming than the preparation of a DNA probe. It is for these reasons that immunological approaches have been utilized far less than DNA approaches for the detection of β-lactamases in clinical isolates. However, recent studies on the rapid detection of TEM β-lactamases have successfully combined the two approaches. One of the drawbacks of PCR for routine use is the requirement for a time-consuming electrophoresis step in order to detect the PCR product. By utilizing specific TEM and control rRNA primers labeled with either dinitro-phenol, digoxigen, or biotin, Curran et al. *(56)* were able to detect PCR product by immunoassay. PCR product was detected by employing an immunoassay detection device consisting of a membrane containing blue latex beads labeled with either anti-dinitrophenol, digoxigen, or biotin antibody. A positive reaction was indicated by the appearance of a blue line and in most cases was obtained within 5 min. Comparison with isoelectric focusing for clinical *E. coli* confirmed the reliability of this method *(56)*. The optimization of this technology could herald the rapid detection of resistance genes direct from clinical specimens in routine laboratories.

References

1. Medeiros, A. A. (1984) β-lactamases. *Brit. Med. Bull.* **40,** 19–27.
2. Payne, D. J. (1993) Metallo-β-lactamases: a new therapeutic challenge. *J. Med. Microbiol.* **39,** 93–99.
3. BSAC Working party. (1991) A guide to sensitivity testing. *J. Antimicrob. Chemother.* **27,** 1–50.
4. Dubois, S. K., Marriott, M. S., and Amyes, S. G. B. (1995) TEM and SHV-derived extended-spectrum β-lactamases, relationship between selection, structure and function. *J. Antimicrob. Chemother.* **35,** 7–22.
5. Thomson, C. J. and Amyes, S. G. B. (1993) Selection of variants of the TEM-1 beta-lactamase, encoded by a plasmid of clinical origin, with increased resistance to beta-lactamase inhibitors. *J. Antimicrob. Chemother.* **31,** 655–664.
6. Payne, D. J., Cramp, R., Winstanley, D. J., and Knowles, D. (1994) Comparative activities of clavulanic acid, sulbactam and tazobactam against clinically important β-lactamases. *Antimicrob. Agents Chemother.* **38,** 767–772
7. Jarlier, V., Nicolas, M. H., Fournier, G., and Philippon, A. (1988) Extended broad-spectrum beta-lactamases conferring transferable resistance to newer beta-lactam agents in Enterobacteriaceae, hospital prevalence and susceptibility patterns. *Rev. Infect. Dis.* **10,** 867–878.
8. Bolstrom, A., Karlsson, A., and Mills, K. (1995) Detection of ESBLs Using a new E-test strip. Abstracts of the Seventh European Congress of Clinical Microbiology and Infectious Diseases. pp. 49–50.

9. Thomson, C. J., Miles, R. S., and Amyes, S. G. B. (1995) Sensitivity testing with clavulanic acid, fixed concentration versus fixed ratio. *Antimicrob. Agents Chemother.* **39,** 2591–2592

10. Towner, K. J. and Cockayne, A. (1993) Identification by nucleic acid hybridization techniques, in *Molecular Methods for Microbial Identification and Typing.* Chapman and Hall, London, pp. 64–92.

11. Thomson, C. J. and Amyes, S. G. B. (1992) TRC-1, emergence of a clavulanic acid resistant TEM beta-lactamase in a clinical strain. *FEMS Microbiol. Lett.* **91,** 113–118.

12. Meinkoth, J. and Wahl, G. (1984) Hybridization of nucleic acids immobilized on solid supports. *Analy. Biochem.* **138,** 267–284.

13. Arlet, G. and Philippon, A. (1991) Construction by polymerase chain reaction and intragenic DNA probes for three main types of transferable β-lactamases (TEM, SHV, CARB). *FEMS Microbiol. Lett.* **81,** 57–60.

14. Arlet, G., Rouveau, M., Fournier, G., Lagrange, P. H., and Philippon, A. (1993) Novel, plasmid encoded, TEM- derived extended -spectrum β-lactamase in *Klebsiella pneumoniae* conferring higher resistance to aztreonam than to extended-spectum cephalosporins. *Antimicrob. Agents Chemother.* **37,** 2020–2023.

15. Ouellette, M., Rossi, J. J., Bazin, R. and Roy, P. H. (1987) Oligonucleotide probes for the detection of TEM-1 and TEM-2 β-lactamase genes and their transposons. *Can. J. Microbiol.* **33,** 205–211.

16. Urban, C., Meyer, K. S., Mariano, N., Rahal, J. J., Flamm, R., Rasmussen, B. A., and Bush, K. (1994) Identification of TEM-26 β-lactamase responsible for a major outbreak of ceftazidime-resistant *Klebsiella pneumoniae. Antimicrob. Agents Chemother.* **38,** 392–395.

17. Bradford, P. A., Urban, C., Jaiswal, A., Mariano, N., Rasmussen, B. A., Projan, S. J., Rahal, J. J., and Bush, K. (1995) SHV-7, a novel cefotaxime-hydrolysing β-lactamase identified in *Escherichia coli* isolates from hospitalised nursing home patients. *Antimicrob. Agents Chemother.* **39,** 899–905.

18. Levesque, R. C., Medeiros, A. A., and Jacoby, G. A. (1987) Molecular cloning and DNA homology of plasmid mediated β-lactamase genes. *Mol. Gen. Genet.* **206,** 252–258.

19. Rice, L. B., Willey, S. H., Papanicolaou, G. A., Medeiros, A. A., Eliopoulos, G. M., Moellering, R. C., and Jacoby, G. A. (1990) Outbreak of ceftazidime resistance caused by extended-spectrum β-lactamases at a Massachusetts chronic-care facility. *Antimicrob. Agents Chemother.* **34,** 2193–2199.

20. Pornull, K. J., Goransson, E., Rytting, A., and Dornbusch, K. (1993) Extended-spectrum β-lactamases in *Escherichia coli* and *Klebsiella* spp. in European septicaemia isolates. *J. Antimicrob. Chemother.* **32,** 559–570.

21. Houovinen, S., Huovinen, P., and Jacoby, G. A. (1988) Detection of plasmid-mediated β-lactamases with DNA probes. *Antimicrob. Agents Chemother.* **32,** 175–179.

22. Tenover, F. C., Huang, M. B., Rasheed. J. K., and Persing D. H. (1994) Development of PCR assays to detect ampicillin resistance genes in cerebrospinal fluid samples containing *Haemophilus influenzae. J. Clin. Microbiol.* **32,** 2729–2737.

23. Arlet, G., Rouveau, M., Bengoufa, D., Nicolas, M. H., and Philippon, A. (1991) Novel transferable extended spectrum β-lactamase (SHV-6) from klebsiella conferring selective resistance to ceftazidime. *FEMS Microbiol. Lett.* **81**, 57–60.

24. M'Zali, F., Gascoyne-Binzi, D. M., Heritage, J., and Hawkey, P. M. (1996) Detection of mutations conferring extended-spectrum activity on SHV β-lactamases using polymerase chain reaction single strand conformational polymorphism (PCR-SSCP). *J. Antimicrob. Chemother.* **37**, 797–802.

25. Shlaes, D. M., Currie-McCumber, C., Hull, A., Behlau, I., and Kron, M. (1990) OHIO-1 β-lactamase is part of the SHV-1 family. *Antimicrob. Agents Chemother.* **34**, 1570–1576.

26. Ouellette, M., Paul, G. C., Philippon, A. M., and Roy, P. H. (1988) Oligonucleotide probes (TEM-1, OXA-1) verus isoelectric focusing in β-lactamase characterisation of 114 resistant strains. *Antimicrob. Agents Chemother.* **32**, 397–399.

27. Juteau, M., Deschaseaux, M. L., Royez, M., Mougin, C., Cooksey, R. C., Michel-Briand, Y., and Adessi, G. L. (1987) Molecular hybridization verus isoelectric focusing to determine TEM-type β-lactamases in gram-negative bacteria. *Antimicrob. Agents Chemother.* **31**, 300–305.

28. Ito, H., Arakawa, Y., Ohsuka, S., Wacharotayankun, R., Kato, N., and Ohta, N. (1995) Plasmid dissemination of the metallo-β-lactamase genebla IMP among clinically isolated Serratia marcescens. *Antimicrob. Agents Chemother.* **39**, 824–829.

29. Podglagen, I., Breuil, J. Bordon, F., Gutmann, L., and Collatz, E. (1992) A silent carbapenemase gene in strains of *Bacteroides fragilis* can be expressed after a one-step mutation. *FEMS Microbiol. Lett.* **91**, 21–30.

30. Tzouvelekis, L. S., Tzelepi, E., Mentis, A. F., and Tsakris, A. (1993) Identification of a novel plasmid mediated β-lactamase with chromosomal cephalosporinase characteristics from *Klebsiella pneumoniae. J. Antimicrob. Chemother.* **31**, 645–654.

31. Sakurai, Y., Tsukamoto, K., and Sawai, T. (1991) Nucleotide sequence and characterisation of a carbacillin hydrolysing penicillinase gene from *Proteus mirabilis. J. Bacteriol.* **173**, 7038–7041.

32. Zwadyk, P., Cooksey, R. C., and Thornsberry, C. (1986) Commercial detection methods for biotinylated gene probes, comparison with 32P-labeled DNA probes. *Curr. Microbiol.* **14**, 95–100.

33. Gallego, L., Umaran, A., Garaizar, J., Colom, K., and Cisterna, R. (1990) Digoxigenin-labelled DNA probe to detect TEM type β-lactamases. *J. Microbiol. Meth.* **11**, 261–267.

34. Adrian, P. V., Thomson, C. J., Klugman, K. P., and Amyes, S. G. B. (1995) Prevalence and genetic location of non-transferable trimethoprim resistant dihydrofolate reductase genes in South African commensal faecal isolates. *Epidemiol. Infect.* **115**, 255–267.

35. Brown, J. C., Thomson, C. J., and Amyes, S. G. B. (1996) Mutations of the *gyrA* gene of clinical isolates of *Salmonella typhimurium* and three other Salmonella species leading to decreased susceptibilities to 4-quinolone drugs. *J. Antimicrob. Chemother.* **37**, 351–356.

36. Vila, J., Ruiz, R., Goni, P., Marcos, A., and Deanta, T. J. (1995) Mutation in the *gyrA* gene of quinolone-resistant clinical isolates of *Acinetobacter baumannii. Antimicrob. Agents Chemother.* **39,** 1201–1203.

37. Persing, D. H. (1993) In vitro nucleic acid amplification techniques, in *Diagnostic Molecular Microbiology. Principles and Applications,* (Persing, D. H., Smith, T. F., Tenover, F. C., and White, T. J., eds.), American Society for Microbiology, Washington, DC, pp. 51–87.

38. Brakstad, O. G., Aasbakk, K., and Maeland, J. A. (1992) Detection of Staphylococcus aureus by polymerase chain reaction amplification of the nuc gene. *J. Clin. Microbiol.* **30,** 1654–1660.

39. Shirai, H., Nishibuchi, M., Ramamurthy,T., Bhattacharya, S. K., Pal, S. C., and Takeda, Y. (1991) Polymerase chain reaction for detection of cholera enterotoxin operon of *Vibrio cholerae. J. Clin. Microbiol.* **29,** 2517–2521.

40. Hunt, J. M., Roberts, G. D., Stockman, L., Felmlee, T. A., and Persing, D. H. (1994) Detection of a genetic locus encoding resistance to rifampicin in mycobacterial cultures and in clinical specimens. *Diagn. Microbial. Infect. Dis.* **18,** 219–227.

41. Telenti, A., Imboden, F., Merchesi, F., Schmidheini, T., and Bodmer, T. (1993) Direct, automated detection of rifampicin-resistant Mycobacterium tuberculosis by polymerase chain reaction and single-strand confirmation polymorphism analysis. *Antimicrob. Agents Chemother.* **37,** 2054–2058.

42. Jones, M., E., Avison, M. B., Damdinsuren, E., MacGowan, A. P., and Bennett, P. M. (1994) Heterogeneity at the β-lactamase structural gene *ampC* amonst Citrobacter spp. assessed by polymerase chain reaction analysis, potential for typing at the molecular level. *J. Med. Microbiol.* **41,** 209–214.

43. Payne, D. J., Marriott, M. S., and Amyes, S. G. B. (1990) Characterisation of a unique ceftazidime hydrolysing β-lactamase, TEM-E2. *J. Med. Microbiol.* **32,** 131–134.

44. Henquell, C., Chanal, C., Sirot, D., Labia, R., and Sirot, J. (1995) Molecular characterisation of nine different types of mutants among 107 inhibitor resistant TEM β-lactamases from clinical isolates of *Escherichia coli. Antimicrob. Agents Chemother.* **39,** 427–430.

45. Chanal, C., Poupart, M-C., Sirot, D., Labia, R., Sirot, J., and Cluzel, R. (1992) Nucleotide sequences of CAZ-2, CAZ-6 and CAZ-7 β-lactamase genes. *Antimicrob. Agents Chemother.* **36,** 1817–1820.

46. Malibat, C. and Courvalin, P. (1990) Development of 'oligotyping' for characterisation and molecular epidemiology of TEM β-lactamases in members of the family *Enterobacteriaceae. Antimicrob. Agents Chemother.* **34,** 2210–2216.

47. Ambler, R. P. (1980) The structure of β-lactamases. *Phil. Trans. Royal Soc. Lon.* **B289,** 321–331.

48. Tham, T. N., Mabilat, C., Courvalin, P., and Guesdon, J-L. (1990) Biotinylated oligonucleotide probes for the detection and the characterisation of TEM-type extended broad spectrum β-lactamases in Enterobacteriaceae. *FEMS Microbiol. Lett.* **69,** 109–116.

49. Hultman, T., Bergh, S., Moks, T., and Uhlen, M (1991) Bidirectional solid-phase sequencing of in vitro-amplified plasmid DNA. *Biotechniques* **10,** 84–93.
50. Dynal. (1995) *Biomagnetic Techniques in Molecular Biology.* Dynal A. S., Oslo, Norway.
51. Jouvenot, M., Deschaseaux, M. L., Royez, M., Mougin, C., Cooksey, R. C., Michel-Briand, Y., and Adessi, G. L. (1987) Molecular hybridization verus iso-electric focusing to determine TEM-type β-lactamases in Gram-negative bacteria. *Antimicrob. Agents Chemother.* **31,** 300–305.
52. Podglagen, I., Breuil, J., and Collatz, E. (1994) Insertion of a novel DNA sequence IS*1186,* upstream of the silent carbapenemase gene cfiA promotes expression of carbapenem resistance in clinical isolates of *Bacteroides fragilis. Molecular Microbiol.* **12,** 105–114.
53. Bush, K., Jacoby, G. A., and Medeiros, A. A. (1995) A functional classification scheme for β-lactamases and its correlation with molecular structure. *Antimicrob. Agents Chemother.* **39,** 1211–1233.
54. Giwercman, B., Rasmussen, J. W., Cioufu, O., Clemmentintsen, I., Schumacher, H., and Hoiby, N. (1994) Antibodies against chromosomal β-lactamase. *Antimicrob. Agents Chemother.* **38,** 2306–2310.
55. Morin, C. J., Patel, P. C., Levesque, R. C., and Letarte, R. (1987) Monoclonal antibodies to TEM-1 plasmid mediated β-lactamase. *Antimicrob. Agents Chemother.* **12,** 461–464.
56. Curran, R., Talbot, D. C. S., and Towner, K. (1996) A rapid immunoassay method for the direct detection of PCR products, application to detection of TEM β-lactamase genes. *J. Med. Microbiol.* **45,** 76–78.

26

Biochemical and Enzyme Kinetic Applications for the Characterization of β-Lactamases

David J. Payne and Tony H. Farmer

1. Introduction

For the last 20 yr thin-layer polyacrylamide isoelectric focusing (IEF) has played a major role in the identification and characterization of β-lactamases. IEF is able to distinguish enzymes that focus only 0.05 pI apart *(1)*, but the exponential increase in the numbers of β-lactamases discovered over the last 10 yr has meant that this method no longer provides sufficient resolution to distinguish the majority of β-lactamases. Today the pI of a β-lactamase is still an essential determinant that must be used *in combination* with a variety of other data. Moreover, the IEF of β-lactamases is now entering a new era as this technique can be adapted to provide important biochemical information on β-lactamases other than simply their pI values. These approaches are discussed in **Subheading 3.1.**

β-lactamases can also be characterized by their interaction with inhibitors and substrates. The most rapid and least complex measurements are I$_{50}$ values (the concentration of inhibitor that inhibits the hydrolytic activity of the β-lactamase by 50% compared with an untreated control) and the rates of hydrolysis of a range of different β-lactam substrates at fixed concentrations. Protocols for performing these assays and approaches for maximizing their accuracy are discussed in **Subheadings 3.2.1** and **3.2.7.** However, because many of the currently identified β-lactamases have similar characteristics (e.g., the TEM and SHV β-lactamases), more discriminatory methods for characterizing β-lactamases are required, and it is becoming increasingly important to determine more complex kinetic data to distinguish different β-lactamases. These approaches are discussed in **Subheadings 3.2.2.–3.2.6.** and **3.2.8.**

From: *Methods in Molecular Medicine, Vol. 15: Molecular Bacteriology: Protocols and Clinical Applications*
Edited by: N. Woodford and A. P. Johnson © Humana Press Inc., Totowa, NJ

2. Materials

2.1. Major Equipment

1. Refrigerated centrifuge capable of spinning from 6000g–32,000g.
2. Grant LTD 6 (Cambridge, UK) circulating water bath.
3. Power pack capable of delivering 800 V, 30 A, and 20 W (e.g., Pharmacia MultiDrive XL, Piscataway, NJ).
4. Horizontal gel apparatus (e.g., Multiphor II Electrophoresis System, Pharmacia).
5. Light box.
6. Hood Camera with fixed focal length (e.g., Model QSP Quickshooter Photosystem IBI [International Biotechnologies, Inc., CT]).
7. Pharmacia PhastSystem.
8. UV/VIS spectrophotometer (e.g., Beckman DU7500, Fullerton, CA).
9. Microtiter plate reader (e.g., Bio-Tek EL312e Microplate Biokinetics Reader with KinetiCalc software [Bio-Tek Instruments, Inc., VT], or SpectraMAX 250 [Molecular Devices, Crawley, West Sussex, UK]).
10. Sonicator (e.g., Soniprep 150 [MSE]).
11. Scientific graphics package (e.g., GraFit [Erithacus Software, Ltd. Staines, UK]).

2.2. Consumables

1. Solution I: 25 mM 1,4-piperazine-diethanesulphonic acid (PIPES), pH 7.0.
2. Solution II: 25 mM 1,4-PIPES, pH 7.0, 0.5 mg/mL nitrocefin (Unipath, Basingstoke, UK). The nitrocefin should be dissolved in a minimum volume of dimethylsulfoxide (1% of final volume) and then made up to the final volume with PIPES buffer. All solutions of nitrocefin should be maintained on ice.
3. Solution III: 25 mM 1,4-PIPES, pH 7.0, 0.5 mg/mL nitrocefin and 1 mM zinc sulfate (added last).
4. Solution IV: 25 mM 1,4-PIPES, pH 7.0, 0.05 mg/mL nitrocefin.
5. Solution V: 25 mM 1,4-PIPES, pH 7.0, 0.5 M EDTA.
6. Dimethylsulfoxide (DMSO).
7. Colored pI markers (e.g., BDH Electran IEF Markers pI Calibration Kit Range 4.7–10.6, Poole, UK [the lyophilized sample should be reconstituted with 100 µL of distilled water]).
8. Quartz cuvets: 1–10 mm pathlength.
9. Microtiter plates, and deep-well plates, 96 well (Beckman).
10. Precast IEF gels (e.g., Pharmacia, various pH ranges available).
11. PhastGels IEF 3-9 (Pharmacia).
12. Filter paper (e.g., Whatman 1541320).
13. Ice.
14. Film: Polaroid Type 665 (positive/negative) or equivalent.

3. Methods

3.1. Isoelectric Focusing

IEF separates β-lactamases on the basis of their overall electric charge. The isoelectric point (pI) is the pH at which the protein bears no net charge, and is, therefore, the pH at which the protein is electrophoretically immobile. This section will summarize the various protocols that can be used for the IEF of β-lactamases, including the use of PhastSystem protocols that have increased the speed of the technique fivefold. The alternative uses of IEF for identifying different β-lactamases will also be discussed.

3.1.1. Preparation of Samples for Isoelectric Focusing

The volume of bacterial culture required for analysis by IEF depends on the quantity of enzyme produced by the isolate. The authors recommend starting with a 100 mL culture. If the β-lactamase produced by an isolate is unknown, then it is advisable to grow both induced and uninduced cultures as the isolate may produce insufficient β-lactamase in the basal state to allow identification by IEF. The optimal conditions for induction of a β-lactamase produced by a clinical isolate will be unknown, but the conditions below have been shown to cause induction of a variety of β-lactamases:

1. Inoculate 2X 100 mL of broth with a 1% inoculum of an overnight broth culture.
2. These cultures should be grown until they have an optical density of approx $A_{500} = 0.7$.
3. Addition of inducer. β-Lactams that are effective inducers include cefoxitin (for most serine β-lactamases) and imipenem (the latter is particularly good for inducing some metallo-β-lactamases). Sufficient inducer should be added to one of the two flasks to give a final concentration of 1/4 the MIC of the particular isolate.
4. Grow the cultures for 2 h more.
5. Harvest cells from each culture by centrifugation. Typical conditions are 6000g, for 15 min at 4°C.
6. Resuspend and wash the pellet in 10 mL of Solution I.
7. Centrifugation at 6000g for 15 min at 4°C.
8. For the best results, resuspend the sample in a minimum volume of Solution I (ideally ≤1% of the original culture volume; more buffer can be added if the pellet is too viscous). The pellet should be maintained on ice.
9. The β-lactamase is then released from the whole cells by sonication while cooled in a mixture of ice and water. A typical treatment for *Escherichia coli* would be 2 × 30 s bursts at 8 μ separated by a 1 min cooling period; longer sonication periods may be required for other organisms (e.g., *Pseudomonas* spp.).
10. Subject the cell-free lysate to high speed centrifugation (32,000g for 30 min at 4°C) to remove the cell debris. The supernatant, containing the β-lactamase, is removed.
11. Estimate the activity of the preparation by mixing 33 μL of the sample with 100 μL of Solution IV, and record the time in seconds taken for the color to change from yellow to red.
12. Store the β-lactamase in aliquots at −20°C; freeze/thawing of the sample should be avoided.

3.1.2. Isoelectric Focusing Methodologies: Precast Gels

The methodology for performing IEF of β-lactamases is based on the protocol devised by Matthew et al. *(2)*. This method examines β-lactamases on thin-layer polyacrylamide gels employing ampholines (multicharged compounds that have different isoelectric points and stack across the gel to form a pH gradient). Once the gels have run, β-lactamases are selectively identified from other bacterial proteins by staining the gel with nitrocefin, a chromogenic cephalosporin. This technique is, therefore, equally effective with crude lysates as with purified β-lactamases. Few β-lactamases focus as discrete bands, and in most cases, a series of satellite bands are observed (*see* **Note 1**).

1. When investigating an enzyme of unknown p*I*, a gel of pH range 3–10 is the most appropriate. If further resolution is required, gels covering narrower pH ranges are available (Pharmacia).
2. Place the precast IEF gel on the gel apparatus (*see* **Note 2**). Cut paper electrode contacts (supplied with Pharmacia gels) to an appropriate size and moisten with distilled water, Remove excess water by rigorously blotting on filter paper.
3. Place the electrode contacts on the edge of the longest sides of each gel.
4. Load samples. The number of µL of β-lactamase sample added to the gel should roughly correspond to the activity of the preparation in seconds (*see* **Subheading 3.1.1., step 11**). For example, if the activity is 5 s then add 5 µL. Volumes of <5 µL may be directly applied onto the gel, whereas 5–15 µL of sample should be applied to a loading tab (supplied with the Pharmacia gels) and 15–30 µL should be added to two loading tabs placed on top of each other. For best results the minimum volume of sample should be added to the gel and β-lactamase samples must be prepared as concentrated as possible. Space the sample loadings at least 0.5 cm apart (**Fig. 1**).
5. Loading position. For best results it is advisable to load β-lactamases known to focus close to the anode (with low p*I*) at the cathode and vice versa. Since the loading site normally causes some distortion of the gel, it is best that the β-lactamase band does not focus in this area. Test samples are best loaded close to the anode.
6. p*I* markers. 5 µL of colored broad range p*I* markers should be loaded in 2–3 tracks on the gel.
7. Focusing. The power pack should be set at 800 V, 30 A, and 20 W, and the temperature of the circulating water bath set to 9°C. Focusing is complete when the p*I* markers have focused as distinct bands (approx 4 h). The high voltage results in better resolution and less drift at the cathode.
8. Staining of gels. Soak filter paper in Solution II and overlay on the gel. After a few minutes, red bands will start to appear where the β-lactamases have focused. For optimum results, fresh overlays should be applied at regular intervals, although the bands will start to diffuse after 10–15 min.
9. Recording gels. Fixed focal length "hood" cameras can provide good photographs of gels illuminated over a suitable light box (orange filter, f/16, 1 s). Alterna-

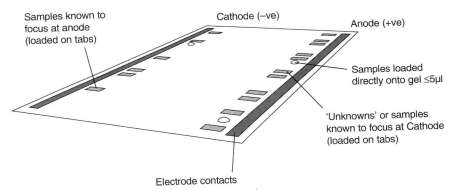

Fig. 1. Loading β-lactamase preparations onto IEF gels.

Table 1
Optimized Method for IEF with PhastGel IEF 3-9

Step	Volts	mA	W	Temp	Volt hours	Function of step
1.1	2000	2.5	3.5	15°C	75	Prefocusing
1.2	200	2.5	3.5	15°C	75	Sample loading
1.3	2000	2.5	3.5	15°C	410	Sample focusing

The run takes 500 Vh or 30 min, prefocusing takes 10 min.

tively, providing the gels can be sealed effectively in plastic film, they can be photocopied to provide a record of the gel. The p*I* of unknown enzymes can be determined by plotting a graph of p*I* against distance migrated for the p*I* markers.

3.1.3. IEF Methodologies: PhastSystem (Pharmacia)

The advantage of the PhastSystem is that it enables β-lactamase IEFs to be run in only 30 min, but it has the disadvantage that the resolution is not as good as with the gels recommended in **Subheading 3.1.2.**

1. β-lactamase samples are loaded onto the applicator comb as described by the manufacturer; the load volume varies with the size of the comb, but generally no more than 2–3 μL of sample can be loaded.
2. Set the PhastSystem to the appropriate program for IEF 3-9 p*I* gels (**Table 1**). This program takes 30 min to run.
3. For best staining results, soak a piece of filter paper in Solution II and then blot onto clean filter paper to remove all the surface liquid. Lay this onto the gel and gently remove air bubbles. The filter paper is removed after 30 s and the process repeated 2–5 times until satisfactory staining is achieved.
4. The gels can then be photographed as described in **Subheading 3.1.2.**

3.1.4. Use of IEF to Distinguish Metallo and Serine β-Lactamases

Protocols have been developed to distinguish the metallo and serine β-lactamases of *Stenotrophomonas maltophilia* (formerly *Xanthomonas maltophilia*) by IEF *(3)*. This methodology enabled the identification of seven different metallo-β-lactamases and at least eight different serine β-lactamases in 17 clinical isolates of *S. maltophilia (3)*. This approach is particularly attractive, since it circumvents the requirement for lengthy purification and biochemical characterization of each of the enzymes. To date, this methodology has only been optimized for the study of β-lactamases produced by *S. maltophilia*, but it is probably appropriate for β-lactamases produced by other species. The protocol is shown below:

1. *Identification of metallo-β-lactamase.* Mix a 10 µL aliquot of the cell free lysate from an induced culture of *S. maltophilia* with 2 µL of Solution V to create a final EDTA concentration of 83 mM (Sample 1). Mix another 10 µL aliquot with 2 µL of Solution I (Sample 2).
2. Load both samples onto the IEF gel, as described previously (*see* **Subheading 3.1.2.**).
3. Focus and stain the gel in the normal way with Solution II. Sample 2 will elaborate both serine and metallo-β-lactamases, but in Sample 1 the metallo-β-lactamases are inhibited by the chelating action of EDTA and only the serine enzymes will be visualized.
4. *Identification of serine β-lactamase.* BRL42715, an experimental serine β-lactamase inhibitor, enabled serine β-lactamases to be distinguished by this method *(3)*. The method employed was as follows: Load two 10 µL aliquots (A and B) of the cell free lysate from an induced culture on to the IEF gel and focus in the normal way.
5. When focusing is complete, overlay Sample B with 100 µM BRL42715 for 10 min. The gel is then stained with Solution III (nitrocefin containing ZnSO$_4$). BRL 42715 is a very potent inhibitor of all serine β-lactamases, but is a substrate for metallo-β-lactamases. Therefore, only the metallo-β-lactamase will be visualized in Sample B, and both the serine and metallo-β-lactamase will be visualized in Sample A. Unfortunately, BRL42715 is no longer available. However, commercial serine β-lactamase inhibitors would probably make suitable replacements, although the appropriate method development studies would have to be performed.

Table 2 summarizes each of these treatments and **Fig. 2** demonstrates the effectiveness of the EDTA treatment. Adjustments to this protocol have been used to visualize metallo-β-lactamases produced by *Bacteroides fragilis (4)*, *see* **Note 3**.

3.2. Kinetic Analysis of β-Lactamases

This section provides protocols for the kinetic characterization of β-lactamases and, more importantly, discusses the principles and limitations of

Table 2
Summary of the Assays Used to Identify Serine and Metallo-β-Lactamases on IEF

Treatment	Result, following overlay with Solution II (nitrocefin)	Indication of enzyme type
Cell free extract mixed with Solution V (EDTA) prior to IEF	Band unaffected Band eradicated	Serine Metallo-
Gel overlaid with 100 μ*M* BRL42715 + 1 m*M* zinc sulfate	Band enhanced/unaffected Band eradicated	Metallo- Serine
Gel overlaid with Solution III (1 m*M* zinc sulfate)	Band activated Band unaffected	Metallo- Serine

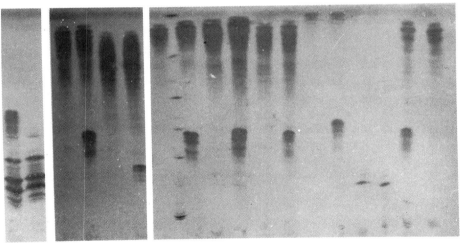

A B C D E F G H I J K L M N O P Q R S

Fig. 2. Effect of 83m*M* EDTA on the IEF of β-lactamases from *S.maltophilia*. **(A)** *S.maltophilia* A37454; **(B)** A37454 + EDTA; **(C)** 37 + EDTA; **(D)** 37; **(E)** 152 + EDTA; **(F)** 152; **(G)** 10258 + EDTA; **(H)** pI markers; **(I)** 10258; **(J)** 10257 + EDTA; **(K)** 10257; **(L)** 00157 + EDTA; **(M)** 00157; **(N)** ED136 + EDTA; **(O)** ED136; **(P)** TEM-1 + EDTA; **(Q)** TEM-1; **(R)** 511 (known metallo-β-lactamase producer); **(S)** 511 + EDTA.

each approach. 25 m*M* sodium phosphate (pH 7.0) is an appropriate buffer for the assay of serine β-lactamases. 25 m*M* 1,4-PIPES, pH 7.0 is recommended for the assay of metallo-β-lactamases *(5)*. A range of ZnSO$_4$ concentrations should be tested to determine the optimal concentration for the particular metallo-β-lactamase being examined. Where possible all assays should be run at 37°C.

3.2.1. Rapid Substrate Profiling of β-Lactamases

Substrate profiling is a widely used tool that enables the rapid comparison of different β-lactamases by the relative hydrolysis rates of a series of β-lactam substrates tested at fixed concentrations. Results are most accurate and reproducible if rates are at or close to V_{max}, and for this it is advisable to use high substrate concentrations (1 mM) where possible (*see* below and **Note 4**). The procedure given below is for the direct spectrophotometric assay *(6,7)*, which is commonly used (*see* **Note 5**).

1. Optimal wavelengths and changes in absorbance ($\Delta\varepsilon$) on substrate hydrolysis can be generally obtained from the literature *(8)*. If such data are not available, proceed as shown below:
2. Scan a solution of the β-lactam to be assayed from 200–400 nm and store the spectrum. If the substrate absorbs in the visible part of the spectrum, extend the range of the scan.
3. Add β-lactamase and record spectra periodically; when no further changes are observed, subtract this spectrum from that of the intact β-lactam to obtain λ_{max} for the *change* in absorbance on hydrolysis. Generally, hydrolysis is accompanied by a fall in absorbance; in some cases, such as nitrocefin (λ_{max} = 482 nm), there is an increase in absorbance on hydrolysis.
4. If the β-lactams are relatively stable, β-lactamase II from *Bacillus cereus* can be used as this enzyme hydrolyses virtually all β-lactams.
5. Set the instrument to the desired wavelength (λ_{max}, for change in absorbance on hydrolysis) and incubate 1 mM substrate at 37°C in a 10 mm pathlength quartz cuvet. If the β-lactam absorbs too strongly at this concentration, shorter pathlength (e.g., 2 mm) cuvets can be used, or, alternatively shift the wavelength to bring the reading onto the scale. In the latter case $\Delta\varepsilon$ at this wavelength will have to be found. The most commonly used cuvets have a capacity of 2.5–3.0 mL, but 10 mm pathlength cells are available with working volumes of 0.8–1.0 mL, if materials are in short supply.
6. Check stability of substrate. Generally β-lactams are not spontaneously hydrolysed, but some carbapenems and nitrocefin at high concentrations may be unstable. It is particularly important that substrate stability is checked when assaying metallo-β-lactamases, as Zn^{2+} ions may accelerate nonenzymatic hydrolysis of β-lactams *(5)*. When measuring the activity of these enzymes it is best to add the metal salt to the substrate from a concentrated stock solution (in distilled water, as concentrated solutions of zinc salts precipitate in buffers), just prior to carrying out the assay. If spontaneous hydrolysis does occur, record the rate; this will have to be subtracted from the enzymatic rate.
7. Check stability of the enzyme. Diluted enzymes often lose activity, particularly when pure. Solutions to be added to reactions should be as concentrated as pos-

sible, and it is advisable to prepare these in buffers containing 5–10 mg/mL bovine serum albumin (BSA). It may also prove necessary to include 50–100 µg/mL BSA in assay solutions.

8. Add the β-lactamase to the substrate and mix by inversion or with a microstirrer. It is best to add a small volume of a concentrated enzyme solution as this will reduce any changes in volume and in temperature (e.g., if enzyme stock solutions are kept on ice). Concentrated enzymes also tend to be more stable.

9. If using very concentrated crude preparations (e.g., if β-lactamase activity is low), a blank should be set up containing enzyme and buffer only and rates measured against this.

10. The rate of absorbance change should be immediately recorded after adding the β-lactamase. Record as many data points as practically possible, but ensure that each reading is measured for a sufficient time to give a stable signal (particularly with cell changers). Generally 10–20 data points should be sufficient over 5 min.

11. Use only the initial linear phase of the reaction to calculate rates (i.e., ΔAbs/time) It is advisable to check the digitalized data to ensure linearity (many modern spectrophotometers will automatically do this, but this should be confirmed by closer inspection). Subtract blank rates (double beam and many single beam instruments do this automatically) and any nonenzymatic hydrolysis.

12. As crude preparations will often be used for substrate profiles, rates cannot be related to mg (or moles) of β-lactamase. The calculations given below, therefore, relate to per mL enzyme. The units for substrate hydrolysis are in µMoles min^{-1} as these tend to give manageable figures. Rates are therefore found as follows:

$$\frac{\Delta Abs\ min^{-1}}{\Delta\varepsilon \times 10^{-6}} \quad * \quad \frac{v_r\ (mL)}{1000} \quad * \quad \frac{1}{v_e(mL)} \quad = \mu M\ \text{substrate}\ min^{-1}\text{enz}\ mL^{-1}$$

(where Δε is the molar extinction coefficient of hydrolysis; v_r is the volume of the reaction; and v_e is the volume of enzyme added).

13. To obtain the substrate and profile, it is necessary to relate all rates back to a standard substrate. This is usually done by setting the value for benzylpenicillin at 100, but sometimes other reference substrates are used such as nitrocefin.

3.2.2. Evaluation of the Kinetic Parameters of Substrate Hydrolysis by β-Lactamases

Substrate profiles provide a rapid characterization of β-lactamases. However, such an approach does have short falls (*see* **Note 4**). Increasingly it is necessary to determine K_m and V_{max} to distinguish different β-lactamases, and the ratio of V_{max}/K_m is especially useful for comparing both β-lactamases and substrates. Discussion of the most appropriate methods for obtaining and processing enzyme kinetic data has generated a vast literature and space does not permit a critical evaluation here. Anyone contemplating kinetic characterization of enzymes should certainly consult Fersht *(9)*, whereas other workers give more comprehensive treatments for obtaining and processing data *(10–12)*. Data for the determination of K_m and V_{max} values can either be determined by recording the initial rate of hydrolysis of different concentrations of β-lactams (**Subheading 3.2.3.**) or by determining the rates of hydrolysis at different points along a single progress curve (**Subheading 3.2.4.**).

3.2.3. Measuring Initial Rates

This is the classical approach to obtaining rate data for finding kinetic constants. Good results are generally obtained for cephalosporins and carbapenems, but for reasons given in **Subheading 3.2.4.** the method is often inappropriate for penicillins. Basic assay procedures for measuring initial rates were given in **Subheading 3.2.1.** (e.g., check for enzyme and substrate stability). Key points pertinent to determining kinetic parameters are given below.

1. Prepare a range of substrate solutions from $0.2–10 \times K_m$ (if the K_m is unknown and preliminary investigations have given no indication, it will be necessary to carry out a "sighting" experiment). At least five, ideally 10–12, concentrations will be needed. The stability of the substrate should be checked at each of the concentrations assayed as described in **Subheading 3.2.1.**
2. Start recording the absorbance of the β-lactam immediately after adding and mixing the enzyme. Preferably, substrate hydrolysis should be followed as a continuous progress curve. However, it is not necessary to follow reactions to completion (provided the $\Delta\varepsilon$ is known).
3. When processing the results, it is absolutely essential that only true initial rates are used. Ideally these should be no more than 5% of the reactions and never more than 10%. For this reason it is advisable that reactions are followed singly, and cell programmers generally avoided. Plate readers can be utilized as long as the enzyme is added to the range of substrate concentrations simultaneously, and all the wells are read with a minimum time delay between the first and last assay well.

Subheadings 3.2.5. and **3.2.6.** detail recommended approaches for calculating V_{max} and K_m data from rate-substrate data pairs.

3.2.4. Use of Single Progress Curves

The $\Delta\varepsilon$s for penicillin are typically approx 10-fold lower than those for cephalosporins, thus reducing assay sensitivity. Furthermore, their K_m values are often rather low (<100 μM). The combination of these two factors make it virtually impossible to obtain linear initial rates, and data obtained by this method will give erroneous kinetic constants, especially for K_m. Consequently, the use of single progress curves to obtain rate-substrate data pairs has been extensively used to find kinetic parameters for β-lactamases *(6,13–15)*. The authors strongly recommend this approach is used for penicillins and cephalosporins with K_ms of <100 μM and <10 μM, respectively. Other advantages of this procedure are that it saves on materials, reduces problems from changes in enzyme activity, and is less time consuming. By cutting the number of pipeting and manipulation steps, potential sources of error are also reduced. Guidelines are given below.

1. Select a substrate concentration that is about 2.5–5.0 times the K_m. This should ensure that sufficient data points are obtained where a noticeable change in rate is seen.
2. It is advisable to carry out assays at two to three different starting substrate concentrations.
3. The starting ($Abs_{t=0}$) and final absorbances (i.e., when the reaction has gone to completion, Abs_∞) must be accurately determined as these will be needed to process the results.
4. While satisfactory estimates of K_m and V_{max} can be obtained from as few as 10 data points, preferably at least 20 should be taken.
5. Checks should be made to make sure that the enzyme is not spontaneously losing activity during the assay or being inhibited by substrate or product. Whereas this may only become apparent after data processing, it is often apparent from premature slowing or stopping of the reaction.
6. The substrate concentration $[S_1]$ at time t is found from $(Abs_t - Abs_\infty)/\Delta\varepsilon$; at t + 1 the concentration $[S_2]$ is $(Abs_{t+1} - Abs_\infty)/\Delta\varepsilon$. The rate for $[S_1]$ is given from ΔAbs between t and t + 1. Although this is not a strictly valid assumption, assuming that the curve is divided into small enough sections, this method provides rate-substrate pairs that give good and reproducible estimates of K_m and V_{max} when fitted to the Michaelis-Menten equation. Excellent results are obtained if the procedure is modified such that the rate between $[S_1]$ and $[S_2]$ (for example) is related to the substrate concentration $[S_1]$ + $[S_2]/2$ (*see* also **ref. 13** for related approaches). **Subheadings 3.2.5.** and **3.2.6.** detail recommended approaches for determining K_m and V_{max} from rate-substrate pairs.

3.2.5. Graphical Determination of Kinetic Data

The two practical approaches detailed in **Subheadings 3.2.3.** and **3.2.4.** give rise to data sets of rate versus substrate concentration. Traditionally, K_m and V_{max} have been found from plotting rate versus substrate concentration using linear transformations of the Michaelis-Menten equation. Fitting of lines can be made by eye or using linear regression. Whereas the former is obviously subjective to some extent, the latter is not necessarily more accurate because rearrangement of the equation also leads to a rearrangement of the error distribution *(16)*. To fit the line accurately to the data requires a weighting scheme for the transformed data *(10,11)*. This can be co;nfusing, difficult, and time consuming, especially for the nonspecialist. However, the authors have found that the following graphical methods give good, reproducible kinetic constants for β-lactamases.

1. The direct linear plot *(17)* is, in many respects, the best method because it has no requirement for weighting. Substrate concentrations are marked on the abscissa (as negative values) and the measured velocities on the ordinate. The individual

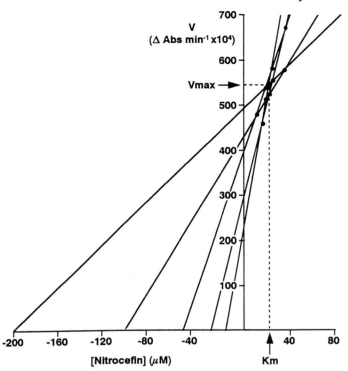

Fig. 3. Determination of kinetic constants for the hydrolysis of nitrocefin by the *B. cereus* II β-lactamase using direct linear plot; data from initial rates. K_m = 22 μ*M*; V_{max}= 0.054 ΔAbs min^{-1}. Data constants are determined from the median value of the intersections (for n data there are [n(n-1)/2] intersections). This is the middle point from an odd numbered set and the mean of the middle pair in an even numbered set.

data pairs are joined by lines that intersect with those from other data pairs. The intersection of the lines gives V_{max} when drawn back to the v axis (ordinate) and K_m when dropped down to the S axis (abscissa). In theory, all lines should intersect at one point, but this is unlikely to happen in practice. When the lines do not meet at one point the kinetic parameters are found from the median (i.e., middle) of the intersections (not the mean) (**Fig. 3**).

2. Although, as indicated above, there are objections to fitting unweighted data to straight line transformations of the Michaelis-Menten equation, we have found that the Hanes plot, where [S]/v is plotted against [S], gives reliable estimates of K_m and V_{max} for β-lactamases (**Fig. 4**). The abscissa intercept gives $-K_m$, the ordinate K_m/V_{max} and the slope $1/V_{max}$. The equation is:

$$\frac{[S]}{v} = \frac{K_m}{V_{max}} + \frac{1}{V_{max}} * [S]$$

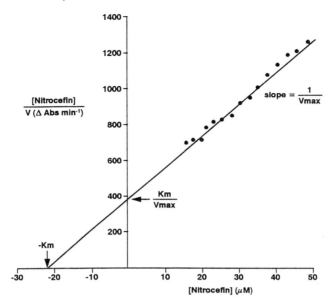

Fig. 4. Determination of kinetic constants for the hydrolysis of nitrocefin by the *B. cereus* II β-lactamase using the Hanes plot; data from continuous progress curve. ($K_m = 22$ μM; $V_{max} = 0.057$ ΔAbs min^{-1}.)

3. Although popular, the Lineweaver-Burk plot is very susceptible to error distribution and should not be used.

3.2.6. Computerized Determination of Kinetic Data

Error distribution becomes much less of a problem if initial rate and substrate data are fitted directly to the Michaelis-Menten equation.

$$v = \frac{V_{max} * [S]}{K_m + [S]}$$

Fortunately, a number of computer programs are available that easily accomplish this (e.g., GraFit or similar programs *[18,19]*). This procedure can be very quick if results are directly transferred from the spectrophotometer to the program. Such programs should not be used uncritically. It is especially important that v against [S] plots are inspected to ensure that data are evenly distributed. Results should not be clustered well below or above K_m.

Many modern spectrophotometers enable rate-substrate data sets to be easily obtained from single progress curves that can then be fitted to the Michaelis-Menten equation as described above. Alternatively, some workers have fitted absorbance data to the integrated form of the equation; in fact, this method was used in one of the first papers describing the spectrophotometric assay *(6)*. The integrated equation can be cast in various forms, of which the following is a common example:

$$V_{max} * t = ([S_0] - [S_t]) + K_m * \ln [S_0]/[S_t]$$

Details of methods used to obtain and process data can be found elsewhere *(6,14,15)*. The review by De Meester et al. *(14)* is a particularly comprehensive treatment of the fitting of both substrate and inhibitor data for β-lactamases using computerized procedures. Data for the integrated rate equation can also be obtained relatively easily from primary results using a calculator and then plotting these manually.

3.2.7. Rapid Inhibitory Profiling of β-Lactamases

Microtiter plate readers with kinetics software packages have enabled high throughput approaches for the determination of I_{50} values. The Biotek Microplate Biokinetics Reader with KinetiCalc software and the SPECTRA$_{max}$ 250 SOFTmax PRO software both enable percent inhibition compared with a control to be calculated for each of the 96 wells in a microtiter plate. The SPECTRAmax has the added advantage of being able to detect substrates that absorb in the UV, which means almost any β-lactam can be used as a reporter substrate. In addition, the SOFTmax PRO software will plot and calculate the I_{50}.

The protocol below will enable I_{50}s of eight compounds to be measured for one or more β-lactamases and is based on the method previously described *(20,21)*.

1. Dissolve the inhibitors in buffer to produce a 3 mM solution. The compounds can be predissolved in a small volume of DMSO, but this must not exceed 5% of the final volume.
2. In a deep 96-well plate add 600 μL of buffer into wells in columns 3–11.
3. Pipet 1 mL of the 3 mM inhibitor solutions into the wells of column 12.
4. Make 1:3 serial dilution of the 3 mM solution down to column 3; take 300 μL of the 3 mM solution from column 12 and dispense into column 11. Mix, and take 300 μL from column 11 and dispense into column 10, and so on, across the columns to column 3, which will then be at 0.15 μM.
5. Transfer 50 μL of the contents from columns 3 to 12 in the deep-well plate to the equivalent positions on the assay microtiter plate. 100 μL and 50 μL of buffer are added to the Blank (column 1) and Control (column 2) wells of the assay microtitre plate, respectively.
6. For I_{50} determinations with preincubation of enzyme and inhibitor, add 50 μL of appropriately diluted enzyme and incubate the plate for 5 min at 37°C.
7. Add 50 μL of a reporter substrate (*see* **Note 8**), such as nitrocefin, to all 96 wells and rapidly place into the microtiter plate reader and initiate absorbance readings. **Table 3** shows the final concentration of inhibitor in each well of the microtiter plate, and **Table 4** lists the contents of each well. Assays performed on

Table 3
Example of an I$_{50}$ Set Up

	Row			Final concentration of inhibitor in assay microtitre plate (μM)									
Inhibitor 1	A	Blank	Control	0.05	0.15	0.46	1.4	4.1	12	37	111	333	1000
Inhibitor 2	B	Blank	Control	0.05	0.15	0.46	1.4	4.1	12	37	111	333	1000
Column no.		1	2	3	4	5	6	7	8	9	10	11	12

Table 4
Composition of Blank, Control, and Test Wells for I_{50} Determination

Reagent	Blank	Control	Test
Buffer (pH 7.0)	100	50	—
Reporter substrate (e.g., nitrocefin)	50	50	50
Appropriately diluted β-lactamase	—	50	50
Diluted inhibitor	—	—	50
Total volume	150	150	150

    plate readers that only read in the visible range are restricted to using chromogenic substrates such as nitrocefin. Alternative β-lactam substrates can be used with plate readers that read in the UV range, such as the SPECTRAmax 250 plate reader.

8. When measuring I_{50} values without preincubation of enzyme and inhibitor, the enzyme solution should be added last.

With the appropriate software, this methodology enables 80 I_{50} values to be determined in 3–4 h using up to 20 times less reagents than standard spectrophotometric assays in 1-mL cuvets. The interpretation of I_{50} values and how their accuracy can be maximized is discussed in **Notes 6–9**.

3.2.8. Evaluation of Kinetic Constants for β-Lactamase Inhibitors

Many of the best known β-lactamase inhibitors such as clavulanic acid, sulbactam, β-bromopenicillanic acid, and the olivanic acids react with β-lactamases in an extremely complex way *(22,23)*. A full discussion of the methods used for the kinetic evaluation of such compounds is beyond the scope of this review. Such investigations may be warranted if inhibitor-resistant enzymes are being studied *(24,25)*. Essentially the following pieces of information should be ascertained:

1. The stoichiometry of inhibition; how much inhibitor is needed to inhibit a given amount of β-lactamase?
2. The stability of inhibition; does the enzyme activity return on removal of the inhibitor?
3. The rate of inhibition; how does the speed of inactivation relate to the inhibitor concentration?

The methods used are described in recent publications *(26,27)*, whereas a useful review is given by Matagne et al. *(28)*. Because inhibition of β-lactamases is complex, classical methods for finding K_i values and interpreting results should be used with caution. The authors have found Dixon plots (1/v against [I], **Fig. 5**) and the method of Cornish-Bowden ([S]/v against [I]) the most useful for

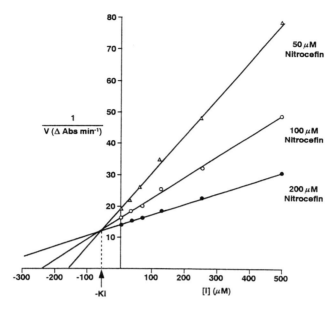

Fig. 5. Determination of K_i for a competitive inhibitor of a β-lactamase using the Dixon plot; nitrocefin is the substrate ($K_i = 62$ μM).

determining K_i values for more straightforward inhibitors *(10,29)*. If such plots indicate noncompetitive inhibition, further studies should be carried out, as many time-dependent inhibitors will give such a pattern. Classical procedures for analyzing results and determining K_i values should never be used for results obtained from assays carried out with preincubation.

4. Notes

1. Satellite bands associated with IEF of β-lactamases. Simpson and Plested *(30)* showed that β-lactamase satellite bands had similar biochemical characteristics to each other and to the main β-lactamase band. Arstila et al. *(31)* illustrated that some satellite bands arose as a consequence of enzyme preparation methodology, with fewer bands observed if the enzyme samples are prepared by osmotic shock. Matagne et al. *(32)* illustrated that satellite bands arose from the loss of different numbers of amino acids from the N-terminus of the enzyme (ragged ends). However, electrospray mass spectrometry can determine the mass of proteins with an accuracy of 99.99% *(33)*, and has been used to illustrate that samples of TEM-1 and P99 β-lactamases, which possessed no ragged ends produced a range of satellite bands, suggesting that these bands may result from more subtle changes to the β-lactamase *(34)*. Concomitant with this observation, it was discovered that a satellite band in the vicinity of p*I* 5.2 was identified when ceftazidime was added to 14 TEM-derived β-lactamases while other β-lactams

gave rise to satellite bands of different pI values. β-lactams were also shown to have an effect on the IEF focusing patterns of other β-lactamases *(35)*. In addition, these "β-lactam-induced" satellite bands could revert to the pI of the main β-lactamase band, illustrating that satellite bands may not always be permanent features of a β-lactamase sample. Another explanation for β-lactam induced satellite bands is that the β-lactam or other molecules in the growth medium bind to a proportion of the β-lactamase molecules, causing an alteration in the overall charge to mass ratio of the protein. This leads to a shift in pI and results in the appearance of a satellite band.

2. The precast gels may contain some unpolymerized acrylamide. This is neurotoxic and protective gloves must be worn throughout the procedures described in **Subheadings 3.1.2.** and **3.1.3.**

3. The CfiA-type metallo-β-lactamase is produced by the majority of imipenem-resistant clinical isolates of *Bacteroides fragilis*. As part of an investigation of six such isolates, it was found that the CfiA β-lactamases were only observed once the IEF gel had been overlaid with Solution III *(4)*. Therefore, when examining the β-lactamases produced by imipenem-resistant strains, it is advisable to stain the gels in the presence of 1 mM zinc sulfate to ensure identification of any metallo-β-lactamases that may have been depleted of zinc during the preparation process.

4. Limitations of substrate profiles. It is important to appreciate that rates of hydrolysis will depend on assay conditions and, ideally, substrate concentrations should be high enough to be approaching V$_{max}$. Where this is not possible because of substrate instability, substrate inhibition of β-lactamase activity, or simply shortage of materials, it may be necessary to use subsaturating β-lactam concentrations. Under these conditions, rates will be very sensitive to even small changes in substrate concentration. It is essential, therefore, that only true (i.e., linear) initial rates are measured. It is also very important that assay conditions are clearly defined, especially when comparisons are to be made with published data. Providing assays are carried out carefully and these points borne in mind, substrate profiles are usually sufficiently accurate to allow unambiguous identification of β-lactamases, especially when used in conjunction with inhibitor profiles and IEF. It is always advisable to check profiles against IEF data when using crude preparations, as some organisms may produce appreciable amounts of more than one enzyme.

   When β-lactamases have rather similar profiles and cannot be readily distinguished by inhibition characteristics or IEF, a more detailed kinetic analysis is called for, as described in **Subheadings 3.2.3.–3.2.6.** This is true for some of the extended-spectrum β-lactamases *(36,37)*. A full kinetic evaluation should always be carried out when a novel enzyme is suspected, particularly when trying to ascertain its role in resistance. For example, some years ago it was proposed that Class I β-lactamases caused resistance to 3rd generation cephalosporins by binding or "trapping" them *(38,39)*. Proper kinetic analysis showed that resistance was, in fact, due to β-lactamase hydrolysis *(40–42)*.

Table 5
Interpretation of I_{50} Values

Effect of preincubation of enzyme and inhibitor	Interpretation
No effect	Reversible inhibition[a]
	Transiently stable E-I complex with short $t_{1/2}$[a]
I_{50} decreased	Progressive inhibition where rate is fast[a, b]
I_{50} increased	Competitive substrate[a]
	Unstable inhibitor

[a]If the I_{50} is affected by changing the substrate concentration or type; then the inhibitor probably competes for the substrate binding site.
[b]If the rate is slow (i.e., half-life >> preincubation time) preincubation will have no effect.
(*See* also Reading and Farmer, **ref. 44**.)

5. Approaches for increasing throughput of substrate profiling. Most spectrophotometers now have at least six cuvet positions. This will enable a blank and five different enzyme preparations to be simultaneously assessed against the same β-lactam substrate. Some spectrophotometers, such as the Beckman DU7500 and Beckman 650, have attachments that enable 12 microcuvets to be assayed. These cuvets are in a microtiter plate spacing format to assist liquid handling. UV/visible microtiter plate readers (e.g., SPECTRAmax 250) significantly increase the number of enzyme preparations (up to 96) that can be assayed against a particular substrate.

6. Interpretation and limitations of I_{50} values. It is important to realize that I_{50} values are not kinetic constants, but relative parameters that depend on, and, therefore, vary with assay conditions. For this reason, their use has been criticized, but, experience has shown that not only are a wide variety of inhibition mechanisms encountered with β-lactamases, but also that inhibition can be complex *(43,44)*. Analysis of results for classical inhibition patterns (e.g., competitive, uncompetitive, and noncompetitive), can, therefore, be at best difficult, and at worst, misleading. Given these factors, the I_{50} assay is probably the most appropriate method for initial characterization of both β-lactamases and inhibitors.

7. Interpretation of results with and without preincubation. When assays are undertaken both with and without preincubation, useful preliminary information can be obtained, that can then be used to direct more detailed studies. Interpretation of results is summarized in the **Table 5**.

8. Effect of substrate concentration. As indicated above (*see* **Note 6**), the I_{50} will depend on both the substrate and its concentration if the inhibitor and substrate compete for the same binding site (generally the case for β-lactamase inhibitors). For a reversible competitive inhibitor, the effect of substrate is described by the following relationship:

$$I_{50} = K_i (1 + [S]/K_m) \text{ where } K_i = \text{inhibitor constant}$$

Similarly, for a progressive inhibitor binding at the active site, the rate of inactivation will be decreased by increasing the substrate concentration.

In practical terms, for a competitive inhibitor assayed with the substrate at K_m, the I_{50} will be $2 \times K_i$; when [S] is at $9 \times K_m$, the I_{50} will be $10 \times K_i$. Therefore, when assays are carried out at $<K_m$, I_{50} values will be little affected by substrate concentration. When materials are in short supply, assaying at low substrate concentrations may be the method of choice. If this is done, however, control (i.e., uninhibited) rates will soon become nonlinear. It is, therefore, essential that care is taken to measure only over the initial linear phase, otherwise misleading and poorly reproducible results will be obtained. As a rule of thumb, substrate hydrolysis should not exceed 10–20% of the total.

9. Use of multiples of K_m for a more standardized procedure. If $[S]>>K_m$, then:
   a. Control rates will be linear for a larger proportion of the reaction.
   b. I_{50} values will be more reproducible.
   c. More complex inhibition patterns (i.e., progressive and slow binding) will be more readily detected.

When sufficient materials are available, and when inhibitors are being compared against a number of β-lactamases, it is advisable to use the same substrate and multiple of its K_m for each of the respective enzymes (suggested concentration = $5 \times K_m$). The I_{50} will then be a strict function of the inhibitory properties of a particular compound against the β-lactamases, and will reflect its potency against the different enzymes. When using inhibitors to characterize and classify β-lactamases, it is essential that procedures are standardized to give reproducible results.

References

1. Payne, D. J., Blakemore, P. H., Drabu, Y., and Amyes, S. G. B. (1989) Comparison of the TEM-E3 and TEM-5 β-lactamases. *J. Antimicrob. Chemother.* **24**, 615–617.

2. Matthew, M., Harris, A. M., Marshall, M. J., and Ross, G. W. (1975) The use of analytical isoelectric focusing for the detection and identification of β-lactamases. *J. Gen. Microbiol.* **88**, 169–178.

3. Payne, D. J., Cramp, R., Bateson, J. H., Neale, J., and Knowles, D. (1994) Rapid identification of serine and metallo-β-lactamases. *Antimicrob. Agents Chemother.* **38**, 991–996.

4. Khushi, T., Payne, D. J., Fosberry, A., and Reading, C. (1996) Production of metal dependent β-lactamases by clinical strains of *B. fragilis* isolated before 1987. *J. Antimicrob. Chemother.* **37**, 345–350.

5. Benitez, M. J., Company, M., Arevalillo, A., and Jimenez, J. S. (1991) Comparative study of various hydrogen ion buffers to assay Zn^{2+}-dependent β-lactamases. *Antimicrob. Agents Chemother.* **35**, 1517–1519.

6. Samuni, A. (1975) A direct spectrophotometric assay and determination of Michaelis constants for the β-lactamase reaction. *Anal. Biochem.* **63**, 17–26.

7. Waley, S. G. (1974) A spectrophotometric assay of β-lactamase action on penicillins. *Biochem. J.* **139**, 789,790.

8. Felici, A. and Amicosante, G. (1995) Kinetic analysis of extension of sub-strate specificity with *Xanthomonas maltophilia, Aeromonas hydrophila* and *Bacillus cereus* metallo-β-lactamase. *Antimicrob. Agents Chemother.* **39,** 192–199.
9. Fersht, A. (1985) *Enzyme Structure and Mechanism,* W. H. Freeman and Company, New York.
10. Cornish-Bowden, A. (1995) *Fundamentals of Enzyme Kinetics,* Portland, London.
11. Henderson, P. J. F. (1993) in *Enzyme Assays: A Practical Approach,* (Eisenthal, R and Danson, M. J., eds.), IRL Press at Oxford University Press, Oxford, New York, pp. 277–316.
12. Segel, I. H. (1993) *Enzyme Kinetics,* John Wiley and Sons, London.
13. Buckwell, S. C., Page, M. I., and Longridge, J. L. (1988) Hydrolysis of 6-alkyl penicillins catalysed by β-lactamase I from *Bacillus cereus* and by hydroxide ion. *J. Chem. Soc. Perkin Trans.* **II,** 1809–1813.
14. De Meester, F., Joris, B., Reckinger, G., Bellefroid-Bourguignon, C., Frere, J.-M., and Waley, S. G. (1987) Automated analysis of enzyme inactivation phenom-ena; application to β-lactamases and DD peptidases. *Biochem. Pharmacol.* **36,** 2393–2403.
15. Fukagawa, Y., Takei, T., and Ishikura, T. (1980) Inhibition of β-lactamase of *Bacillus licheniformis* 749/C by compound PS-5: a new β-lactam antibiotic. *Biochem. J.* **185,** 177–188.
16. Leatherbarrow, R. J. (1990) Using linear and non-linear regression to fit biochemi-cal data. *TIBS* **15,** 455–458.
17. Eisenthal, R. and Cornish-Bowden, A. (1974) The direct linear plot. A new graphi-cal procedure for estimating enzyme kinetic parameters. *Biochem. J.* **139,** 715–720.
18. Beynon, R. J. *Curvefit.* IRL Press Software.
19. Leatherbarrow, R. J. (1992) *Grafit,* Erithacus Software.
20. Payne, D. J., Coleman, K., and Cramp, R. (1991) The automated in-vitro assess-ment of β-lactamase inhibitors. *J. Antimicrob. Chemother.* **28,** 773–780.
21. Payne, D. J., Cramp, R., Winstanley, D., and Knowles, D. J. (1994) Comparative activities of clavulanic acid, sulbactam, and tazobactam against clinically impor-tant β-lactamases. *Antimicrob. Agents Chemother.* **38,** 767–772.
22. Cartwright, S. J. and Waley, S. G. (1983) β-lactamase inhibitors. *Med. Res. Rev.* **3,** 341–382.
23. Knowles, J. R. (1983) Anti-β-lactamase agents, in *Antibiotics: Volume 6: Modes and Mechanisms of Microbial Growth Inhibitors,* (Hahn, F., ed.), Springer-Verlag, Berlin, Heidelberg, New York and Tokyo, pp. 90–107.
24. Bonomo, R. A., Currie-McCumber, C., and Shlaes, D. M. (1992) OHIO-1 β-lactamase resistant to mechanism-based inactivators. *FEMS Microbiol. Lett.* **92,** 79–82.
25. Henquell, C., Chanal, C., Sirot, D., Labia, R., and Sirot, J. (1995) Molecular characterisation of nine different types of mutants among 107 inhibitor-resis-tant TEM β-lactamases from clinical isolates of *E. coli. Antimicrob. Agents Chemother.* **39,** 427–430.

26. Farmer, T. H., Page, J. W. J, Payne, D. J., and Knowles, D. J. C. (1994) Kinetic and physical studies of β-lactamase inhibition by a novel penem, BRL 42715. *Biochem. J.* **303,** 825–830.

27. Matagne, A., Ledent, P., Monnaie, D., Felici, A., Jamin, M., Raquet, X., Galleni, M., Klein, D., Francois, I., and Frere, J.-M. (1995) Kinetic study of interaction between BRL 42715, β-lactamases and D-alanyl-D-alanine peptidases. *Antimicrob. Agents Chemother.* **39,** 227–231.

28. Matagne, A., Ghuysen, M.-F., and Frere, J.-M. (1993) Interactions between active site serine β-lactamases and mechanism based inactivators: a kinetic study and overview. *Biochem. J.* **295,** 705–711.

29. Dixon, M. (1953) The determination of enzyme inhibitor constants. *Biochem. J.* **55,** 170–171.

30. Simpson, I. N. and Plested, S. J. (1983) The origin and properties of β-lactamase satellite bands seen in isoelectric focusing. *J. Antimicrob. Chemother.* **12,** 127–131.

31. Arstila, T., Jacoby, G. A., and Huovinen, P. (1993) Evaluation of five different methods to prepare bacterial extracts for the identification of β-lactamases by isoelectric focusing. *J. Antimicrob. Chemother.* **32,** 809–816.

32. Matagne, A., Joris, B., Van Beeuman, J., and Frere, J.-M. (1991) Ragged N-termini and other variants of Class A β-lactamases analysed by chromatofocusing. *Biochem. J.* **273,** 503–510.

33. Smith, R. G., Loo, J. A., Edmonds, C. G., Barinaga, C. J., and Udseth, H. R. (1990) New developments in biochemical mass spectrometry: electrospray ionization. *Anal. Chem.* **62,** 882–899.

34. Payne, D. J., Skett, P., Aplin, R. T., Robinson, C. and Knowles, D. (1994) β-lactamase ragged ends detected by electrospray mass spectrometry correlates poorly with multiple banding on isoelectric focusing. *J. Biol. Mass Spectrometry* **23,** 159–164.

35. Payne, D. J. and Amyes, S. G. B. (1994) The effects of β-lactams on the isoelectric focusing of β-lactamases. *J. App. Bacteriol.* **76,** 500–505.

36. Bush, K., Jacoby, G. A., and Medeiros, A. A. (1995) A functional classification scheme for β-lactamases and its correlation with molecular structure. *Antimicrob. Agents Chemother.* **39,** 1211–1233.

37. Philippon, A., Labia, R., and Jacoby, G. (1989) Extended-spectrum β-lactamases. *Antimicrob. Agents Chemother.* **33,** 1131–1136.

38. Sanders, C. C. (1984) Inducible β-lactamases and non-hydrolytic resistance mechanisms. *J. Antimicrob. Chemother.* **13,** 1–3.

39. Then, R. L. and Angehrn P. (1982) Trapping of nonhydrolyzable cephalosporins by cephalosporinases in *Enterobacter cloacae* and *Pseudomonas aeruginosa* as a possible resistance mechanism. *Antimicrob. Agents Chemother.* **21,** 711–717.

40. Farmer, T. H. and Reading, C. (unpublished results)

41. Livermore, D. M., Riddle, S. J., and Davy, K. W. M. (1986) Hydrolytic model for cefotaxime and ceftriaxone resistance in β-lactamase-derepressed *Enterobacter cloacae. J. Infect. Dis.* **153,** 619–622.

42. Vu, H. and Nikaido, H. (1985) Role of β-lactam hydrolysis in the mechanism of resistance of a β-lactamase-constitutive *Enterobacter cloacae* strain to expanded-spectrum β-lactams. *Antimicrob. Agents Chemother.* **27,** 393–398.

43. Bush, K. and Sykes, R. B. (1983) β-lactamase inhibitors in perspective. *J. Antimicrob. Chemother.* **11,** 97–107.

44. Reading, C. and Farmer, T. H. (1983) The biochemical evaluation of β-lactamase inhibitors, in *Antibiotics: Assessment of Antimicrobial Activity and Resistance,* (Denver-Russell, A. and Quesnel, L. B., eds.), Academic, New York, pp. 141–159.

β-Lactam Resistance Mediated
by Changes in Penicillin-Binding Proteins

Christopher G. Dowson and Tracey J. Coffey

1. Introduction
1.1. Mechanism of Resistance

The widespread use, or perhaps overuse, of penicillin during the past 50 yr has driven the evolution of resistance to penicillin in numerous different species of bacteria. Typically, resistance has arisen as a result of the acquisition of β-lactamases that inactivate the antibiotic (*see* Chapter 25). Alternatively, in some Gram-negative bacteria, resistance may have arisen by a reduction in the ability of the antibiotic to access its target. However, in a number of clinically important Gram-negative and Gram-positive bacteria, resistance has arisen by alteration of the targets for penicillin and other β-lactam antibiotics, namely, the penicillin-binding proteins (PBPs).

PBPs are involved in the final stages of peptidoglycan biosynthesis and each PBP possesses a penicillin-sensitive transpeptidase domain that catalyses the crosslinking of peptides within the cell wall. Penicillin and other β-lactam antibiotics inhibit these enzymes by acting as structural analogs of these cell wall peptides, and forming an essentially stable acyl-enzyme complex, rather than the transient complex that occurs during the normal substrate-enzyme reaction. Inhibition of PBPs that are essential for growth results in cell death. PBP-mediated resistance occurs either when the resident PBPs are altered to forms where they have a substantially reduced affinity for penicillin or other β-lactams, as in the evolution of penicillin-resistant strains of *Streptococcus pneumoniae*, viridans streptococci, *Haemophilus* spp., and *Neisseria* spp., or when these essential PBPs are made redundant by the acquisition of a novel, low-affinity PBP, such as PBP2' in the evolution of methicillin-resistant

From: *Methods in Molecular Medicine, Vol. 15: Molecular Bacteriology: Protocols and Clinical Applications*
Edited by: N. Woodford and A. P. Johnson © Humana Press Inc., Totowa, NJ

Staphylococcus aureus (MRSA). This chapter will focus upon the mechanism of PBP alterations and the detection of altered PBPs in clinical isolates of the naturally transformable species of *Streptococcus* and *Neisseria.*

1.2. Evolution of Resistance in S. pneumoniae and the Pathogenic Neisseria spp.

To date, one of the largest clinical impacts resulting from PBP-mediated resistance has occurred in clinical isolates of *S. pneumoniae.* Penicillin resistance in *S. pneumoniae* was first detected in the late 1960s *(1)* and has now spread worldwide *(2).* PBP-mediated resistance in pneumococci can result in up to a 2000-fold increase in resistance to penicillin or cephalosporins. Although PBP-mediated resistance has also arisen in the pathogenic members of *Haemophilus* and *Neisseria,* the role of PBPs in resistance is in part augmented by alterations in the access of the antibiotics to the PBPs.

In the late 1980s, it was found that low affinity PBPs in penicillin-resistant clinical isolates of *S. pneumoniae, N. gonorrhoeae,* and *N. meningitidis* had evolved as the result of interspecies gene transfer *(3,4)* rather than as a result of point mutation. Genes encoding low-affinity PBPs had been acquired by one species from another, presumably by transformation and homologous recombination.

Altered PBP genes from penicillin-resistant strains were found to be "mosaics," composed of blocks of nucleotides derived from the recipient and at least one other species. In the case of some penicillin-resistant strains of *N. meningitidis,* it is quite clear that alterations are as a result of the acquisition of *pbp 2* genes from *N. flavescens (4).* For altered *pbp2b* and *pbp2x* genes of *S. pneumoniae,* it has been found that the oral streptococci *S. mitis* and *S. oralis* have donated the homologous DNA *(5,6),* and that these homologs, as with *N. flavescens,* differ in nucleotide sequence from the susceptible recipient by around 20%. Mosaic PBP genes from resistant isolates are, therefore, substantially different from those found in typical penicillin-susceptible isolates, which, for *S. pneumoniae* and *Neisseria* spp., are very uniform within each species and generally differ by less than 1% in nucleotide sequence *(3,4).* Using restriction fragment length polymorphism (RFLP) analysis it is, therefore, possible both to distinguish mosaic PBP genes from typical PBP genes, and to compare different mosaics that have arisen as independent recombinational events.

As penicillin resistance continues to spread, and PBP genes continue to move within species and from one species to another, PBP gene fingerprints can be used in helping to identify specific resistant clones or to show evidence of horizontal transfer of PBP genes from one strain to another *(7,8).*

2. Diagnosis

2.1. PBP Genes Involved in Resistance

The definition of bacterial resistance is relative and frequently arbitrary. The absolute criterion of resistance is failure of the clinical response to therapy, although such failure often is difficult to define given the many variables affecting clinical outcome. Few clinical studies of pneumococcal disease have attempted to correlate measures of in vitro susceptibility with clinical responsiveness to penicillin. With the appearance of resistant pneumococci, it has become necessary to test not only for penicillin susceptibility, but also to ascertain the susceptibility patterns for alternative agents. A truly penicillin-susceptible strain of *S. pneumoniae* has a MIC of penicillin of ≤ 0.016 µg/mL, although strains with MICs of ≤ 0.06 µg/mL are regarded as susceptible. Reviews of penicillin-resistant pneumococci *(2,9)* classify resistance to penicillin into two distinct categories; intermediate-level resistance (MICs of 0.1–1.0 µg/mL), or high-level resistance (MICs of ≥ 2.0 µg/mL). However, in laboratory studies, it is useful to distinguish a further category, low-level penicillin resistance, which includes those isolates that are not truly susceptible, exhibiting MICs greater than 0.016 µg/mL, but that do not fall into the penicillin-resistant categories.

A general correlation between increased MICs of penicillin and increased MICs of cephalosporins and other β-lactam antibiotics has been found *(10)*. However, exceptions can occur in this general picture of cross resistance and it is, therefore, important to be clear about the underlying basis of PBP-mediated resistance to different β-lactams. Low-, intermediate-, or high-level resistance to β-lactams in *S. pneumoniae* can arise by the alteration of one or more PBPs (1A, 2X, and 2B) *(11)*. Different combinations of alterations to one or more of these three PBPs can result in resistance to one β-lactam, but not another. For example, a 1000-fold increase in resistance to third-generation cephalosporins can be obtained by alterations to PBPs 1A and 2X, yet this increases resistance to benzylpenicillin less than tenfold to 0.06 µg/mL, a level that is regarded clinically as penicillin-susceptible *(12)*. Likewise, the acquisition of a low-affinity PBP 2X alone may confer resistance to isoxazolylpenicillins, such as oxacillin (MIC 0.5 to 1.0 µg/mL), but not penicillin *(13)*. However, the further acquisition of a low affinity PBP2B would not only increase oxacillin resistance two- to eightfold, but also penicillin resistance from 0.03–0.06 µg/mL to 0.25 µg/mL. High-level penicillin resistance in clinical isolates of *S. pneumoniae* requires alterations to each of PBPs 1A, 2X, and 2B *(11)*.

In PBP-mediated penicillin resistance in *N. gonorrhoeae*, PBPs 1 and 2 have been found to have a reduced affinity for penicillin *(14,15)*, and among the few reports of PBP-mediated resistance in *N. meningitidis,* resistance is the result, at least in part, of an altered PBP2 *(16,17)*. Typically, meningococci have penicllin MICs of <0.04 μg/mL. However, acquisition of a low-affinity PBP2 from the naturally more penicillin-resistant *N. flavescens* can result in a ten-fold increase in resistance.

2.2. Detection of Resistance to Penicillin and Other β-Lactams in S. pneumoniae and Pathogenic Neisseria spp.

The upward trend of β-lactam resistance in *S. pneumoniae* and, to a lesser degree, the pathogenic *Neisseria* emphasizes the need for a reliable and convenient method for their detection.

For the routine testing of penicillin susceptibility of pneumococci, the use of 1 μg oxacillin discs is thought to be a reliable method, and is currently recommended by the National Committee for Clinical Laboratory Standards *(18)* and the Working Party on Antibiotic Sensitivity Testing of the British Society for Antimicrobial Chemotherapy *(19)*. This method does not give a truly quantitative susceptibility determination, and it is difficult to distinguish between the transitions from low-level penicillin resistance, to intermediate-level resistance, and to high-level resistance. Such distinctions are particularly important in meningeal infections with pneumococci, as the failure of penicillin therapy in cases of meningitis caused by pneumococci with intermediate-level resistance has been reported *(20)*. In addition, the recommended therapy for pneumococcal infections differs for infections caused by highly penicillin-resistant strains and those caused by intermediately-resistant strains *(20,21)*. All pneumococcal isolates that show zones of inhibition around an oxacillin disk of less than 19 mm must be further investigated to ascertain their degree of resistance *(22)*. The importance of this recommendation is highlighted by the recent finding that some oxacillin-resistant pneumococci remain susceptible to penicillin *(13)*. The further investigation of the degree of resistance to penicillin in such cases may alleviate the need for an alternative to penicillin in therapy. The E-test (AB Biodisk, Solna, Sweden) is thought to be one of the most accurate of the commercial tests that are available *(23)*, and a reliable alternative to agar or broth dilution methods for the determination of penicillin, cefotaxime, and ceftriaxone MICs for *S. pneumoniae* *(23,24)*.

Classification for penicillin resistance in *Neisseria*, like pneumococci, falls into three categories. Fully susceptible isolates have MICs for penicillin ≤0.06 μg/mL, MICs for moderately susceptible isolates are 2- to 20-fold higher than those for fully susceptible ones, and isolates with MICs of penicillin >1 μg/mL are considered to be resistant. The laboratory detection of moderately penicillin-susceptible isolates by currently recommended methods is unreliable and results in false-positives *(25)*. In testing levels of penicillin resistance in *N. meningitidis* isolates, the use of the 1 μg oxacillin disk has been shown to discriminate between fully susceptible and moderately suscept- ible ones *(26)*. Such a distinction is clinically important as the failure of penicillin treatment in meningococcal meningitis caused by a moderately susceptible men- ingococcus has been reported *(27)*. A comparison of the E-test with agar dilution for the antimicrobial susceptibility testing of *N. gonorrhoeae* and *N. meningitidis* also found no significant difference between the methods *(28,29)*.

2.3. Detection of PBPs and their Affinity for β-Lactams

PBPs can be detected by incubating whole cells or membrane preparations either directly with radiolabeled penicillin, or in competition with unlabeled compounds, then resolving the proteins by sodium dodecyl sulfate-gel electro- phoresis (SDS-PAGE) and visualizing them by fluorography *(30)*. Fluores- cein-coupled penicillins are now available that may, in some instances, supercede the use of radiolabeled compounds as this new method appears to be both more rapid and more sensitive than classical fluorography *(31)*. PBPs dif- fer in their natural affinity for different β-lactams; those with the highest affin- ity are inactivated by the lowest concentration of antibiotic. These PBPs may reduce their affinities for a β-lactam by 2000-fold. This obviously presents problems when trying to visualize both high-affinity and low-affinity PBPs within a particular strain. Usually PBP affinity gels use a single concentration of radiolabeled penicillin and report either the presence or absence of PBPs, depending upon whether they have either a relatively high or low affinity for the compound used. Reproducibility between laboratories requires a strict adherence to protocol as even a change in the manufacturer of the SDS used for PAGE can alter the final fluorogram. Apart from this, altered low-affinity PBPs can migrate either fast or slow to give an apparently altered molecular weight when compared with their high-affinity forms *(32)*. Nevertheless, PBP profiles have been used to help identify penicillin-resistant clones of *S. pneumoniae* *(7,33)*. The methodology associated with PBP gels has been extensively reported elsewhere *(34)*, and although this method may be important in the initial identification of low-affinity PBPs, restriction fragment length polymor- phism (RFLP) analysis of PBP genes is a simpler, and more informative method, that can be readily used to identify and track the presence of a particu- lar altered PBP in epidemiologic analyses of penicillin-resistant isolates with mosaic PBP genes. High resolution PBP fingerprinting has recently been used very successfully in a number of studies and is the main focus of methodology relating to this chapter (*see* **Subheading 6.**).

2.3.1. RFLP Analysis of Mosaic PBP Genes

The method of gene fingerprinting or RFLP analysis of amplified DNA pro- vides a convenient high resolution method for the examination of the genetic diversity within a particular locus. It can be applied in the analysis of *pbp* genes from both *Neisseria* spp. and *S. pneumoniae* where it is able to distinguish between both susceptible and resistant altered forms of the genes and also between different mosaic structures of resistant forms. Ideally this involves the use of a single restriction enzyme that cuts frequently along the entire length of the gene. In some instances it is necessary to use a combination of restriction enzymes to obtain appropriately-sized fragments for resolution within a single gel.

Several studies of β-lactam-resistant *Neisseria* spp. and pneumococci have used this technique in the analysis of their epidemiology *(7,8,35–38)*. A study of *pbp 2* genes of penicillin-resistant meningococci *(35)* showed that a high resolution of allelic variation was achieved by the use of a single restriction enzyme. For pneumococci, however, a comparable degree of resolution for the analysis of the *pbp* genes is best achieved by the use of two separate digests.

3. Epidemiologic Investigation: *Bacterial Population Structure*

Before embarking upon an epidemiologic analysis of penicillin resistance in *S. pneumoniae* or *Neisseria* spp., it is important to appreciate something of the population structure of these organisms as this will help in the interpretation of results obtained from PBP analysis, whether by affinity SDS-PAGE gels or by RFLPs.

Early work, examining the population structure of *Escherichia coli (39)* and *Salmonella (40)*, indicated that bacterial populations were essentially composed of independently evolving clones, each clone being derived from a common ancestor and with little recombination occurring between clones. However, more recently it has been found that for the naturally transformable *Neisseria* spp. *(41)* and for *S. pneumoniae (42)* there appear to be substantial levels of recombination within species. This results in populations that should be pictured as being composed of individuals that are linked by a mesh of interconnections as opposed to a bifurcating tree (*see* Chapter 14). In *N. gonorrhoeae*, the level of recombination is such that of nine loci examined by their relative mobility during starch gel electrophoresis (multilocus enzyme electrophoresis: MLEE), none were linked to each other. This is in contrast to *E. coli* where the electrophoretic mobilities of all of 20 loci were identical in strains isolated widely in space and time *(43)*.

N. gonorrhoeae is extremely recombinogenic. However, in other species, such as *N. meningitidis* and *S. pneumoniae*, recombination appears to be less frequent, and the population might be better described as epidemic, where occasionally from the recombining masses a successful clone arises and spreads—perhaps because of a recently acquired trait such as antibiotic resistance or a novel virulence determinant. The ET-V complex of *N. meningitidis* (*see* Chapter 14) or the multiply resistant serotype 23F clone of *S. pneumoniae* (*see* Chapter 9) are probable examples of this *(44,45)*. Although prolific, neither the ET-V complex nor the multiply resistant serotype 23F clone possess the inherent stability of truly clonal species, such as *E. coli* or *Salmonella*. In both instances, within a decade or so of their emergence, these two epidemic clones started to lose their integrity following recombination with other strains that introduced new traits, and changed those markers originally used to identify each clone.

In species where recombination is common, one has to be careful in attributing inappropriate epidemiologic weight to traditional markers such as serotype, auxotype, or antibiotic resistance when trying to determine whether resistance has spread clonally or by horizontal gene transfer. This distinction can be achieved using a combination of methods such as MLEE, ribotyping (*see* Chapter 2), and gene fingerprinting or partial gene sequencing that are able to index either the overall genetic relatedness between isolates or the relatedness of their resistance determinants. Resistant pneumococci, that are not closely related, can possess identical altered *pbp* genes and isolates that are apparently genetically identical and can possess different *pbp* genes *(7,8,36,38)*.

4. Future Prospects and Goals
4.1. PCR Detection of Penicillin Resistance in Pneumococci

In severe pneumococcal infections, it is important to determine as soon as possible if the infecting strain is resistant to penicillin or third-generation cephalosporins to aid choice of antibiotic therapy. Attention has, therefore, turned to finding a rapid method for detecting such resistance. A recent study by Ubukata et al. *(46)* used polymerase chain reaction (PCR) to identify penicillin resistance in clinical isolates of *S. pneumoniae*. Three sets of primers were used, two of which were designed to amplify previously identified altered blocks within the *pbp2b* gene *(3)*. The third primer set was designed for the amplification of part of the *pbp2b* gene of penicillin-susceptible pneumococci, *(7)*. The third primer set produced an amplification product from 98.9% of the isolates with a MIC for penicillin of <0.06 μg/mL. The drawback of this technique is that the primers were designed to amplify known altered blocks from previously sequenced *pbp2b* genes. As shown in the study by Ubukata et al. *(46)*, the primers designed gave no amplification product with 27.9% of the penicillin-resistant pneumococci studied. As penicillin resistance is thought to have arisen on numerous occasions around the world by recombination with several different species of streptococci, the existence of *pbp* genes dissimilar to those previously reported is to be expected. Further characterization of alterations to PBPs from resistant pneumococci has to be undertaken if PCR detection of PBP-mediated resistance is to be more successful than previously reported. As *pbp* genes from penicillin-susceptible strains of *S. pneumoniae* and *Neisseria* spp. are highly uniform, it may be more useful to determine susceptibility, or lack of susceptibility, rather than resistance.

5. Materials

5.1. Growth of Bacteria and Preparation of Chromosomal DNA

1. Brain heart infusion (BHI) agar (per 100 mL). 2 g agar, 3.7 g BHI (Difco, Detroit, MI) in H_2O; autoclave at 15 psi for 15 min.
2. BHI blood agar: as above, but add 1 mL of defibrinated sheep blood per 20 mL molten agar (cooled to 55°C) immediately prior to pouring.
3. GC agar base (per 100 mL). 3.7 g GC agar base (Oxoid, Basingstoke, UK) in H_2O; autoclave at 15 psi for 15 min.
4. GC supplements (per 100 mL). 40 g glucose, 0.5 g L-glutamine, 0.05 g ferric nitrate, 1 mL 0.2% thiamine pyrophosphate. Add 1 mL of supplements per 100 mL GC agar.
5. 50 mM Tris-HCl, 20 mM EDTA (per L). 6.055 g Tris-HCl, 7.5 g EDTA, pH 7.5.
6. Sodium deoxycholate. 5% (w/v).
7. Proteinase K. 2.5 mg/mL.
8. Phenol:chloroform:isoamyl alcohol (25:24:1).
9. 3 M ammonium acetate.
10. 100% ethanol.
11. Lysozyme. 10 mg/mL.
12. Triton X. 2% in 50 mM Tris-HCl, 20 mM EDTA.
13. RNase. 0.5 mg/mL.
14. TE buffer. 10 mM Tris-HCl, 1 mM EDTA, pH 8.0.

5.2. Amplification of pbp Genes and PBP Gene Fingerprinting

1. PCR buffer (5X). 250 mM KCl, 50 mM Tris-HCl, pH 8.5, 12.5 mM $MgCl_2$, 1 mg/mL gelatin, 1 mM dATP, 1 mM dCTP, 1 mM dGTP, 1 mM dTT
2. Labeling buffer (2X): 11 mM Tris-HCl, pH 7.4, 11 mM $MgCl_2$, 6.5 mM dithiothreitol (DTT), 0.1 mM dATP, 0.1 mM dGTP, 0.1 mM dCTP or dTTP (add CTP to buffer when labeling with α-[^{32}PdTTP] or dTTP when labeling with α-[^{32}PdCTP].
3. TE buffer: 10 mM Tris-HCl, 1 mM EDTA, pH 8.0.
4. Loading buffer: 0.6 g/mL sucrose, 0.375 mg/mL bromophenol blue in TE buffer.
5. Ethidium bromide (EB). 10 mg/mL.
6. Formamide loading buffer. 98% formamide, 1 mg/mL bromophenol blue, 1 mg/mL xylene cyanol.
7. TBE buffer (per L). 121.1 g Tris, 61.8 g boric acid, 7.44 g EDTA; autoclave at 15 psi for 15 min.
8. 6% (or 8%) nondenaturing polyacrylamide gel (per 100 mL). 15 mL (or 20 mL) of 40% acrylamide, 1 mL of 10% ammonium persulfate, 100 μL TEMED in TBE buffer.

6. Methods

6.1. Growth of Bacteria and Preparation of Chromosomal DNA

6.1.1. Streptococcus pneumoniae

1. Grow pneumococci on BHI agar plus catalase (1000 U/mL) at 37°C in a 5% CO_2 atmosphere. Growth on BHI blood agar can sometimes results in poor lysis,

especially following incubation for more than 24 h. However, genomic DNA can be routinely prepared from cells grown on BHI blood plates.

2. Resuspend the bacterial growth from a single BHI/catalase plate in 400 μL of 50 mM Tris-HCl20 mM EDTA (pH 7.5) and lyse the cells by the addition of 80 μL of sodium deoxycholate and 100 μg of proteinase K.
3. Following lysis, treat the sample with 480 μL (an equal volume) of phenol: chloroform:isoamyl alcohol.
4. Precipitate the DNA with 1000 μL ethanol plus 100 μL 3 M ammonium acetate and dry the pellet. Resuspend in 50 μL of water.

6.1.2. Neisseria *spp.*

1. Meningococci and gonococci are routinely grown on gonococcal (GC) agar base plus GC supplements at 37°C in a 5% CO_2 atmosphere.
2. Harvest the bacterial growth from a single agar plate using sterile cotton wool swabs in 1.5 mL of 50 mM Tris-HCl, 20 mM EDTA (pH 7.5), and pellet the cells in a microfuge at 2400g for 15 min.
3. Resuspend the pellet in 250 μL of 50 mM Tris-HCl, 20 mM EDTA, and lyse the cells by the adding 50 μL of freshly prepared lysozyme solution and incubating at room temperature for 15 min.
4. Add an equal volume of 2% Triton X to complete lysis.
5. Mix the DNA suspension and add 10 μL of proteinase K.
6. After 30 min at room temperature (during which time occasionally mix by inversion), add 400 μL of phenol:chloroform:isoamyl alcohol (25:24:1).
7. Mix the lysate and phenol by inversion, vortex for 30 s and centrifuge at 16000g for 5 min.
8. Extract the aqueous layer to a fresh tube, treat the lysate with 1 μL of RNase, and incubate at 37°C for 10 min.
9. Re-extract the lysate with 400 μL of the phenol mix, repeat **step 7** and remove the aqueous upper phase to a clean tube.
10. Precipitate the DNA with two volumes of ethanol, dry the pellet and resuspend it in 100 μL of TE buffer.

6.2. Amplification of pbp *Genes*

6.2.1. Streptococcus pneumoniae

For pneumococci, three of the five high molecular weight *pbp* genes, *pbp2b* **(47)**, *pbp2x* **(48)**, and *pbp1a* **(6)** are routinely amplified by PCR. The 1.5-kb amplified region of the *pbp2b* gene includes the penicillin-binding transpeptidase domain. The 2-kb amplified region of the *pbp2x* gene includes all but the first 80 bp of the coding region. For the *pbp1a* gene, the entire coding region is amplified resulting in a 2.1-kb fragment. The primers used in the amplification of these genes are shown in **Table 1**.

1. Make a 100 μL reaction mixture of 1X PCR buffer containing 2 μL of chromosomal DNA and 500 ng of each primer.

Table 1
Primers Used to Amplify *pbp* Genes
from *N. meningitidis*, *N. gonorrhoeae*, and *S. pneumoniae*

Species	Primer	Sequence (5'-3')[a]
Neisseria spp.	GC11	GCC TGT GTG CCG GAA TCG
	GCdown 3	TCG TGA ATT CGG GGA TAT AAC TGC GGC CGT C
S. pneumoniae	*pbp2b* up	CTA CGG ATC CTC TAA ATG ATT CTC AGG TGG
	pbp2b down	CAA TTA GCT TAG CAA TAG GTG TTG G
	pbp2x up	CGT GGG ACT ATT TAT GAC CGA AAT GG
	pbp 2x down	AAT TCC AGC ACT GAT GGA AAT AAA CAT ATT A
	pbp1a up	GGT AAA ACA TGA AYA ARC C
	pbp1a down	TGG ATG ATA AAT GTT ATG GTT G

[a]IUB codes: Y = C or T; R = A or G

2. The PCR program typically consists of an initial denaturation step at 95°C for 5–10 min, followed by cooling (1°C/10 s) down to the annealing temperature (usually 58°C for the *pbp2b* gene and 55°C for the *pbp2x* and *pbp1a* genes).
3. The reaction is held at the annealing temperature for 2 min (for a "hot start" reaction), during which *AmpliTaq* (Perkin Elmer, Norwalk, CT; 2 U) is added (unless the polymerase was added prior to the start of the PCR program, e.g., if using *Amplitaq* Gold). An initial extension at 72°C for 2 min precedes the start of the cycling step.
4. Amplification is then carried out over 25 cycles using a denaturation step (95°C) for 1 min, annealing for 2 min, and extension (72°C) for 30–60 s/kb. A final extension at 72°C for 5 min follows the cycling step and the samples are then cooled to 4°C.
5. A 5 μL aliquot is applied to a 0.7% agarose gel together with a known size marker. This allows a rough quantitation of the product to be made.

6.2.2. Neisseria *spp.*

Amplification of the *pbp 2* gene from gonococci and meningococci uses a reaction mix very similar to that of pneumococci. However, for these *Neisseria* spp. it is possible to use the crude cell lysate directly in PCR reactions, a 2 μL aliquot of which is added to a mixture containing 1 μ*M* of primers GC11 and GCdown3 (*see* **Table 1**). The amplification parameters are identical to those used with pneumococci except that annealing is routinely carried out at 52°C.

6.3. PBP Gene Fingerprinting (see Notes 1–3)

6.3.1. Radioactive Labeling

1. Purify the amplified fragments (gel purified if necessary) with GeneClean (BIO 101, La Jolla, CA) and resuspended at 20 ng/μL in water. Alternatively, aliquots can be taken directly from "clean" PCR.

2. Digest 20 ng of PCR product with suitable restriction enzymes (*Sty*I and *Hin*fI for the *pbp2b* gene, *Hin*fI and *Mse*I + *Dde*I for the *pbp2x* and *pbp1a* genes, and *Hpa*II for the *pbp2* genes from *Neisseria* spp.) in a final volume of 20 µL.

3. Label 2.5 µL of each digest with a suitable α-[^{32}P] deoxynucleoside triphosphate using Klenow DNA polymerase with a mixture of "hot" and "cold" dNTPs depending upon the nucleotide sequence of the "sticky" end generated by each restriction enzyme; make a master mix of labeling buffer containing 10 µCi/sample of a suitable α-[^{32}P] deoxynucleoside triphosphate (use α-[^{32}dCTP] for *Sty*I and *Hpa*II digests and α-[^{32}dTTP] for *Hin*fI and *Dde*I +*Mse*I digests) and 0.4 U/sample of Klenow polymerase.

4. Add 2.5 µL of this labeling mix to 2.5 µL aliquots of the digests.

5. Add 1 µL of a loading buffer to each sample prior to loading on a 6% non-denaturing polyacrylamide gel (30 × 40 cm). Molecular size markers are run alongside (e.g., pBR322 digested with *Hpa*II and end-labeled as described above using α-[^{32}dCTP]).

6. Run the gels at a constant voltage of 600 V for 30 min and then alter to a constant current of 35 mA for the remainder of the run. Electrophoresis is continued until the dye front reaches 10 cm from the bottom of the gel.

7. Following electrophoresis, dry the gel under vacuum for 60 min.

8. Visualize the gene fingerprints by autoradiography. **Figure 1** shows an example of the resulting fingerprints following digestion of *pbp* genes from a penicillin-susceptible isolate and three penicillin-resistant isolates of *S. pneumoniae*.

9. Fingerprint patterns on the X-ray films can be digitized and compared by using a Summasketch II Plus digitizer tablet (Summagraphics, Seymour, CT) and Molmatch software (Ultra-Violet Products, Cambridge, UK) or one of a number of other commercially available gel analysis packages.

6.3.2. Ethidium Bromide Staining

A nonradioactive alternative to this technique has been used more recently.

1. Digest 200 ng of "clean" or purified PCR product in a total volume of 15 µL.

2. Following digestion, add formamide loading buffer to each sample, apply to a 8% nondenaturing polyacrylamide gel (16 × 18 cm) and run on a water cooled Hoeffer SE 600 series electrophoresis unit.

3. Following electrophoresis, stain the gel in a 1000 mL bath of TBE buffer containing 100 µL of EB (final concentration 1 µg/mL) for 20 min, and the gene fingerprint is then visualized on a UV transilluminator. **Figure 2** shows an example of *pbp* gene fingerprints visualized following EB staining.

7. Notes

Having used both radioactive labeling and EB techniques (*see* **Subheadings 6.3.1.** and **6.3.2.**) in the analysis and comparison of *pbp* gene fingerprints, it is worthwhile to highlight the benefits and drawbacks of both to the reader.

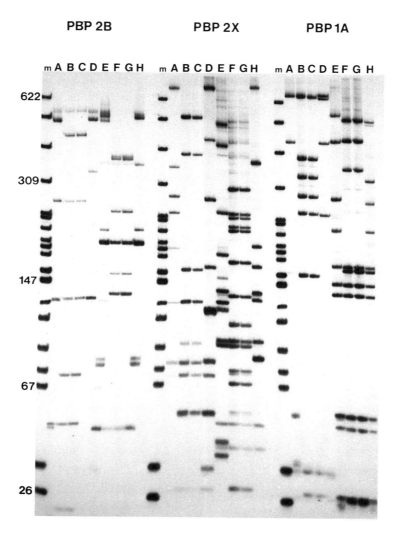

Fig. 1. *pbp* gene fingerprints of *S. pneumoniae* visualized by autoradiography. *pbp2b* gene fingerprints are shown following digestion with *Sty*I and *Hin*fI. A penicillin-susceptible isolate (lane A and E, respectively), two members of the multiply-resistant serotype 23F clone (lanes B, C, and F, G), and a third unrelated penicillin-resistant isolate (lanes D and H). The *pbp2x* and *pbp1a* gene fingerprints for the same strains are shown in lanes A, B, C, D, following digestion with *Hin*fI, and lanes E, F, G, H, following digestion with *Mse*I + *Dde*I. pBR322 DNA digested with *Hpa*II was used as a marker, the size of some of some of the resulting bands are shown in bp.

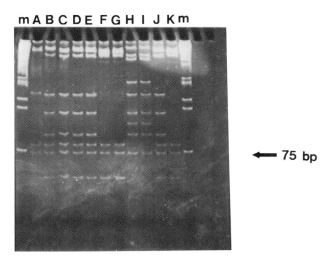

mA B C DE F G H I J Km

← 75 bp

Fig. 2. *pbp2x* gene fingerprints of *S. pneumoniae* visualized following EB staining. Lanes A–K are the fingerprints resulting from the *Hin*fI digestion of the *pbp2x* gene from penicillin-resistant pneumonococci insolated in Poland. The outside lanes are Kb-ladder (Gibco-BRL, Gaithersburg, MD) included as a marker.

1. The EB staining method has several advantages over the ^{32}P method. There is the obvious avoidance of using ^{32}P, in terms of both hazard and cost. The smaller gels used in this method are easier to assemble and run, and the results are visualized much quicker.
2. The ^{32}P method, however, yields a far greater resolution than the EB staining method, which is important when fragment sizes within one sample can range from twenty to several hundred base pairs.
3. Whichever method the reader may choose, it is important to keep the gels cool, particularly the larger ones, to avoid separation of the strands of the smaller fragments on heating. The inclusion of a size standard when running gels does allow a size approximation to be made, and enables the comparison of the gene fingerprints from separate gels, either manually or using commercially available software (*see* Chapter 3).

Acknowledgments

Basic research leading up to the authors' general understanding of PBP-mediated alterations in the evolution of β-lactam resistance has been funded by the Lister Institute, Royal Society, Medical Research Council and the Wellcome Trust to CGD and the Wellcome Trust to TJC and Brian Spratt.

References

1. Hansman, D. and. Bullen, M. M. (1967) A resistant pneumococcus. *Lancet* **ii,** 179–184.
2. Klugman, K. P. (1990) Pneumococcal resistance to antibiotics. *Clin. Microbiol. Rev.* **3,** 171–96.
3. Dowson, C. G., Hutchison, A., Brannigan, J. A., George, R. C., Hansman, D., Liñares, J., Tomasz, A., Smith, J. M., and Spratt, B. G. (1989) Horizontal gene transfer of penicillin-binding protein genes in penicillin-resistant clinical isolates of *Streptococcus pneumoniae. Proc. Natl. Acad. Sci. USA* **86,** 8842–8846.
4. Spratt, B. G., Zhang, Q., Jones, D., Hutchison, A., Brannigan, J. A., and Dowson, C. G. (1989) Recruitment of a penicillin-binding protein gene from *Neisseria flavescens* during the emergence of penicillin resistance in *Neisseria meningitidis. Proc. Natl. Acad. Sci. USA* **86,** 8988–8992.
5. Dowson, C. G., Coffey, T. J., Kell, C., and Whiley, R. A. (1993) Evolution of penicillin resistance in *Streptococcus pneumoniae*; the role of *Streptococcus mitis* in the formation of a low affinity PBP2B in *S. pneumoniae. Mol. Microbiol.* **9,** 635–643.
6. Martin, C., Briese, T., and Hakenbeck, R. (1992) Nucleotide sequences of genes encoding penicillin-binding proteins from Streptococcus pneumoniae and Streptococcus oralis with high homology to *Escherichia coli* penicillin-binding protein 1A and 1B. *J. Bacteriol.* **174,** 4517–4523.
7. Munoz, R., Coffey, T. J., Daniels, M., Dowson, C. G., Laible, G., Casal, J., Hakenbeck , R., Jacobs, M., Musser, J., Spratt, B. G., and Tomasz, A. (1991) Intercontinental spread of a multiresistant clone of serotype 23F *Streptococcus pneumoniae. J. Infect. Dis.* **164,** 302–306.
8. Coffey, T. J., Dowson, C. G., Daniels, M., Zhou, J., Martin, C., Spratt, B. G., and Musser, J. M. (1991) Horizontal transfer of multiple penicillin-binding protein genes, and capsular biosynthetic genes, in natural populations of *Streptococcus pneumoniae. Mol. Microbiol.* **5,** 2255–2260.
9. Ward, J. (1981) Antibiotic-resistant *Streptococcus pneumoniae*: clinical and epidemiological aspects. *Rev. Infect. Dis.* **3,** 254–266.
10. Liñares, J., Alonso, T., Perez, J. L., Ayats, J., Dominguez, M. A., Pallares, R., and Martin, R. (1992) Decreased susceptibility of penicillin-resistant pneumococci to twenty-four β-lactam antibiotics. *J. Antimicrob. Chemother.* **30,** 279–288.
11. Barcus, V. A., Ghanekar, K., Yeo, M., Coffey, T. J., and Dowson, C. G. (1995) Genetics of high-level penicillin resistance in clinical isolates of *Streptococcus pneumoniae. FEMS Microbiol. Lett.* **126,** 299–304.
12. Coffey, T. J., Daniels, M., McDougal, L. K., Dowson, C. G., Tenover, F. C., and Spratt, B. G. (1995) Genetics analysis of clinical isolates of *Streptococcus pneumoniae* with high-level resistance to expanded-spectrum cephalosporins. *Antimicrob. Agents Chemother.* **39,** 1306–1313.
13. Dowson, C. G., Johnson, A. P., Cercenado, E., and George, R. C. (1994) Genetics of oxacillin resistance in clinical isolates of *Streptococcus pneumoniae* that are oxacillin resistant and penicillin susceptible. *Antimicrob. Agents Chemother.* **38,** 49–53.

14. Dougherty, T. J., Koller, A. E., and Tomasz, A. (1980) Penicillin binding proteins of penicillin-susceptible and intrinsically resistant *Neisseria gonorrhoeae*. *Antimicrob. Agents Chemother.* **18**, 730–737.

15. Faruki, H. and Sparling, P. F. (1986) Genetics of resistance in a non-β-lactamase-producing gonococcus with a relatively high level penicillin resistance. *Antimicrob. Agents Chemother.* **30**, 856–860.

16. Mendelman, P. M., Caugant, D. A., Kalaitzoglou, G., Wedege, E., Chaffin, D. O., Serfass, D. A., Smith, A. L., Saez-Nieto, J. A., Vinas, M., and Selander, R. K. (1989) Genetic diversity of penicillin G-resistant *Neisseria meningitidis* from Spain. *Infect. Immun.* **57**, 1025–1029.

17. Guichard, D., Aubert, G., and Michel, V. P. (1994) What about penicillin resistance in *Neisseria meningitidis*? *Med. Mal. Infect.* **24**, 252–254.

18. National Committee for Clinical Laboratory Standards. (1990) Performance standards for antimicrobial disk susceptibility tests. Approved standard M2–A4. NCCLS, Villanova PA.

19. Report of the working party on antibiotic sensitivity testing of the British Society for Antimicrobial Chemotherapy. (1991) A guide to sensitivity testing. *J. Antimicrob. Chemother.* **27 Suppl. D,** 37.

20. Paris, M. M., Ramilo, O., and McCracken Jr., G. H. (1995) Management of meningitis caused by penicillin-resistant *Streptococcus pneumoniae*. *Antimicrob. Agents Chemother.* **39**, 2171–2175.

21. Viladrich, P. F., Gudiol, F., Liñares, J., Rufí, G., Ariza, J., and Pallares, R. (1988) Characteristics and antibiotic therapy of adult meningitis due to penicillin-resistant pneumococci. *Am. J. Med.* **84**, 839–846.

22. Jorgensen, J. H., Ferraro, M. J., McElmeel, M. L., Spargo, J., Swenson, J. M., and Tenover, F. C. (1994) Detection of penicillin and extended-spectrum cephalosporin resistance among *Streptococcus pneumoniae* clinical isolates by use of the E-test. *J. Clin. Microbiol.* **32**, 159–163.

23. Tenover, F. C., Baker, C. N., and Swenson, J. M. (1996) Evaluation of commercial methods for determining antimicrobial susceptibility of *Streptococcus pneumoniae*. *J. Clin Microbiol.* **34**, 10–14.

24. Skulnick, M., Small, G. W., Lo, P., Patel, M-P., Porter, C. R., Low, D. E., Matsumura, S., and Mazzulli, T. (1995) Evaluation of accuracy and reproducibility of E-test for susceptibility testing of *Streptococcus pneumoniae* to penicillin, cefotaxime, and ceftriaxone. *J. Clin. Microbiol.* **33**, 2334–2337.

25. Campos, J., Mendelman, P. M., Sako, M. V., Chaffin, D. O., Smith A. L., and Saez-Nieto, J. A. (1987) Detection of relatively penicillin G-resistant *Neisseria meningitidis* by disc susceptibility testing. *Antimicrob. Agents Chemother.* **31**, 1478–1482.

26. Campos, J., Trujillo G., Seuba T., and Rodriguez A. (1992) Discriminative criteria for *Neisseria meningitidis* isolates that are moderately susceptible to penicillin and ampicillin. *Antimicrob. Agents Chemother.* **36**, 1028–1031.

27. Bardi, L., Badolati, A., Corso, A., and Rossi, M. A. (1994) Failure of penicillin treatment of *Neisseria meningitidis* meningitis. *Medicina-Buenos Aires* **54**, 427–430.

28. VanDyck, E., Smet, H., and Piot, P. (1994) Comparison of E-test with agar dilution for antimicrobial susceptibility testing of *Neisseria gonorrhoeae. J. Clin. Microbiol.* **32,** 1586–1588.

29. Gomezherruz, P., Gonzalezpalacios, R., Romanyk, J., Cuadros, J. A., and Ena, J. (1995) Evaluation of the E-test for penicillin susceptibility testing of *Neisseria meningitidis. Diag. Microbiol. Infect. Dis.* **21,** 115–117.

30. Spratt, B. G. (1977) Properties of the penicillin-binding proteins of *Escherichia coli* K12. *Eur. J. Biochem.* **72,** 341–352.

31. Galleni, M., Lakaye, B., Lepage, S., Jamin, M., Thamm, I., Joris, B., and Frere, J-M. (1993) A new highly sensitive method for the detection and quantification of penicillin-binding proteins. *Biochem. J.* **291,** 19–21.

32. Markiewicz, Z. and Tomasz, A. (1989) Variation in penicillin-binding protein patterns of penicillin-resistant clinical isolates of pneumococci. *J. Clin. Microbiol.* **27,** 405–410.

33. Jabes, D., Nachman, S., and Tomasz, A. (1989) Penicillin-binding protein families: evidence for the clonal nature of penicillin resistance in clinical isolates of pneumococci. *J. Infect. Dis.* **159,** 16–25.

34. Hakenbeck, R., Briese, T., Ellerbrok, H., Laible, G., Martin, C., Metelmann, C., Schier, H. -M., and Tornette, S. (1988) Targets of β-lactams, in *Streptococcus pneumoniae,* in *Antibiotic Inhibition of Cell Surface Assembly and Function,* (Actor, P., Daneo-Moore, L., Higgins, M. L., Salton, M. R. J., and Shockman, G. D., eds.), American Society for Microbiology, Washington, D.C., pp. 390–399.

35. Zhang, Q-Y., Jones D. M., Saez Nieto, J. A., Pérez Trallero, E., and Spratt, B. G. (1990) Genetic diversity of penicillin-binding protein 2 genes of penicillin-resistant strains of *Neisseria meningitidis* revealed by fingerprinting of amplified DNA. *Antimicrob. Agents Chemother.* **34,** 1523–1528.

36. Kell, C. M., Jordens, J. Z., Daniels, M., Coffey, T. J., Bates J. Paul, J., Gilks, C., and Spratt, B. G. (1993) Molecular epidemiology of penicillin-resistant pneumococci isolated in Nairobi, Kenya. *Infect. Immun.* **61,** 4382–4391.

37. Klugman, K. P., Coffey, T. J., Smith, A., Wasas, A., Meyers, M., and Spratt, B. G. (1994) Cluster of an erythromycin-resistant variant of the Spanish multiply-resistant 23F clone of *Streptococcus pneumoniae* in South Africa. *Euro J. Clin. Microbiol. Infect. Dis.* **13,** 171–174.

38. Smith, A. M. and Klugman, K. P. (1995) Alterations in penicillin–binding protein 2B from penicillin-resistant wild-type strains of *Streptococcus pneumoniae. Antimicrob. Agents Chemother.* **39,** 859–867.

39. Selander, R. K., Caugant D. A., and Whittam T. S. (1987) Genetic structure and variation in natural populations of *Escherichia coli,* in *Cellular and Molecular Biology* vol. 2., (Neidhardt, F. C., ed), American Society of Microbiology, Washington D.C., pp. 1625–1648.

40. Selander, R. K. and Smith, N. H. (1990) Molecular population genetics of *Salmonella. Med. Rev. Microbiol.* **1,** 219–228.

41. Spratt, B. G., Smith, N. H., Zhou, J., O'Rourke, M., and Feil, E. (1995) The population genetics of the pathogenic *Neisseria*, in *The Population Genetics of Bacteria,* (Baumberg, S., Young, J. P. W., Wellington, E. M. H., and Saunders, J. R., eds.), SGM Symposium No. 52, Cambridge University Press, Cambridge, pp. 143–160.

42. Hall, L. M. C., Whiley, R. A., Duke, B., George, R. C., and Efstratiou, A. (1996) Genetic relatedness within and between serotypes of *Streptococcus pneumoniae* from the United-Kingdom—analysis of multilocus enzyme electrophoresis, pulsed-field gel electrophoresis, and antimicrobial resistance patterns. *J. Clin. Microbiol.* **34,** 853–859.

43. Selander, R. K. and Levin, B. R. (1980) Genetic diversity and structure *in Escherichia coli* populations. *Science* **210,** 545–547.

44. Caugant, D. A., Froholm, L. O., Bovre, K., Holten, E., Frasch, C. E., Mocca, L. F., Zollinger, W. D., and Selander, R. K. (1987) Intercontinental spread of *Neisseria meningitidis* clones of the ET-5 complex. Antonie van Leeuwenhoek *J. Microbiol.* **53,** 389–394.

45. Dowson, C. G., Coffey, T. J., and Spratt, B. G. (1994) Origin and molecular epidemiology of penicillin-binding-protein-mediated resistance to β-lactam antibiotics. *Trends Microbiol.* **2,** 361–366.

46. Ubukata, K., Asahi, Y., Yamane A., and Konno, M. (1996) Combinational detection of autolysin and penicillin-binding protein 2B genes of *Streptococcus pneumoniae* by PCR. *J. Clin Microbiol.* **34,** 592–596.

47. Dowson, C. G., Hutchison, A., and Spratt, B. G. (1989) Nucleotide sequence of the penicillin-binding protein 2B gene of *Streptococcus pneumoniae* strain R6. *Nucleic Acid Res.* **17,** 7518.

48. Laible, G., Spratt, B. G., and Hakenbeck, R. (1991) Interspecies recombinational events during the evolution of altered PBP 2X genes in penicillin-resistant clinical isolates of *Streptococcus pneumoniae. Mol. Microbiol.* **5,** 1993–2002.

28

The Application of Molecular Techniques for the Study of Aminoglycoside Resistance

Karen J. Shaw, Frank J. Sabatelli, Linda Naples, Paul Mann, Roberta S. Hare, and George H. Miller

1. Introduction

Aminoglycosides have been clinically used since 1944. Although this class of antibacterial agents has some nephrotoxicity and ototoxicity issues, they continue to be part of the hospital armamentarium because of their rapid bactericidal activity, especially in combination with β-lactams. Bacterial resistance to aminoglycosides can be caused by modifying enzymes, changes in cell permeability, and changes in the cellular target. The clinical observation of high levels of aminoglycoside resistance most often results from the acquisition of genes that encode modifying enzymes and are often plasmid-borne. Aminoglycosides are inactivated by three classes of enzymes:

1. Acetyltransferases [AAC], which use acetyl Co-A to modify specific amino groups on the aminoglycoside;
2. Adenylyltransferases [ANT], which use ATP to adenylylate specific hydroxyl groups; and
3. Phosphotransferases [APH], which use ATP to phosphorylate specific hydroxyl groups.

The common sites of modification are the 3-, 2'-, and 6'-amino groups and the 2"-, 3'-, and 4'-hydroxyl groups. The acquisition of resistance to the gentamicin (G) and kanamycin (K) families of aminoglycosides is usually mediated by specific enzymatic inactivation *(1–3)*. In contrast, a variety of factors may be responsible for a more generalized mechanism referred to as permeability resistance. These include changes in outer membrane proteins, alterations in active transport systems, or decreased ribosomal binding affinity.

From: *Methods in Molecular Medicine, Vol. 15: Molecular Bacteriology: Protocols and Clinical Applications*
Edited by: N. Woodford and A. P. Johnson © Humana Press Inc., Totowa, NJ

Whatever the cause of the decreased uptake of aminoglycosides by bacteria, permeability resistance is not structure-specific, and applies equally to all aminoglycosides of the G and K families. However, the level of resistance (low-level or high-level) may vary markedly from strain to strain.

Over 50 genes encoding aminoglycoside-modifying enzymes have been cloned and their DNA sequences determined *(3)*. Some modifying enzymes confer identical resistance profiles even though they show differences in amino acid sequence, whereas others have single changes in amino acid sequences that result in phenotypic differences. Thus, it is important to know both the genotype and phenotype of the genes encoding the aminoglycoside-modifying enzymes in order to understand clinical outbreaks and follow the spread of resistant organisms.

The acquisition of plasmid-borne genes encoding aminoglycoside-modifying enzymes has resulted in the rapid dissemination of aminoglycoside resistance across species boundaries. Changes in the frequency of specific aminoglycoside-modifying enzymes in clinical isolates have been correlated with aminoglycoside usage patterns *(4)*. Furthermore, surveillance of the frequency of the genes encoding these enzymes in specific hospital or geographical settings provides information on the origin, evolution, and dissemination of these genes. The authors have, therefore, developed a method in which information on the phenotype (aminoglycoside resistance patterns) and genotype (DNA hybridization or PCR) are obtained for individual organisms in order to determine which genes and which mechanisms are present.

1.1. Detection of Resistance

A variety of methods are used in clinical practice for the detection of aminoglycoside-resistant isolates, and in principle all are suitable. However, there are certain pitfalls that may result in the failure of clinical laboratories to detect isolates with specific aminoglycoside-modifying enzymes. These include screening only a limited number of aminoglycosides, such as gentamicin (G) and amikacin (A), and/or the use of susceptibility criteria that are in part based on clinical pharmacokinetics, and are, therefore, too high to detect low levels of resistance.

Laboratories that screen isolates for aminoglycoside resistance using tobramycin (T), but not G, will fail to detect those that carry AAC(3)-I since this protein modifies G, but not T *(see below)*. Laboratories that screen with G and A, but not T or netilmicin (N) will often fail to detect AAC(6')-I, which does not modify gentamicin C1 and usually causes only low level resistance to A. Screening with G, T, or N, but not A or isepamicin I would result in the failure to detect APH(3")-VI, a frequent cause of resistance to A and I in strains of *Acinetobacter*. Screening with G, T, or N, and A or I would be

Table 1
Midpoint of Zone Size Diameter (mm) Distribution
for Aminoglycoside-Susceptible Isolates

Pathogen	Aminoglycoside disks (disk content)				
	G (10 µg)	T (10 µg)	N (30 µg)	A (30 µg)	I (30 µg)
CEK[a]	21	21	24	22	24
EMPSS[b]	21	20	24	20	22
Providencia	23	22	24	27	29
Serratia	23	20	25	24	26
Pseudomonas	19	24	21	23	23
Acinetobacter	21	21	23	22	23
Staphylococcus	22	22	25	21	21

[a]*Citrobacter, Enterobacter, Klebsiella.*
[b]*Escherichia, Morganella, Proteus, Salmonella, Shigella.*

expected to detect most aminoglycoside-modifying enzymes in Gram-negative bacteria (GNB).

Susceptibility criteria in the clinic are usually based on three factors:

1. The range of zone sizes or MICs seen with normal isolates;
2. Serum levels achievable at the usual clinical doses; and
3. Clinical outcome of patients with infections caused by susceptible organisms at the usual clinical doses.

The detection of low levels of aminoglycoside-modifying enzymes in clinical isolates would be enhanced if changes in the relative activities of aminoglycosides were monitored, rather than absolute breakpoints. For example in *Citrobacter, Enterobacter, Klebsiella, Escherichia, Morganella, Proteus, Salmonella,* and *Shigella,* 10 µg disks of G and T, and 30 µg disks of A should produce very similar zone sizes, whereas 30 µg disks of N and I will produce zone sizes generally 2–3 mm larger in diameter. Typical zone size diameters observed in susceptible isolates lacking aminoglycoside-modifying enzymes are shown in **Table 1** (*see* **Subheading 3.1.1.**). Any change in these ratios is sufficient cause to suspect the presence of a modifying enzyme.

2. Materials
2.1. Phenotypic Methods

1. Mueller-Hinton agar.
2. Mueller-Hinton broth.
3. Sterile distilled water.
3. Aminoglycoside disks (*see* **Subheading 3.1.** and **Note 1**).

2.2. Genotypic Methods

2.2.1. DNA Hybridization for Large-Scale Screening of Isolates

1. GeneScreen Plus membranes (NEF-976, NEN Research Products, Boston, MA).
2. Whatman 3MM paper.
3. Shallow cafeteria trays.
4. 1 M NaOH.
5. 1 M Tris-HCl, pH 7.0.
6. Oligonucleotide labeling kit (Pharmacia, Piscataway, NJ) and ^{32}P.
7. Sephadex G-50 columns (5 Prime → 3 Prime, West Chester, PA).
8. Hybridization bags.
9. Prehybridization solution. 50% formamide, 1% sodium dodecyl sulfate (SDS), 1 M NaCl, 10% dextran sulfate, containing sonicated salmon sperm DNA at a final concentration of 100 µg/mL.
10. 20X SSC. 3 M NaCl, 0.3 M sodium citrate, pH 7.0.
11. Wash buffer I. 2X SSC, 0.1% SDS.
12. Wash buffer II. 0.1X SSC, 0.1% SDS.
13. X-ray film and cassettes.

3. Methods

3.1. Phenotypic Methods: The Use of Aminoglycoside Resistance Profiles for the Determination of Aminoglycoside Resistance Mechanisms

Different aminoglycoside-modifying enzymes have distinct substrate and resistance profiles. If susceptibility determinations are made using 12 specific aminoglycosides, it is often possible to determine which of the known classes of aminoglycoside-modifying enzymes are present within the strain. The 12 aminoglycosides that have been used for the phenotypic characterization (*see* **Tables 2** and **3** and **Note 1**) are apramycin (Apra), fortimicin (astromicin-Astm), SCH21562 6'-*N*-ethylnetilmicin (6'-Net), SCH21561 2'-*N*-ethylnetilmicin (2'-Net), gentamicin (G), tobramycin (T), amikacin (A), isepamicin (I), netilmicin (N), SCH22591 5-episisomicin (5-epi), kanamycin (K), and neomycin (Neo). The two netilmicin derivatives were chosen because they can differentiate between AAC(2'), AAC(6'), and AAC(3) mechanisms. One complication of only using susceptibility determinations to assign resistance mechanisms is the frequent presence of multiple mechanisms with overlapping resistance profiles in a single strain. In fact, individual strains with up to six aminoglycoside-modifying enzymes have been documented *(5)*. Therefore, it has been necessary to combine the use of resistance profiling with that of DNA hybridization or PCR techniques in order to characterize fully aminoglycoside resistant strains (*see* below).

Table 2
Determination of Aminoglycoside Resistance mechanisms in GNB by Phenotype

Observed Zone Sizes (mm) [a]

Ag (Disk Content µg)	1 Apra resistance AAC(3)-IV	1 Perm.	1 Normal potency relationships [b]	2 Apra res. AAC(3)-I	3 2'-Net = 6'-Net AAC(3)-II	3 AAC(3)-?	3 AAC(3)-?	4 6'-Net << 2'-Net AAC(2')-Neo	4 AAC(3)-K	4 AAC(6')-VI	5 2'-Net << 6'-Net AAC(6')-I	5 AAC(6')-II	5 AAC(6')-III	5 AAC(6')-I	5 AAC(6')-I+ ANT(2')-I	5 AAC(6')-I+ AAC(3)-I	5 AAC(6')-IV	6 2'-Net ≤ 6'-Net AAC(3)-II	6 AAC(6')-I+ APH(2')-I+ AAC(3)-III	6 AAC(6')-I+ AAC(6')-III	7 G res. ANT(2')-I	8 T res. AAC(3)-II	9 T res. ANT(4')-II	10 A, I res. APH(3')-VI
Apra (100)	6	15		25	24	20	25	28	24	25	27	26	24	25	24	26	25	23	19	22	24	25	23	24
Astm (100)	26	11		6	26	18	28	25	26	26	25	13	26	26	26	25	24	26	22	24	24	21	25	24
6'-Net (100)	15	11		26	11	6	15	24	11	8	20	6	6	8	6	6	11	6	11	6	23	13	27	23
2'-Net (100)	13	11		25	12	6	24	25	12	20	16	6	19	14	14	14	7	6	6	6	8	10	20	22
G (10)	11	11		13	6	11	14	9	6	10	21	6	8	9	9	9	7	15	7	7	9	20	11	21
T (10)	6	15		21	11	19	13	15	11	16	12	20	12	15	14	14	14	22	14	11	22	20	20	22
A (30)	22	14		23	22	20	23	22	14	21	13	21	13	22	22	21	21	21	20	14	23	21	23	9
I (30)	25	15		24	24	20	24	24	18	11	11	6	11	11	8	9	9	9	9	10	22	22	26	9
N (30)	11	14		25	11	11	15	19	14	12	11	6	11	8	8	7	7	9	6	6	20	17	21	22
5-epi (10)	15	14		22	18	18	23	22	18	11	12	6	11	6	6	6	6	6	6	6	8	6	21	23
K (30)	20	6		22	27	19	24	17	19	20	17	6	11	6	6	6	6	6	6	6	8	6	17	6
Neo (30)	18	9		21	20	17	16	18	17	19	18	14	18	17	17	17	17	12	12	12	18	15	19	14
Potency Relationships →	Apra<T <G<N <6'-Net= 2'-Net= 5-epi	Normal potency relation- ships [b]		Astm < G	G<T≤ N= 6'-Net< 2'-Net< 5-epi=K	6'-Net= 2'-Net= G≤N	G< 6'-Net≤ T<N K	G=T≤ N= 6'-Net≤ Neo	T<N= 6'-Net≤ K	G< 6'-Net≤ T<N K	T=N= 2'-Net= 5-epi=K <A<G= I<6'-Net	T=N= =K<G= A<I< 6'-Net	T=T=N= 2'-Net= 5-epi=K G≤6'-Net	T=K< A<I< 6'-Net	T=K< G=N< A<I< 6'-Net	T=K< G=N< A<I< 6'-Net	2'-Net= G=T≤ N=5-epi ≤6'-Net =K<A <I	T=N= 2'Net= 5-epi=K =G=A< 6'-Net= I=Neo			G=T= K	G=T= K≤5-epi	T≤K< A=I	K<A=I ≤Neo

Legend: ■ most affected; ▨ affected; ▨ least affected; □ unaffected.

[b] Normal potency relationships: Apra < Astm except in *Pseudonomas*, G = T except in *Pseudonomas* and *Serratia*, N > 5-epi except in *Pseudonomas* and *Providencia*, 2'-Net = 6'-Net except in *Providencia* and *Serratia*, I > A except in *Pseudonomas* and *Staphylococcus*, and K = Neo except in *Pseudonomas* and *Providencia*.

Table 3
Determination of Aminoglycoside Resistance Mechanisms in GPB by Phenotype

	Observed Zone Sizes (mm) [a]				
Steps →	**1**			**2**	**3**
	Astm = 2'-Net = 6'-Net			T res.	K, A res.
	AAC(6')-V + APH(2")-I				
Ag (Disc Content)	Alone	+ ANT(4')-I	+ APH(3')-III	ANT(4')-I	APH(3')-III
Apra (100)	27	29	28	31	31
Astm (100)	6	6	6	31	35
6'-Net (100)	10	9	13	34	34
2'-Net (100)	9	7	12	33	34
G (10)	6	6	6	29	30
T (10)	6	6	6	6	30
A (30)	17	14	18	22	27
I (30)	18	16	19	22	28
N (30)	19	20	21	35	31
5-epi (10)	10	9	12	30	28
K (30)	6	6	6	11	6
Neo (30)	18	13	7	17	14
Potency Relationships →	K = G ≤ T = Astm ≤ 6'-Net= 2'-Net = 5-epi < A = I = N = Neo	K = G ≤ T = Astm ≤ 6'-Net = 2'-Net = 5-epi < A = I = Neo < N	K = G ≤ T = Astm ≤ Neo ≤ 6'-Net = 2'-Net = 5-epi < A = I < N	T ≤ K ≤ Neo < A = I	K ≤ Neo < A = I < All other Ags

[a]Shadings are as in **Table 2**.

■ most affected; ▨ affected; ▢ least affected; □ unaffected.

3.1.1. Performing Antimicrobial Disk Susceptibility Tests

1. Mueller-Hinton broth and agar are prepared as directed by the manufacturer. Sixty to seventy milliliters of agar medium are used for 15-cm Petri dishes.
2. A single colony of an isolate to be tested is inoculated into 10-mL Mueller-Hinton broth, and grown overnight at 37°C. One milliliter of the overnight culture is diluted into 9 mL of sterile water. The remainder of the overnight culture is saved for preparing GeneScreen Plus filters for DNA hybridization (*see* **Subheading 3.2.1.1.**).
3. A sterile cotton swab on an applicator is dipped into the diluted suspension and then wiped against the inside wall of the tube above the fluid level to remove excess inoculum from the swab. The swab is then streaked onto the entire surface of a Mueller-Hinton agar plate. The plate is rotated 60° and streaked, and then rotated again 60° and streaked, in order to ensure an even distribution of the inoculum.

4. Aminoglycoside disks are placed on the surface of the dried plate. The plates are then incubated at 37°C. Zone diameters are recorded after overnight growth (18 h). The endpoint of the zone is the area showing no obvious visable growth.

3.1.2. Identification of Aminoglycoside Resistance Mechanisms in Gram-Negative Bacteria

The following procedure distinguishes the 15 aminoglycoside resistance mechanisms that occur in GNB, as well as the four most common combinations of resistance mechanisms. Data are analyzed by examining the relationships between the sizes of the zones of inhibition obtained with disks of the 12 aminoglycosides listed above (*see* below). When a changed zone of inhibition is encountered, one looks at all of the remaining zones in order to distinguish among the possibilities that can occur in each given step.

1. Compare the zone of inhibition around the Apra disk to the other zones. The only known causes of changes in Apra susceptibility in GNB are the AAC(3)-IV mechanism or permeability resistance (**Table 2**, columns 2 and 3). Equal changes to 6'-Net and 2'-Net suggest the presence of an AAC(3) enzyme. If equal changes in resistance are observed to 2'-Net, 6'-Net, and 5-epi, then the pattern is AAC(3)-IV. If resistance to all of the aminoglycosides is observed, then the mechanism is permeability resistance.

2. If the Apra zone is unchanged, but the Astm zone is changed, the aminoglycoside resistance mechanism must be either AAC(3)-I, alone or in combination with other aminoglycoside-modifying enzymes (**Table 2**, column 4). AAC(3)-I is further distinguished by changes in G zones without changes in any other tested aminoglycoside. AAC(3)-I is frequently found in combination with other aminoglycoside resistance mechanisms. These can be identified by continuing with **steps 3–10**. One combination that could be identified at this step is a very rare combination, AAC(3)-I + AAC(3)-IV. This combination differs from permeability since A, I, K, and Neo zones are unchanged. One might expect to see greater changes in G zones than in the other aminoglycosides, since it can be modified by both enzymes.

3. If Apra/Astm zones are unchanged, but there are changes in either 2'-Net and/or 6'-Net zones, then five patterns may occur (*see* also **steps 4–7**). If the changes to 2'-Net and 6'-Net zones are equal, then the aminoglycoside resistance mechanism must be either AAC(3)-II or AAC(3)-? They can be distinguished since AAC(3)-II leads to additional resistance to G, T, N, 5-epi, and K, where resistance to G is much greater than T, and resistance to T is greater than to N and 5-epi (**Table 2**, column 5). AAC(3)-II is the most common aminoglycoside resistance mechanism in GNB, especially in genera other than *Pseudomonas*. AAC(3)-? leads to additional resistance to G and N, but not to T (**Table 2**, column 6). It is frequently observed in *Acinetobacter*. The gene conferring this profile has not been cloned. It is possible that this phenotype may be conferred by subtle changes in permeability, therefore the ? is used rather than a roman numeral to describe this AAC(3)-like phenotype.

4. If there are changes to 6'-Net, but not to 2'-Net zones, then the aminoglycoside resistance mechanism must be either AAC(2')-I or AAC(3)-VI. AAC(2')-I is chromosomally-encoded in *Providencia stuartii* and is quite commonly expressed in this species. It has also been reported in Mycobacteria. It is very rare or nonexistent in other GNB. The presence of AAC(2')-I leads to additional resistance to G, T, N, and Neo, where the changes in the zone sizes to G and T are identical, and the change in the zone size to N is slightly less (**Table 2**, column 7). AAC(3)-VI leads to additional resistance to G, T, N, and K, where the change in the zone size of G is much greater than the change to T, and N (**Table 2**, column 8). This mechanism is currently rare, and occurs in *Enterobacteriaceae coli* and *Pseudomonas*.

5. Changes to 2'-Net, but not to 6'-Net zones, are characteristic of an AAC(6') resistance mechanism. Although all AAC(6') enzymes cause reistance to T, N, K, the four types of AAC(6') activity in GNB can be distinguished as follows: AAC(6')-I confers resistance to A; AAC(6')-II confers resistance to G (but not A); AAC(6')-III confers resistance to A and I; AAC(6')-IV confers resistance to A and G. A fifth type of AAC(6'), AAC(6')-V, occurs only in GPB as a fusion protein with APH(2")-Ia. AAC(6')-V confers resistance to A, I, and Astm and will be discussed below (**Subheading 3.1.3, step 1**). For strains with an AAC(6')-I profile (**Table 2**, column 9), resistance to T and N are usually similar but may be slightly greater to T except in *Pseudomonas*. Both T and N zone sizes are changed more than A, which often remains susceptible by normal clinical standards. Zone sizes for G and I are very similar (if they are not similar, additional resistance genes are present) and both have only slightly reduced zones. Although AAC(6')-I is very common in Enterobacteriaceae, it is not found very frequently in *Pseudomonas*. As in the case of AAC(6')-I, zone sizes for 2'-Net, T, N, 5-epi, and K are similar in AAC(6')-III (**Table 2**, column 10); however, AAC(6')-III also shows reduced zone sizes for A and I. The AAC(6')-III profile, first observed in Venezuela, is uncommon but has been increasing since 1993. In *Pseudomonas*, AAC(6')-II is the most common aminoglycoside resistance mechanism *(4,6)*. Resistance to G, T, and N is usually at a very high level (**Table 2**, column 11). Although T may be a better substrate for this enzyme than G, zones of inhibition of G and T are nearly always equal due to the compensating differences in potency observed against *Pseudomonas*. Zones around N are usually the smallest. Zone sizes around Apra, Astm, 6'-Net, A, and I disks are unchanged. The AAC(6')-IV mechanism (**Table 2**, column 14), whose resistance profile is identical to the combination of AAC(6')-I plus ANT(2")-I, will be discussed below along with the discussion of AAC(6')-I plus ANT(2")-I.

In addition to the three types of enzymes AAC(6')-I, II, and III, combinations of other aminoglycoside resistance mechanisms with AAC(6')-I are more frequent than any other type of combination in GNB. One of the more frequent combinations of aminoglycoside resistance mechanisms observed in Enterobacteriaceae is AAC(6')-I + AAC(3)-I, which can be distinguished from AAC(6')-I by the addition of Astm and G resistance to the profile (**Table 2**, column 12). A second very frequent combination, AAC(6')-I + ANT(2")-I (**Table 2**, column 13), is the exact summation of the two individual resistance mechanisms. ANT(2")-I confers resistance to G, T, K (*see* **step 8**), and AAC(6')-I confers resistance to T, N, 2'-Net, 5-epi, K, and A. As in AAC(6')-I, zones around 6'-Net are larger than those around I. Zones for T and K are usually the smallest, since both enzymes cause resistance to T and K. The combination of AAC(6')-I + ANT(2")-I (**Table 2**, column 13) is phenotypically indistinguishable from the fourth AAC(6') mechanism, AAC(6')-IV (**Table 2**, column 14). Therefore, assignment of a resistance mechanism relies heavily on genotypic methods to detect the presence of the gene encoding an ANT(2")-I mechanism (see **Subheading 3.2.**). The most frequent combination of mechanisms in *Pseudomonas* is AAC(6')-II + permeability *(4)*. When the permeability resistance level is low to moderate, this combination can be distinguished from AAC(6')-II by high-level resistance to the AAC(6')-II substrates (G, T, N, K, 5-epi, 2'-Net) and small but equal changes to A, I, Apra, Astm, and 6'-Net due to permeability.

6. If there are changes to 2'-Net and 6'-Net zones, but the changes for 2'-Net are greater, then the pattern of resistance suggests the presence of both AAC(3), conferring equal changes to both 2'-Net and 6'-Net (**step 3**, above), and AAC(6')-I (or II in *Pseudomonas*) conferring resistance only to 2'-Net. AAC(6')-I + AAC(3)-II is the most frequent combination of aminoglycoside resistance mechanisms in Enterobacteriaceae, and in fact, the second overall most common aminoglycoside resistance mechanism found in a recent survey *(4)*. The combination of these two enzymes results in resistance to most commonly available aminoglycosides. Changes in zone sizes for G, T, N, 2'-Net, and 5-epi are usually very large, whereas those for 6'-Net are smaller. However, if AAC(3)-II expression is at a very high level, this differential between 2'-Net and 6'-Net may be obscured (**Table 2**, column 15). As expected, the presence of AAC(6')-I results in reduced zones of inhibition for A relative to I. Zone sizes with I are usually diminished by 3–5 mm relative to Apra and Astm.

   Combinations of AAC(6')-I with other AAC(3) enzymes such as AAC(3)-IV or AAC(3)-? can occur. The combination of AAC(6')-I with AAC(3)-IV, which is very rare, can be seen by the diminished Apra zone sizes relative to Astm and I. The combination of AAC(3)-? with AAC(6')-I occurs more frequently in *Acinetobacter*. Whereas AAC(3)-? can be distinguished from AAC(3)-II by hybridization with probes for AAC(3)-II, it may often be detected by smaller changes in T and 5-epi zone sizes than occur in the combination with AAC(3)-II. In a recent survey of over 10,000 Enterobacteriaceae *(4)* the four aminoglycoside resistance mechanisms, AAC(6')-I, AAC(3)-II, AAC(3)-I, and ANT(2")-I occurred in all six possible double combinations, all four possible triple combinations, and in the quadruple combination. However, the three double combinations AAC(6')-I + AAC(3)-II, AAC(6')-I + AAC(3)-I, and AAC(6')-I + ANT(2")-I, were by far the most common.

   If there are changes to 2'-Net and 6'-Net zones and the changes for 2'-Net are greater, then the pattern of resistance can alternatively be the combination of

APH(2") and AAC(6')-III (**Table 2,** column 16). This combination is very rare, and can be distinguished from AAC(6')–I + AAC(3)–II (**Table 2,** column 15) by greater effects on A and I, an dby moderate resistance to neomycin. Only Apra and Astm are unaffected when APH(2") + AAC(6')-III mechanism is present.

7. If there are changes to both 6'-Net and 2'-Net zone sizes, but the changes to 6'-Net are greater than those to 2'-Net, then it is likely that the mechanism is a combination of AAC(3) and AAC(2')-I. Since this combination is confined to *Providencia* and is very rare at this time, it will not be discussed further.

8. If Apra, Astm, 2'-Net and 6'-Net zone sizes are unchanged but G and T are changed, the aminoglycoside resistance mechanism can be either ANT(2")-I or AAC(3)-III. These two mechanisms can be distinguished since AAC(3)-III leads to resistance to 5-epi (**Table 2,** column 18) whereas ANT(2")-I does not (**Table 2,** column 17). ANT(2")-I is one of the most common aminoglycoside resistance mechanisms in GNB, whereas AAC(3)-III is relatively rare. The limited occurrence of AAC(3)-III has been most often associated with specific outbreaks in individual hospitals. The rare combination of AAC(3)-III and AAC(6')-I cannot be distinguished from the very frequently occurring combination of ANT(2")-I + AAC(6')-I, except by DNA hybridization or PCR. The phenotype (**Table 2,** column 13) resulting from the either of these two combinations could also result from AAC(6')-IV, such that both amikacin and gentamicin are acetylated.

9. If Apra, Astm, 2'-Net, 6'-Net, and G zones sizes are unchanged, but T, A, I and K are changed, the aminoglycoside resistance mechanism must be ANT(4')-II (**Table 2,** column 19). ANT(4')-II is a very rare aminoglycoside resistance mechanism, and one gene from a localized outbreak has been sequenced *(7)*. Zones of inhibition around T are usually much smaller than those around A, I, and K. In fact, A and I may appear normal with low level expression of this aminoglycoside resistance mechanism. Because of this, assignment of the ANT(4')-II mechanism on the basis of phenotype must be done carefully.

10. If Apra, Astm, 2'-Net, 6'-Net, G, and T zone sizes are unchanged but A is changed, the aminoglycoside resistance mechanism can only be APH(3')-VI (**Table 2,** column 20). This is the only enzyme in GNB known to cause resistance to A and I without affecting Apra, Astm, 2'-Net, 6'-Net, G, or T. APH(3')-VI is a very common aminoglycoside resistance mechanism in *Acinetobacter* from which one gene, *aph(3')-VIa*, has been cloned *(8)*. It rarely occurs alone; rather it is frequently found with ANT(2")-I (G, T, K); AAC(3)-I (G, Astm), AAC(6')-I (T, N, A, 5-epi, K), and AAC(3)-? [G, N, 2'-Net, 6'-Net). APH(3')-VI has also been observed in GNB other than *Acinetobacter,* although this is still quite rare *(4)*. In genera other than *Acinetobacter*, it has always been found combined with other aminoglycoside resistance mechanisms—never alone. However, this finding may be associated with the manner in which most isolates are screened for aminoglycoside resistance, i.e., with G.

Recently, isolates of GNB other than *Acinetobacter* have been observed with resistance to A and I that did not hybridize to an *aph(3')-VIa* probe. This profile has been referred to previously as ? R ? A/I or APH(3')-VIII *(4,6)*. Since these publications, most but not all of the strains with this resistance profile have been shown to have a gene conferring AAC(6')-III (unpublished data). In strains obtained from Venezuela, both *aac(6')-IL* and *aac(6')-In* have been shown to be responsible for this phenotype. In isolates obtained from other countries, this phenotype was always associated with *aac(6')-IL*. Several explanations for the observations of this resistance phenotype in Gram-negative bacteria other than AAC(6')-III or APH(3')-IV are possible:

a. New and as yet unidentified genes encoding AAC(6')-III or APH(3')-VI activity may exist, or

b. APH(2") + AAC(6')-III also cause this same phenotype. None of the unexplained strains form the previous publications *(4,6)* were due to *aph(2")-Ic* + *aac(6')-Im*, since they failed to hybridize with an *aac(6')-Im* probe (*see* **Subheading 3.2.2., step 3**). However, as with APH(3')-VI or AAC(6')-III, new and as yet unidentified genes encoding this activity was exist.

3.1.3. Identification of Aminoglycoside Resistance Mechanisms in Gram-Positive Bacteria (GPB)

The following procedure is intended to distinguish the three single aminoglycoside resistance mechanisms that occur in GPB, as well as the most common combinations. Disk susceptibilities are determined as described in **Subheading 3.1.1.** above.

1. Compare the zone sizes of Astm, 2'-Net and 6'-Net to those of the other aminoglycosides. If Astm, 2'-Net, and 6'-Net zones are changed, the aminoglycoside resistance mechanism must be AAC(6') + APH(2") (bifunctional enzyme) alone (**Table 3**, column 2) or in combination with either ANT(4')-I (**Table 3**, column 3) or APH(3')-III (**Table 3**, column 4). Combinations of APH(2") + AAC(6') with either ANT(4')-I or APH(3')-III frequently occur and are discussed in **steps 2** and **3** below. In strains that express the AAC(6') + APH(2") mechanism, resistance to G, T, K, and 6'-Net is caused by phosphorylation of the 2"-hydroxyl of the aminoglycoside by the APH(2") domain. This resistance is almost always at a high level. The AAC(6')-V domain activity causes high levels of resistance to Astm, 2'-Net and K, but resistance to A, I, N, 5-epi and Neo can be quite variable.

   In a recent survey *(4)*, APH(2") + AAC(6') was the most common aminoglycoside resistance mechanism in GPB and frequently occurred (41.7% of aminoglycoside-resistant staphylococci). It is also the most common aminoglycoside resistance mechanism in *Enterococcus faecalis* (*see* **Note 2**).

2. If Astm, 2'-Net and 6'-Net are unchanged, but T is changed, the aminoglycoside resistance mechanism can be ANT(4')-I (**Table 3**, column 5), or very rarely ANT(4')-I + APH(3')-III. Resistance to T is significantly greater than to A, I, K, or Neo. ANT(4')-I is the second most common aminoglycoside resistance mechanism in GPB and occurs in both *Staphylococcus* and *Enterococcus*.

   The most common combination of aminoglycoside resistance mechanisms in GPB is AAC(6') + APH(2") with ANT(4')-I. Distinguishing features of this combination are:

   a. Smaller zone sizes for T than G are sometimes observed;

   b. A and I zone sizes are smaller than zones for N;

   c. The Neo zone size is smaller than the zone for N (**Table 3**, column 2). Hybridization with aminoglycoside resistance gene probes or PCR *(9)* have been used to confirm this phenotype.

3. If Astm, 2'-Net, 6'-Net, and T zone sizes are unchanged, but A and I are changed, then the aminoglycoside resistance mechanism can only be APH(3')-III (**Table 3**, column 6). The greatest affects of APH(3')-III are on K and Neo. The occurrence of APH(3')-III in GPB had been thought to be relatively rare. However, recent studies using hybridization to *aph(3')-IIIa* or PCR *(9)* have shown an incidence of this gene as high as 15% in aminoglycoside-resistant *Staphylococcus* and *Enterococcus*. The reason for this underestimation is that resistance to A and I is usually marginal. Laboratories that screen GPB for aminoglycoside resistance using K or Neo can, however, detect this aminoglycoside resistance mechanism quite readily.

   In isolates with APH(3')-III, resistance to Neo relative to A and I is greater than that seen in isolates with ANT(4')-I. This difference can be helpful in distinguishing the combination of AAC(6') + APH(2") with either APH(3')-III or ANT(4')-I. The combination of APH(3')-III with AAC(6') + APH(2") occurs frequently (13%) in *Staphylococcus* and *Enterococcus*. It can be detected by resistance phenotype, but is best detected by DNA hybridization or PCR (*see* **Subheading 3.2**).

   In strains carrying the combination of AAC(6') + APH(2") with APH(3')-III, zone sizes of G and T are affected similarly; A and I are usually more affected than N; Neo will be more affected than A and I and, in particular, Neo will have reduced zones relative to N. These changes may not always be noticeable in individual strains since it may be obscured by a background of aminoglycoside permeability.

3.2. Genotypic Methods

3.2.1. DNA Hybridization for Large-Scale Screening of Isolates

The procedure described here is a modification of the protocol described in Shaw et al. *(5)*.

3.2.1.1. PREPARATION OF FILTERS FOR HYBRIDIZATION

1. This technique utilizes GeneScreen Plus hybridization membranes (*see* **Note 3**). To avoid contamination with nucleases, the membranes are always handled while wearing gloves and are placed on absorbant paper (such as Whatman 3MM paper) for spotting, labeling, or drying.

2. Grids with 0.5×0.5 in squares are drawn or stamped onto the membranes. Grids generally contain 50 squares for convenience in handling sets of strains.

3. Twenty replicates of each set of isolates are made so that the isolates can be simultaneously probed for the presence of several genes encoding aminoglycoside-modifying enzymes (*see* **Table 4** and **Note 4**).

4. Approximately 10 µL of a 48 h culture of GNB is spotted onto the squares (*see* **Notes 5** and **6** for GPB). A 1-mL pipet is convenient for spotting the replicate

Table 4
Commonly Used Probes for Genes Endocing Aminoglycoside Modifying Enzymes in GNA and GPB

Resistance Mechanism	Cloned Gene	GenBank Accession Number	Plasmid	Cloned Fragment (Vector, site)a	Fragment Size (Restriction Digest)	PCR Product Size (PCR Primers)b	ATCC Number	References
AAC(2')-I	aac(2')-Ia	L06156	pSCH4500	1306 EcoRV/XhoI (pBluescript SK-)	375 (BclI)		87412	(10)
AAC(3)-I	aac(3)-Ia	X15852	pSCH2006	343 EcoRI/HindIII (pBluescript KS-)	343 (EcoRI/HindIII)	496 (T3*/T7*)	87413	This work
AAC(3)-I	aac(3)-Ib	L06157	pSCH6008	226 SacII/HindIII (pBluescript KS-)	226 (SacII/HindIII)	334 (T3*/T7*)	87414	This work
AAC(3)-I	aac(3)-IIa	X13543	pSCH2007	517 ClaI/SalI (pBluescript KS-, HincII)	538 (XhoI/HindIII)	682 (T3*/T7*)	87415	This work
AAC(3)-I	aac(3)-IIb	M97172	pSCH4203	1572 Sau3AI (pBluescript KS-)	485 (Asp718/EcoRV)		87416	This work
AAC(3)-III	aac(3)-IIIa	X55652	pJV305	2336 KpnI (M13mp18)				(11)
AAC(3)-IV	aac(3)-IVa	X01385	pWP701	~1650 Pst I (pBR322)	754 (SstI)			(12)
AAC(3)-VI	aac(3)-VIa	M88012	pSCH4101	2077 Sau3AI (pBluescript KS-, BamHI)	400 (ClaI/EcoRV)		87417	(13)
AAC(6)-I	aac(6')-Ia	M18967	pSCH2008	405 SspI (pBluescript KS-, HincII)	426 (HindIII/XhoI)	570 (T3*/T7*)	87418	This work
AAC(6)-I	aac(6')-Ib	M21682	pSCH2009	331 AvaI/DdeI (pBluescript KS-, HincII)	352 (HindIII/XhoI)	496 (T3*/T7*)	87419	This work
AAC(6)-I	aac(6')-Ic	M94066	pSCH2014	358 BamHI/EcoRV (pBluescript KS-)	358 (BamHI/EcoRV)	497 (T3*/T7*)	87420	This work
AAC(6)-I	aac(6')-Id	X12618						(14)
AAC(6)-V + APH(2")-I	aac(6')-Ie	M18086	pSCH3019	1558 EcoRI/BamHI (pBluescript KS+)	616 (HpaI/ScaI)	450 (DL/DR)	87421	This work

(continued)

567

Table 4 (continued)

AAC(6')-I	aac(6')-If	M55353	pSCH5410	282 HindIII/RsaI (pBluescript KS-, HindIII/SmaI)	288 (HindIII/BamHI)	421 (T3*/T7*)	87422	This work
AAC(6')-I	aac(6')-Ig	L09246	pAT475	554 FokI (pUC19)	188 (HindIII/AvaI)	438 (GL/GR)[c]		(14)
AAC(6')-I	aac(6')-Ih	L29044	pAT479	~2300 HindIII (pUC18)	172 (SspI/HindIII)	400 (HL/HR)		(14, 15)
AAC(6')-I	aac(6')-Ii	L12710	pAT434	391 Sau3AI/EcoRI (pUC18)	391 (Sau3AI/EcoRI)			(16)
AAC(6')-I	aac(6')-Ij	L29045	pAT481	~2900 PvuII (pUC19)	177 (SspI/ClaI)	768 (JL/JR)		(15)
AAC(6')-I	aac(6')-Ik	L29510	pAT477	~3200 EcoRI/PstI (pUC19)		438 (GL/GR)		(17)
AAC(6')-III	aac(6')-Il	U13880	pSCH129	263 PCR fragment (PCRScript)		428 (T3*/T7*)	87423	(18), this work
AAC(6')-III	aac(6')-Im		pSCH130	302 PCR fragment (PCRScript)		467 (T3*/T7*)	87424	This work
AAC(6')-III	aac(6')-In		pSCH131	297 PCR fragment (PCRScript)		462 (T3*/T7*)	87425	This work
AAC(6')-II	aac(6')-IIa[c]	M29695			273 (RsaI)			This work
AAC(6')-II	aac(6')-IIb	L06163	pSCH5102	1499 XbaI/PstI (pBluescript SK+- ?)			87426	This work
ANT(2")-I	ant(2")-Ia	X04555	pSCH2003	305 AvaI (pBluescript KS-, HincII)	326 (HindIII/XhoI)	491 (T3*/T7*)	87427	This work
ANT(3")-I	ant(3")-Ia	M10241	pSCH2004	483 RsaI (pBluescript KS-, HincII)	504 (HindIII/XhoI)	648 (T3*/T7*)	87428	This work
ANT(4')-I	ant(4')-Ia	V01282	pSCH2005	605 SacI/BamHI (pBluescript KS-)	317 (SacI/BglII)		87429	This work
ANT(4')-II	ant(4')-IIa	M98270	pSCH2012	204 EcoRV/SspI (pBluescript KS-, EcoRV)	222 (HindIII/PstI)	369 (T3*/T7*)	87430	This work
ANT(6)-I	ant(6)-Ia	J01839	pJH1	467 (HpaII)				(19)
APH(3')-I	aph(3')-Ia		pSCH2001	520 XhoI/HindIII (pBluescript KS-)	520 (XhoI/HindIII)	664 (T3*/T7*)	87431	This work

						1088 (T3*/T7*)	87432	This work
APH(3')-II	aph(3')-IIa	V00618	pSCH2002	923 PstI (pBluescript KS-)[d]	923 (PstI)			
APH(3')-III	aph(3')-IIIa	V01547	pSCH2011	563 EcoRI/HindIII (pBluescript KS-)	563 (EcoRI/HindIII)	716 (T3*/T7*)	87433	This work
APH(3')-VI	aph(3')-VIa	X07753	pSCH2010	377 EcoRI/AccI (pBluescript KS-)	377 (EcoRI/AccI)	515 (T3*/T7*)	87434	This work
Control 16S RNA		V00348	pSCH038	737 PCR fragment (pCRII)	742 (EcoRI)		87435	This work

[a]The site(s) of fragment insertion in the vector is the same as the site(s) used for digesting the fragment, except where noted. Therefore the designation "2077 Sau3AI (pBluescript KS-, BamHI)" refers to a 2077 Sau3AI fragment ligated in the BamHI site of pBluescript KS-.

[b]Primers: T3*, 5'-CCCTCACTAAAGGGAACAAAAGCTG-3'; T7*, 5'-GCGCGTAATACGACTCACTATAGGGCGAA-3'; DL, 5'-ATGATCGAAGCGGTGTCACT-3'; DR, 5'-TCATTCTGGCGCAAGCAT-3'; GL, 5'-TCATTCTGGCGCAAGCAT-3'; GR,5'-TTAATCTAT TTTTTACT-3'; HL, 5'-TCTGAATCACAATTATCA-3'; HR, 5'-CACCACAGTTCAGTTTC-3'; JL, 5'-CTCTCGGACCCATGCAGT-3'; JR, 5'-GATGTTAAATTTAGCTT-3'.

[c]Due to the DNA sequence similarity between the aac(6')-Ib and aac(6')-IIa genes, the aac(6')-Ib probes is used to detect both genes. Similarly, the primers for aac(6')-Ig are used for detecting both aac(6')-Ig and aac(6')-Ik.

[d]This fragment encodes the last 617 bp of the aph(3')-IIa gene plus the adjacent 3' flanking sequence of Tn5. This includes 290 bp of sequence from a gene which encodes bleomycin resistance. A better probe to use would be a 383 bp by Pst I/Nco I fragment which is only contains coding sequences of the aph(3')-IIa gene.

569

filters. Culture 1 is spotted on square 1 of all 20 replicate filters, culture 2 is spotted on square 2 of all 20 replicate filters, continuing in this manner to culture 50. If a culture is not dense enough, the cells can be concentrated by centrifugation.

5. The filters are dried at room temperature, but are not stacked until after they have been "fixed" (*see* **step 9** below).
6. Shallow "cafeteria" trays (approx 18" × 14" inches) are used in the following procedure. Absorbant paper (Whatman 3 MM) is placed on the first tray and saturated with 1 *M* NaOH (the excess solution is poured off). The GeneScreen filters, which have been spotted with bacterial cultures, are placed grid-side-up on the saturated paper for 5 min.
7. A second sheet of absorbant paper is placed on a second tray and is saturated with 1 *M* Tris, pH 7.0. The membrane filters are transferred with forceps to the second tray, and placed grid side up on the saturated paper.
8. After 5 min, all filters are placed into a container with 500 mL of 1 *M* Tris pH 7.0, for 3 min.
9. The filters are then placed on fresh absorbant paper and allowed to dry at room temperature (no baking). At this point the filters are "fixed" and can be indefinitely stored in plastic bags.

3.2.1.2. DNA HYBRIDIZATION

1. DNA probes generated from restriction digests or by polymerase chain reaction (PCR) (**Table 4**) are labeled with ^{32}P by using the Pharmacia oligonucleotide labeling kit according to the manufacturer's instructions (*see* **Notes 7** and **8**). Unbound label is removed by column chromatography on prepackaged Sephadex G-50 columns.
2. The filters from **Subheading 3.2.1.1.** are prehybridized for about 4 h at 42°C in Ziploc bags containing a solution of 50% formamide, 1% SDS, 1 *M* NaCl, 10% dextran sulfate, and 100 µg/mL sonicated salmon sperm DNA.
3. After prehybridization, the excess solution is drained off and the oligo-labeled probe (1 × 10^6 dpm/mL), which had been denatured and diluted into 5 mL of prehybridization mix, is added.
4. The filters are hybridized overnight at 42°C with shaking.
5. The filters are rinsed in 2X SSC, 0.1% SDS (prewarmed to 55°C), washed twice in 2X SSC, 0.1% SDS for 30 min (55°C), and once in 0.1X SSC, 0.1% SDS for 30 min (55°C). All washes are done with mild agitation.
6. The filters are then dried at room temperature, followed by exposure for 1–2 d to X-ray film (with an intensifying screen).

3.2.2. The Use of PCR for Detection of Aminoglycoside Resistance Genes

The proliferation of PCR technology has permitted the rapid detection of genes encoding aminoglycoside-modifying enzymes in bacterial strains. Specific pairs of oligonucleotide primers have been used to determine the distribution of the *aac(6')-Ic* gene *(20)*, the *ant(2")-Ia* gene (21), and the *aac(6')-Ia, aac(6')-Ib, aac(6')-If,* and *aac(6')-IIa* genes *(22)* in collections of clinical isolates. Additional studies have utilized up to five sets of primers (including a set that amplifies a fragment of the 16S rRNA gene) in a single PCR reaction in order to detect as many as four aminoglycoside-modifying genes in a single strain *(23–25)*. For isolation of template DNA, these studies have relied upon various modifications of the alkaline lysis technique, or upon rapid one-step techniques *(21,24)*. The simplest of the techniques requires suspending the pellet of 1.5 mL overnight culture in 500 μL of distilled water, heating for 10 min at 94°C, centrifuging for 5 min at 16,000g and using 10 μL of the supernatant as a template in a 50 μL PCR reaction *(24)*. **Table 4** lists the primer pairs and the sizes of the probes that have been successfully used in PCR to generate specific probes.

Several issues still have to be resolved before PCR can be used for high through-put, wide-scale screening of resistant organisms:

1. Whether a single lysis technique will be usable on a large scale for an extensive number of different GNB and GPB;
2. Whether mutant resistance genes, which no longer precisely match the chosen primer set, will be detected in the PCR reaction. How much divergence exists in the coding sequence of these globally spread genes has never been established. This issue becomes increasingly important for genes that encode AAC(6') enzymes. At least 17 genes have been identified, and several of these genes show significant DNA sequence homology.
3. A large number of PCR primer sets have to be developed, since over 50 genes encoding aminoglycoside-modifying enzymes have been identified. It is possible to use subsets of these genes for most screening, since many of the genes are now only found in GNB, GPB or actinomycetes. Thus, it is currently only necessary to use probes for the AAC(6')-APH(2"), ANT(4')-I, APH(3')-III for GPB. However, this distinction may become increasingly obsolete as genes or their homologs cross over these division boundaries. An example of this is the recent detection of homologs to the two domains of the Gram-positive bifunctional AAC(6')-APH(2") enzyme in GNB *(26)*.
4. The use of multiple primer sets in PCR reactions must be further explored. Although five PCR primer sets (including an internal control set for the 16S rRNA gene) have been used successfully in a single reaction *(24)*, it is likely that multiple reactions will be needed to test for the approx 20 other common genes that are found.
5. The use of PCR without a phenotypic test may be misleading. Genes may be present in strains, but for a variety of reasons may not lead to resistance to aminoglycosides. These reasons include poor expression or lack of expression because of a distal location in an integron (e.g., too far from the promoter—*see* **Subheading 3.2.4.**) *(27,28)*, poor recognition of a promoter in a heterologous host *(29)*, lack of a complete gene, and repression by endogenous regulators *(30,31)*. If the presence of these genes is detected solely by PCR or hybridization, the appropriate treatment regimen may not be chosen, since the organisms can

possess these genes, but still be susceptible to the predicted drugs. Thus, it is important that the expression of these genes is also monitored by resistance profiling. The detection of these genes by PCR or hybridization, however, may have implications for predictions of patient failures, since mutants that express these genes may arise and be selected for in a patient undergoing aminoglycoside treatment.

3.2.3. The Use of PCR Products as Probes
for Aminoglycoside Resistance Genes in DNA Hybridization Experiments

PCR is useful for generating DNA probes for large-scale hybridization experiments. Whereas it was previously necessary to isolate restriction fragments to generate probes *(32)*, it is now possible to prepare rapidly large quantities of probes by PCR, requiring very little starting DNA (*see* **Note 9**). In order to expedite the generation of probes for DNA hybridization, the authors have developed a set of clones in pBluescript II vectors (Stratagene, La Jolla, CA) that contain internal fragments of genes encoding aminoglycoside-modifying enzymes (*see* **Table 4**). These clones have been deposited with the American Type Culture Collection (ATCC) (Rockville, MD). Since these fragments have been cloned into the polylinker region, one set of vector primers can be used to obtain all of the cloned probes by PCR. The primers are based on the T3 and T7 primers, but are longer (5'-CCCCTCACTAAAGGGAACAA-AAGCTG-3' and 5'-CGCGTAATACGACTCACTATAGGGCGAA-3'). The authors have found that the shorter T3 and T7 primers failed to yield reliably PCR products. Oligonucleotides were obtained from Research Genetics (Huntsville, AL).

A variety of enzymes and conditions have been used for generating probes by PCR. Generally, plasmid DNA is prepared by a suitable methodology (e.g., Qiagen Plasmid Kit, Qiagen, Studio City, CA). The plasmid DNA is diluted 1:50–1:100 with water, and 2 μL of the diluted DNA (~20 ng) is used in a PCR reaction. Several enzymes have been successfully used, including *Taq* polymerase, Vent polymerase, and *Pfu* polymerase. The following is an example of a typical PCR reaction: reactions are carried out in 100 μL vol containing 10 μL of 10X buffer [200 mM Tris-HCl, pH 8.75, 100 mM KCl, 100 mM (NH$_4$)$_2$SO$_4$, 20 mM MgCl$_2$, 1 mg/mL BSA, 1% Triton X-100], 600 ng of each primer, 400 μM each dATP, dCTP, dGTP, and dTTP, 2.5 U of cloned *Pfu* DNA polymerase enzyme (Stratagene), and 2–20 ng of template DNA. The reaction mix is overlaid with mineral oil. Cycle times in the thermal cycler are: 2 min of denaturation at 95°C, 1 min of annealing at 55°C, and 1 min of extension at 72°C for a total of 30 cycles. The amplification products are gel-purified to ensure a low background, and are labeled for use in probing clinical isolates by DNA hybridization. One must be cautioned against using the PCR product for probing against other genes cloned into pBluescript II vectors as the presence of polylinker sequence on the ends of the probe will give false-positive results. DNA fragments are labeled and used in hybridization experiments, as described above (*see* **Subheading 3.2.1.2.** and **Notes 7** and **8**).

3.2.4. Detection of Integron Structure by PCR

Genes that encode aminoglycoside-modifying enzymes are often found associated with a structure known as an integron. These genetic elements have several distinct features, including a site-specific recombinase or integrase *(intI)* responsible for the insertion of resistance gene cassettes at a specific site *(attI)*, and a strong promoter that allows expression of the distally inserted resistance genes. Gene cassettes contain a single gene and a recombination site known as the 59-base element, which is located downstream of the resistance gene. The class 1 integron is often associated with the *sulI* gene (sulfonamide resistance) and *qacEΔ1* (resistance to quaternary ammonium compounds). Oligonucleotides that derive from the *intI*, *qacEΔ1*, *sulI*, and portions of the 5' and 3' conserved elements are useful for detecting the presence of the integron in clinical isolates and for determining the number of inserts and their order in the integron *(33,34)*. Some of the primers that have been shown to be useful in integron mapping are: 5'-CTACCTCTCACTAGTGAGGGGCGG-3', 5'-GCCCTTGCCCTCCCGCACCATG-3' and 5'-GCCTCGACTTCGCTGCTGCCC-3' for *intI;* 5'-AAGCTTTTGCCCATGAAGCAACCA-3' for *qacEΔ1;* 5'-GAATGCCGAACACCGTCACC-3' and 5'-TGAAGGTTCGACAGCAC-3' for *sulI;* and 5'-GGCATCCAAGCAGCAAG-3' and 5'-AAGCAGACT TGACCTGA-3' for the 5' and 3' conserved segments, respectively *(33,34)*. Pairs of these primers (such as an *intI* and a *sulI* primer) can be used to determine the sizes of the inserted gene cassettes, and to estimate the numbers of genes present within the integron. A single primer from *intI* or *sulI* can also be used in combination with a resistance gene-specific primer in order to map the distance from the gene to the boundaries of the integron.

3.3. Future Goals and Prospects

It is clear that resistance to aminoglycosides continues to be an increasingly complex problem. Because of the selection for and prevalence of multiply resistant strains, it is now even more important to use both phenotypic and genotypic methods to characterize these organisms fully. The utility of DNA hybridization and PCR for the detection of genes that encode aminoglycoside-modifying enzymes has been demonstrated, and the use of these techniques in clinical laboratories will increase. The combination of these techniques with susceptibility testing to an expanded spectrum of aminoglycosides (including at least G, T, N, A, I) allows resistance genes to be rapidly identified and assessed for expression in clinical isolates. Results from both detection studies

(PCR and DNA hybridization) and expression studies (susceptibility profiling) are important for predicting clinical outcome.

There are still several unexplained resistance phenotypes, and the gene(s) responsible need to be identified. Additional new genes are still being discovered, and additional probes to these genes will have to be developed. In addition, it is clear that amino acid substitutions in currently-known enzymes have led to alterations in resistance phenotypes. The identification of these changes by DNA sequence analysis will allow us to understand why isolates carrying these genes hybridize to known probes, yet express a new phenotype.

4. Notes

1. The usual potency disks are available to investigators free of charge by contacting the authors. Many of the strains containing cloned internal fragments of aminoglycoside modifying enzymes are available from the ATCC (Rockville, MD).

2. Because of the poor permeability of aminoglycosides in enterococci, the disks used for GNB cannot be conveniently used to interpret aminoglycoside resistance mechanisms in *Enterococcus*. Greater potency disks (3–10X higher μg/disk) must be used to distinguish the presence of aminoglycoside-modifying enzymes in enterococci. Many of the strains have the bifunctional AAC(6')-APH(2") in combination with ANT(4')-I and/or APH(3')-III. Therefore, in this genus, PCR or DNA hybridization are important in determining which aminoglycoside resistance genes are present in clinical isolates.

3. Although nitrocellulose membranes can be used for hybridization, their fragility and the necessity to fix the DNA to the membrane makes them unsuitable for large-scale screening.

4. Although it is possible to prepare a single hybridization filter and reuse it many times by washing off the probe, the authors have found this method to be too time consuming for analyzing many isolates with multiple probes.

5. Modifications for *Staphylococcus*: cultures are grown for 48 h. Then 150 μL of sterile water is added to a 1 mg vial of lyophilized lysostaphin (Sigma, St. Louis, MO); and 15 μL of this 6.67 mg/mL lysostaphin solution is added to 200 μL of cells. After incubation at 37°C for 30 min, with shaking, 200 μL of 1 M NaOH is added. Then 10 μL of the lysed cell suspension is spotted on each of the Gene Screen grids, and the filters are air dried at room temperature.

6. Modifications for *Enterococcus:* cultures are grown for 48 h, 15 μL of a 7.5 mg/mL solution of lysozyme (Sigma, St. Louis, MO) is added to 200 μL cells. After incubation at 37°C for 30 min, with shaking, 200 μL of 1 M NaOH is then added. Then 10 μL of the lysed cell suspension is spotted on each of the Gene Screen grids, and the filters are air dried at room temperature.

7. Hybridization kits that utilize streptavidin/biotin detection systems have not been used successfully in dot blots because of a high background problem. This is presumably caused by high levels of biotin in lysed cells. The advantages and limitations of nonradioactive labeling are discussed elsewhere in this volume (e.g., *see* Chapter 31).

8. Not all of the clones listed in **Table 4** are subcloned internal gene fragments. Some probes are isolated by purifying a restriction fragment from a plasmid that contains

the whole gene plus flanking sequences, whereas other plasmids are PCR-ready using a common set of primers. Single gel purification of PCR fragments and two rounds of gel purification of restriction fragments are necessary before using these probes in hybridization experiments. Failure to gel-purify these fragments often results in a high background and false-positive hybridization with clinical *E. coli* strains.

9. Alternatively, it is possible to use labeled oligonucleotide probes; however, in the authors' experience the signal obtained using short probes is significantly less than with larger DNA fragments and the results in large-scale screening are more often difficult to interpret.

Notes Added in Proof

1. The web site http://warn.utia.cas.cz contains additional information about the frequency and distribution of the aminoglycoside modifying enzymes described here.
2. **References *4* and *5*** can be difficult to obtain. These data, as well as additional data can be found in **ref. *35***.

References

1. Davies, J. and Smith, D. I. (1978) Plasmid-determined resistance to antimicrobial agents. *Ann. Rev. Microbiol.* **32,** 469–518.
2. Foster, T. J. (1983) Plasmid-determined resistance to antimicrobial drugs and toxic metal ions in bacteria. *Microbiol. Rev.* **47,** 361–409.
3. Shaw, K. J., Rather, P. N., Hare, R. S., and Miller, G. H. (1993) Molecular genetics of aminoglycoside resistance genes, and familial relationships of the aminoglycoside-modifying enzymes. *Microbiol. Rev.* **57,** 138–163.
4. Miller, G. H., Sabatelli, F. J., Naples, L., Hare, R. S., Shaw, K. J., and The Aminoglycoside Resistance Groups (1995) The most frequently occuring aminoglycoside resistance mechanisms-combined results of surveys in eight regions of the world. *J. Chemother.* **7 suppl. 2,** 17–30.
5. Shaw, K. J., Hare, R. S., Sabatelli, F. J., Rizzo, M, Cramer, C. A., Naples, L., Kocsi, S., Munayyer, H., Mann, P., Miller, G. H., Verbist, L., Landuyt, H. V., Glupczynski, Y., Catalano, M., and Woloj, M. (1991) Correlation between aminoglycoside resistance profiles and DNA hybridization of clinical isolates. *Antimicrob. Agents Chemother.* **35,** 2253–2261.
6. Miller, G. H., Sabatelli, F. J., Naples, L., Hare, R. S., Shaw, K. J., and The Aminoglycoside Resistance Study Groups (1995) The changing nature of aminoglycoside resistance mechanisms and the role of isepamicin—a new broad-spectrum aminoglycoside. *J. Chemother.* **7 suppl. 2,** 31–44.
7. Shaw, K. J., Munayyer, H., Rather, P. N., Hare, R. S., and Miller, G. H. (1993) Nucleotide sequence analysis and DNA hybridization studies of the *ant(4')-IIa* gene from *Pseudomonas aeruginosa. Antimicrob. Agents Chemother.* **37,** 708–714.
8. Martin, P., Jullien, E., and Courvalin, P. (1988) Nucleotide sequence of *Acinetobacter baumannii aphA-6* gene: evolutionary and functional implications of sequence homologies with nucleotide-binding proteins, kinases and other aminoglycoside-modifying enzymes. *Mol. Microbiol.* **2,** 615–625.

9. Van de Klundert, J. A. M., Gestal, M. H. V., Vliegenthart, J. S., and Ketelaar-Van Gaalen, P. A. G. (1992) Aminoglycoside-resistance genes in coagulase-negative staphylocci as detected by PCR. *Abstr. Ann. Meeting*, ICAAC #700.

10. Rather, P. N., Orosz, E., Shaw, K. J., Hare, R., and Miller, G. (1992) Genetic analysis of the 2'-*N*-acetyltransferase from *Providencia stuartii*. Abstr. Ann. Meeting, ICAAC #436.

11. Vliegenthart, J. S., Ketelaar-van Gaalen, P. A. G., and van de Klundert, J. A. M. (1991) Nucleotide sequence of the *aacC3* gene, a gentamicin resistance determinant encoding aminoglycoside-(3)-*N*-acetyltransferase III expressed in *Pseudomonas aeruginosa*, but not in *Escherichia coli*. *Antimicrob. Agents Chemother.* **35,** 892–897.

12. Chaslus-Dancla, E., Glupczynski, Y., Gerbaud, G., Lagorce, M., Lafont, J. P., and Courvalin, P. (1989) Detection of apramycin resistant Enterobacteriaceae in hospital isolates. *FEMS Microbiol. Lett.* **61,** 261–266.

13. Rather, P. N., Mann, P., Mierzwa, R., Hare, R. S., Miller, G. H., and Shaw, K. J. (1993) Analysis of the *aac(3)-VIa* gene encoding a novel 3-*N*-acetyltransferase. *Antimicrob. Agents Chemother.* **37,** 2074–2079.

14. Ploy, M. C., Giamarellou, H., Bourlioux, P., Courvalin, P., and Lambert, T. (1994) Detection of *aac(6')-I* genes in amikacin-resistant *Acinetobacter* spp. by PCR. *Antimicrob. Agents Chemother.* **38,** 2925–2928.

15. Lambert, T., Gerbaud, G., and Courvalin, P. (1994) Characterization of the chromosomal *aac(6')-Ij* gene of *Acinetobacter* sp. 13 and the *aac(6')-Ih* plasmid gene of *Acinetobacter baumannii*. *Antimicrob. Agents Chemother.* **38,** 1883–1889.

16. Costa, Y., Galimand, M., Leclercq, R., Duval, J., and Courvalin, P. (1993) Characterization of the chromosomal *aac(6')-Ii* gene specific for *Enterococcus faecium*. *Antimicrob. Agents Chemother.* **37,** 1896–1903.

17. Rudant, E., Bourlioux, P., Courvalin, P., and Lambert, T. (1994) Characterization of the *aac(6')-Ik* gene of *Acinetobacter* sp. 6. *FEMS Microbiol. Lett.* **124,** 49–54.

18. Bunny, K. L., Hall, R. M., and Stokes, H. W. (1995) New mobile gene cassettes containing an aminoglycoside resistance gene, *aacA7,* and a chloramphenicol resistance gene, *catB3,* in an integron in pBW301. *Antimicrob. Agents Chemother.* **39,** 686–693.

19. Ounissi, H. and Courvalin, P. (1987) Nucleotide sequence of the gene *aadE* encoding the 6-streptomycin adenylyltransferase in *Enterococcus faecalis*, in *Streptococcal Genetics,* (Ferretti, J. J. and Curtiss, R. III, eds.), American Society for Microbiology, Washington D.C., p. 275.

20. Snelling, A. M., Hawkey, P. M., Heritage, J., Downey, P., Bennett, P. M., and Holmes, B. (1993) The use of a DNA probe and PCR to examine the distribution of the *aac(6')-Ic* gene in *Serratia marcescens* and other Gram-negative bacteria. *J. Antimicrob. Chemother.* **31,** 841–854.

21. Vanhoof, R., Content, J., Bossuyt, E. V., Dewit, L., and Hannecart-Pokorni, E. (1992) Identification of the *aadB* gene coding for the aminoglycoside-2"-O-nucleotidyltransferase, ANT(2"), by means of the polymerase chain reaction. *J. Antimicrob. Chemother.* **29,** 365–374.

22. Vanhoof, R., Content, J., Bossuyt, E. V., Nulens, E., Sonck, P., Depuydt, F., Hubrechts, J. M., Maes, P., and Hannecart-Pokorni, E. (1993) Use of the polymerase

chain reaction (PCR) for the detection of *aacA* genes encoding aminoglycoside-6'-N-acetyltransferases in reference strains and Gram-negative clinical isolates from two Belgium hospitals. *J. Antimicrob. Chemother.* **32,** 23–35.

23. Vliegenthart, J. S., Ketelaar-Van Gaalen, P. A. G., and van de Klundert, J. A. M. (1990) Identification of three genes coding for aminoglycoside-modifying enzymes by means of the polymerase chain reaction. *J. Antimicrob. Chemother.* **25,** 759–765.

24. Van de Klundert, J. A. M., and Vliegenthart, J. S. (1993) PCR detection of genes coding for aminoglycoside-modifying enzymes, in *Diagnostic Molecular Microbiology—Principles and Applications,* (Persing, D. H., Smith, T. F., Tenover, F. C., and White, T. J., eds.) ASM, Washington, D.C., pp. 547–552.

25. Van Asselt, G. J., Vliegenthart, J. S., Petit, P. L. C., van de Klundert, J. A. M., and Mouton, R. P. (1992) High-level aminoglycoside resistance among enterococci and group A streptococci. *J. Antimicrob. Chemother.* **30,** 651–659.

26. Petrin, J., Kuvelkar, R., Munayyer, H., Kettner, M., Hare, R. S., Miller, G. H., and Shaw, K. J. (1995) Cloning of two novel genes leading to aminoglycoside resistance. Abstr. Cold Spring Harbor Molecular Genetics of Bacteria and Phages #197.

27. Collis, C. M. and Hall, R. M. (1995) Expression of antibiotic resistance genes in the integrated cassettes of integrons. *Antimicrob. Agents Chemother.* **39,** 155–162.

28. Centron, C. and Roy, P. H. (1996) Analysis of a *Serratia marcescens* strain that harbors three integrons. Program Abst. Molecular Basis for Drug Resistance in Bacteria, Parasites and Fungi #403.

29. Elisha, B. G., Segal, H., and Steyn, L. M. (1996) The promoter of a cryptic, integron associated *ant(2″)* gene from *Acinetobacter baumannii* is recognized in *A. calcoaceticus.* Program Abst. Molecular Basis for Drug Resistance in Bacteria, Parasites and Fungi #406.

30. Shaw, K. J., Rather, P. N., Sabatelli, F. J., Mann, P., Munayyer, H., Mierzwa, R., Petrikkos, G. L., Hare, R. S., and Miller, G. H. (1992) Characterization of the chromosomal *aac(6')-Ic* gene from *Serratia marcescens. Antimicrob. Agents Chemother.* **36,** 1447–1455.

31. Rather, P. N., Orosz, E., Shaw, K. J., Hare, R., and Miller, G. (1993) Characterization and transcriptional regulation of the 2'-*N*-acetyltransferase gene from *Providencia stuartii. J. Bacteriol.* **175,** 6492–6498.

32. Tenover, F. C. (1988) Diagnostic deoxyribonucleic acid probes for infectious diseases. *Clin. Microbiol. Rev.* **1,** 82–101.

33. Schwocho, L. R., Schaffner, C., Miller, G. H., Hare, R. S., and Shaw, K. J. (1995) Cloning and characterization of a 3-*N*-aminoglycoside acetyltransferase gene, *aac(3)-Ib,* from *Pseudomonas aeruginoasa. Antimicrob. Agents Chemother.* **39,** 1790–1796.

34. Lévesque, C., Piché, L., Larose, C., and Roy, P. H. (1995) PCR mapping of integrons reveals several novel combinations of resistance genes. *Antimicrob. Agents Chemother.* **39,** 185–191.

35. Miller, G. H., Sabatelli, F. J., Hare, R. S., Glupczynski, Y., Mackay, P., Shlaes, D., Shimizu, K., Shaw, K. J., and the Aminoglycoside Resistance Study Groups. (1997) The most frequent aminoglycoside resistance mechanism—changes with time and geographic area—a reflection of aminoglycoside usage patters? *Clin. Infect. Dis.* **24(suppl. 1),** S46–62.

29

Molecular Investigation of Glycopeptide Resistance in Gram-Positive Bacteria

Neil Woodford and Jill M. Stigter

Editor's Note

Our original intention was for this chapter to be written by Sandra Handwerger (Rockefeller University, New York), but tragically she died in April 1996 before she could undertake the task. In our opinion, Sandra was one of the leading research workers on genetic aspects of glycopeptide resistance in enterococci. Although we never had the opportunity to meet her personally, her many publications on this subject have been and will continue to be a valuable source of both information and inspiration. This chapter has been based in part on the outline prepared by Sandra shortly before her death and is dedicated to her memory.

Neil Woodford and Alan Johnson

1. Introduction

Glycopeptide antibiotics have been used for the treatment and prophylaxis of serious disease caused by Gram-positive bacteria since the introduction of vancomycin into clinical use in the late 1950s. Throughout the 1980s and 1990s, the increasing clinical problems posed by multiresistant Gram-positive bacteria, such as staphylococci, JK coryneforms and so forth, have led to a corresponding increase in the use of vancomycin and, to a lesser extent, teicoplanin.

Glycopeptides inhibit cell wall synthesis in Gram-positive bacteria. They bind to the D-alanyl-D-alanine (D-ala-D-ala) residue of peptidoglycan precursors, and consequently cause steric inhibition of the transglycosylase and transpeptidase enzyme activities responsible for the incorporation of these

From: *Methods in Molecular Medicine, Vol. 15: Molecular Bacteriology: Protocols and Clinical Applications*
Edited by: N. Woodford and A. P. Johnson © Humana Press Inc., Totowa, NJ

precursors into the growing peptidoglycan chains. As the D-ala-D-ala residue is highly conserved among Gram-positive and Gram-negative bacterial genera and represents a fundamental component of cell wall synthesis, the emergence of glycopeptide resistance was not anticipated, and for many years clinical laboratories did not screen large numbers of Gram-positive bacteria for resistance. However, transmissible high-level glycopeptide resistance was first recognized in 1986 in enterococci. As glycopeptides frequently represent the "last line" of defense against severe Gram-positive infection, the subsequent dissemination of enterococcal glycopeptide resistance, and the recognition of distinct forms of resistance in staphylococci has caused great concern *(1)*.

1.1. Vancomycin Resistance in Enterococci

1.1.1. Phenotypic Resistance Classes

Three phenotypes of glycopeptide resistance are now recognized in enterococci, designated VanA, VanB, and VanC *(1)*. The VanA phenotype was first described in 1988 in the UK and France *(2,3)* and is still the most widely reported form of resistance. It is characterized by high-level resistance to vancomycin (minimum inhibitory concentration [MIC] >128 µg/mL) and cross-resistance to teicoplanin (MIC ≤ 8 µg/mL). In contrast, the VanB phenotype is associated with variable levels of vancomycin resistance (MICs 8 to >1024 µg/mL), although most VanB isolates remain sensitive to teicoplanin in vitro (MIC ≤ 4 µg/mL). The VanC resistance phenotype confers low-level resistance to vancomycin (MICs ≤ 32 µg/mL) and isolates are sensitive to teicoplanin (MIC ≤ 4 µg/mL). As can be seen, there is some overlap between the levels of resistance conferred by these three phenotypes, and increased awareness and interest in glycopeptide resistance has inevitably resulted in "graying" of definitions that were set before large numbers of glycopeptide-resistant enterococci had been isolated.

The three classes differ in their distribution among the various species of enterococci, and also in their expression. Both VanA and VanB classes are examples of acquired resistance traits and result from the acquisition of exogenous DNA. In addition, both are regulated at the genetic level so that expression of resistance is inducible (i.e., resistance genes are only expressed after exposure of the resistant bacterium to glycopeptides). Expression of the genes responsible for the VanA phenotype may be induced by both vancomycin or teicoplanin, whereas expression of the VanB phenotype can be induced only by vancomycin. The VanB phenotype has been found almost exclusively in isolates of *E. faecalis* and *E. faecium*, but VanA is more widely distributed and has been found in at least seven species of enterococci *(1)*. In contrast to VanA and VanB, the VanC phenotype is an intrinsic property of some motile species of enterococci, namely *E. gallinarum* and *E. casseliflavus*, and is usually expressed constitutively *(1)*. As will be discussed later in this chapter, investigations of glycopeptide resistance using a variety of molecular techniques have provided explanations for many of the phenotypic and epidemiologic differences observed between the three phenotypes.

1.1.2. Mechanisms of Resistance

In enterococci of the VanA and VanB phenotypes, resistance to glycopeptide antibiotics results from the production of altered peptidoglycan precursors in which the usual terminal D-ala-D-ala dipeptide of the pentapeptide sidechain is replaced, after induction of resistance (e.g., by exposure to vancomycin), by a D-ala-D-lactate depsipeptide moiety. In contrast, enterococci with the VanC phenotype have precursors that terminate in the modified dipeptide D-ala-D-serine *(1)*. These altered precursors have a marked reduction in their binding affinity for vancomycin and those terminating in D-ala-D-lactate show similar reduced binding for teicoplanin. Hence, all VanA strains, induced VanB strains, and constitutive VanB-expression mutants are resistant to teicoplanin. As VanC enterococci are sensitive to teicoplanin in vitro, it has been proposed that the binding site of the teicoplanin molecule may be less specific for D-ala-D-ala than that of vancomycin *(4)*.

Phenotypic expression of acquired glycopeptide resistance (VanA or VanB phenotypes) requires the expression of several genes *(1,5,6)*. The eponymous *vanA* and *vanB* genes are diagnostic of these resistance phenotypes and both encode ligases that produce the depsipeptide D-ala-D-lactate. The *vanH* and *vanH$_B$* genes encode dehydrogenases that convert pyruvate to D-lactate (in VanA and VanB strains, respectively), whereas the *vanX* and *vanX$_B$* genes each encode a dipeptidase that cleaves D-ala-D-ala to prevent its incorporation into peptidoglycan precursors. Both phenotypes also have penicillin-insensitive carboxypeptidases (encoded by *vanY* and *vanY$_B$*) that are able to cleave the terminal D-ala (and possibly D-lactate) residues from peptidoglycan precursors to give tetrapeptides (*see* **Subheading 4.3.**). Thus, after induction of resistance, gene expression results in the production of glycopeptide-resistant precursors and the simultaneous degradation of the glycopeptide-sensitive precursors and D-ala-D-ala dipeptide. In both phenotypes, induction involves sensing the presence of environmental glycopeptide by transmembrane peptides encoded by the *vanS* (which responds to vancomycin or teicoplanin) or *vanS$_B$* (which responds only to vancomycin) genes, and the subsequent activation of a cytoplasmic peptide (encoded by the *vanR* and *vanR$_B$* genes) that binds to DNA and allows transcription of resistance genes (*see* **Subheading 4.2.**). In VanC enterococci, a D-ala-D-serine ligase is encoded by the genes *vanC-1* and *vanC-2/3* in *E. gallinarum* and in *E. casseliflavus,* respectively. For a detailed description of the genetic basis of glycopeptide resistance, readers are referred to ref. *(1)*.

1.2. Vancomycin Resistance in Other Gram-Positive Bacteria

1.2.1. Intrinsic Resistance

Resistance to vancomycin and teicoplanin, often at high levels, is an intrinsic characteristic of lactic acid bacteria (including *Leuconostoc* spp., *Pediococcus* spp., and many *Lactobacillus* spp.), *Erysipelothrix rhusiopathiae*, and some *Nocardia* spp. *(1)*. The molecular basis of resistance has only been investigated in the lactic acid bacteria, where it results from production of peptidoglycan precursors that terminate in the depsipeptide D-ala-D-lactate *(1,7,8)*. The similarity of the resistance mechanism in these species and that identified in VanA and VanB enterococci led to the suggestion that the latter may have acquired resistance from lactic acid bacteria. However, transfer of glycopeptide resistance from lactic acid bacteria has not been demonstrated to date. Furthermore, comparison of the D-ala-D-lactate ligases from these organisms with those of enterococci does not support a common ancestry *(9)*.

1.2.2. Staphylococci

At the time of writing, there are no substantiated, published reports of vancomycin resistance in clinical isolates of *Staphylococcus aureus*. The enterococcal *vanA* transposon, Tn*1546* (*see* **Subheading 3.2.**) has been transferred in vitro from an isolate of *E. faecalis* to a methicillin-sensitive laboratory strain of *S. aureus* *(10)*. Most of the genes on this element were expressed in the transconjugants (Reynolds, P.E., University of Cambridge, UK, personal communication), resulting in phenotypic resistance. This suggests that such transfer events could happen naturally, and that vancomycin-resistant *S. aureus* may eventually emerge. Vancomycin resistance has been documented in clinical isolates of coagulase-negative staphylococci *(1)*, but this resistance is unrelated to the VanA, VanB, or VanC mechanisms found in enterococci, and has not been fully characterized. The most frequently reported glycopeptide resistance phenotype among clinical isolates of staphylococci, including rare isolates of *S. aureus*, is teicoplanin resistance combined with vancomycin sensitivity *(1)*. This phenotype has not been reported in any other Gram-positive genera.

2. Detection of Resistance

2.1. Phenotypic Methods

Detection of some types of glycopeptide resistance by phenotypic methods may be difficult. Determination of the minimum inhibitory concentration using an agar incorporation or broth dilution technique provides the most accurate method for reliably detecting all forms of resistance. High-level glycopeptide resistance, as displayed by intrinsically resistant lactic acid bacteria and VanA enterococci, is usually reliably detected by all available methods. However, methods based upon diffusion of antibiotic through agar (i.e., disk diffusion tests and the E-test) have often proved less able to detect isolates displaying low levels of vancomycin resistance. Similarly, some automated systems have difficulty detecting low-level resistance. This subject has been recently reviewed *(1)*.

For enterococci, there is often a requirement to determine the resistance phenotype of an isolate, particularly for epidemiologic purposes. A working "ruleset," based on phenotypic methods, is used in the authors' (NW) laboratory and assumes prior recognition of a vancomycin-resistant enterococcus,

1. If the isolate is also resistant to teicoplanin, it almost certainly has the VanA phenotype, irrespective of the particular species;
2. If the isolate is sensitive to teicoplanin, it must be speciated;
3. If it is then identified as *E. gallinarum* or *E. casseliflavus*, it almost certainly has the VanC phenotype; and
4. However, if it is identified as any other species, it will almost certainly have the VanB phenotype.

There are isolates reported in the literature that provide exceptions to each of these "rules." For example, constitutive VanB-expression mutants are indistinguishable phenotypically from VanA isolates (exception to **step 1** above; **ref. *11***), rare isolates of *E. gallinarum* and *E. casseliflavus* contain genes required to confer both the VanB and VanC phenotypes (exception to **step 3** above; authors' unpublished observations) and some isolates with *vanA* genes express resistance only at low levels and display a VanB phenotype (exception to **step 4** above; **ref. *12***). Despite these exceptions, the aforementioned rules will provide laboratories with a guide to resistance phenotypes. However, if definitive characterization is required, or if the occasional exceptions are required for study, then molecular methods must be used.

2.2. Molecular Methods for Enterococci

2.2.1. Polymerase Chain Reaction (PCR)

Despite a requirement for many genes before resistance is expressed, detection of the VanA, VanB, and VanC phenotypes by PCR is not difficult; as mentioned above, the presence of the eponymous *vanA* or *vanB* genes is diagnostic for the two forms of acquired glycopeptide resistance. Many groups have devised PCR assays that amplify fragments of these genes (**Table 1**) and all appear to work well. The two multiplex assays described at the end of this chapter (*see* **Subheadings 7.1.** and **7.2.**) are those with which the authors are most familiar.

2.2.2. DNA Hybridization

A number of groups have used probes to detect the *vanA, vanB, vanC-1,* and *vanC-2/3* genes in clinical isolates. However, as with many probes for antibi-

Table 1
PCR Primers Suitable for Detecting Enterococcal Glycopeptide Resistance in the Clinical Laboratory

Target gene	Primer sequences (5'–3')	Product, bp	Uses	Refs.
vanA	1. TCTGCAATAGAGATAGCCGC	377	Detection	*13*
	2. GGAGTAGCTATCCCAGCATT			
	3. ATGGCAAGTCAGGTGAAGATGG	399	Detection and product as probe	*14*
	4. TCCACCTCGCCAACAACTAACG			
	5. AATGTGCGAAAAACCTTGC	535	Detection	*15*
	6. AACAACTAACGCGGCACT			
	7. GGGAAAACGACAATTGC	732	Detection	*16*
	8. GTACAATGCGGCCGTTA			
	9. GCTATTCAGCTGTACTC	783	Detection	*17*
	10. CAGCGGCCATCATACGG			
	11. GAGCATGACGTATCG	978	Product as probe	*18*
	12. AAGCGGTCAATCAGT			
	13. ATGAATAGAATAAAAGTTGCAATAC	1029	Detection and product as probe	*19*
	14. CCCCTTTAACGCTAATACGAT			
	15. CATGAATAGAATAAAAGTTGCAATA	1030	Detection	*12*
	16. CCCCTTTAACGCTAATACGATCAA			
vanB	17. CATCGCCGTCCCGAATTTCAAA	298	Detection	*17*
	18. GATGCGGAAGATACCGTGGCT			
	19. GTGACAAACCGGAGGCGAGGA	433	Detection and product as probe	*12*
	20. CCGCCATCCTCCTGCAAAAAA			
	21. CCCGAATTTCAAATGATTGAAAA	457	Detection and product as probe	*19*
	22. CGCCATCCTCCTGCAAAA			

584

Gene		Size	Purpose	Ref.
vanB	23. GCTCCGCAGCTTGCATGGACA	529	Detection and RFLP	*20*
	24. ACGATGCCGCCATCCTCCTGC			
	25. TCTGTTTGAATTGTCTGGTAT	589	Detection and product as probe	*21*
	26. GACCTCGTTTAGAACGATG			
	27. ATGGGAAGCCGATAGTC	635	Detection	*16*
	28. GATTTCGTTCCTCGACC			
vanB2	29. ATTGTCTGGATCCCCTATG	528	Detection	*22*
	30. GCAAGCCCTCTGCATCAAG			
vanC-1	31. GACCCGCTGAAATATGAAG	438	Detection and speciation	*17*
	32. CGGCTTGATAAAGATCGGG			
	33. GAAAGACAACAGGAAGACCGC	796	Detection, speciation, and product as probe	*12*
	34. ATCGCATCACAAGCACCAATC			
	35. GCTGAAATATGAAGTAATGACCA	811	Detection, speciation, and product as probe	*19*
	36. CGGCATGGTGTTGATTTCGTT			
	37. GGTATCAAGGAAACCTC	822	Detection and speciation	*16*
	38. CTTCCGCCATCATAGCT			
vanC-2/3	39. CTCCTACGATTCTCTTG	439	Detection and speciation	*16*
	40. CGAGCAAGACCTTTAAG			
	41. CTCCTACGATTCTCTTG	431	Detection and speciation	*17*
	42. GAATTTCCAGAACGAGC			
*ddl*_{E.faecalis}	43. ATCAAGTACAGTTAGTCTT	941	Speciation	*16*
	44. ACGATTCAAAGCTAACTG			
*ddl*_{E.faecium}	45. GCAAGGCTTCTTAGAGA	550	Speciation	*16*
	46. CATCGTGTAAGCTAACTTC			

585

otic resistance genes, there has been no standardized approach between groups of researchers and most use "in-house" probes. Many of these probes consist of intragenic restriction fragments of the resistance genes excised from appropriate recombinant plasmids, with the subsequent incorporation of an isotopic or nonisotopic label. However, this method of manufacturing probes is time-consuming, relatively technically demanding, and does not generate high yields. It is not, therefore, recommended as a convenient way of producing probes for glycopeptide resistance genes. Furthermore, labeling these restriction fragment probes with ^{32}P is obviously associated with numerous safety and administrative issues, and the probes are short-lived. In the authors' experience, the incorporation of a nonradioactive label into these fragments (e.g., by random priming) is not adequately reproducible between batches, although the probes have a far longer shelf-life (months or years) in comparison with isotopically-labeled DNA.

With the widespread availability of PCR, most groups now generate probes by incorporating a suitably-labeled deoxynucleotide (such as digoxigenin-11-dUTP) during the actual amplification reaction. The PCR primers listed in **Table 1** include several pairs that have been used to generate labeled products that can be used as gene-specific probes. In the authors' own laboratory, they routinely label the products obtained with primers 3 and 4 (for the *vanA* gene) or 25 and 26 (for the *vanB* gene) using two successive rounds of PCR (*see* **Subheading 7.1.**).

The use of hybridization as a means of detecting the presence of resistance genes has been largely superseded by PCR detection. A PCR-based approach allows rapid analysis of large numbers of isolates, but has the disadvantage of indicating only the presence or absence of a particular gene; it does not reveal the location of that gene. In addition, two groups have reported isolates of enterococci that hybridized with *vanA*-specific probes (i.e., probes that have proved reliable in all previous and subsequent uses), but that had the VanB phenotype and were confirmed to contain the *vanB* gene by PCR *(12,21)*. Thus, despite the limited numbers reported to date, the occurrence of these anomalous isolates suggests that PCR detection of resistance genes may be less likely to assign an incorrect genotype than DNA hybridization. However, hybridization of Southern blots is still widely used to determine whether acquired glycopeptide resistance genes are carried on plasmids or the chromosome, as is often required in epidemiologic investigations.

3. Epidemiologic Investigation
3.1. Spread of Glycopeptide Resistant Bacteria

Investigation of the degree of relatedness among isolates of glycopeptide-resistant bacteria (e.g., those causing hospital outbreaks of infection or colonizations) has been attempted by each of the methods detailed earlier in this volume (*see* Chapters 2–7). For specific guidance on the suitability of these methods for staphylococci and enterococci, readers are also referred to Chapters 23 and 24, respectively. PCR primers that have been used for the epidemiologic investigation of enterococci are listed in **Table 2** (primers 47–53). These various methods could also be applied to intrinsically resistant lactic acid bacteria, if epidemiologic investigation is considered necessary.

3.2. Dissemination of Glycopeptide Resistance Genes Among Enterococci

In contrast with the other forms of glycopeptide resistance described to date, the VanA and VanB phenotypes of enterococci can be transferable in vitro and, epidemiologic evidence suggests, in vivo. Hence, in addition to characterizing collections of resistant isolates in order to define strains (*see* **Subheading 3.1.**), detailed epidemiologic investigations of these phenotypes must also consider the possibility of horizontal transfer of resistance genes between strains.

3.2.1. Plasmid-Mediated Resistance

Most clinical isolates of enterococci contain one, or usually more, plasmids. As with plasmids in other bacteria, those in enterococci may be broadly classified as conjugative (capable of self-transfer), or nonconjugative (although some of these plasmids may be mobilized by coresident conjugative plasmids) (*see* Chapter 4). In the first reported examples of VanA glycopeptide-resistant enterococci, resistance was mediated by nontransferable (plasmid pIP816) *(3)* or transferable plasmids *(2,14,30)*. Transfer of plasmids between strains and between different species of enterococci is important, but the recognition of glycopeptide resistance transposon Tn*1546* on pIP816 *(5)* also emphasized the importance of transposition in the dissemination of VanA resistance. It is likely that both processes occur naturally, and that they are responsible for the occurrence of the VanA resistance phenotype in heterogeneous strains of multiple species of enterococci from widely differing geographic areas. Recently, the VanB phenotype has also been associated with both transposable elements *(31,32)* and transferable plasmids *(21,33)* and so dissemination of this resistance appears to involve similar mechanisms.

3.2.1.1. Plasmid Transfer

Exchange of plasmids in enterococci occurs by several mechanisms, but the high efficiency conjugative system mediated by the production of sex pheromones in *E. faecalis* has been studied most widely. Potential "recipient" strains produce and secrete short peptide pheromones that are recognized by "donor" cells containing plasmids responsive to a particular pheromone. As a result of

Table 2
PCR Primers that have been Applied to the Epidemiologic Investigation of Glycopeptide Resistance

Target Gene	Primer sequences (5'–3')	Product	Uses	Ref.
Arbitrary[a] (RAPD)	47. AGCAGCGTGG	Various	Typing	23
	48. TCACGCTGCA	Various	Typing	24
	49. AAGTAAGTGACTGGGGTGAGCG (ERIC2)			
	50. GGAGGGTGTT	Various	Typing	25
	51. IIICGICGICATCIGGC (REP1R)	Various	Typing	26
	52. GAGGGTGGCGGTTCT (M13)			
	53. TACCTTGTTACGACTTCACCCCA (16S rRNA-2)			
prgA	54. ATGAAGTACAGGCAGCAGAAC	701 bp	Detection and product as probe	27
	55. ACTAGTTGTCACAACGGTTTG			
prgB	56. ATACAAAGCCAATGTCG	427 bp	Detection and product as probe	27
	57. TACAAACGGCAAGACAAG			
prgX	58. ATGTTTAAGATAGGTTCTGTCC	561 bp	Detection and product as probe	27
	59. ATCGTAATCTTTACCAAAGG			
Tn1546	60. GGAAAATGCGGATTTACAACGCTAAG	ca. 11 kb	Long PCR	28
vanH	61. GGCAGATGCATTCCATGC	480 bp	Product as probe	29
	62. CTCTATACTTCGGCTGCG			
	63. ATCGGCATTACTGTTTATGGAT	943 bp	Detection and product as probe	19
	64. TCCTTTCAAAATCCAAACAGTTT			
vanR	65. ACAAGTCTGAGATTGACC	500 bp	Product as probe	29
	66. GGATTATCAATGGTGTCG			
	67. AGCGATAAAATACTTATTGTGGA	645 bp	Detection and product as probe	19
	68. CGGATTATCAATGGTGTCGTT			

vanS	69. AACGACTATTCCAAACTAGAAC	1094 bp	Detection and product as probe	19
	70. GCTGGAAGCTCACCCTAAA			
vanY_A	71. ACTTAGGTTATGACTACGTTAAT	866 bp	Detection and product as probe	19
	72. CCTCCTTGAATTAGTATGTGTT			
orf-1	73. TAGATCCGTCTCATGATG	740 bp	Product as probe	29
	74. GATACATGGAATCAATCG			
	75. AGGGCGACATATGGTGTAACA	844 bp	Detection	19
	76. GGGCGACGGTACAACATCTT			
	77. TGGTGGCTCCTTTTCCCAGTTC	1007 bp	Detection	19
	78. CGTCCTGCCGACTATGATTATTT			
	79. ACCGTTTTGCAGTAAGTCTAAAT	1086 bp	Detection	19
	80. AAACGGGATTTAGAAATAGTTAAT			
orf-2	81. TTGCGGAAAATCGGTTATATTC	540 bp	Detection	19
	82. AGCCCTAGATACATTAGTAATT			
	83. CCATTTCTGTATTTCAATTTATTA	925 bp	Detection	19
	84. CATAGTTATCACCCTTTCACATA			
IS1251	85. CTCAGCTTTCCCACATGG	1400 bp	Cloning and sequencing	29
	86. TGTCAGCGGGACCTTTGTCG			

---

^{a}I denotes inosine.

this recognition, a plasmid-encoded protein termed "aggregation substance," is produced by the "donor" cells and establishes mating aggregate formation with the "recipient" cells. This event is followed by replicative exchange of plasmids from the "donor" to the "recipient" cells *(34)*. **Table 2** lists three primer pairs that have been used to investigate the distribution of pheromone response genes among enterococci *(27)*. The *prgA* (primers 54 and 55) and *prgB* (primers 56 and 57) genes are highly conserved among pheromone response plasmids and encode an entry exclusion protein and the aggregation substance, respectively. However, *prgX* (primers 58 and 59) is a negative regulatory gene that has been found on a specific response plasmid (designated pCF10) and so can be used to identify plasmids highly related to this *(27)*.

Although pheromone-responsive plasmids transfer at high efficiency between strains of *E. faecalis* in vitro, the role of the pheromone system in the dissemination of antibiotic resistance has not been extensively studied. Limited numbers of such plasmids have been identified that encode resistance to tetracycline, erythromycin, chloramphenicol, or to high levels of aminoglycosides. In one instance, high-level aminoglycoside resistance was linked with β-lactamase production *(34)*. The prevalence of pheromone-responsive plasmids in enterococcal species other than *E. faecalis* is not known, but a few studies, undertaken with individual strains of *E. faecium,* indicate first that pheromone-responsive plasmids do occur in this species and, second, that they may play a role in the dissemination of glycopeptide resistance; a pheromone-responsive, conjugative plasmid encoding VanA resistance was described in one isolate of *E. faecium (35)*, whereas in a second strain of *E. faecium*, a nonconjugative VanA resistance plasmid was mobilized by a coresident pheromone-responsive conjugative plasmid *(27,36)*. Despite these two specific examples, generalizations about the importance of the pheromone system in dissemination of resistance to glycopeptides in *E. faecalis* or other species are not possible with available data.

3.2.1.2. THE IMPORTANCE OF OTHER PLASMID-ENCODED MARKERS

The clinical use of glycopeptides will select both for resistant enterococci and for maintenance of plasmids encoding resistance to these agents. However, in hospitals, and especially on the high-dependency units on which glycopeptide-resistant enterococci tend to be a problem, numerous antibiotic classes are used. Hence, genes that confer resistance to nonglycopeptides, but that are linked with glycopeptide resistance, must be understood and considered. The use of these other agents may select for the dissemination of resistance plasmids that also carry genes conferring glycopeptide resistance. In the authors' experience, erythromycin resistance is often linked to VanA glycopeptide resistance in vitro *(30)*, and so could allow transmission of resistance between enterococci under the pressure of antibiotic administration in vivo. There are few data for other antibiotics, but the authors have recently identified a plasmid that carried genes for the VanB glycopeptide resistance phenotype, and also a gene conferring resistance to all aminoglycosides, except streptomycin *(33)*. The importance of linked resistance genes is highlighted by the in vitro transfer of VanA glycopeptide resistance from *E. faecalis* to *S. aureus*, which was achieved, not by glycopeptide selection, but by screening transconjugants obtained by transfer of erythromycin resistance *(10)*.

3.2.1.3. USE IN INVESTIGATIONS

When investigating resistant isolates of the VanA or VanB phenotypes from nosocomial outbreaks, it is often beneficial in the first instance to define strains (e.g., by pulsed-field gel electrophoresis [PFGE]; *see* Chapters 3 and 24). Second, representatives of those strains may be subjected to plasmid analysis (*see* Chapter 4) to determine whether the various strains contain the same or distinct resistance plasmids. Digestion of plasmids with restriction endonucleases is useful, but the presence of multiple plasmids in clinical isolates often necessitates the use of transconjugants generated in vitro (this approach is also recommended if linked resistances are to be investigated). Southern blotting (*see* Chapter 2) of agarose gels of intact or digested plasmids, followed by hybridization of the blots with probes for the *vanA* or *vanB* genes may allow further comparison of plasmids from different strains. Using this approach the authors were able to show that various strains isolated during a nosocomial episode all carried *vanB*-encoding plasmids, but that the plasmids were heterogeneous *(21)*. Despite the heterogeneity observed among VanB plasmids in the UK, the *vanB* gene has been located on a conserved 2.1-kb *Eco*RV restriction fragment on all examined to date. This observation is consistent with predictions made from sequence data generated from a strain in which the *vanB* gene cluster was located on the chromosome *(6)*. This supports a transposition event at some point in the past causing the *vanB* cluster to move from the chromosome to become part of a plasmid.

3.2.2. Chromosomally Encoded Resistance

When first described, the VanB phenotype was thought to be nontransferable and conferred by genes located on the chromosome *(1)*. Although plasmid-mediated VanB resistance has been documented in the UK *(21,33)*, the prevalence of this form of resistance in other countries is not clear. Investigation of chromosomally encoded VanB (or VanA) resistance is best accomplished using an approach similar to that of Quintiliani and Courvalin *(31)*. The location of the gene in the clinical isolate should be determined by hybridization of a gene-specific probe with a Southern blot of DNA fragments separated (PFGE) (*see* Chapter 3). If resistance is transferable, then the PFGE pattern of the recipient strain and the transconjugants can be compared to identify fragments that have altered in size because of acquisition and chromosomal integration of the resistance element. Again, hybridization with a specific probe can confirm the location of the resistance genes. Using *Sfi*I digests of chromosomal DNA separated by zero-integrated gel electrophoresis (*see* Chapter 3), the *vanB* gene cluster was found to be located on large insertions of 90–250 kb *(31)*. Further analysis led to characterization of a 64-kb composite *vanB* transposon, designated Tn*1547 (32)*.

There is one report that describes chromosomally-encoded VanA resistance in detail; a Tn*1546* transposon containing an IS*1251* insert (*see* **Subheading 3.2.3.**) was integrated into the chromosome of an *E. faecium* clinical isolate as part of a larger element of at least 25 kb. This larger element was tentatively designated Tn*5482 (37)*. The clinical isolate carried plasmids, but these did not hybridize with probes for genes in the *vanA* cluster. Furthermore, transfer of resistance was observed, but did not correlate with the presence of plasmids in the transconjugants. Similar observations have been previously reported, including 12 of 13 *E. faecalis* from the first nosocomial episode caused by VanA enterococci *(14,30)*. Therefore, it is possible that chromosomal VanA is more widespread than current thinking might suggest.

3.2.3. Comparison of Mobile Elements Conferring VanA Resistance

The transposable nature of the VanA and VanB resistance phenotypes may, in some instances, limit the usefulness of direct comparison of resistance plasmids, so alternative molecular techniques are required to investigate detailed aspects of the epidemiology of glycopeptide-resistant enterococci. Increasingly, various groups have begun to examine the actual resistance elements in VanA enterococci.

In the prototype strain, the VanA phenotype is conferred by transposon Tn*1546 (5)* that carries two genes that encode transposase and resolvase enzymes (*orf1* and *orf2*), five genes (*vanR, vanS, vanH, vanA,* and *vanX*) necessary for the expression of inducible glycopeptide resistance, and two genes (*vanY* and *vanZ)* that are not essential for the expression of resistance. Other VanA enterococci have been shown to contain elements indistinguishable from, or highly related to, Tn*1546 (5,19,28,29,38)*, but despite extensive similarity in their gross structure, genetic variation can be detected between elements conferring the VanA phenotype. The methods used in these studies have usually involved either hybridization of digested genomic DNA (*see* Chapters 2 and 24) with a variety of probes specific for individual genes or parts of the transposon (*19,29)* (primers given in **Tables 1** and **2**) or amplifying overlapping fragments of the transposon, and then comparing the sizes or sequences of

Table 3
PCR Primers for Amplifying Overlapping Fragments
of VanA Transposon Tn*1546* (5)

Primer sequences (5'–3')	Product, bp	Nucleotides
87. GGATTTACAACGCTAAG (P1)	1309	0022–1330
88. GCCTTTATCAGATGCTA (P2)		
89. GGTTTTCGATTATTGGA (P3)	1132	1222–2353
90. AAATAATAGAACGACTC (P4)		
91. CGGAATGCATACGGCTC (P5)	1299	2227–3525
92. AGCCATTACAGTAATTA (P6)		
93. GGATGGACTAACACCAA (P7)	1274	2769–4042
94. TTAAGTATAATTCAACC (P8)		
95. GTGAAGGGATTGAATTG (P9)	1225	3569–4793
96. TCCAATCCCCAAGTTTC (P10)		
97. AAACGACTATTCCAAAC (P11)	1679	4675–6353
98. CATAGTATAATCGGCAA (P12)		
99. GTGTGAAATATATTTCT (P13)	1793	6229–8021
100. (AAGCATGC)TTATCACCCCTTTAACG (P14)[a]		
101. (TTGTCGAC)TTTGGATTTTGAAAGG (P15)[b]	1965	6956–8920
102. GGATTTACTATTATCAC (P16)		
103. ATTCATCTACATTGGTG (P17)	1584	8890–10,473
104. TCAGTCCAAGAAAGCCT (P18)		
105. TATCTTCGCTATTGGAG (P19)	428	10,403–10,830
106. GGATTTACAACGCTAAG (P1)		

[a]Primer P14 is tagged with a restriction site for *Sph*I.
[b]Primer P15 is tagged with a restriction site for *Sal*I.

the products with those amplified from Tn*1546* *(5,19,38)*. **Table 3** lists the primers used by Arthur et al. *(5)* to amplify 10 overlapping fragments of Tn*1546*-related elements. An alternative approach taken by the author (NW) is to amplify the entire element by Long PCR (L-PCR), and then to compare the restriction fragment length polymorphisms (RFLPs) following digestion of the L-PCR products from different isolates *(28)* (*see* **Subheading 7.3.**).

In two studies, *vanA* gene clusters of Tn*1546*-related elements have been divided into four groups *(19,28)*, but, because of the different methodologies used, further work is necessary before it will be possible to compare these groupings. Available data suggest that the intergenic regions between the *vanS* and *vanH* genes and also between the *vanX* and *vanY* genes may also be useful indicators of variation between elements *(5,29)*. In one study, some VanA enterococci from the Northeastern US had a 1.5-kb insertion sequence (designated IS*1251*) between the *vanS* and *vanH* genes *(29)*, increasing the overall size of the VanA element to 12.3 kb. Interestingly, this study also identified strains in which the *vanA* cluster was present on more than one plasmid. Furthermore, the configuration of the gene cluster on these plasmids differed, with some plasmids carrying elements in a truncated form (29).

These various approaches are beginning to generate interesting data that will hopefully provide a fuller understanding of the evolution and dissemination of the VanA resistance mechanism. A major issue currently debated is the extent to which the use of the glycopeptide avoparcin as a growth promoter for meat production, especially in pigs and poultry, has contributed to this process (39,40). In Europe, this agent is widely used (it has never been licensed for use in the US), and enterococci with the VanA phenotype have been isolated from treated animals and raw meat; these strains carry vanA and other genes on Tn1546 (13,15,28,41–43). The potential for avoparcin use to increase the reservoir of resistance genes in animal enterococci is fairly obvious, and cannot be readily contested. However, the ability of these resistant isolates to colonize humans subsequently is not known, nor is their ability to transfer resistance to resident enterococci during passage through the human gut. It is to be hoped that direct comparison of the resistance elements of VanA enerococci from human and nonhuman sources will provide answers to these questions.

To date, no studies have investigated whether similar variation occurs in the elements responsible for VanB resistance. However, these elements are again complex, which makes such variation likely; the gene cluster in the prototype VanB strain carries adjacent $vanR_B$, $vanS_B$, $vanY_B$, $vanW$, $vanH_B$, $vanB$, and $vanX_B$ genes (6). Variation in the chromosomally encoded or plasmid-mediated VanB resistance elements could also be explored as a potential epidemiologic tool.

4. Research Uses

4.1. Identification
of Genes Encoding D-ala-D-ala Related Enzymes

Two groups have published primers designed to amplify genes encoding D-ala-D-ala ligases and related enzymes (44,45). These so-called universal primers (Table 4, primer pairs 111 and 112, and 113 and 114) generate products of approx 600 bp and their use has greatly facilitated our understanding of the genes associated with glycopeptide resistance in both enterococci and intrinsically resistant bacterial genera. These primers have allowed amplification and subsequent characterization of intragenic fragments of vanB (47,48), vanB2 (45), vanC-1 (44), vanC-2/3 (49), genes encoding the D-ala-D-lac ligases of Leuconostoc mesenteroides, Lactobacillus spp. (9), and the genes (ddl) that encode the D-ala-D-ala ligases of many enterococcal species. The amplification protocol used with 113 and 114 consisted of 30 cycles of 94°C for 30 s, 55°C for 30 s, and 72°C for 2 min (45). The PCR products obtained have been used for sequencing and have been labeled for use as DNA probes.

Table 4
PCR Primers for Specific Research Applications

Gene	Primer sequences (5'–3')	Product size, bp	Uses	Ref.
ddl[a]	111. GGIGA(AG)GA(TC)GGI(TA)(CG)I(TCA)TICA(AG)GG 112. GT(AG)AAICCIGGI(TAG)TIGT(AG)TT	approx 600	Universal primers and product as probe	44
	113. GA(AG)GATGGIT(CG)CAT(AC)CA(AG)GG(AT) 114. (AC)GT(AG)AAICCIGGCA(GT)(AG)GT(AG)TT	approx 600	Universal primers and product as probe	45
vanH promoter	115. CGCGGATCCGCGGGGATGCCAATGGT 116. CGGAATTCCGGAAAGCAATGATAACTAT	254	DNA/protein interactions and gel shift assays	46
	117. CGAATTCAAGAACACTG 118. AAGCTTTCTGTGAAAGGC	440	DNA/protein interactions and gel shift assays	46
vanR promoter	119. GAATTCTGTATTCCGCTA 120. CGCGGATCCGCGCACATACGTTTGCCCTTA	197	DNA/protein interactions and gel shift assays	46

[a]I denotes inosine.
[b]Bases in parentheses indicate degenerate positions.

The reader should be aware of a potential problem that may arise when these universal primers are used with bacteria, such as VanA, VanB and VanC isolates of enterococci. These isolates have two D-ala-D-ala related genes; the chromosomal *ddl* gene and the gene encoding the D-ala-D-lactate or D-ala-D-serine ligase. Therefore, when "universal" PCR products are analyzed on agarose gels, although they will appear as single bands of approx 600 bp, the amplified DNA may actually be a mixture of amplicons derived from the *ddl* gene and, for example, from the *vanA* or *vanB* gene. For this reason, products amplified with these primers should not be labeled and used directly as probes, they should be cloned into suitable vectors first. The addition of extra nucleotides, that include target sites for restriction enzymes, onto the primers may facilitate cloning *(45)*. To illustrate the need for this precaution, when the primers were used with *E. faecalis* V583 (*vanB* genotype), preparation of labeled probes from the amplicons resulted in one probe (derived from the *E. faecalis ddl* gene) that hybridized with all isolates of *E. faecalis*, irrespective of their resistance or susceptibility to glycopeptides, and a second probe (derived from the *vanB* gene) that hybridized with all enterococci with the VanB resistance phenotype *(47)*. Caution should, therefore, be exercised when using these universal primers and, if probes for specific genes are required, it is advisable to use the primers listed in **Table 1**.

4.2. Induction of Glycopeptide Resistance in Enterococci

Expression of the VanA resistance phenotype in enterococci is induced following exposure to vancomycin or teicoplanin in the growth medium. This induction is regulated by the *vanS* and *vanR* genes that encode a two-component, signal-transducing system *(50)*. On exposure to glycopeptides, the membrane-spanning VanS sensory peptide becomes phosphorylated and passes this phosphoryl group to the cytoplasmic VanR peptide. The phosphorylated form of VanR (phospho-VanR) binds to DNA upstream of the *vanH* gene and allows transcription of the *van* operon and results in expression of phenotypic VanA resistance. Three DNA fragments, generated by PCR with primers 115–120 (**Table 4**), have been used to determine the sequence recognized by phospho-VanR *(46)*. The primers allow amplification of fragments that include the promoters for the *vanH* (primers 115 and 116, and 117 and 118) and *vanR* (primers 119 and 120) genes. Gel mobility shift studies have confirmed that purified phospho-VanR binds to these fragments (probably as oligomers). DNase I footprinting of the promoter regions indicated that phospho-VanR protects a ca. 80-bp region of the *vanH* promoter and a ca. 40-bp fragment of the *vanR* promoter. Further analysis of these regions identified a 12-bp consensus sequence, T(TC)TTA(GA)GAAA(TA)T, which is thought to be the recognition site for VanR binding. This sequence is present in a single copy within the *vanR* promoter, but in two copies in the *vanH* promoter. Bacteria contain other two-component, signal-transducing systems. However, gene inactivation studies have shown that induction of the VanA phenotype appears to be dependent upon the presence of functional VanR peptide, although phosphorylation of this peptide can be accomplished by *trans*-acting sensory peptides other than VanS *(50,51)*. This VanR-specificity suggests that the primers listed here may prove useful for studies of the mechanism of induction of the VanA resistance phenotype.

Recently, the sequences of the *vanR$_B$* and *vanS$_B$* genes that regulate induction of the VanB enterococcal glycopeptide resistance phenotype have been determined *(6)*. Similar approaches to those outlined above will enable the recognition site of the phospho-VanR$_B$ peptide to be determined. It is thought that the gene cluster is probably transcribed from a promoter downstream of the *vanS$_B$* gene, but the 12-bp consensus sequence of VanA enterococci is not present.

4.3. Analysis of Peptidoglycan Precursors

In the pathway of peptidoglycan synthesis that is inhibited by glycopeptide antibiotics, the precursor, UDP-N-acetyl muramyl-L-Ala-D-Glu-g-L-Lys-D-Ala-D-Ala (pentapeptide) is synthesized. When glycopeptides bind to acyl-D-Ala-D-Ala, the transglycosylation reaction that links the nascent glycan chain to the new wall-subunit, and which occurs in the cytoplasmic membrane, is the first step to be inhibited as a consequence of steric hindrance, and the subsequent transpeptidation is also inhibited *(52)*. This prevents cell-wall synthesis and results in cell death. VanA and VanB glycopeptide-resistant enterococci are capable of synthesizing peptidoglycan in the presence of vancomycin by producing modified peptidoglycan precursors. In the resistant pathway (controlled by *vanR/vanS* or *vanR$_B$/vanS$_B$*, *see* **Subheading 4.2.** above) modified peptidoglycan precursors terminating in D-lactate (i.e., UDP-N-acetyl muramyl-L-Ala-D-Glu-g-L-Lys-D-Ala-D-lactate [pentadepsipeptide]) are the main product of the cytoplasmic stage of peptidoglycan biosynthesis *(7,53–57)*.Glycopeptides have a very low affinity for acyl D-Ala-D-lactate (approx 1000-fold less that that for acyl-D-Ala-D-Ala), and do not effectively bind to it. Hence, resistant bacteria are able to evade the action of glycopeptides *(58,59)*.

As explained previously, various genes have a major role in glycopeptide resistance (*vanH, vanA, vanX,* and *vanY* in VanA enterococci and *vanH$_B$, vanB, vanX$_B$,* and *vanY$_B$* in VanB enterococci). The *vanH* and *vanH$_B$* genes encode dehydrogenases that reduce pyruvate to D-lactate *(59)* and this, rather than D-alanine as in the glycopeptide-sensitive pathway, is ligated to D-alanine by the ligase of altered specificity encoded by *vanA* (or *vanB*) *(59)*. The *vanX* and *vanX$_B$* genes encode D,D-dipeptidases that hydrolyze D-ala-D-ala *(54)*, and consequently reduce the amount of dipeptide available, which favors the production of glycopeptide-resistant precursors. The *vanY* and *vanY_B* genes encode D,D-carboxypeptidases that hydrolyze the terminal D-ala from UDP-MurNAc-pentapeptide, thereby reducing the availability of pentapeptide precursor for peptidoglycan biosynthesis *(60,61)*. The *vanY* gene of VanA enterococci has been referred to as an accessory gene, but *vanY_B* is likely to have a more significant role in expression of the VanB phenotype.

Investigation of the resistant pathway of peptidoglycan biosynthesis has been performed by employing various biochemical techniques involving the extraction of cell-wall precursors, partial purification by trichloroacetic acid (TCA) precipitation, desalting of neutralized extracts by gel-filtration chromatography, and subsequent analysis by high performance liquid chromatography (HPLC) (*see* **Subheading 7.4.**). The precursor profile of a particular strain is dependent upon the relative activities of the genes of the resistance operon.

5. Future Prospects and Goals

The application of molecular biological methods has provided a detailed insight into the mechanisms of glycopeptide resistance in gram-positive bacteria, particularly enterococci. However, mechanistic, epidemiologic, and clinical questions remain unanswered. Future goals include detailed characterization of the form(s) of resistance to teicoplanin (and, less commonly, to vancomycin) observed in clinical isolatea of coagulase-negative staphylococci and occasional isolates of *S. aureus*. Within enterococci, there are still many unanswered questions. In particular, the origins of the *vanA* and *vanB* gene clusters remain to be discovered. The answer to this question is likely to be intimately associated with the question concerning the relative importance of the clinical use of vancomycin and teicoplanin versus the ergotropic use of avoparcin as factors promoting the evolution, selection, and dissemination of glycopeptide resistance. To date, glycopeptide-resistant enterococci (excluding intrinsically resistant species) isolated outside the hospital environment have all contained the *vanA* gene cluster. The apparent absence of the *vanB* cluster outside hospitals is puzzling, but could indicate major differences in the epidemiologies of these two acquired resistance phenotypes. Since the first reports of glycopeptide resistance in enterococci, the VanB phenotype has been documented less frequently than VanA. Differences in the ability of clinical laboratories to detect the two types of resistance could explain this *(1)*, but it is also possible that this variation reflects a true difference in prevalence; existence of an environmental reservoir of VanA enterococci and, perhaps more importantly, of the *vanA* gene cluster, but no similar reservoir of VanB enterococci or *vanB* genes could also significantly contribute to the different isolation rates.

A major aim should also be the design of new antimicrobials active against glycopeptide-resistant enterococci. A knowledge of the genetic and biochemical bases of resistance encoded by the *vanA* and *vanB* clusters offers several choices that may be considered. For example,

1. Is it possible to synthesize agents that are able to bind to D-ala-D-lactate or D-ala-D-serine in peptidoglycan precursors and so inhibit bacterial growth?
2. If resistant enterococci continue to produce D-ala-D-ala, and must prevent formation of glycopeptide-susceptible peptidoglycan precursors, would inhibition of the products of the *vanX/vanX$_B$* or *vanY/vanY$_B$* genes be possible and, if so, would these offer effective antimicrobial targets?
3. As expression of the *vanA* and *vanB* clusters is usually inducible, would inhibition of the two-component regulatory system be possible? As the VanR peptide appears to be essential for induction of resistance (at least in VanA enterococci) this intracytoplasmic peptide may be a possible target. However, the pitfalls of developing new antimicrobials are numerous and include the cost involved and the unfailing ability of bacteria to become resistant. It is, therefore, perhaps preferable to avoid exploring agents that act on targets for which a theoretical resistance mechanism already exists; constitutive expression of glycopeptide resistance may bypass the need for VanR or VanR$_B$, although this has not been proven to date *(62)*.

6. Materials

6.1. Detection of vanA and vanB Genes in Enterococci

1. Pure growth of bacteria. Discrete colonies.
2. Sterile distilled water. Tissue culture grade or equivalent (e.g., Sigma W-3500, St. Louis, MO).
3. dATP, dGTP, dCTP, dTTP solutions: each at 10 mM.
4. Primers 3, 4, 25, and 26 (**Table 1**). A diluted solution containing 50 ng/µL of each primer in sterile distilled water.
5. *Taq* polymerase (Life Technologies, Bethesda, MD): supplied with 10X reaction buffer and buffer W1.
6. Mineral oil.
7. Loading buffer. 50 mM EDTA, pH 8.0 containing 25% Ficoll and 0.25% bromophenol blue.
8. 0.5X TBE buffer. 44.5 mM Tris base, 44.5 mM boric acid, 1 mM EDTA, pH 8.0.
9. 2% agarose gel in 0.5X TBE buffer.
10. Ethidium bromide (EB). 1 µg/mL in distilled water.

6.2. Speciation of Enterococci by PCR

1. As in **Subheading 6.1.** above, except;
2. Primers 37–40 and 43–46 (**Table 1**; i.e., eight primers): diluted solution containing 50 ng/µL of each primer in sterile distilled water.

6.3. Amplification of Tn1546-Related Elements by Long PCR

1. Genomic DNA (e.g., extracted using guanidium thiocyanate, *see* Chapter 2).
2. Expand™ Long Template PCR System (Boehringer Mannheim, Germany).
3. Primer 60 (**Table 2**).
4. Sterile distilled water. Tissue culture grade or equivalent (e.g., Sigma W-3500).
5. dATP, dGTP, dCTP, dTTP solutions: each at 10 m*M*.
6. Mineral oil.
7. Loading buffer. 50 m*M* EDTA, pH 8.0 containing 25% Ficoll and 0.25% bromophenol blue.
8. 0.5X TBE buffer. 44.5 m*M* Tris base, 44.5 m*M* boric acid, 1 m*M* EDTA, pH 8.0.
9. 0.8% agarose gels in 0.5 x TBE buffer (1% agarose gels for RFLP analysis).
10. EB. 1 µg/mL in distilled water.

6.4. Analysis of Peptidoglycan Precursors in Glycopeptide-Resistant Enterococci

6.4.1. Extraction of Precursors

1. Brain heart infusion broth (3.4% w/v) supplemented with yeast extract (0.5% w/v) (BHY).
2. BHY agar. As in **step 1** above, but containing 1.6% (w/v) agar.
3. Sterile deionized water.
4. Stock vancomycin solution. 2 mg/mL in H_2O.
5. Stock ramoplanin solution. 2 mg/mL in H_2O.
6. Trichloroacetic acid (TCA). 20%.
7. Test tubes containing approx 70 mg sodium bicarbonate $NaHCO_3$.
8. Sephadex G10 column (approx dimensions 25 × 1 cm).

6.4.2. HPLC Analysis of Precursors

1. Prepare the following buffers for HPLC using Milli-Q grade deionized water and degas prior to use.
2. Buffer A. 50 m*M* ammonium acetate, pH 5.1.
3. Buffer B. 10% methanol in 50 m*M* ammonium acetate, pH 5.1. N.B. Use HPLC grade methanol.
4. HPLC grade acetonitrile.

7. Methods

7.1. Detection of vanA and vanB Genes in Enterococci

The authors use a multiplex PCR assay (with primers 3 and 4, and 25 and 26; **Table 1**) to detect the *vanA* and *vanB* genes in clinical isolates of enterococci. These primers have the added advantage of generating fragments suitable, after addition of a label, for use as DNA probes (*see* **Subheading 2.2.2.**).

1. Resuspend two colonies of the enterococcus in 100 µL of sterile distilled water. Vortex and pulse on a microfuge for 5–10 s.

2. Remove 2 μL of the supernatant to a suitable PCR tube (e.g., a 500 μL microfuge tube). This is used as template DNA (*see* **Note 1**).
3. Prepare a 100 μL diluted mix containing primers 3, 4, 25, and 26 (**Table 1**) each at a concentration of 50 ng/μL (*see* **Note 2**).
4. Prepare master mix (sufficient for 40 isolates) by adding to a 1.5-mL microfuge tube on ice, 700 μL of sterile distilled water, 100 μL of 10X PCR buffer, 50 μL of buffer W1 (supplied with Life Technologies *Taq* polymerase), 40 μL of 50 mM MgCl$_2$, 25 μL of each of dATP, dGTP, dTTP, and dCTP (each at 10 mM stock concentrations), 40 μL of diluted primer mix and 25 U of *Taq* polymerase. Vortex and pulse down on a microfuge.
5. Add 23 μL of mix to each tube of template DNA, vortex and pulse down on a microfuge. Add one drop of mineral oil and pulse down again. Transfer the tubes to a thermal cycler.
6. In the authors' laboratory, amplification is carried out using an Omnigene thermal cycler (Hybaid, Teddington, UK) with the following protocol:
   a. 94°C for 5 min (1 cycle).
   b. 94°C for 60 s, 52°C for 30 s, 72°C for 30 s (30 cycles)
   c. 72°C for 10 min (1 cycle)
7. After amplification, remove 4 μL of PCR product, add 4 μL of gel-loading buffer to it, and electrophorese 3 μL on a 2% agarose gel at 90 V for 1–2 h in 0.5X TBE buffer.
8. The gel is stained in 1 μg/mL EB for 15 min and visualized under UV light.
9. The genotypes of unknown isolates are determined by comparing the migration of the amplicons through the gel with those derived from known *vanA* and *vanB* control strains. These controls are amplified in every batch.
10. The specific amplicons generated by this protocol may be used as template DNA for 'second round' PCR reactions to generate probes for hybridization studies (*see* **Notes 3** to **6**).

7.2. Speciation of Enterococci by PCR (64)

For routine identification of clinically relevant enterococci, the multiplex assay described by Dutka-Malen et al. *(16)* is used. The amplification products are derived from the genes encoding the D-ala-D-ala ligases (of *E.faecalis* and *E. faecium*) or the D-ala-D-serine ligases (of *E. gallinarum* and *E.casseliflavus*). These authors also described primers for the *vanA* (**Table 1**, primers 7 and 8) and *vanB* (primers 27 and 28) genes that could also be included in the assay described below. The authors do not use their own primers for the *vanA* (primers 3 and 4) and *vanB* genes (primers 25 and 26) in this assay as they have observed false amplification products when this multiplex is performed.

1. Resuspend two colonies of the enterococcus in 100 μL of sterile distilled water. Vortex and pulse on a microfuge for 5–10 s.
2. Remove 2 μL of the supernatant to a suitable PCR tube (e.g., a 500 μL microfuge tube). This is used as template DNA (*see* **Note 1**).

3. Prepare a 100 µL diluted mix containing primers 37–40 inclusive and 43–46 inclusive (**Table 1**; i.e., eight primers) each at a concentration of 50 ng/µL (*see* **Note 2**).

4. Prepare master mix (sufficient for 40 isolates) by adding to a 1.5 mL microfuge tube on ice, 700 µL of sterile distilled water, 100 µL of 10X PCR buffer, 50 µL of buffer W1 (supplied with Life Technologies *Taq* polymerase), 40 µL of 50 m*M* MgCl$_2$, 25 µL of each of dATP, dGTP, dTTP, and dCTP (each at 10 m*M* stock concentrations), 40 µL of diluted primer mix and 25 U of *Taq* polymerase. Vortex and pulse down on a microfuge.

5. Add 23 µL of mix to each tube of template DNA, vortex, and pulse down on a microfuge. Add one drop of mineral oil and pulse down again. Transfer the tubes to a thermal cycler.

6. In the authors' laboratory, amplification is carried out using an Omnigene thermal cycler (Hybaid) with the following protocol:
   a. 94°C for 2 min (1 cycle).
   b. 94°C for 60 s, 54°C for 60 s, 72°C for 60 s (30 cycles).
   c. 72°C for 10 min (1 cycle).

7. After amplification, remove 4 µL of PCR product, add 4 µL of gel-loading buffer to it and electrophorese 3 µL on a 2% agarose gel at 90 V for 1–2 h in 0.5X TBE buffer.

8. The gel is stained in 1 µg/mL EB for 15 min and visualized under UV light.

9. The identification of unknown isolates is determined by comparing the migration of the amplicons through the gel with those derived from control strains of *E. faecalis*, *E. faecium*, *E. gallinarum*, and *E. casseliflavus*. These controls are amplified in every batch.

7.3. Amplification of Tn1546-Related Elements by Long PCR (L-PCR) (28)

The L-PCR procedure allows amplification of DNA fragments that, because of their length (>5 kb), would not be amplified reliably under normal conditions by *Taq* polymerase. The procedure uses a mixture of two polymerase enzymes, *Taq* polymerase plus an additional enzyme with proofreading (3'-exonuclease) activity that, when used with modified PCR buffers and cycling conditions, is able to amplify long target sequences *(65)*. It is, therefore, possible to amplify the entire length of Tn*1546*-related glycopeptide resistance elements as single amplicons. As it is PCR-based, this approach allows large numbers of isolates to be processed and, furthermore, the amplicons from different isolates can be compared directly by RFLP analysis.

Several commercial kits are now available for L-PCR, but the authors describe below the use of the Expand™ Long Template PCR System (Boehringer Mannheim) as this is the kit they have used. This system uses the combination of *Taq* polymerase with *Pwo* polymerase (from *Pyrococcus woesei*), and is supplied with three buffers optimized for amplification of fragments of phage λ DNA of 0.5–25 kb, 25–30 kb, and more than 30 kb, respectively.

1. Extract genomic DNA from enterococci to be investigated using guanidium thiocyanate *(66)* (*see* Chapter 2 and **Note 7**). *E. faecium* BM4147 (the host of Tn*1546* on plasmid pIP816) and its glycopeptide-susceptible derivative, BM4147-1 *(3)* are suitable positive and negative controls, respectively.

2. Prepare master mix A (sufficient for 20 isolates) by adding to a 1.5-mL microfuge tube on ice, 35 µL each of dATP, dGTP, dTTP, and dCTP (each at 10 m*M* stock concentrations), 600 pmol of primer 60 (**Table 2**, *see* **Note 8**), and sterile distilled water to a final volume of 500 µL.

3. Mix B is separately prepared for each isolate tested by adding to a 0.5-mL microfuge tube on ice, "x" µL of genomic DNA (approx 250 ng), 5 µL of 10X concentrated buffer 1 (supplied with the kit), 0.75 µL of polymerase enzyme mixture (supplied with the kit), and sterile distilled water to a final volume of 25 µL.

4. Add 25 µL of master mix A to each tube of mix B to give the final reaction volume of 50 µL, mix well and overlay with two drops (approx 30 µL) of mineral oil. Transfer the tubes to a thermal cycler.

5. The concentrations of each component in the final 50 µL reaction will be 50 m*M* of Tris-HCl, pH 9.2, 16 m*M* of $(NH_4)_2SO_4$, 1.75 m*M* of $MgCl_2$, 350 µ*M* of each dNTP, and 600 n*M* of primer.

6. In the authors' laboratory, amplification is carried out using an Omnigene thermal cycler (Hybaid) with the following protocol:
   a. 94°C for 2 min (1 cycle).
   b. 94°C for 10 s, 65°C for 30 s, 68°C for 10 min (10 cycles).
   c. 94°C for 10 s, 65°C for 30 s, 68°C for 10 min (increase elongation time by 20 s per cycle for a further 20 cycles).
   d. 68°C for 7 min.

7. After amplification, dilute 5 µL of PCR product in 45 µL sterile distilled water (*see* **Note 9**).

8. Add 4 µL of gel-loading buffer to 5 µL of the diluted products (i.e., containing 0.5 µL of amplified DNA) and electrophorese at 90 V for 2–3 h through a 0.8% agarose gel in 0.5X TBE buffer.

9. Stain the gel in 1 µg/mL EB for 30 min and visualize under UV light (*see* **Note 10**).

10. For RFLP analysis of the amplicons, 5 µL of the diluted product (i.e., containing 0.5 µL of amplified DNA) may be digested overnight with 20 U of restriction enzyme (*see* **Note 11**) plus the appropriate reaction buffer in a final volume of 20 µL.

11. After digestion, add 4 µL of gel-loading buffer and electrophorese at 90 V for 2–3 h through a 1% agarose gel in 0.5X TBE buffer.

12. Stain the gel in 1 µg/mL EB for 30–60 min (to allow staining of small fragments) and visualize under UV light.

13. The sizes of fragments obtained by digestion of a Tn*1546* control are compared with those predicted from DNA sequence data (*see* **Note 11**). If there is agreement between the predicted and observed RFLP pattern for Tn*1546*, other banding patterns are compared. This comparison may be done visually or by computer analysis (*see* Chapter 3).

7.4. Analysis of Peptidoglycan Precursors in Glycopeptide-Resistant Enterococci

7.4.1. Extraction of Precursors

1. Two overnight 6 mL cultures of enterococci grown at 30°C, are used to inoculate two fresh 120-mL culture flasks. One culture is grown (shake slowly in an orbital shaker) and diluted in BHY medium (prewarmed to 37°C; uninduced control), whereas the other is grown and diluted in BHY containing 4 µg/mL vancomycin (to induce resistance).
2. Monitor the growth of the 120 mL cultures at 37°C by measuring absorbance at 600 nm. When the absorbance reaches approx 1.0, add ramoplanin to the culture to give a final concentration of 10 µg/mL.
3. Incubate the cultures with ramoplanin for 50% of the mean generation time (previously determined from the growth curve) to block late stages of peptidoglycan synthesis and cause the accumulation of peptidoglycan precursors.
4. After incubation, harvest the bacteria by centrifugation at 15,000g at 4°C. Discard the culture supernatant and resuspend the cell pellet in 1.2 mL of sterile H_2O. Add 0.7 mL of 20% trichloroacetic acid (TCA) to give a final concentration of 5.75% (including the pellet volume in the calculation) and incubate on ice for 15 min to precipitate proteins.
5. Centrifuge for 30 s at 40,000g at 4°C. Transfer the supernatant, containing the cell wall precursors, to a tube containing approx 70 mg $NaHCO_3$ to neutralize the TCA.
6. Desalt the neutralized samples on a Sephadex G10 gel filtration column. Elute with Milli Q H_2O at a flow rate of 0.5 mL per min and collect 2-mL fractions.
7. Monitor the eluate at wavelength 260 nm (the wavelength at which UDP-linked peptidoglycan precursors absorb maximally). The precursor peak elutes immediately after the void volume and this fraction should be retained for subsequent HPLC analysis. The salt elutes later and is discarded.

7.4.2. HPLC Analysis of Precursors (see **Note 12**)

Analysis of precursors is performed on a Hewlett Packard 1050 Series HPLC using a reverse-phase ODS column (220 × 4.6 mm, Brownlee Labs [Santa Clara, CA] C_{18} reverse-phase) with a matching NewGuard™ guard column cartridge. Apparatus of similar specifications may be used. Integration software may be used to determine the volume under each peak eluted.

1. Prepare and degas buffers A and B (*see* **Subheading 6.4.2.**).
2. Set the flow rate to 0.2 mL per min and the chart speed to 5 mm per min. Equilibrate the HPLC column with buffer A.
3. Inject 20 µL samples of precursors, extracted as described above, into the sample loop of the HPLC apparatus, and analyze.
4. Program the HPLC apparatus with the following elution protocol:
5. From start time to 5 min time point elute with 50 mM ammonium acetate pH 5.1 (i.e., 100% buffer A, 0% buffer B). At the 5 min time-point commence elution

with a methanol gradient starting at 0% methanol in 50 m*M* ammonium acetate pH 5.1, increasing to 2% methanol in 50 m*M* ammonium acetate pH 5.1 at 45 min (i.e., 0% buffer B at 5 min, increasing to 20% buffer B at 45 min). Continue eluting at 2% methanol in 50 m*M* ammonium acetate pH 5.1 until 55 min (i.e., the end of the elution protocol). Re-equilibrate the column between sample analyses with 5 mL of 50 m*M* ammonium acetate pH 5.1 (100% buffer A).

6. After completion of analyses wash the column with 5 mL of 100% acetonitrile to remove any impurities still bound to the column for storage purposes.

7.4.3. Interpreting HPLC Chromatograms

The different peaks obtained by HPLC were originally identified by collecting fractions for further analysis by mass spectrometry or amino acid analysis. However, this would be both costly and extremely time-consuming to do regularly. Hence, to identify the individual peaks on a chromatogram, a standard mixture of purified peptidoglycan precursors is analyzed by HPLC; unknowns are then identified by comparison of their retention times with those of the standard peaks (*see* **Figs. 1** and **2**). Biochemical techniques may be used to confirm the identities of the peaks if necessary.

8. Notes

1. This very crude method of preparing PCR template works very well in the authors' experience and avoids the need for more complex protocols, such as those involving boiling or actual DNA extraction. In their laboratory this method is used for generating PCR templates from a variety of bacteria, including *Enterococcus* spp., *Staphylococcus aureus*, coagulase-negative staphylococci, *Escherichia coli*, *Klebsiella* spp. However, it is only used for amplifying specific products, e.g., antibiotic-resistance genes. This method is not suitable for generating template for use in RAPD/AP-PCR (*see* Chapter 6) or long PCR protocols.

2. Diluted primer mix (sufficient for 100 PCR tests) is prepared by mixing together the volume containing 5 µg of each primer and then adjusting the final volume to 100 µL with sterile distilled water. This diluted stock may be stored at –20°C for reuse.

3. This labeling is accomplished in a "second-round" PCR reaction that uses 1 µL of a specific *vanA* or *vanB* PCR product as template DNA. The second round reaction mix is identical to that described in **Section 7.1**, except that

   a. Only the pair of specific primers are used.

   b. The concentration of dTTP is reduced to 166 µ*M* (instead of 250 µ*M*), whereas

   c. Digoxigenin-11-dUTP (Boehringer Mannheim) is added to a final concentration of 83 µ*M*. The 2:1 ratio of dTTP to substituted-dUTP results in efficient incorporation of label *(63)*.

4. A sample (2–3 µL) of the labeled "second-round" product is analyzed on an agarose gel. The labeled "second round" PCR product is run on a gel next to a lane loaded with the unlabeled "first-round" product. Incorporation of a nonradioactive label, such as digoxigenin, increases the molecular weight and, hence,

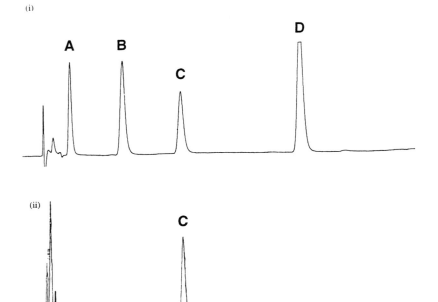

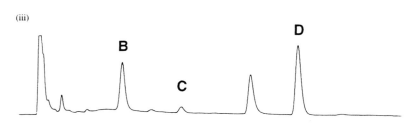

Fig. 1

(i)   HPLC Trace of Standard Cocktail

Precursor	Retention time, min	Area, %
(A) UDP-MurNAc-tripeptide	6.77	17.01
(B) UDP-MurNAc-tetrapeptide	14.15	19.51
(C) UDP-MurNAc-pentapeptide	22.27	14.31
(D) UDP-MurNAc-tetrapeptide-D-lactate	39.03	32.73

decreases the mobility of the labeled product. Thus the labeled product should be slightly higher up the gel than the unlabeled product.

5. The gel is then blotted to nylon membrane (*see* Chapter 2) and incorporation of the label is confirmed. The resulting Southern blot should contain labeled DNA. Detection, therefore, does not require hybridization and can be done directly as recommended by the supplier of the labeled nucleotide. Typically, this involves sequentially washing the blot in blocking solution, an antibody conjugated to an enzyme (e.g., alkaline phosphatase), washing solution, and then adding substrates for the enzyme; the authors use nitroblue tetrazolium and 5-bromo-4-chloro-3-indolylphosphate (NBT/BCIP) colorimetric detection.

6. When a batch of probe is required, several 3 µL aliquots of the labeled PCR product are run on an agarose gel. In order to reduce the risk of nonspecific reactions in hybridization experiments, the authors purify the labeled amplicon from a gel prior to using it as a probe, rather than using the labeled PCR product

---

Fig. 1 *(continued)*

(ii) Uninduced VanB Strain

Precursor	Retention time, min	Area, %
(C) UDP-MurNAc-pentapeptide	22.30	48.32

In the uninduced state, UDP-MurNAc-pentapeptide is synthesised almost exclusively. The absence of UDP-MurNAc-tetrapeptide and UDP-MurNAc-tetrapeptide-D-lactate (pentadepsipeptide) indicate that there is no activity of the resistance genes prior to induction by vancomycin.

(iii) Induced VanB Strain

Precursor	Retention time, min	Area, %
(B) UDP-MurNAc-tetrapeptide	14.93	16.84
(C) UDP-MurNAc-pentapeptide	23.40	1.96
(D) UDP-MurNAc-tetrapeptide-D-lactate	40.15	24.66

After induction of resistance by vancomycin, the modified precursors UDP-MurNAc-tetrapeptide-D-lactate and UDP-MurNAc-tetrapeptide were produced in significant amounts, whereas the production of UDP-MurNAc-pentapeptide (normal precursor) was greatly reduced. This reflects the activities of the resistance genes encoding the enzymes, VanH$_B$ (dehydrogenase), VanB (ligase), VanX$_B$ (D,D-dipeptidase), and VanY$_B$ (D,D-carboxypeptidase). The unlabeled peak that elutes just before pentadepsipeptide (peak D) has not been identified and is of unknown significance.

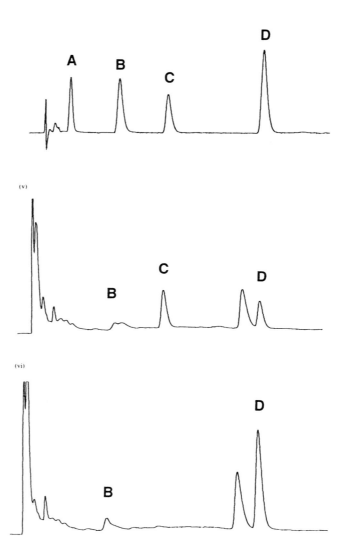

Fig. 2

(iv) HPLC Trace of Standard Cocktail

Precursor	Retention time, min	Area, %
(A) UDP-MurNAc-tripeptide	7.40	15.88
(B) UDP-MurNAc-tetrapeptide	16.03	19.57
(C) UDP-MurNAc-pentapeptide	24.65	14.89
(D) UDP-MurNAc-tetrapeptide-D-lactate	41.59	34.41

directly. After staining in EB, the required bands are carefully cut from the gel using a scalpel and placed into microfuge tubes. A commercial kit (e.g., a Prep-a-Gene kit; Bio-Rad, Hercules, CA, or a Recovery kit; Hybaid) is used to dissolve the agarose and recover the labeled DNA. The DNA from each band is used as a "batch" of probe and is usually sufficiently labeled to allow its use with five to six blots.

7. Unlike PCR for short, specific products, but in common with AP-PCR/RAPD applications (*see* Chapter 6), the quality of the template DNA has a marked effect on the efficiency of L-PCR reactions. The authors have successfully used genomic DNA prepared from enterococci with guanidium thiocyanate. However, to date they have not been successful consistently when using DNA extracted by shorter protocols.

Fig. 2 (*continued*)

(v)   Uninduced VanA Strain

Precursor	Retention time, min	Area, %
(**B**) UDP-MurNAc-tetrapeptide	17.60	1.92
(**C**) UDP-MurNAc-pentapeptide	26.40	11.07
(**D**) UDP-MurNAc-tetrapeptide-D-lactate	43.90	9.26

In the uninduced state not only is the pentapeptide precursor produced, but some UDP-MurNAc-tetrapeptide and UDP-MurNAc-tetrapeptide-D-lactate (pentadepsipeptide) are also synthesized, indicating that the VanA resistance genes are not completely switched off in the absence of inducer. This is quite different from the situation in VanB resistance. The unlabeled peak that elutes just before pentadepsipeptide (peak D) has not been identified and is of unknown significance.

(vi)   Induced VanA Strain

Precursor	Retention time, min	Area, %
(**B**) UDP-MurNAc-tetrapeptide	16.96	3.87
(**D**) UDP-MurNAc-tetrapeptide-D-lactate	42.98	24.99

After induction of resistance by vancomycin, the modified precursors UDP-MurNAc-tetrapeptide-D-lactate and UDP-MurNAc-tetrapeptide were produced. Pentapeptide was not detected. This reflects the activities of the resistance genes encoding the enzymes, VanH (dehydrogenase), VanA (ligase), VanX (D,D-dipeptidase), and VanY (D,D-carboxypeptidase). The unlabeled peak that elutes just before pentadepsipeptide (peak D) has not been identified and is of unknown significance.

8. Primers suitable for L-PCR tend to be longer than those for other PCR studies. In the authors' initial studies of this technique we have used a 26 mer (primer 60 in **Table 2**) that is complementary to sequences in the 38-bp terminal inverted repeats of Tn*1546*.

9. The products of L-PCR reactions (and the 1 in 10 working dilutions) may be stored at 4°C for extended periods (weeks or months). In the authors' experience, RFLP analysis of the stored products generates patterns indistinguishable from those obtained with freshly amplified material.

10. For Tn*1546* and related elements, products should be approx 11 kb in size. Repeated failure to obtain products with this protocol from isolates known to have a *vanA* genotype only indicates that they significantly differ from Tn*1546* in at least one of the terminal inverted repeats. Further studies are required to confirm the presence of a novel genetic element. In the authors' initial studies, 15 of 18 VanA strains tested gave products with primer 60 *(28)*.

11. The authors have used *Eco*RI, *Eco*RV, *Hin*dIII, and *Xba*I. The predicted sizes of fragments obtained from Tn*1546* (GenBank M97297) are
    a. *Eco*RI, 5564 bp, 4113 bp, and 1174 bp;
    b. *Eco*RV, 4879 bp, 2701 bp, 1726 bp, 961 bp, and 584 bp;
    c. *Hin*dIII, 7116 bp, 2454 bp, 1163 bp, and 118 bp; and
    d. *Xba*I, 9002 bp, 1129 bp, and 720 bp.

12. Basic principles of HPLC. A mixture of molecules in an aqueous solution (buffer A) binds to the hydrophobic solid phase via hydrophobic interactions with the C_{18} column (a silica-based beaded matrix to which alkyl chains 18 carbon atoms long are attached). The separation of the different molecules is achieved by exploitation of both hydrophobicity and molecular size. Molecules are eluted from the column by applying a gradient of increasing concentration of a water-miscible organic solvent (in this case methanol). The increasing hydrophobic character of the mobile phase begins to compete for the binding sites of the molecules that are bound to the stationary phase. At a particular concentration of organic solvent, a particular molecule will be displaced from the solid phase and will move into the mobile phase and, thus, be eluted from the column at a particular retention time and is detected by absorbance, using a variable wavelength detector. For mixtures of molecules with similar binding characteristics, larger molecules elute later, however, differences in hydrophobicity will influence retention times with more hydrophobic molecules eluting later than those that are less hydrophobic. The C_{18} reverse-phase column is hydrophobic and, therefore, more hydrophobic compounds elute later as the methanol gradient increases, displacing the solute from the nonionic stationary phase into the mobile phase. The retention times of the different precursors are determined by eluting a standard sample containing a known mixture of purified precursors. This standard mixture of precursors is analyzed by HPLC whenever a new batch of sample analyses are performed.

Acknowledgment

The authors would like to thank Prof. Michel Arthur (Institut Pasteur, Paris) for providing confirmation of the primer sequences shown in **Table 3**.

Addendum in Proof

With regard to glycopeptide resistance in staphylococci (*see* **Subheading 1.2.2.**), readers should be aware that low-level vancomycin resistance (MIC 8 μg/mL) has recently been confirmed in a clinical isolate of *S. aureus* from Japan (Hiramatsu, K., Hanaki H., Ino, T., Yabuta, K., Oguri, T., and Tenover, F. C. [1997] Methicilin-resistant *Staphylococcus aureus* clinical strain with reduced vancomycin susceptibility. *J. Antimicrob. Chemother.* **40**, 135,136). The mechanism of resistance in this strain was unrelated to those found in enterococci.

References

1. Woodford, N., Johnson, A. P., Morrison, D., and Speller, D. C. E. (1995) Current perspectives on glycopeptide resistance. *Clin. Microbiol. Rev.* **8**, 585–615.
2. Uttley, A. H. C., Collins, C. H., Naidoo, J., and George, R. C. (1988) Vancomycin-resistant enterococci. *Lancet* **i**, 57,58.
3. Leclercq, R., Derlot, E., Duval, J., and Courvalin, P. (1988) Plasmid-mediated resistance to vancomycin and teicoplanin in *Enterococcus faecium*. *N. Eng. J. Med.* **319**, 157–161.
4. Reynolds, P. E., Snaith, H. A., Maguire, A. J., Dutka-Malen, S., and Courvalin, P. (1994) Analysis of peptidoglycan precursors in vancomycin resistant *Enterococcus gallinarum* BM4174. *Biochem. J.* **301**, 5–8.
5. Arthur, M., Molinas, C., Depardieu, F., and Courvalin, P. (1993) Characterization of Tn*1546*, a Tn*3*-related transposon conferring glycopeptide resistance by synthesis of depsipeptide peptidoglycan precursors in *Enterococcus faecium* BM4147. *J. Bacteriol.* **175**, 117–127.
6. Evers, S. and Courvalin, P. (1996) Regulation of VanB-type vancomycin resistance gene expression by the VanS$_B$-VanR$_B$ two-component regulatory system in *Enterococcus faecalis* V583. *J. Bacteriol.* **178**, 1302–1309.
7. Handwerger, S., Pucci, M. J., Volk, K. J., Liu, J., and Lee, M. S. (1994) Vancomycin-resistant *Leuconostoc mesenteroides* and *Lactobacillus casei* synthesize cytoplasmic peptidoglycan precursors that terminate in lactate. *J. Bacteriol.* **176**, 260–264.
8. Billot-Klein, D., Gutmann, L., Sable, S., Guittet, E., and van Heijenoort, J. (1994) Modification of peptidoglycan precursors is a common feature of the low-level vancomycin-resistant VANB-type *Enterococcus* D366 and of the naturally glycopeptide-resistant species *Lactobacillus casei, Pediococcus pentosaceus, Leuconostoc mesenteroides,* and *Enterococcus gallinarum*. *J. Bacteriol.* **176**, 2398–2405.
9. Elisha, B. G. and Courvalin, P. (1995) Analysis of genes encoding D-alanine:D-alanine ligase-related enzymes in *Leuconostoc mesenteroides* and *Lactobacillus* spp. *Gene* **152**, 79–83.
10. Noble, W. C., Virani, Z., and Cree, R. G. A. (1992) Co-transfer of vancomycin and other resistance genes from *Enterococcus faecalis* NCTC 12201 to *Staphylococcus aureus*. *FEMS Microbiol. Lett.* **93**, 195–198.

11. Hayden, M. K., Trenholme, G. M., Schultz, J. E., and Sahm, D. F. (1993) In vivo development of teicoplanin resistance in a VanB *Enterococcus faecium* isolate. *J. Infect. Dis.* **167,** 1224–1227.

12. Clark, N. C., Cooksey, R. C., Hill, B. C., Swenson, J. M., and Tenover, F. C. (1993) Characterization of glycopeptide-resistant enterococci from U. S. hospitals. *Antimicrob. Agents Chemother.* **37,** 2311–2317.

13. Klare, I., Heier, H., Claus, H., Reissbrodt, R., and Witte, W. (1995) *vanA*-mediated high-level glycopeptide resistance in *Enterococcus faecium* from animal husbandry. *FEMS Microbiol. Lett.* **125,** 165–172.

14. Woodford, N., Morrison, D., Johnson, A. P., Briant, V., George, R. C., and Cookson, B. D. (1993) Application of DNA probes for rRNA and *vanA* genes to investigation of a nosocomial cluster of vancomycin-resistant enterococci. *J. Clin. Microbiol.* **31,** 653–658.

15. Aarestrup, F. M. (1995) Occurrence of glycopeptide resistance in *Enterococcus faecium* isolates from conventional and ecological poultry farms. *Microbial Drug Resist.* **1,** 255–257.

16. Dutka-Malen, S., Evers, S., and Courvalin, P. (1995) Detection of glycopeptide resistance genotypes and identification to the species level of clinically relevant enterococci by PCR. *J. Clin. Microbiol.* **33,** 24–27 (Erratum in *J. Clin. Microbiol.* **33,** 1434).

17. Sahm, D. F., Free, L., and Handwerger, S. (1995) Inducible and constitutive expression of *vanC-1* encoded resistance to vancomycin in *Enterococcus gallinarum. Antimicrob. Agents Chemother.* **39,** 1480–1484.

18. Rosato, A., Pierre, J., Billot-Klein, D., Buu-Hoi, A., and Gutmann, L. (1995) Inducible and constitutive expression of resistance to glycopeptides amd vancomycin dependence in glycopeptide-resistant *Enterococcus avium. Antimicrob. Agents Chemother.* **39,** 830–833.

19. Miele, A., Bandera, M., and Goldstein, B. (1995) Use of primers selective for vancomycin resistance genes to determine *van* genotype in enterococci and to study gene organization in VanA isolates. *Antimicrob. Agents Chemother.* **39,** 1772–1778.

20. Fraimow, H. S., Jungkind, D. L., Lander, D. W., Delso, D. R., and Dean, J. L. (1994) Urinary tract infection with an *Enterococcus faecalis* isolate that requires vancomycin for growth. *Ann. Intern. Med.* **121,** 22–26.

21. Woodford, N., Morrison, D., Johnson, A. P., Bateman, A. C., Hastings, J. G. M., Elliott, T. S. J., and Cookson, B. D. (1995) Plasmid-mediated *vanB* glycopeptide resistance in enterococci. *Microbial Drug Resist.* **1,** 235–240.

22. Plessis, P., Lamy, T, Donnio, P. Y., Autuly, F., Grulois, I., Le Prise, P. Y., and Avril, J. L. (1995) Epidemiologic analysis of glycopeptide resistant *Enterococcus* strains in neutropenic patients receiving prolonged vancomycin administration. *Eur. J. Clin. Microbiol. Infect. Dis.* **14,** 959–963.

23. Cocconcelli, P. S., Porro, D., Galandini, S., and Senini, L. (1995) Development of RAPD protocol for typing strains of lactic acid bacteria and enterococci. *Lett. Appl. Microbiol.* **21,** 376–379.

24. Barbier, N., Saulnier, P., Chachaty, E., Dumontier, S., and Andremont, A. (1996) Random amplified polymorphic DNA typic versus pulsed-field gel electrophoresis for epidemiological typing of vancomycin-resistant enterococci. *J. Clin. Microbiol.* **34,** 1096–1099.
25. Issack, M. I., Power E. G. M., and French G. L. (1996) Investigation of an outbreak of vancomycin-resistant *Enterococcus faecium* by random amplified polymorphic DNA (RAPD) assay. *J. Hosp. Infect.* **33,** 191–200.
26. Morrison, D., Jones, B., Egelton, C., and Cookson B. D. (1997) PCR typing of *Enterococcus faecium*: An evaluation, in *Streptococci and the Host,* (Horaud, T., Bouvet, A., Leclerq, R., de Montclos, H., and Sicard, M., eds.), Plenum, New York, 387–391.
27. Heaton, M. P. and Handwerger, S. (1995) Conjugative mobilization of a vancomycin resistance plasmid by a putative *Enterococcus faecium* sex pheromone response plasmid. *Microbial Drug Resist.* **1,** 177–183.
28. Woodford, N., Watson, A. P., and Chadwick, P. R. (1997) Investigation by long PCR of the genetic elements mediating VanA glycopeptide resistance in enterococci from uncooked meat in South Manchester, in *Streptococci and the Host* (Horaud, T., Bouvet, A., Leclerq, R., de Montclos, H., and Sicard, M., eds.), Plenum, New York, 409–412.
29. Handwerger, S., Skoble, J., Discotto, L. F., and Pucci, M. J. (1995) Heterogeneity of the *vanA* gene cluster in clinical isolates of enterococci from the Northeastern United States. *Antimicrob. Agents Chemother.* **39,** 362–368.
30. Uttley, A. H. C., George, R. C., Naidoo, J., Woodford, N., Johnson, A. P., Collins, C. H., Morrison, D., Gilfillan, A. J., Fitch, L. E., and Heptonstall, J. (1989) High-level vancomycin-resistant enterococci causing hospital infection. *Epid. Infect.* **103,** 173–181.
31. Quintiliani, R., Jr. and Courvalin, P. (1994) Conjugal transfer of the vancomycin resistance determinant *vanB* between enterococci involves the movement of large genetic elements from chromosome to chromosome. *FEMS Microbiol. Lett.* **119,** 359–364.
32. Quintiliani, R. Jr. and Courvalin, P. (1996) Characterization of Tn*1547*, a composite transposon flanked by IS*16* and IS*256*-like elements, that confers vancomycin resistance in *Enterococcus faecalis* BM4281. *Gene,* **172,** 1–8.
33. Woodford, N., Jones, B. L., Baccus, Z., Ludlam, H. A., and Brown, D. F. J. (1995) Linkage of vancomycin and high-level gentamicin resistance genes on the same plasmid in a clinical isolate of *Enterococcus faecalis. J. Antimicrob. Chemother.* **35,** 179–184.
34. Wirth, R. (1994) The sex pheromone system of *Enterococcus faecalis*: more than just a plasmid-collection system? *Eur. J. Biochem.* **222,** 235–246.
35. Handwerger, S., Pucci, M. J., and Kolokathis, A. (1990) Vancomycin resistance is encoded on a pheromone response plasmid in *Enterococcus faecium* 228. *Antimicrob. Agents Chemother.* **34,** 358–360.
36. Heaton, M. P., Discotto, L. F., Pucci, M. J., and Handwerger, S. (1996) Mobilization of vancomycin resistance by transposon-mediated fusion of a VanA plasmid with an *Enterococcus faecium* sex pheromone-response plasmid. *Gene* **171,** 9–17.

37. Handwerger, S. and Skoble, J. (1995) Identification of chromosomal mobile element conferring high-level vancomycin resistance in *Enterococcus faecium*. *Antimicrob. Agents Chemother.* **39,** 2446–2453.

38. Van de Auwera, P., Pensart, N., Korten, V., Murray, B. E., and Leclercq, R. (1996) Influence of oral glycopeptides on the fecal flora of human volunteers: selection of highly glycopeptide-resistant enterococci. *J. Infect. Dis.* **173,** 1129–1136.

39. Mudd, A. (1996) Vancomycin resistance and avoparcin. *Lancet* **347,** 1412.

40. Wise, R. (1996) Avoparcin and animal feedstuff. *Lancet* **347,** 1835.

41. Bates, J., Jordens, J. Z., and Griffiths, D. T. (1994) Farm animals as a putative reservoir for vancomycin-resistant enterococcal infection in man. *J. Antimicrob. Chemother.* **34,** 507–514.

42. Chadwick, P. R., Woodford, N., Kaczmarski, E. B., Gray, S., Barrell, R. A., and Oppenheim, B. A. (1996) Glycopeptide-resistant enterococci from uncooked meat. *J. Antimicrob. Chemother.* **38,** 908,909.

43. Aarestrup, F. M., Ahrens, P., Madsen, M., Pallesen, L. V., Poulsen, R. L., and Westh, H. (1996) Glycopeptide susceptibility among Danish *Enterococcus faecium* and *Enterococcus faecalis* isolates of animal and human origin and PCR identification of genes within the VanA cluster. *Antimicrob. Agents Chemother.* **40,** 1938–1940.

44. Dutka-Malen, S., Molinas, C., Arthur, M., and Courvalin, P. (1992) Sequence of the *vanC* gene of *Enterococcus gallinarum* BM4174 encoding a D-alanine:D-alanine ligase-related protein necessary for vancomycin resistance. *Gene* **112,** 53–58.

45. Gold, H. S., Unal, S., Cercenado, E., Thauvin-Eliopoulos, C., Eliopoulos, G. M., and Moellering, R. C., Jr. (1993) A gene conferring resistance to vancomycin but not teicoplanin in isolates of *Enterococcus faecalis* and *Enterococcus faecium* demonstrates homology with *vanB, vanA,* and *vanC* genes of enterococci. *Antimicrob. Agents Chemother.* **37,** 1604–1609.

46. Holman, T. R., Wu, Z., Wanner, B. L., and Walsh, C. T. (1994) Identification of the DNA-binding site for the phosphorylated VanR protein required for vancomycin resistance in *Enterococcus faecium*. *Biochemistry* **33,** 4625–4631.

47. Evers, S., Reynolds, P. E., and Courvalin, P. (1994) Sequence of the *vanB* and *ddl* genes encoding D-alanine:D-lactate and D-alanine:D-alanine ligases in vancomycin-resistant *Enterococcus faecalis* V583. *Gene* **140,** 97–102.

48. Evers, S., Sahm, D. F., and Courvalin, P. (1993) The *vanB* gene of vancomycin-resistant *Enterococcus faecalis* V583 is structurally related to genes encoding D-Ala:D-Ala ligases and glycopeptide-resistance proteins VanA and VanC. *Gene* **124,** 143–144.

49. Navarro, F. and Courvalin, P. (1994) Analysis of genes encoding D-alanine:D-alanine ligase related enzymes in *Enterococcus casseliflavus* and *Enterococcus flavescens*. *Antimicrob. Agents Chemother.* **38,** 1788–1793.

50. Arthur, M., Molinas, C., and Courvalin, P. (1992) The VanS-VanR two-component regulatory system controls synthesis of depsipeptide peptidoglycan precursors in *Enterococcus faecium* BM4147. *J. Bacteriol.* **174,** 2582–2591.

51. Handwerger, S., Discotto, L., Thanassi, J., and Pucci, M. J. (1992) Insertional inactivation of a gene which controls expression of vancomycin resistance on plasmid pHKK100. *FEMS Microbiol. Lett.* **92,** 11–14.

52. Reynolds, P. E. (1989) Structure, biochemistry and mechanism of action of glycopeptide antibiotics. *Eur. J. Clin. Microbiol. Infect. Dis.* **8,** 943–950.

53. Messer, J. and Reynolds, P. E. (1992) Modified peptidoglycan precursors produced by glycopeptide-resistant enterococci. *FEMS Microbiol. Lett.* **94,** 195–200.

54. Reynolds, P. E., Snaith, H. A., Maguire, A. J., Dutka-Malen, S., and Courvalin, P. (1994) Analysis of peptidoglycan precursors in vancomycin-resistant *Enterococcus gallinarum* BM4174. *Biochem. J.* **301,** 5–8.

55. Allen, N. E., Hobbs, J. N., Richardson, J. M., and Riggin, R. M. (1992) Biosynthesis of modified peptidoglycan precursors by vancomycin-resistant *Enterococcus faecium. FEMS Microbiol. Lett.* **98,** 109–115.

56. Handwerger, S., Pucci, M. J., Volk, K. J., Liu, J. P., and Lee, M. S. (1992) The cytoplasmic peptidoglycan precursor of vancomycin-resistant *Enterococcus faecalis* terminates in lactate. *J. Bacteriol.* **174,** 5982–5984.

57. Billot-Klein, D., Gutmann, L., Collatz, E., and van Heijenoort, J. (1992) Analysis of peptidoglycan precursors in vancomycin-resistant enterococci. *Antimicrob. Agents Chemother.* **36,** 1487–1490.

58. Arthur, M. and Courvalin, P. (1993) Genetics and mechanisms of glycopeptide resistance in enterococci. *Antimicrob. Agents Chemother.* **37,** 1563–1571.

59. Bugg, T. D. H., Wright, G. D., Dutka-Malen, S., Arthur, M., Courvalin, P., and Walsh, C. T. (1991) Molecular basis for vancomycin resistance in *Enterococcus faecium* BM4147: biosynthesis of a depsipeptide peptidoglycan precursor by vancomycin resistance proteins VanH and VanA. *Biochemistry* **30,** 10,408–10,415.

60. Arthur, M., Molinas, C., and Courvalin, P. (1992) Sequence of the *vanY* gene required for production of a vancomycin-inducible D,D-carboxypeptidase in *Enterococcus faecium* BM4147. *Gene* **120,** 111–114.

61. Wright, G. D., Molinas, C., Arthur, M., Courvalin, P., and Walsh, C. T. (1992) Characterization of VanY, a DD-carboxypeptidase from vancomycin-resistant *Enterococcus faecium* BM4147. *Antimicrob. Agents Chemother.* **36,** 1514–1518.

62. Baptista, M., Depardieu, F., Courvalin, P., and Arthur, M. (1996) Specificity of induction of glycopeptide resistance genes in *Enterococcus faecalis. Antimicrob. Agents Chemother.* **40,** 2291–2295.

63. Emanuel, J. R. (1991) Simple and efficient system for synthesis of nonradioactive nucleic acid hybridization probes using PCR. *Nucleic Acids Res.* **19,** 2790.

64. Woodford, N., Egelton, C. M., and Morrison, D. Comparison of PCR with phenotypic methods for the speciation of enterococci, in *Streptococci and the Host,* (Horaud, T., Bouvet, A., Leclerq, R., de Montclos, H., and Sicard, M., eds.), Plenum, New York, 387–391.

65. Barnes, W. M. (1994) PCR amplification of up to 35-kb DNA with high fidelity and high yield from λ bacteriophage templates. *Proc. Natl. Acad. Sci. USA* **91,** 2216–2220.

66. Pitcher, D. G., Saunders N. A., and Owen, R. J. (1989) Rapid extraction of bacterial genomic DNA with guanidium thiocyanate. *Letts. Appl. Microbiol.* **8,** 151–156.

30

Quinolone Resistance

Janice C. Brown and Sebastian G. B. Amyes*

1. Introduction
1.1. Quinolone Antibacterials: An Overview

Quinolone antibacterial agents were first introduced into the clinical environment in the early 1960s. The first quinolone to be clinically used was nalidixic acid, which was used for the treatment of enteric and urinary tract infections. As a result of increased clinical resistance to this drug, its use has declined. However, the development of other chemically related antimicrobials with activities approaching one thousand times that of nalidixic acid has meant that bacteria resistant to this early nonfluorinated quinolone are susceptible to the action of the newer fluoroquinolones. The fluoroquinolones, such as ciprofloxacin and ofloxacin, have proved to be potent antimicrobials and are used throughout the world in the treatment of bacterial infections, ranging from urinary tract infections to life-threatening septicemia. The clinical success of these agents can be attributed to their broad spectrum of activity, unique mechanism of action, good tissue distribution, and absorption from the gastrointestinal tract after oral administration *(1)*.

1.2. Mechanism of Action

Quinolones have been shown to target primarily DNA gyrase, the bacterial type II topoisomerase. DNA gyrase is responsible for the negative supercoiling of chromosomal DNA, and accomplishes this by nicking the DNA strands at sites 4 bp apart, supercoiling the DNA, and then resealing the nick. It is this resealing step that is inhibited by quinolones. The diversity between DNA gyrase and eukaryotic type II topoisomerases makes the enzyme a suitable antibacterial target. Quinolones bind to DNA and this complex is then bound by DNA gyrase. This results in the DNA gyrase being "poisoned," and perma-

*Corresponding author.

From: *Methods in Molecular Medicine, Vol. 15: Molecular Bacteriology: Protocols and Clinical Applications*
Edited by: N. Woodford and A. P. Johnson © Humana Press Inc., Totowa, NJ

nently attached to the bacterial chromosome via a covalent bond *(2)*. Through a combination of double-stranded DNA breaks and induction of the bacterial SOS response, the bacteria die rapidly.

1.3. Resistance to Quinolones at a Molecular Level

1.3.1. Mutations in the Genes Encoding DNA Gyrase

As a consequence of the extensive and, in some cases, indiscriminate use of the fluoroquinolones, clinical resistance to these agents has been widely reported *(3)*. At a molecular level, bacteria have evolved a number of mechanisms to counteract the effects of the quinolones. Clinical pathogens with a decreased susceptibility to quinolones commonly have alterations in the *gyrA* gene that encodes the GyrA subunit of the DNA gyrase enzyme. Point mutations that correlate with decreased susceptibility to quinolones have been located in a region termed the "quinolone resistance determining region" (QRDR) of the GyrA subunit, which, in *Escherichia coli*, includes amino acids 67–106 (encoded by nucleotides 201–320). DNA gyrase with mutations in this region has been found to bind less well to the DNA and quinolone in vitro *(4)*. This seems to be the predominant mechanism of resistance to quinolones, certainly in the Enterobacteriaceae. There have also been reports of mutations in the *gyrB* gene of bacteria with decreased susceptibilities to quinolones, however these seem to be of minor importance compared with the *gyrA* mutations. Nucleotide changes occurring in response to quinolone resistance in various micro-organisms are shown in **Table 1**.

1.3.2. Mutations in parC

E. coli, as well as other bacteria, is known to possess another quinolone target, topoisomerase IV (encoded by the *parC* and *parE* genes). The role of the inhibition of topoisomerase IV became apparent only when resistance was examined *(24)*. Quinolones cause accumulation of replication catenanes, which is diagnostic of a loss of topoisomerase IV activity. Mutant forms of topoisomerase IV can provide an additional 10-fold resistance to quinolones by the prevention of drug-induced catenane accumulation. It should be noted that quinolone inhibition of topoisomerase IV differs from that of either of the component subunits of DNA gyrase in that

1. Wild-type topoisomerase IV is not dominant over the resistant allele;
2. Inhibition of topoisomerase IV leads to only a slow stop in replication; and
3. Inhibition of topoisomerase IV is primarily bacteriostatic.

These differences suggest that it is a secondary target to DNA gyrase. However, nucleotide sequencing results suggest that it shows some considerable homology with the *gyrA* gene of the host organism *(25)*, and that it can be

Table 1
Mutations in Gyrase Genes Conferring Resistance to Quinolones (taken from ref. 5)

Species	Subunit	Amino acid no.	Quinolone mutation	Resistance[a]	Ref.
E. coli	GyrA	67	Ala→Ser	low-level	6
E. coli	GyrA	81	Gly→Cys	medium-level	7
E. coli	GyrA	83	Ser→Ala	low-level	2
E. coli	GyrA	83	Ser→Leu	high-level	6
E. coli	GyrA	83	Ser→Trp	high-level	6
E. coli	GyrA	84	Ala→Pro	medium-level	7
E. coli	GyrA	87	Ser→Leu	high-level	8
			Asp→Gly[†]		
E. coli	GyrA	87	Asp→Asn	medium-/high- level	7
E. coli	GyrA	87	Asp→Val	low-level	9
E. coli	GyrA	106	Gln→Arg	low-level	2
E. coli	GyrA	106	Gln→His	low-level	6
E. coli	GyrB	426	Asp→Asn	low/medium-level	10
E. coli	GyrB	447	Lys→Glu	low-level	10
P. aeruginosa	GyrA	87	Asp→Asn	medium-level	11
P. aeruginosa	GyrA	87	Asp→Tyr	medium/high- level	
P. aeruginosa	GyrA	83	Thr→Ile	high-level	
N. gonorrhoeae	GyrA	83	Ser→Phe	low-level	12
N. gonorrhoeae	GyrA	87	Asp→Asn	medium-level †	
Helicobacter pylori	GyrA	87 [83]	Asn→Lys	medium/high	13
	GyrA	88 [84]	Ala→Val	medium-level	
	GyrA	91 [87]	Asp→ Gly/Asn/Tyr	medium/high	
S. aureus	GyrA	84 [83]	Ser→Leu	medium-level	14
S. aureus	GyrA	84 [83]	Ser→Ala	medium-level	15
S. aureus	GyrA	85 [84]	Ser→Pro	high-level	14
S. aureus	GyrA	88 [87]	Glu→Lys	medium-level	15
S. aureus	GyrA	84 [83]	Ser→Leu	medium-level	16
		88 [7]	Glu→Lys	medium/high- level	
C. jejuni	GyrA	70 [67]	Ala→Thr	low/medium-level	17
C. jejuni	GyrA	86 [83]	Thr→Ile	high-level	
C. jejuni	GyrA	90 [87]	Asp→Ala	medium/low-level	
Mycobacterium avium	GyrA	83	Ala→Val	high-level	18
M. smegmatis	GyrA	83 87	Ala→Val	low-level	19
			Asp→Gly	high-level †	
M. tuberculosis	GyrA	90 [83]	Ala→Val	high-level	20
Salmonella	GyrA	83	Ser→Phe/Tyr	medium/low-level	21
typhimurium		87	Asp→Tyr/Asn	medium/low-level	
A. baumannii	GyrA	83	Ser→Leu	medium-level	22
		81	Gly→ Val	low-level	
A. salmonicida	GyrA	83	Ser→Ile	low-level	23
		67	Ala→Gly	medium-level†	

[ ] denotes equivalent amino acid position in *E. coli;* † denotes had both mutations.
[a]Low-level resistance up to 1 mg/L of quinolones; medium-level up to 32 mg/L; high-level above 32 mg/L.

studied in a similar manner to *gyrA*, except that the "plasmid dominance test" (*see* **Subheading 5.1.**) cannot be performed. In particular, mutations in residues glycine-78 to aspartate, serine-80 to arginine or isoleucine, and glutamate-84 to glycine or lysine, may contribute to decreased quinolone susceptibility *(26,27)*. The authors are just beginning to examine the role of *parC* mutations in quinolone resistance, and any specific recommendations made here will almost immediately become out-of-date.

1.3.3. Mutations in the Outer Membrane Proteins (OMPs)

Resistance to quinolones via altered drug permeability has been documented for *E. coli (28,29)*, *Klebsiella pneumoniae (30)*, *Serratia* spp. *(31)*, *Pseudomonas aeruginosa (32,33)*, *Proteus vulgaris (34)*, and *Citrobacter* spp. *(35)*. Most reports conclude that this resistance is brought about by changes in outer membrane proteins, or porins, although some reports have also implicated lipopolysaccharide (LPS) changes in reduced uptake *(29,36)*. The resistance of *E. coli* to quinolones via altered permeability mechanisms has been well-documented, and for this reason *E. coli* will be concentrated upon here.

Quinolone resistance brought about by altered permeability is generally characterized by low-level resistance to quinolones, and also to structurally unrelated drugs. High-level resistance linked to permeability changes, however, has been documented in *Citrobacter freundii* and *Bacillus fragilis (37)*. It has been speculated that drug permeability was decreased as a result of a decreased expression of the outer membrane porin OmpF *(38)*. The importance of OmpF expression in quinolone resistance was established by Hirai and colleagues *(29)*, who documented that spontaneous single-step *E.coli* mutants deficient in OmpF were fourfold less susceptible to quinolones than the parent strain and also accumulated norfloxacin at a slower rate. Decreases in OmpF proteins have been implicated in fleroxacin resistance *(39)*, norfloxacin resistance *(29)*, and ciprofloxacin resistance *(40)* in *E. coli*. The *norB* mutation (decreased expression of OmpF protein) results in low-level resistance to quinolones, cefoxitin, tetracycline, and chloramphenicol *(29)*. The *norC* mutation (decreased expression of OmpF, defective LPS core, and hypersusceptibility to detergents) encodes hypersusceptibility to hydrophobic quinolones, such as nalidixic acid. Resistance brought about by outer membrane protein mutations is thought to have less effect on nalidixic acid than modern fluorinated quinolones, as nalidixic acid is more hydrophobic and may, therefore, permeate into the bacterial cell through the lipid bilayer *(28)*.

1.3.4. The Multiple Antibiotic Resistance Locus (Mar)

Another nonspecific mechanism of resistance to quinolones involves the multiple antibiotic resistance locus or *mar* gene. This can be exemplified by laboratory-derived, quinolone-resistant mutants created by serial passage of *P. aeruginosa* on quinolone-containing agar to obtain high-level resistance (e.g., ciprofloxacin MIC of 1024 mg/L). After 4- to 32-fold increases in quinolone MICs (during sequential passage on quinolone-containing agar), *P. aeruginosa* alters its morphology and acquires a multiple-antibiotic-resistant (Mar) phenotype (including cross-resistance to β-lactams, chloramphenicol, and tetracycline). This is manifested by reduced uptake of quinolones and the other agents, as well as by altered outer membrane proteins (reductions in the 25 and 38 kDa bands as well as of several bands in the 43–66 kDa region) *(41)*.

1.3.5. Efflux Pumps

There is evidence that quinolone resistance can be mediated by an efflux system. This is epitomized by the *norA* gene. The NorA protein has been well-studied in *Staphylococcus aureus* where it mediates the active efflux of hydrophilic fluoroquinolones from the cell, conferring low-level resistance upon the organism. This is an efflux pump similar to those responsible for tetracycline resistance (*see* Chapter 31). This efflux pump is not specific to the quinolones and can transport structurally diverse substrates. Increased transcription of the *norA* gene, leading to a greater quantity of the NorA protein within the cytoplasmic membrane, is considered to be the mechanism of quinolone insusceptibility *(42)*. As this mechanism is so similar to that of the plasmid-encoded tetracycline resistance genes, it is possible that the gene could migrate to a plasmid location. Indeed, if transferable quinolone resistance emerges in the future, this is the most likely mechanism.

2. Diagnosis
2.1. Phenotypic Detection of Resistance

Unlike the aminoglycosides (*see* Chapter 28) and even the β-lactams (*see* Chapter 25), it is not possible to examine the resistance pattern to a cohort of quinolones and predict the resistance mechanism. Resistance to one of the group is almost always accompanied by a proportional increase in resistance to the others, which means that distinction cannot be made by observing differential increases in resistance.

2.2. Molecular Characterisation of Resistance Mechanisms

Quinolone resistance found in clinical isolates of bacteria is rarely the result of a single mutation, instead it arises from a sequential build-up of alterations in permeability and in structural genes, usually *gyrA* and *parC*. It is very unlikely that permeability mutations will produce a high level of resistance, except in certain species. Thus MICs of greater than 10-times that of the sensitive strain indicate that mutations are present in the structural genes of DNA gyrase. If the MICs are high (over 100-times the sensitive MIC), then this would suggest alterations in the *gyrA* gene, most probably in the amino acid that corresponds to amino acid 83, which for many of the Enterobacteriaceae is a serine residue. At that point, it may be suitable to set up the "plasmid dominance test" (*see* **Subheading 5.1.**) to look specifically for alterations in *gyrA* structure.

In this day and age, one should be looking for direct molecular distinction of quinolone resistance genes. This is only feasible at the moment for the *gyrA* gene and perhaps soon also for the *gyrB* gene, but there are many problems associated with this approach. As this is not a plasmid-encoded gene, one is dealing with different genes in each species studied. There is often surprisingly limited similarity between *gyrA* genes of apparently closely related species. Thus to employ a genetic approach to distinguish *gyrA* alterations makes the presumption that the species identification is unflawed. This may be extremely difficult to guarantee as the techniques that are used to distinguish species may not be sensitive enough. This may result in *gyrA* genes from strains of apparently the same species appearing different, as has already been shown to be the case with *gyrB* genes of *Acinetobacter* spp. *(43)*.

This diversity in the DNA gyrase structural genes makes diagnosis of quinolone resistance particularly difficult and may also cause problems with direct polymerase chain reaction (PCR) amplification. With some other resistance genes, it is possible to design oligonucleotides that straddle the region of alteration and that will therefore identify changes in the nucleotide sequence by amplifying only those sequences that, for instance, decrease susceptibility. These are usually plasmid-encoded genes and are present in greater numbers than the DNA gyrase structural genes. If we make the assumption that we can distinguish bacterial species sufficiently, then the problems still faced by a molecular characterization scheme are great.

With some plasmid-encoded genes, it has been possible to use oligonucleotide probes to distinguish changes, particularly with the extended-spectrum β-lactamases (*see* Chapter 25). This approach could cause problems with *gyrA* alterations because of the wide diversity of alterations that could be present and the limited number of *gyrA* genes for the probe to adhere to. Look simply at the possibilities of alterations in one site of the *gyrA* gene in *E. coli*, the alteration of the codon for serine-83 (**Fig. 1**). Three single-step mutations give three different amino acid alterations, each capable of raising the MIC of quinolones. Therefore, to diagnose an alteration, the loss of binding to an oligonucleotide straddling the sequence TCG would have to be identified, but this is a negative result. To demonstrate that an alteration had occurred would require positive hybridization to another oligonucleotide. In this case, three

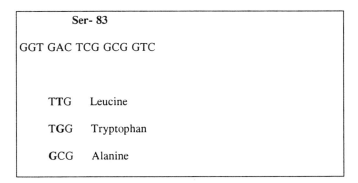

Fig. 1. DNA sequence surrounding serine-83 of *E.coli.*

new oligonucleotides would be required, and even then it would not be certain that all possibilities had been covered. The use of oligonucleotides for distinguishing alterations in resistance genes can be extremely difficult, and has not yet adapted well to a diagnostic environment; it would not be appropriate at the moment for quinolone resistance.

More recently, PCR has been used to amplify the QRDR of the *gyrA* gene and, more latterly, the *parC* gene. This takes oligonucleotides that match conserved sequences and part of the gene, of a known and measurable size, is amplified. If one of the oligonucleotides is labeled with biotin, it is possible to separate the DNA strands and directly sequence them. This is not feasible in a diagnostic laboratory as, although fairly rapid nowadays, the use of a radiolabel (for sequencing) and the time taken for analysis makes it impractical. If a suitable restriction site is found, then it is possible to distinguish alterations in a specific codon. This is much more practicable, and it is quite possible to prepare amplified DNA and examine the restriction fragment length polymorphisms within one day. It may also be possible to use the diversity on the *gyrA* genes to distinguish the type of organism as well. For instance, the oligonucleotides 5' ATGAGCGACCTTGCGAGAGAAATTACACCG 3' and 5' TTCCATCAGCCCTTCAATGCTGATGTCTTC 3' will amplify the QRDR of *gyrA* from most of the Enterobacteriaceae. Variation of the oligonucleotide can be used to make the distinction more specific as discussed later.

3. Epidemiologic Investigation

In the study of resistance epidemiology, the authors are particularly concerned to know whether resistance is emerging as a spontaneous response to therapy in different hospitals, or whether there is clonal spread of a resistant variant. This is not only important in studying the epidemiology of plasmid-encoded resistance determinants, but also with those determined by the chromosome. In the past, techniques such as bacteriophage typing, serotyping, plasmid fingerprinting (*see* Chapter 4), and ribotyping (*see* Chapter 2) have been extensively used for this purpose. However, the recent development of more accurate molecular techniques, such as PCR-based methods and analysis of chromosomal DNA restriction patterns by pulsed-field gel electrophoresis (PFGE) (*see* Chapter 3) has meant that these early techniques have been superseded. Indeed, for the investigation of quinolone-resistant outbreaks, the use of PFGE has featured strongly in the recent literature *(44,45)*.

In India, there has recently been an epidemic of multiresistant *Salmonella typhi.* The isolates are identifiable by the presence of a 180 kb *incH* (H incompatibility group) plasmid that confers trimethoprim, chloramphenicol, and ampicillin resistance. Pulsed-field characterization of the isolates revealed that, as far as could be determined, most of the host strains were virtually identical, no matter which part of India they derived from. Thus, clonal spread appeared to be the cause of the epidemic. However, some strains showed decreased susceptibility to ciprofloxacin and sequencing of the *gyrA* genes revealed diversity in the mutations amongst the isolates. This suggests that spontaneous emergence of quinolone resistance has occurred against a background of clonal spread. On the other hand, isolates in a recent *Acinetobacter baumannii* outbreak in Edinburgh showed rapid acquisition of quinolone resistance, which was probably too rapid to be explained by spontaneous emergence. The difference in Edinburgh, as distinct from the outbreak in India, is that the quinolone resistance is being maintained by the antibiotic blanket of the intensive care units of Edinburgh Royal Infirmary. In India, the selective pressure is immediately removed on the cessation of treatment and, unlike plasmid-encoded genes, it seems improbable that mutations in the *gyrA*, *gyrB*, *parC*, or outer membrane proteins will remain competitive once the quinolone environment is lost.

3.1. Molecular Typing of Clinical Isolates by PCR Fingerprinting

Molecular typing by PCR fingerprinting can accurately detect whether resistance has spontaneously emerged throughout a bacterial population, or if clonal spread of one resistant organism has occurred. Amplification of the bacterial DNA with an arbitrary primer (e.g., the M13 core primer) will generate PCR "fingerprints" of the resistant isolates in such a way that if the emergence of spontaneous resistance has occurred, different "fingerprints" will be observed and if clonal spread has occurred, the same "fingerprints" will be generated for all the investigated isolates. For further details, the reader is referred to Chapters 6 and 7.

3.2. Molecular Typing of Clinical Isolates by PFGE

PFGE (*see* Chapter 3) stands out as a superior technique for investigating a clinically important outbreak of resistant bacteria. Quinolone-resistant bacteria derived from the same parent strain can be distinguished from those that are probably not part of the same outbreak. The presence or absence of various bands for a particular isolate on a PFGE gel compared with other isolates, enables the clinical microbiologist to reach a quick and usually accurate conclusion. The accuracy of this technique can be improved by repeating the digestion of chromosomal DNA with a different enzyme. For a protocol, the reader is referred to Chapter 3.

4. Materials

4.1. The Plasmid Dominance Test

4.1.1. Conjugation

1. Donor. *E. coli* that carries one of the plasmids listed in **Table 2**.
2. ML broth. 10 g Bacto tryptone, 5 g Bacto yeast extract, 5 g NaCl, and 2 g K_2HPO_4 per liter.
3. Recipient strain. The quinolone-resistant isolate under investigation.
4. 0.15 M NaCl.
5. ML agar. ML broth plus 1.5% Bacteriological Agar No. 1 and appropriate antibiotic selection.

4.1.2. Transformation

1. TSS buffer. 1% (w/v) Bacto tryptone, 0.5% (w/v) yeast extract, 1% (w/v) NaCl, 10% (w/v) PEG3350, 20 mM $MgSO_4$, 10 mM PIPES pH 6.5.
2. Dimethyl sulfoxide (DMSO).
3. LBG broth. LB broth containing 20 mM glucose.

4.2. Analysis of Outer Membrane Proteins of Gram-Negative Bacteria

1. Medium A. 7 g nutrient broth, 1 g yeast extract, 2 g glycerol, 3.7 g K_2HPO_4, and 1.3 g KH_2PO_4 per liter, supplemented with different concentrations of sucrose.
2. 10 mM sodium phosphate pH 7.2.
3. 2% Triton X-100.
4. 2X SDS-PAGE sample buffer: 0.125 M Tris-HCl, pH 6.8, 4% SDS, 20% glycerol, 10% 2-mercaptoethanol, 0.004% bromophenol blue.
5. SDS-PAGE gel. A complete description of SDS-PAGE is given by Sambrook et al. *(46)*.

4.3. Amplification of the QRDR-Containing Region of gyrA in Clinical Isolates

4.3.1. Preparation of Chromosomal DNA

1. Nutrient broth.
2. HTE buffer. 50 mM Tris-HCl, pH 8.0, 20 mM EDTA.
3. 2% sarcosyl in HTE buffer.

4. 10 mg/mL RNase in TE buffer
5. 10 mg/mL pronase in 10 mM Tris-HCl, pH 8.0, 10 mM NaCl, 0.1 mM EDTA buffer.
6. Phenol:chloroform:isoamyl alcohol (25:24:1).
7. 3 M sodium acetate pH 5.5.
8. Isopropanol.
9. 70% ethanol.

4.3.2. Amplification of the QRDR of gyrA by PCR

1. PCR buffer (10X stock).
2. 50 mM MgCl$_2$.
3. 1 mM stock dNTPs.
4. Appropriate oligonucleotide primers (*see* **Note 1**).
5. DNA sample (*see* **Note 2**).
6. Diluted *Taq* polymerase (*see* **Note 3**).

4.4. DNA Sequencing of the gyrA QRDR

4.4.1. Sequencing Single-Stranded DNA Using Dynabeads

1. Appropriate primer (*see* **Table 3**). The primer used in the annealing reaction is always the opposite one which will extend the opposite strand to the template. By using both templates in separate reactions, it is possible to sequence both strands of DNA.
2. Annealing reaction buffer (Amersham, Arlington Heights, IL).
3. DNA attached to Dynabeads (Dynal A.S., N-0212, Oslo, Norway; *see* **Note 4**).
4. Sequenase 2.0 and other sequencing components (*see* **Note 5**)
5. ^{35}S-dATP radiolabel (Amersham).

4.4.2. Enzymatic Degradation of Unwanted dNTPs and Primers

1. Exonuclease I (Amersham).
2. Shrimp alkaline phosphatase (Amersham).

4.5. Restriction Fragment Length Polymorphisms in the gyrA Gene

1. Appropriate restriction enzyme (*see* **Table 4**) along with appropriate buffer.
2. PCR-amplified DNA.
3. Agarose gel and running buffer (*see* Chapter 4).
4. Standard DNA loading buffer (*see* Chapter 4).

4.6. Single-Stranded Conformational Polymorphism Analysis

1. Purifed PCR-amplified DNA (*see* **Note 6**).
2. Loading solution. 95% deionized formamide, 20 mM EDTA, 0.05% xylene cyanol, 0.05% bromophenol blue.
3. MDE gel matrix solution (FMC Bioproducts).

5. Methods

Quinolone resistance in clinical isolates of bacteria results from either a mutation in the DNA encoding *gyrA,* a mutation in the DNA encoding *gyrB,* a change in the OMP profile, or a mutation in the DNA encoding *parC.* In order to determine the cause of quinolone resistance in a particular clinical isolate, a number of procedures can be carried out as detailed in **Fig. 2**. Each method displayed in the flow diagram will be described separately.

Although not so well-documented, the investigation of mutations in *parC* can be carried out following these protocols since these are general protocols for detecting point mutations in DNA. Restriction enzymes and primer sequences would of course have to be adapted.

5.1. The Plasmid Dominance Test

The "plasmid dominance test" is a well-documented technique for the investigation of the nature of quinolone resistance *(47)*. Transformation of a quinolone-resistant isolate with a plasmid containing the sensitive allele of *gyrA* will result in a complete reversion to quinolone sensitivity if one or more point mutations have occurred in the chromosomal copy of the *gyrA* gene. Likewise, transformation of a quinolone-resistant isolate with a plasmid containing the sensitive allele of *gyrB* will result in a complete reversion to quinolone sensitivity if one or more point mutations have occurred in the chromosomal copy of the *gyrB* gene. This "genetic dominance" results from the quinolone-sensitive GyrA and GyrB sub-units produced by the plasmids competing with the mutated quinolone-resistant chromosomal GyrA and GyrB subunits resulting in an increased susceptibility of the cell to quinolones. However, if the isolate under investigation also has OMP mutations, a much less pronounced decrease in the MIC of quinolones will be observed. Various cosmid and plasmid constructs have been used for these types of investigations. Important information in the use of such constructs is displayed in **Table 2**.

1. Transfer one of the constructs listed in **Table 2** into the cell to be tested by the appropriate method (*see* **Subheadings 5.1.1.** and **5.1.2.**).
2. Plate out the mixtures onto a series of agar plates containing doubling dilutions of quinolone as well as the appropriate selective antibiotic and incubate overnight at 37°C.
3. Read the MIC of the transferred strain. A drop in MIC of the isolate being tested signifies a mutation in the same gene that is carried on the plasmid.

5.1.1. Conjugation

1. Grow donor and recipient strains statically overnight at 37°C.
2. Mix 0.1 mL of donor culture with 1 mL of recipient culture in 4.5 mL of ML broth.

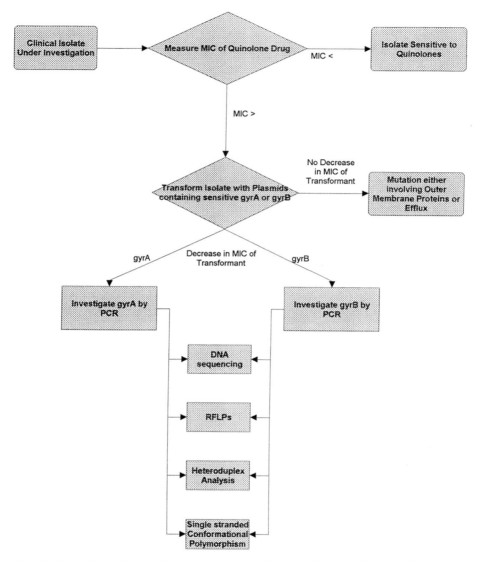

Fig. 2. Detection of mutations associated with quinolone resistance: An over-view of the stepwise progression of experimental procedures used to investigate mutations in *gyrA* and *gyrB* genes (this may also be used to detect mutations in *parC*).

3. Incubate mixtures at 37°C for 6 h.
4. Plate out mating mixtures on appropriate antibiotic-containing plates.

Table 2
Plasmid and Cosmids Used for the Plasmid Dominance Test

Name	Selective markers	Size	Mode of transfer into cell	Host range	Ref.
pNJR3-2 (*gyrA*)	Tetracycline	28kb	Conjugation	*E. coli* *P. aeruginosa*	*48*
pBP515 (*gyrA*)	Kanamycin Gentamicin	11kb	Transformation	Gram-negative bacteria	*47*
pBP547 (*gyrB*)	Kanamycin	11kb	Transformation	Gram-negative bacteria	*47*

5.1.2. Transformation

1. Grow up an overnight culture of recipient cells.
2. Dilute the culture 1 in 100 in fresh LB broth.
3. Grow up to an OD_{600} of between 0.3 and 0.4. Place the cells on ice, then pellet in a benchtop centrifuge at 2060g for 10 min.
4. Drain the broth from the cells and resuspend the cells in ice-cold TSS buffer (a volume equivalent to 0.1X original volume of cells is used). Vortex the cells to resuspend thoroughly and then add a volume of DMSO equivalent to 0.05X original volume of cells.
5. Add 1 to 100 ng of plasmid DNA (in a volume of less than 10 μL) to 100 μL of cells.
6. Leave the cells on ice for 30 min, then add 400 μL of LBG broth.
7. Incubate the cells at 37°C for 1 h.
8. Plate samples of the mixture on appropriate antibiotic-containing plates and incubate overnight.

5.2. Analysis of Outer Membrane Proteins of Gram-Negative Bacteria

If it is suspected that changes in permeability are the cause or part of the cause of quinolone resistance in gram-negative clinical isolates, the following protocol can be used to extract OMPs. The stained proteins can be visualized after gel electrophoresis and changes compared.

1. Inoculate a single colony into 10 mL of Medium A and grow overnight at 37°C in a shaking incubator.
2. Spin the culture in a bench top centrifuge at 2060g for 10 min.
3. Discard the broth and wash the cell pellet in 2 mL of phosphate buffer and resuspend the pellet in 0.5 mL of the same buffer.
4. Sonicate, spin down at 1400g for 10 min to remove unbroken cells and debris.
5. Centrifuge at 100,000g for 30 min and resuspend the cell envelopes in 4 mL of 2% Triton X-100 at 37°C for 15 min.
6. Centrifuge at 100,000g to recover insoluble fractions.

7. Wash once with 2 mL of phosphate buffer.
8. Standardize the protein concentrations by comparing to a standard of known concentration.
9. Preheat preps at 100°C in an equal volume of 2X SDS-PAGE sample buffer.
10. Analyze on a SDS-PAGE gel.

5.3. Amplification of the QRDR-Containing Region of gyrA in Clinical Isolates

The DNA encoding the QRDR of *gyrA* can be amplified from chromosomal DNA by PCR as a starting point in the molecular investigation of quinolone resistance in clinical bacteria.

5.3.1. Preparation of Chromosomal DNA

The following method has been used successfully for *E.coli* and several other types of Gram-negative bacteria, including *Rhizobium*, *Serratia*, and *Vibrio* species *(49)*. For Gram-positive organisms, the lysis step has to be modified *(50)* (*see* also **Note 2**).

1. Inoculate a single colony into 5 mL of nutrient broth and grow overnight at 37°C in a shaking incubator.
2. Spin 1.2 mL of this culture in a microcentrifuge for 15 s.
3. Discard the broth and resuspend the cell pellet in 0.31 mL of HTE buffer by briefly vortexing.
4. Add 0.35 mL of 2% sarcosyl in HTE buffer with brief inversion of the tube, followed by 5 µL of RNase.
5. Add 35 µL of pronase and incubate the tubes in a 50°C water bath until lysis occurs.
6. Extract the DNA three times with phenol:chloroform:isoamyl alcohol.
7. Add a 1/10 vol of 3 *M* sodium acetate and 1 vol of isopropanol (stored at –20°C). DNA precipitation is achieved after placing the tubes at –70°C for 15 min followed by centrifugation at 11,600*g* in a microcentrifuge for 20 min.
8. Wash the resulting pellet with 70% ethanol and then air-dry for 15 min.
9. Resuspend the pellet in 100 µL of sterile distilled water.

5.3.2. Amplification of the QRDR of gyrA by PCR

For investigation of the *gyrA* QRDR of clinical isolates, amplification of a 300 bp region by PCR will generate a fragment of DNA suitable for subsequent manipulations. Published primer sequences corresponding to part of the *gyrA* gene of various bacterial species are detailed in **Table 3**. Careful choice of primers will permit usage in other species. For example, the *Salmonella* spp. primers listed above were designed to amplify the QRDR of *gyrA* of *Salmonella* spp., but have since also been found to amplify the QRDR of *gyrA* of *E. coli*, *K. pneumoniae*, *C. freundii*, and *Enterobacter sakazakii*.

Table 3
Primer Sequences Used to Amplify the QRDR of *gyrA* in Various Species

Bacterial species	Primer sequence (5'–3')	Ref.
Mycobacterium spp.	GGTAAGTACCACCCGCCACGGC	*19*
	CTCGGTGTAACGCATGGCGGC	
Escherichia coli	TACACCGGTCAACATTGAGG	*9*
	TTAATGATTGCCGCCGTCGG	
Salmonella spp.	ATGAGCGACCTTGCGAGAGAAATTACACCG	*51*
	TTCCATCAGCCCTTCAATGCTGATGTCTTC	
Acinetobacter baumannii	AAATCTGCCCGTGTCGTTGGT	*22*
	GCCATACCTACGGCGATACC	
Shigella dysenteriae	TACACCGGTCAACATTGAGG	*52*
	TTAATGATTGCCGCCGTCGG	
Staphylococcus aureus	ATGAACAAGGTATGACACCGG	*53*
	CACGTATCGTTGGTGACGTA	
Neisseria gonorrhoeae	CGCGATGCACGAGCTGAAAAA	*12*
	ATTTCGGTATAGCGCATGGCTG	
Helicobacter pylori	AATTAGGCCTTACTTCCAAAGTCGCTTACA	*13*
	TCTTCACTCGCCTTAGTCATTCTGGC	
Enterococcus faecalis	CGGGATGAACGAATTGGGTGTGA	*54*
	AATTTTACTCATACGTGCTTCGG	
Aeromonas hydrophila	TCTGTTTGCTATGAACGAGTTG	*5*
	GTGGAACCACCCGTACTGTCG	
Pseudomonas aeruginosa	GAGCTGGGCAACGACTGGAACAAGCCC	*55*
	GATACCGCTGGAACCGTTGACCAGCAG	

1. Prepare the PCR mix by adding together 75 µL of sterile distilled water, 10 µL of 10X PCR buffer, 5 µL of 50 m*M* MgCl$_2$, 5 µL of 1 m*M* stock dNTPs, 1 µL of 5' oligo (10 pmoles), 1 µL of 3' oligo (10 pmoles), 1 µL of DNA (*see* **Subheading 5.3.1.** and **Note 2**) and 2 µL of diluted *Taq* polymerase.
2. A standard thermocycler is used for DNA amplification. The program includes a preliminary stage of 96°C for 180 s followed by 24 cycles at 96°C for 15 s, 50°C for 30 s and 70°C for 90 s. A final extension step runs at 70°C for 5 min. A complete description of the choice of conditions for PCR is given in the Recombinant DNA Laboratory Manual *(49)* (*see* also Chapter 5)
3. The above standard conditions can be employed initially and modified later if required. In general, it may be necessary to adjust the melting and annealing temperature depending on the characteristics of the primers.

5.4. DNA Sequencing of the gyrA *QRDR*

DNA sequencing of the *gyrA* QRDR can be carried out in a number of ways, two of which are described below. The advantage of sequencing a approx 300

Table 4
Modifications to Primers Required
for Subsequent Manipulation of PCR Products

Subsequent manipulation	Modification to primer(s)
Sequencing of single-stranded DNA by Dynabeads	A biotin label is attached to one of the primers
Cloning of PCR fragment and site-directed mutagenesis	Two different restriction enzyme "handles" are attached to each end of the primer. This allows ligation of the PCR fragment directly into a suitably digested vector in the correct orientation

bp fragment of DNA is that in most cases, only one sequencing gel is needed to visualize the entire region of interest. Most commonly, in DNA sequencing of the QRDR of clinical isolates, direct sequencing of PCR fragment is carried out. It is only when further manipulations of the DNA (such as site-directed mutagenesis) are planned that it is necessary to clone the region of interest into a cloning vector (*see* **Table 4**).

5.4.1. Sequencing Single-Stranded DNA Using Dynabeads

Dynabeads consist of streptavidin-coated beads that bind strongly to the biotin label incorporated in one of the primers. This method is simple to use and only involves the preparation of a limited number of basic solutions. Primer is annealed to the DNA template as follows.

1. Mix together 1 µL of primer, 1 µL of reaction buffer, 5 µL of DNA attached to Dynabeads, and 3 µL of sterile distilled water.
2. Warm the tube to 65°C for 2 min.
3. Allow the temperature of the tube to cool slowly to room temperature over a period of about 30 min, then place on ice.
4. Direct sequencing can then be carried out on either strand as detailed in **Subheading 5.4.3.**
5. Electrophoresis of the sequencing reactions is carried out as described in the Recombinant DNA Laboratory Manual *(49)*.

5.4.2. Enzymatic Degradation of Unwanted dNTPs and Primers

This sequencing method is convenient as it eliminates all DNA purification steps after PCR. The double-stranded template is sequenced directly without having to change the buffer. Two enzymes, exonuclease I and shrimp alkaline phosphatase, are added to the amplified PCR product, incubated at 37°C and excess dNTPs and primers are thus degraded. The enzymes themselves are heat-labile and can be removed from the reaction mixture by heating at 80°C. The DNA template is denatured and the primer annealed by the use of a heating and snap-cooling procedure. The PCR product can subsequently be sequenced as described in **Subheading 5.4.3.**

1. Mix 1 µL of primer (5–10 pmol) with 1–9 µL PCR-amplified DNA (0.5 pmol) and make up to 10 µL with sterile distilled water.
2. Denature by heating to 100°C for 2 min, then allow the temperature of the tube to cool rapidly by immersing in an ice/water bath for 5 min.
3. Centrifuge briefly, then place on ice. Direct sequencing can then be carried out on either strand as detailed in **Subheading 5.4.3.**

5.4.3. Sequencing Reaction

1. Add 10 µL of template-primer (from above) to 1 µL of 0.1 M DTT, 2 µL of diluted labeling mix, 0.5 µL of ^{35}S-dATP, and 2 µL of diluted Sequenase. This is the labeling reaction.
2. Mix thoroughly, avoiding bubbles, and incubate for 2–5 min at room temperature.
3. Label four tubes "G," "A," "T" and "C". Place 2.5 µL of the ddGTP termination mix in the tube labeled "G." Similarly fill the "A," "T," and "C" tubes with 2.5 µL of the ddATP, ddTTP and ddCTP respectively. Prewarm the tubes for at least 1 min at 37°C. When the labeling incubation is complete, remove 3.5 µL and transfer it to the tube labeled "G." Mix and continue incubation of the "G" tube at 37°C. Similarly transfer 3.5 µL of the labeling reaction to the "A," "T," and "C" tubes, mixing and returning them to the 37°C water bath.
4. Continue the incubations for 3 min. Add 4 µL of stop solution to each, mix and quench on ice. Prior to electrophoresis, denature by heating at 80°C for 4 min.

5.5. Restriction Fragment Length Polymorphisms in the gyrA Gene

As certain common *gyrA* mutations (e.g., some found in *E. coli*) often cause the abolition of a *Hinf*I restriction site within a region of the enzyme centred around serine-83, *Hinf*I restriction can be used as a screen for amino acid substitutions in the QRDR. The sequence recognized and cleaved by this enzyme is G/ANTC, thus a *gyrA* PCR fragment of *E.coli* amplified from a strain with no amino acid substitution at serine-83 (nucleotide sequence encoding amino acids 82 and 83 being GACTCG) would be cleaved by the enzyme whereas a *gyrA* PCR fragment amplified from a quinolone-resistant strain with a mutation at serine-83 would not be cut since this would change the nucleotide sequence in this region. As many other prokaryotes have the sequence GANTC centered around serine-83, *Hinf*I RFLPs can be conveniently used to screen point mutations in *gyrA* occurring in response to a decreased susceptibility to ciprofloxacin in species other than *E. coli* **(56)**. Details of restriction enzymes used to detect point mutations in *gyrA* of a range of organisms are given in **Table 5**.

**Table 5
Restriction Enzymes Used
to Detect Point Mutations in *gyrA* in a Range of Organisms**

Bacterial species	Restriction enzyme	Sequence recognized	Reference
Escherichia coli	*Hin*fI	GANTC	*9*
Neisseria gonorrhoeae	*Hin*fI	GANTC	*12*
Staphylococcus aureus	*Hin*fI	GANTC	*53*
Shigella dysenteriae	*Hin*fI	GANTC	*52*
Salmonella spp.	*Hin*fI	GANTC	*56*
Acinetobacter baumannii	*Hin*fI	GANTC	*22*
Pseudomonas aeruginosa	*Sst*II	CCGCGG	*57*
Aeromonas hydrophila	*Bsr*DI	GCAATGNN	*5*

1. Add 40 µL of PCR-amplified DNA to 1 µL of *Hin*fI (or appropriate enzyme from **Table 4**) along with 5 µL of 10X buffer and 4 µL of sterile distilled water, to make a total volume of 50 µL.
2. Incubate for 3 h at 37°C.
3. Run fragments out on a 4% agarose gel after addition of 10 µL of standard DNA loading buffer.

5.6. Single-Stranded Conformational Polymorphism Analysis

Single stranded conformational polymorphism (SSCP) is a sensitive procedure for detecting polymorphisms by analyzing mobility shifts of single-stranded DNAs on neutral polyacrylamide gels. The incorporation of a point mutation into a fragment of DNA will usually result in a mobility shift with respect to an identical DNA fragment with no mutation. In this way, PCR products can be rapidly screened for mutations simply after a brief period of denaturation followed by electrophoresis. To identify band shifts, either the PCR product can be radiolabeled (e.g., using ^{32}P-dATP) and detected by autoradiography, or the DNA bands can be detected by staining either with ethidium bromide (EB), or silver stain. The optimum length of PCR-amplified DNA to use is 300 bp. Although not a commonly used technique in the detection of quinolone resistance, SSCP has been used successfully to distinguish between isolates with and without mutations in *gyrA* (*53*).

1. Amplify a suitable section of chromosomal DNA by PCR (as in **Subheading 5.3.**).
2. Clean the DNA by the use of a PCR-purification kit.
3. Add 1 µL of PCR product to 10 µL of loading solution.
4. After denaturation at 80°C for 5 min, analyze 10 µL by electrophoresis. Electrophoresis of the SSCP reactions is carried out by the use of specialized MDE gel matrix solution (FMC Bioproducts) on standard sequencing electrophoresis apparatus.

6. Notes

1. When designing the primers, the following should be considered. Depending on what one intends to do with the PCR-amplified DNA, various modifications to the primers can be carried out during their manufacture to aid subsequent manipulations. These are briefly detailed in **Table 4** (*see* also Chapter 5).

2. For PCR, either 1 μL chromosomal DNA prepared as in **Subheading 5.3.1.** can be used or 1 μL of a bacterial suspension prepared by adding a single bacterial colony in 100 μL of sterile distilled water, boiling for 15 min and removing the cell debris by briefly centrifuging in a microcentrifuge.

3. *Taq* polymerase, 10X PCR buffer and MgCl$_2$ are readily available from common molecular biology reagent suppliers such as Gibco-BRL. Dilute *Taq* polymerase by adding 2 μL of *Taq* polymerase to 6.5 μL of dH$_2$O, 1 μL of 10X PCR buffer, and 0.5 μL of MgCl$_2$. Too much *Taq* polymerase will result in mispriming of the template.

4. The procedure used to separate the DNA strands takes about 30 min and uses common laboratory solutions. By retaining the supernatant after DNA duplex denaturation, both strands can be sequenced.

5. A Sequenase 2.0 kit can be obtained from Amersham. This contains all necessary reagents for sequencing, except the radiolabel.

6. PCR products are purified using a suitable kit (e.g., a Wizard prep kit from Promega, Madison, WI).

References

1. Smith, J. T. and Lewin, C. S. (1988) Chemistry and mechanisms of action of the quinolone antibacterials. in *The Quinolones,* (Andriole, V. T., ed.), Academic, pp. 23–82.

2. Hallett, P. and Maxwell, A. (1991) Novel quinolone resistance mutations of the *Escherichia coli* DNA gyrase A protein: enzymatic analysis of the mutant proteins. *Antimicrob. Agents Chemother.* **35**, 335–340.

3. Wiedemann, B. and Heisig, P. (1994) Mechanisms of quinolone resistance. *Infection* **22**, S73–S79.

4. Willmott, C. J. R. and Maxwell, A. (1993) A single point mutation in the DNA gyrase A protein greatly reduces binding of fluoroquinolones to the gyrase-DNA complex. *Antimicrob. Agents Chemother.* **37**, 126–127.

5. Hayes, M. V. (1995) Beta lactam and quinolone resistance in *Aeromonas* spp. PhD thesis. Edinburgh University.

6. Yoshida, H., Kojima, T., Yamagishi, J., and Nakamura, S. (1988) Quinolone-resistant mutations of the *gyrA* gene of *Escherichia coli. Mol. Gen. Genet.* **211**, 1–7.

7. Yoshida, H., Bogaki, M., Nakamura, M., and Nakamura, S. (1990) Quinolone resistance-determining region in the DNA gyrase *gyrA* gene of *Escherichia coli. Antimicrob. Agents Chemother.* **34**, 1271–1272.

8. Heisig, P., Schedletzky, H., and Fackensteinpaul, H. (1993) Mutations in the *gyrA* gene of a highly fluoroquinolone-resistant clinical isolate of *Escherichia coli. Antimicrob. Agents Chemother.* **37**, 696–701.

9. Oram, M. and Fisher, L. M. (1991) 4-Quinolone resistance mutations in the DNA gyrase of *Escherichia coli* clinical isolates identified by using the polymerase chain reaction. *Antimicrob. Agents Chemother.* **35**, 387–389.

10. Yamagishi, J., Yoshida, H., Yamayoshi, M., and Nakamura, S. (1986) Nalidixic acid-resistant mutations of the *gyrB* gene of *Escherichia coli*. *Mol. Gen. Genet.* **204**, 367–373.

11. Kureishi, A., Diver, J. M., Beckthold, B., Schollaardt, T., and Bryan, L. E. (1994) Cloning and nucleotide sequence of *Pseudomonas aeruginosa* DNA gyrase *gyrA* gene from strain PAO1 and quinolone-resistant clinical isolates. *Antimicrob. Agents Chemother.* **38**, 1944–1952.

12. Deguchi, T., Yasuda, M., Asano, M., Tada, K., Iwata, H., Komeda, H., Ezaki, T., Saito, I., and Kawada, Y. (1995) DNA gyrase mutations in quinolone-resistant clinical isolates of *Neisseria gonorrhoeae*. *Antimicrob. Agents Chemother.* **39**, 561–563.

13. Moore, R. A., Beckthold, B., Wong, S., Kureishi, A., and Bryan, L. E. (1995) Nucleotide sequence of the *gyrA* gene and characterization of ciprofloxacin-resistant mutants of *Helicobacter pylori*. *Antimicrob. Agents Chemother.* **39**, 107–111.

14. Sreedharan, S., Oram, M., Jensen, B., Peterson, L. R., and Fisher, L. M. (1990) DNA gyrase *gyrA* mutations in ciprofloxacin-resistant strains of *Staphylococcus aureus*: close similarity with quinolone resistance mutations in *Escherichia coli*. *J. Bacteriol.* **172**, 7260–7262.

15. Goswitz, J. J., Willard, K. E., Fasching, C. E., and Peterson, L. R. (1992) Detection of *gyrA* gene mutations associated with ciprofloxacin resistance in meticillin-resistant *Staphylococcus aureus*: analysis by polymerase chain reaction and automated direct DNA sequencing. *Antimicrob. Agents Chemother.* **36**, 1166–1169.

16. Tanaka, M., Zhang, Y. X., Ishida, H., Akasaka, T., Sato, K., and Hayakawa, I. (1995) Mechanisms of 4-quinolone resistance in quinolone-resistant and methicillin-resistant *Staphylococcus aureus* isolates from Japan and China. *J. Med. Microbiol.* **42**, 214–219.

17. Wang, Y., Huang, W. M., and Taylor, D. E. (1993) Cloning and nucleotide sequence of the *Campylobacter jejuni gyrA* gene and characterization of quinolone resistance mutations. *Antimicrob. Agents Chemother.* **37**, 457–463.

18. Cambau, E., Sougakoff, W., and Jarlier, V. (1994) Amplification and nucleotide sequence of the quinolone resistance-determining region in the *gyrA* gene of mycobacteria. *FEMS Microbiol. Lett.* **116**, 49–54.

19. Revel, V., Cambau, E., Jarlier, V., and Sougakoff, W. (1994) Characterisation of mutations in *Mycobacterium smegmatis* involved in resistance to fluoroquinolones. *Antimicrob. Agents Chemother.* **38**, 1991–1996.

20. Cambau, E., Sougakoff, W., Besson, M., Truffotpernot, C., Grosset, J., and Jarlier, V. (1994) Selection of a *gyrA* mutant of *Mycobacterium tuberculosis* resistant to fluoroquinolones during treatment with ofloxacin. *J. Infect. Dis.* **170**, 479–483.

21. Reyna, F., Huesca, M., Gonzalez, V., and Fuchs, L. Y. (1995) *Salmonella typhimurium gyrA* mutations associated with fluoroquinolone resistance. *Antimicrob. Agents Chemother.* **39**, 1621–1623.

22. Vila, J., Ruiz, J., Goni, P., Marcos, A., and Deanta, T. J. (1995) Mutation in the *gyrA* gene of quinolone-resistant clinical isolates of *Acinetobacter baumannii*. *Antimicrob. Agents Chemother.* **39**, 1201–1203.

23. Oppegaard, H. and Sorum, H. (1994) *gyrA* mutations in quinolone-resistant isolates of the fish pathogen *Aeromonas salmonicida. Antimicrob. Agents Chemother.* **38**, 2460–2464.

24. Khodursky, A. B., Zechiedrich, E. L., and Cozzarelli, N. R. (1995) Topoisomerase IV is a target of quinolones in *Escherichia coli. Proc. Natl. Acad. Sci. USA* **92**, 11,801–11,805.

25. Deguchi, T., Yasuda, M., Nakano, M., Ozeki, S., Ezaki, T., Saito, I., and Kawada, Y. (1996) Quinolone-resistant *Neisseria gonorrhoeae*: Correlation of alterations in the GyrA subunit of DNA gyrase and the ParC subunit of topoisomerase IV with antimicrobial susceptibility profiles. *Antimicrob. Agents Chemother.* **40**, 1020–1023.

26. Vila, J., Ruiz, J., Goni, P., and Deanta, M. T. J. (1996) Detection of mutations in *parC* in quinolone-resistant clinical isolates of *Escherichia coli. Antimicrob. Agents Chemother.* **40**, 491–493.

27. Heisig, P. (1996) Genetic evidence for a role of *parC* mutations in development of high-level fluoroquinolone resistance in *Escherichia coli. Antimicrob. Agents Chemother.* **40**, 879–885.

28. Hirai, K., Aoyama, H., Irikura, T., Iyobe, S., and Mitsuhashi, S. (1986) Differences in susceptibility to quinolones of outer membrane mutants of *Salmonella typhimurium* and *Escherichia coli. Antimicrob. Agents Chemother.* **29**, 535–538.

29. Hirai, K., Aoyama, H., Suzue, S., Irikura, T., Iyobe, S., and Mitsuhashi, S. (1986) Isolation and characterization of norfloxacin-resistant mutants of *Escherichia coli* K-12. *Antimicrob. Agents Chemother.* **30**, 248–253.

30. Sanders, C. C., Sanders, W. E., Goering, R. V., and Werner, V. (1984) Selection of multiple antibiotic resistance by Quinolones, Beta-lactams and Aminoglycosides with special reference to cross-resistance between unrelated drug classes. *Antimicrob. Agents Chemother.* **26**, 797–801.

31. Gutmann, L., Williamson, R., Moreau, N., Kitzis, M.-D., Collatz, E., Acar, J. F., and Goldstein, F. W. (1985) Cross-resistance to nalidixic acid, trimethoprim and chloramphenicol associated with alterations in outer membrane proteins of *Klebsiella, Enterobacter* and *Serratia. J. Infect. Dis.* **151**, 501–507.

32. Hirai, K., Suzue, S., Irikura, T., Iyobe, S., and Mitsuhashi, S. (1987) Mutations producing resistance to norfloxacin in *Pseudomonas aeruginosa. Antimicrob. Agents Chemother.* **31**, 582–586.

33. Inoue, Y., Sato, K., Fujii, T., Hirai, K., Inoue, M., Iyobe, S., and Mitsuhashi, S. (1987) Some properties of subunits of DNA gyrase from *Pseudomonas aeruginosa* PAO1 and its nalidixic acid-resistant mutant. *J. Bacteriol.* **169**, 2322–2325.

34. Ishii, H., Sato, K., Hoshino, K., Sato, M., Yamaguchi, A., Sawai, T., and Osada, Y. (1991) Active efflux of ofloxacin by a highly quinolone-resistant strain of *Proteus vulgaris. J. Antimicrob. Chemother.* **28**, 827–836.

35. Aoyama, H., Sato, K., Fujii, T., Fujimaki, K., Inoue, M., and Mitsuhashi, S. (1988) Purification of *Citrobacter freundii* DNA gyrase and inhibition by quinolones. *Antimicrob. Agents Chemother.* **32**, 104–109.

36. Moniot-Ville, N., Guibert, J., Moreau, N., Acar, J. F., Collatz, E., and Gutmann, L. (1991) Mechanisms of quinolone resistance in a clinical isolate of *Escherichia coli* highly resistant to fluoroquinolones but susceptible to nalidixic acid. *Antimicrob. Agents Chemother.* **35**, 519–523.

37. Kato, N., Miyauchi, M., Muto, Y., Watanabe, K., and Ueno, K. (1988) Emergence of fluoroquinolone resistance in *Bacteroides fragilis* accompanied by resistance to beta-lactam antibiotics. *Antimicrob. Agents Chemother.* **32**, 1437,1438.
38. Cohen, S. P., McMurray, L. M., Hooper, D. C., Wolfson, J. S., and Levy, S. B. (1989) Cross resistance to fluoroquinolones in multiple-antibiotic-resistanct (Mar) *Escherichia coli* selected by tetracycline or chloramphenicol: Decreased drug accumulation associated with membrane changes in addition to OmpF reduction. *Antimicrob. Agents Chemother.* **33**, 1318–1325.
39. Chapman, J. S., Bertasso, A., and Georgopapadakou, N. H. (1989) Fleroxacin resistance in *Escherichia coli*. *Antimicrob. Agents Chemother.* **33**, 239–241.
40. Bedard, J., Chamberland, S., Wong, S., Schollaardt, T., and Bryan, L. E. (1989) Contribution of permeability and sensitivity to inhibition of DNA synthesis in determining susceptibilities of *Escherichia coli, Pseudomonas aeruginosa and Alcaligenes faecalis* to ciprofloxacin. *Antimicrob. Agents Chemother.* **33**, 1457–1464.
41. Zhanel, G. G., Karlowsky, J. A., Saunders, M. H., Davidson, R. J., Hoban, D. J., Hancock, R. E. W., Mclean, I., and Nicolle, L. E. (1995) Development of multiple-antibiotic-resistant (Mar) mutants of *Pseudomonas aeruginosa* after serial exposure to fluoroquinolones. *Antimicrob. Agents Chemother.* **39**, 489–495.
42. Kaatz, G. W. and Seo, S. M. (1995) Inducible NorA-mediated multidrug resistance in *Staphylococcus aureus. Antimicrob. Agents Chemother.* **39**, 2650–2655.
43. Yamamoto, S. and Harayama, S. (1996) Phylogenetic analysis of *Acinetobacter* strains based on the nucleotide sequences of *gyrB* genes and on the amino acid sequences of their products. *Int. J. Syst. Bacteriol.* **46**, 506–511.
44. Lehn, W., Stowerhoffman, J., Kott, T., Strassner, C., Wagner, H., Kronke, M., and Schneiderbrachert, W. (1996) Characterisation of clinical isolates of *E.coli* showing high levels of fluoroquinolone resistance. *J. Clin. Microbiol.* **34**, 597–602.
45. Oethinger, M., Conrad, S., Kaitel, K., Cometta, A., Bille, J., Klotz, G., Glauser, M. P., Marre, R., and Kern, W. V. (1996) Molecular epidemiology of fluroquinolone-resistant *Escherichia coli* bloodstream isolates from patients admitted to European cancer centres. *J. Clin. Microbiol.* **40**, 387–392.
46. Sambrook, J., Fritsch, E. F., and Maniatis, T. (1989) *Molecular Cloning: A Laboratory Manual*, Cold Spring Harbor Laboratory Press, Cold Spring Harbor, New York.
47. Heisig, P. (1993) High-level fluoroquinolone resistance in a *Salmonella typhimurium* isolate due to alterations in both *gyrA* and *gyrB* genes. *J. Antimicrob. Chemother.* **32**, 367–377.
48. Robillard, N. J. (1990) Broad-host-range gyrase A gene probe. *Antimicrob. Agents Chemother.*, **34**, 1889–1894.
49. Zyskind, J. W. and Bernstein, S. I. (1992) Isolation and purification of *E.coli* and Drosophila chromosomal DNA, in *Recombinant DNA Laboratory Manual,* 2nd ed. (Zyskind, J. W. and Bernstein, S. I., eds.), Academic, New York, pp. 12–15.

50. Del Vecchio, V. G., Petroziello, J. M., Gress, M. J., McCleskey, F. K., Melcher, G. P., Crouch, H. K., and Lupski, J. R. (1995) Molecular genotyping of methicillin-resistant *Staphylococcus aureus* via fluorophore-enhanced repetitive-sequence PCR. *J. Clin. Microbiol.* **33**, 2141–2144.

51. Brown, J. C., Thomson, C. J., and Amyes, S. G. B. (1996) Mutations of the *gyrA* gene of clinical isolates of *Salmonella typhimurium* and three other *Salmonella* species leading to decreased susceptibilities to 4-quinolone drugs. *J. Antimicrob. Chemother.* **37**, 359–366.

52. Rahman, M., Mauff, G., Levy, J., Couturier, M., Pulverer, G., Glasdorff, N., and Butzler, J. P. (1994) Detection of 4-quinolone resistance mutation in *gyrA* gene of *Shigella dysenteriae* type 1 by PCR. *Antimicrob. Agents Chemother.* **38**, 2488–2491.

53. Tokue, Y., Sugano, K., Saito, D., Noda, T., Ohkura, H., Shimosato, Y., and Sekiya, T. (1994) Detection of novel mutations in the *gyrA* gene of *Staphylococcus aureus* by nonradioisotopic single-strand conformation polymorphism analysis and direct DNA sequencing. *Antimicrob. Agents Chemother.* **38**, 428–431.

54. Korten, V., Huang, W. M., and Murray, B. E. (1994) Analysis by PCR and direct DNA sequencing of *gyrA* mutations associated with fluoroquinolone resistance in *Enterococcus faecalis*. *Antimicrob. Agents Chemother.* **38**, 2091–2094.

55. Yonezawa, M., Takahata, M., Matsubara, N., Watanabe, Y., and Narita, H. (1995) DNA gyrase *gyrA* mutations in quinolone-resistant clinical isolates of *Pseudomonas aeruginosa*. *Antimicrob. Agents Chemother.* **39**, 1970–1972.

56. Brown, J. C., Shanahan, P. M. A., Jesudason, M. V., Thomson, C. J., and Amyes, S. G. B. (1996) Mutations responsible for reduced susceptibilities to 4-quinolones in clinical isolates of multi-resistant *Salmonella typhi* in India. *J. Antimicrob. Chemother.* **37**, 891–900.

57. Quibell, K. J. (1993) Mechanisms of resistance of *Pseudomonas aeruginosa* to the 4-quinolones. PhD thesis. Edinburgh University.

31

Resistance to Tetracyclines, Macrolides, Trimethoprim, and Sulfonamides

Marilyn C. Roberts

1. Introduction
1.1. Tetracycline

Tetracyclines are antimicrobial agents that interact with bacterial ribosomes and block protein synthesis. They have activity against a wide range of gram-positive, gram-negative, anaerobic and aerobic bacteria, cell-wall free myco-plasmas, chlamydiae, mycobacteria, rickettsia, *Helicobacter*, *Listeria*, and protozoan parasites, such as *Entamoeba histolytica, Giardia lamblia*, and *Plas-modium falciparum (1–6)*. Tetracyclines have been extensively used in humans for therapy of bacterial respiratory and urogenital tract diseases, as well as periodontal, Lyme, and rickettsial diseases. They are also used for therapy in animals and as growth promoters for animal food production *(1)*. Tetracyclines were the first major group of antibiotics to which the term "broad-spectrum" was applied. Because of this spectrum of activity, their relative safety and low cost, tetracyclines have been widely used throughout the world, and are second after the penicillins in world consumption *(7)*.

The first multiresistant *Shigella dysenteriae*, which causes bacterial dysen-tery, was isolated in 1955 and was resistant to tetracycline, streptomycin, and chloramphenicol *(8–10)* and represented 0.02% of the isolates tested. By 1960, multiresistant *Shigella* represented almost 10% of the strains tested in Japan *(8–10)*, a dramatic increase in 5 yr. The increase in multidrug resistant *Shigella* has continued to the present as illustrated by a recent study *(11)* where over 60% of the *S. flexneri* isolated between 1988 and 1993 were resistant to tetra-cycline, streptomycin, and chloramphenicol. This is the same combination of antibiotic resistance determinants found in 1955 in *S. dysenteriae*. Tetracy-

From: *Methods in Molecular Medicine, Vol. 15: Molecular Bacteriology: Protocols and Clinical Applications*
Edited by: N. Woodford and A. P. Johnson © Humana Press Inc., Totowa, NJ

cline resistance has also been seen with increasing frequencies in Gram-positive and cell-wall free species *(12,13)*.

Mendez et al. *(14)* first examined the genetic heterogeneity of tetracycline resistance determinants on plasmids of the Enterobacteriaceae and Pseudomonadaceae using restriction enzyme analysis, DNA-DNA hybridization, and expression of resistance to tetracycline and its various analogs. DNA-DNA hybridization with the structural genes as probes is now the standard method used to distinguish different *tet* genes *(15)*. A new gene is identified by its inability to hybridize with probes for any of the known *tet* genes under stringent conditions; this criterion indicates less than 80% sequence identity. If this is demonstrated for a new gene, then a letter designation is given in accordance with nomenclature standards *(15)*. However, many of the different *tet* genes share significant homology, as illustrated by *tetM*, *tetO*, and *tetS*, which share 75% sequence identity, but do not cross-hybridize under stringent conditions and are thus considered separate genes. There have been 17 different tetracycline resistance (Tet) determinants characterized. This includes *tetU* recently found in *Enterococcus faecium* *(16)*. Two other Gram-negative *tet* genes have been identified. One has been cloned from *Pseudomonas aeruginosa* *(17)*, and the author has shown that it does not hybridize with either TetG or TetH determinants. It is called TetI since it meets the current criteria for a new determinant. It most likely encodes for an efflux protein. To date, nothing has been done with the second gene.

Sixteen *tet* genes confer resistance by coding either for an efflux protein that pumps tetracycline out of the cell, or for a ribosomal protection protein *(12)*. However, other mechanisms have been described including *tetX* which codes for an enzyme that alters the tetracycline molecule *(12)*. Recently, TetU has been described. It has limited homology with the ribosomal protection protein TetM, and the author has suggested that it may have a different mechanism of resistance *(16)*. Both types of genes are found in gram-positive, gram-negative, aerobic, and anaerobic species (**Table 1**). The *tetB* gene confers resistance to tetracycline, doxycycline, and minocycline while all the other efflux genes confer resistance to only tetracycline and doxycycline *(12,14)*. The ribosomal protection proteins confer resistance to tetracycline, doxycycline and minocycline. Enteric species carry one or more of the *tet* efflux genes *tetA-tetE*, *tetI*, and *tetG*. The *tetB* has the broadest host range in Gram-negative species, and the *tetM* has the broadest host range found in both Gram-negative and Gram-positive species. gram-positive species often carry *tetK*, *tetL*, *tetM*, or *tetO*, whereas few studies have investigated the distribution of *tetS* or *tetQ* *(6,12,18,19)*. A detailed distribution can be found in a recent review *(20)*.

Table 1
Mechanisms of Tetracycline Resistance

Efflux		Ribosomal	
Gram-negative	Gram-positive	Gram-negative	Gram-positive
Tet A-E		Tet M	Tet M
Tet G-H		Tet O	Tet O
Tet I	Tet A(P)	Tet S	
Tet K (rare)	Tet K	Tet Q	Tet Q
Tet L (rare)	Tet L		Tet B(P)

1.2. Erythromycin

Erythromycin is active against gram-positive bacteria and has been used for a range of Gram-positive infections *(21)*. Interest in the macrolides has increased in recent years because of new derivatives, such as azithromycin and clarithromycin *(22)*. These newer analogs have a broader spectrum of activity than erythromycin and are being suggested for use in a number of different infections *(22)*. Macrolides inhibit protein synthesis by binding to the 50S bacterial ribosomal subunit *(21)*. There are a number of different mechanisms of resistance to macrolides, including efflux, mutation that replaces a specific adenine residue at nucleotide position 2058 (or neighboring nucleotides) in the 23S ribosomal RNA (rRNA) and drug inactivation. However, the most common mechanism of resistance results from the presence of rRNA methylases that modify a single adenine residue at position 2058 in the 23S rRNA. This prevents macrolides, lincosamides, and type B streptogramins (MLS) from binding to the 50S ribosomal subunit *(21)*. Thus, one enzyme gives rise to cross-resistance to three structurally unrelated groups of antibiotics that have overlapping binding sites on the ribosomal subunit *(23)*. Lincosamides, like clindamycin, have been extensively used for treatment of anaerobic infections, whereas lincosamide has also been used for animals. Streptogramins have been used in Europe, but are not licensed in the United States *(24)*.

The rRNA methylase genes can protect the cell against the newer macrolides as well as erythromycin. These genes can be expressed constitutively or, inducibly regulated by translational attenuation *(21)*. Most, if not all, of the rRNA methylases seem to have originated in gram-positive species, even those found in gram-negative species. Erythromycin is usually a good inducer for these genes whereas lincosamides are generally not. As a consequence, when an isolate carrying an inducibly regulated gene is tested for antibiotic susceptibility it may appear either susceptible or more often borderline intermediate to erythromycin, but susceptible to lincosamides like clindamycin. However, if

Table 2
Groups of Related rRNA Methylases

Group	Gene	Isolated from[a]
Erm B	*ermAM*	*Streptococcus sanguis, Streptococcus* spp.
	ermBC	*Escherichia coli*
	ermB	*Enterococcus faecalis, Streptococcus* spp.
	ermBP	*Clostridium perfringens*
	ermZ	*Clostridium difficile, E. faecalis*
	ermB-like	*Actinobacillus actinomycetemcomitans,*
		Campylobacter rectus, Peptostreptococcus
		Fusobacterium, Bacteroides, Prevotella
Erm C	*ermC*	*Staphylococcus aureus, Bacillus subtilis*
	ermM	*Staphylococcus epidermidis*
	ermGT	*Lactobacillus reuteri*
	ermC-like	*Peptostreptococcus, Streptococcus*
ErmD	*ermD*	*Bacillus licheniformis*
	ermJ	*Bacillus anthracis*
	ermK	*Bacillus licheniformis*
ErmF	*ermF*	*Bacteroides ovatus*
	ermFS	*Bacteroides fragilis*
	erm*FU*	*Bacteroides fragilis*
	ermF-like	*A. actinomycetemcomitans, Prevotella*
		Peptostreptococcus, Streptococcus spp.

[a]This list of host species is not complete, but the selected representatives illustrate the wide host-range of these rRNA methylase genes *(21,35,47)*.

the same isolate is grown in a low dose of erythromycin (0.5 μg) for 4 h and retested, it will appear resistant to both erythromycin and clindamycin. Thus, when susceptibility testing, if a gram-positive species is borderline intermediate or resistant to erythromycin and/or clindamycin, it should be assumed that it most likely carries a rRNA methylase gene until proven otherwise.

Erythromycin was discovered in 1952 *(1)* and the first macrolide-resistant staphylococci were described in 1953 from a variety of countries *(21)*. The number of species and isolates resistant to the macrolides has increased over time. Some gram-negative enteric species also carry rRNA methylase genes in addition to a variety of anaerobic gram-negative species *(21)* (**Table 2**). There are over 30 different erm genes identified as well as a number of related genes from *Streptomyces* spp. *(21)*. Unfortunately, no nomenclature has been established for naming *erm* genes and as a result, most *erm* genes isolated have been given a separate letter designation, even if they differ from a previously named gene by only a few base pairs *(21,25)*. However, the different *erm* genes can be grouped into families based on their high DNA and amino acid homology. For example, what is called the Erm B group includes *ermAM, ermB, ermBC, ermBP*, and *ermZ*, and similarly *ermD, ermJ*, and *ermK* are nearly identical and comprise the Erm D group *(21)*. Some of the genes that are highly related are grouped in **Table 2**.

1.3. Trimethoprim and Sulfonamides

Trimethoprim and sulfonamides are synthetic drugs unlike either tetracycline or erythromycin. Trimethoprim interacts with the dihydrofolate reductase (DHFR), whereas sulfonamides interact with dihydropteroate synthase (DHPS). These two enzymes are part of the folate biosynthetic pathway required for thymine production and bacterial cell growth *(26)*. Both drugs have a wide spectrum of activity against gram-positive and gram-negative bacteria and have been used to treat urinary tract, respiratory, and skin diseases as well as certain enteric diseases *(26)*. Sulfonamides was first used in 1932, and trimethoprim was available 30 yr later. Since 1968, the combination of the two drugs (trimethoprim and sulfamethoxazole) has been used because of its low cost and the fact that it reduces selection of antibiotic resistance to either drug *(26)*. In contrast, sulfonamides are used much more extensively in veterinary practice than in human medicine *(26)*.

In the 1970s trimethoprim-resistant (Tmp^r) *Escherichia coli* were found in approx 10% of the outpatients and by the 1980s this had increased to 15–20% *(26)*. Increased trimethoprim and combined trimethoprim and sulfamethoxazole resistance has continued to increase in a variety of enteric pathogens isolated from both humans and animals *(26)*. More recently, there has been an increase in trimethoprim and sulfonamide resistance in respiratory pathogens, such as *Streptococcus pneumoniae*. Some of these isolates are multiresistant *(26)*. Resistance has also been found in other species, such as *Neisseria gonorrhoeae (26)*. Plasmid-mediated trimethoprim resistance (*dfrA*) is found in both *Staphylococcus aureus* and *S. epidermidis (27)*, whereas nonmobile trimethoprim resistance that differs from the staphylococcal *dfrA* has recently been described in enterococci *(28)*. Trimethoprim-resistant *Enterococcus faecium* may also carry mobile non-*dfrA* genes linked to high-level gentamicin resistance *(29)*.

Trimethoprim resistance can result either from changes in chromosomal genes or by acquisitions of new genes *(26)*. Spontaneous mutations that block the ability of the cell to methylate deoxyuridylic acid to thymidylic acid have been isolated. These mutations make the cell dependent on external supplies of thymine. Trimethoprim resistance may also result from mutational changes that lead to overproduction of the bacterial dihydrofolate reductase *(26)*. Similar mutations have been found in laboratory-created *S. pneumoniae* strains.

Table 3
Transferable Genes Encoding Trimethoprim
and Sulfonamide Resistance

Gram-negative				Gram-positive	
DHFRs					
Group 1	Group 2	Other	DHPS	DHFR	DHPS
dhfrI	*dhfrIIa*	*dhfrIII*	*sulI*	*dfrA*	*sulII*[la]
dhfrV	*dhfrIIb*	*dhfrVIII*	*sulII*		
dhfrVI	*dhfrIIc*	*dhfrIV*			
dhfrVII		*dhfrIX*			
dhfrIb		*dhfrX*			
		dhfrIIIb			
		dhfrIIIc			

[a]A variant found in *Mycobacterium fortuitum* that is highly related to *sulI (26)*.

Plasmid-mediated Tmp^r in gram-negative bacteria was first described in 1972. It resulted from a DHFR that was not susceptible to the actions of trimethoprim. This new DHFR was unrelated to the chromosomal DHFR, and the origin of it and most of the other acquired gram-negative DHFRs is not known *(26)*. There are now 16 different types of resistant DHFRs in gram-negative facultative rods. Many are associated with transposons or transferable cassettes *(26)*. Most have been defined by amino acid and/or nucleotide sequence analysis *(26)*. The different DHFRs show different biochemical characteristics and, more importantly, confer different levels of resistance to trimethoprim. Two families of DHFR genes have been characterized in the gram-negative species *(26)*. Family 1, with five different genes, share 64–88% identity with each other (**Table 3**). These encode 157 amino acids and mediate high levels of trimethoprim resistance. Family 2, with three different genes, share 78–86% amino acid identity and encode 78 amino acids. The other DHFR genes listed in **Table 3** are less related to each other with 20–50% amino acid sequence identity (**Table 3**).

One transferable DHFR has been characterized in *S. aureus*. It is part of a transposon (Tn*4003*) and has 80% identity to the susceptible chromosomal DHFR from *S. aureus*. The transposon has been sequenced and is thought to be ancestrally related to the chromosomal DHFR *(27)*. Trimethoprim-resistant enterococci do not carry sequences related to Tn*4003 (28)*. In surveys, other Tmp^r bacteria also appear to carry unknown genes *(26)*.

Sulfonamides have been used for prophylaxis and treatment of meningococcal disease and resistance is common. Different types of resistance have been found. The most common type resulted from *dhps* genes that differed at the nucleotide level by approx 10% from the susceptible *dhps*, and had a 6 bp insertion that corresponds to an insertion of two additional amino acids *(26,30)*. Two different insertions have now been found that result in increased resistance. Mosaic genes are thought to have developed by recombination with genes from other *Neisseria* species. Changes by recombination have also been suggested for the *N. meningitidis* penicillin-binding proteins with reduced susceptibility to penicillin *(31)*.

2. Detection of Resistance

2.1. DNA Hybridization

The use of molecular tests to determine specific antibiotic resistance genes began over 20 yr ago *(14)*. Initially the entire plasmid or a fragment carrying the gene of interest was radiolabeled and used as a DNA probe in hybridization assays. This technique has also been used to distinguish different *tet* genes from each other *(13–16,32–34)*. The probes used for hybridization have been refined over time. First, intragenic fragments containing internal segments of the structural gene were used. These were made by cutting the desired band out of a recombinant plasmid after it had been cleaved with restriction enzyme(s) and the fragments separated on an agarose gel. Currently, when the gene sequence is known, a short oligonucleotide is used as the probe. In many cases, radiolabeled probes have been replaced by nonradiolabeled systems of which many are currently available commercially. More recently, the author has used PCR assays for detection of specific antibiotic resistance genes from purified bacteria or direct patient material *(18,19,35)* as have others *(36–38)*.

Southern blot hybridization for detection of specific *tet* genes is still considered to be the standard for detection of genes in new species or genera. For screening of large numbers of isolates, the use of whole chromosomal dot blots where the bacteria are spotted onto a support, such as nitrocellulose or nylon, and then lysed *in situ* has been used *(32)*. A radiolabeled probe is then hybridized; either a specific DNA fragment or oligonucleotide probe may be used *(32,33)*. To increase the sensitivity of the dot blot assay, the author has used purified DNA instead of lysed bacteria. However, in her hands the dot blot assay is less sensitive for detection of antibiotic resistance genes than Southern blot hybridization. The other advantage of the Southern blot over dot blots is that the hybridization pattern indicates the location of the antibiotic resistance gene (chromosome vs plasmid) *(34)*. The author uses the Southern blot hybridization as the standard method for confirmation of dot blot and PCR assays (*see* **Note 1**).

Each method for labeling probes has advantages and disadvantages. In the author's laboratory, they routinely use a nonradiolabeled system for most of their work. They find it very convenient because we do not have to worry about the short shelf-life inherent in ^{32}P-labeled probes. Also, nonradiolabeled probes do not have disposal problems. The nonradiolabeled probes can be reused at least five to seven times. A disadvantage of the nonradiolabeled method is that the probes are unsatisfactory when using anything other than purified DNA because of the high level of false-positives generated. Thus, nonradiolabled probes are not used in the author's laboratory for whole-bacterial dot blots, because of the high background and the nonspecificity of binding with this type of target sample. In contrast, ^{32}P-labeled probes work well with whole-bacterial dot blots. The result is that the author routinely uses either purified DNA or PCR products when using nonradiolabeled probes (see **Note 2**).

Another disadvantage of the nonradiolabeled probes is that they fail to be adequately labeled more often than radiolabeled probes in the author's laboratory. Some of the newer systems are reducing this problem. Also the author believes that nonradiolabeled systems tend to be more expensive in materials than radiolabeled probes, but save in time. The big advantage of the nonradiolabeled probes is that all the regulations on handling and disposing of radioactive material no longer has to be dealt with. The probes are always available and hybridizations can be done as needed.

Originally, all the probes for hybridizations were DNA fragments cut from plasmids. The preferred fragment was a relatively small (<1 kb) piece cut within the structural gene. However, the author has used large fragments (>1 kb) with no problems *(32–34)*. As the cost of oligonucleotide manufacturing has fallen, they are becoming more affordable and are replacing the fragment probes. The advantage is that oligonucleotide probes do not have vector contamination, and thus should not nonspecifically bind in the assay (see **Note 3**). Oligonucleotide probes that have been reported in the literature for the detection of tetracycline, rRNA methylase, and trimethoprim resistance genes are listed in **Table 4**. Not all genes have had oligonucleotide probes determined. This can be easily done if the DNA sequence of the desired gene is available. New oligonucleotides probes will then need to be tested to determine their specificity and optimal hybridization conditions.

Hybridization conditions for either Southern or dot blots are set so that *tet* genes that share less than 80% DNA sequence identity do not show significant hybridization with each other *(18)*. The hybridization conditions will vary depending on the type of probe used (fragment vs oligonucleotide), the type of label, and the kit used for the detection. Often optimal conditions will be obtained only after several sets of experiments.

2.2. Polymerase Chain Reaction (PCR)

A variety of PCR assays have been developed in the last 5 yr for detection of specific antibiotic resistance genes *(18,19,35–38)*. They have the advantage of

Table 4
Oligonucleotides Used for Detection of Specific Genes (*see* Note 4)

Gene	Oligonucleotide (5'-3')	Ref.
tetA	GCC TCC TGC GCG ATC TGG	*50*
tetB	CAG TGC TGT TGT TGT CAT TAA	*50*
	GCT TGG AAT ACT GAG TGT AA	*35,50*
tetC	TTG CAT GCA CCA TTC CTT GCG	*50*
tetK, tetL	CCT GTT CCC TCT GAT AAA	*19*
tetK, tetL	CAA ACT GGG TGA ACA CAG	*19*
tetM, tetO, tetS	GTT TAT CAC GGA AGT/C GCA/T A	*18,19*
tetM, tetO, tetS	GAA GCC CAG AAA GGA TTC/T GG	*18*
dhfrI	TTT AGG CCA GTT TTT ACC CAA GAC TTC GC	*26*
dhfrV	GAG AAA CAT TGC CCA TGG CCT CTA CGC TC	*26*
	CCT GGA CGG CCG ATA ATG ACA ACG TAA TAG	*26*
dhfrVII	AGT GTC GAG GAA AGG AAT TTC AAG CTC A	*26*
dhfrVIb	GTT GGA CAT CAA ATG ATG ACA ATG TAG TTG	*26*
dhfrIIa	GAT CGC GTG CGC AAG AAA T	*26*
dhfrVIII	CTA ACG GCG CTA TCT TCG TGA ACA ACG	*26*
dhfrIX	AGT GTC GAG GAA AGG AAT TTC AAG CTC	*26*

being able to use whole proteinase K-lysed bacteria or purified DNA as templates *(18,19)*. The PCR products can then be confirmed using nonradiolabeled probes. Thus, PCR assays can more quickly identify which antibiotic resistance genes a particular group of bacteria carry. However, the PCR assays are known to give both false-negative and false-positive results, especially when proteinase K-treated bacteria are used as a template for the assay. The PCR assay can also use direct patient material as the source of the template to determine whether a particular antibiotic resistance gene is present in the sample without growing the bacteria *(18)*. This is because the PCR does not require growth of the organisms for detection, unlike either the dot blot or Southern hybridization assays.

The PCR assay is becoming more common and, as its use increases, it may make its way into the clinical laboratory as a method for detection of specific antibiotic resistance genes *(19,35,36)*. Some advantages include the ability to use proteinase K-treated whole bacteria as the source of the template, the relative ease in testing large number of isolates, and the ability to look for genes directly in patient material without growing the bacteria. More PCR assays have been developed for the detection of *tetM* or *tetM/tetO* than for other antibiotic resistance genes *(18,37,38)*. As PCR becomes less expensive, it will be used more frequently for surveying distributions of specific antibiotic resistance genes.

The author, as well as others, has often used the close genetic relationship between groups of antibiotic resistance genes to develop a single PCR assay that can identify two or three different but related genes *(18,19,36)*. The author has then used specific labeled DNA probes to determine which of the possible genes the isolate carries. The advantage of this is that if there is no PCR product, one can assume that the two or three genes looked for in the assay are not likely to be present. Also, it has been used to amplify more distantly related antibiotic resistance genes that then can be used to isolate the entire gene (*36*, author's unpublished observations). However, with the gram-negative efflux genes the author has found that it is often difficult to develop a PCR assay that will distinguish one gene from another; in these cases single PCR assays are preferable, but not always easy to develop (author's unpublished observations).

The disadvantage of this type of PCR assay is that if one has an amplified product, additional tests are needed to determine which gene is present. However, the PCR product produced is adequate to make two Southern blots that will allow testing for two different genes at the same time. For example, with the *tetK/tetL* PCR assay the total volume of sample is 30 μL. This can be divided and 10–15 μL run on two different gels that can then be probed *(19)*. This can reduce time, but still requires an extra set of hybridization experiments as compared with having a PCR assay for each different gene. However, the author has found that when looking for rare genes, a PCR that detects multiple genes saves time and reagents.

Because of the reduced time needed for PCR detection of antibiotic resistance genes, this has often replaced whole-bacterial dot blots as the preferred method of screening large numbers of bacteria. Thus, instead of screening dot blots with a variety of antibiotic resistance gene probes, the author screens the bacteria with a variety of PCR assays that will detect one or more antibiotic resistance genes. If PCR products are found, then they are hybridized with specific labeled probes to verify which gene has been amplified. When new species have been shown to carry a specific antibiotic resistance gene, selected isolates are checked by Southern blot hybridization for verification of the presence of that particular gene. Similarly, when large surveys are being done, selected isolates are checked by Southern blot hybridization. This helps verify that the PCR assays are giving accurate results. If a particular isolate gives a questionable Southern blot hybridization, the author does a confirmatory PCR assay as a check. Thus, when something unusual is found therefore different methods to confirm the presence of these genes *(39)*.

There has been discussion of whether a genetic test for the presence of particular antibiotic resistance genes should replace the phenotypic testing of the ability of bacteria to grow in the presence of particular antibiotics, which currently is the standard in clinical settings *(40,41)*. Before this can happen, the costs of any of these tests needs to be reduced, and comparisons with pheno-typic methods done to determine whether the genotypic methods will provide the clinicians with the information they need to prescribe appropriate antibiotic therapy.

3. Epidemiologic Uses

The detection of specific antibiotic resistance genes for epidemiologic stud-ies began with the introduction of the molecular assays described in **Subhead-ings 1.1.**, **2.1.**, and **2.2.** These methods allowed investigators to track specific genes and/or transposons through bacterial populations, such as those in hospi-tals, communities, and different continents. As specific genes were followed, it became clear that these genes were spread from the host organism to indig-enous species sharing the same environment, whether it is the human body, a marine environment, or a farm *(1,3,6,12,13,18,32–34)*. Unique genes and plas-mids, found in host-specific bacteria, are moving to other ecosystems and other host species. How these genes are moved is not clear, but the sharing of the same plasmid or transposon in these separate environments clearly shows that there is gene exchange and movement through the bacterial population.

These molecular techniques have enabled the author to demonstrate that genes from very different species are related, and indicates that gene exchange is occurring in nature, which on occasion, can be duplicated in the laboratory *(6,35)*. An understanding of the extent that genes may move within the bacterial com-munity is important for surveillance and ultimately, for antibiotic therapy.

Benveniste and Davies *(42)* in 1973 suggested that many of the antibiotic resistance genes found in bacteria originated in antibiotic-producing micro-organisms. With the genetic tools available, the evidence now shows that some of the same genes found in resistant bacteria are also found in the antibiotic-producing micro-organism, suggesting that an ancestral relationship may exist for a variety of different antibiotic resistance genes *(16,43,44)*. Data now sug-gest that tetracycline efflux, ribosomal protection genes, and rRNA methylase genes all have counterparts in the antibiotic producers *(43,44)*. In contrast, the origin of most of the DHFRs is not clear *(26)*. It has also allowed us to docu-ment gene exchange across genetic and biochemical barriers *(6,12)*. The au-thor can show that gram-positive genes are being introduced, maintained, and expressed in gram-negative species *(6,12)*.

In the early 1970s, ampicillin-resistant β-lactamase-producing *Haemophilus influenzae* were isolated, and this together with previous observations that most *N. gonorrhoeae* carried indigenous plasmids, led to speculation that plasmid-mediated penicillin resistance in gonococci was a distinct possibility. Shortly thereafter, plasmid-mediated β-lactamase-producing *N. gonorrhoeae* were iso-lated *(45)*. The small β-lactamase plasmids from *N. gonorrhoeae* were related

(70% DNA sequence homology) to a previously characterized β-lactamase plasmid from *H. parainfluenzae*. In addition, both types of plasmids carried the TEM-type β-lactamase *(45)*. Thus, as early as the 1970s, molecular tools were being used to make accurate predictions about the future spread of antibiotic resistance genes.

The detection of specific antibiotic resistance genes allows for surveillance over time and space *(46)*. It will enable us to begin to understand what influences the spread of particular antibiotic resistance genes, to devise measures to reduce or prevent this from continuing at its current pace, and to make recommendations for antibiotic therapy changes as they become necessary *(46)*. Thus, continual surveillance is needed to monitor the spread of antibiotic resistance genes and to determine their potential impact on antibiotic therapy.

3.1. Characterization of Tetracycline Resistance Genes

3.1.1. DNA Hybridization

The detection of specific genes has primarily been done with either specific labeled DNA fragments or oligonucleotide probes. The commonly used DNA fragments are listed in **Table 5**. In a few cases, more than one plasmid has been used as the source of the fragment. The majority of the plasmids listed are vectors that carry the antibiotic resistance gene as an insert. To release the particular fragment, the relevant plasmid is digested with the appropriate enzyme(s), run on an agarose gel, and then purified using the molecular weight as a guide to the proper fragment. This protocol is time-consuming and often has vector contamination that can lead to false-positive results (*see* **Note 3**). In the last 5 yr, there has been a move to replace DNA fragment probes with oligonucleotide probes for hybridization. Many of the oligonucleotide probes described in the literature are listed in **Table 4**.

The author has developed oligonucleotide probes that detect several *tet* genes as part of developing PCR assays for detection of different genes (*see* **Subheading 3.1.2.**). Using a single probe to detect two or three different, but related *tet* genes is very cost-effective when working with bacterial populations that do not have a high prevalence of these genes.

3.1.2. PCR (see **Note 5**)

The author has developed three PCR assays for the detection of *tetB, tetK,* and *tetL,* and *tetM, tetO,* and *tetS* genes *(18,19,35)*. Each PCR assay uses only one pair of primers regardless of the number of determinants detected. Using one PCR assay to screen for multiple *tet* genes is cost-effective. The author has even found that one can combine two of the PCR assays to screen for *tetK, tetL tetM, tetO,* and *tetS* in a single assay *(19)*. This was useful when screening direct patient material where the amount of material was limited. It can reduce time and reagents needed for screening. The advantage of using the PCR assays

Table 5
Selected Fragments for Detection of Specific Genes

Gene(s)	Plasmid	Probe Fragment	Reference
tetA	pSL18	0.75 kb *Sma*I	*16*
tetA	pRU1000	0.3 kb *Nru*I-*Sma*I	*16*
tetB	pRT11, pRT29	1.27 kb *Hinc*II	*16*
tetC	pBR322	0.93 kb *Bst*N1	*16*
tetD	pSL106	3.1 kb *Hind*III-*Pst*I	*16*
tetE	pSL1504	2.5 kb *Cla*I-*Pvu*I	*16*
tetH	pVM112	1.44 kb *Pst*I-*Xho*II	*51*
tetK	pT181	0.87 kb *Hinc*II	*52*
tetK	pAT102	0.87 kb *Hinc*II	*53*
tetL	pBC16	0.31 kb *Cla*I-*Hpa*II	*52*
tetL	pMV158	1.1 kb *Hha*I	*19*
tetM	Tn*1545*	0.85 kb *Eco*RI-*Hind*III	*52*
tetM	pUW-JKB1, pJI3	1.8 kb *Hinc*II	*54*
tetO	pIP1433	1.46 kb *Hind*III-*Nde*I	*52*
tetO	pAT121	0.87 kb *Hinc*II	*55*
tetO	pUOA4	1.8 kb *Hinc*II	*56*
tetA(P)	pJI71	0.9 kb *Sph*I-*Eco*RI	*57*
tetB(P)	pCW3	1.1 kb *Pst*I-*Eco*RI	*57*
tetQ	pNFD13-2	0.9 kb *Eco*RI-*Eco*RV	*58*
tetS	pIP811	0.50 kb *Hind*III-*Eco*RI	*59*
ermA	pEM9592, Tn*554*	0.7 kb *Ssp*I, 0.42 kb *Mbo*I	*60*
ermB[a]	pAD2	1.3 kb *Kpn*I-*Hpa*I	*36*
ermBP[a]	pJIR229	0.8 kb *Pst*I-*Eco*RI	*60*
ermAM[a]	Tn1545	0.56 kb *Ssp*I	*36*
ermB[a]	pIP1527	0.35 kb *Rsa*I	*36*
ermC[b]	pE194	0.47 kb *Hae*III-*Hinc*II	*36*
ermGT[b]	pBR328:33RV	0.9 kb *Hpa*I	*61*
ermD	pBD90	0.28 kb *Sau*3A	*62*
ermF	pFD292	0.75 kb *Pvu*II	*60*
ermF	pBF4	0.45 kb *Eco*RI-*Hind*III,	*60*
		0.29 kb *Hind*III-*Taq*I	*60*
ermQ	pJIR745	1.0 kb *Eco*RI-*Hind*III	*60*
dhfrI	pLMO150	0.49 kb *Hpa*I	*26*
dhfrV	pLMO20	0.49 kb *Hpa*I	*26*
dhfrVII	Tn*5086*	0.31 kb *Eco*RV	*26*
dhfrVIb	Tn*4132*	0.49 kb *Hpa*I	*26*
dhfrIIc	Tn*5090*	0.28 kb *Nhe*I-*Eco*RI	*26*
dhfrIII	pFE1242	0.85 kb *Eco*RI-*Hind*III	*26*
dfrA	pSK1, Tn*4003*	0.9 kb *Eco*RV,	*48*
		0.5 kb*Hind*III-*Eco*RI	
sulI	Tn*21*, R388	0.66 *Sac*II-*Bgl*II	*26*
sulII	RSF1010, pGS05	0.78 kb *Hinc*II	*26*

[a]All part of the ErmB group of genes.
[b]Part of the ErmC group of genes.

over Southern blots is that purified DNA is not required, and thus, more isolates can be screened in any one time period. PCR does not allow determination
of the location of the gene as Southern hybridization does. The disadvantage is
that one may get false results. However, these are usually false-negative rather
than false-positive results (author's unpublished observations).

All PCR products are hybridized with the appropriate DNA fragments or
internal oligonucleotide probes to verify the gene present. This is because in
the author's assays the size of the PCR product is the same for either *tetK* and
tetL, or *tetM, tetO,* and *tetS (18,19,35)*. However, either stripping and reprobing
or making two Southern blots from the single agarose gel eliminates the need
to run more than a single PCR for each sample. Such double Southern blots are
made by placing a membrane on each side of the agarose gel, and then sheets
of absorbant paper as done with a single Southern blot (*see* Chapter 2). This
should sit no more than 24 h, otherwise it is very difficult to remove the gel
from the membrane. No liquid reservoir or weights are used with this procedure. The basic assays for each PCR are given below.

Amplification was performed using purified DNA from bacteria, proteinase
K-treated whole bacteria (*see* **Note 6**), proteinase K-treated patient material, or
boiled bacteria. For the *tetK/tetL* assay a reaction volume of 30 µL contained 2 U
of *Taq* polymerase (Perkin Elmer-Cetus, Norwalk, CT), 200 mM dNTP, 1X PCR
buffer II (50 mM KCl, 10 mM Tris-HCl, pH 8.3, 0.001% w/v gelatin), 2.5 mM
MgCl$_2$, and 100 ng of each oligonucleotide. Reactions were conducted using a
Perkin-Elmer DNA thermal cycler and the program consisted of 3 min at 95°C,
then 35 cycles of 20 s at 95°C, 30 s at 45°C, and 2 min at 72°C. Cloned control
Tet K and Tet L plasmids were used at 5–10 ng, whereas 20 ng of purified total
genomic DNA, or 1 µL of proteinase K-treated bacteria and patient materials
was used. A negative control that lacked DNA was used in each run. The
amplified DNA products were electrophoresed on 1.5% agarose gels using 0.5X
TBE buffer (44.5 mM Tris base, 44.5 mM boric acid, and 1.25 mM EDTA) at
4 V/cm for 45 min. The DNA was visualized by ethidium bromide staining *(19)*.

The *tetM/tetO/tetS* PCR assay used a reaction volume of 100 µL consisting
of purified DNA, proteinase K-treated whole bacteria (*see* **Note 6**), or boiled
whole bacteria as the template, 100 mM dNTPs, 1X PCR buffer (50 mM KCl,
10 mM Tris-HCl, 0.001% w/v gelatin, pH 8.3) 1.5 mM MgCl$_2$, 200 ng of each
primer, and 2 U of *Taq* polymerase (Perkin Elmer-Cetus). This is a slight modification of the conditions described in **ref. *18***. Controls of the cloned determinants were used and visualized as described in **Table 4**. In **ref. *18*** the
oligonucleotide primer M4 has a misprint and should read 5'GAA GCC CAG
AAA GGA TTC/T GGT 3' as in **Table 4**.

The author used the *tetK/tetL* and *tetM/tetO/tetS* primers combined in one
assay using the *tetK/tetL* PCR conditions and found that one could amplify

both groups of Tet determinants equally well *(19)*. The different PCR products could be distinguished from each other by the differences in their molecular weight; the *tetK* product (1048 bp) comigrates with the *tetL* product (1028 bp), but both are easily distinguished from the *tetM/tetO/tetS* products (686 bp).

Two other laboratories have devised PCR assays using different primers and conditions for detection of *tetM*, or *tetM*, and *tetO*. The specifics of these assays can be found in the following references *(37,38)*, but are not significantly different than the ones we developed.

3.2 Characterization of Macrolide Resistance Genes

3.2.1 DNA Hybridization

A number of DNA fragment probes have been used for detection of different *erm* genes (**Table 5**). The Erm B class has had the most different probes used, presumably because it is widely distributed and found in a large number of different species and genera as illustrated in **Table 2**. In general, less work on epidemiology and distribution in bacterial populations has occurred with the *erm* genes. This may change as the newer macrolides are used and as important pathogens, such as *S. pneumoniae,* become resistant to macrolides.

The lack of research into the distribution of the genes is also reflected by the relatively few oligonucleotide probes listed in **Table 4**. Oligonucleotide primers previously developed by Arthur et al. *(36)* for PCR (*see* **Subheading 3.2.2.**) were specifically stated not to be used as oligonucleotide probes because they cross-hybridized with other enteric genes *(36)*. The result is that this area has not advanced as quickly as the study of the *tet* genes and the majority of the literature uses DNA fragment probes for detection.

3.2.2. PCR

A PCR assay for the detection of *erm* genes has been published *(36)* that detects a number of different Erm groups. The common motif of *ermA, ermB, ermC*, and *ermG* genes was used to develop a single assay that would detect all four groups of genes. This assay also gives PCR products with cloned *ermQ* about 50% of the time, but does not give PCR products with cloned *ermF* or *ermCD* genes *(47)*. The original article does not show a gel of the PCR products. Using the original degenerate primers 5' GAA/G ATI GGI III GGI AAA/G GGI CA 3' and 5' AA GTT/C TTA/G TTC/T TGI AAA/G 3', (where A/G means both of these nucleotides are added in equal molarities at this position, while I means deoxyinosine) many PCR assays gave multiple PCR products, including cloned controls. To reduce this the author, in collaboration with Dr. Gherardini (from the University of Georgia), has modified the PCR primers so that, instead of an inosine, all four bases are added at the same positions. In contrast, where two bases are added at one position these have been left as originally described. This modification results in cleaner PCR products. The author has also recently modified the conditions as follows: 1.5 m*M* MgCl₂, 50 m*M* KCl, 10 m*M* Tris-HCl, 0.2 m*M* dNTPs, 2.5 U *Taq*, 100 ng of each primer and template to a volume of 100 µL/reaction. The cycles are 95°C for 1 min, 37°C for 2 min and 72°C for 2 min for a total of 45 cycles. The last cycle is 72°C for 5 min.

Using either set of primers, PCR products may be obtained that look the appropriate size, but do not hybridize to any of the four *erm* gene groups. Recently, the author has cloned and sequenced one of these PCR products from *Borrelia burgdorferi* and found that the amino acid motif is conserved and associated with erythromycin resistance (manuscript in preparation). This suggests that a novel rRNA methylase gene is present. This assay has allowed us to identify other organisms carrying rRNA methylase genes that are not identical to known characterized genes. The advantage of PCR is that the product can be labeled directly or cloned, and then used as a probe to examine the complete restricted genome of the isolate in question. This will allow one to identify the size of the chromosomal DNA fragment carrying the complete gene. This size fragment can be isolated and used for cloning. This procedure greatly increases the chances of cloning the desired fragment than does using the entire restricted genome for cloning. The labeled PCR product or labeled cloned PCR fragment can then be used to screen a cloned library for the gene. Thus, the PCR product will be a very useful tool for cloning the complete gene from the bacterial host's chromosome. If the gene is located on a plasmid, the cloning is much easier since the plasmid can be isolated from the chromosome, which reduces the unimportant DNA, prior to cloning.

3.3. Characterization of Trimethoprim and Sulfonamide Resistance Genes

3.3.1. DNA Hybridization

Much of the work that has been done on acquired trimethoprim and sulfonamide resistance has been with genes found in the gram-negative enteric species *(26)*. The staphylococcal Tn*4003* has also been studied *(48)*. In many cases specific DNA fragment probes have been used as illustrated in **Table 5**. In fewer cases, oligonucleotide probes have been developed and are listed in **Table 4**.

3.3.2. PCR

One PCR assay for amplification of the chromosomal *dhps* gene has been described *(49)*. The primers NM1 5' GGG TCG ACG GTT TCA GAC GGC ATA TAA 3' and NM2 5' AAG GTA CCG CCC CGT CCT TTT CAG ACG 3' were used. The reaction contained 100–200 ng of purified chromosomal DNA, 50 m*M* KCl, 10 m*M* Tris-HCl, pH 8.4, 1.5 m*M* MgCl₂, 100 mg of gelatin per mL,

1 mM of each primer, 200 mM of dNTPs and 2.5 U of *Taq* polymerase in a volume of 100 µL. The cycle consists of 94°C for 1 min, 55°C for 1 min, and 72°C for 2 min for 28 cycles. A final incubation at 72°C for 7 min is then performed. However, depending on the bacteria used, the author varied the MgCl$_2$ concentration between 1 and 2 mM, and the annealing temperature between 45 and 60°C. Samples (10 µL) are run on a 1% agarose gel.

4. Research Uses

The detection of specific antibiotic resistance genes has allowed the author to follow the spread of particular genes into different bacterial populations and ecosystems. It has provided evidence that gram-positive genes are now found in gram-negative species in nature, and for the spread of these genes from animal to human bacteria, and in the reverse direction *(6,12)*. It has also allowed the author to understand how antibiotic resistance is developing in different species and has provided tools for detection of particular genes. The author has been able to track specific genes over time and location and to see that they are continuing to move into new genera and species *(6,12)*.

The author can show that the mechanisms of resistance in many classes of bacteria are the same as those in antibiotic-producing species. This lends support to the hypothesis that many of these genes originated from the antibiotic-producing species.

5. Future Prospects and Goals

Currently, the standard for determining antibiotic resistance pattens of pathogens involves culture and traditional susceptibility testing methods *(40,41)*. This looks at phenotypic expression of resistance and cannot generally distinguish between the various genes a cell might carry. However, as health care reform continues to spread, the number of laboratories performing susceptibility testing is expected to decrease. Under these circumstances, it is likely that molecular methods for detecting resistance determinants will play a larger role in the future of clinical laboratory. Before widespread application of molecular technology for the detection of antibiotic resistance genes occurs, there will need to be simplification, standardization, and validation of these alternative methods in clinical laboratory settings. How the detection of antibiotic resistance genes will provide the information needed to guide the clinician in antimicrobial therapy choices and what information molecular tests can provide, have not yet been adequately addressed. The molecular methods also have some advantages over traditional susceptibility testing especially when looking at antibiotic resistance that is inducibly regulated. For example, it is possible to miss that a particular bacterium carries an inducible rRNA methylase gene by standard susceptibility testing, since the organism may look susceptible, in this type of test, to macrolides and clindamycin. Alternatively, it may also look resistant or intermediate to macrolides, but susceptible to clindamycin even though, once induced, it is resistant to both classes of antibiotics. However, neither antibiotic would be appropriate therapy under this situation. In contrast, showing that an rRNA methylase gene is present, by molecular techniques, is straightforward and provides the information to the clinician that there is a potential for treatment failure with either macrolides or clindamycin. In addition, molecular methods can be considerably faster than traditional methods, especially when working with slow growing pathogens. Therefore, the author would hypothesize that as these molecular tests are simplified and kits become available, they will be used with increasing frequency in the clinical laboratory. They will continue to be used in the research setting to monitor changes in gene carriage and characterization of new antibiotic-resistant species and genera.

6. Notes

1. When detecting specific genes by hybridization with dot blots of whole chromosomal DNA, both false-positives and false-negatives can occur more frequently than with Southern blots. The author has always read the whole-cell dot blot assay very conservatively so false-positive results are not obtained, but, on occasion, some false-negative results have been obtained. Confirmation of questionable results is done by Southern blot hybridization and/or by PCR testing.
2. When using whole chromosomal dots the author has found that radiolabeled probes work best because nonradiolabeled probes have a tendency to give nonspecific false-positive results owing to binding of the antibody used as part of the detection system. Thus, nonradiolabeled probes are used only when purified DNA is used as the template in either dot or Southern blots.
3. The disadvantage of using DNA fragments purified from cloned genes is the potential of vector contamination in the fragment preparation. This can be tested before use by determining if the fragment will hybridize with the vector DNA. If it does then there is significant vector contamination and the fragment should not be used as probe. This is less of an issue if the antibiotic resistance gene has been cloned in a gram-negative vector, and will be used to probe gram-positive bacteria since there is less of a chance that the gram-positive species will carry DNA homologous to any contaminating vector DNA. Another alternative would be to use PCR-generated fragments as probes for hybridization. The label can be incorporated during PCR synthesis or end-labeling can be done after the PCR product is generated. These types of probes should not have vector contamination.
4. The author found with the *tetB* oligonucleotides that labeling each of the two listed in **Table 4** then combining them for a single labeled probe worked the best.
5. With detection of specific genes by PCR, the author finds that the most reproducible results are obtained when the number of samples is limited to between 20 to 25 per run. Positive and negative controls are run with each assay and both are

run on every gel. Physically separating the areas in which PCR assays are mixed, run in the thermal cycler, and where the PCR products are run on agarose gels helps to reduce contamination problems. The biggest factor in reducing contamination in the author's hands has been the use of dedicated positive-displacement pipets. When these are not available, the use of tips with barrier filters is a possible, but not a long-term substitute. The author has also found that even when a PCR assay is up and running it still takes time for a new person to learn how to run the assays. With the published *erm* PCR assay the author found that, even with the modified primers, the diluted working stock cannot be reused more than two to three times. If used more, the assay either gives multiple bands or no bands.

6. Proteinase K-treatment for bacteria can use either plate-grown bacteria or broth-grown bacteria. The bacteria are suspended in proteinase K solution containing 0.5% Tween-20, 100 mM Tris-HCl, pH 8.5, and 0.05 mg/mL proteinase K in sterile water and incubated at 60°C for 1–6 h. Longer times are needed for gram-positive bacteria than gram-negative species. To inactivate the enzyme, samples are boiled for 10 min before being used as templates for PCR. Samples are aliquoted and stored at −20°C. One microliter usually contains between 10^2 to 10^4 bacteria and 1–2 µL are used as the template in the PCR assay. In some gram-positive bacteria additional treatment may be required or proteinase K incubation prolonged for 24–72 h to sufficiently lyse the cells (*18,19,35*, author's unpublished observations). Proteinase K-treated biopsy material and direct patient material have been used as templates for a variety of PCR assays (*18*). The biopsies consisted of 10–50 mg of wet tissue placed in 50 mM KCl, 20 mM Tris-HCl, pH 8.3, 2.5 mM MgCl$_2$ containing 0.5 mg/mL of freshly added proteinase K. The tissue is incubated for 2 h at 55°C with vortexing at 1 and 1.5 h. The proteinase K is then inactivated by heating at 95°C for 20 min. Only 1–2 µL of this material can be used as template without inhibiting the PCR reaction. A similar procedure is used with direct patient material obtained from swabs (*18*).

References

1. Levy, S. B. (1992) *The Antibiotic Paradox: How Miracle Drugs Are Destroying the Miracle*. Plenum, New York.
2. Chopra, I., Hawkey, P. M., and Hinton, M. (1992) Tetracyclines, molecular and clinical aspects. *J. Antimicrob. Chemother.* **29,** 245–277.
3. Levy, S. B. (1988) Tetracycline resistance determinants are widespread. *ASM News* **54,** 418–421.
4. Klein, N. C. and Cunha, B. A. (1995) Tetracyclines. *Med. Clin. N. Am.* **79,** 789–801.
5. Freeman, C. D., Nightingale, C. H., and Quintiliani, R. (1994) Minocycline: old and new therapeutic uses. *Int J. Antimicrob. Agents* **4,** 325–335.
6. Roberts, M. C. (1989). Gene transfer in the urogenital and respiratory tract, in *Gene Transfer in the Environment* (Levy, S. and Miller, R. V., eds.) McGraw-Hill, New York, pp 347–375.
7. Col, N. F. and O'Connor, R. W. (1987) Estimating worldwide current antibiotic usage: Report of task force 1. *Rev Infect. Dis.* **9,** S232–243.

8. Akiba, T., Koyama, K., Ishiki, Y., Kimura, S., and Fukushima, T. (1960) On the mechanism of the development of multiple-drug-resistant clones of Shigella. *Jpn. J. Microbiol.* **4**, 219.

9. Falkow, S. (1975) *Infectious Multiple Drug Resistance*, Pion Limited, London England.

10. Watanabe, T. (1963) Infectious heredity of multiple drug resistance in bacteria. *Bacteriol. Rev.* **27**, 87–115.

11. Lima, A. A. M., Lima, N. L., Pinho, M. C, Barros, Jr., E. A., Teixeria, M. J., Marins, M. C. V., and Guerrant, R. L. (1995) High frequency of strains multiply resistant to ampicillin, trimethoprim-sulfamethoxazole, streptomycin, chloramphenicol, and tetracycline isolated from patients with shigellosis in northeastern Brazil during the period 1988 to 1993. *Antimicrob. Agents Chemother.* **39**, 256–259.

12. Roberts, M. C. (1994) Epidemiology of tetracycline resistance determinants. *Trends Microbiol.* **2**, 353–357.

13. Roberts, M. C. and Kenny, G. E. (1986) Dissemination of the *tetM* tetracycline resistance determinant to *Ureaplasma urealyticum*. *Antimicrob. Agents Chemother.* **29**, 350–352.

14. Mendez, B., Tachibana, C., and Levy, S. B. (1980) Heterogeneity of tetracycline resistance determinants. *Plasmid* **3**, 99–108.

15. Levy, S. B., McMurry, L. M., Burdett, V., Courvalin, P., Hillen, W., Roberts, M. C., and Taylor, D. E. (1989) Nomenclature for tetracycline resistance determinants. *Antimicrob. Agents Chemother.* **33**, 1373,1374.

16. Rienhour, M. B., Fletcher, H. M., Mortensen, J. E., and Daneo-Moore, L. (1996) A novel tetracycline-resistant determinant, *tet(U)*, is encoded on the plasmid pKQ10 in *Enterococcus faecium*. *Plasmid* **35**, 71–80.

17. Jones, C. S., Osborne, D. J., and Stanley, J. (1992) Cloning of a probe for a previously undescribed enterobacterial tetracycline resistance gene. *Lett. Appl. Microb.* **15**, 106–108.

18. Roberts, M. C., Pang, Y., Riley, D. E., Hillier, S. L., Berger, R. C., and Krieger, J. N. (1993) Detection of Tet M and Tet O tetracycline resistance genes by polymerase chain reaction. *Molec. Cell. Probes.* **7**, 387–393.

19. Pang, Y., Bosch, T., and Roberts, M. C. (1994) Single polymerase chain reaction for the detection of tetracycline resistant determinants Tet K and Tet L. *Molec. Cell. Probes.* **8**, 417–422.

20. Roberts, M. C. (1996). Mechanism, regulation, mobility and distribution of tetracycline-resistant determinants. *FEMS Microbiol. Rev.* **19**, 1–24.

21. Weisblum, B. (1995) Erythromycin resistance by ribosome modification. *Antimicrob. Agents Chemother.* **39**, 577–585.

22. Kirst, H. A. (1991) New macrolides: expanded horizons for an old class of antibiotics. *J. Antimicrob. Chemother.* **28**, 787–790.

23. Leclercq, R. and Courvalin, P. (1991) Bacterial resistance to macrolide, lincosamide, and streptogramin antibiotics by target modification. *Antimicrob. Agents Chemother.* **35**, 1267–1272.

24. Barriere, J. C., Bouanchaud, D. H., Paris, J. M., Rolin, O., Harris, N. V., and Smith, C. (1992) Antimicrobial activity against *Staphylococcus aureus* of semisynthetic injectable streptogramins: RP 59500 and related compounds. *J. Antimicrob. Chemother.* **30, Suppl. A** 1–8.

25. Halula, M. C., Manning, S., and Macrina, F. L. (1991) Nucleotide sequence of *ermFU*, a macrolide-lincosamide-streptogramin (MLS) resistance gene encoding an RNA methylase from the conjugal element of *Bacteroides fragilis* V503. *Nucleic Acids Res.* **19**, 3454.

26. Huovinen, P., Sundstrom, L., Swedberg, G., and Skold, O. (1995) Trimethoprim and sulfonamide resistance. *Antimicrob. Agents Chemother.* **39**, 279–289.

27. Rouch, D. A., Messerotti, L. J., Loo, S. L., Jackson, C. A., and Skurray, R. A. (1989) Trimethoprim resistance transposon Tn*4003* from *Staphylococcus aureus* encodes genes for a dihydrolfolate reductase and thymidylate synthetase flanked by three copies of IS*257*. *Molec. Microbiol.* **3**, 161–175.

28. Frosolono, M., Hodel-Christian, S. L., and Murray, B. E. (1991) Lack of homology of enterococci which have high-level resistance to trimethoprim with the *dfrA* gene of *Staphylococcus aureus*. *Antimicrob. Agents Chemother.* **35**, 1928–1930.

29. Woodford, N. and George, R. C. (1992) Trimethoprim resistance linked to high-level gentamicin resistance in *Enterococcus faecium*. *Med. Microbiol. Lett.* **1**, 11–16.

30. Fermer, C., Kristiansen, B-E., Skold, O., and Swedberg, G. (1995) Sulfonamide resistance in *Neisseria meningitidis* as defined by site-directed mutagenesis could have its origin in other species. *J. Bacteriol.* **177**, 4669–4675.

31. Dowson, D. G., Coffey, T. J., and Spratt, B. G. (1994) Origin and molecular epidemiology of penicillin-binding-protein-mediated resistance to β-lactam antibiotics. *Trends Microbiol.* **2**, 361–366.

32. Roberts, M. C., Hillier, S. L., Hale, J., Holmes, K. K., and Kenny, G. E. (1986) Tetracycline resistance and *tetM* in pathogenic urogenital bacteria. *Antimicrob. Agents Chemother.* **30**, 810–812.

33. Roberts, M. C. and Kenny, G. E. (1986) Dissemination of the *tetM* tetracycline resistance determinant to *Ureaplasma urealyticum*. *Antimicrob. Agents Chemother.* **29**, 350–352.

34. Roberts, M. C. (1989) Plasmid-mediated Tet M in *Haemophilus ducreyi*. *Antimicrob. Agents Chemother.* **33**, 1611–1613.

35. Roe, D. E., Roberts, M. C., Braham, P., and Weinberg, A. (1995) Characterization of tetracycline resistance in *Actinobacillus actinomycetemcomitans*. *Oral Microbiol. Immun.* **10**, 227–232.

36. Arthur, M., Molinas, C., Mabilat, C., and Courvalin, P. (1990) Detection of erythromycin resistance by the polymerase chain reaction using primers in conserved regions of *erm* rRNA methylase genes. *Antimicrob. Agents Chemother.* **34**, 2024–2026.

37. Lacrox, J-M. and Walker, C. B. (1995) Detection and incidence of the tetracycline resistance determinant *tet(M)* in the microflora associated with adult periodontitis. *J. Perio.* **66**, 102–108.

38. Olsvik, B., Olsen, I., and Tenover, F. C (1995) Detection *tet(M)* and *tet(O)* using the polymerase chain reaction in bacteria isolated from patients with periodontal disease. *Oral Microbiol. Immun.* **10**, 87–92.

39. Pang, Y., Brown, B. A., Steingrube, V. A., and Wallace, R. J. Jr., and Roberts, M. C. (1994) Acquisition of gram-positive tetracycline resistance genes in *Mycobacterium* and *Streptomyces* species. *Antimicrob. Agents Chemother.* **38,** 1408–1412.

40. National Committee for Clinical Laboratory Standards. (1995) Suggested modifications of standard methods for susceptibility testing of some fastidious and special problem bacteria. Approved standard M100 M7-A3. Villanova, PA.

41. National Committee for Clinical Laboratory Standards. (1995) Zone diameter interpretive standards and equivalent minimum inhibitory concentrations (MIC) breakpoints for streptococci, including *Streptococcus pneumoniae*. Approved standard M100-S6 M2–A5. Villanova, PA.

42. Benveniste, R. and Davies, J. (1973) Aminoglycoside antibiotic-inactivation enzymes in actinomycetes similar to those present in clinical isolates of antibiotic resistant bacteria. *Proc. Natl. Acad Sci. USA* **172,** 3628–3632.

43. Pernodet, J-L., Fish, S., Blondelet-Rouault, M-H., and Cundliffe, E. (1996) The macrolide-lincosamide-streptogramin B resistance phenotypes characterized by using a specifically deleted, antibiotic-sensitive strain of *Streptomyces lividans*. *Antimicrob. Agents Chemother.* **40,** 581–585.

44. Dittrich, W. and Schrempf, H. (1992) The unstable tetracycline resistance gene of *Streptomyces lividans* 1326 encodes a putative protein with similarities to translational elongation factors and Tet (M) and Tet (O) proteins. *Antimicrob. Agents Chemother.* **36,** 1119–1124.

45. Roberts, M., Elwell, L. P., and Falkow, S. (1997) Molecular characterization of two beta-lactamase-specifying plasmids isolated from *Neisseria gonorrhoeae*. *J. Bacteriol.* **131,** 557–563.

46. Schwarz, S. K., Zenilman, J. M., Schnell, D., Knapp, J. S., Hook, E. W. III, Thompson, S., Judson, F. N., and Homes, K. K. (1990) National surveillance of antimicrobial resistance in *Neisseria gonorrhoeae*. *JAMA* **264,** 1413–1417.

47. Roberts, M. C., and Brown, M. B. (1994) Macrolide-lincosamide resistance determinants in streptococcal species isolated from the bovine mammary gland. *Vet. Microbiol.* **40,** 253–261.

48. Coughter, J. P., Johnston, J. L., and Archer, G. L. (1987) Characterization of a staphylococcal trimethoprim resistance gene and its product. *Antimicrob. Agents Chemother.* **31,** 1027–1032.

49. Radstrom, P., Fermer, C., Kristiansen, B.-E., Jenkins, A., Skold, O., and Swedberg, G. (1992) Transformational exchanges in the dihydropteroate synthase gene of *Neisseria meningitidis*: a novel mechanism for acquisition of sulfonamide resistance. *J. Bacteriol.* **174,** 6386–6393.

50. Nikolich, M. P., Shoemaker, N. B., and Salyers, A. A. (1992) A *Bacteroides* tetracycline resistance gene represents a new class of ribosome protection tetracycline resistance. *Antimicrob. Agents Chemother.* **36,** 1005–1012.

51. Hansen, L. M, McMurry, L. M., Levy, S. B., and Hirsch, D. C. (1993) A new tetracycline resistance determinant, Tet H, from *Pasterurella multocida* specifying active efflux of tetracycline. *Antimicrob. Agents Chemother.* **37,** 2699–2705.

52. Chaslus-Dancla, E., Lesage-Descauses, M-C., Leroy-Setrin, S., Martel, J-L., and Lafont, J-P. (1995) Tetracycline resistance determinants, Tet B and Tet M, detected in *Pasteurella haemolytica* and *Pasteurella multocida* from bovine herds. *J. Antimicrob. Chemother.* **36**, 815–819.

53. Zilhao, R., Papadopoulou, B., and Courvalin, P. (1988) Occurrence of *Campylobacter* resistance gene *tetO* in *Enterococcus* and *Streptococcus* spp. *Antimicrob. Agents Chemother.* **32**, 1793–1796.

54. Brown, J. T. and Roberts, M. C. (1987) Cloning and characterization of *tetM* from a *Ureaplasma urealyticum* strain. *Antimicrob. Agents Chemother.* **31**, 1852–1854.

55. Sougakoff, W., Papadopoulou, B., Nordmann, P., and Courvalin, P. (1987) Nucleotide sequence and distribution of gene *tetO* encoding tetracycline resistance in *Campylobacter coli. FEMS Microbiol. Lett.* **44**, 153–159.

56. Ng, L-K., Stiles, M. E., and Taylor, D. E. (1987) DNA probes for identification of tetracycline resistance genes in *Campylobacter* species isolated from swine and cattle. *Antimicrob. Agents Chemother.* **31**, 1669–1674.

57. Sloan, J., McMurry, L. M., Lyras, D., Kevy, S. B., and Rood, J. I. (1994) The *Clostridium perfringens* Tet P determinant comprises two overlapping genes: *tetA(P),* which mediates active tetracycline efflux, and *tetB(P),* which is related to the ribosomal protection family of tetracycline-resistance determinants. *Molec. Microbiol.* **11**, 403–415.

58. Shoemaker, N. B., Barber, R. D., and Salyers, A. A., (1989) Cloning and characterization of a *Bacteroides* conjugal tetracycline-erythromycin resistance element by using a shuttle cosmid vector. *J. Bacteriol.* **171**, 1294–1302.

59. Charpentier, E., Gerbaud, G., and Courvalin, P. (1993) Characterization of a new class of tetracycline-resistance gene *tet(S)* in *Listeria monocytogenes* BM4210. *Gene* **131**, 27–34.

60. Roe, D. E., Weinberg, A., and Roberts, M. C. (1996) Mobile rRNA methylase genes in *Actinobacillus actinomycetemcomitans. J. Antimicrob. Chemother.* **37**, 457–464.

61. Rinckel, L. A. and Savage, D. C., (1990) Characterization of plasmid-borne macrolide resistance from *Lactobacillus* sp. strain. 100–33. *Plasmid* **23**, 119–123.

62. Gryczan, T., Grandi, G., Hanh, J., and Dubnau, D. (1980) DNA sequence and regulation of *ermD*, a macrolide-lincosamide-streptogramin B resistance element from *Bacillus licheniformis. Mol. Gen. Genet.* **194**, 349–356.

Index

A

Accessory cholera enterotoxin, 378
AccuProbe Culture Identification Test, 140
AccuProbe Enterococcus, 470
Accuprobe system, 160, 161
Acrylamide gel electrophoresis
    PCR end-points, 69, 70
Agarose gel electrophoresis
    GDD and ribotyping, 21, 25
    PCR end-points, 68
    plasmid analysis, 54, 58–59
Algorithms
    PCR primer design, 64
Aminoglycoside resistance, 555–575
    detection, 556–557
        Gram-negative bacteria, 561–565
        Gram-positive bacteria, 565–566
    Enterococci, 480
    future, 573–574
    genotypic methods, 558, 566–573
    materials, 557–558
    methods, 558–573
    phenotypic methods, 557, 558–566
Aminoglycoside resistance genes
    PCR, 570–573
Amplicor, 63, 164
Amplified *Mycobacterium Tuberculosis*
    Direct Test, 164
Antibiograms
    Enterococci, 475
    Salmonella, 357
Antibiotic resistance genes
    detection
        research, 657
    future, 657–658
    Mycobacterium tuberculosis
        detection, 8–9

Antigen detection
    Legionellae, 216
    Neisseria gonorrhoeae, 295
Antigenic variation
    Vibrio cholera, 371–372
Antimicrobial disk susceptibility tests
    aminoglycoside resistance, 560–561
Antitoxin
    diphtheria, 193
AP-PCR, 85
    bacterial diseases, 83–97
    fingerprinting RNA, 95–97
    fingerprinting vs RFLP, 90–92
    fingerprinting with oligonucleotides, 92–95
    Helicobacter pylori detection, 425
    materials, 85–86
    methods, 86–97
    Mycobacterium tuberculosis, 173–175
    pneumococcal diseases, 150–151
    sensitivity, 88–90
    Staphylococci, 452–455
    technical considerations, 88–92
    template preparation, 86–87
Arbitrarily primed PCR, *see* AP-PCR
Aspergillosis
    AP-PCR, 94

B

B-lactam resistance
    epidemiologic investigation, 542–543
    future, 543
    Neisseria, 538, 540–541
    PBP, 537–549
        detection, 541–542
        diagnosis, 539–542
        materials, 544
        methods, 544–547
        resistance mechanism, 537–538

PBP genes, 539–540
Streptococcus pneumoniae, 538,
     540–541
B-lactamase, 495–508
   characterization, 513–532
      consumables, 514
      equipment, 514
      materials, 514
      methods, 514–529
   diagnosis, 496
   direct sequence analysis, 504–505
   epidemiologic uses, 496–504
   future, 505–506
   isoelectric focusing, 514–518
      metallo vs serine, 518
      PhastSystem, 516–517
      precast gels, 516–517
      sample preparation, 515
   kinetic analysis, 518–529
      computerized determination,
         525–526
      graphical determination, 523–525
      initial rate measurement, 522
      kinetic constants evaluation,
         528–529
      rapid inhibitory profiling, 526–528
      rapid substrate profiling, 520–522
      single process curves, 522–523
   research, 504–505
B-lactamase genes
   PCR detection, 500–501
B-lactamase probes
   labeling, 500
BACTEC
   Mycobacterium tuberculosis, 159–160
Bacterial diseases
   AP-PCR methods, 83–97
Bacterial isolates
   epidemiologically related
      PFGE comparison, 38
   identification, 3–4
Bacterial lysis
   PFGE, 39, 41–42
Bacterial lysis buffers
   PFGE, 40
Bacteriocin typing
   Corynebacterium diphtheriae, 198–200

Enterococci, 475
Bacteriophage typing
   Enterococci, 475
Banding patterns interpretation
   PFGE, 37, 43–44
bDNA assay, 77
Biotin-labeled probes
   Enterococci, 471
Biotyping
   Corynebacterium diphtheriae, 198
   Enterococci, 475
   Helicobacter pylori detection, 422
   Lior
      Campylobacter, 408
   Preston
      Campylobacter, 409
   Vibrio cholera, 374
Birnboim and Doly method
   plasmid analysis, 52
BOX-fingerprinting
   pneumococcal diseases, 148
Branched DNA assay, 77
BRENDA, *see* REA
Brij lysis method
   plasmid analysis, 52

C

Campylobacter
   epidemiologic investigation, 408–414
   phage typing, 409
   Preston biotyping scheme, 409
   speciation, 408
      goals, 414
Campylobacter infection, 407–415
Campylobacterosis
   cost, 207
Candida albicans isolates
   PFGE comparison, 39
Candida septicemia
   AP-PCR, 94
Capillary blotting
   GDD and ribotyping, 22, 25–26
Capsular genes
   meningococcal disease, 270–271
Capsular genotyping
   Haemophilus influenzae, 252–254
Cerebral nocardia infection

PCR misdiagnosis, 4–5
Chancroid, 309–327
  diagnosis, 309–311
  epidemiologic investigation, 311–312
  incidence, 309
  pathogenesis, 313–314
  research uses, 313–314
CHEF, 34, 35
  Staphylococci, 449
CHEF Mapper, 35
Chemiluminescent detection
  GDD and ribotyping, 23–24, 27
Cholera, 369–381
  diagnosis, 370–374
  epidemiologic investigation, 374–378
  future, 379–380
    improved diagnostic tests, 379
    typing method standardization, 379,
      380
  history, 369–370
  research, 378–379
    epidemic origins, 378–379
    pathogenesis, 378
Cholera toxin gene
  detection, 372–373
Chromosomal DNA analysis
  Enterococci, 476–479
  Salmonella, 359–362
Chromosomally mediated resistance
  Enterococci, 591–592
  Neisseria gonorrhoeae, 299
Chromosome mapping
  physical
    PFGE, 38
Ciprofloxacin
  tuberculosis, 167
Clinical bacteriology
  molecular methods impact, 1–14
Clinical bacteriology laboratory
  functions, 1–2
Clostridium difficile
  AP-PCR, 93–94
Colitis
  AP-PCR, 93–94
Commercial kits
  DNA probes
    Pneumococci, 148

ELISA
  Legionellae, 216
  Helicobacter pylori, 420
  Mycobacterium tuberculosis, 164–165
  Mycoplasma detection, 348
  Neisseria gonorrhoeae, 295–296
  nucleic acid probes
    Legionellae, 216, 221
  PCR, 63
  plasmid DNA extraction, 52–53
Conjugative plasmids, 51
Contour-clamped homogenous electric
  field, 34–35
Conventional typing
  Campylobacter, 408–409
  diphtheria, 198–204
  Enterococci, 475
  Helicobacter pylori, 422–423
  microbial
    advantages, 10–11
    disadvantages, 11
Cornybacterium diphtheriae
  biotyping, 198
Corynebacterium diphtheriae, 191
  PCR typing, 203–204
  phage typing, 200
  toxin, 193–194
Cost
  campylobacterosis, 407
  isolated bacteria identification, 3–4
  Mycobacterium tuberculosis
    antibiotic resistance genes, 9
  nested PCR, 68
  plasmid analysis, 52–53
Cross contamination
  isolated bacteria identification, 3
  PCR assays, 64
Cryptic plasmids, 51
Cryptococcus neoformans
  AP-PCR, 93
ctx
  DNA sequencing, 377

D

D-ala-D-ala genes
  identification, 594–596
DAEC, 395–396

Dendron, 37
Detection
    aminoglycoside resistance, 556–557
        Gram-negative bacteria, 561–565
        Gram-positive bacteria, 565–566
    antibiotic resistance genes
        research, 657
    antigen
        Legionellae, 216
        Neisseria gonorrhoeae, 295
    chemiluminescent
        GDD and ribotyping, 23–24, 27
    cholera toxin gene, 372–373
    glycopeptide resistance
        Gram-positive bacteria, 582–586
    Legionellae, 214–224
        antigen, 216
        conventional methods, 215–216
        culture, 215–216
        molecular approaches, 216–224
        serology, 215
    mecA, 8
    MRSA, 7–8
    Neisseria
        B-lactam resistance, 540–541
    PCR
        penicillin resistance
            pneumococcal infection, 543
        Pneumococci, 146
    rapid
        isolated bacteria detection, 3
        Mycobacterium tuberculosis
            antibiotic resistance genes,
                8–9
        pathogen detection, 4
        therapy, 6
    tetracycline resistance, 647–651
    vanA gene
        Enterococci, 599, 600–601
    vanB gene
        Enterococci, 599, 600–601
DFA test
    Legionellae, 221
dhps gene
    meningococcal disease, 272
Diagnosis
    B-lactam resistance, 539–542

B-lactamase, 496
chancroid, 309–311
cholera, 370–374
diarrheagenic *Escherichia coli*,
    396–397
diphtheria, 192–193, 195–197
    laboratory, 196
    molecular, 196–197
gonorrhea, 294–298
    conventional methods, 294–295
    molecular methods, 295–298
Haemophilus influenzae, 244–248
Helicobacter pylori infection, 419–422
    conventional methods, 419–420
    molecular methods, 420–422
meningococcal disease, 267–273
molecular methods impact, 3–6
Mycobacterium, 158–169
    conventional, 158–160
    molecular, 160–165
Mycobacterium tuberculosis, 177
    targets and primer sequences, 162
Mycoplasma, 339–342
pneumococcal infection, 139–145
quinolone resistance, 621–623
tuberculosis, 158–169
Vibrio cholerae, 371–374
Diagnostic probes
    Haemophilus influenzae, 244
Diagnostic System, 169
Diarrhea
    AP-PCR, 93–94
Diarrheagenic *Escherichia coli*, 387–400
    categories, 387–388
    clinical diagnosis, 396–397
    epidemiologic investigation, 397
    future, 398
    molecular diagnostic tests, 387–396
        criteria, 387–388
    research uses, 397–398
Dice coefficient formula, 44
Diffusely adherent *Escherichia coli*,
    395–396
Dihydropterate synthesis gene
    meningococcal disease, 272
Diphtheria, 191–208
    diagnosis, 192–193, 195–197

epidemiologic investigation, 197–198
epidemiology, 194–195, 205
future, 206
history, 191–192
microbiologic surveillance, 205–206
molecular typing, 200–205
traditional typing methods, 198–204
treatment, 193
typing methods
    rationale, 205
Diphtheria toxin, 193–194
Diphtheria toxin gene
    Corynebacterium diphtheriae, 201
Direct fluorescent antibody test
    Legionellae, 221
DNA amplification, 63–79
    Haemophilus influenzae, 244–248
    PCR assay design, 63–68
        heminested PCR, 66–68
        hot-start PCR, 65
        nested PCR, 66–68
        PCR optimization, 65
        PCR process, 63–64
        primer design, 64–65
        touch-down PCR, 65–66
    PCR end-point evaluation, 68–75
        gel electrophoresis, 68–69
        internal probes, 70–74
        RFLP, 69–70
        sequencing, 74–75
    signal amplification methods, 76–77
    target amplification methods, 75–76
DNA extraction
    plasmid analysis, 53, 54–55
DNA fingerprint analysis
    Pneumococci, 147
DNA fragment separation
    PFGE, 41, 42–43
DNA gyrase gene mutations
    quinolone resistance, 618–619
DNA hybridization
    aminoglycoside isolate screening,
        558, 566–570
    Enterococcal glycopeptide
        resistance, 583–586
    macrolide resistance genes, 655
    Staphylococci, 435

sulfonamide resistance genes, 656
tetracycline resistance, 647–648
tetracycline resistance genes, 652
trimethoprim resistance genes, 656
DNA isolation
    guanidinium thiocyanate lysis
        Streptococci identification, 118,
            120, 121
    long-PCR and electrophoresis
        Streptococci identification, 119, 121
    SDS lysis
        Streptococci identification, 118, 120
DNA probes
    aminoglycoside, 567–569, 572–573
    B-lactamase, 496–500
    Corynebacterium diphtheriae,
        200–201
    diarrheagenic *Escherichia coli*, 396
    Enterococci, 470–474
    Haemophilus ducreyi, 309–310
    Haemophilus influenzae, 251
    Helicobacter pylori detection, 421
    Mycoplasma, 340
    Neisseria gonorrhoeae, 295–296
    pneumococcal diseases, 139
    Staphylococci, 434–437
    tox, 206
DNA probing kit
    Pneumococci, 140
DNA sequencing
    ctx, 377
    gyrA QRDR, 626, 631–633
    Vibrio cholera
        future, 380
DNA–DNA hybridization
    Helicobacter pylori detection, 420
    Streptococci identification, 131–132
DNASIS, 69
Dot blotting
    PCR product identification, 72–73
DT, 193–194

E

EAEC, 394
EAF probe, 391
Efflux pumps
    quinolone resistance, 621

EHEC, 387, 392–394
EIEC, 394–395
Electrophoresis
  DNA isolation
    Streptococci identification, 119, 121
    Enterococci, 476–477
  gel
    PCR end-points evaluation, 68–69
Electroporation
  Haemophilus ducreyi, 314, 317,
    322–325
Elek immunoprecipitation test, 196
ELISA
  Helicobacter pylori, 420
  Legionellae, 216
ELWGD, 205–206
emm gene identification
  oligonucleotide probes, 130
Entamoeba histolytica
  AP-PCR, 94
Enteroaggregative *Escherichia coli*, 394
Enterobacter cloacea
  AP-PCR, 94
Enterobacterial repetitive intergenic
  consensus sequences, 103
Enterococcal glycopeptide resistance
  DNA hybridization, 583–586
  PCR, 583
  PCR primers, 584–585, 588–589
Enterococcal nosocomial infection,
  469–483
Enterococci
  biotyping, 475
  epidemiologic investigation, 475–481
  identification, 470–475
    nucleic acid techniques, 470–474
    protein analysis, 474
  plasmid analysis, 54, 56–57
  research
    mutant analysis, 481
    virulence marker detection, 481
  vancomycin resistance, 580–582
Enterococci speciation
  PCR, 599, 601–602
Enterohemorrhagic *Escherichia coli*,
  387, 392–394
Enteroinvasive *Escherichia coli*, 394–395

Enteropathogenic *Escherichia coli*,
  391–392
Enterotoxigenic *Escherichia coli*,
  388–391
EPEC, 391–392
Epidemiologic investigation
  B-lactam resistance, 542–543
  Campylobacter, 408–414
    conventional typing methods,
    408–409
  chancroid, 311–312
  cholera, 374–378
  diarrheagenic *Escherichia coli*, 397
  diphtheria, 197–198
  Enterococci, 475–481
    conventional typing, 475
    nucleic acid analysis, 475–481
  gonorrhea, 299–303
  Haemophilus influenzae, 249–256
  Helicobacter pylori, 422–426
  Legionellae, 227–231
    conventional approaches, 228
    molecular methods, 229–231
    typing rationale, 227–228
  meningococcal disease, 273–276
  Mycobacterium tuberculosis, 169–175
  penicillin resistance, 542–543
  pneumococcal infection, 145–151
  quinolone resistance, 623–625
  Staphylococci, 441–455
    criteria, 441
  Vibrio cholerae, 374–378
Epidemiologic typing
  isolated bacteria, 10–12
Epidemiology
  molecular methods impact, 10–12
ERIC, 103–104
  Campylobacter, 413
  Haemophilus influenzae, 255
  Pneumococci, 149
ERIC-PCR
  Salmonella, 362–363
Erythromycin resistance, 643–645
Escherichia coli
  trimethoprim resistance, 645
Escherichia coli K-12
  AP-PCR primer PCR fingerprinting, 89

ET-37 complex
  Meningococci, 278
ET-5 complex
  Meningococci, 279
ETEC, 388–391
European Laboratory Working Group
  on Diphtheria, 205–206

F

Field-inversion gel electrophoresis, 34,
  449
FIGE, 34
  Staphylococci, 449
Fingerprint analysis
  meningococcal disease, 276
Fingerprinting
  genomic
    rep-PCR, 103–113
FLA-typing
  Campylobacter, 412–413
Fluorescein-labeled B-lactamase
  probes, 500–501

G

GDD, *see* also REA
  Campylobacter, 409–411
  problems, 18
  ribotyping, 17–30
GDD and ribotyping
  agarose gel electrophoresis, 21, 25
  capillary blotting, 22, 25–26
  chemiluminescent detection, 23–24,
    27
  DNA digestion, 21, 24
  DNA rapid extraction, 21, 23–24
  hybridization, 22–23, 26–27
  materials, 21–23
  methods, 23–27
  probe preparation, 22, 26
  Southern blotting, 22, 25–26
  UV spectroscopy, 21, 24
Gel electrophoresis
  PCR end-points evaluation, 68–69
GelCompar, 37
GelManager, 37
Gen Probe system, 160
Gen-Probe PACE-2 assay
  Neisseria gonorrhoeae, 295–296

Gene Navigator, 35
Gene probes
  Mycobacterium tuberculosis, 160–161
GeneJockey, 64
Genomic DNA digestion, *see* GDD
Genomic DNA rapid extraction
  GDD and ribotyping, 21, 24
Genomic fingerprinting
  rep-PCR, 103–113
Genomic typing
  Campylobacter, 409–414
Glycopeptide resistance
  Enterococci, 480–481
    induction, 596–597
    materials, 599–600
    methods, 600–605
    molecular methods, 583–586
    vancomycin resistance, 580–582
  Gram-positive bacteria, 579–611
    phenotypic detection, 582–583
    resistance detection, 582–586
Glycopeptide resistant bacteria
  spread, 586–587
Glycopeptide resistant Enterococci
  peptidoglycan precursors analysis,
    600, 605–606
Glycopeptide resistant genes
  dissemination
    Enterococci, 587–594
Gonorrhea, 293–303
  antibiotic resistance, 294
  diagnosis, 294–298
  epidemiologic investigation, 299–303
  prevalence, 293–294
Gram-negative bacteria
  plasmid analysis, 53, 56
Gram-positive bacteria
  glycopeptide resistance, 579–611
  vancomycin resistance, 582
Guanidinium thiocyanate lysis
  DNA isolation
    Streptococci identification, 118,
      120–121
gyrA gene
  RFLP, 626, 633–634
gyrA QRDR
  DNA sequencing, 626, 633–634

gyrA QRDR amplification
  quinolone isolates, 625, 630–631

H

Haemophilus aegyptius, 257
Haemophilus ducreyi, 309–327
  electroporation, 314, 317, 322–325
  Escherichia coli, 314
  materials, 314–317
  methods, 318–328
  PCR, 314–315, 318–319
  ribotyping, 315–317, 320–322
Haemophilus ducreyi genes
  complementation, 327–328
  directed mutagenesis, 327
  transpositional mutagenesis,
    325–327
Haemophilus influenzae, 243–258
  diagnosis, 244–248
  epidemiologic investigation, 249–256
  future, 257–258
  research, 256–257
Helicobacter pylori, 419–427
  conventional typing, 422–423
  epidemiologic investigation, 422–426
  future, 426
  molecular typing, 423–426
Helicobacter pylori detection
  biotyping, 374
Helicobacter pylori infection
  diagnosis, 419–422
  epidemiology, 419
Heminested PCR, 66–68
Hemolytic uremic syndrome, 387
High performance liquid
    chromatography, 159–160, 600,
    605–606
Hot-start PCR, 65
HPLC
  Mycobacterium tuberculosis, 159–160
  peptidoglycan precursors, 600,
    605–606
Hybridization
  GDD and ribotyping, 22–23,
    26–27
  Legionellae, 221
  Streptococci identification, 127–131

I

Identification
  bacterial isolates, 3–4
  D-ala-D-ala genes, 594–596
  emm gene, 130
  Enterococci, 470–475
    nucleic acid techniques, 470–474
    protein analysis, 474
  isolated bacteria, 3–4
    cost, 3–4
    cross contamination, 3
    rapid detection, 3
  Legionellae, 224–227
    conventional methods, 224–225
    molecular methods, 225–227
  Salmonella, 359–362
  Staphylococci, 432–441
    criteria, 432
  Streptococci, 117–136
    DNA extraction, 119–121
    DNA–DNA hybridization,
      131–132
    hybridization analysis, 127–131
    materials, 118–119
    methods, 119–133
    PCR amplification techniques,
      121–131
      PCR product analysis, 124–125
      PCR reaction setup, 122–124
    PCR primers, 123
    PFGE, 132
    RFLP, 125–126
IEF
  B-lactamase, 514–518
IEF gels
  B-lactamase, 516–517
Immunofluorescence
  Legionellae, 216
Infection
  nosocomial
    Enterococcal, 469–483
    Staphylococcal, 431–458
Insertion sequence elements, *see* IS
    elements
Integron structure
  aminoglycosides
    PCR, 573

Internal probes
  PCR product identification, 71
    hybridization assay formats, 72
    internal sequence selection, 71
    oligonucleotide labeling, 71
IS elements
  Corynebacterium diphtheriae, 201, 207
  Mycobacterium tuberculosis, 171
IS1106
  meningococcal disease, 272–273
IS200 fingerprinting
  Salmonella, 361–362
IS6110-RFLP
  Mycobacterium tuberculosis, 171–173
IS6770 typing
  Enterococci, 479
Isoelectric focusing
  B-lactamase, 514–518
Isoenzyme typing
  Legionellae, 229
Isolated bacteria
  identification, 3–4
    cost, 3–4
    cross contamination, 3
    rapid detection, 3
    molecular fingerprinting, 11–12
Izuma fever
  AP-PCR, 93

J

Jaccard coefficient formula, 44

K

Kado and Liu method
  plasmid analysis, 52, 229

L

Laboratory methods
  molecular methods impact, 3–6
LaneManager, 37
LAR, 76
Lasergene, 64, 65, 69
LCR, 76, 165
  Neisseria gonorrhoeae, 296–298
LDR, 76
Legionellae, 213–231
  detection, 214–224
    antigen, 216

conventional methods, 215–216
  culture, 215–216
  molecular approaches, 216–224
  serology, 215
epidemiologic investigation, 227–231
future, 231
Identification, 224–227
  conventional methods, 224–225
  molecular methods, 225–227
identification, 224–227
PCR typing, 231
Legionnaires' disease, 213
Leptospira
  AP-PCR fingerprinting, 89–90
Ligase chain reaction, 76, 165
Ligation amplification reaction, 76
Ligation detection reaction, 76
Linear cycle sequencing
  PCR products, 74
Lior biotyping scheme
  Campylobacter, 408
Lipase chain reaction
  Neisseria gonorrhoeae, 296–298
Listeria monocytogenes
  AP-PCR, 92
  rep-PCR profiles, 110
Lysis buffers
  bacterial
    PFGE, 40
lytA gene
  Pneumococci identification, 143–144

M

Macrolide resistance, 641–649
Macrolide resistance genes
  characterization, 655–656
  DNA hybridization, 655
  PCR, 655–656
MacVector, 64
Mar
  quinolone resistance, 620–621
mecA detection, 8
Medical microbiology
  molecular methods impact, 1–14
Meningococcal disease, 265–283
  diagnosis, 267–273
  epidemiologic investigation, 273–276

future, 279–280
research, 276–279
Meningococcus
population biology, 276–279
Methicillin resistant staphylococcus
aureus
AP-PCR, 93
MGIT
Mycobacterium tuberculosis, 159
Microbial typing
traditional techniques
advantages, 10–11
disadvantages, 11
MIP gene
Legionellae, 223
MLEE
Corynebacterium diphtheriae, 204
Enterococci, 475
Haemophilus influenzae, 249, 254
meningococcal disease, 281
Mycobacterium tuberculosis, 170
pneumococcal diseases, 145–147
Vibrio cholera, 375–376
MLST
Meningococci, 283
Modified Elek test, 206
Molecular fingerprinting, 10
isolated bacteria, 11–12
Molecular methods
impact on clinical bacteriology, 1–14
impact on diagnosis, 3–6
impact on epidemiology, 10–12
impact on therapy, 6–9
isolated bacteria identification, 3–4
MRSA detection, 7–8
Mycobacterium tuberculosis
detection, 8–9
pathogen detection, 4–6
quality assurance, 12–14
Molecular typing, 10
diphtheria, 200–205
Helicobacter pylori, 423–426
Mollicutes
taxonomy, 335
Monoclonal antibody subgrouping
Legionellae, 228
Mosaic PBP genes

RFLP analysis, 541–542
MRSA
AP-PCR, 93
MRSA detection, 7–8
Multilocus enzyme electrophoresis,
see MLEE
Multilocus sequence typing
Meningococci, 283
Multiple antibiotic resistance locus
quinolone resistance, 620–621
MVSP, 37, 43
Mycobacteria growth indicator tube
Mycobacterium tuberculosis, 159
Mycobacterium
AP-PCR, 173–175
BACTEC, 159–160
diagnosis, 158–169
gene probes, 160–161
HPLC, 159–160
IS elements, 171
IS6110-RFLP, 171–173
MGIT, 159
microbiology, 158
molecular approaches, 157–179
PCR, 161–164
PFGE, 173
RAPD, 173
rapid detection, 159–160
repetitive elements, 171
spoligotyping, 175
target amplification, 161–165
Mycobacterium tuberculosis
antibiotic resistance genes
detection, 8–9
diagnosis, 177
drug resistance, 165–169
molecular basis, 166–167
resistant isolate identification,
167–168
drug susceptibility, 177
epidemiologic investigation, 169–175
materials, 177–178
methods, 178–179
molecular approaches, 157–179
phage typing, 169
Q-beta replicase-directed
amplification, 77

Mycoplamsa pneumoniae, 342
Mycoplasma
  colonization, 338
  diagnosis, 339–342
  epidemiologic investigation,
    342–345
  future, 348
  PCRs, 343–344
  research uses, 345–348
    animal models, 347
    cultivation, 346–347
    genetics, 347–348
    genome sequencing, 345–346
    pathogenicity, 347
  STDs, 346
  taxonomy, 336
  vs Eubacteria, 337
Mycoplasma genitalium, 341, 345
Mycoplasma infection, 335–350
  taxonomy, 336
Mycoplasma PCR ELISA, 348
Mycoplasma PCR Primer Set, 348

N

NASBA, 75–76
Neisseria
  B-lactam resistance, 538
    detection, 540–541
  population structure, 542–543
Neisseria gonorrhoeae, 293
  antigen detection tests, 295
  antimicrobial susceptibility, 298
  chromosomally mediated resistance,
    299
  plasmid-mediated resistance, 298–299
  trimethoprim and sulfonamide
    resistance, 645, 647
Neisseria meningitidis
  AP-PCR, 92–93
  population genetics, 277
  serologic typing scheme, 266
Nested PCR, 66–68
  cost, 68
  disadvantages, 67–68
Nested primer PCR
  pneumococcal DNA detection,
    140–141

Nosocomial infection
  Enterococcal, 469–483
  Staphylococcal, 431–458
Nucleic acid analysis
  Enterococci, 475–481
Nucleic acid probes
  Legionellae, 216–221
Nucleic acid sequence based
    amplification, 75–76
Nucleic acid techniques
  Enterococci, 470–474

O

OLIGO, 64
Oligonucleotide probes
  advantages, 70
  emm gene identification, 130
  Enterococci, 470–473
  PCR diagnosis
    Pneumococci, 142
  tetracycline resistance, 648–649
Oligonucleotide sequence
  Neisseria gonorrhoeae, 297
Oligonucleotide sequences
  diarrheagenic *Escherichia coli*, 390
  Pneumococci, 148
Oligotyping
  TEM B-lactamase, 502–504
OMP analysis
  Gram-negative bacteria
    quinolone resistance, 625, 629–630
OMP mutations
  quinolone resistance, 620
ompP2 gene, 251–252
Opa-typing
  Neisseria gonorrhoeae, 302–303
Otitis media
  PCR
    Haemophilus influenzae, 248
  REA
    Haemophilus influenzae,
      249–251
Outer membrane protein analysis
  Gram-negative bacteria
    quinolone resistance, 625, 629–630
Outer membrane protein mutations
  quinolone resistance, 620

P
PACE, 35
PAGE
  PCR end-points, 69–70
Palindromic units, 103
parC mutations
  quinolone resistance, 618–619
Pathogen detection
  rapid detection, 4
  sensitivity, 4
PAUP, 89
PBP
  B-lactam resistance, 537–549
    detection, 541–542
    diagnosis, 539–542
    materials, 544
    resistance mechanism, 537–538
  detection
    B-lactam resistance, 541–542
PBP 2B gene
  Pneumococci identification, 141–143
PBP gene fingerprinting
  B-lactam resistance, 544, 546–547
  pneumococcal diseases, 147
PBP genes
  B-lactam resistance, 539–540
  mosaic
    RFLP analysis, 541–542
PCR
  aminoglycoside resistance genes,
    570–573
  B-lactamase gene detection, 500–501
  cerebral nocardia infection
    misdiagnosis, 4–5
  Corynebacterium diphtheriae, 196–197
  DNA isolation
    Streptococci identification, 119, 121
  DNA preparation
    Streptococci identification,
      118, 121
  Enterococcal glycopeptide resistance,
    583
  Enterococci speciation, 599, 601–602
  Haemophilus ducreyi, 310–311,
    314–315, 318–319
  Haemophilus influenzae, 244–248,
    251–252

  ampicillin resistant, 248
  Helicobacter pylori detection, 422
  integron structure
    aminoglycosides, 573
  kits, 63
  Legionellae, 221–223
  macrolide resistance genes, 655–656
  meningococcal disease, 268
  Mycobacterium tuberculosis, 161–165
  Mycoplasma, 341–342
  Neisseria gonorrhoeae, 296, 301–302
  optimization, 65
  otitis media
    Haemophilus influenzae, 248
  pneumococcal diseases, 140–145
  quinolone resistance, 622–623
  Salmonella, 362–363
  tetracycline resistance, 648–651
  tetracycline resistance genes, 652–655
PCR amplification
  Staphylococci, 437–441
    methicillin resistance detection,
      438–440
    species identification, 437, 441
  Streptococci identification, 121–131
  Tn1546-related elements, 600,
    602–603
PCR assays
  cross-contamination, 64
  design, 63–68
  diarrheagenic *Escherichia coli*, 396
  DNA amplification, 63–68
  Helicobacter pylori detection, 420–422
  problems, 64, 66–68
PCR detection
  penicillin resistance
    pneumococcal infection, 543
  Pneumococci, 146
PCR DNA fingerprinting
  Campylobacter, 412–413
PCR end-points evaluation, 68–75
  gel electrophoresis, 68–69
  internal probes, 71–74
  RFLP, 69–70
PCR fingerprinting
  Enterococci, 479
  quinolone resistance, 624

Staphylococci, 452–455
PCR primers
  Enterococcal glycopeptide
    resistance, 584–585, 588–589
PCR product identification
  sequencing, 74–75
PCR sequencing
  TEM-related enzymes, 502
PCR typing
  Corynebacterium diphtheriae, 203–204
  Legionellae, 231
PCR-RFLP analysis
  Helicobacter pylori detection, 425–426
Pearson's correlation coefficient, 44, 45
Penicillin resistance
  epidemiologic investigation, 542–543
  Neisseria
    detection, 540–541
  Neisseria gonorrhoeae, 298–299
  pneumococcal infection
    PCR detection, 543
  Streptococcus pneumoniae
    detection, 540–541
Penicillin-binding proteins, *see* PBP
Peptidoglycan precursors
  analysis, 597–598
  glycopeptide-resistant Enterococci,
    600, 605–606
  HPLC analysis, 600, 605–606
PFGE, 33–49, 91–92
  advances, 35
  applications, 38–39
  banding pattern interpretation, 37,
    43–44
    computer analysis, 43–44
    visual, 43
  Campylobacter, 410–411
  Corynebacterium diphtheria, 203,
    207
  description, 33
  DNA restriction, 39, 42
  DNA separation, 35–37, 41, 42–43
    agarose concentration, 36
    buffer strength and temperature, 37
    DNA quality, 36
    voltage and pulse times, 36–37
  Enterococci, 478–479

equipment, 34–35
  Legionellae, 230
  materials, 39–41
  meningococcal disease, 276, 281
  methods, 41–47
  Mycobacterium tuberculosis, 173
  Neisseria gonorrhoeae, 302
  pattern interpretation, 19–20
  pneumococcal diseases, 148
  quinolone resistance, 624–625
  Salmonella, 359–360
  Staphylococci, 449–451
  steps, 34
  Streptococci identification, 132
  troubleshooting, 44–47
  unsheared DNA preparation, 39,
    41–42
  Vibrio cholera, 376–377
Phage typing
  Campylobacter, 409
  Corynebacterium diphtheriae, 200
  Mycobacterium tuberculosis, 169
PhastSystem
  B-lactamase, 516–517
Phenotypic detection
  glycopeptide resistance
    Gram-positive bacteria, 582–583
Phoretix 1D & 2D, 37
PHYLIP, 89
Physical chromosome mapping
  PFGE, 38
Plasmid analysis, 51–61
  agarose gel electrophoresis, 54, 58–59
  cost, 52–53
  Enterococci, 475–476
  history, 52
  Legionellae, 229
  materials, 53–54
  methods, 54–58
  Neisseria gonorrhoeae, 300
  plasmid DNA extraction, 53, 54–55
  restriction endonucleases, 53–54,
    56–57
  Salmonella, 358–359
  Staphylococci, 441–443
Plasmid DNA extraction
  kits, 52–53

plasmid analysis, 53, 54–55
Plasmid dominance test
   quinolone resistance, 625, 627–629
Plasmid fingerprinting
   Salmonella, 358–359
Plasmid profile typing
   Salmonella, 358–359
Plasmid profiling
   Campylobacter, 413–414
   Mycobacterium tuberculosis, 169–170
   Vibrio cholera, 377
Plasmid typing, 51
Plasmid-mediated resistance
   Enterococci, 487–591
   Neisseria gonorrhoeae, 298–299
Plasmids
   conjugative, 51
   cryptic, 51
   selftransferable, 51
   separation and sizing
      PFGE, 38
pMC5, 340
Pneumococcal infection, 139–151
   AP-PCR, 150–151
   BOX-fingerprinting, 148
   diagnosis, 139–145
   DNA probing, 139
   epidemiologic investigation, 145–151
   MLEE, 145–147
   molecular diagnosis
      future, 151
   PBP gene fingerprinting, 147
   PCR, 139–140
   penicillin resistance
      PCR detection, 543
   PFGE, 148
   rep-PCR, 148–150
   ribotyping, 147
Pneumococci
   ERIC, 149
   oligonucleotide sequences, 148
   PCR diagnosis, 146
   REP, 149
Polyacrylamide gel electrophoresis,
   see PAGE
Polynucleotides
   diarrheagenic *Escherichia coli*, 389

Por typing
   Neisseria gonorrhoeae, 303
porA gene
   meningococcal disease, 271–272
Porphyromonas gingivalis
   AP-PCR, 94–95
Preston biotyping scheme
   Campylobacter, 409
Primer Premier, 64
Primers
   B-lactamase, 498–500
   Haemophilus influenzae, 246–247
   PCR assays, 64–65
   PCR-based typing
      Mycobacterium tuberculosis, 174
   streptococcal DNA, 122–123
Probe preparation
   GDD and ribotyping, 22, 26
Probes
   B-lactamase
      labeling, 500
   biotin-labeled
      Enterococci, 471
   diagnostic
      Haemophilus influenzae, 244
   fluorescein-labeled B-lactamase, 500
   gene
      Mycobacterium tuberculosis,
         160–161
   internal
      PCR product identification, 72
   nucleic acid
      Legionellae, 216–221
   oligonucleotide
      advantages, 70
      emm gene identification, 130
      Enterococci, 470–473
      tetracycline resistance, 648–649
   urease gene
      Helicobacter pylori detection,
         24–425
Programmable autonomously controlled
   electrodes, *see* PACE
Protein analysis
   Enterococci, 474–475
Pulsed-field gel electrophoresis, *see*
   PFGE

Q

Q-beta replicase-directed amplification,
77
Quality assurance
molecular methods, 12–14
Quinolone antibacterials, 617–618
mechanism of action, 617–618
Quinolone resistance, 617–635
diagnosis, 621–623
epidemiologic investigation, 623–625
materials, 625–626
methods, 627–634
molecular characterization, 621–623
molecular level, 618–621
mutation detection, 628

R

Random amplified polymorphic DNA,
see RAPD
Random cloned chromosomal sequence
typing
Salmonella, 360–361
RAP-PCR, 96
RAPD, 85
Campylobacter, 413
Haemophilus influenzae, 255
meningococcal disease, 276
Mycobacterium tuberculosis, 173
Rapid detection
isolated bacteria detection, 3
Mycobacterium tuberculosis
antibiotic resistance genes, 8–9
pathogen detection, 4
therapy, 6
RCCS typing
Salmonella, 360–361
REA
Campylobacter, 409–410
Enterococci, 475–477
Haemophilus influenzae, 249–251
Helicobacter pylori detection,
423–424
otitis media
Haemophilus influenzae, 249–251
Pneumococci, 147
Staphylococci, 443–444
REP, 103

Campylobacter, 413
Haemophilus influenzae, 255
Helicobacter pylori detection, 425
Mycobacterium tuberculosis, 171
Pneumococci, 149
rep-PCR
buffers and stock solutions, 106–107
cell lysates preparation, 105, 107
defined, 104–105
DNA template preparation, 105–106,
107–108
Enterococci, 479
fingerprints computer-assisted
analysis, 110–111
genomic fingerprinting, 103–113
materials, 105–107
methods, 107–111
oligonucleotide primers, 106
PCR product analysis, 109
pneumococcal diseases, 148–150
reaction mixes preparation, 108
results, 109–111
Salmonella, 362–363
Staphylococci, 452
thermal cycling, 109
Repetitive extragenic palindrome
elements, 103
RES
Enterococci, 479–480
Resisotyping
Campylobacter, 409
Resistance detection
glycopeptide resistance
Gram-positive bacteria, 582–586
Resistance genes
Haemophilus influenzae, 255–256
Restriction endonuclease DNA
digestion
GDD and ribotyping, 21, 24
Restriction endonuclease fingerprinting
Neisseria gonorrhoeae, 300–301
Restriction endonucleases
plasmid analysis, 53–54, 56–57
Restriction enzyme analysis,
see REA
Restriction fragment length
polymorphism, *see* RFLP

RFLP
  gyrA gene, 626, 633–634
  Haemophilus influenzae, 251
  Legionellae, 230
  mosaic PBP genes, 541–542
  PCR end-points evaluation, 69–70
    electrophoretic analysis, 69–70
    enzyme selection, 69
  Staphylococci, 444–449
  Streptococci identification, 119,
    125–126
  vs AP-PCR fingerprinting, 90–92
Ribosomal RNA genes
  Legionellae, 223–224
Ribotyping
  Campylobacter, 411–412
  Corynebacterium diphtheriae,
    202–203, 207
  Enterococci, 479
  GDD, 17–30
  Haemophilus ducreyi, 311, 315–317,
    320–322
  Haemophilus influenzae, 254–255
  Helicobacter pylori detection, 422
  Neisseria gonorrhoeae, 300–301
  pneumococcal diseases, 147
  probes, 18
  Salmonella, 360
  Staphylococci, 432–434, 445–446
  Vibrio cholera, 376
RNA fingerprinting
  AP-PCR, 96
rrn gene
  meningococcal disease, 268–270

S

Salmonella, 355–364
  identification
    chromosomal analysis methods,
      359, 362
    DNA-based methods, 358–359
    PCR-based methods, 362–363
  RCCS typing, 360–361
  strain identification, 356–357
  taxonomy, 356–357
Salmonella typhimurium
  AP-PCR fingerprinting, 96

Salmonellosis, 355
  subspecies, 355–357
SDA, 76, 164–165
SDS lysis
  DNA isolation
    Streptococci identification, 118, 120
Self-sustained sequence (synthetic)
  reaction, 75–76
Selftransferable plasmids, 51
Sensitivity
  AP-PCR methods, 88–90
  pathogen detection, 4
Sequencing
  linear cycle
    PCR products, 74
  PCR product identification, 74–75
Serogroup A meningococci, 277–278
Serogroup B meningococci, 279
Serogroup C meningococci, 278–279
Serologic typing scheme
  Neisseria meningitidis, 266
Serosubtype genes
  meningococcal disease, 271–272
Serotype genes
  meningococcal disease, 271–272
Serotyping
  Campylobacter, 409
  Corynebacterium diphtheriae, 198
  Enterococci, 475
  Helicobacter pylori detection, 423
  Mycobacterium tuberculosis, 169
  Neisseria meningitidis, 266
  Vibrio cholera, 371
Signal amplification, 76–77
Signature Tagged Mutagenesis
  Haemophilus influenzae, 256
Single-strand conformational
    polymorphism
  Corynebacterium diphtheriae, 204
Single-stranded conformational
    polymorphism analysis
  quinolone resistance, 626, 634
SNAP system, 160
Southern blot
  GDD and ribotyping, 22, 25–26
  Helicobacter pylori detection, 424
  meningococcal disease, 276

PCR product identification, 72
Staphylococci, 444–449
viriophages, 377–378
Spoligotyping
  Mycobacterium tuberculosis, 175
SSCP
  Corynebacterium diphtheriae, 204
  quinolone resistance, 626, 634
ST polynucleotide probe, 389
Staphylococcal nosocomial infection,
  431–458
Staphylococci
  epidemiologic investigation,
    441–455
  identification, 432–441
    criteria, 432
  methicillin resistance detection,
    432–441
  molecular methods future, 456–457
    research, 455–456
  species identification, 432–441
  vancomycin resistance, 582
Staphylococcus aureus
  plasmid analysis, 54, 57
Stenotrophomonas maltophilia
  AP-PCR, 94
Strand Displacement Assay, 76, 164
Streptococci identification, 117–136
  DNA extraction, 119–121
  DNA–DNA hybridization, 131–132
  hybridization analysis, 127–131
  materials, 118–119
  methods, 119–133
  PCR amplification techniques,
    121–131
    PCR product analysis, 124–125
    PCR reaction setup, 122–124
  PCR primers, 123
  PFGE, 132
  RFLP, 125–126
Streptococcus pneumoniae
  B-lactam resistance, 538
    detection, 540–541
  population structure, 542–543
  trimethoprim and sulfonamide
    resistance, 645
Sulfonamide resistance, 641–649

Sulfonamide resistance genes
  characterization, 656–657
  DNA hybridization, 656

T

Taq-man system
  PCR product identification, 74
Target amplification, 75–76
TAS, 75–76
TCP, 378
TEM B-lactamase
  characterization, 501–504
  oligotyping, 502–504
TEM-related enzymes
  PCR sequencing, 502
Tetracycline resistance, 641–649
  detection, 647–651
  epidemiologic uses, 651–657
  Neisseria gonorrhoeae, 299
Tetracycline resistance genes
  DNA hybridization, 652
Therapy
  molecular methods impact, 6–9
Tn1546-related elements
  PCR amplification, 600, 602–603
Touch-down PCR, 65–66
tox gene
  Corynebacterium diphtheriae, 201,
    206
Toxin coregulated pili, 378
Toxoplasma gondii
  AP-PCR, 94
Traditional typing
  Campylobacter, 408–409
  diphtheria, 198–204
  Enterococci, 475
  Helicobacter pylori, 422–423
  microbial
    advantages, 10–11
    disadvantages, 11
Transcription-based amplification
  system, 75–76
Traveler's diarrhea, 387
Trimethoprim resistance, 641–649
Trimethoprim resistance genes
  characterization, 656–657
  DNA hybridization, 656

Tuberculosis
    ciprofloxacin, 167
    diagnosis, 158–169
    prevalence, 157–158
Typing
    bacteriocin
        Corynebacterium diphtheriae,
            198–200
        Enterococci, 475
    bacteriophage
        Enterococci, 475
    conventional
        Campylobacter, 408–409
        diphtheria, 198–204
        Enterococci, 475
        Helicobacter pylori, 422–423
        microbial, 10–11
    epidemiologic
        isolated bacteria, 10–12
    genomic
        Campylobacter, 409–414
    IS6770
        Enterococci, 479
    isoenzyme
        Legionellae, 229
    molecular, 10
        diphtheria, 200–205
        Helicobacter pylori, 423–426
    multilocus sequence
        Meningococci, 283
    PCR
        Corynebacterium diphtheriae,
            203–204
        Legionellae, 231
    phage
        Campylobacter, 409
        Corynebacterium diphtheriae,
            200
        Mycobacterium tuberculosis, 169
    plasmid, 51
    plasmid profile
        Salmonella, 358–359
    por
        Neisseria gonorrhoeae, 303
    random cloned chromosomal
            sequence
        Salmonella, 360–361

    rationale
        Legionellae, 227–228
    RCCS
        Salmonella, 360–361
    serologic
        Neisseria meningitidis, 266
    standardization
        cholera, 379, 380

U

Unsheared DNA preparation
    PFGE, 39, 41–42
Ureaplasma infection, 335–350
Urease gene probes
    Helicobacter pylori detection, 424
UV spectroscopy
    GDD and ribotyping, 21, 24

V

vanA gene detection
    Enterococci, 599, 600–601
vanA resistance
    mobile elements, 592–594
vanB gene detection
    Enterococci, 599, 600–601
Vancomycin resistance
    Enterococci
        phenotypic resistance classes, 580
        resistance mechanisms, 581
    Gram-positive bacteria, 582
Vibrio cholerae, 369–381
    AP-PCR, 92
    biotyping, 374
    diagnosis, 371–374
    epidemiologic investigation, 374–378
    microbiology, 370–371
    research, 378–379

Y

Yeast isolates
    PFGE comparison, 39
Yersinia pseudotuberculosis
    AP-PCR, 93

Z

Zero-integrated field electrophoresis, 34
ZIFE, 34
Zonula occludens toxin, 378